PHYSIOLOGY OF SPORT AND EXERCISE

THIRD EDITION

Jack H. Wilmore, PhD

Margie Gurley Seay Centennial Professor Emeritus
The University of Texas at Austin

David L. Costill, PhD

John and Janice Fisher Chair in Exercise Science
Human Performance Laboratory
Ball State University

Human Kinetics

Library of Congress Cataloging-in-Publication Data

Wilmore, Jack H., 1938-
    Physiology of sport and exercise / Jack H. Wilmore, David L.
Costill.-- 3rd ed.
       p. ; cm.
Includes bibliographical references and index.
   ISBN 0-7360-4489-2 (hard cover)
 1.   Exercise-Physiological aspects. 2.   Sports--Physiological aspects.

   [DNLM: 1.   Exercise--physiology. 2.   Sports--physiology. 3.   Physical
Endurance. 4.   Physical Fitness.   QT 260 W744p 2004]   I. Costill, David
L. II. Title
   QP301.W6749 2004
   612' .044--dc21

                                  2003011842
ISBN-10: 0-7360-4489-2
ISBN-13: 978-0-7360-4489-9

The Web addresses cited in this text were current as of September 15, 2003, unless otherwise noted.

For instructor and student resources, visit the *Physiology of Sport and Exercise* Web site at www.humankinetics.com/physiologyofsportandexercise.

Acquisitions Editor: Michael Bahrke; **Editor of Original Edition:** Lori Garrett; **Developmental Editor:** Julie Rhoda; **Assistant Editor:** Carla Zych; **Copyeditor:** Julie A. Anderson; **Proofreader:** Sarah Wiseman; **Indexer:** Michael Ferreira; **Permission Manager:** Dalene Reeder; **Graphic Designer:** Andrew Tietz; **Graphic Artist:** Dawn Sills; **Photo Manager:** Kareema McLendon; **Cover Designer:** Keith Blomberg; **Photographer (cover):** Adam Pretty/Getty Images; **Photographer (interior):** see page xiv for a full listing; **Art Manager:** Kelly H. Hendren; **Illustrators of Graphs and Computer-Generated Art:** Mic Greenberg and Chuck Nivens; **Medical Illustrators:** Mic Greenberg and Kristin Mount.

Printed in Hong Kong
10 9 8 7 6

Human Kinetics
Web site: www.HumanKinetics.com

United States: Human Kinetics
P.O. Box 5076
Champaign, IL 61825-5076
800-747-4457
e-mail: humank@hkusa.com

Canada: Human Kinetics
475 Devonshire Road Unit 100
Windsor, ON N8Y 2L5
800-465-7301 (in Canada only)
e-mail: orders@hkcanada.com

Europe: Human Kinetics
107 Bradford Road
Stanningley
Leeds LS28 6AT, United Kingdom
+44 (0) 113 255 5665
e-mail: hk@hkeurope.com

Australia: Human Kinetics
57A Price Avenue
Lower Mitcham, South Australia 5062
08 8277 1555
e-mail: liaw@hkaustralia.com

New Zealand: Human Kinetics
Division of Sports Distributors NZ Ltd.
P.O. Box 300 226 Albany
North Shore City
Auckland
0064 9 448 1207
e-mail: info@humankinetics.co.nz

To those who have had the greatest impact on my life: to my loving wife, Dottie, and our three wonderful daughters, Wendy, Kristi, and Melissa, for their patience, understanding, and love; to my sons-in-law, Craig, Brian, and Randall, for being good husbands, fathers, and friends; to my grandchildren, who are a constant source of joy and amazement; to Mom and Dad for their love, sacrifice, direction, and encouragement; to my former students, who continue to be my friends and inspiration; and to my Lord, Jesus Christ, who provides for every one of my needs.

Jack H. Wilmore

To my wife, Judy, who has tolerated my professional, sports, and hobby obsessions for nearly four decades; to my wonderful daughters, Jill and Holly, who mean everything to me; and to my former students, who taught me more than I taught them—their subsequent successes have been the highlight of my career.

David L. Costill

CONTENTS

PART I   ESSENTIALS OF MOVEMENT

PART II   ENERGY FOR MOVEMENT

PART III   CARDIOVASCULAR AND RESPIRATORY FUNCTION AND PERFORMANCE

PART IV   ENVIRONMENTAL INFLUENCES ON PERFORMANCE

PART V   OPTIMIZING PERFORMANCE IN SPORT

PART VI   AGE AND SEX CONSIDERATIONS IN SPORT AND EXERCISE

PREFACE

Your body is an amazingly complex machine. All of its various cells and tissues communicate with each other, and their activities are precisely coordinated. When you think of the numerous processes occurring within your body at any given time, it is truly remarkable that all the body systems function so well together. Even as you sit reading this, your heart pumps blood throughout your body, your intestines digest and absorb nutrients, your kidneys clear waste products, your lungs bring in oxygen, and your muscles hold this book while your brain concentrates on reading. Although you may feel at rest, your body is physiologically quite active. Imagine, then, how much more active all your body systems become when you engage in active movement. As your physical activity increases, so does your muscles' physiological activity. Active muscles require more nutrients, more oxygen, more metabolic activity, and thus more efficient clearance of waste products. How does your body respond to the high physiological demands of physical activity?

That is the key question when you study the physiology of sport and exercise, and we answer it in this book. *Physiology of Sport and Exercise, Third Edition*, introduces you to the fields of sport and exercise physiology. Our goal is to build on the foundation of knowledge that you have constructed through basic coursework in human anatomy and physiology by applying the principles you've learned to how the body performs and responds to physical activity.

We begin in the introduction with a historical overview of sport and exercise physiology as they have emerged from the parent disciplines of anatomy and physiology, and we explain basic principles that will be used throughout the text. In parts I through III, we review selected physiological systems, focusing on their responses to acute bouts of exercise, and then we consider how these systems adapt to long-term exposure to exercise—training. In part I, we focus on how the muscular and nervous systems coordinate to produce body movement. In part II, we address how the basic energy systems provide the energy needed for movement and the role of the endocrine system in regulating metabolism. In part III, we look at how the cardiovascular and respiratory systems transport nutrients and oxygen to the active muscles and waste products away from them during physical activity.

We change perspective in part IV to examine the impact of the external environment on physical performance. We consider the body's response to heat and cold, and then we examine the impact of low atmospheric pressure experienced at altitude and high atmospheric pressure experienced during diving. We conclude by considering the effects of a unique

environment—one of little gravity—experienced during space travel.

In part V, we shift our attention to how athletes can optimize physical performance. We evaluate the effects of different amounts of training. We examine athletes' special dietary needs and how nutrition can be used to enhance performance. We consider the importance of appropriate body composition for performance. Finally, we explore the use of ergogenic aids: substances purported to improve athletic ability.

In part VI, we examine unique considerations for specific populations of athletes. We look first at the processes of growth and development and how they affect the performance capabilities of young athletes. We evaluate changes that occur in physical performance as we age and explore the ways that physical activity can prolong our youthfulness. Finally, we examine issues and special physiological concerns of female athletes.

In the final part of the book, part VII, we turn our attention to the application of sport and exercise physiology to prevent and treat various diseases and the use of exercise for rehabilitation. We look at prescribing exercise to maintain health and fitness, and then we close the book with a discussion of cardiovascular disease, obesity, and diabetes.

This third edition of *Physiology of Sport and Exercise* offers a novel approach to the study of sport and exercise physiology. It was designed for the student reader with the goal of making learning easy and enjoyable. This text is comprehensive, but we don't want you to be overwhelmed by either its size or its scope. We have included special features to help you progress through the book.

Key point

Key terms

Review box

neuron, and transmitting it to the next cell. We continue the story with what happens after the neurotransmitter binds with the postsynaptic receptors.

Once the neurotransmitter binds to the receptors, the chemical signal that traversed the synaptic cleft once again becomes an electrical signal. The binding causes a graded potential in the postsynaptic membrane. An incoming impulse may be either excitatory or inhibitory. An excitatory impulse causes a hypopolarization, or depolarization, known as an excitatory postsynaptic potential (EPSP). An inhibitory impulse causes a hyperpolarization, known as an inhibitory postsynaptic potential (IPSP).

The discharge of a single presynaptic terminal generally changes the postsynaptic potential less than 1 mV. Clearly this is not sufficient to generate an action potential, because reaching threshold requires a change of at least 15 to 20 mV. But when a neuron transmits an impulse, several presynaptic terminals typically release their neurotransmitters so that they can diffuse to the postsynaptic receptors. Also, presynaptic terminals from numerous axons can converge on the dendrites and cell body of a single neuron. When multiple presynaptic terminals discharge at the same time, or when only a few fire in rapid succession, more neurotransmitter is released. With an excitatory neurotransmitter, the more that is bound, the greater the EPSP will be.

Triggering an action potential at the postsynaptic neuron depends on the combined effects of all incoming impulses from these various presynaptic terminals. A number of impulses are needed to cause sufficient depolarization to generate an action potential. Specifically, the sum of all changes in the membrane potential must equal or exceed the threshold. This summing of the individual impulses' effects is called summation.

For summation, the postsynaptic cell must keep a running total of the neuron's responses, both EPSPs and IPSPs, to all incoming impulses. This task is done at the axon hillock, which lies on the axon just past the cell body. Only when the sum of all individual graded potentials meets or exceeds threshold can an action potential occur.

The process of summation is very important and has great relevance to muscle function. Human muscle is capable of generating considerable force, far greater than we normally see even in highly trained weightlifters or bodybuilders. Under extreme, life-threatening circumstances, such as an airplane or automobile crash, individuals have been able to exert superhuman levels of force, breaking bones and tearing muscles from bones in the process. The IPSPs protect the muscle–tendon–bone complex under normal conditions. As we show in chapter 3, a reduction in IPSPs most likely is a major factor explaining the gains in strength we experience following a period of resistance training.

Summation refers to the cumulative effect of all individual graded potentials as processed by the axon hillock.

Now that we have carefully considered the function of the most basic units of the nervous system, the neurons, we are ready to examine how these cells work together. Individual neurons are grouped together into bundles. In the central nervous system (brain and spinal cord), these bundles are referred to as tracts or pathways. Neuron bundles in the peripheral nervous system are referred to as nerves.

- ▶ EPSPs are hypopolarizations of the postsynaptic membrane. IPSPs are hyperpolarizations of that membrane.

- ▶ A single presynaptic terminal cannot generate enough of a depolarization to generate an action potential. Multiple signals are needed. These may come from numerous neurons or from a single neuron with numerous axon terminals releasing neurotransmitters repeatedly and rapidly.

- ▶ The axon hillock keeps a running total of all EPSPs and IPSPs. When the total meets or exceeds the threshold for depolarization, an action potential occurs. This process of accumulating incoming signals is known as summation.

At the beginning of each part, you will find brief text that describes the contents of that part's chapters. Each chapter begins with a brief overview and a chapter outline with page numbers to help you locate material. Within a chapter, key points are placed in violet boxes for quick reference. Key terms are highlighted in the text in red, are listed at the end of the chapter, and are defined in the glossary at the end of the book. Review boxes scattered throughout each chapter summarize the major points presented. Boxes marked "fyi" provide interesting physiological facts related to the text.

At each chapter's end, the key terms are listed so you can check your vocabulary comprehension. Study questions allow you to test your knowledge of the chapter's contents.

Central Nervous System

To comprehend how even the most basic stimulus can cause muscle activity, we must next consider the complexity of the central nervous system (CNS). In this section, we present an overview of the components of the central nervous system and their functions.

> fyi The central nervous system houses more than 100 billion neurons.

Brain

The brain is composed of numerous parts. For our purposes, we subdivide it into the four major regions illustrated in figure 2.5: the cerebrum, diencephalon, cerebellum, and brain stem. Let's discuss each briefly.

Cerebrum

The cerebrum is composed of the right and left cerebral hemispheres. These are connected to each other by fiber bundles (tracts) referred to as the *corpus callosum*, which allows the two hemispheres to communicate with each other. The cerebral cortex forms the outer portion of the cerebral hemispheres and has been referred to as the site of the mind and intellect. It is also called the gray matter, which simply reflects its distinctive color resulting from lack of myelin on the cell bodies located in this area. The cerebral cortex is your conscious brain. It allows you to think, to be aware of sensory stimuli, and to voluntarily control your movements.

The cerebrum consists of five lobes—four outer lobes and the central insula, which we will not discuss. Its four outer lobes (see figure 2.5) have the following general functions:

- Frontal lobe: general intellect and motor control
- Temporal lobe: auditory input and its interpretation
- Parietal lobe: general sensory input and its interpretation
- Occipital lobe: visual input and its interpretation

The three areas in the cerebrum that are of primary concern to our discussion and which we discuss later in this chapter are the primary motor cortex, in the frontal lobe; the basal ganglia, in the white matter below the cerebral cortex; and the primary sensory cortex, in the parietal lobe.

Diencephalon

This region of the brain is composed mostly of the thalamus and the hypothalamus. The thalamus is an important sensory integration center. All sensory input (except smell) enters the thalamus and is relayed to the appropriate area of the cortex. The thalamus regulates what sensory input reaches your conscious brain and thus is very important for motor control.

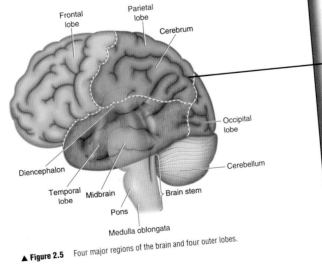

▲ **Figure 2.5** Four major regions of the brain and four outer lobes.

FYI-
For your information

Full-color
anatomical art

References, numbered throughout each chapter, also appear here, as do selected readings, which provide additional information about topics of special interest within that chapter. Finally, at the end of the book you will find a comprehensive glossary that includes definitions of all key terms, a thorough index, and a list of common abbreviations as well as a table of conversions and metric equivalents for easy reference.

Many of you will read this book only because it is a required text for a course. But we hope that the information will entice you to continue to study this relatively new and exciting area. We hope at the very least to further your interest and understanding of your body's marvelous abilities to perform physical work, to adapt to stressful situations, and to improve its physiological capacities. What you learn here is useful and practical not only for anyone who pursues a career in exercise or sport science but also for anyone who wants to be active, healthy, and fit.

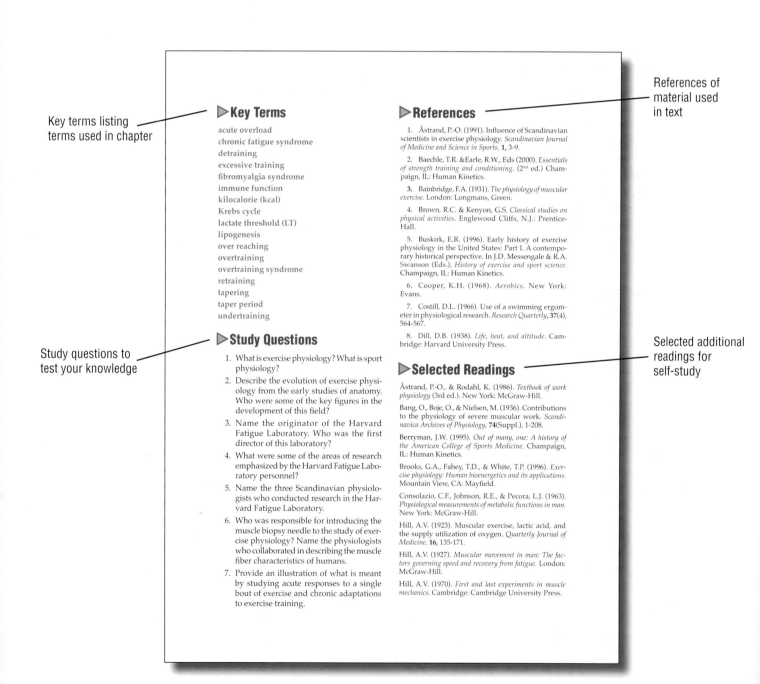

Key terms listing terms used in chapter

▷ **Key Terms**

acute overload
chronic fatigue syndrome
detraining
excessive training
fibromyalgia syndrome
immune function
kilocalorie (kcal)
Krebs cycle
lactate threshold (LT)
lipogenesis
over reaching
overtraining
overtraining syndrome
retraining
tapering
taper period
undertraining

Study questions to test your knowledge

▷ **Study Questions**

1. What is exercise physiology? What is sport physiology?
2. Describe the evolution of exercise physiology from the early studies of anatomy. Who were some of the key figures in the development of this field?
3. Name the originator of the Harvard Fatigue Laboratory. Who was the first director of this laboratory?
4. What were some of the areas of research emphasized by the Harvard Fatigue Laboratory personnel?
5. Name the three Scandinavian physiologists who conducted research in the Harvard Fatigue Laboratory.
6. Who was responsible for introducing the muscle biopsy needle to the study of exercise physiology? Name the physiologists who collaborated in describing the muscle fiber characteristics of humans.
7. Provide an illustration of what is meant by studying acute responses to a single bout of exercise and chronic adaptations to exercise training.

References of material used in text

▷ **References**

1. Åstrand, P.-O. (1991). Influence of Scandinavian scientists in exercise physiology. *Scandinavian Journal of Medicine and Science in Sports*, **1**, 3-9.

2. Baechle, T.R. &Earle, R.W., Eds (2000). *Essentials of strength training and conditioning*. (2nd ed.) Champaign, IL: Human Kinetics.

3. Bainbridge, F.A. (1931). *The physiology of muscular exercise*. London: Longmans, Green.

4. Brown, R.C. & Kenyon, G.S. *Classical studies on physical activities*. Englewood Cliffs, N.J.: Prentice-Hall.

5. Buskirk, E.R. (1996). Early history of exercise physiology in the United States: Part I. A contemporary historical perspective. In J.D. Messengale & R.A. Swanson (Eds.), *History of exercise and sport science*. Champaign, IL: Human Kinetics.

6. Cooper, K.H. (1968). *Aerobics*. New York: Evans.

7. Costill, D.L. (1966). Use of a swimming ergometer in physiological research. *Research Quarterly*, **37**(4), 564-567.

8. Dill, D.B. (1938). *Life, heat, and altitude*. Cambridge: Harvard University Press.

Selected additional readings for self-study

▷ **Selected Readings**

Åstrand, P.-O., & Rodahl, K. (1986). *Textbook of work physiology* (3rd ed.). New York: McGraw-Hill.

Bang, O., Boje, O., & Nielsen, M. (1936). Contributions to the physiology of severe muscular work. *Scandinavica Archives of Physiology*, **74**(Suppl.), 1-208.

Berryman, J.W. (1995). *Out of many, one: A history of the American College of Sports Medicine*. Champaign, IL: Human Kinetics.

Brooks, G.A., Fahey, T.D., & White, T.P. (1996). *Exercise physiology: Human bioenergetics and its applications*. Mountain View, CA: Mayfield.

Consolazio, C.F., Johnson, R.E., & Pecora, L.J. (1963). *Physiological measurements of metabolic functions in man*. New York: McGraw-Hill.

Hill, A.V. (1923). Muscular exercise, lactic acid, and the supply utilization of oxygen. *Quarterly Journal of Medicine*, **16**, 135-171.

Hill, A.V. (1927). *Muscular movement in man: The factors governing speed and recovery from fatigue*. London: McGraw-Hill.

Hill, A.V. (1970). *First and last experiments in muscle mechanics*. Cambridge: Cambridge University Press.

STUDENT AND INSTRUCTOR RESOURCES

Student Resources

Students, visit the **Online Study Guide** at www.humankinetics.com/physiologyofsport andexercise/osg for dynamic and interactive learning activities, all of which can be conducted outside the lab or classroom. You'll be able to apply the key concepts learned by conducting self-made experiments and recording your own physiological responses to exercise. The guide also includes study questions and activities to test your knowledge as you prepare for quizzes or tests. You'll also have access to links to professional journals, organizations, and career information. The Web site format conveniently allows you to practice, review, and develop knowledge and skills about the physiology of sport and exercise via the Internet.

Instructor Resources

Instructor Guide. Specifically developed for instructors of *Physiology of Sport and Exercise, Third Edition,* the Instructor Guide includes sample course syllabi, sample lecture outlines, suggestions for class projects and student assignments, key points, references for the presentation package, laboratory experiences for case studies, and direct links to detailed sources on the Internet for every chapter in the text.

Test Package. The Test Package, created with Respondus 2.0, includes a bank of over 700 questions created especially for *Physiology of Sport and Exercise, Third Edition.* A variety of question types are included, such as true-false,

fill-in-the-blank, essay/short answer, and multiple-choice. With Respondus LE, a free version of the Respondus software, instructors can

create print versions of their own tests by selecting from the question pool;

create, store, and retrieve their own questions; and

select their own test forms, save them for later editing or printing, or

export them into a word processing program.

Respondus also offers the capability to create and manage exams that can be published directly to Blackboard, eCollege, WebCT, and other course management systems. Instructions for downloading a free version of Respondus and for acquiring the test package are at www.humankinetics.com/physiologyofsport andexercise/osg.

The Instructor Guide and Test Package are free to course adopters.

Presentation Package. The CD-ROM Presentation Package includes a comprehensive series of PowerPoint® slides for each chapter. Learning objective slides present the major topics covered in each chapter, text slides list key points, and illustration and photo slides contain the outstanding graphics found in the text. The presentation package has more than 930 slides that can be used directly with PowerPoint® and used to print transparencies, slides, or make copies for distribution to students. Instructors can easily add, modify, or rearrange the order of the slides, as well as search for images based on keywords. You may order the CD-ROM Presentation Package by visiting www.humankinetics.com/ physiologyofsportandexercise.

▶▶▶ ACKNOWLEDGMENTS

We would like to thank Rainer Martens for his continued support of this third edition of *Physiology of Sport and Exercise.* Rainer and the staff at Human Kinetics have been supportive of our many demands and have been extremely dedicated to publishing a quality product. Special recognition must go to Julie Rhoda, our developmental editor for both the second and third editions of this book, who has worked very hard to bring this edition to completion. She has spent countless hours during evenings and on weekends on top of her normal work week to keep the book on schedule while maintaining a quality product. She has also had to work with two "grumpy old men" who were not always considerate of her special needs and the conditions under which she had to work. A very sincere thank you to Julie!

We especially appreciate the efforts of Lawrence Armstrong, Calvin Caplan, James Gale, Dan Halvorsen, Peter Iltis, Anthony Mahon, Donald Michielli, Robert Ruhling, Frans Verstappen, Lawrence Weiss, and Ben B. Yaspelkis, III, who provided us important feedback on the first edition as well as the efforts of Scott Trappe, Howard G. Knuttgen, and the anonymous colleagues who reviewed our more recent work.

Finally, we thank our families, who put up with our many long hours of isolation while we were writing, rewriting, editing, and finally proofing this book. Their patience and support can never be repaid.

Jack H. Wilmore
David L. Costill

PHOTO CREDITS

Figure 0.0s (sidebar on p. 5), 0.6a, 0.7a Photo courtesy of American College of Sports Medicine Archives.

Figure 0.1, 0.2, 0.3a-b, 0.4a-b, 0.5a-c, 0.6b, 0.7b-c, 0.8a-c, 0.9a-b, 0.12, 1.1a-c, 1.10a-b, 1.11, 1.12a-c, 1.13a-b, 3.4a-b, 4.20a-b, 6.1, 6.3a-b, 6.6, 10.3a-b, 17.6, 19.1, 19.2, 19.4a-b, 20.7, 21.9a-c, and back cover author photos: Photo provided courtesy of the authors.

Figure 0.10, 0.11, 3.2, 4.12, 9.4a-b, 12.11, 12.13, and 19.3 © Tom Roberts

Figure 1.4, 2.2a, 7.4a, and 20.3a-b © Custom Medical Stock Photo

Figure 3.1, 3.15, 3.16a-c, 14.05, 14.06, 18.12, 19.5, and 19.8 © Human Kinetics

Figure 3.10, and 3.11a-b From F.C. Hagerman, R.S. Hikida, R.S. Staron, W.M. Sherman, and D.L. Costill, 1984, "Muscle damage in marathon runners," *Physician and Sportsmedicine 12: 39-48.*

Figure 08.0s (sidebar on p. 267) Photographer/Bruce Coleman, Inc.

Figure 09.0s (sidebar on p. 272) and 19.9 © Empics

Figure 11.9 Photo courtesy of Dennis Graver

Figure 11.14 Courtesy of NASA

Figure 14.2a Tom Pantages

Figure 14.3a-b Photo courtesy of Hologic, Inc.

Figure 14.4 Photo courtesy of Life Measurement, Inc.

Figure 18.10a-b Reproduced from J Bone Miner Res 1986, 1:15-21 with permission of the American Society for Bone and Mineral Research. Photos provided courtesy of D.W. Dempster.

Figure 19.10 © 2002 iPhotoNews.com

Introduction: Photo courtesy of Cosmed, Srl; © Empics; Photo provided courtesy of the authors

Chapter 1: © Custom Medical Stock Photo; © Human Kinetics; Photographer/Bruce Coleman, Inc.

Chapter 2: © Custom Medical Stock; © Human Kinetics; © Empics

Chapter 3: © Human Kinetics; © Empics; © Human Kinetics

Chapter 4: © Human Kinetics; © Tom Roberts; © Human Kinetics

Chapter 5: © Empics; © Human Kinetics; © Empics

Chapter 6: © Human Kinetics; © Human Kinetics; Photo courtesy of the authors

Chapter 7: Custom Medical Stock Photo; © Human Kinetics; © Human Kinetics

Chapter 8: Photographer/Bruce Coleman, Inc.; © Human Kinetics; © Human Kinetics

Chapter 9: © Empics; © Human Kinetics; © Human Kinetics

Chapter 10: Photo courtesy of the authors; Photographer/Bruce Coleman, Inc.; © Human Kinetics

Chapter 11: © Empics; © Digital Vision; Courtesy of NASA

Chapter 12: Photographer/Bruce Coleman, Inc.; © Empics; © Human Kinetics

Chapter 13: © Human Kinetics; © Human Kinetics; © Human Kinetics

Chapter 14: Tom Pantages; © Human Kinetics; © Human Kinetics

Chapter 15: © Human Kinetics; © Empics; © Human Kinetics

Chapter 16: © Photographer/Bruce Coleman, Inc.; © Human Kinetics; © Human Kinetics

Chapter 17: © Human Kinetics; © Human Kinetics; © Human Kinetics

Chapter 18: © Human Kinetics ; © Empics; © Human Kinetics

Chapter 19: © Human Kinetics; © 2002 iPhotoNews.com; © Human Kinetics

Chapter 20: © Human Kinetics; © Custom Medical Stock; Jim Whitmer

Chapter 21: © Human Kinetics; © Human Kinetics; Courtesy of Lifescan by Johnson and Johnson

AN INTRODUCTION TO EXERCISE AND SPORT PHYSIOLOGY

overview

The human body is an amazing machine. As you sit reading this introduction, countless perfectly coordinated events are occurring simultaneously in your body. These events allow complex functions, such as hearing, seeing, breathing, and information processing, to continue without your conscious effort. If you stand up, walk out the door, and jog around the block, almost all your body's systems will be called into action, enabling you to successfully shift from rest to exercise. If you continue this routine daily for weeks or months and gradually increase the duration and intensity of your jogging, your body will adapt so that you can perform better.

For centuries, scientists have studied how the human body works. During the past several centuries, a small but growing group of scientists have focused their studies on how the body's functioning, or physiology, is altered during physical activity and sport. This introduction presents a historical overview of exercise and sport physiology and then explains some basic concepts that form the foundation for the chapters that follow.

outline

Much of the history of exercise physiology in the United States can be traced to the effort of a Kansas farm boy, David Bruce ("D.B.") Dill, whose interest in physiology first led him to study the composition of crocodile blood. Fortunately for us, this young scientist redirected his research to humans when he became the first director of the Harvard Fatigue Laboratory, established in 1927. Throughout his life he was intrigued by the physiology and adaptability of many animals who survive extreme environmental conditions; but he is best remembered for his research on human responses to exercise, heat, high altitude, and other environmental factors. Dr. Dill always served as one of the human "guinea pigs" in these studies. During the Harvard Fatigue Laboratory's 20-year existence, he and his coworkers produced 330 scientific papers along with a classic book titled *Life, Heat, and Altitude*.[8]

After the Harvard Fatigue Laboratory closed its doors in 1947, he began a second career as deputy director of medical research for the Army Chemical Corps, a position he held until his retirement from that post in 1961. Dr. Dill was then 70 years old—an age he considered too young for retirement—so he moved his exercise research to Indiana University, where he served as a senior physiologist until 1966. In 1967 he obtained funding to establish the Desert Research Laboratory at the University of Nevada at Las Vegas. Dr. Dill used this laboratory as a base for his studies on human tolerance to exercise in the desert and at high altitude. He continued his research and writing until his final retirement at age 93, the same year he produced his last publication, a book titled *The Hot Life of Man and Beast*.[10] Dr. Dill once bragged to me (DLC) that he was the only scientist to have retired four times.

As the busy executive goes out for his morning jog and the point guard directs her team down the basketball court on a fast break, their bodies must make many adjustments that require a series of complex interactions involving many body systems. Consider a few examples:

- The skeletal system provides the basic framework through which muscles act.
- The cardiovascular system delivers nutrients to the body's various cells and removes waste products.
- The cardiovascular and respiratory systems together provide oxygen to the cells and remove carbon dioxide.
- The integumentary system (skin) helps maintain body temperature by allowing the exchange of heat between the body and its surroundings.
- The urinary system helps maintain fluid and electrolyte balance and assists in the long-term regulation of blood pressure.
- The nervous and endocrine systems coordinate and direct all this activity to meet the body's needs.

Adjustments occur even at the cellular and molecular levels. For example, to enable your biceps muscle to contract to lift a 20-kg (44-lb) weight, nerve cells from the brain, referred to as motor neurons, conduct electrical impulses down the spinal cord to the arm. On reaching the biceps muscle, these neurons release chemical messengers that cross the gap between the nerve and muscle, each neuron exciting a number of individual muscle cells or fibers. Once the nerve impulses cross this gap, they spread along the length of each muscle fiber, entering into the muscle fiber through small pores. Once inside the muscle fiber, the impulse activates the muscle fiber's contraction processes, which involve specific protein molecules—actin and myosin—and an elaborate energy system to provide the fuel necessary to sustain a single contraction and subsequent contractions. It is at this level that other molecules, such as adenosine triphosphate (ATP) and phosphocreatine (PCr), become critical for providing the energy necessary to fuel contraction.

Physical activity is a complicated process. Scientists must examine each adjustment that the body makes by observing these events both singly and collectively. Here we discuss how scientists approach this task.

Focus of Exercise and Sport Physiology

Exercise and sport physiology have evolved from anatomy and physiology. Anatomy is the study of an organism's structure, or morphology. From anatomy, we learn the basic structure of various body parts and their interrelationships. Physiology is the study of body function. In physiology, we study how our organ systems, tissues, cells, and molecules within cells work and how their functions are integrated to regulate our internal environments. Because physiology focuses on the functions of structures, we can't easily discuss physiology without understanding anatomy.

Exercise physiology is the study of how our bodies' structures and functions are altered when we are exposed to acute and chronic bouts of exercise. **Sport physiology** further applies the concepts of exercise physiology to training the athlete and enhancing the athlete's sport performance. Thus, sport physiology is derived from exercise physiology.

Exercise physiology has evolved from its parent discipline, physiology. It is concerned with the study of how the body adapts physiologically to the acute stress of exercise, or physical activity, and the chronic stress of physical training. Although a number of physiological studies were performed in the mid-1800s, it might be argued that exercise physiology as we know it today was an outgrowth of sport physiology, because many of the earliest physiological studies focused on athletes and sport performance.

Let's consider an example to help us distinguish between these two closely related branches of physiology. Through considerable research in exercise physiology, we have come to better understand how our bodies derive energy from the foods we eat to permit muscle actions to initiate and sustain movement. We have learned

that fat is our major energy source when we are at rest and during low-intensity exercise but that our bodies use proportionately more carbohydrate as exercise intensity increases, until carbohydrate becomes our primary energy source. Prolonged high-intensity exercise can substantially reduce our bodies' carbohydrate stores, which can contribute to exhaustion.

Recognizing that the body has limited carbohydrate energy stores, sport physiology uses this information about energy to find ways to

- increase the body's carbohydrate storage capacity (carbohydrate loading),
- decrease the rate at which the body uses carbohydrate during physical performance (carbohydrate sparing), and
- improve the athlete's diet both before and during competition to minimize the risk of depleting carbohydrate stores.

The area of sport nutrition, a subdiscipline of sport physiology, is one of the most rapidly growing areas of research in this field.

Similarly, exercise physiology has uncovered an important sequence of events that occur when the body is trained beyond its ability to adapt, a condition known as overtraining. Sport physiology has applied this information to both the design and the evaluation of training programs to reduce the risk of overtraining.

But sport physiology is not merely applied exercise physiology. Because exercise physiology also has its own applications, it is often hard to clearly distinguish the two. For this reason, exercise and sport physiology are often considered together, as they are in this text. Now let's look at how exercise physiology, the parent discipline of sport physiology, has evolved through the years.

Historical Events

It is tempting to think that the information in this book is new and contains the final word on each topic. Contributions of contemporary exercise physiologists might seem to present new ideas never before treated to the rigors of science, but this is not the case. Rather, the information we'll explore represents the lifelong efforts of many outstanding scientists who have helped piece together the puzzle of human movement. The thoughts and theories of today's physiological detectives often were shaped by the efforts of scientists who may be long forgotten. What we consider as original or new is most often an assimilation of previous findings or the application of basic science to problems in exercise physiology. There are, of course, a number of pivotal scientific findings and investigators who have made significant leaps in our knowledge of the physiological and biochemical responses to physical activity. To help you appreciate this, let's briefly reflect on the history and people who have shaped the field of exercise physiology.

Beginnings of Anatomy and Physiology

One of the earliest attempts to explain human anatomy and physiology was Claudius Galen's Greek text *De fascius*, published in the first century. As a physician to the gladiators, Galen had ample opportunity to study and experiment on human anatomy. His theories of anatomy and physiology were so widely accepted that they remained unchallenged for nearly 1,400 years. Not until the 1500s were any truly significant contributions made to the understanding of both the structure and function of the human body. A landmark text by Andreas Vesalius, titled *Fabrica Humani Corporis* [Structure of the Human Body], presented his findings on human anatomy in 1543. Although Vesalius' book focused primarily on anatomical descriptions of various organs, the book occasionally attempted to explain their functions as well. British historian Sir Michael Foster said, "This book is the beginning, not only of modern anatomy, but of modern physiology. It ended, for all time, the long reign of fourteen centuries of precedent and began in a true sense the renaissance of medicine" (p. 354).[13]

Most early attempts at explaining physiology were either incorrect or so vague that they could be considered no more than speculation. Attempts to explain how a muscle generates force, for example, were usually limited to a description of its change in size and shape during action because observations were limited to what could be seen with the eye. From such observations, Hieronymus Fabricius (ca. 1574) suggested that a muscle's contractile power resided in its fibrous tendons, not in its "flesh." Anatomists didn't discover the existence of individual muscle fibers until Dutch scientist Anton van Leeuwenhoek introduced the microscope (ca. 1660). But how these fibers shortened and created force remained a mystery until the middle of the 20th century, when the intricate workings of muscle proteins could be studied by electron microscopy.

Historical Aspects of Exercise Physiology

Exercise physiology is a relative newcomer to the world of science, although muscular activity played an interesting role in a physiological study as early as 1793. A celebrated paper by Séguin and Lavoisier describes the oxygen consumption of a young man as measured in the resting state and while he lifted a weight of 7.3 kg numerous times to a total height of 200 m in 15 min. At rest the man used 24 L of oxygen per hour (L/h), which increased to 63 L/h during exercise. Lavoisier believed that the site of oxygen utilization and carbon dioxide production was in the lungs. This belief was doubted by other physiologists but remained accepted doctrine until the middle of the 1800s, when several German physiologists demonstrated that combustion occurred in tissues throughout the entire body.

Although advances were made in the understanding of circulation and respiration during the 1800s, few efforts were made to focus on the physiology of physical activity. However, in 1888, an apparatus was described that enabled scientists to study subjects during mountain climbing, even though the subjects had to carry a 7-kg gasometer on their backs.[22]

The first published textbook on exercise physiology, *Physiology of Bodily Exercise*, was written by Fernand LaGrange in 1889.[16] Considering the small amount of research on exercise that had been conducted at that time, it is intriguing to read the author's accounts of such topics as "Muscular Work," "Fatigue,"

"Habituation to Work," and "The Office of the Brain in Exercise." This early attempt to explain the body's response to exercise was, in many ways, limited to a lot of rambling theory and little fact. Although some basic concepts of exercise biochemistry were emerging at that time, LaGrange was quick to admit that many details were still in the formative stages. For example, he stated that "vital combustion [energy metabolism] has become very complicated of late; we may say that it is somewhat perplexed, and that it is difficult to give in a few words a clear and concise summary of it. It is a chapter of physiology which is being rewritten, and we cannot at this moment formulate our conclusions" (p. 395).[16]

Since the early text by LaGrange offered only limited physiological insights regarding bodily functions during physical activity, it might be argued that the third edition of a text by F.A. Bainbridge entitled *The Physiology of Muscular Exercise* be considered the earliest scientific text on this subject.[3] Interestingly, that third edition was written by A.V. Bock and D.B. Dill, at the request of A.V. Hill, three men we discuss in this chapter.

> The first published textbook on exercise physiology was an 1889 work by Fernand LaGrange titled *Physiology of Bodily Exercise*.

A.V. Hill

October 16, 1923, was a significant milestone in the history of exercise physiology. A.V. Hill (figure 0.1) was inaugurated that day as Joddrell Professor of Physiology at University College, London. In his inaugural address he stated the principles that subsequently influenced the field of exercise physiology:

▲ **Figure 0.1** 1921 Nobel Prize winner Archibald Hill (1927).

It is strange how often a physiological truth discovered on an animal may be developed and amplified, and its bearings more truly found, by attempting to work it out on man. Man has proved, for example, far the best subject for experiments on respiration and on the carriage of gases by the blood, and an excellent subject for the study of kidney, muscular, cardiac and metabolic function . . . . Experiment on man is a special craft requiring a special understanding and skill, and "human physiology," as it may be called, deserves an equal place in the list of those main roads which are leading to the physiology of the future. The methods, of course, are those of biochemistry, of biophysics, of experimental physiology; but there is a special kind of art and knowledge required of those who wish to make experiments on themselves and their friends, the kind of skill that the athlete and the mountaineer must possess in realizing the limits to which it is wise and expedient to go.

Quite apart from direct physiological research on man, the study of instruments and methods applicable to man, their standardization, their description, their reduction to routine, together with the setting up of standards of normality in man are bound to prove of great advantage to medicine; and not only to medicine but to all those activities and arts where normal man is the object of study. Athletics, physical training, flying, working, submarines, or coalmines, all require a knowledge of the physiology of man, as does also the study of conditions in factories. The observation of sick men in hospitals is not the best training for the study of normal man at work. It is necessary to build up a sound body of trained scientific opinion versed in the study of normal man, for such trained opinion is likely to prove of the greatest service, not merely to medicine, but in our ordinary social and industrial life. Haldane's unsurpassed knowledge of the human physiology of respiration has often rendered immeasurable service to the nation in such activities as coal-mining or diving; and what is true of the human physiology of respiration is likely also to be true of many other normal human functions.

During the late 1800s, many theories were proposed to explain the source of energy for muscle contraction. Muscles were known to generate much heat during exercise, so some theories suggested that this heat was used directly or indirectly to cause muscle fibers to shorten. After the turn of the century, Walter Fletcher and Sir Frederick Gowland Hopkins observed a close relationship between muscle action and lactate formation.[11] This observation led to the realization that energy for muscle action is derived from the breakdown of muscle glycogen to lactic acid (chapter 4), although the details of this reaction remained obscure.

Because the energy demands for muscle action are high, this tissue served as an ideal model to help unravel the mysteries of cellular metabolism. In 1921, Archibald V. (A.V.) Hill was awarded the Nobel Prize for his findings on energy metabolism. At that time, biochemistry was in its infancy, although it was rapidly gaining recognition through the research efforts of such Nobel laureates as Albert Szent Gorgyi, Otto Meyerhof, August Krogh, and Hans Krebs, who were all actively studying how living cells generate energy.

Although much of Hill's research was conducted with isolated frog muscle, he also conducted some of the first physiological studies of runners. Such studies were possible through the technical contributions of John S. Haldane, who developed the methods and equipment needed to measure oxygen use during exercise. These and other investigators provided the basic framework for our understanding of whole-body energy production, which became the focus of considerable research during the middle of this century and is incorporated into computer-based systems used to measure oxygen uptake in exercise physiology laboratories today.

Harvard Fatigue Laboratory

No laboratory has had more impact on the field of exercise physiology than the Harvard Fatigue Laboratory (HFL). A visit by A.V. Hill to Harvard University in 1926 appeared to have a significant impact on the founding and activities of the HFL, which was initiated a year later. Creation of this laboratory is credited to the insightful planning of world-famous biochemist and Nobel Laureate, Lawrence J. Henderson. Because he was not interested in heading the laboratory's research program, he chose a young biochemist from Stanford University, David Bruce "D.B." Dill (see figure 0.2) as the first director.

As noted earlier, Dill aided A.V. Bock in writing the third edition of Bainbridge's text on exercise physiology. Later in his career he was quoted as crediting the writing of that textbook with "shaping the program of the Fatigue Laboratory." Although he had little experience in applied human physiology, Dill's creative thinking and ability to surround himself with young, talented scientists created an environment that would lay the foundation for modern exercise and environmental physiology. For example, HFL personnel examined the physiology of endurance exercise and described the physical requirements for success in events such as distance running. Some of the most outstanding HFL investigations were conducted not in the laboratory but in the Nevada desert, on the Mississippi Delta, and on White Mountain in California (altitude 3,962 m, or 13,000 ft). These and other studies provided the foundation for future investigations on the effects of the environment on physical performance and human physiology.

In its earlier years, the HFL focused primarily on general problems of exercise, nutrition, and health. For example, studies on exercise

▲ **Figure 0.2**   David Bruce (D.B.) Dill as director of the Harvard Fatigue Laboratory.

and aging were first conducted in 1939 by Sid Robinson (see figure 0.3), a student at the HFL. Based on his studies of subjects from 6 to 91 years old, Robinson described the effect of aging on maximal heart rate and oxygen uptake.[19] But with the onset of World War II, the HFL took a different direction. Henderson and Dill realized the HFL's potential contribution to the war effort. They and other HFL personnel were instrumental in forming new laboratories for the Army, Navy, and Army Air Corps (what is now the Air Force). They also published the methodologies necessary for relevant military research; those methods are still in use throughout the world.

Today's exercise physiology students would be amazed by the technology used in the early days of the HFL and by the time and energy committed to early research. What is now accomplished in mere milliseconds with the aid of computers and automatic analyzers demanded days of effort by HFL personnel. Measurements of oxygen uptake during exercise, for example, required collecting expired air in Douglas bags and analyzing it for oxygen and carbon dioxide by using a manually operated chemical analyzer, without the help of a computer, of course (see figure 0.4 on page 8). The analysis of a single 1-min sample of air required 20 to 30 min of effort by one or more laboratory personnel. Today, laboratory personnel make such measurements almost instantaneously with little effort. We must marvel at the dedication of the HFL's pioneers of knowledge.

The HFL was an intellectual center that attracted young physiologists from many places. Scholars from 15 countries worked in the HFL between 1927 and its closure in 1947. Most went on to develop their own laboratories and to become noteworthy international figures in exercise physiology. Thus, the HFL planted seeds of intellect around the world that resulted in an explosion of knowledge and interest in this new field.

Founded by biochemist L.J. Henderson in 1927 and directed by D.B. Dill through its closure in 1947, the Harvard Fatigue Laboratory (HFL) trained most of those who became world leaders in exercise physiology during the 1950s and 1960s. Most contemporary exercise physiologists can trace their roots back to the HFL.

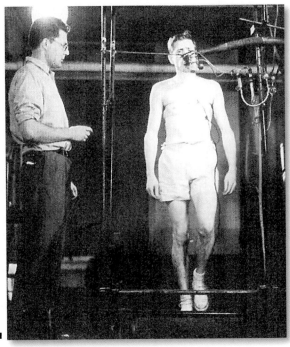

▲ **Figure 0.3** Sid Robinson being tested by R.E. Johnson on the treadmill in the Harvard Fatigue Laboratory (left) and as a Harvard Student and athlete (right) in (1938).

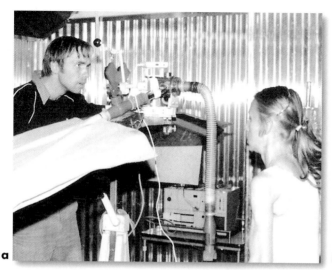

▲ **Figure 0.4**   Early measurements of metabolic responses to exercise required the collection of expired air in a sealed bag known as a Douglas bag (*a*). A sample of that gas then was measured for oxygen and carbon dioxide using a chemical gas analyzer, as illustrated by this photo (*b*) of Nobel Laureate August Krogh.

Scandinavian Influence

As a result of contacts between D.B. Dill and August Krogh, a Danish Nobel Prize winner, three young Danish physiologists came to the HFL in the 1930s. Krogh encouraged Erik Hohwü-Christensen, Erling Asmussen, and Marius Nielsen to spend time at Harvard studying exercise in the heat and at high altitude. After returning to Scandinavia, each man established a separate line of research. Asmussen and Nielsen became professors at the University of Copenhagen, where Asmussen studied the mechanical properties of muscle and Nielsen conducted studies on body temperature control. Both remained active at the University of Copenhagen's August Krogh Institute until their retirements.

In 1941, Hohwü-Christensen (see figure 0.5) moved to Stockholm to become the first physiology professor at the College of Physical Education at Gymnastik-och Idrottshögskolan (GIH). In the late 1930s, he teamed with Ole Hansen to conduct and publish a series of five studies of carbohydrate and fat metabolism during exercise. These studies still are cited frequently and are considered among the first and most important sport nutrition studies. Hohwü-Christensen introduced Per-Olof (P.-

O.) Åstrand to the field of exercise physiology. Åstrand, who conducted numerous studies related to physical fitness and endurance capacity during the 1950s and 1960s, became the director of GIH after Hohwü-Christensen retired in 1960. During his tenure at GIH, Hohwü-Christensen mentored a number of outstanding scientists, including Bengt Saltin, who was the 2002 Olympic Prize winner for his many contributions to the field of exercise and clinical physiology (see figure 0.5*b*).

In addition to their work at GIH, both Hohwü-Christensen and Åstrand interacted with physiologists at the Karolinska Institute in Stockholm who studied clinical applications of exercise. It is hard to single out the most exceptional contributions from this institute, but Jonas Bergstrom's reintroduction of the biopsy needle (ca. 1966) to sample muscle tissue was a pivotal point in the study of human muscle biochemistry and muscle nutrition (figure 0.5*c*). This technique, which involves withdrawing a tiny sample of muscle tissue through a small incision, was first introduced in the early 1900s to study muscular dystrophy. The needle biopsy enabled physiologists to conduct histological and biochemical studies of human muscle before, during, and after exercise.

▲ **Figure 0.5**   Erik Hohwü-Christensen (*a*) was the first physiology professor at the College of Physical Education at Gymnastik-och Idrottshögskolan in Stockholm, Sweden. Bengt Saltin (*b*), winner of the 2002 Olympic Prize. (*c*) Jonas Bergstrom (left) and Eric Hultman (right) were the first to use the muscle biopsy to study muscle glycogen use and restoration before, during, and after exercise.

Other invasive studies of blood circulation were subsequently conducted by physiologists at GIH and at the Karolinska Institute. Just as the HFL had been the mecca of exercise physiology research between 1927 and 1947, the Scandinavian laboratories have been equally noteworthy since the late 1940s. Many leading investigations during the past 35 years were collaborations between American and Scandinavian exercise physiologists. (For a more detailed listing of the Scandinavian contributions to exercise physiology, consult Åstrand's review.[1])

Exercise Physiology and Other Fields

Physiology has always been the basis for clinical medicine. In the same manner, exercise physiology has provided essential knowledge for many other areas, such as physical education, physical fitness, physical therapy, and health promotion. In the late 1800s and early 1900s, physicians such as Edward Hitchcock, Jr. (Amherst College) and Dudley Sargent (Harvard University) studied body proportions (anthropometry) and the effects of physical training on strength and endurance. Although a number of physical educators introduced science to the undergraduate physical education curriculum, Peter Karpovich, a Russian immigrant who had been briefly associated with the HFL, played a major role in introducing physiology to physical education. Karpovich (see figure 0.6a) established his own research facility and taught physiology at Springfield College (Massachusetts) from 1927 until his death in 1968. Although he made numerous contributions to physical education and exercise physiology research, he is best remembered for the outstanding students he advised, including Charles Tipton and Loring Rowell, both recipients of the American College of Sports Medicine Honor and Citation Awards.

Another Springfield faculty member, swim coach T.K. Cureton (see figure 0.6b), created an exercise physiology laboratory at the University of Illinois in 1941. He continued his research and taught many of today's leaders in physical fitness and exercise physiology until his retirement in 1971. Physical fitness programs developed by Cureton and his students and Kenneth Cooper's 1968 book, *Aerobics*, established a physiological rationale for using exercise to promote a healthy lifestyle.[6]

▲ **Figure 0.6** Peter Karpovich (a) introduced the field of exercise physiology during his tenure at Springfield College. Thomas K. Cureton (b) directed the exercise physiology laboratory at the University of Illinois at Urbana–Champaign from 1941 to 1971.

Although there was some awareness as early as the mid-1800s of a need for regular physical activity to maintain optimal health, this idea did not gain popular acceptance until the late 1960s. Subsequent research continues to support the importance of exercise in resisting the physical decline associated with aging.

Awareness of the need for physical activity has alerted the public to the importance of preventive medicine and the establishment of wellness programs. Although exercise physiology cannot be given credit for the current wellness movement, it provided the basic knowledge and justification for including exercise as an integral component of a healthy lifestyle and laid the foundation for the science of exercise prescription in both sickness and health.

Contemporary Exercise and Sport Physiology

Much advancement in exercise physiology must be credited to improvements in technology. For example, in the 1960s, development of electronic analyzers to measure respiratory gases made studying energy metabolism much easier and more productive than before. This technology and radiotelemetry (which uses radio-transmitted signals), used to monitor heart rate and body temperature during exercise, were developed as a result of the U.S. space program. Although such instruments took the labor out of research, they did not alter the direction of scientific inquiry. Until the late 1960s, most exercise physiology studies focused on the whole body's response to exercise. The majority of investigations involved measurements of such variables as oxygen uptake, heart rate, body temperature, and sweat rate. Cellular responses to exercise received little attention.

In the mid-1960s, three biochemists emerged who were to have a major impact on the field of exercise physiology. John Holloszy (figure 0.7a) at Washington University (St. Louis), Charles "Tip" Tipton (figure 0.7b) at the University of Iowa, and Phil Gollnick (figure 0.7c) at Washington State University first used rats and mice to study muscle metabolism and to examine factors related to fatigue. Their publications and training of graduate and postdoctoral students have resulted in a more biochemical approach to exercise physiology research. Holloszy was ultimately awarded the 2000 Olympic Prize for his contributions to exercise physiology and health.

At about the time that Bergstrom reintroduced the needle biopsy procedure, exercise physiologists who were well trained as biochemists emerged. In Stockholm, Bengt Saltin realized the value of this procedure for studying human muscle structure and biochemistry. He first collaborated with Bergstrom in the late 1960s to study the effects of diet on muscle endurance and muscle nutrition. About the same time, Reggie Edgerton (University of California, Los Angeles) and Phil Gollnick were using rats to study the characteristics of individual muscle fibers and their responses to training. Saltin subsequently combined his knowledge of the biopsy procedure with Gollnick's biochemical talents. These researchers were responsible for many early studies on human muscle fiber characteristics and use during exercise. Although many biochemists have used exercise to study metabolism, few have had more impact on the current direction of human exercise physiology than Bergstrom, Saltin, Tipton, Holloszy, and Gollnick.

For more than 100 years athletes have served as subjects to study the upper limits of human endurance. Perhaps the first physiological studies on athletes occurred in 1871. Austin Flint studied one of the most celebrated athletes of that era, Edward Payson Weston, an endurance runner/walker. Flint's investigation involved measuring Weston's energy balance (i.e., food intake vs. energy expenditure) during Weston's attempt to walk 400 mi in 5 days. Although the study resolved few questions about muscle metabolism during exercise, it did demonstrate

▲ **Figure 0.7** John Holloszy (*a*), winner of the 2000 Olympic Prize for scientific contributions in the field of exercise science. Charles Tipton (*b*) was a professor at the University of Iowa and the University of Arizona, and mentor to many students who have become the leaders in molecular biology and genomics. Phil Gollnick (*c*) conducted muscle and biochemical research at Washington State University.

Women in Exercise Physiology

As in many areas of science, the contributions of female exercise physiologists have been slow to gain recognition. In 1954, Irma Rhyming collaborated with her future husband, P.-O. Åstrand, to publish a classic study that provided a means to predict aerobic capacity from submaximal heart rate. Although this indirect method of assessing physical fitness has been challenged over the years, its basic concept is still in use today.

In the 1970s, two Swedish women, Birgitta Essen and Karen Piehl, gained international attention for their research on human muscle fiber composition and function. Essen (figure 0.8a), who collaborated with Bengt Saltin, was instrumental in adapting microbiochemical methods to study the small amounts of tissue obtained with the needle biopsy procedure. Her efforts enabled others to conduct studies on the muscle's use of carbohydrates and fats and to identify different muscle fiber types. Piehl (see figure 0.8b) published a number of studies that illustrated which muscle fiber types were activated during both aerobic and anaerobic exercise.

In the 1970s and 1980s, a third Scandinavian female physiologist, Bodil Nielsen, daughter of Marius Nielsen, actively conducted studies on human responses to environmental heat stress and dehydration. Her studies even encompassed measurements of body temperature during immersion in water. Interestingly, at about the same time an American female exercise physiologist, Barbara Drinkwater (see figure 0.8c), was doing similar work at the University of California, Santa Barbara. Her studies were often conducted in collaboration with Steven Horvath, D.B. Dill's son-in-law and director of the Environmental Physiology Laboratory. Drinkwater's contributions to environmental physiology and the physiological problems confronting the female athlete gained international recognition. In addition to their scientific contributions, the legacy of these and other women in physiology is the credibility they earned and roles they played in attracting other young women to the field of exercise physiology and medicine.

▲ **Figure 0.8** Birgitta Essen (*a*) collaborated with Bengt Saltin and Phil Gollnick in publishing the earliest studies on muscle fiber types in human muscle. Karen Piehl (*b*) was among the first physiologists to demonstrate that the nervous system selectively recruits slow- and fast-twitch fibers during exercise of differing intensities. Barbara Drinkwater (*c*) was among the first to conduct studies on female athletes and to address issues specifically related to the female athlete.

that some body protein is lost during prolonged heavy exercise.[12]

Throughout the 20th century, athletes were used repeatedly to assess the physiological capabilities of human strength and endurance and to ascertain the characteristics needed for record-setting performances. Some attempts have been made to use the technology and knowledge derived from exercise physiology to predict performance, to prescribe training, or to identify athletes with exceptional potential. In most cases, however, these applications of physiological testing are of little more than academic interest because few laboratory or field tests can accurately assess all the qualities required to become a champion.

The intent of this section has been to provide you with an overview of the personalities and technologies that have helped to shape the field of exercise physiology. Naturally, a comprehensive review of all the scientists and research associated with this field is unwarranted for a text intended to introduce you to exercise physiology, but for those students who wish to gain an in-depth look at the historical background in exercise physiology, there are several good sources.

In 1968, D.B. Dill wrote a chapter, "History of the Physiology of Exercise," which detailed many of the events and scientists who contributed to this field before the founding of the HFL[9]. In that same year, Roscoe Brown, Jr., the first African-American exercise physiologist, coauthored *Classical Studies on Physical Activity* [4]. Although the authors subjectively selected those scientific studies they considered worthy of publication, the edited book provides an excellent sampling of important exercise physiological research from the early 1900s.

In the early 1970s, D.B. Dill's son-in-law and daughter (Steven and Betty Horvath) published a detailed history of HFL's history, which included the laboratory and field studies conducted by the key scientists of that era.[15] Although others have written different versions of exercise physiology history,[5, 21] most tend to provide the authors' views of important scientists and events, perhaps as we have done here. Finally, McArdle, Katch, and Katch[18] published one of the most comprehensive reviews

of the evolution of exercise physiology. Their description of the early anatomists, physiologists, and exercise physiologists clearly illustrates the complexity and diversity of this field of science.

Now that we understand the historical basis for the discipline of exercise physiology, from which sport physiology emerged, we can explore the scope of exercise and sport physiology.

Acute Physiological Responses to Exercise

When you begin your study of exercise and sport physiology, you must first learn how the body responds to an individual bout of exercise, such as running on a treadmill. This response is called an **acute response.** You can then better understand the **chronic adaptations** that the body makes when it is challenged with repeated bouts of exercise, such as changes in cardiovascular function after 6 months of endurance training. In the following sections, we address basic concerns, concepts, and principles associated with both acute responses to exercise and chronic adaptations to training. This material is vital to your understanding of much of the material in later chapters.

How are physiological responses to exercise determined? Neither the highly skilled athlete nor the recreational jogger performs her daily runs in environments that permit extensive physiological monitoring. Few selected physiological variables can be monitored during exercise in the field, but some can be assessed accurately without disrupting performance. For example, radiotelemetry and miniature tape recorders can be used during activity to monitor

- heart activity (heart rate and electrocardiogram),
- respiration rate,
- ventilation,
- oxygen consumption,
- skin and deep body temperature, and
- muscle activity (electromyogram).

Recent developments even allow direct monitoring of oxygen consumption during free-ranging activity outside the confines of the research laboratory. Unfortunately, participants most often must be evaluated in the laboratory, where they can be studied more thoroughly and under tightly controlled conditions.

Factors to Consider During Monitoring

Many factors can alter the body's acute response to a bout of exercise. For example, environmental conditions must be carefully controlled. Factors such as the temperature and humidity of the laboratory and the amount of light and noise in the test area can markedly affect the body's response, both at rest and during exercise. Even the time and size of the subject's last meal and the quantity and quality of sleep the night before must be controlled.

To illustrate this, table 0.1 shows how varying environmental and behavioral factors can alter heart rate at rest and during running on a treadmill at 14 km/h (9 mph). The subject's heart rate response during exercise differed by 25 beats/min when the temperature was increased from 21° to 35° C (70°-95° F). Most physiological variables normally assessed during exercise are similarly influenced by environmental fluctuations. Whether you are comparing a person's test results from different days or comparing one person's results to another's, these factors must be carefully controlled.

Physiological responses, both at rest and during exercise, also vary throughout the day. The term **diurnal variation** refers to fluctuations that occur during a 24-h day. Table 0.2 illustrates this variation for heart rate at rest, during various levels of exercise, and during recovery. Deep body (rectal) temperature shows similar fluctuations throughout the day. As seen in table 0.2, tests of the same person in the morning on one day and in the afternoon on the next can produce quite different results. Test times must be standardized to control for this diurnal effect.

At least one other cycle must also be considered. The normal 28-day menstrual cycle often involves considerable variations in

- body weight,
- total body water,
- body temperature,
- metabolic rate,
- heart rate,
- stroke volume (the amount of blood leaving the heart with each contraction), and
- mood state.

The scientist must control for these variables when testing women. Testing should always be conducted at the same point in the menstrual cycle when possible.

> Conditions under which research participants are monitored, at rest and during exercise, must be carefully controlled. Environmental factors, such as temperature, humidity, altitude, and noise, can affect the magnitude of response of all basic physiological systems, as can behavioral factors such as eating patterns and sleep. Likewise, monitoring must be controlled for diurnal and menstrual cycle variations.

Use of Ergometers

When physiological responses to exercise are assessed in a laboratory setting, the participant's physical effort must be controlled to provide a constant and known rate of work. This is generally accomplished by using ergometers. An **ergometer** (*ergo* = work, *meter* = measure) is an exercise device that allows the amount and rate of a person's physical work to be controlled (standardized) and measured. Let's consider some examples.

Cycle Ergometers

For many years, the **cycle ergometer** was the primary testing device in use. It is still used extensively in both research and clinical settings, although the trend in the United States

Table 0.1

Variations in Heart Rate Response to Running at 14 km/h on a Treadmill With Environmental and Behavioral Alterations

Environmental and behavioral factors	Heart rate (beats/min)	
	Rest	Exercise
Temperature (50% humidity)		
21° C (70° F)	60	165
35° C (95° F)	70	190
Humidity (21° C)		
50%	60	165
90%	65	175
Noise level (21° C, 50% humidity)		
Low	60	165
High	70	165
Food intake (21° C, 50% humidity)		
Small meal 3 h before exercising	60	165
Large meal 30 min before exercising	70	175
Sleep (21° C, 50% humidity)		
8 h or more	60	165
6 h or less	65	175

Table 0.2

An Example of Diurnal Variations in Heart Rate at Rest and During Exercise

	Time of day					
	2 A.M.	6 A.M.	10 A.M.	2 P.M.	6 P.M.	10 P.M.
Condition	Heart rate (beats/min)					
Resting	65	69	73	74	72	69
Light exercise	100	103	109	109	105	104
Moderate exercise	130	131	138	139	135	134
Maximal exercise	179	179	183	184	181	181
Recovery, 3 min	118	122	129	128	128	125

Data from T. Reilly, and G.A. Brooks, 1990, "Selective persistence of circadian rhythms in physiological responses to exercise," *Chronobiology International* 7: 59-67.

Tools of Exercise Physiology Research

The history of exercise physiology has, in some ways, been driven by advancements in technologies adapted from basic sciences. The early studies of energy metabolism during exercise were made possible by the invention of gas-collecting equipment and chemical analysis of oxygen and carbon dioxide. Chemical determination of blood lactic acid seemed to provide some insights regarding the aerobic and anaerobic aspects of muscular activity, but these data told us little regarding the production and removal of this by-product of exercise. Likewise, blood glucose measurements taken before, during, and after exhaustive exercise proved to be interesting data but were of limited value in understanding the energy exchange at the cellular level.

Prior to the 1960s, there were few biochemical studies on the adaptations of muscle to training. Although the field of biochemistry can be traced to the early part of the 20th century, this special area of chemistry was not applied to human muscle until Bergstrom and Hultman reintroduced and popularized the needle biopsy procedure in 1966. Initially, this procedure was used to examine glycogen depletion during exhaustive exercise and its resynthesis during recovery. In the early 1970s, as noted earlier, a number of exercise physiologists used the muscle biopsy method, histochemical staining, and the light microscope to determine human muscle fiber types.

Over the last 30 years, muscle physiologists have used various chemical procedures to understand how muscles generate energy and adapt to training. Test tube experiments (in vitro) with muscle biopsy samples have been used to measure muscle proteins (enzymes) and to determine the muscle fiber's capacity to use oxygen. Although these studies provided a snapshot of the fiber's potential to generate energy, they often left more questions than answers. It was natural, therefore, that the sciences of cell biology move to an even deeper level. It was apparent that the answers to those questions must lie within the fiber's molecular makeup.

Although not a new science, molecular biology has become a useful tool for exercise physiologists who wish to delve deeper into the cellular regulation of metabolism and adaptations to the stress of exercise. Physiologists like Frank Booth and Ken Baldwin (see figure 0.9) have dedicated their careers to understanding the molecular regulation of muscle fiber characteristics and function and have laid the groundwork for our current understanding of the genetic controls of muscle growth and atrophy. The use of molecular biological techniques to study the contractile characteristics of single muscle fibers is discussed in chapter 1.

▲ **Figure 0.9**　Frank Booth (*a*) and Ken Baldwin (*b*).

Well before James Watson and Francis Crick unraveled the structure of DNA (1953), scientists already appreciated the importance of genetics in predetermining the structure and function of all living organisms. The newest frontier in exercise physiology combines the study of molecular biology and genetics. Since the early 1990s, scientists have attempted to explain how exercise emits signals that affect the expression of genes within skeletal muscle.

In retrospect, it is apparent that since the beginning of the 20th century, the field of exercise physiology has evolved from measuring whole-body function (i.e., oxygen consumption, respiration, and heart rate) to molecular studies of muscle fiber genetic expression. There is little doubt that exercise physiologists of the future will need to be well grounded in biochemistry, molecular biology, and genetics.

has been toward more widespread use of treadmills. Cycle ergometers can be used by a person either in the normal upright position (see figure 0.10) or in the supine position.

Cycle ergometers in a research setting generally use either mechanical friction or electrical resistance. With mechanical friction devices, a belt encompassing a flywheel is tightened or loosened to adjust the resistance against which you pedal. Your power output depends on your pedal rate—the faster you pedal, the greater your power output. To maintain the same power output throughout the test, you must maintain the same pedal rate, so pedal rate must be constantly monitored.

With electrical resistance devices (shown in figure 0.10), also known as electrically braked cycle ergometers, resistance is provided by an electrical conductor that moves through a magnetic or electromagnetic field. The strength of the magnetic field determines the resistance to pedaling. These ergometers can be controlled so that the resistance increases automatically as pedal rate decreases, and decreases as pedal rate increases, to provide a constant power output.

Cycle ergometers offer some advantages over other ergometric devices. Your upper body remains relatively immobile when you use a cycle ergometer, allowing more accurate determination of blood pressure and easier blood sampling during exercise. Furthermore, the rate of work when you pedal does not depend on your body weight. This is important when one is investigating physiological responses to a standard rate of work (power output). As an example, if you lost 5 kg (11 lb), data derived from treadmill testing could not be compared with data obtained before your weight loss because physiological responses to a set speed and grade on the treadmill vary with body weight. After the weight loss, your rate of work at the same speed and grade would be less than before. With the cycle ergometer, weight loss does not have as great an effect on physiological response to a standardized power output.

> Cycle ergometers are the most appropriate devices for evaluating changes in submaximal physiological function before and after training in people whose weights have changed. Resistance on a cycle ergometer is independent of body weight, but the rate of work on a treadmill is directly related to body weight.

Cycle ergometers also have disadvantages. If you don't cycle regularly, your leg muscles will likely fatigue before your body does. In addition, peak (maximal) values for some physiological variables obtained on a cycle ergometer are frequently lower than comparable values obtained on a treadmill. This may be attributable to local leg fatigue, blood pooling in the legs (less blood returns to the heart), or the use of less muscle mass during cycling than during treadmill exercise. Trained cyclists, however, tend to achieve their peak values on the cycle ergometer.

Treadmills

Treadmills are the ergometers of choice for an increasing number of researchers and clinicians, particularly in the United States. With these devices, a motor-and-pulley system

▲ **Figure 0.10**   An electrical resistance cycle ergometer.

drives a large belt on which you can either walk or run (see figure 0.11). Belt length and width must accommodate your body size and stride length. It is nearly impossible to test elite athletes on treadmills that are too narrow or too short.

Treadmills offer a number of advantages. Unlike most cycle ergometers, the rate of work on a treadmill need not be monitored closely—if you don't maintain the belt speed, you are carried off the back of the device. Treadmill walking is a very natural activity, so individuals normally adjust to the skill required within 1 to 2 min. Also, average people almost always achieve their highest physiological values on the treadmill, although some athletes achieve higher values on ergometers that more closely match their mode of training or competition.

> Treadmills generally produce higher peak values than other ergometric devices for almost all assessed physiological variables, such as heart rate, ventilation, and oxygen uptake.

▲ **Figure 0.11**   A motor-driven treadmill.

Treadmills also have certain disadvantages. They are generally more expensive than cycle ergometers. They are also bulky, require electrical power, and are not very portable. Accurate measurement of blood pressure during treadmill exercise can be difficult because the noise associated with normal treadmill operation makes hearing through a stethoscope difficult. Also, obtaining accurate blood pressure measurements becomes difficult when treadmill speed requires jogging. Obtaining blood samples from a person who is on a treadmill is also difficult.

Other Ergometers

Other ergometers allow athletes who compete in specific sports or events to be tested in a manner that more closely approximates their training and competition. For example, an arm ergometer is used to test athletes or nonathletes who use primarily their arms and shoulders in physical activity (such as swimmers). The rowing ergometer was devised to test competitive rowers.

Valuable research data have been obtained by instrumenting swimmers and monitoring them during swimming in a pool. However, the problems associated with turns and constant movement led investigators to try **tethered swimming,** in which the swimmer is attached to a harness connected to a rope, a series of pulleys, and a pan containing weights.[7] The swimmer swims at an effort that maintains a constant body position in the pool. As weights are added to the pan, the swimmer must swim faster to maintain position.

Although important data have come from tethered swimming, the swimmer's technique is not at all similar to that used in free swimming. The **swimming flume** (see figure 0.12) allows swimmers to more closely simulate their natural swimming strokes. The swimming flume operates by propeller pumps that circulate water past the swimmer, who attempts to maintain body position in the flume. The pump circulation can be increased or decreased to vary the speed at which the swimmer must swim. The swimming flume, which unfortunately is very expensive, has at least partially resolved the problems with tethered swimming and has created new

▲ **Figure 0.12**   The swimming treadmill, or flume, enables researchers to instrument and study subjects while they swim in a constant flow of water.

opportunities to investigate the sport of swimming.

Specificity of Exercise Testing

When one is choosing an ergometer for testing, the concept of **test specificity** is particularly important with highly trained athletes, as has been illustrated by two research studies.[14, 17] In both studies, improvements in aerobic endurance after 10 weeks of swim training were monitored. Subjects performed maximal treadmill running and tethered swimming, both before and after training. Endurance as measured by the tethered swimming test increased by 11% to 18%, but treadmill endurance capacity did not change. If the treadmill alone had been used for testing, the researchers would have concluded that swim training had no influence on cardiorespiratory endurance capacity! These two studies point to the importance of testing athletes on ergometers that most closely approximate their sport or selecting ergometers

that closely mimic the mode or type of training used in training studies. For running, cycling, and swimming, the choice is easy. However, for sports such as ice hockey and tennis, the decision is not as clear.

▶ When researchers assess acute responses to exercise, environmental conditions—such as temperature, humidity, light, and noise—must be carefully controlled so that variations in these factors do not alter the body's responses.

▶ Diurnal (daily) cycles, the menstrual cycle, and sleep patterns also must be considered. Testing should be conducted at the same time of day and at the same point in the menstrual cycle, and subjects must have adequate sleep the night before any testing.

▶ An ergometer is a device that allows the amount and rate of physical work to be measured under standardized conditions.

▶ Cycle ergometers allow easier blood pressure assessment and blood sampling because the upper body remains relatively immobile. Results on these devices are not greatly affected by changes in body weight.

▶ Treadmills ensure that the rate of work remains relatively constant because a person cannot stay on the treadmill if the work rate is not maintained. Treadmill walking is also a very natural activity. But the results are affected by body weight; thus, measuring physiological change is more difficult than on a cycle ergometer. Maximal physiological values are generally higher when tested on the treadmill.

▶ The swimming flume allows free swimming and provides results that can be applied to competitive and recreational swimming.

▶ To assess the body's responses to exercise, it is essential that the mode of testing be carefully matched to the type of activity that a person usually performs.

Chronic Physiological Adaptations to Training

When examining the acute response to exercise, we are concerned with the body's immediate response to a single exercise bout. The other major area of interest in exercise and sport physiology is how the body responds over time to the stress of repeated exercise bouts. When you perform regular exercise over a period of weeks, your body adapts. The physiological adaptations that occur with chronic exposure to exercise or training improve both your exercise capacity and your efficiency. With resistance training, your muscles become stronger. With aerobic training, your heart and lungs become more efficient, and your endurance capacity increases. These adaptations are highly specific to the type of training you do.

In the following chapters, we discuss in detail specific physiological adaptations that result from chronic exercise, or training. Several principles can be applied to all forms of physical training. Each is discussed as it applies to specific types of training later in this text. For now, let's examine basic training principles.

Principle of Individuality

We were not all created with the same capacity to adapt to exercise training. Heredity plays a major role in determining how quickly and to what degree your body adapts to a training program. Except for identical twins, no two people have exactly the same genetic characteristics, so individuals are unlikely to show precisely the same adaptations to a given training program. Variations in cellular growth rates, metabolism, and neural and endocrine regulation also lead to tremendous individual variation. Such individual variation might explain why some people show great improvement after participating in a given program (responders) whereas others experience little or no change after following the same program (nonresponders). We discuss this phenomenon of responders and nonresponders in more detail in chapter 9. For these reasons, any training program must take into account the specific needs and abilities of the individuals for whom it is designed. This is the **principle of individuality.**

Principle of Specificity

Training adaptations are highly specific to the type of activity and to the volume and intensity of the exercise performed. To improve muscular power, for example, the shot-putter would not emphasize distance running or slow, low-intensity resistance training. Similarly, the distance runner wouldn't concentrate on sprint-type interval training. This is likely the reason that athletes who train for strength and power, such as weightlifters, often have great strength but don't have better aerobic endurance than untrained people. According to the **principle of specificity,** the training program must stress the physiological systems that are critical for optimal performance in the given sport in order to achieve specific training adaptations.

Principle of Disuse

Most athletes would agree that regular physical exercise improves your muscles' capacity to generate more energy and to resist fatigue. Likewise, endurance training improves your ability to perform more work for longer periods. But if you stop training, your state of fitness will decrease to a level that meets only the demands of daily use. Any gains you achieved with training will be lost. This **principle of disuse** leads to the popular phrase, "Use it or lose it." A training program must include a maintenance plan. In chapter 12 we examine specific physiological changes that occur when the training stimulus stops.

Principle of Progressive Overload

Two important concepts, overload and progressive training, form the foundation of all training. According to the **principle of progressive overload,** all training programs must include these components. For example, to gain strength, the muscles must be overloaded, which means they must be loaded beyond the point to which they are normally loaded. Progressive resistance training implies that as the muscles become stronger, a proportionately greater resistance is required to stimulate further strength increases.

As an example, consider a young man who can perform only 10 repetitions of a bench press before reaching fatigue, using 68 kg (150 lb) of weight. With a week or two of resistance training, he should be able to increase to 14 or 15 repetitions with the same weight. He then adds 2.3 kg (5 lb) to the bar, and his repetitions decrease to 8 or 10. As he continues to train, the repetitions continue to increase, and within another week or two, he is ready to add an additional 2.3 kg (5 lb). Thus, there is a progressive increase in the amount of weight lifted. Similarly, with anaerobic and aerobic training, training volume (intensity and duration) can be increased progressively.

Principle of Hard/Easy

Serious athletes and fitness buffs tend to train hard most days, if not every day, of the week for months or even years at a time. Training hard includes training each day at high intensities or for long durations or both, leading to little variation in the total volume of training. We are now realizing that the body begins to maladapt under these conditions (see chapter 12). Bill Bowerman, former track coach at the University of Oregon, former U.S. Olympic track coach, and cofounder of NIKE, Inc., was particularly recognized for his development of world-class distance runners. Part of his success was attributable to his understanding that hard workouts break the body down and that you need a day or two to recover to achieve your optimal training adaptation. He developed a training strategy for his distance runners that became known as the principle of hard/easy training. As an example, on a day following a high-intensity interval training workout (hard), you might prescribe an easy day of training consisting of a slow, 8-km (5-mi) run in the country. The easy day provides a period of active recovery so that the body is ready for the next day of hard training.

Principle of Periodization

Closely related to the principle of hard/easy, the principle of periodization has become very popular over the past 30 years in the area of resistance training. Periodization is the gradual cycling of specificity, intensity, and volume of training to achieve peak levels of fitness for competition.[2] With periodization, the volume and intensity of training are varied over a macrocycle, which is generally a year of training. A macrocycle is composed of two or more mesocycles that are dictated by the dates of major competitions. Each mesocycle is subdivided into periods of preparation, competition, and transition. This principle is discussed in more detail in chapter 3.

Research: Foundation of Understanding

Exercise and sport scientists actively engage in research to better understand the mechanisms that regulate the body's physiological responses to acute bouts of exercise as well as its adaptations to extended periods of training and reduced training. Most of this research is conducted at major research universities and at specialized centers and institutes. As you read through this book you will find numerous references to specific research studies that have made a major impact on our understanding within a given area. Once a group of scientists complete a research project, they submit the results of their research to one of the many research journals in sport and exercise physiology. Some of the more widely used research journals can be seen in the selected readings and references at the end of each chapter of this book as well as on the Web site www.humankinetics.com/physiologyofsport andexercise/osg.

Most research findings published in these journals are presented in the form of tables and graphs. Thus, this textbook has included tables and graphs from selected research studies to help you become familiar with actual research data and how they are interpreted.

- A major area of concern for exercise physiologists is how the body adapts to chronic exposure to exercise or training.

- According to the principle of individuality, each person must be recognized as unique, and training programs must be designed to allow for individual variation. Different people respond to a given training program in different ways.

- According to the principle of specificity, to maximize the benefits, training must be specifically matched to the type of activity the person normally engages in. An athlete involved in a sport that requires tremendous strength, such as weightlifting, would not expect great strength gains from endurance running.

- According to the principle of disuse, training benefits are lost if training is either discontinued or reduced too abruptly. To avoid this, all training programs must include a maintenance program.

- According to the principle of progressive overload, training must involve working the body (muscles, cardiovascular system) harder than normal; as the body adapts, training progresses to a higher work level.

- According to the principle of hard/easy, one or two days of hard training should be followed by one day of easy training, allowing the body and mind to fully recover before the next hard day of training.

- According to the principle of periodization, a macrocycle (a full year of resistance training, for example) is divided into two or more mesocycles, with variations in the intensity and volume of training as well as the specific form or type of training.

Reading and Interpreting Tables and Graphs

Tables and graphs provide an efficient way for researchers to communicate the results of their studies to other scientists. Tables and graphs are also an effective way for us to transmit this information to you, providing you understand how to read and interpret them. Before going any further, let's review how to read and interpret tables and graphs.

With respect to tables, let's use table 0.2 as an example. It is important to first look at the title of the table, which tells you what information is being presented. In this case, we are demonstrating how heart rate varies throughout a 24-h day, at rest and during exercise and recovery from exercise. The left-hand column specifies the conditions under which the heart rate measurement was taken, and the times across the top of the table are the specific times these measurements were taken. You can see from the table that heart rate is expressed in "beats/min" or beats per minute. It is important to know the units of measure used when interpreting a table or graph. From this table we see that heart rate is at its lowest early in the morning and highest midafternoon, both at rest and during exercise. These data illustrate the importance of standardizing the time of day that measurement is taken for a given variable, either when you are comparing the same person across different days or when comparing two or more individuals on the same day.

Graphs can be more difficult to read and interpret. First of all, each graph has a horizontal or x-axis for the independent variable and one or two vertical or y-axes for the dependent variable or variables. If we used the data from table 0.2, the x-axis would represent the time of day and the y-axis would represent heart rate, or the heart rate during moderate exercise as illustrated in figure 0.13a. The unit of measure is also displayed on the graph; again, heart rate is expressed in beats per minute. These data can also be plotted in the format of a bar graph (figure 0.13b).

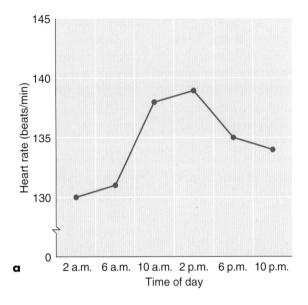

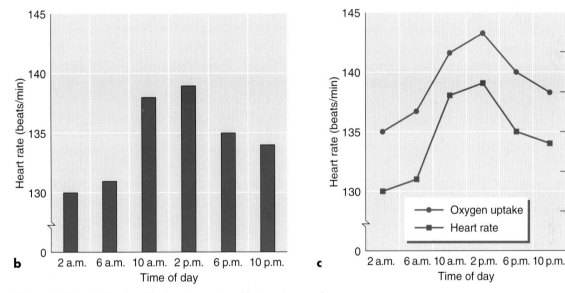

▲ **Figure 0.13**   Understanding how to read and interpret a graph.

If you were interested in plotting both heart rate and the metabolic rate (oxygen uptake) during light exercise, the left-hand *y*-axis would show heart rate and the right-hand *y*-axis would show oxygen uptake, both during light exercise (figure 0.13*c*). From figure 0.13*c*, we can see that oxygen consumption and heart rate change in the same direction over the course of the day, in that when heart rate is increased oxygen uptake is increased, and when heart rate is decreased oxygen uptake is decreased.

Research Design

In this text, we refer primarily to two basic types of research design: cross-sectional and longitudinal. With a **cross-sectional research design,** a large cross-section of the population is tested at one specific time, and the differences between individual groups within that population are used to estimate change in any given physiological variable across time. With a **longitudinal research design,** participants are tested one or more times after initial testing to measure their changes over time.

The differences between these two approaches are best understood through an example. Let's say that you want to determine if distance running increases the concentration of high-density–lipoprotein cholesterol (HDL-C) in the blood. HDL-C is the desirable form of cholesterol; increased concentrations are associated with reduced risk for heart disease. Using the cross-sectional approach, you could, for example, test a large number of people who fall into the following categories:

- No training
- Run 24 km (15 mi) per week
- Run 48 km (30 mi) per week
- Run 72 km (45 mi) per week
- Run 96 km (60 mi) per week

You would then compare the results from each group, basing your conclusions on how much running was done. This approach has been used in past studies, and scientists found that the greater the weekly distance, the higher the HDL-C level; this suggests a positive

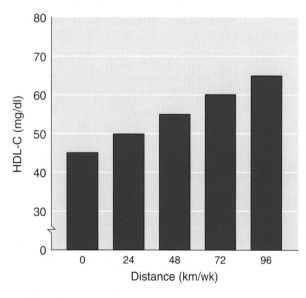

▲ **Figure 0.14**   The relationship between distance run per week and average high-density–lipoprotein cholesterol (HDL-C) concentrations across five groups: nontraining control (0 km/week), 24 km/week, 48 km/week, 72 km/week, and 96 km/week. This illustrates a cross-sectional study design.

health benefit related to running distance, as illustrated in figure 0.14. In this figure there is clearly a higher concentration of HDL-C with an increase in the distance run per week. It is important to remember, however, that these are different groups of runners, not the same runners at different training volumes.

Using the longitudinal approach to test the same question, you could design a study in which untrained people would be recruited to participate in a 12-month distance running program. You could, for example, recruit 40 people willing to begin running and then randomly assign 20 to a training group and the remaining 20 to a control group. Both groups would be followed for 12 months. Blood samples would be tested at the beginning of the study, then at 3-month intervals, concluding at 12 months when the program ends.

With this design, both the running group and the control group are followed over the entire period of the study, and you can determine changes in their HDL-C levels across each period. The control group acts as a comparison group to make certain that any changes observed in the running group are attributable solely to the training program and not to any other factors, such as the time of the year or aging. Actual studies using this design to study changes in HDL-C with distance running have been conducted, but their results have not been as clear as the results of the cross-sectional studies. See figure 0.15 as an example. Note that in this figure, in contrast to figure 0.14, there is only a small increase in HDL-C in the subjects who are training. The control group stays relatively stable with only minor fluctuations in their HDL-C from one 3-month period to the next.

A longitudinal research design is best suited to studying this problem. Too many factors that may taint results can influence cross-sectional designs. For example, genetic factors might interact so that those who run long distances are also those who have high HDL-C levels. Also, different populations might follow different diets, but in a longitudinal study, diet and other variables can be more easily controlled.

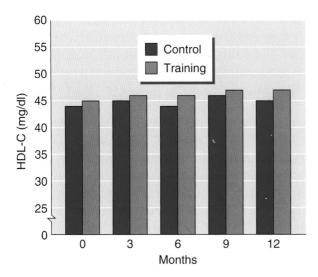

▲ **Figure 0.15** The relationship between months of distance-running training and average high-density–lipoprotein cholesterol (HDL-C) concentrations in an experimental group (20 subjects, distance training) and a sedentary (20 subjects) control group. This illustrates a longitudinal study design.

However, longitudinal studies are not always possible, and cross-sectional studies provide some insight into these questions.

> Longitudinal research studies are generally the most accurate for studying a given problem or issue. However, it is not always possible to use a longitudinal study, so cross-sectional studies must sometimes be used.

Research Controls

When we conduct research, it is important to be as careful as possible when designing the study and collecting the data. We have seen from figure 0.15 that changes in a variable over time resulting from an intervention such as exercise can be very small. Yet, even small changes in a variable such as HDL-C can mean a substantial reduction in risk for heart disease. Recognizing this, scientists design studies that attempt to provide results which are both accurate and reproducible. This requires that studies be carefully controlled. What does this mean?

Research controls are applied at various levels. Starting with the design of the research project, the scientist must determine how to control for variation in the subjects used in the study. The scientist must determine if it is important to control for the subjects' sex, age, or body size. To use age as an example, for certain variables, the response to an exercise training program might be different for a child or an aged person compared with a young or middle-aged adult. Is it important to control for the subject's smoking or dietary status? Considerable thought and discussion are needed to make sure that the subjects used in a study are appropriate for the specific research question being asked. The results of some past research have been contaminated because researchers did not properly control for variable factors.

For most studies, it is important to have a control group. However, it might also be necessary to have a **placebo group.** Thus, in a study in which a subject might expect to have a benefit from the proposed intervention, such as the use of a specific food or drug, a scientist might decide to use three groups of subjects: an intervention group that receives the actual substance, a placebo group that receives an inert substance, and a control group that receives nothing. (The placebo group would be told that they were receiving the specific intervention, e.g., food or drug, yet would be given an inert substance that has no known physiological effects.) If the intervention and placebo groups both improve their performance to the same level and the control group does not improve performance, then the improvement is likely the result of the "placebo effect," or the expectation that the substance will improve performance. If the intervention group improves performance and the placebo and control groups do not, then we can conclude that the intervention does improve performance.

One other way of controlling for the placebo effect is to conduct a study that uses a "crossover" design. In this case, you have two groups. One group is administered the intervention for the first half of the study (e.g., 6 months of a

12-month study) and serves as a control the last half of the study. The second group serves as a control during the first half of the study and receives the intervention during the second half. In some cases, a placebo can be used in the "control" phase of the study. A further discussion of placebo groups is found in chapter 15, Ergogenic Aids and Sport.

It is equally important to control data collection. The equipment must be calibrated so you know that the numbers generated by a given piece of equipment are accurate, and the procedures used in collecting data must be standardized. For example, when using a scale to measure the weight of subjects, you need to calibrate that scale by using a set of calibrated weights (e.g., 10 kg, 20 kg, 30 kg, and 40 kg) that have been measured on a precision scale. These weights are placed on the weighing scale to be used in the study, individually and in combination, at least once a week to be certain that the scale is measuring the weights accurately. As another example, electronic analyzers used to measure respiratory gases need to be calibrated frequently with gases of known concentration to ensure the accuracy of these analyses.

Finally, it is important to know that all test results are reproducible. Let's take the example illustrated in figure 0.15, where HDL-C of a person on an exercise training program is monitored every 3 months. If you test that person 5 days in a row before he or she starts the training program, you would expect the HDL-C results to be similar across all 5 days, providing diet, exercise, sleep, and time of day for testing remained the same. In figure 0.15, the values for the control group across 12 months varied from about 44 to 45 mg/dl, whereas the exercise group increased from 45 to 47 mg/dl. Over 5 consecutive days, the measurements should not vary by more than 1 mg/dl for any one person if you are going to pick up this small change over time. To control for reproducibility of results, scientists generally take several measurements, each on different days, and then average the results, before, during and at the end of an intervention.

A scientist will spend considerable time before starting an experiment to try to identify all of those sources that can cause variation when measuring these variables so these sources can be controlled. A series of experiments in our laboratory (JHW) were directed at identifying the effect of exercise training on resting metabolic rate (RMR). RMR is very difficult to measure accurately because it varies considerably with movement, changes in light and temperature, and duration and quality of sleep the night before the measurement. To measure RMR takes between 45 and 60 min. We spent several months perfecting our technique by controlling these extraneous conditions, but we were still seeing considerable variation. We allowed subjects to wear headphones to listen to their favorite music during the period of measurement, which we believed would keep them quiet and reduce the variation in measurement. We finally discovered that the variations we were seeing were attributable to the different songs that were playing, with RMR increasing with fast, loud music and decreasing with slow, soft music! No more headphones and no more music!

Research Settings

Research can be conducted either in the laboratory or in the field. Laboratory tests are usually more accurate because more specialized equipment can be used and conditions can be carefully controlled. As an example, the direct laboratory measurement of maximal oxygen uptake ($\dot{V}O_2$max) is considered the most accurate estimate of cardiorespiratory endurance capacity. However, some field tests, such as the 1.5-mi (2.4-km) run, are used to predict or estimate $\dot{V}O_2$max. The field test is not totally accurate, but it provides a reasonable estimate of $\dot{V}O_2$max, it is inexpensive to conduct, and you can test many people in a short time. To have your $\dot{V}O_2$max measured directly, you would need to go to a university or clinical facility such as a hospital, but you could easily have your $\dot{V}O_2$max estimated from your time for the 1.5-mi run.

Sometimes the field test is the most appropriate. For example, during early research on blood doping—in which blood is removed from an athlete, stored, and later reinfused (see chapter 15)—all research was conducted in the laboratory. The resulting data were very accurate because they were collected under tightly controlled conditions. But they really didn't determine whether blood doping improves performance. Only later, when studies were designed that combined laboratory tests with field tests of actual racing performance, were we able to see the influence of blood doping.

▶ Cross-sectional research involves collecting data once from a diverse population and then making comparisons based on groups in that population.

▶ Longitudinal research involves following participants over an extended period of time, collecting data at intervals to note individual changes with time. It is the more accurate design but cannot always be done, in which case a cross-sectional design can provide some insight.

▶ To obtain meaningful results, research studies must be carefully controlled. Consideration must be given to the characteristics of the subjects used, the design of the experiment (use of control and/or placebo groups), standardization of procedures, calibration of equipment, and reproducibility of results.

▶ Research can be conducted in the laboratory or in the field. Laboratory research allows careful control of most variables and use of elaborate and very accurate equipment. Field research is less controllable and fewer equipment options are available, but participants' activity is often more natural in the field than in a laboratory. Each setting has advantages and drawbacks, and often research is conducted in both settings to get a more accurate picture.

In Closing . . .

In this introduction we uncovered the roots of exercise and sport physiology. We learned that the current state of knowledge in these fields builds on the past and is merely a bridge to the future—many questions remain unanswered. We examined the two primary areas of concern for today's researcher: acute responses to exercise bouts and chronic adaptations to long-term training. We discussed basic training principles and concluded with an overview of the most common types of research.

In part I, we begin examining physical activity the way exercise physiologists do as we explore the essentials of movement. In the next chapter, we examine the structure and function of skeletal muscle, how it produces movement, and how it responds during exercise.

▶ Key Terms

acute response
chronic adaptation
cross-sectional research design
cycle ergometer
diurnal variation
ergometer
exercise physiology
longitudinal research design
placebo group
principle of disuse
principle of hard/easy
principle of individuality
principle of periodization
principle of progressive overload
principle of specificity
sport physiology
swimming flume
test specificity
tethered swimming
treadmill

▶ Study Questions

1. What is exercise physiology? What is sport physiology?

2. Describe the evolution of exercise physiology from the early studies of anatomy. Who were some of the key figures in the development of this field?

3. Name the originator of the Harvard Fatigue Laboratory. Who was the first director of this laboratory?

4. What were some of the areas of research emphasized by the Harvard Fatigue Laboratory personnel?

5. Name the three Scandinavian physiologists who conducted research in the Harvard Fatigue Laboratory.

6. Who was responsible for introducing the muscle biopsy needle to the study of exercise physiology? Name the physiologists who collaborated in describing the muscle fiber characteristics of humans.

7. Provide an example of what is meant by studying acute responses to a single bout of exercise and chronic adaptations to exercise training.

8. What are several environmental conditions that could affect your response to an acute bout of exercise? What is meant by diurnal variation? Why is it important when testing to control for environmental and diurnal conditions? Give several examples.

9. What is an ergometer? Give examples of two ergometers and explain their principle of operation.

10. What would be the appropriate ergometer to use when testing a cyclist? A distance runner? A swimmer?

11. Define the principle of individuality and provide an example.

12. Define the principle of specificity and provide an example.

13. Define the principle of disuse and provide an example.

14. Define the principle of progressive overload and provide an example.

15. Define the principles of hard/easy and periodization and discuss how they are similar.

16. What factors must you consider when designing a research study to ensure that you get accurate and reproducible results?

▶ References

1. Åstrand, P.-O. (1991). Influence of Scandinavian scientists in exercise physiology. *Scandinavian Journal of Medicine and Science in Sports,* **1,** 3-9.

2. Baechle, T.R., & Earle, R.W. (Eds.). (2000). *Essentials of strength training and conditioning* (2nd ed.). Champaign, IL: Human Kinetics.

3. Bainbridge, F.A. (1931). *The physiology of muscular exercise.* London: Longmans, Green.

4. Brown, R.C., & Kenyon, G.S. (1968). *Classical studies on physical activity.* Englewood Cliffs, NJ: Prentice Hall.

5. Buskirk, E.R. (1996). Early history of exercise physiology in the United States: Part I. A contemporary historical perspective. In J.D. Messengale & R.A. Swanson (Eds.), *History of exercise and sport science* (pp. 55-74). Champaign, IL: Human Kinetics.

6. Cooper, K.H. (1968). *Aerobics.* New York: Evans.

7. Costill, D.L. (1966). Use of a swimming ergometer in physiological research. *Research Quarterly,* **37**(4), 564-567.

8. Dill, D.B. (1938). *Life, heat, and altitude.* Cambridge, MA: Harvard University Press.

9. Dill, D.B. (1968). Historical review of exercise physiology science. In R. Warren & R.E. Johnson (Eds.), *Science and medicine of exercise and sports* (2nd ed., pp. 42-48). New York: Harper

10. Dill, D.B. (1985). *The hot life of man and beast.* Springfield, IL: Charles C Thomas.

11. Fletcher, W.M., & Hopkins, F.G. (1907). Lactic acid in amphibian muscle. *Journal of Physiology,* **35,** 247-254.

12. Flint, A., Jr. (1871). On the physiological effects of severe and protracted muscular exercise; with special

reference to the influence of exercise upon the excretion of nitrogen. *New York Medical Journal, 13,* 609-697.

13. Foster, M. (1970). *Lectures on the history of physiology.* New York: Dover.

14. Gergley, T.J., McArdle, W.D., DeJesus, P., Toner, M.M., Jacobowitz, S., & Spina, R.J. (1984). Specificity of arm training on aerobic power during swimming and running. *Medicine and Science in Sports and Exercise, 16,* 349-354.

15. Horvath, S.M., & Horvath, E.C. (1973). *The Harvard Fatigue Laboratory: Its history and contributors.* Englewood Cliffs, NJ: Prentice Hall.

16. LaGrange, F. (1889). *Physiology of bodily exercise.* London: Kegan Paul International.

17. Magel, J.R., Foglia, G.F., McArdle, W.D., Gutin, B., Pechar, G.S., & Katch, F.I. (1975). Specificity of swim training on maximum oxygen uptake. *Journal of Applied Physiology, 38,* 151-155.

18. McArdle, W.D., Katch, F.I., & Katch, V.L. (2001). *Exercise physiology: Energy, nutrition, and human performance* (5th ed.). Baltimore: Williams & Wilkins.

19. Robinson, S. (1938). Experimental studies of physical fitness in relation to age. *Arbeitsphysiologie, 10,* 251-327.

20. Séguin, A., & Lavoisier, A. (1793). Premier mémoire sur la respiration des animaux. Dans *Histoire et Mémoires de l'Academie Royale des Sciences. 92,* 566-584.

21. Tipton, C.M. (1997). Early history of exercise physiology in the United States: Part II. A contemporary historical perspective. In J.D. Massengale & R.A. Swanson (Eds.), *History of exercise and sport science* (pp. 396-438). Champaign, IL: Human Kinetics.

22. Zuntz, N., & Schumberg, N.A.E.F. (1901). Studien Zur Physiologie des Marches. Berlin: A. Hirschwald. p. 211.

▷ Selected Readings

Åstrand, P.-O., & Rodahl, K. (1986). *Textbook of work physiology* (3rd ed.). New York: McGraw-Hill.

Bang, O., Boje, O., & Nielsen, M. (1936). Contributions to the physiology of severe muscular work. *Scandinavica Archives of Physiology, 74*(Suppl.), 1-208.

Beecher, C.E. (1858). *Physiology and calisthenics.* New York: Harper & Brothers.

Bergstrom, J. (1962). Muscle electrolytes in man. *Scandinavian Journal of Clinical Investigation, 14*(Suppl.), 1-110.

Berryman, J.W. (1995). *Out of many, one: A history of the American College of Sports Medicine.* Champaign, IL: Human Kinetics.

Brooks, G.A., Fahey, T.D., & White, T.P. (1996). *Exercise physiology: Human bioenergetics and its applications.* Mountain View, CA: Mayfield.

Consolazio, C.F., Johnson, R.E., & Pecora, L.J. (1963). *Physiological measurements of metabolic functions in man.* New York: McGraw-Hill.

Costill, D.L. (1985). Practical problems in exercise physiology research. *Research Quarterly, 56,* 378-384.

Fox, E.L., Bowers, R.W., & Foss, M.L. (1993). *The physiological basis for exercise and sport* (5th ed.). Madison, WI: Brown & Benchmark.

Hill, A.V. (1923). Muscular exercise, lactic acid, and the supply utilization of oxygen. *Quarterly Journal of Medicine, 16,* 135-171.

Hill, A.V. (1927). *Muscular movement in man: The factors governing speed and recovery from fatigue.* London: McGraw-Hill.

Hill, A.V. (1970). *First and last experiments in muscle mechanics.* Cambridge, UK: Cambridge University Press.

MacDougall, J.D., Wenger, H.A., & Green, H.J. (1991). *Physiological testing of the high performance athlete* (2nd ed.). Champaign, IL: Human Kinetics.

Maud, P.J., & Foster, C. (Eds.). (1995). *Physiological assessment of human fitness.* Champaign, IL: Human Kinetics.

McCurdy, J.H., & Larson, L. (1939). *The physiology of exercise.* Philadelphia: Lea & Febiger.

Park, R.J. (1995). History of research on physical activity and health: Selected topics, 1867 to the 1950s. *Quest, 47,* 274-287.

Powers, S.K., & Howley, E.T. (2001). *Exercise physiology: Theory and application to fitness and performance* (4th ed.). Madison, WI: McGraw-Hill.

Reilly, T., & Brooks, G.A. (1990). Selective persistence of circadian rhythms in physiological responses to exercise. *Chronobiology International, 7,* 59-67.

Robergs, R.A., & Roberts, S.O. (2000). *Exercise physiology for fitness, performance, and health.* Dubuque, IA: McGraw-Hill.

Sargent, D.A. (1906). *Physical education.* Boston: Ginn.

Sargent, D.A. (1921). The physical test of a man. *American Physical Education Review, 26,* 188-194.

Strømme, S.B., Ingjer, F., & Meen, H.D. (1977). Assessment of maximal aerobic power in specifically trained athletes. *Journal of Applied Physiology, 42,* 833-837.

Trine, M.R., & Morgan, W.P. (1995). Influence of time of day on psychological responses to exercise. *Sports Medicine, 20,* 328-337.

PART I

Essentials of Movement

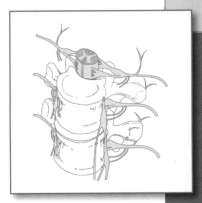

In the introduction, we explored the foundations of exercise and sport physiology. We defined these two fields of study, gained a historical perspective of their development, and established basic concepts that underlie the remainder of this book. With this foundation, we can begin our quest to understand how the human body performs physical activity. We start our journey in chapter 1, "Muscles and How They Move," where we focus on skeletal muscle, examining the structure and function of muscles and muscle fibers and how they produce body movement. We will learn how muscle fiber types differ and why these differences are important to specific types of activity. In chapter 2, "Neurological Control of Movement," we discuss how the nervous system coordinates muscle action by integrating sensory information coming from all parts of the body and then signaling the appropriate muscles to act. Finally, in chapter 3, "Neuromuscular Adaptations to Resistance Training," we learn how the muscular system and nervous system adapt when they are subjected to several weeks or months of resistance training. We examine changes in individual muscle fibers and neural control, how these changes improve performance in strength activities, and how to design a resistance training program to maximize the benefits.

CHAPTER 1

MUSCLES AND HOW THEY MOVE

overview

All human movement, from the blinking of an eye to the running of a marathon, depends on the proper functioning of skeletal muscle. Whether it is the strained effort of a sumo wrestler or the graceful pirouette of a ballet dancer, physical activity can be accomplished only through muscle force.

The title for this chapter, "Muscles and How They Move," was first used by A.V. Hill in his 1927 text *Living Machinery.* Although our knowledge of how muscles function has grown significantly since then, Hill's description of the basic actions of whole muscle remains unchanged. Technological developments since the 1950s have enabled us to understand the molecular design and biochemical reactions that enable muscles to create movement.

In this chapter we examine the structure and function of skeletal muscle. We begin by reviewing basic anatomy and physiology, examining muscle at the gross, microscopic, and molecular levels. Then we discuss how muscle functions during exercise and how the force needed to create movement is generated.

outline

Nine-year-old Jeremy Schill weighed only 29.5 kg (65 lb), but that didn't stop him from lifting the rear of the family's 1,860-kg (4,100-lb) car off his father's chest! The car had slipped off the jack while Rique Schill was working underneath it, pinning him under the rear axle. When Jeremy realized that his father was slowly suffocating, the third-grader lifted the car enough to enable his father to breathe and to allow his mother to place another jack under the rear bumper. How could a small boy lift such a great weight? In this chapter we consider how his muscles might have generated the force that enabled him to save his father's life.

When our hearts beat, when a meal we've eaten moves through our intestines, and when we move any part of our bodies, muscle is involved. The myriad functions of the muscular system are performed by only three types of muscle (see figure 1.1): skeletal, cardiac, and smooth.

Smooth muscle is called involuntary muscle, because it is not directly under our conscious control. It is found in the walls of most blood vessels, allowing them to constrict or dilate to regulate blood flow. It is also found in the walls of most internal organs, allowing them to contract and relax, perhaps to move food along the digestive tract, to expel urine, or to give birth.

Cardiac muscle is found only in the heart, composing most of the heart's structure. It shares some characteristics with skeletal muscle but, like smooth muscle, is not under conscious control. Cardiac muscle controls itself, with some mere fine-tuning by the nervous and endocrine systems. Cardiac muscle is discussed fully in chapter 7.

We usually pay attention only to those muscles we can consciously control. These are the skeletal, or voluntary, muscles, so named because most attach to and move the skeleton. We know many of these muscles by their names—such as deltoid, pectorals, and biceps—but the human body contains more than 600 skeletal muscles. The thumb alone is controlled by nine separate muscles!

Exercise requires movement of the body, which is accomplished through the action of skeletal muscles. Because this is an exercise and sport physiology book, our primary interest is the structure and function of skeletal muscle. Although the anatomical structures of smooth, cardiac, and skeletal muscle differ somewhat, their principles of action are similar.

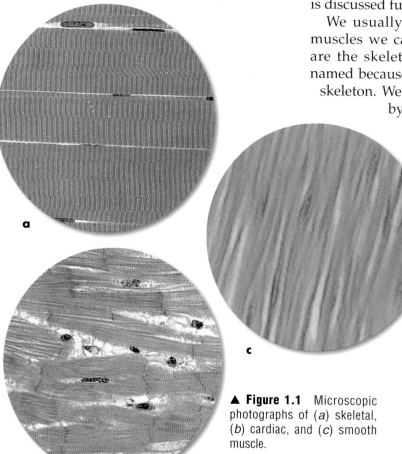

▲ **Figure 1.1** Microscopic photographs of (*a*) skeletal, (*b*) cardiac, and (*c*) smooth muscle.

Structure and Function of Skeletal Muscle

When we think of muscles, we visualize each muscle as a single unit. This is natural because a skeletal muscle seems to act as a single entity. But skeletal muscles are far more complex than that (see figure 1.2).

If you were to dissect a muscle, you would first cut through the outer connective tissue covering. This is the **epimysium.** It surrounds the entire muscle, holding it together. Once you cut through the epimysium, you would see small bundles of fibers wrapped in a connective tissue sheath. These bundles are called fasciculi. The connective tissue sheath surrounding each **fasciculus** is the **perimysium.**

Finally, by cutting through the perimysium and using a magnifier, you could see the **muscle fibers,** which are the individual muscle cells. A sheath of connective tissue, called the **endomy-**sium, also covers each muscle fiber. It has often been suggested that muscle fibers extend from one end of the muscle to the other, but under the microscope one often divides the muscle bellies into compartments or more transverse fibrous bands (inscriptions). Because of this compartmentalization, the longest human muscle fibers are about 12 cm (4.7 in.), which corresponds to about 500,000 sarcomeres, the basic functional unit of the myofibril. The number of fibers in different muscles ranges from 10,000 (e.g., first lumbrical muscle) to more than a million (e.g., gastrocnemius muscle).[8]

Now that we know how muscle fibers fit into the whole muscle, let's look at them more closely.

> A single muscle cell is known as a muscle fiber.

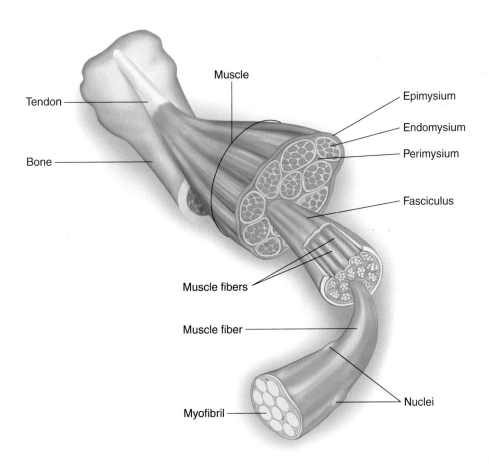

▲ **Figure 1.2** The basic structure of muscle.

Muscle Fiber

Muscle fibers range in diameter from 10 to 80 μm, so they are nearly invisible to the naked eye. Let's take a look at the structure of the individual muscle fiber.

Sarcolemma

If you looked closely at an individual muscle fiber, you would see that it is surrounded by a plasma membrane, called the sarcolemma (figure 1.3). At the end of each muscle fiber, its sarcolemma fuses with the tendon, which inserts into the bone. Tendons are made of fibrous cords of connective tissue that transmit the force generated by muscle fibers to the bones, thereby creating motion. So typically, each individual muscle fiber is ultimately attached to bone via the tendon.

Sarcoplasm

Inside the sarcolemma, with the aid of a microscope, you could see that a muscle fiber contains successively smaller subunits, as shown in figure 1.3. The largest of these are myofibrils, which we discuss separately. For now, consider myofibrils to be rodlike structures running the length of the muscle fibers. A gelatin-like substance fills the spaces between the myofibrils. This is the sarcoplasm. It is the fluid part of the muscle fiber—its cytoplasm. The sarcoplasm contains mainly dissolved proteins, minerals, glycogen, fats, and the necessary organelles. It differs from the cytoplasm of most cells because it contains a large quantity of stored glycogen as well as the oxygen-binding compound myoglobin, which is quite similar to hemoglobin.

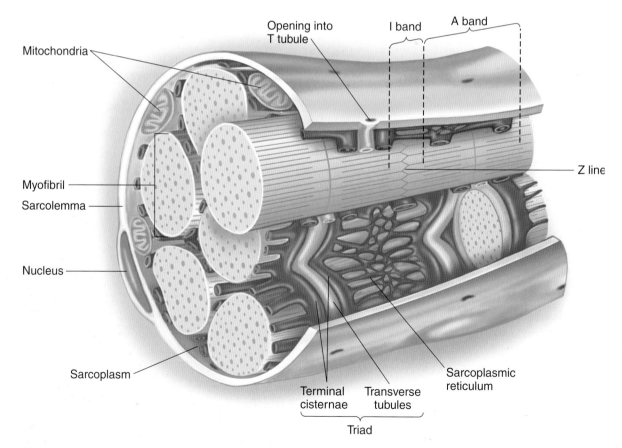

▲ **Figure 1.3** The structure of a single muscle fiber.

The Transverse Tubules The sarcoplasm also houses an extensive network of **transverse tubules** (T tubules), which are extensions of the sarcolemma (plasma membrane) that pass laterally through the muscle fiber. These tubules are interconnected as they pass among the myofibrils, allowing nerve impulses received by the sarcolemma to be transmitted rapidly to individual myofibrils. Tubules also provide pathways from outside the fiber to its interior, enabling substances to enter the cell and waste products to leave the fibers.

The Sarcoplasmic Reticulum A longitudinal network of tubules, known as the **sarcoplasmic reticulum** (SR), is also found within the muscle fiber. These membranous channels parallel the myofibrils and loop around them. The sarcoplasmic reticulum serves as a storage site for calcium, which is essential for muscle contraction. Figure 1.3 depicts the T tubules and the sarcoplasmic reticulum. We discuss their functions in more detail later in this chapter, when we outline the process of muscle action.

> ▶ An individual muscle cell is called a muscle fiber.
>
> ▶ A muscle fiber is enclosed by a plasma membrane called the sarcolemma.
>
> ▶ The cytoplasm of a muscle fiber is called the sarcoplasm.
>
> ▶ The extensive tubule network found in the sarcoplasm includes T tubules, which allow communication and transport of substances throughout the muscle fiber, and the sarcoplasmic reticulum, which stores calcium.

Myofibril

Each muscle fiber contains several hundred to several thousand **myofibrils.** These are the contractile elements of skeletal muscle. Myofibrils appear as long strands of still smaller subunits—the sarcomeres.

Striations and the Sarcomere

Under a light microscope, skeletal muscle fibers have a distinctive striped appearance. Because of these markings, or striations, skeletal muscle is also called striated muscle. This striation also is seen in cardiac muscle, so it too can be considered striated muscle.

Refer to figure 1.4, showing a single muscle fiber and its myofibrils. You can see the striations. Note that dark regions, known as A bands, alternate with light regions, known as I bands. Each dark A band has a lighter region in its center, the H zone, which is visible only when the myofibril is relaxed. Now look at the light I bands. These are interrupted by a dark stripe known as the Z disk.

> The sarcomere is the smallest functional unit of a muscle.

A **sarcomere** is the basic functional unit of a myofibril. Each myofibril is composed of numerous sarcomeres joined end to end at the Z disks. Each sarcomere includes what is found between each pair of Z disks, in this sequence:

- An I band (light zone)
- An A band (dark zone)
- An H zone (in the middle of the A band)
- The rest of the A band
- A second I band

▲ **Figure 1.4** An electron micrograph of myofibrils. Note the presence of striations (the thick, purple bands).

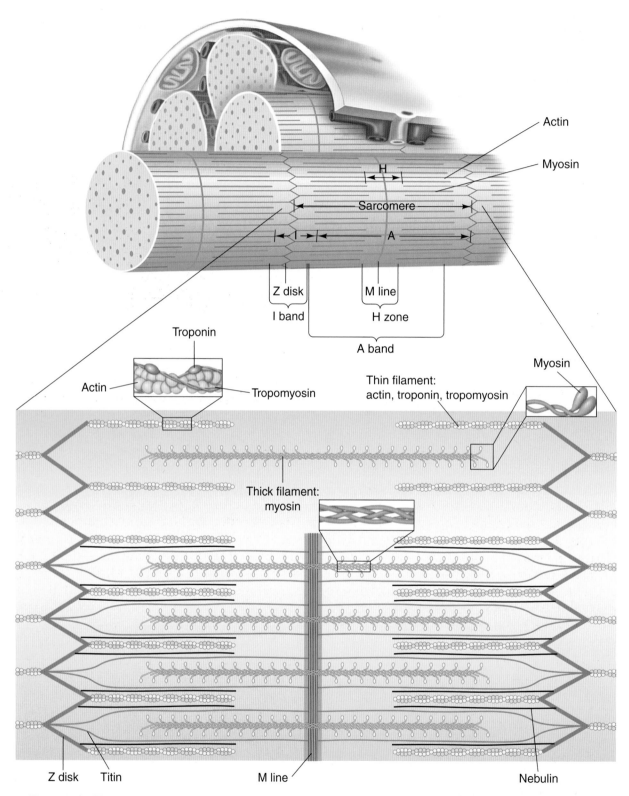

Actin

Myosin

H

Sarcomere

I A

Z disk

M line

I band

H zone

A band

Troponin

Actin

Tropomyosin

Thin filament:
actin, troponin, tropomyosin

Myosin

Thick filament:
myosin

Z disk   Titin

M line

Nebulin

▲ **Figure 1.5** The basic functional unit of a myofibril is the sarcomere, which contains a specialized arrangement of actin and myosin filaments. The role of titin is to position the myosin filament to maintain equal spacing between the actin filaments. Nebulin is often referred to as an "anchoring protein," because it provides a framework that helps stabilize the position of actin.

If you look at an individual myofibril through an electron microscope, you can differentiate two types of small protein filaments that are responsible for muscle action. The thinner filaments are **actin,** and the thicker ones are **myosin.** Approximately 3,000 actin and 1,500 myosin filaments lie side by side within each myofibril. The striations seen in muscle fibers result from alignment of these filaments, as illustrated in figure 1.4. The light I band indicates the region of the sarcomere where there are only thin actin filaments. The dark A band represents the regions that contain both thick myosin filaments and thin actin filaments. The H zone is the central portion of the A band that appears only when the sarcomere is in a resting state. Only the thick myosin filaments occupy it. The absence of the actin filaments causes the H zone to appear lighter than the adjacent A band. The H zone is visible only when the sarcomere is relaxed because the sarcomere shortens during contraction and the actin filaments are pulled into this zone, giving it the same appearance as the rest of the A band.

Myosin Filaments   Although we just told you that each myofibril contains about 3,000 actin (thin) filaments and 1,500 myosin (thick) filaments, those numbers are misleading. About two thirds of all skeletal muscle protein is myosin. Recall that myosin filaments are thick. Each myosin filament typically is formed by about 200 myosin molecules lined up end to end and side by side.

Each myosin molecule is composed of two protein strands twisted together (see figure 1.5). One end of each strand is folded into a globular head, called the myosin head. Each filament contains several such heads, which protrude from the myosin filament to form cross-bridges that interact during muscle action with specialized active sites on the actin filaments. There is an array of fine filaments, composed of **titin,** that stabilizes the myosin filaments in the longitudinal axis (see figure 1.5). **Nebulin,** an anchoring protein for actin, coextends with actin and appears to play a regulatory role in mediating actin and myosin interactions. These filaments are approximately 5 nm in diameter and 1 µm in length.

Actin Filaments   Each actin filament has one end inserted into a Z disk, with the opposite end extending toward the center of the sarcomere, lying in the space between the myosin filaments. Each actin filament contains active sites to which a myosin head can bind.

Each thin filament, although referred to simply as an actin filament, is actually composed of three different protein molecules—actin, **tropomyosin,** and **troponin.**

Actin forms the backbone of the filament. Individual actin molecules are globular and join together to form strands of actin molecules. Two strands then twist into a helical pattern, much like two strands of pearls twisted together. Each actin molecule has active binding sites that serve as a point of contact with the myosin heads.

Tropomyosin is a tube-shaped protein that twists around the actin strands, fitting in the groove between them. Troponin is a more complex protein that is attached at regular intervals to both the actin strands and the tropomyosin. This arrangement is depicted in figure 1.6. Tropomyosin and troponin work together in an intricate manner along with calcium ions to maintain relaxation or initiate action of the myofibril, which we discuss later in this chapter.

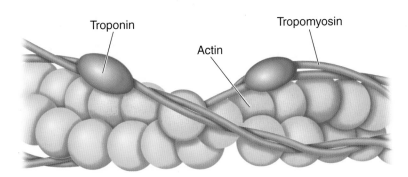

Troponin   Actin   Tropomyosin

◀ **Figure 1.6**   An actin filament, composed of molecules of actin, tropomyosin, and troponin. At rest, the active binding sites on the actin molecules are covered by strands of tropomyosin.

> ▶ Myofibrils are composed of sarcomeres, the smallest functional units of a muscle.
>
> ▶ A sarcomere is composed of filaments of two proteins, myosin and actin, which are responsible for muscle contraction.
>
> ▶ Myosin is a thick filament, composed of two protein strands, each folded into a globular head at one end.
>
> ▶ An actin filament is composed of actin, tropomyosin, and troponin. One end of each actin filament is attached to a Z disk.

Muscle Fiber Action

A motor nerve may connect with and innervate many muscle fibers. A motor nerve and all the muscle fibers it supplies are collectively termed a **motor unit** (see figure 1.7). The synapse or gap between a motor nerve and a muscle fiber is referred to as a neuromuscular junction. This is where communication between the nervous and muscular systems occurs. Let's examine this process.

> A motor unit consists of a single motor neuron and all the muscle fibers it supplies.

Motor Impulse

The events that trigger a muscle fiber to act are complex. The process, depicted in figure 1.8, is initiated by a motor nerve impulse from the brain or spinal cord. The neural impulse arrives at the nerve's endings, called axon terminals, which are located very close to the sarcolemma. When the impulse arrives, these nerve endings secrete a neurotransmitter substance called acetylcholine (ACh), which binds to receptors on the sarcolemma (see figure 1.8a). If enough ACh binds to the receptors, an electrical charge will be transmitted the full length of the muscle fiber as ion gates open in

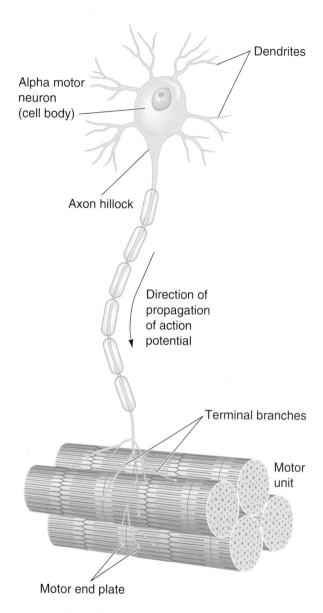

▲ **Figure 1.7**   Motor units include the alpha motor neurons and the muscle fibers they innervate.

the muscle cell membrane and allow sodium to enter. This process is referred to as depolarization and results in firing, or generating, an action potential. An action potential must be generated in the muscle cell before the muscle cell can act. These neural events are discussed more fully in chapter 2.

Role of Calcium in the Muscle Fiber

In addition to depolarizing the fiber membrane, the electrical impulse travels through the fiber's network of tubules (T tubules and SR) to the interior of the cell. The arrival of an electrical charge causes the SR to release large quantities of stored calcium ions (Ca^{2+}) into the sarcoplasm (see figure 1.8b).

In the resting state, tropomyosin molecules are believed to lie on top of the active sites on the actin filaments, preventing or weakening the binding of the myosin heads. Once calcium ions are released from the sarcoplasmic reticulum, they bind with the troponin on the actin filaments. Troponin, with its strong affinity for calcium ions, is believed to then initiate the action process by lifting the tropomyosin molecules off the active sites on the actin filaments. This is shown in figure 1.8c. Because tropomyosin normally hides the active sites,

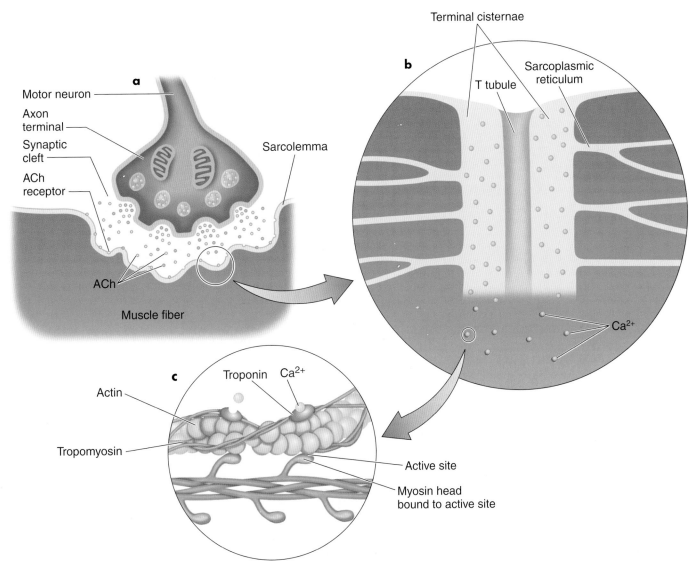

▲ Figure 1.8 The sequence of events leading to muscle action. (*a*) A motor neuron releases acetylcholine (ACh), which binds to receptors on the sarcolemma. If enough ACh binds, an action potential is generated in the muscle fiber. (*b*) The action potential triggers release of calcium ions (Ca^{2+}) from the sarcoplasmic reticulum into the sarcoplasm. (*c*) The Ca^{2+} binds to troponin on the actin filament, and the troponin pulls tropomyosin off the active sites, allowing myosin heads to attach to the actin filament.

it blocks the attraction between the **myosin cross-bridge** and actin filament. However, once the tropomyosin has been lifted off the active sites by troponin and calcium, the myosin heads can attach to the active sites on the actin filaments.

Sliding Filament Theory: How Muscle Creates Movement

How do muscle fibers shorten? The explanation for this phenomenon is termed the **sliding filament theory.** When the myosin cross-bridges are activated, they bind strongly with actin, resulting in a conformational change in the cross-bridge, which causes the myosin head to tilt toward the arm of the cross-bridge and to drag the actin and myosin filaments in opposite directions (see figure 1.9). This tilting of the head is referred to as the **power stroke.** The pulling of the actin filament past the myosin shortens the muscle and generates force. When the fibers are not contracting, the myosin head remains in contact with the actin binding site, but the molecular bonding

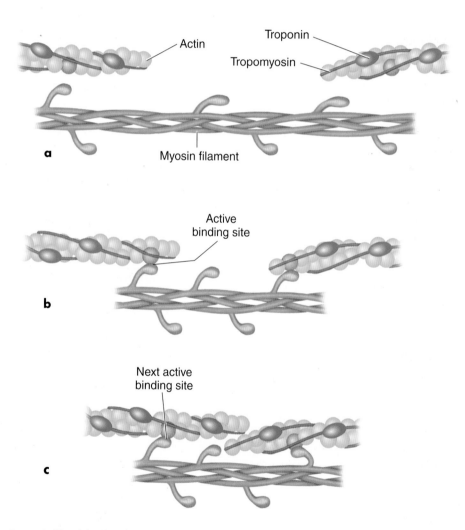

▲ **Figure 1.9**   A muscle fiber (*a*) relaxed, (*b*) contracting, and (*c*) fully contracted, illustrating the ratchetlike action responsible for the sliding of actin and myosin filaments.

at the site is weakened or blocked by tropomyosin.

Immediately after the myosin head tilts, it breaks away from the active site, rotates back to its original position, and attaches to a new active site further along the actin filament. Repeated attachments and power strokes cause the filaments to slide past one another, giving rise to the term *sliding filament theory.* This process continues until the ends of the myosin filaments reach the Z disks. During this sliding (contraction), the actin filaments are brought closer to each other and protrude into the H zone, ultimately overlapping. When this occurs, the H zone is no longer visible.

Energy for Muscle Action

Muscle action is an active process requiring energy. In addition to the binding site for actin, a myosin head contains a binding site for **adenosine triphosphate (ATP)**. The myosin molecule must bind with ATP for muscle action to occur because ATP supplies the needed energy.

The enzyme **adenosine triphosphatase (ATPase)**, which is located on the myosin head, splits the ATP to yield adenosine diphosphate (ADP), inorganic phosphate (P_i), and energy. The energy released from this breakdown of ATP is used to bind the myosin head to the actin filament. Thus, ATP is the chemical source of energy for muscle action. We discuss this in much more detail in chapter 4.

End of Muscle Action

Muscle action continues until the calcium is depleted. Calcium then is pumped back into the SR, where it is stored until a new nerve impulse arrives at the muscle fiber membrane. Calcium is returned to the SR by an active calcium-pumping system. This is another energy-demanding process that also relies on ATP. Thus, energy is required for both the action and relaxation phases.

When the calcium is removed, troponin and tropomyosin are deactivated. This blocks the linking of the myosin cross-bridges and actin filaments and stops the use of ATP. As a result, the myosin and actin filaments return to their original relaxed state.

▶ Muscle action is initiated by a motor nerve impulse. The motor nerve releases ACh, which opens up ion gates in the muscle cell membrane, allowing sodium to enter the muscle cell (depolarization). If the cell is sufficiently depolarized, an action potential is fired and muscle action occurs.

▶ The action potential travels along the sarcolemma, then moves through the tubule system, and eventually causes stored calcium ions to be released from the sarcoplasmic reticulum.

▶ Calcium ions bind with troponin, and then troponin lifts the tropomyosin molecules off of the active sites on the actin filament, opening these sites to allow the myosin heads to bind strongly with them.

▶ Once a strong binding state is established with actin, the myosin head tilts, pulling the actin filament so that the two slide across each other. The tilting of the myosin head is the power stroke.

▶ Energy is required before muscle action can occur. The myosin head binds to ATP, and ATPase found on the head splits ATP into ADP and Pi, releasing energy to fuel the contraction.

▶ Muscle action ends when calcium is actively pumped out of the sarcoplasm back into the sarcoplasmic reticulum for storage. This process, leading to relaxation and the creation of a weak binding state between the myosin heads and the active sites, also requires energy supplied by ATP.

Skeletal Muscle and Exercise

Now that we have reviewed the overall structure of muscles and the process by which myofibrils act, we are ready to look more specifically at muscle function during exercise. Your endurance and speed during exercise depend largely on your muscles' ability to produce energy and force. Let's examine how muscles accomplish this task.

Slow-Twitch and Fast-Twitch Muscle Fibers

Not all muscle fibers are alike. A single skeletal muscle contains fibers having different speeds of shortening and strength: slow-twitch (ST) and fast-twitch (FT). **Slow-twitch fibers** take approximately 110 ms to reach peak tension when stimulated. **Fast-twitch fibers**, on the other hand, can reach peak tension in about 50 ms and can create more force than an ST fiber.

The Muscle Biopsy Needle

It was once difficult to examine human muscle tissue from a living person. Most early muscle research used muscle from laboratory animals or open-incision surgery to obtain muscle from humans. In the early 1900s, a needle biopsy procedure was developed to study muscular dystrophy. In the 1960s, this technique was adapted to sample muscle for studies in exercise physiology.

Samples are removed by muscle biopsy, which involves removing a very small piece of muscle from the muscle belly for analysis. The area from which the sample is taken is first deadened with a local anesthetic, and then a small incision (approximately 1 cm, or 0.4 in.) is made with a scalpel through the skin, subcutaneous tissue, and fascia. A hollow needle is then inserted to the appropriate depth into the belly of the muscle (see figure 1.10a). A small plunger is pushed through the center of the needle to snip off a very small sample of muscle.

The biopsy needle is withdrawn, and the sample, weighing 10 to 100 mg, is removed (see figure 1.10b), cleaned of blood, mounted, and quickly frozen. It is then thinly sliced, stained, and examined under a microscope. Figure 1.10a illustrates the use of a biopsy needle to obtain a sample from the leg muscle of an elite female runner; figure 1.10b shows a close-up view of a muscle biopsy needle and a small piece of muscle tissue.

This method allows us to study muscle fibers and gauge the effects of acute exercise and chronic training on fiber composition. Microscopic and biochemical analyses of the samples aid our understanding of the muscles' machinery for energy production.

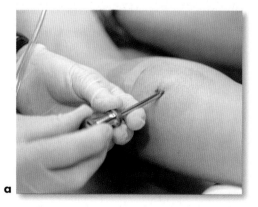
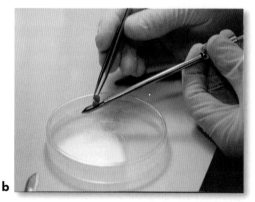

a   b

▲ **Figure 1.10**   (*a*) A muscle biopsy needle is inserted into the belly of the muscle to remove a sample of muscle tissue. (*b*) The removed muscle sample then can be studied.

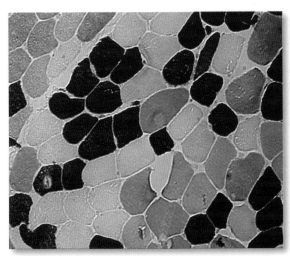

▲ **Figure 1.11**   A photomicrograph showing slow-twitch (ST, black), fast-twitch type a (FT$_a$, white), and fast-twitch type b (FT$_b$, gray) muscle fibers.

Although only one type of ST fiber has been identified, FT fibers can be further classified. The two major types of FT fibers are fast-twitch type a (FT$_a$) and fast-twitch type b (FT$_b$). Figure 1.11 is a micrograph of human muscle in which thinly sliced (10 μm) cross sections of a muscle sample have been chemically stained to differentiate the fiber types. The ST fibers are stained black, FT$_a$ fibers are unstained, and FT$_b$ fibers appear gray. Although not apparent in this figure, a third subtype of FT fibers has also been identified: type c (FT$_c$).

The differences among the FT$_a$, FT$_b$, and FT$_c$ fibers are not fully understood, but FT$_a$ fibers are believed to be the most frequently recruited. Only ST fibers are recruited more frequently than FT$_a$ fibers. FT$_c$ fibers are the least often used. On the average, most muscles are composed of roughly 50% ST fibers and 25% FT$_a$ fibers. The remaining 25% are mostly FT$_b$, with FT$_c$ fibers making up only 1% to 3% of the muscle. Because knowledge about FT$_c$ fibers is limited, we will not discuss them further. The exact percentages of these fiber types vary greatly in various muscles and among individuals, so the numbers listed here are only averages.

Characteristics of ST and FT Fibers

We need to understand the significance of different muscle fiber types. What roles do they play in physical activity? To answer this, let's first examine how the fiber types differ.

ATPase   The ST and FT fiber types derive their names from the difference in their speed of action. This difference results primarily from different forms of myosin ATPase. Recall that myosin ATPase is the enzyme that splits ATP to release energy to drive contraction or allow relaxation. ST fibers have a slow form of myosin ATPase, whereas FT fibers have a fast form. In response to neural stimulation, ATP is split more rapidly in FT fibers than in ST fibers. As a result, the FT fibers have energy for contraction available more quickly than do the ST fibers.

One of the methods used to classify muscle fibers employs a chemical staining procedure applied to a thin slice of tissue. This staining technique acts on the ATPase in the fibers. Thus, the ST, FT$_a$, and FT$_b$ fibers stain differently, as we saw in figure 1.11. This technique makes it appear that each muscle fiber has only one type of ATPase, but fibers can have a mixture of ATPase types. Some have a predominance of ST-ATPase, but others have mostly FT-ATPase. Their appearance in a stained slide preparation should be viewed as a continuum rather than as absolutely distinct types.

A newer method for identifying fiber types is to chemically separate the different types of myosin molecules (isoforms) by using a process called gel electrophoresis. As shown in figure 1.12 on page 46, the isoforms are separated and stained to show the bands of protein (i.e., myosin) that characterize ST, FT$_a$, and FT$_b$ fibers. Although our discussion here categorizes fiber types simply as slow (ST) and fast twitch (FT$_a$ and FT$_b$), other scientists have subdivided these fiber types. The use of electrophoretic technology has led to the detection of myosin hybrids or fibers that possess two or more forms of myosin. With this method of analysis, the fibers are classified as I, IIa, IIx, I/IIa, I/IIa/IIx, IIa/IIx, and I/IIx. So, you can see why we prefer to use the histochemical method of identifying fibers by their primary isoforms, type I (ST), IIa (FT$_a$), and IIb (FT$_b$).

Table 1.1 on page 46 summarizes the characteristics of the different muscle fiber types. The table also includes alternative names that are used in other classification systems to refer to the muscle fiber types.

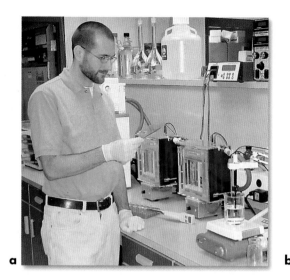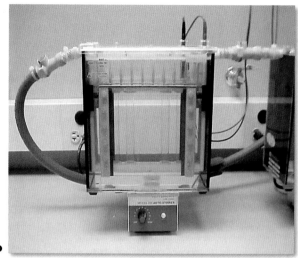

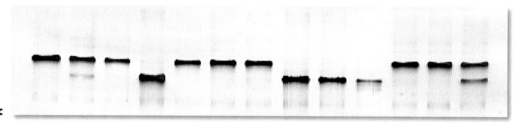

▲ **Figure 1.12**   Electrophoretic separation of myosin isoforms to identify type I (slow-twitch, or ST), IIa (fast-twitch type a, or FT$_a$), and IIb (fast-twitch type b, FT$_b$) fibers. (*a*) Single fibers are isolated under a dissecting microscope. (*b*) The myosin isoforms are separated for each fiber using electrophoretic techniques. (c) The isoforms are then stained to show the myosin that indicates the fiber type.

Table 1.1			
Classification of Muscle Fiber Types			
	Fiber classification		
System 1	Slow twitch (ST)	Fast twitch a (FT$_a$)	Fast twitch b (FT$_b$)
System 2	Type I	Type IIa	Type IIb
System 3	SO	FOG	FG
	Characteristics of fiber types		
Oxidative capacity	High	Moderately high	Low
Glycolytic capacity	Low	High	Highest
Contractile speed	Slow	Fast	Fast
Fatigue resistance	High	Moderate	Low
Motor unit strength	Low	High	High

Note. In this text we use System 1 to classify muscle fiber types. Other terminologies (Systems 2 and 3) have also been used to identify the three primary fiber types. SO = slow oxidative; FOG = fast oxidative glycolytic; FG = fast glycolytic.

Sarcoplasmic Reticulum FT fibers have a more highly developed sarcoplasmic reticulum than do ST fibers. Thus, FT fibers are more adept at delivering calcium into the muscle cell when stimulated. This ability is thought to contribute to the faster speed of action (V_o) of FT fibers. On average, human FT fibers have a V_o that is five to six times faster than ST fibers. Although the amount of force (P_o) generated by FT and ST fibers having the same diameter is about the same, the calculated power (μN · fiber length^{-1} · s^{-1}) of an FT fiber is three to five times greater than that of an ST fiber. This may explain in part why individuals who have a predominance of FT fibers in their leg muscles tend to be better sprinters than individuals who have a high percentage of ST fibers.

Motor Units Recall that a motor unit is a single motor neuron and the muscle fibers it innervates. The neuron appears to determine whether the fibers are ST or FT. The motor neuron in an ST motor unit has a small cell body and innervates a cluster of 10 to 180 muscle fibers. In contrast, the motor neuron in an FT motor unit has a larger cell body and more axons and innervates from 300 to 800 muscle fibers.

This difference in the arrangement of motor units means that when a single ST motor neuron stimulates its fibers, far fewer muscle fibers contract than when a single FT motor neuron stimulates its fibers. Consequently, FT motor fibers reach peak tension faster and collectively generate more force than ST fibers do.[1]

Single Muscle Fiber Physiology

One of the most advanced methods for the study of human muscle fibers is to dissect fibers out of a muscle biopsy sample, suspend a single fiber between force transducers, and measure its strength and **single fiber contractile velocity (V$_o$)** (see figure 1.13). Figure 1.14 illustrates the difference in the rate for ST and FT fibers to reach their peak tension, as measured by the single muscle fiber technology.

a

b

▲ **Figure 1.13**   (*a*) The dissection and (*b*) suspension of a single muscle fiber to study the physiology of different fiber types.

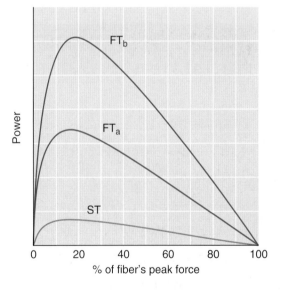

▲ **Figure 1.14**   Differences in peak power generated by each fiber type at various percentages of maximal force. Note that all of the fibers tend to reach their peak power when the fibers are generating only about 20% of their peak force. It is quite clear that the peak power of the FT fibers is considerably higher than that of the ST fibers.

Distribution of Fiber Types

As mentioned earlier, the percentages of ST and FT fibers are not the same in all the muscles of the body. Generally, a person's arm and leg muscles have similar fiber compositions. Studies have shown that people with a predominance of ST fibers in their leg muscles will likely have a high percentage of ST fibers in their arm muscles as well. A similar relationship exists for FT fibers. There are some exceptions, however. The soleus muscle (beneath the gastrocnemius in the calf), for example, is composed of a very high percentage of ST fibers in everyone.

Fiber Type and Exercise

We've looked at various ways in which ST and FT fibers differ. Because of these differences, you might expect that these fiber types would also have different functions when you are physically active. Indeed, this is the case.

ST Fibers   In general, ST muscle fibers have a high level of aerobic endurance. Aerobic means "in the presence of oxygen," so oxidation is an aerobic process. ST fibers are very efficient at producing ATP from the oxidation of carbohydrate and fat.

Recall that ATP is required to produce the energy needed for muscle fiber action and relaxation. As long as oxidation occurs, ST fibers continue producing ATP, allowing the fibers to remain active. The ability to maintain muscular activity for a prolonged period is known as muscular endurance, so ST fibers have high aerobic endurance. Because of this, they are recruited most often during low-intensity endurance events (e.g., marathon running) and during most daily activities where the muscle force requirements are low (e.g., walking).

FT Fibers   FT muscle fibers, on the other hand, have relatively poor aerobic endurance. They are better suited to perform anaerobically (without oxygen) than the ST fibers. This means that in the absence of adequate oxygen, ATP is formed through anaerobic pathways, not oxidative pathways. (We discuss these pathways in detail in chapter 4.)

FT_a motor units generate considerably more force than do ST motor units, but FT_a motor units fatigue easily because of their limited endurance. Thus, FT_a fibers appear to be used mainly during shorter, higher intensity endur-

Table 1.2

Structural and Functional Characteristics of Muscle Fiber Types

Characteristic	Fiber type		
	ST	FT_a	FT_b
Fibers per motor neuron	10-180	300-800	300-800
Motor neuron size	Small	Large	Large
Nerve conduction velocity	Slow	Fast	Fast
Contraction speed (ms)	110	50	50
Type of myosin ATPase	Slow	Fast	Fast
Sarcoplasmic reticulum development	Low	High	High
Motor unit force	Low	High	High
Aerobic capacity (oxidative)	High	Moderate	Low
Anaerobic capacity (glycolytic)	Low	High	High

Adapted, by permission, from R. Close, 1967, "Properties of motor units in fast and slow skeletal muscles of the rat," *Journal of Physiology* 193: 45-55.[2]

The difference in maximal isometric force development between FT and ST motor units is attributable to the number of muscle fibers per motor unit and the difference in strength of FT and ST fibers. ST and FT fibers of the same diameter generate about the same force. On average, however, FT fibers tend to be larger than ST fibers. So, when stimulated, FT motor units generate more force because the fibers are larger and have more muscle fibers per motor unit than do the ST motor units.

ance events, such as the mile run or the 400-m swim.

Although the significance of the FT_b fibers is not fully understood, they apparently are not easily activated by the nervous system. Because of this, they are used rather infrequently in normal, low-intensity activity but are predominantly used in highly explosive events such as the 100-m dash and the 50-m sprint swim. Characteristics of the various fiber types are summarized in table 1.2.

Determination of Fiber Type

The ST and FT characteristics of muscle fibers appear to be determined early in life, perhaps within the first few years. Studies with identical twins have shown that muscle fiber composition, for the most part, is genetically determined, changing little from childhood to middle age. These studies reveal that identical twins have nearly identical fiber compositions, whereas fraternal twins differ in their fiber profiles. The genes we inherit from our parents determine which motor neurons innervate our individual muscle fibers. After innervation is established, our muscle fibers differentiate (become specialized) according to the type of neuron that stimulates them. Some recent evidence, however, suggests that endurance training, strength training, and muscular inactivity may cause a shift in the myosin isoforms.[1] Consequently, training may induce a small change, perhaps less than 10%, in the percentage of ST and FT fibers. As well, endurance training has been shown to reduce the percentage of FT_b fibers while increasing the fraction of FT_a fibers.

Studies of older men and women have shown that aging may alter the distribution of ST and FT fibers. As we grow older, our muscles tend to lose FT motor units, which increases the percentage of ST fibers.

▶ Most skeletal muscles contain both ST and FT fibers.

▶ The different fiber types have different ATPase. The ATPase in the FT fibers acts faster, providing energy for muscle action more quickly than the ATPase in ST fibers.

▶ FT fibers have a more highly developed sarcoplasmic reticulum, enhancing the delivery of calcium needed for muscle action.

▶ Motor neurons supplying FT motor units are larger and supply more fibers than do neurons for ST motor units. Thus, FT motor units have more fibers to contract and can produce more force than ST motor units.

▶ The proportions of ST and FT fibers in an individual's arm and leg muscles are usually quite similar.

▶ ST fibers have high aerobic endurance and are well suited to low-intensity endurance activities.

▶ FT fibers are better for anaerobic activity. FT_a fibers play a major role in explosive bouts of exercise. The use of FT_b fibers is not well understood, but it appears that they are not easily activated unless the force demanded of the muscle is high.

Muscle Fiber Recruitment

When a motor neuron stimulates a muscle fiber, a minimum amount of stimulation, called the threshold, is required to elicit a response. If the stimulation is less than this threshold, no muscle action occurs. But with any stimulation equal to or exceeding the threshold, maximal action occurs in the muscle fiber. This is known as the *all-or-none response*. Because all muscle fibers in a single motor unit receive the same neural stimulation, all the fibers in the motor unit act maximally any time the threshold is met. Thus, the motor unit also exhibits an all-or-none response.

Activating more muscle fibers produces more force. When little force is needed, only a few motor units are stimulated to act. Recall from our earlier discussion that FT_a and FT_b motor units contain more muscle fibers than ST motor units do. Skeletal muscle action involves selective recruitment of ST or FT muscle fibers, depending on the requirements of the activity being performed. In the early 1970s, Gollnick and coworkers demonstrated that, indeed, this orderly recruitment is determined not by the speed of action but by the level of force demanded of the muscle.[6,7]

Orderly Recruitment of Muscle Fibers and the Size Principle

Most researchers agree that motor units are generally activated on the basis of a fixed order of fiber recruitment. This is known as the principle of orderly recruitment, in which the motor units within a given muscle appear to be ranked. Let's use the biceps brachii as an example: Assume a total of 200 motor units, which are ranked on a scale from 1 to 200. For an extremely fine muscle action requiring very little force production, the motor unit ranked number 1 would be recruited. As the requirements for force production increase, numbers 2, 3, 4, and so on would be recruited, up to a maximal muscle action that would activate 50% to 70% of the motor units. For the production of a given force, the same motor units are recruited each time.

A mechanism that may partially explain the principle of orderly recruitment is the size principle, which states that the order of recruitment of motor units is directly related to their motor neuron size. Motor units with smaller motor neurons will be recruited first. Because the ST motor units have smaller motor neurons, they are the first units recruited in graded movement (going from very low to very high rates of force production). The FT motor units then are recruited as the force needed to perform the movement increases. Some questions remain about how the size principle relates to most athletic movements because it has been

examined only in graded movements that represent muscle actions of less than 25% relative intensity.

Figure 1.15 illustrates the relationship between force development and the recruitment of ST, FT_a, and FT_b fibers. During low-intensity exercise, such as walking, most of the muscle force is generated by ST fibers. As the muscle tension requirements increase at higher exercise intensities, such as jogging, FT_a fibers are added to the work force. Finally, in events where maximal strength is needed, such as sprint running, FT_b fibers are also activated.

However, even during maximal efforts, the nervous system does not recruit 100% of the available fibers. Despite your desire to produce more force, only a fraction of your muscle fibers are stimulated at any specific time. This prevents damage to your muscles and tendons. If you could contract all the fibers in your muscle at the same instant, the force generated would likely tear the muscle or its tendon.

During events that last several hours, you must exercise at a submaximal pace, and the tension in your muscles is relatively low. As a result, the nervous system tends to recruit

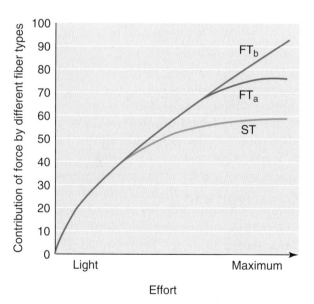

▲ **Figure 1.15** The ramplike recruitment of slow-twitch (ST) and fast-twitch (FT) muscle fibers. Note that ST fibers are responsible for most of the force development at the lighter loads, whereas FT_a and FT_b fibers contribute progressively more to the force production as the force requirement increases. All fiber types are recruited when near-maximal force is required.

those muscle fibers best adapted to endurance activity: the ST and some FT_a fibers. As the exercise continues, these fibers become depleted of their primary fuel supply (glycogen), and the nervous system must recruit more FT_a fibers to maintain muscle tension. Finally, when the ST and FT_a fibers become exhausted, the FT_b fibers may be recruited to continue exercising.

This may explain why fatigue seems to come in stages during events such as the marathon, a 42-km (26.2-mi) run. It also may explain why it takes great conscious effort to maintain a given pace near the finish of the event. This conscious effort results in the activation of muscle fibers that are not easily recruited. Such information is of practical importance to our understanding of the specific requirements of training and performance. We discuss this further in chapters 3 and 6.

> ▶ Motor units give all-or-none responses. For a unit to be recruited into activity, the motor nerve impulse must meet or exceed the threshold. When this occurs, all muscle fibers in the motor unit act maximally. If the threshold is not met, no fibers in that unit act.
>
> ▶ Activating more motor units and thus more muscle fibers produces more force.
>
> ▶ In low-intensity activity, most muscle force is generated by ST fibers. As the intensity increases, FT_a fibers are recruited, and at the higher intensities, the FT_b fibers are activated. The same pattern of recruitment is followed during events of long duration.

Fiber Type and Athletic Success

Knowledge of the composition and use of muscle fibers suggests that athletes who have a high percentage of ST fibers might have an advantage in prolonged endurance events, whereas those with a predominance of FT fibers could be better suited for short-term and explosive activities. Can it be that the proportions of an athlete's various muscle fiber types determine athletic success?

The muscle fiber makeup of successful athletes from a variety of athletic events and of nonathletes is shown in table 1.3 on page 52. As anticipated, the leg muscles of distance runners, who rely on endurance, have a predominance of ST fibers.[3] Studies of elite male and female distance runners revealed that many of these athletes' gastrocnemius (calf) muscles contain more than 90% ST fibers. Also, although muscle fiber cross-sectional area varies markedly among elite distance runners, ST fibers in their leg muscles average about 22% more cross-sectional area than FT fibers.

In contrast, the gastrocnemius muscles are composed principally of FT fibers in sprint runners, who rely on speed and strength. Although swimmers tend to have higher percentages of ST fibers (60-65%) in their arm muscles than untrained subjects (45-55%), fiber-type differences between good and elite swimmers are not apparent.[4, 5]

> World champions in the marathon are reported to possess 93% to 99% ST fibers in their gastrocnemius muscles. World-class sprinters, on the other hand, have only about 25% ST fibers in this muscle.

The fiber composition of muscles in distance runners and sprinters is markedly different. However, it may be a bit risky to think we can select champion distance runners and sprinters solely on the basis of predominant muscle fiber type. Other factors, such as cardiovascular function, motivation, training, and muscle size, also contribute to success in such events of endurance, speed, and strength. Thus, fiber composition alone is not a reliable predictor of athletic success.

Use of Muscles

We have examined the different muscle fiber types. We understand that all fibers in a motor unit, when stimulated, act at the same time and that different fiber types are recruited in stages, depending on the force required to perform an activity. Now we can move back to the gross level, turning our attention to how whole muscles work to produce movement.

The more than 600 skeletal muscles in the body vary widely in size, shape, and use. Every

Table 1.3

Percentages and Cross-Sectional Areas of Slow-Twitch (ST) and Fast-Twitch (FT) Fibers in Selected Muscles of Male and Female Athletes

Athlete	Sex	Muscle	% ST	% FT	Cross-sectional area ($\mu m2$) ST	FT
Sprint runners	M	Gastrocnemius	24	76	5,878	6,034
	F	Gastrocnemius	27	73	3,752	3,930
Distance runners	M	Gastrocnemius	79	21	8,342	6,485
	F	Gastrocnemius	69	31	4,441	4,128
Cyclists	M	Vastus lateralis	57	43	6,333	6,116
	F	Vastus lateralis	51	49	5,487	5,216
Swimmers	M	Posterior deltoid	67	33	—	—
Weightlifters	M	Gastrocnemius	44	56	5,060	8,910
	M	Deltoid	53	47	5,010	8,450
Triathletes	M	Posterior deltoid	60	40	—	—
	M	Vastus lateralis	63	37	—	—
	M	Gastrocnemius	59	41	—	—
Canoeists	M	Posterior deltoid	71	29	4,920	7,040
Shot-putters	M	Gastrocnemius	38	62	6,367	6,441
Nonathletes	M	Vastus lateralis	47	53	4,722	4,709
	F	Gastrocnemius	52	48	3,501	3,141

coordinated movement requires the application of muscle force. This is accomplished by

- agonists, or prime movers, muscles primarily responsible for the movement;
- antagonists, muscles that oppose the prime movers; and
- synergists, muscles that assist the prime movers.

As illustrated in figure 1.16, the smooth flexion of the elbow requires shortening of the brachialis and biceps brachii muscles (agonists) and relaxation of the triceps brachii (antagonist). The brachioradialis muscle (synergist) assists the brachialis and biceps brachii in their flexion of the joint.

The agonists produce most of the force needed for any particular movement. The muscles act on the bones to which they are attached, pulling them toward each other. Synergists assist this action and sometimes are involved in fine-tuning the direction of movement. The antagonists play a protective role. Think of the quadriceps (front) and hamstrings (back) in your thigh. When your hamstrings (agonists) contract forcefully, your quadriceps (antagonists) also contract slightly, opposing the motion of the hamstrings. This prevents overstretching of your quadriceps by the strong contraction of the hamstrings and

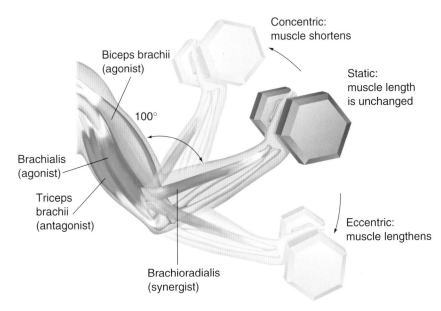

Concentric:
muscle shortens

Biceps brachii
(agonist)

100°

Static:
muscle length
is unchanged

Brachialis
(agonist)

Triceps
brachii
(antagonist)

Eccentric:
muscle lengthens

Brachioradialis
(synergist)

▲ **Figure 1.16**   The actions of agonistic, antagonistic, and synergistic muscles during elbow flexion.

allows more controlled thigh movement. This opposing action between agonists and antagonists also produces muscle tone.

Types of Muscle Action

Muscle movement generally can be categorized into three types of actions—concentric, static, and eccentric. In many activities, such as running and jumping, all three types of actions may occur in the execution of a smooth, coordinated movement. For the sake of clarity, though, we will examine each type separately.

Concentric action. A muscle's principal action, shortening, is referred to as a concentric action, shown in figure 1.16. We are most familiar with this type of action. To understand muscle shortening, recall our earlier discussion of how the actin and myosin filaments slide across each other. In a concentric action, the actin filaments are pulled closer together. Because joint movement is produced, concentric actions are considered **dynamic actions.**

Static isometric muscle action. Muscles can also act without moving. When this happens, the muscle generates force, but its length remains static (unchanged). This is called a static action, or an isometric action, because the joint angle does not change (see figure 1.16) . A static action occurs, for example, when you try to lift an object that is heavier than the force generated by your muscle or when you sup-

port the weight of an object by holding it steady with your elbow flexed. In both cases, you feel your muscles tense, but they can't move the weight so they don't shorten. In a static action, the myosin cross-bridges form and are recycled, producing force, but the external force is too great for the actin filaments to be moved. They remain in their normal position, so shortening can't occur. If enough motor units can be recruited to produce sufficient force to overcome the resistance, a static action can become a dynamic one.

Eccentric action. Muscles can exert force even while lengthening. This movement is an eccentric action. Because joint movement occurs, this is also a dynamic action. An example of an eccentric action is the action of the biceps brachii when you extend your elbow to lower a heavy weight. In this case, the actin filaments are pulled farther away from the center of the sarcomere, essentially stretching it.

Generation of Force

Your muscles' strength reflects their ability to produce force. If you have the strength to bench-press 150 kg (330 lb), your muscles are capable of producing enough force to overcome a load of 150 kg. Even when unloaded (not trying to lift a weight), your muscles must still generate enough force to move the bones to which they are attached. The development of this muscle force depends on the number and type of motor units activated, the size of the muscle, the muscle's initial length when activated, the angle of the joint, and the muscle's speed of action. Let's examine these components.

Motor Units and Muscle Size   We previously discussed motor units. To recap, more force can be generated when more motor units are activated. FT motor units generate more force than ST motor units because each FT unit has more muscle fibers than an ST unit. In a similar manner, larger muscles, having more muscle fibers, can produce more force than smaller muscles.

Muscle Length   Muscles and their connective tissues (fasciae and tendons) have the property of elasticity. When tissues are stretched, this elasticity results in stored energy. During subsequent muscle activity, this stored energy is released, increasing the amount of force.

In the intact body, the anatomical arrangement and attachment of muscle to bone restricts muscle length. When attached to the skeleton, a muscle at resting length is normally under slight tension because it is moderately stretched. If a muscle were freed from its attachments, it would assume a relaxed, somewhat shorter length.

Measurements indicate that maximal force can be generated in a muscle when the muscle is first stretched to a length approximately 20% greater than its resting length. When the muscle is stretched to this length, the combination of stored energy and the force of muscle action is optimized, resulting in maximal force production.

Increasing or decreasing the muscle length beyond 20% reduces force development. For example, if the muscle is stretched to twice its resting length, the force it produces will be nearly zero. Energy is still stored in the muscle because of the stretching. In fact, more stretching means more stored energy. But another factor must be considered. The force created by muscle fibers during muscle action depends on the number of cross-bridges in contact with the actin filaments at any given time. The more cross-bridges that are in contact at once, the more forceful the muscle action. When muscle fibers are overstretched, the actin and myosin filaments are pulled farther apart. The decreased overlap of these filaments results in fewer cross-bridges binding to create force.

Angle of the Joint   Because muscles exert their force through skeletal levers, understanding the physical arrangement of these muscle pulleys and bone levers is crucial to understanding movement. Consider the biceps brachii. The tendon attachment for the biceps is only one tenth the distance from the elbow fulcrum to the weighted resistance held in the hand. Thus, to hold a 5-kg (11-lb) weight, the muscle must exert 10 times as much force (50 kg, or 110 lb).

The force generated in the muscle is transferred to the bone through the muscle's insertion (tendon). As with muscle length, an optimal joint angle will maximize the amount of force transmitted to the bone. This angle depends on the relative position of the tendons insertion on the bone and the load being moved. In our example of the biceps brachii, the best joint angle for application of the needed 50 kg (110 lb) of force is 100°. More or less flexion of the elbow joint will alter the angle at which the force is applied, reducing the amount of force transferred to the bone.

Speed of Action   The ability to develop force also depends on the speed of muscle action. During concentric (shortening) actions, maximal force development decreases progressively at higher speeds. Think of when you try to lift a very heavy object. You tend to do it slowly, maximizing the force you can apply to it. If you grab it and quickly try to lift it, you will likely fail, if not injure yourself. However, with eccentric (lengthening) actions, the opposite is true. Fast eccentric actions allow maximal application of force. These relationships are depicted in figure 1.17. Eccentric actions are shown on the left and concentric on the right.

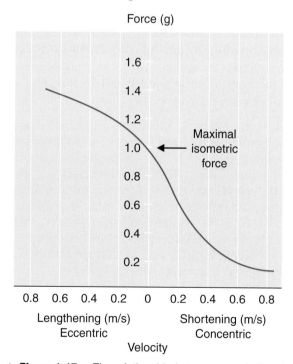

▲ **Figure 1.17**   The relationship between muscle lengthening and shortening velocity and force production. Note that the capacity for the muscle to generate force is greater during eccentric (lengthening) actions than during concentric (shortening) actions.

▶ Muscles involved in a movement can be classed as agonists (prime movers), antagonists (opponents), or synergists (assistants).

▶ The three main types of muscle action are concentric, in which the muscle shortens; static, in which the muscle acts but the joint angle is unchanged; and eccentric, in which the muscle lengthens.

▶ Force production can be increased by recruiting more motor units.

▶ Force production can be maximized if the muscle is stretched 20% before action. At this length, the amount of energy stored and the number of linked actin–myosin cross-bridges are optimal.

▶ All joints have an optimal angle at which the muscles crossing the joint produce maximal force. This angle varies with the relative position of the muscle's insertion on the bone and the load placed on the muscle.

▶ Speed of action also affects the amount of force produced. For concentric action, maximal force can be achieved with slower contractions. The closer you get to zero velocity (static), the more force can be generated. With eccentric actions, however, faster movement allows more force production.

In Closing . . .

In this chapter, we reviewed the components of skeletal muscle. We considered the differences in fiber types and their impact on physical performance. We learned how muscles generate force and produce movement by pulling on bones. Now that we understand how movement is produced, it is time to turn our attention to how movement is coordinated. In the next chapter, we focus on the neurological control of movement.

▶ Key Terms

actin
adenosine triphosphatase (ATPase)

adenosine triphosphate (ATP)
concentric action
dynamic action
eccentric action
endomysium
epimysium
fasciculus
fast-twitch (FT) fiber
motor unit
muscle fiber
myofibril
myosin
myosin cross-bridge
nebulin
perimysium
power stroke
principle of orderly recruitment
sarcolemma
sarcomere
sarcoplasm
sarcoplasmic reticulum (SR)
single fiber contractile velocity (V_o)
size principle
sliding filament theory
slow-twitch (ST) fiber
static (isometric) action
titin
transverse tubules (T tubules)
tropomyosin
troponin

▶ Study Questions

1. List and define the components of a muscle fiber.

2. List the components of a motor unit.

3. What is the role of calcium in muscle action?

4. Describe the sliding filament theory. How do muscle fibers shorten?

5. What are the basic characteristics of slow- and fast-twitch muscle fibers?

6. What is the role of genetics in determining the proportions of muscle fiber types

and the potential for success in selected activities?

7. Describe the relationship between muscle force development and the recruitment of slow- and fast-twitch motor units.

8. What is the pattern of muscle fiber recruitment during (a) high jumping, (b) running a 10-km race, and (c) running a marathon?

9. Differentiate and give examples of concentric, static, and eccentric actions.

10. What is the optimal length of a muscle for maximal force development?

11. What is the relationship between maximal force development and the speed of shortening (concentric) and lengthening (eccentric) actions?

▶ References

1. Bouchard, C., Dionne, F.T., Simoneau, J.-A., & Boulay, M.R. (1992). Genetics of aerobic and anaerobic performance. *Exercise and Sport Sciences Reviews*, **20**, 27-58.

2. Close, R. (1967). Properties of motor units in fast and slow skeletal muscles of the rat. *Journal of Physiology* (London), **193**, 45-55.

3. Costill, D.L., Daniels, J., Evans, W., Fink, W., Krahenbuhl, G., & Saltin, B. (1976). Skeletal muscle enzymes and fiber composition in male and female track athletes. *Journal of Applied Physiology*, **40**, 149-154.

4. Costill, D.L., Fink, W.J., Flynn, M., & Kirwan, J. (1987). Muscle fiber composition and enzyme activities in elite female distance runners. *International Journal of Sports Medicine*, **8**, 103-106.

5. Costill, D.L., Fink, W.J., & Pollock, M.L. (1976). Muscle fiber composition and enzyme activities of elite distance runners. *Medicine and Science in Sports*, **8**, 96-100.

6. Gollnick, P.D., & Hodgson, D.R. (1986). The identification of fiber types in skeletal muscle: A continual dilemma. *Exercise and Sport Sciences Reviews*, **14**, 81-104.

7. Gollnick, P.D., Piehl, K., & Saltin, B. (1974). Selective glycogen depletion pattern in human muscle fibers after exercise of varying intensity and at varying pedal rates. *Journal of Physiology*, **241**, 45-47.

8. McComas, A.J. (1996). *Skeletal muscle form and function*. Champaign, IL: Human Kinetics.

▶ Selected Readings

Brobeck, J.R. (Ed.). (1979). *Best and Taylor's physiological basis of medical practice* (10th ed.). Baltimore: Williams & Wilkins.

Bouchard, C., Simoneau, J.A., Lortie, G., Boulay, M.R., Marcotte, M., & Thibault, M.C. (1986). Genetic effects in human skeletal muscle fiber type distribution and enzyme activities. *Canadian Journal of Applied Physiology and Pharmacology*, **64**, 1245-1251.

Buchthal, F., & Schmalbruch, H. (1970). Contraction times and fiber types in intact muscle. *Acta Physiologica Scandinavica*, **79**, 435-452.

Burke, R.E., & Edgerton, V.R. (1975). Motor unit properties and selective involvement in movement. *Exercise and Sport Sciences Reviews*, **3**, 31-81.

Coggan, A.R., Spina, R.J., Rogers, M.A., King, D.S., Brown, M., Nemeth, P.M., & Holloszy, J.O. (1992). Characteristics of skeletal muscle in master athletes. *Journal of Applied Physiology*, **68**, 1896-1901.

Essen-Gustavsson, B., & Borges, O. (1986). Histochemical and metabolic characteristics of human skeletal muscle in relation to age. *Acta Physiologica Scandinavica*, **126**, 107-114.

Gordon, T., & Pattullo, M.C. (1993). Plasticity of muscle fiber and motor unit types. *Exercise and Sport Sciences Reviews*, **21**, 331-362.

Heckman, C.J., & Sandercock, T.G. (1996). From motor unit to whole muscle properties during locomotor movements. *Exercise and Sport Sciences Reviews*, **24**, 109.

Hultman, E. (1995). Fuel selection, muscle fibre. *Proceedings of the Nutrition Society*, **54**, 107-121.

Karlsson, J. (1977). Skeletal muscle fibres and muscle enzyme activities in monozygous and dizygous twins of both sexes. *Acta Physiologica Scandinavica*, **100**, 385-392.

Komi, P.V., & Karlsson, J. (1979). Physical performance, skeletal muscle enzyme activities, and fibre types in monozygous and dizygous twins of both sexes. *Acta Physiologica Scandinavica*, **462**(Suppl.), 1-28.

Roy, R.R., Baldwin, K.M., & Edgerton, V.R. (1991). The plasticity of skeletal muscle: Effects of neuromuscular activity. *Exercise and Sport Sciences Reviews*, **19**, 269-312.

Staron, R.S. (1997). Human skeletal muscle fiber types: Delineation, development and distribution. *Canadian Journal of Applied Physiology*, **22**, 307-327.

Staron, R.S., Hagerman, F.C., Hikida, R.S., Murray, T.F., Hostler, D.P., Crill, M.T., Ragg, K.E., & Toma, K. (2000). *Journal of Histochemistry and Cytochemistry*, **48**, 623-629.

Trappe, S.W., Williamson, D.L., Godard, M.P., Porter, D.A., Rowden, G., & Costill, D.L. (2000). Effect of resistance training on single muscle fiber contractile function in older men. *Journal of Applied Physiology,* **89,** 143-152.

Trappe, S.W. (2001). Master athletes. *International Journal of Sport Nutrition and Exercise Metabolism,* **11,** S194-S205.

Williamson, D.L., Godard, M.P., Porter, D.A., Costill, D.L., & Trappe, S.W. (2000). Progressive resistance training reduces myosin heavy chain coexpression in single muscle fibers from older men. *Journal of Applied Physiology,* **88,** 627-633.

CHAPTER 2

NEUROLOGICAL CONTROL OF MOVEMENT

overview

In chapter 1 we discussed how muscles, by generating force, pull on the bones to which they are attached to produce movement. This movement would not be possible without the nervous system. Just as the skeleton remains motionless without the application of force by the muscles, the muscles themselves cannot move unless activated by the nervous system. The nervous system plans, initiates, and coordinates all human movement. But not all movement is planned, and the nervous system must deal with incoming as well as outgoing signals. We shall see that it is a very complex system, still not fully understood by those scientists researching this area. This role of the nervous system—controlling body movement—is our focus in this chapter.

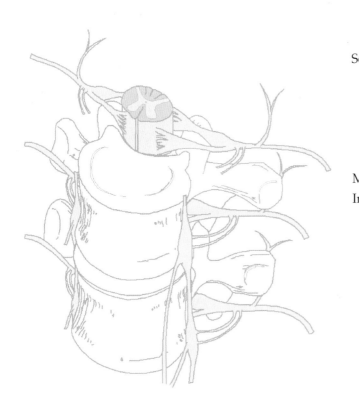

outline

In 1959, at age 15, Jimmie Heuga was the youngest male ever to make the U.S. Ski Team. He raced internationally for 10 years, competing on the 1964 and 1968 Olympic teams and on the 1962 and 1966 World Championship teams. In 1964, Heuga and teammate Billy Kidd made history by winning the first U.S. Olympic medals in men's alpine skiing. In 1967, Heuga won a World Cup in the giant slalom, finishing third in the world for the entire season. He remains the only American male to win the Arlberg-Kandahar at Garmisch, Germany, one of the oldest and most prestigious alpine ski races.

After competing in the 1968 Olympics, troubled by unknown physical ailments, Heuga retired from the U.S. Ski Team. In 1970, he was diagnosed with multiple sclerosis (MS), a neurological disorder. At that time, people with MS were told that physical activity would exacerbate their condition, so he was advised to live a quiet and tranquil life. Heuga followed that advice and began feeling unhealthy, unmotivated, and less energetic. He began to deteriorate physically and mentally.

Six years later, Heuga decided to defy medical convention. He developed a cardiovascular endurance exercise program and began stretching and strengthening exercises. He established realistic goals for his personal wellness program. With this program, Heuga regained his health within the constraints of MS. In 1984, inspired by his own success, he created the Jimmie Heuga Center, a nonprofit organization based in Edwards, Colorado. Since then, several thousand people with MS have gone through the center's medical program. Furthermore, the center has been able to fund research studies and provide financial support for both doctoral and postdoctoral students. The center's most important contribution to research on MS, published in the *Annals of Neurology* in 1996,[4] demonstrated that a physical activity training program enhances physiological and psychological function and general quality of life in MS patients, countering the medical wisdom at that time, which dictated a life of restricted activity.

Source: Dr. Richard W. Hicks, Executive Director; Jimmie Heuga Center; Edwards, Colorado; June 1997.

All physiological activity in the human body can be influenced by the nervous system. Nerves provide the wiring through which electrical impulses are received from and sent to virtually all parts of the body. The brain acts as a computer, integrating all incoming information, selecting an appropriate response, and then instructing the involved body parts to take appropriate action. Thus, the nervous system forms a vital link, allowing communication and coordination of interaction among the various tissues in the body as well as with the outside world.

The nervous system is one of the body's most complex systems. Many of its functions are not yet fully understood. For these reasons, and because this book is concerned only with neural control of voluntary movement, we will not go into as much detail about the nervous system as a whole as you did in introductory anatomy and physiology. Rather, we first look at an overview of the structure and function of the nervous system and then focus on specific topics relevant to sport and exercise.

Overview of the Nervous System

Before we delve into the intricate details of the nervous system, it is important to first step back and look at the big picture, how the nervous system is organized. This should help you better understand how the pieces of the nervous system come together as a whole—in the form of nervous system function—to integrate the body's movement. First, the nervous system as a whole is composed of two components: the **central nervous system (CNS)** and the **peripheral nervous system (PNS)**. The central nervous system is composed of the brain and spinal cord, whereas the peripheral nervous system is composed of two major divisions, the **sensory division** (or **afferent division**) and the motor division (or efferent division). The sensory division is responsible for informing the central nervous system what is going on within and outside the body. The motor division is responsible for sending information from the central nervous system to the various parts of the body in response to the signals coming in from the sensory division. The motor division is composed of two parts, the autonomic nervous system and the somatic nervous system. Figure 2.1 provides a schematic of these relationships. Much more detail concerning each of these individual units of the nervous system is presented later in this chapter.

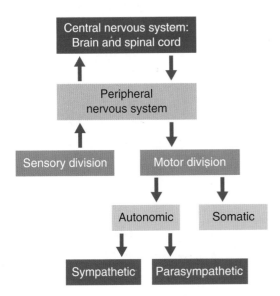

▲ **Figure 2.1**   Organization of the nervous system.

Structure and Function of the Nervous System

The **neuron** is the structural unit of the nervous system. We first review the anatomy of the neuron and then look at how it functions—allowing electrical impulses to be transmitted throughout the body.

Neuron

Individual nerve fibers (nerve cells), depicted in figure 2.2 on page 62, are called neurons. A typical neuron is composed of three regions:

- The cell body, or soma
- The dendrites
- The axon

The cell body contains the nucleus. Radiating out from the cell body are the cell processes: the dendrites and the axon. On the side toward the axon, the cell body tapers into a cone-shaped region known as the **axon hillock.** It has an important role in impulse conduction, which we discuss later.

Most neurons contain many dendrites. These are the neuron's receivers. Most impulses coming into the nerve, from sensory stimuli or from adjacent neurons, typically enter the neuron via the dendrites. These processes then carry the impulses toward the cell body.

In contrast, most neurons have only one axon. The axon is the neuron's transmitter. It conducts impulses away from the cell body. Near its end, an axon splits into numerous **end branches.** The tips of these branches are dilated into tiny bulbs known as **axon terminals** or synaptic knobs. These terminals or knobs house numerous vesicles (sacs) filled with chemicals known as **neurotransmitters** that are used for communication between a neuron and another cell. (This is discussed later in this chapter in more detail.) The structure of the neuron allows nerve impulses to enter the neuron through the dendrites, and to a lesser extent through the cell body, and to travel through the cell body and axon hillock, down the axon, and out through the end branches to the axon terminals. We next explain how this happens in somewhat more detail, including how these impulses travel from one neuron to another and from a motor neuron to muscle fibers.

Nerve Impulse

A **nerve impulse**—an electrical charge—is the signal that passes from one neuron to the next and finally to an end organ, such as a group of muscle fibers, or back to the central nervous system. For simplicity, you can think of the nerve impulse traveling through a neuron much as electricity travels through the electrical wires in your home. Let's look at how this electrical impulse is generated and how it travels through a neuron.

Resting Membrane Potential

The cell membrane of a neuron at rest has a negative electrical potential of about –70 mV. This means that if you were to insert a voltmeter probe inside the cell, the electrical charges found there and the charges found outside the cell would differ by 70 mV, and the inside would be negative relative to the outside. This potential difference is known as the **resting membrane potential (RMP)**. It is caused by a separation of charges across the membrane. When the charges across the membrane differ, the membrane is said to be polarized.

The neuron has a high concentration of potassium ions (K^+) on the inside and a high concentration of sodium ions (Na^+) on the outside. The imbalance in the number of ions inside and outside

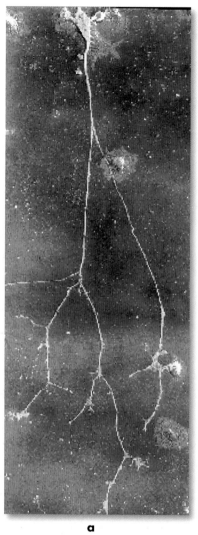

a

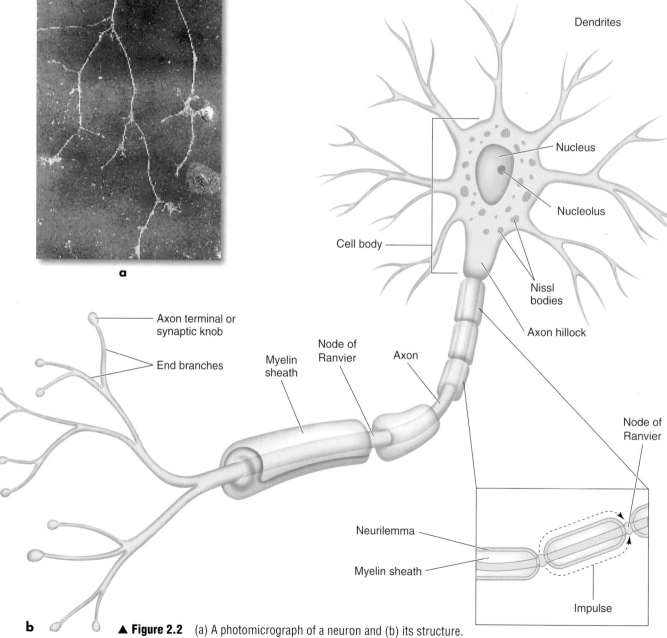

b ▲ **Figure 2.2** (a) A photomicrograph of a neuron and (b) its structure.

the cell causes the resting membrane potential. This imbalance is maintained in two ways. First, the cell membrane is much more permeable to K^+ than to Na^+, so the K^+ can move more freely. Because ions tend to move to establish equilibrium, some of the K^+ will move to an area where it is less concentrated, outside the cell. The Na^+ cannot move in this manner. Second, the neuron's **sodium–potassium pump**, which is actually an enzyme called Na^+–K^+ adenosine triphosphatase (ATPase), maintains the imbalance on each side of the membrane by actively transporting potassium and sodium ions. The sodium–potassium pump moves three Na^+ out of the cell for each two K^+ it brings in. The end result is that more positively charged ions are outside the cell than inside, creating the potential difference across the membrane. Maintenance of a constant resting membrane potential of –70 mV is primarily a function of the sodium–potassium pump.

Depolarization and Hyperpolarization

If the inside of the cell becomes less negative relative to the outside, the potential difference across the membrane decreases. The membrane will be less polarized. When this happens, the membrane is said to be depolarized. Thus, **depolarization** occurs any time the charge difference becomes less than the RMP of –70 mV, moving closer to zero. This typically results from a change in the membrane's Na^+ permeability.

The opposite can also occur. If the charge difference across the membrane increases, moving from the RMP to an even more negative number, then the membrane becomes more polarized. This is known as **hyperpolarization.**

Changes in the membrane potential are actually signals used to receive, transmit, and integrate information within and between cells. These signals are of two types, graded potentials and action potentials. Both are electrical currents created by the movement of ions. Let's look at them now.

Graded Potentials

Graded potentials are localized changes in the membrane potential. These changes can be either depolarizations or hyperpolarizations. The membrane contains ion channels with ion gates that act as doorways into and out of the neuron. These gates are usually closed, preventing ion flow, but they open with stimulation, allowing ions to move from the outside to the inside or vice versa. This ion flow alters the charge separation, changing the polarization of the membrane.

Graded potentials are triggered by a change in the neuron's local environment. Depending on the location and type of neuron involved, the ion gates may open in response to the transmission of an impulse from another neuron or in response to sensory stimuli such as changes in chemical concentrations, temperature, or pressure.

Recall that most neuron receptors are located on the dendrites (although some are on the cell body), yet the impulse is always transmitted from the axon terminals at the opposite end of the cell. For a neuron to transmit an impulse, the impulse must travel almost the entire length of the neuron. Although a graded potential may result in depolarization of the entire cell membrane, it is usually just a local event, and the depolarization does not spread very far along the neuron. To travel the full distance, an impulse must generate an action potential.

> Nerve impulses typically pass from the dendrites to the cell body and from the cell body along the length of the axon to its terminal fibrils.

Action Potentials

An **action potential** is a rapid and substantial depolarization of the neuron's membrane. It usually lasts only about 1 ms. Typically, the membrane potential changes from the RMP of –70 mV to a value of +30 mV and then rapidly returns to its resting value. How does this marked change in membrane potential occur?

All action potentials begin as graded potentials. When enough stimulation occurs to cause a depolarization of at least 15 to 20 mV, an action potential results. In other words, if the membrane depolarizes from the RMP of –70 mV to a value of –50 to –55 mV, the cell will experience an action potential. The minimum depolarization required to produce an action potential is called the **threshold.** Any depolarization less than the threshold value of 15 to 20 mV will not result in an action potential. For example, if the membrane potential changes from the RMP of –70 mV to –60 mV, the change is only 10 mV and doesn't meet the threshold; thus, no action potential occurs. But any time depolarization reaches or exceeds the threshold, an action potential will result. This is the all-or-none principle.

When a given segment of an axon is generating an action potential and its sodium gates are open, it is unable to respond to another stimulus. This is referred to as the *absolute refractory period.* When the sodium gates are closed, the potassium gates are open, and repolarization is occurring, the segment of the axon can then respond to a new stimulus, but it must be of substantially greater magnitude to evoke an action potential. This is referred to as the *relative refractory period.*

Propagation of the Action Potential

Now that we understand how a neural impulse, in the form of an action potential, is generated, we can look at how the impulse is propagated, or how it travels through the neuron. Two characteristics of the neuron become particularly important when we consider how quickly an impulse can pass through the axon: myelination and diameter.

Myelin Sheath   The axons of most motor neurons are myelinated, meaning they are covered with a sheath formed by myelin, a fatty substance that insulates the cell membrane. In the peripheral nervous system, this **myelin sheath** (see figure 2.2) is formed by specialized cells called Schwann cells.

The sheath is not continuous. As it spans the length of the axon, the myelin sheath exhibits gaps between adjacent Schwann cells, leaving the axon uninsulated at those points. These gaps are referred to as nodes of Ranvier (see figure 2.2). The action potential appears to jump from one node to the next as it traverses a myelinated fiber. This is referred to as **saltatory conduction,** a much faster type of conduction than occurs in unmyelinated fibers.

> **fyi**   The velocity of nerve impulse transmission in large myelinated fibers can be as high as 100 m/s, or 5 to 150 times faster than in unmyelinated fibers of the same size.

Myelination of motor neurons occurs over the first several years of life, partly explaining why children need time to develop coordinated movement. Individuals affected by certain neurological diseases, such as MS as discussed in our chapter opening, experience degeneration of the myelin sheath and a subsequent loss of coordination.

▶ A neuron's RMP of –70 mV results from the separation of sodium and potassium ions maintained primarily by the sodium–potassium pump, coupled with low sodium permeability and high potassium permeability of the neuron membrane.

▶ Any change that makes the membrane potential less negative beyond a given threshold results in a positive potential and a depolarization. Any change making this potential more negative is a hyperpolarization. These changes occur when ion gates in the membrane open, permitting ions to move from one side to the other.

▶ If the membrane is depolarized by 15 to 20 mV, the threshold is reached and an action potential results. Action potentials are not generated if threshold is not met.

▶ In myelinated neurons, the impulse travels through the axon by jumping between nodes of Ranvier (gaps between the cells that form the myelin sheath). This process, saltatory conduction, is 5 to 150 times faster than in unmyelinated fibers of the same size.

▶ Impulses also travel faster in neurons of larger diameters.

Diameter of the Neuron The velocity of nerve impulse transmission is also determined by the neuron's size. Neurons of larger diameter conduct nerve impulses faster than neurons of smaller diameter because larger neurons present less resistance to local current flow. This is consistent with our discussion in chapter 1, where we established that fast-twitch motor units have large motor neurons and slow-twitch motor units have small motor neurons.

Synapse

For a neuron to communicate with another neuron, an action potential must occur. Once the action potential is fired, the nerve impulse travels the full length of the axon, ultimately reaching the axon terminals. How does the nerve impulse move from the neuron in which it starts to another neuron?

Neurons communicate with each other across synapses. A **synapse** is the site of impulse

transmission from one neuron to another. The most common type of synapse is the chemical synapse, which is our focus.

As seen in figure 2.3, a synapse between two neurons includes

- the axon terminals of the neuron carrying the impulse,
- receptors on the second neuron, and
- the space between these structures.

The neuron sending the impulse across the synapse is called the presynaptic neuron, so axon terminals are presynaptic terminals. Similarly, the neuron receiving the impulse on the opposite side of the synapse is called the postsynaptic neuron, and it has postsynaptic receptors. The axon terminals and postsynaptic receptors are not physically in contact with each other. A narrow gap, the synaptic cleft, separates them.

A nerve impulse can be transmitted across a synapse in only one direction: from the axon terminals of the presynaptic neuron to the postsynaptic receptors, usually on the dendrites, of the postsynaptic neuron. Impulses also can go directly to receptors on the cell body: About 5% to 20% of the axon terminals are adjacent to the cell body instead of the dendrites.[2] Why can the nerve impulse go in only one direction?

The presynaptic terminals of the axon contain a large number of saclike structures, called synaptic vesicles. These sacs contain neurotransmitter chemicals. When the impulse reaches the presynaptic terminals, the synaptic vesicles respond by dumping their chemicals into the

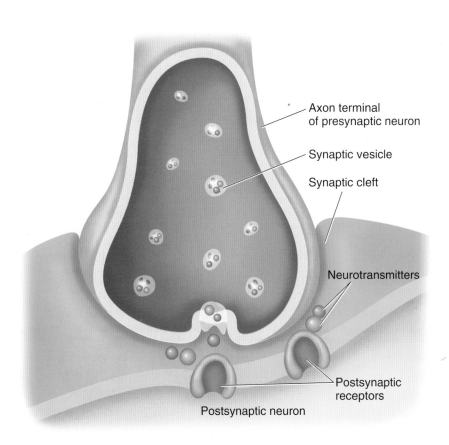

Axon terminal of presynaptic neuron

Synaptic vesicle

Synaptic cleft

Neurotransmitters

Postsynaptic receptors

Postsynaptic neuron

▲ **Figure 2.3**   A chemical synapse between two neurons, showing the synaptic vesicles.

synaptic cleft. These neurotransmitters then diffuse across the synaptic cleft to the postsynaptic neuron's receptors. The postsynaptic receptors bind the neurotransmitter once it diffuses across the synaptic cleft. When this binding occurs, the impulse has been transmitted successfully to the next neuron and can be transmitted onward.

> Neurons communicate with one another across sites of impulse transmission called synapses.

Neuromuscular Junction

Whereas neurons communicate with other neurons at synapses, a motor neuron communicates with a muscle fiber at a site known as a neuromuscular junction. The function of the neuromuscular junction is essentially the same as a synapse. In fact, the proximal part of the neuromuscular junction is the same: It starts with the axon terminals of the motor neuron, which release neurotransmitters into the space between the motor nerve and the muscle fiber in response to an action potential. However,

in the neuromuscular junction, the axon terminals protrude into motor endplates, which are troughlike segments on the sarcolemma membrane (see figure 2.4).

The motor endplate is invaginated (folded to form cavities). The cavity thus formed is called the synaptic gutter. As with synapses, the space between the neuron and the muscle fiber is the synaptic cleft.

Neurotransmitters released from the motor axon terminals diffuse across the synaptic cleft and bind to receptors on the muscle fiber's sarcolemma (membrane). This binding typically causes depolarization by opening sodium ion channels, allowing more sodium to enter the muscle fiber. As always, if the depolarization reaches the threshold, an action potential is fired. It spreads across the sarcolemma into the T tubules, initiating muscle fiber contraction. Like the neuron, the sarcolemma, once depolarized, must undergo repolarization. During the period of repolarization, the sodium gates are closed and the potassium gates are open; thus, like the neuron, the muscle fiber is unable to respond to any further stimulation. This is referred to as the refractory period. Once the

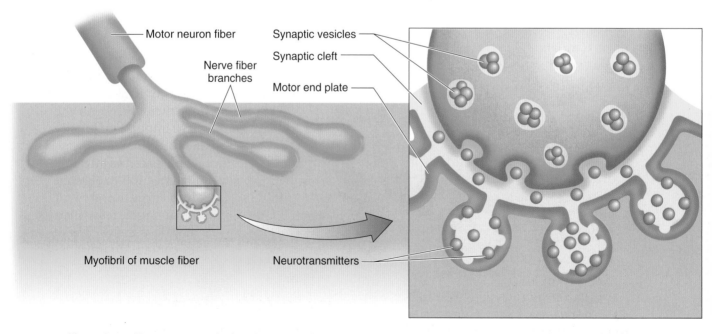

▲ **Figure 2.4**   The neuromuscular junction, illustrating the interaction between the motor neuron and the sarcolemma of a single muscle fiber.

electrical conditions of the muscle fiber are restored to resting levels, the fiber can respond to another stimulus. Thus, the refractory period limits the motor unit's firing frequency.

Now we know how the impulse is transmitted between two cells. But to understand what happens once the impulse is transmitted, we must first examine the chemical signals that accomplish transmission. Let's turn our attention to neurotransmitters.

Neurotransmitters

More than 50 neurotransmitters have been positively identified or are suspected as potential candidates. These can be categorized as either (a) small-molecule, rapid-acting neurotransmitters or (b) neuropeptide, slow-acting neurotransmitters. The small-molecule, rapid-acting transmitters, which are responsible for most neural transmissions, are our main concern.

Acetylcholine and norepinephrine are the two major neurotransmitters involved in regulating our physiological responses to exercise. Acetylcholine is the primary neurotransmitter for the motor neurons that innervate skeletal muscle and for most parasympathetic neurons. It is generally an excitatory neurotransmitter, but it can have inhibitory effects at some parasympathetic nerve endings, such as in the heart. Norepinephrine is the neurotransmitter for most sympathetic neurons, and it too can be either excitatory or inhibitory, depending on the receptors involved. The sympathetic and parasympathetic nervous systems are discussed later in this chapter.

> Acetylcholine and norepinephrine are the two major neurotransmitters—chemical substances that transmit nerve impulses across synapses and synaptic clefts.

Once the neurotransmitter binds to the postsynaptic receptor, the nerve impulse has been successfully transmitted. The neurotransmitter is then either destroyed by enzymes or actively transported back into the presynaptic terminals for reuse when the next impulse arrives.

▶ Neurons communicate with each other across synapses.

▶ A synapse involves
  1. the axon terminals of the presynaptic neuron,
  2. the postsynaptic receptors on the dendrite or cell body of the postsynaptic neuron, and
  3. the space (the synaptic cleft) between the two neurons.

▶ A nerve impulse causes chemicals called neurotransmitters to be released from the presynaptic axon terminals into the synaptic cleft.

▶ Neurotransmitters diffuse across the cleft and are bound to the postsynaptic receptors.

▶ Once neurotransmitters are bound, the impulse has been successfully transmitted and the neurotransmitter is then either destroyed by enzymes or actively returned to the presynaptic neuron for future use.

▶ Neurotransmitter binding at the postsynaptic receptors opens ion gates in that membrane and can cause depolarization (excitation) or hyperpolarization (inhibition), depending on the specific neurotransmitter and the receptors to which it binds.

▶ Neurons communicate with muscle cells at neuromuscular junctions. A neuromuscular junction involves presynaptic axon terminals, the synaptic cleft, and motor endplate receptors on the sarcolemma of the muscle fiber. The neuromuscular junction functions much like a neural synapse.

▶ The neurotransmitters most important in regulating exercise are acetylcholine and norepinephrine.

Postsynaptic Response

We have discussed generating an action potential, conducting the impulse the length of the neuron, and transmitting it to the next cell. We continue the story with what happens after the neurotransmitter binds with the postsynaptic receptors.

Once the neurotransmitter binds to the receptors, the chemical signal that traversed the synaptic cleft once again becomes an electrical signal. The binding causes a graded potential in the postsynaptic membrane. An incoming impulse may be either excitatory or inhibitory. An excitatory impulse causes a hypopolarization, or depolarization, known as an **excitatory postsynaptic potential (EPSP)**. An inhibitory impulse causes a hyperpolarization, known as an **inhibitory postsynaptic potential (IPSP)**.

The discharge of a single presynaptic terminal generally changes the postsynaptic potential less than 1 mV. Clearly this is not sufficient to generate an action potential, because reaching threshold requires a change of at least 15 to 20 mV. But when a neuron transmits an impulse, several presynaptic terminals typically release their neurotransmitters so that they can diffuse to the postsynaptic receptors. Also, presynaptic terminals from numerous axons can converge on the dendrites and cell body of a single neuron. When multiple presynaptic terminals discharge at the same time, or when only a few fire in rapid succession, more neurotransmitter is released. With an excitatory neurotransmitter, the more that is bound, the greater the EPSP will be.

Triggering an action potential at the postsynaptic neuron depends on the combined effects of all incoming impulses from these various presynaptic terminals. A number of impulses are needed to cause sufficient depolarization to generate an action potential. Specifically, the sum of all changes in the membrane potential must equal or exceed the threshold. This summing of the individual impulses' effects is called **summation.**

For summation, the postsynaptic cell must keep a running total of the neuron's responses, both EPSPs and IPSPs, to all incoming impulses. This task is done at the axon hillock, which lies on the axon just past the cell body. Only when the sum of all individual graded potentials meets or exceeds threshold can an action potential occur.

The process of summation is very important and has great relevance to muscle function. Human muscle is capable of generating considerable force, far greater than we normally see even in highly trained weightlifters or bodybuilders. Under extreme, life-threatening circumstances, such as an airplane or automobile crash, individuals have been able to exert superhuman levels of force, breaking bones and tearing muscles from bones in the process. The IPSPs protect the muscle–tendon–bone complex under normal conditions. As we show in chapter 3, a reduction in IPSPs most likely is a major factor explaining the gains in strength we experience following a period of resistance training.

> Summation refers to the cumulative effect of all individual graded potentials as processed by the axon hillock.

Now that we have carefully considered the function of the most basic units of the nervous system, the neurons, we are ready to examine how these cells work together. Individual neurons are grouped together into bundles. In the central nervous system (brain and spinal cord), these bundles are referred to as tracts, or pathways. Neuron bundles in the peripheral nervous system are referred to as nerves.

▶ EPSPs are hypopolarizations of the postsynaptic membrane. IPSPs are hyperpolarizations of that membrane.

▶ A single presynaptic terminal cannot generate enough of a depolarization to fire an action potential. Multiple signals are needed. These may come from numerous neurons or from a single neuron when numerous axon terminals release neurotransmitters repeatedly and rapidly.

▶ The axon hillock keeps a running total of all EPSPs and IPSPs. When their sum meets or exceeds the threshold for depolarization, an action potential occurs. This process of accumulating incoming signals is known as summation.

Central Nervous System

To comprehend how even the most basic stimulus can cause muscle activity, we must next consider the complexity of the central nervous system (CNS). In this section, we present an overview of the components of the central nervous system and their functions.

> **fyi** The central nervous system houses more than 100 billion neurons.

Brain

The brain is composed of numerous parts. For our purposes, we subdivide it into the four major regions illustrated in figure 2.5: the cerebrum, diencephalon, cerebellum, and brain stem. Let's discuss each briefly.

Cerebrum

The cerebrum is composed of the right and left cerebral hemispheres. These are connected to each other by fiber bundles (tracts) referred to as the *corpus callosum,* which allows the two hemispheres to communicate with each other. The cerebral cortex forms the outer portion of the cerebral hemispheres and has been referred to as the site of the mind and intellect. It is also called the gray matter, which simply reflects its distinctive color resulting from lack of myelin on the cell bodies located in this area. The cerebral cortex is your conscious brain. It allows you to think, to be aware of sensory stimuli, and to voluntarily control your movements.

The cerebrum consists of five lobes—four outer lobes and the central insula, which we will not discuss. Its four outer lobes (see figure 2.5) have the following general functions:

- Frontal lobe: general intellect and motor control
- Temporal lobe: auditory input and its interpretation
- Parietal lobe: general sensory input and its interpretation
- Occipital lobe: visual input and its interpretation

The three areas in the cerebrum that are of primary concern to our discussion and which we discuss later in this chapter are the primary motor cortex, in the frontal lobe; the basal ganglia, in the white matter below the cerebral cortex; and the primary sensory cortex, in the parietal lobe.

Diencephalon

This region of the brain is composed mostly of the thalamus and the hypothalamus. The thalamus is an important sensory integration center. All sensory input (except smell) enters the thalamus and is relayed to the appropriate area of the cortex. The thalamus regulates what sensory input reaches your conscious brain and thus is very important for motor control.

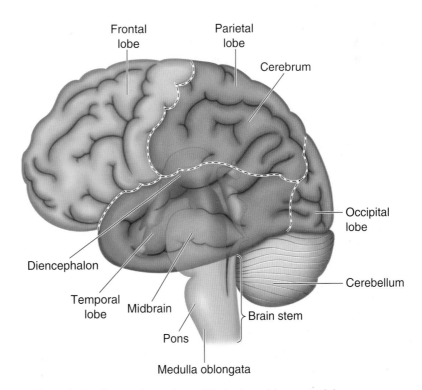

▲ **Figure 2.5** Four major regions of the brain and four outer lobes.

The hypothalamus, directly below the thalamus, is responsible for maintaining homeostasis by regulating almost all processes that affect the body's internal environment. Neural centers here regulate

- the autonomic nervous system (and through it, blood pressure, heart rate and contractility, respiration, and digestion),
- body temperature,
- fluid balance,
- neuroendocrine control,
- emotions,
- thirst,
- food intake, and
- sleep–wake cycles.

Cerebellum

The cerebellum is located behind the brain stem. It is connected to numerous parts of the brain and has a crucial role in coordinating movement, as we see later in this chapter.

Brain Stem

The brain stem, composed of the midbrain, the pons, and the medulla oblongata (see figure 2.5) is the stalk of your brain, connecting the brain and the spinal cord. All sensory and motor nerves pass through the brain stem as they relay information between the brain and the spinal cord. This is the site of origin for 10 of the 12 pairs of cranial nerves. The brain stem also contains the major autonomic regulatory centers that control the respiratory and cardiovascular systems.

A specialized collection of neurons running the entire length of the brain stem, known as the *reticular formation*, are influenced by and have an influence on nearly all areas of the central nervous system. These neurons help

- coordinate skeletal muscle function,
- maintain muscle tone,
- control cardiovascular and respiratory functions, and
- determine our state of consciousness (both arousal and sleep).

The brain has a pain control system, called an analgesia system. The enkephalins and β-endorphin are important opiate substances that act on the opiate receptors in the analgesia system to help reduce pain. Research has demonstrated that exercise of long duration increases the natural levels of these opiate substances.

Spinal Cord

The lowest part of the brain stem, the medulla oblongata, is continuous with the spinal cord below. The spinal cord is composed of tracts of nerve fibers that allow two-way conduction of nerve impulses. The sensory (afferent) fibers carry neural signals from sensory receptors, such as those in the muscles and joints, to the upper levels of the CNS. Motor (efferent) fibers from the brain and upper spinal cord travel down to end organs (e.g., muscles, glands).

- ▶ The central nervous system comprises the brain and the spinal cord.
- ▶ The four major divisions of the brain are the cerebrum, the diencephalon, the cerebellum, and the brain stem.
- ▶ The cerebral cortex is your conscious brain.
- ▶ The diencephalon includes the thalamus, which receives all sensory input entering the brain, and the hypothalamus, which is a major control center for homeostasis.
- ▶ The cerebellum, which is connected to numerous parts of the brain, is critical for coordinating movement.
- ▶ The brain stem is composed of the midbrain, the pons, and the medulla oblongata.
- ▶ The spinal cord carries both sensory and motor fibers between the brain and the periphery.

Peripheral Nervous System

The peripheral nervous system (PNS) contains 43 pairs of nerves: 12 pairs of cranial nerves that connect with the brain and 31 pairs of spinal nerves that connect with the spinal cord. Spinal nerves directly supply the skeletal muscles. Functionally, the peripheral nervous system has two major divisions: the sensory division and the motor division. Let's examine each briefly.

Sensory Division

The sensory division of your peripheral nervous system carries sensory information toward your central nervous system. Sensory (afferent) neurons originate in such areas as

- blood and lymph vessels,
- internal organs,
- special sense organs (taste, touch, smell, hearing, vision),
- the skin, and
- muscles and tendons.

Sensory neurons in the PNS end either in the spinal cord or in the brain, and they continuously convey information to the CNS concerning your body's constantly changing status. By relaying this information, these neurons allow your brain to sense what is going on in all parts of your body and in your immediate environment. Sensory neurons within your CNS carry the sensory input to appropriate areas, where the information can be processed and integrated with other incoming information.

The sensory division receives information from five primary types of receptors:

1. Mechanoreceptors that respond to mechanical forces such as pressure, touch, vibrations, or stretch
2. Thermoreceptors that respond to changes in temperature
3. Nociceptors that respond to painful stimuli

4. Photoreceptors that respond to electromagnetic radiation (light) to allow vision
5. Chemoreceptors that respond to chemical stimuli, such as from foods, odors, or changes in blood concentrations of substances such as oxygen, carbon dioxide, glucose, and electrolytes

Several of these receptors are important in exercise and sport. Let's consider just a few. Free nerve endings detect crude touch, pressure, pain, heat, and cold. Thus, they function as mechanoreceptors, nociceptors, and thermoreceptors. These nerve endings are important for preventing injury during athletic performance.

Special muscle and joint nerve endings are of many types and functions, and each type is sensitive to a specific stimulus. Here are some important examples:

- Joint kinesthetic receptors located in your joint capsules are sensitive to joint angles and rates of change in these angles. Thus, they sense the position and any movement of your joints.
- Muscle spindles sense how much a muscle is stretched.
- Golgi tendon organs detect the tension applied by a muscle to its tendon, providing information about the strength of muscle contraction.

Muscle spindles and Golgi tendon organs are discussed later in this chapter.

Motor Division

Your central nervous system transmits information to various parts of your body through the motor, or efferent, division of your peripheral nervous system. Once your CNS has processed the information it receives from the sensory division, it decides how your body should respond to that input. From your brain and spinal cord, intricate networks of neurons go out to all parts of your body, providing detailed instructions to the target areas—for our purposes, muscles.

Autonomic Nervous System

The autonomic nervous system, often considered part of the motor division of the peripheral nervous system, controls your body's involuntary internal functions. Some of these functions that are important to sport and activity include heart rate, blood pressure, blood distribution, and respiration.

The autonomic nervous system has two major divisions: the sympathetic nervous system and the parasympathetic nervous system. These originate from different sections of the spinal cord and from the base of the brain. The effects of the two systems are often antagonistic, but both systems always function together. Remember this as we examine each separately.

Sympathetic Nervous System

The sympathetic nervous system is your fight-or-flight system: It prepares your body to face a crisis. When you're excited, your sympathetic nervous system produces a massive discharge throughout your body, preparing you for action. A sudden loud noise, a life-threatening situation, or those last few seconds before starting an athletic competition are examples of when you would experience this massive sympathetic discharge. The effects of sympathetic stimulation are important to the athlete:

- Heart rate and strength of cardiac contraction increase.
- Coronary vessels dilate, increasing the blood supply to the heart muscle to meet its increased demands.
- Peripheral vasodilation allows more blood to enter the active skeletal muscles.
- Vasoconstriction in most other tissues diverts blood away from them and to the active muscles.
- Blood pressure increases, allowing better perfusion of the muscles and improving the return of venous blood to the heart.
- Bronchodilation improves gas exchange.
- Metabolic rate increases, reflecting the body's effort to meet the increased demands of physical activity.

- Mental activity increases, allowing better perception of sensory stimuli and more concentration on performance.
- Glucose is released from the liver into the blood as an energy source.
- Functions not directly needed are slowed (e.g., renal function, digestion), conserving energy so that it can be used for action.

These basic alterations in bodily function facilitate your motor response, demonstrating the importance of the autonomic nervous system in preparing you for acute stress or physical activity.

Parasympathetic Nervous System

The parasympathetic nervous system is your body's housekeeping system. It has a major role in carrying out such processes as digestion, urination, glandular secretion, and conservation of energy. This system is more active when you are calm and at rest. Its effects tend to oppose those of the sympathetic system. The parasympathetic division causes decreased heart rate, constriction of coronary vessels, and bronchoconstriction.

▶ The peripheral nervous system contains 43 pairs of nerves: 12 cranial and 31 spinal.

▶ The PNS can be subdivided into the sensory and motor divisions. The motor division also includes the autonomic nervous system.

▶ The sensory division carries information from sensory receptors to the central nervous system so that the CNS is constantly aware of your current status and environment.

▶ The motor division carries motor impulses from the CNS to the muscles, organs, and other tissues.

▶ The autonomic nervous system includes the sympathetic nervous system, which is your fight-or-flight system, and the parasympathetic system, which is your housekeeping system. Although these systems often oppose each other, they always function together.

Table 2.1

Effects of the Sympathetic and Parasympathetic Nervous Systems on Various Organs

Target organ/system	Sympathetic effects	Parasympathetic effects
Heart muscle	Increase rate and force of contraction	Decrease rate of contraction
Heart: coronary blood vessels	Cause vasodilation	Cause vasoconstriction
Lungs	Cause bronchodilation; mildly constrict blood vessels	Cause bronchoconstriction
Blood vessels	Increase blood pressure; cause vasoconstriction in abdominal viscera and skin to divert blood when necessary; cause vasodilation in the skeletal muscles and heart during exercise	Little or no effect
Liver	Stimulate glucose release	No effect
Cellular metabolism	Increase metabolic rate	No effect
Adipose tissue	Stimulate lipolysis[a]	No effect
Sweat glands	Increase sweating	No effect
Adrenal glands	Stimulate secretion of epinephrine and norepinephrine	No effect
Digestive system	Decrease activity of glands and muscles; constrict sphincters	Increase peristalsis and glandular secretion; relax sphincters
Kidney	Cause vasoconstriction; decrease urine formation	No effect

[a]Recall from anatomy that lipolysis is the process of breaking down triglyceride to its basic units to be used for energy.

The various effects of the sympathetic and parasympathetic divisions of the autonomic nervous system are summarized in table 2.1.

Sensory–Motor Integration

Now that we have discussed the components and divisions of the nervous system, we are ready to discuss how a sensory stimulus gives rise to a motor response. How, for example, do the muscles in your hand know to pull your finger away from a hot stove? When you decide to run, how do the muscles in your legs coordinate while supporting your weight and propelling you forward? To accomplish these tasks, the sensory and motor systems must communicate with each other.

This process is called **sensory–motor integration**, and it is depicted in figure 2.6 on page 74. For your body to respond to sensory stimuli, the sensory and motor divisions of your nervous system must function together in the sequence of events enumerated on page 74.

1. A sensory stimulus is received by sensory receptors.

2. The sensory impulse is transmitted along sensory neurons to the CNS.

3. The CNS interprets the incoming sensory information and determines which response is most appropriate.

4. The signals for the response are transmitted from the CNS along motor neurons.

5. The motor impulse is transmitted to a muscle, and the response occurs.

Sensory Input

Recall that sensations and physiological status are detected by sensory receptors throughout your body. The impulses resulting from sensory stimulation are transmitted via the sensory nerves to the spinal cord. When they reach the spinal cord, they can trigger a local reflex at that level, or they can travel to the upper regions of the spinal cord or to the brain. Sensory pathways to the brain can terminate in sensory areas of the brain stem, the cerebellum, the thalamus, or the cerebral cortex. An

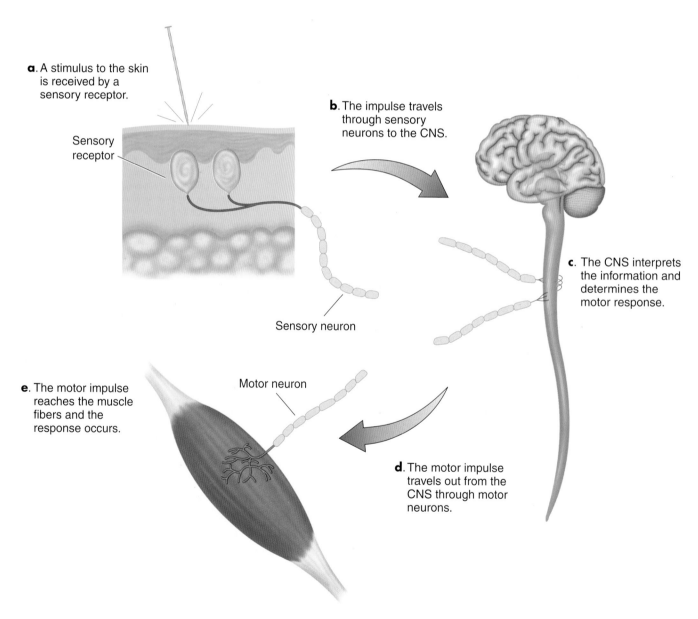

a. A stimulus to the skin is received by a sensory receptor.

Sensory receptor

b. The impulse travels through sensory neurons to the CNS.

Sensory neuron

c. The CNS interprets the information and determines the motor response.

e. The motor impulse reaches the muscle fibers and the response occurs.

Motor neuron

d. The motor impulse travels out from the CNS through motor neurons.

▲ **Figure 2.6**   The sequence of events in sensory–motor integration.

area in which the sensory impulses terminate is referred to as an integration center. This is where the sensory input is interpreted and linked to the motor system. Figure 2.7 illustrates various sensory receptors and their nerve pathways back to the spinal cord and up into various areas of the brain. The integration centers vary in function:

- Sensory impulses that terminate in the spinal cord are integrated there. The response is typically a simple motor reflex, which is the simplest type of integration. We discuss it later.

- Sensory signals that terminate in the lower brain stem result in subconscious motor reactions of a higher and more complex nature than simple spinal cord reflexes. Postural control when sitting, standing, or moving is an example of this level of sensory input.

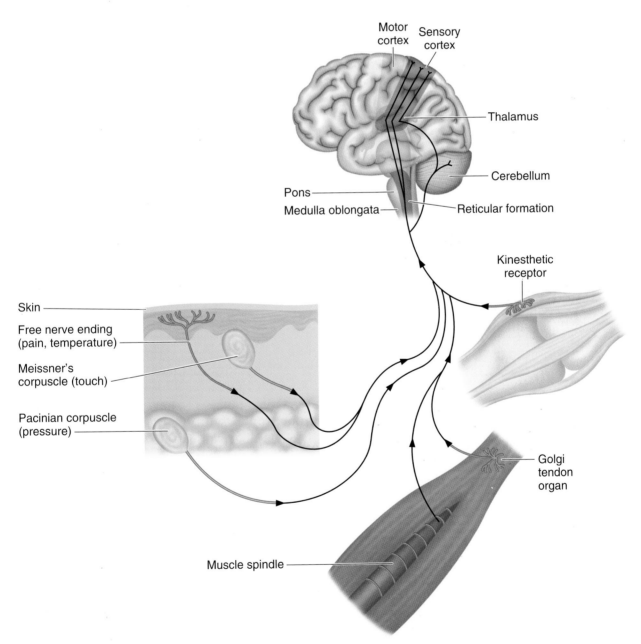

▲ **Figure 2.7** The sensory receptors and their pathways back to the spinal cord and brain.

- Sensory signals that terminate in the cerebellum also result in subconscious control of movement. The cerebellum appears to be the center of coordination, smoothing out movements by coordinating the actions of the various contracting muscle groups to perform the desired movement. Both fine and gross motor movements appear to be coordinated by the cerebellum in concert with the basal ganglia. Without the control exerted by the cerebellum, all movement would be uncontrolled and uncoordinated.

- Sensory signals that terminate at the thalamus begin to enter the level of consciousness, and you begin to distinguish various sensations.

- Only when sensory signals enter the cerebral cortex can you discretely localize the signal. The primary sensory cortex, located in the postcentral gyrus (in the parietal lobe), receives general sensory input from receptors in the skin and from proprioceptors in the muscles, tendons, and joints. This area has a map of the body. Stimulation in a specific area of the body is recognized, and its exact location is known instantly. Thus, this part of our conscious brain allows us to be constantly aware of our surroundings and our relationship to them.

Motor Control

Once a sensory impulse is received, it typically evokes a response through a motor neuron, regardless of the level at which the sensory impulse stops. Skeletal muscles are controlled by impulses conducted by motor (efferent) neurons that originate from any of three levels:

- The spinal cord
- The lower regions of the brain
- The motor area of the cerebral cortex

As the level of control moves from the spinal cord to the motor cortex, the degree of movement complexity increases from simple reflex control to complicated movements requiring

basic thought processes. Motor responses for more complex movement patterns typically originate in the motor cortex of the brain. Figure 2.8 depicts some motor pathways.

At last, we are ready to tie the two systems together through sensory–motor integration. The simplest form of this is the reflex, so we consider it first.

Reflex Activity

What happens when you unknowingly put your hand on a hot stove? First, the stimuli of heat and pain are received by the thermoreceptors and nociceptors in your hand, and then sensory impulses travel to the spinal cord, terminating at the level of entry. Once in the spinal cord, these impulses are integrated instantly by interneurons that connect the sensory and motor neurons. The impulse moves to the motor neurons and travels to the effectors, the muscles controlling the withdrawal of your hand. The result is that you reflexively withdraw your hand from the hot stove without giving the action any thought.

A **motor reflex** is a preprogrammed response; any time your sensory nerves transmit certain impulses, your body responds instantly and identically. In our example, whether you touch something that is too hot or too cold, thermoreceptors will elicit a reflex for you to withdraw your hand. Whether the pain arises from heat or from a sharp object, the nociceptors will also cause a withdrawal reflex. By the time you are consciously aware of the specific stimulus (after sensory impulses also have been transmitted to your primary sensory cortex), the reflex activity is well underway, if not completed. All neural activity occurs extremely rapidly, but a reflex is the fastest mode of response because you don't need time to make a conscious decision. Only one response is possible—no options need to be considered.

The level of nervous system control in response to sensory input varies according to the complexity of movement necessary. Simple reflexes are handled by the spinal cord, whereas complex reactions require involvement of the brain.

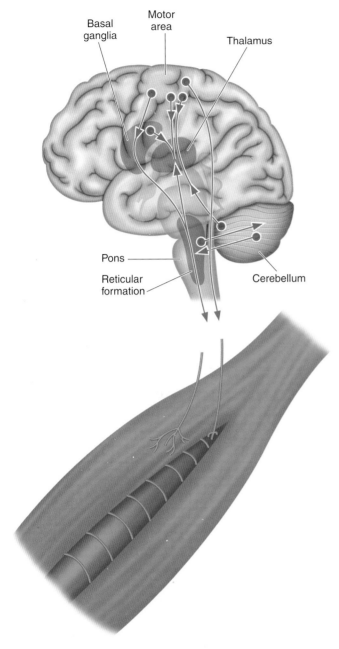

▲ Figure 2.8   Motor pathways of the nervous system.

Basal ganglia

Motor area

Thalamus

Pons

Reticular formation

Cerebellum

Muscle Spindles

Now that you understand the basics of reflex activity, we can look more closely at two reflexes that help control muscle function. The first involves a special structure: the muscle spindle.

The **muscle spindle** lies between regular skeletal muscle fibers, referred to as *extrafusal*

(outside the spindle) fibers. A muscle spindle consists of 4 to 20 small, specialized muscle fibers called *intrafusal* (inside the spindle) fibers and the nerve endings, sensory and motor, associated with these fibers. A connective tissue sheath surrounds the muscle spindle and attaches to the endomysium of the extrafusal fibers. The intrafusal fibers are controlled by specialized motor neurons, referred to as γ-*motor neurons*. In contrast, extrafusal fibers (the regular fibers) are controlled by α-*motor neurons*.

The central region of an intrafusal fiber cannot contract because it contains no or only a few actin and myosin filaments. Thus, the central region can only stretch. Because the muscle spindle is attached to the extrafusal fibers, any time those fibers are stretched, the central region of the muscle spindle is also stretched.

Sensory nerve endings wrapped around this central region of the muscle spindle transmit information to the spinal cord when this region is stretched, informing the CNS of the muscle's length. In the spinal cord, the sensory neuron synapses with an α-motor neuron, which triggers reflexive muscle contraction (in the extrafusal fibers) to resist further stretching.

Let's illustrate this action with an example. Your arm is bent at the elbow, and your hand is extended, palm up. Suddenly someone places a heavy weight in your palm. Your forearm starts to drop, which stretches the muscle fibers in your arm (biceps brachii), which in turn stretch the muscle spindle. In response to that stretch, the sensory neurons send impulses to the spinal cord, which then excites the α-motor neurons. These cause the biceps to increase its force production, overcoming the stretch.

γ-motor neurons excite the intrafusal fibers, prestretching them slightly. Although the midsection of the intrafusal fibers cannot contract, the ends can. The γ-motor neurons cause slight contraction of the ends of these fibers, which stretches the central region slightly. This prestretch makes the muscle spindle highly sensitive to even small degrees of stretch.

The muscle spindle also assists normal muscle action. It appears that when the α-motor neurons are stimulated to contract the extrafusal muscle fibers, the γ-motor neurons are also activated, contracting the ends of the intrafusal fibers. This stretches the central region of the muscle spindle, giving rise to sensory impulses that travel to the spinal cord and then to the motor neurons. In response, the muscle increases its force production. Thus, neural muscle contraction is enhanced through this function of the muscle spindles.

Information brought into the spinal cord from the sensory neurons associated with muscle spindles does not merely end at that level. Impulses are also sent up to higher parts of the CNS, supplying the brain with information on the exact length and contractile state of the muscle, as well as the rate at which those states are changing. This information is essential for maintaining muscle tone and posture and for executing movements. Before the brain can tell a muscle what to do next, the brain must know what the muscle is currently doing.

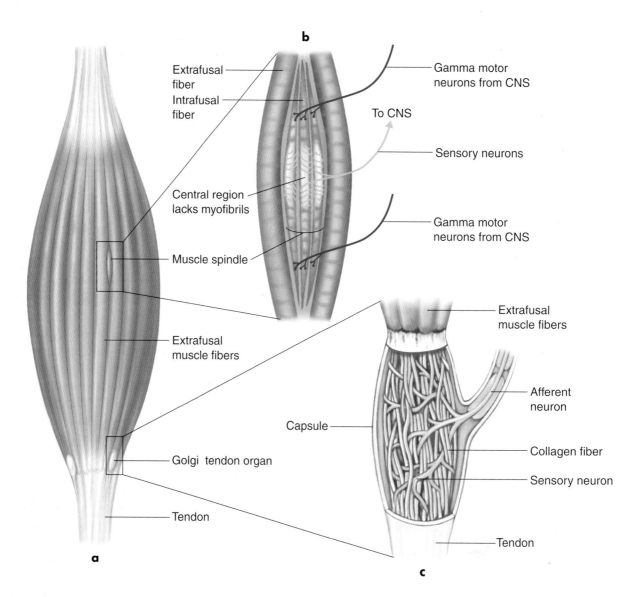

▲ **Figure 2.9**   A muscle belly (a), a muscle spindle (b), and a Golgi tendon organ (c).

Golgi Tendon Organs

Golgi tendon organs are encapsulated sensory receptors through which a small bundle of muscle tendon fibers pass. These organs are located just proximal to the tendon fibers' attachment to the muscle fibers, as shown in figure 2.9. Approximately 5 to 25 muscle fibers are usually connected with each Golgi tendon organ. Whereas muscle spindles monitor the length of a muscle, Golgi tendon organs are sensitive to tension in the muscle–tendon complex and operate like a strain gauge, a device that senses changes in tension. Their sensitivity is so great that they can respond to the contraction of a single muscle fiber. These sensory receptors are inhibitory in nature, performing a protective function by reducing the potential for injury. When stimulated, these receptors inhibit the contracting (agonist) muscles and excite the antagonist muscles.

Some researchers speculate that reducing the influence of Golgi tendon organs disinhibits the active muscles, allowing a more forceful muscle action. This mechanism may explain at least part of the gains in muscular strength that accompany strength training.

Higher Brain Centers

Reflexes involve the simplest form of neural integration. But most movements used in sport activities involve control and coordination through the higher brain centers, specifically the primary motor cortex, the basal ganglia, and the cerebellum. Let's discuss some functions of each of these centers.

Primary Motor Cortex

The primary motor cortex is responsible for the control of fine and discrete muscle movements. It is located in the frontal lobe, specifically within the precentral gyrus. Neurons here, known as *pyramidal cells*, let us consciously control movement of our skeletal muscles. Think of the primary motor cortex as the part of the brain that decides what movement you want to make. For example, if you are sitting in a chair and you want to stand up, the decision to do so is made in your primary motor cortex, where the entire body is carefully mapped out. The areas that require the finest motor control have a greater representation in the motor cortex; thus, more neural control is provided to them.

The cell bodies of the pyramidal cells are housed in the primary motor cortex, and their axons form the extrapyramidal tracts. These are also known as the corticospinal tracts because the nerve processes extend from the cerebral cortex down to the spinal cord. These tracts provide the major voluntary control of our skeletal muscles.

In addition to the primary motor cortex, there is a premotor cortex just anterior to the precentral gyrus in the frontal lobe. Learned motor skills of a repetitious or patterned nature are stored here. This region can be thought of as the memory bank for skilled motor activities.[3]

Basal Ganglia

The basal ganglia (nuclei) are not part of the cerebral cortex. Rather, they are in the cerebral white matter, deep in the cortex. These ganglia are clusters of nerve cell bodies. The complex functions of the basal ganglia are not well understood, but the ganglia are known to be important in initiating movements of a sustained and repetitive nature (such as arm swinging while walking), and thus they control complex semivoluntary movements such as walking and running. These cells also are involved in maintaining posture and muscle tone.

Cerebellum

The cerebellum is crucial to the control of all rapid and complex muscular activities. It helps coordinate the timing of motor activities and the rapid progression from one movement to the next by monitoring and making corrective adjustments in the motor activities that are elicited by other parts of the brain. The cerebellum assists the functions of both the primary motor cortex and the basal ganglia. It facilitates movement patterns by smoothing out the movement, which would otherwise be jerky and uncontrolled.

The cerebellum acts as an integration system, comparing your programmed or intended activity with the actual changes occurring in your body and then initiating corrective adjustments through the motor system. It receives information from the cerebrum and other parts of your brain and also from sensory receptors (proprioceptors) in your muscles and joints that keep your cerebellum informed about your body's current position. The cerebellum also receives visual and equilibrium input. Thus, it notes all incoming information about the exact tension and position of all muscles, joints, and tendons and your body's current position relative to your surroundings, and then it determines the best plan of action to produce the desired movement.

Consider our previous example in which you are sitting but want to stand. Your primary motor cortex is the part of your brain that makes the decision to stand. This decision is relayed to the cerebellum. The cerebellum notes the desired action and then considers the current status of your body, based on all the sensory input it receives. Then the cerebellum decides, based on that input, what is the best plan of action to accomplish your desired movement, standing.

Engrams

When you learn a new motor skill, your initial periods of practice require intense concentration. As you become more familiar with the skill, you find that you don't need to concentrate as much. Finally, once you perfect the skill, it can be recalled with little or no conscious effort. How do you reach this point?

Specific learned motor patterns appear to be stored in the brain, to be replayed on request. These memorized motor patterns are referred to as motor programs, or **engrams.** Engrams are apparently stored in both the sensory and motor portions of the brain, such as the premotor cortex. Those in the sensory portion of the brain are for slower motor patterns, and those in the motor portion are for rapid move-

ments. Little is known about engrams and the mechanisms of their action, so these are important areas for future research.

- ▶ Sensory–motor integration is the process by which your PNS relays sensory input to your CNS and your CNS interprets this information and then sends out the appropriate motor signal to elicit the desired motor response.

- ▶ Sensory input can terminate at various levels of the CNS. Not all of this information reaches the brain.

- ▶ Reflexes are the simplest form of motor control. These are not conscious responses. For a given sensory stimulus, the motor response is always identical and instantaneous.

- ▶ Muscle spindles trigger reflexive muscle action when the muscle spindle is stretched.

- ▶ Golgi tendon organs trigger a reflex that inhibits contraction if the tendon fibers are overstretched.

- ▶ The primary motor cortex, located in the frontal lobe, is the center of conscious motor control.

- ▶ The basal ganglia, in the cerebral white matter, help initiate some movements (sustained and repetitive ones) and help control posture and muscle tone.

- ▶ The cerebellum is involved in all rapid and complex movement processes and assists the primary motor cortex and the basal ganglia in coordinating the response. It is an integration center that decides how to best execute the desired movement, given your body's current position and your muscles' current status.

- ▶ Although not well understood, engrams are memorized motor patterns, stored in both the sensory and motor areas of the brain, that are called on as needed.

Motor Response

Now that we have discussed how sensory input is integrated to determine the appropriate motor response, the last step in the process to consider is how muscles respond to motor impulses once they reach the muscle fibers.

Once an electrical impulse reaches a motor neuron, the impulse travels the length of the neuron to the neuromuscular junction. From there, the impulse spreads to all muscle fibers innervated by that particular motor neuron. Recall that the motor neuron and all muscle fibers it innervates form a single motor unit. Each muscle fiber is innervated by only one motor neuron, but each motor neuron innervates up to several thousand muscle fibers, depending on the function of the muscle. Muscles controlling fine movements, such as those controlling the eyes, have only a small number of muscle fibers per motor neuron. Muscles with more general functions have many fibers per motor neuron.

The muscles that control eye movements (the extraocular muscles) have an innervation ratio of 1:15, meaning that one motor neuron serves only 15 muscle fibers. In contrast, the gastrocnemius and tibialis anterior muscles of the lower leg have innervation ratios of almost 1:2,000.

The muscle fibers in a specific motor unit are homogeneous with respect to fiber type. Thus, you won't find a motor unit that has both FT and ST fibers. In fact, as mentioned in chapter 1, it is generally believed that the characteristics of the motor neuron actually determine the fiber type in that motor unit.[1,5]

> ▶ Each muscle fiber is innervated by only one neuron, but each neuron may innervate up to several thousand muscle fibers.
>
> ▶ All muscle fibers within a single motor unit are of the same fiber type.

In Closing . . .

What we have covered in this chapter is merely a tiny fragment of the complex role that the nervous system plays in regulating movement. All divisions of the nervous system are involved:

- The sensory division of the peripheral nervous system always keeps your central nervous system informed of what is happening in and around your body.
- The CNS interprets all incoming sensory information and decides how you should respond.
- The motor division of the PNS tells your muscles exactly when and how much to act.
- The autonomic division of the PNS adjusts physiological functions throughout your body to ensure that the needs of your active tissues are met.

We have seen how your muscles respond to neural stimulation, whether through reflexes or under complex control of the higher brain centers. We discussed how the individual motor units respond and how they are recruited in an orderly manner depending on the required force. Thus, we have learned how your body functions to allow you to move. In the next chapter, we examine the effects that training has on this neuromuscular control.

▶ Key Terms

acetylcholine
action potential
afferent division
axon hillock
axon terminal
central nervous system (CNS)
depolarization
efferent division
end branches
engram
excitatory postsynaptic potential (EPSP)
Golgi tendon organ
graded potential
hyperpolarization
inhibitory postsynaptic potential (IPSP)
motor division

motor reflex

muscle spindle

myelin sheath

nerve impulse

neuromuscular junction

neuron

neurotransmitter

norepinephrine

peripheral nervous system (PNS)

resting membrane potential (RMP)

saltatory conduction

sensory division

sensory–motor integration

sodium–potassium pump

summation

synapse

threshold

▶ Study Questions

1. Name the different regions of a neuron.

2. Explain the resting membrane potential. What causes it? How is it maintained?

3. Describe an action potential. What is required before an action potential is fired?

4. Explain how an electrical impulse is transmitted from a presynaptic neuron to a postsynaptic neuron. Describe a synapse and a neuromuscular junction.

5. How is an action potential generated in a postsynaptic neuron?

6. What are the major divisions of the nervous system? What are their major functions?

7. What brain centers have major roles in controlling movement, and what are these roles?

8. How do the sympathetic and parasympathetic systems differ? What is their significance in performing physical activity?

9. Explain how movement occurs in response to touching a hot object.

10. Describe the role of the muscle spindle in controlling muscle action.

11. Describe the role of the Golgi tendon organ in controlling muscle action.

12. What is a motor unit, and how are motor units recruited?

References

1. Edstrom, L., & Grimby, L. (1986). Effect of exercise on the motor unit. *Muscle and Nerve, 9,* 104-126.

2. Guyton, A.C., & Hall, J.E. (2000). *Textbook of medical physiology* (10th ed.). Philadelphia: Saunders.

3. Marieb, E.N. (1995). *Human anatomy and physiology* (3rd ed.). New York: Benjamin/Cummings.

4. Petajan, J.H., Gappmaier, E., White, A.T., Spencer, M.K., Mino, L., & Hicks, R.W. (1996). Impact of aerobic training on fitness and quality of life in multiple sclerosis. *Annals of Neurology, 39,* 432-441.

5. Pette, D., & Vrbova, G. (1985). Neural control of phenotypic expression in mammalian muscle fibers. *Muscle and Nerve, 8,* 676-689.

▶ Selected Readings

Bawa, P. (2002). Neural control of motor output: Can training change it? *Exercise and Sport Sciences Reviews, 30,* 59-63.

Binder, M.D., Heckman, C.J., & Powers, R.K. (1996). The physiological control of motoneuron activity. In L.B. Rowell & J.T. Shepherd (Eds.), *Handbook of physiology: Section 12. Exercise: Regulation and integration of multiple systems* (pp. 3-53). New York: Oxford University Press.

Christensen, N.J., & Galbo, H. (1983). Sympathetic nervous activity during exercise. *Annual Review of Physiology, 45,* 139-153.

Cope, T.C., & Pinter, M.J. (1995). The size principle: Still working after all these years. *News in Physiological Sciences, 10,* 280-286.

Edgerton, V.R., Bodine-Fowler, S., Roy, R.R., Ishihara, A., & Hodgson, J.A. (1996). Neuromuscular adaptation. In L.B. Rowell & J.T. Shepherd (Eds.), *Handbook of physiology: Section 12. Exercise: Regulation and integration of multiple systems* (pp. 54-88). New York: Oxford University Press.

Emonet-Denand, F., Hunt, C.C., & Laporte, Y. (1988). How muscle spindles signal changes of muscle length. *News in Physiological Sciences, 3,* 105-109.

Enoka, R.M. (2001). *Neuromechanics of human movement* (3rd ed.). Champaign, IL: Human Kinetics.

Enoka, R.M., & Stuart, D.G. (1984). Henneman's "size principle": Current issues. *Trends in Neurosciences, 7,* 226-227.

Gandevia, S.C. (1996). Kinesthesia: Roles for afferent signals and motor commands. In L.B. Rowell & J.T. Shepherd (Eds.), *Handbook of physiology: Section 12. Exercise: Regulation and integration of multiple systems* (pp. 128-172). New York: Oxford University Press.

Gielen, C.C.A.M., & Denier van der Gon, J.J. (1990). The activation of motor units in coordinated arm movements in humans. *News in Physiological Sciences,* **5,** 159-163.

Hasan, Z., Enoka, R.M., & Stuart, D.G. (1985). The interface between biomechanics and neurophysiology in the study of movement: Some recent approaches. *Exercise and Sport Sciences Reviews,* **13,** 169-234.

Henneman, E., & Mendell, L.M. (1981). Functional organization of motoneuron pool and its inputs. In V.B. Brooks (Ed.), *Handbook of physiology. Section I, Volume II, The nervous system: Motor control, Part I* (pp. 423-507). Bethesda, MD: American Physiological Society.

McArdle, W.D., Katch, F.I., & Katch, V.L. (2001). *Exercise physiology: Energy, nutrition, and human performance* (5th ed.). Baltimore: Williams & Wilkins.

O'Donovan, M.J. (1985). Developmental regulation of motor function: An uncharted sea. *Medicine and Science in Sports and Exercise,* **17,** 35-43.

Powers, S.K., & Howley, E.T. (2001). *Exercise physiology: Theory and application to fitness and performance* (4th ed.). Madison, WI: McGraw-Hill.

Prochazka, A. (1996). Proprioceptive feedback and movement regulation. In L.B. Rowell & J.T. Shepherd (Eds.), *Handbook of physiology: Section 12. Exercise: Regulation and integration of multiple systems* (pp. 89-127). New York: Oxford University Press.

Proske, U. (1997). The mammalian muscle spindle. *News in Physiological Sciences,* **12,** 37-42.

Roy, R.R., Baldwin, K.M., & Edgerton, V.R. (1991). The plasticity of skeletal muscle: Effects of neuromuscular activity. *Exercise and Sport Sciences Reviews,* **19,** 269-312.

Sale, D.G. (1987). Influence of exercise and training on motor unit activation. *Exercise and Sport Sciences Reviews,* **15,** 95-151.

Seals, D.R., & Victor, R.G. (1991). Regulation of muscle sympathetic nerve activity during exercise in humans. *Exercise and Sport Sciences Reviews,* **19,** 313-349.

CHAPTER 3

NEUROMUSCULAR ADAPTATIONS TO RESISTANCE TRAINING

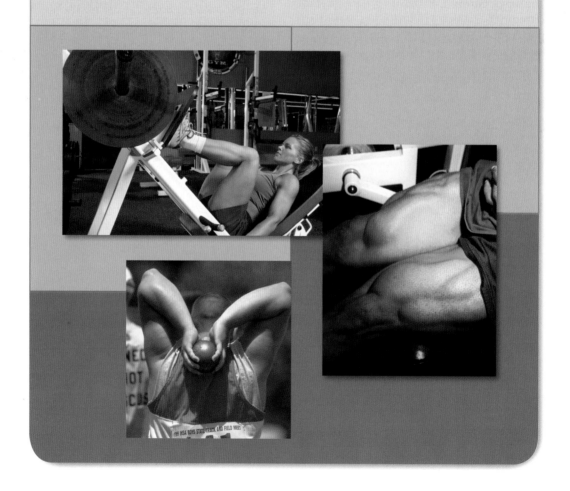

overview

In the preceding chapters, we discussed the functions of the muscular and nervous systems during exercise. But how do we account for the differences between the classic 90-lb (41-kg) weakling and an Olympic weightlifter? What enables a 9-year-old boy to lift a 2-ton car? Why do athletes engage in resistance training even in sports that don't require tremendous strength? Does no pain indeed equal no gain?

We cannot all be Arnold Schwarzenegger, but nearly everyone can improve strength. In this chapter, we examine how strength is gained through resistance training, noting changes that occur in the muscles themselves and in the neural mechanisms controlling them. We explore the phenomenon of muscle soreness and how to prevent it. Finally, we discuss the major concerns in designing a resistance training program and the importance of tailoring it to the specific needs of the individual.

outline

We know that athletes who use **resistance training** become much stronger. For a number of years, Dr. William Gonyea and his colleagues at the University of Texas Health Sciences Center Dallas have been trying to determine how the athlete's muscle gets stronger with resistance training. But Dr. Gonyea and his colleagues have been working with a different type of athlete—cats! The cats receive food rewards for their daily workouts that entice them to work very hard "pumping iron." Like their human counterparts, these cats experience substantial increases in strength and in muscle size. Dogs, beware—these are not cats that you want to mess around with! This chapter discusses the results of Dr. Gonyea's studies as he and his colleagues have challenged traditional thinking on how a muscle increases its size.

With chronic exercise, many adaptations occur in the neuromuscular system. The extent of the adaptations depends on the type of training program followed: Aerobic training, such as jogging or swimming, results in little or no gain in muscular strength and power, but major neuromuscular adaptations occur with resistance training.

Resistance training was once considered inappropriate for athletes except those in competitive weightlifting, weight events in track and field, and—on a limited basis—football, wrestling, and boxing. But in the late 1960s and early 1970s, coaches and researchers discovered that strength and power training are beneficial for almost all sports and activities.

Most athletes now include strength and power training as important components of their overall training programs, including female athletes, who were traditionally excluded from such training. Much of this attitude change is attributable to research that has proven the performance benefits of resistance training and to innovations in training techniques and equipment. Resistance training is now recognized as important even for nonathletes who seek the health-related benefits of exercise (figure 3.1).

▲ **Figure 3.1**   Resistance training is recognized as important for athletes and nonathletes alike.

Terminology

Before discussing the neuromuscular changes that result from resistance training, we'll first define the measurable components of muscular fitness.

Muscular Strength

The maximal force that a muscle or muscle group can generate is termed **strength**. Someone with a maximal capacity to bench-press 150 kg (330 lb) has twice the strength of someone who can bench-press 75 kg (165 lb). In this example, maximal capacity, or strength, is defined as the maximal weight the individual can lift just once. This is referred to as the **1-repetition maximum**, or the **1RM**. To determine your 1RM, select a weight that you know you can lift at least one time. After a proper warm-up, try to execute several repetitions. If you can perform more than one repetition, add weight and try again to execute several repetitions. Continue doing this until you are unable to lift the weight more than a single repetition. This last weight that you are able to lift only once is your 1RM.

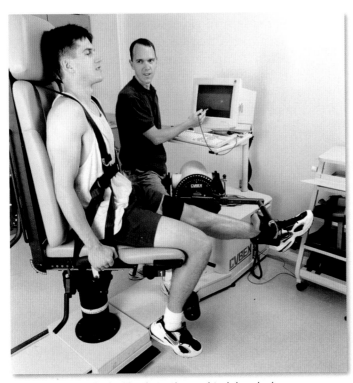

▲ **Figure 3.2**   An isokinetic testing and training device.

Muscle strength can be measured very accurately in the research laboratory by using specialized equipment that allows quantification of static strength and dynamic strength at various speeds and at various angles in the joint's range of motion (see figure 3.2).

Muscular Power

Power, the explosive aspect of strength, is the product of strength and speed of movement

power = force × distance/time,

where force = strength and distance/time = speed. Consider an example. Two individuals can each bench-press 200 kg (441 lb), moving the weight the same distance. But the one who can do it in half the time has twice the power of the slower individual. This is illustrated in table 3.1.

> Power is the functional application of both strength and speed. It is the key component for most athletic performances.

Although absolute strength is an important component of performance, power is even more important for most activities. In football, for example, an offensive lineman with a bench-press 1RM of 200 kg may be unable to control a defensive lineman with a bench-press 1RM of only 150 kg if the defensive lineman can move his 1RM at a much faster speed. The offensive lineman is 50 kg stronger, but the defensive lineman's faster speed coupled with good strength gives him the performance edge. Although field tests are available to estimate power, these tests are generally not very specific to power, in that their results are affected by other factors besides power. Power can be measured, however, by using an electronic device as pictured in figure 3.2.

In this chapter, we primarily concern ourselves with issues of muscular strength, with only brief mention of muscular power. Recall that power has

Table 3.1

Strength, Power, and Muscular Endurance of Three Resistance-Trained Athletes While Performing the Bench Press

Component	Athlete A	Athlete B	Athlete C
Strength[a]	100 kg	200 kg	200 kg
Power[b]	100 kg lifted 0.6 m in 0.5 s = 120 kg · m/s = 1,177 J/s or 1,177 W	200 kg lifted 0.6 m in 2.0 s = 60 kg · m/s = 588 J/s or 588 W	200 kg lifted 0.6 m in 1.0 s = 120 kg · m/s = 1,177 J/s or 1,177 W
Muscular endurance[c]	10 repetitions with 75 kg	10 repetitions with 150 kg	5 repetitions with 150 kg

[a]Strength was determined by the maximum amount of weight the athlete could bench-press just once (i.e., the 1RM).

[b]Power was determined by performing the 1RM test as explosively as possible. Power was calculated as the product of force (weight lifted) times the distance lifted from the chest to full arm extension (0.6 m or about 2 ft), divided by the time it took to complete the lift.

[c]Muscular endurance was determined by the greatest number of repetitions that could be completed using 75% of the 1RM.

two components: strength and speed. Speed is a more innate quality that changes little with training. Thus, power is increased almost exclusively through gains in strength.

Muscular Endurance

Although this chapter focuses on maximal strength and power development, many sporting activities depend on your muscles' ability to repeatedly develop and sustain near-maximal or maximal forces. This capacity to sustain repeated muscle actions, such as when you are performing sit-ups or push-ups, or to sustain fixed or static muscle actions for an extended period of time, such as when attempting to pin an opponent in wrestling, is termed **muscular endurance.** Although there are several excellent laboratory techniques available to directly measure muscular endurance, it can be simply estimated by assessing the maximum number of repetitions you can perform at a given percentage of your 1RM. For example, if you can bench-press 100 kg (220 lb), your muscular endurance could be evaluated independently of your muscular strength by noting how many

repetitions you could perform at, say, 75% of that load (75 kg, or 165 lb). Your muscular endurance is increased through gains in muscular strength and through changes in local metabolic and circulatory function. Metabolic adaptations that occur with training are discussed in chapter 6, and circulatory adaptations are discussed in chapter 9.

Table 3.1 illustrates the functional differences between strength, power, and muscular endurance in three athletes. The actual values have been greatly exaggerated for the purpose of illustration. From this illustration we can see that although Athlete A has half the strength of Athletes B and C, he has twice the power of Athlete B and is equal in power to Athlete C. Therefore, his lack of strength does not seriously limit his power output because of his fast speed of movement. Also, for purposes of designing training programs, the analysis of these three athletes indicates that Athlete A should focus training on developing strength, without losing speed; athlete B should focus training on developing speed of movement, although this is unlikely to change much; and athlete C should focus training on developing muscular endurance. These recommendations are made assuming that each athlete needs to optimize performance in each of these three areas.

> ▶ Muscular strength is the maximum amount of force a muscle or muscle group can generate.
>
> ▶ Muscular power is the product of strength and the speed of a movement. Although two individuals may have the same strength, if one requires less time than the other to move an identical load the same distance, the first individual has more power.
>
> ▶ Muscular endurance is the ability of the muscles to sustain repeated muscle actions or a single static action.

Resistance Training and Gains in Muscular Fitness

Throughout this book we see how important muscular fitness is to athletic performance as well as general health. How do we get stronger, and how do we increase muscle power and muscle endurance? Maintaining an active lifestyle is important in maintaining muscular fitness, but resistance training programs are necessary to increase strength, power, and endurance. In this section, we briefly discuss basic resistance training principles and then review the changes that result from resistance training. We focus on strength, with only a brief mention of power and muscular endurance—topics that are discussed in more detail later in this book.

Basic Resistance Training Principles

In the Introduction chapter of this book, we briefly discussed basic training principles (pages 20-22). Although most of these apply to resistance training, the most important

principle is that of progressive overload. For muscles to get stronger, they must be loaded beyond that point to which they are normally loaded. As a muscle gets stronger, it will require greater resistance to stimulate further increases in strength.

With resistance training, you need to consider the following factors:

• The muscles or muscle groups that you want to condition
• The intensity of training
• The number of repetitions per set
• The number of sets per workout

Two of these factors, intensity and number of repetitions per set, are closely interrelated. As you increase the intensity of training, the number of repetitions per set must decrease. Conversely, as you decrease the intensity of training, you can increase the number of repetitions per set.

To train the muscle for gains in strength, the emphasis is on exercising at higher resistances (high intensity) and lower repetitions. To develop muscular endurance, the emphasis is on lower resistances and higher repetitions. To increase power, part of the workout must emphasize the speed of movement. The number of sets prescribed for resistance training traditionally has been set at three sets. However, more recent research indicates that similar gains in strength can be achieved with only one set. This, as well as how to train muscle, is discussed in more detail later in this chapter in the section "Designing Resistance Training Programs" on page 105.

Now that we have discussed basic resistance training principles, let's look at how the muscles respond to this form of training.

Gains in Strength With Resistance Training

The neuromuscular system is one of the most responsive systems in the body to training. Resistance training programs can produce substantial strength gains. Within 3 to 6 months, you can see from 25% to 100% improvement, sometimes even more. These gains appear to be similar when we compare women to men, children to adults, and the elderly to young and middle-aged adults. Muscle is very plastic, increasing rapidly in size and strength with exercise training and decreasing in size and strength when immobilized. The remainder of this chapter details how these changes occur. How do you become stronger? What physiological adaptations occur that allow you to exert greater levels of strength?

Mechanisms of Gains in Muscle Strength

For many years, strength gains were assumed to result directly from increases in muscle size (hypertrophy). This assumption was logical because most who strength-trained regularly were men, and they often developed large, bulky muscles. Also, muscles associated with a limb immobilized in a cast for weeks or months start to decrease in size (atrophy) and lose strength almost immediately. Gains in muscle size are generally paralleled by gains in strength, and losses in muscle size correlate highly with losses in strength. Thus, we are tempted to conclude that a cause-and-effect relationship exists between muscle size and muscle strength. However, muscle strength involves far more than mere muscle size.

Numerous media reports indicate that people can perform superhuman feats of strength during great psychological stress. Straitjackets were designed specifically to control patients in mental hospitals who suddenly go berserk and are impossible to restrain. Even the world of sport boasts isolated examples of superhuman athletic performances, such as Bob Beamon's long jump of 29 ft 2-1/2 in. (8.9 m) at the 1968 Olympic Games, a jump that exceeded the previous world record by nearly 2 ft (0.6 m)! World records are usually broken by inches or centimeters or, more often, mere fractions of inches or centimeters. Beamon's record stood unbroken until 1991.

Women experience similar strength gains compared with men who participate in the same training program, but the women do not experience as much hypertrophy (see chapter 18). In fact, some women have doubled their strength without any observable change in their muscle size. Similar findings have been reported in children. Thus, strength gains don't require hypertrophy in women, children, or even in men.

This doesn't mean that muscle size is unimportant in the ultimate strength potential of the muscle. Size is extremely important, as revealed by the existing men's and women's world records for competitive weightlifting, shown in figure 3.3. As weight classification increases (implying increased muscle size), so does the record for the total weight lifted. However, examples of superhuman strength and studies on women and children indicate that the mechanisms associated with strength gains are very complex and are not completely understood at this time. How, then, can we explain strength gains with training? Obviously, increased muscle size is important, but there is increasing evidence that the neural control of the trained muscle is also altered, allowing a greater force production from the muscle. Let's look first at the neural control component.

Neural Control of Strength Gains

An important neural component explains at least some of the strength gains that result from resistance training. Enoka has made a convincing argument that strength gains can be achieved without structural changes in muscle but not without neural adaptations.[16] Thus, strength is not solely a property of the muscle. Rather, it is a property of the motor system. Motor unit recruitment is quite important to strength gains. It may well explain most, if not all, strength gains that occur in the absence of hypertrophy, as well as episodic superhuman feats of strength.[40]

Synchronization and Recruitment of Additional Motor Units

Motor units are generally recruited asynchronously; they are not all called on at the same instant. They are controlled by a number of different neurons that can transmit either excitatory or inhibitory impulses (see chapter

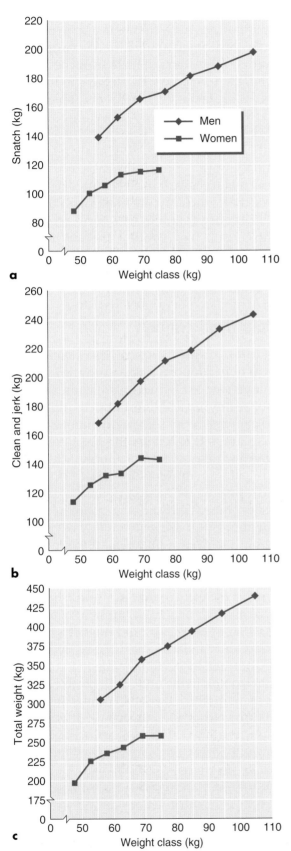

a

b

c

▲ **Figure 3.3**   World records for (*a*) the snatch, (*b*) the clean and jerk, and (*c*) total weight for men and women through February 2002.

2). Whether the muscle fibers contract or stay relaxed depends on the summation of the many impulses received by that motor unit at any one time. The motor unit is activated and its muscle fibers contract only when the incoming excitatory impulses exceed the inhibitory impulses and threshold is met.

Strength gains may result from changes in the connections between motor neurons located in the spinal cord, allowing motor units to act more synchronously, facilitating contraction and increasing the muscle's ability to generate force. There is good evidence to support increased motor-unit synchronization with resistance training, but there is still controversy as to whether synchronization of motor unit activation produces a more forceful contraction.[48, 49, 61] It is clear, however, that synchronization does improve the rate of force development and the capability to exert steady forces.[12]

An alternate possibility is simply that more motor units are recruited to perform the given task, independent of whether these motor units act in unison. Such improvement in recruitment patterns could result from an increase in neural drive to a muscle during maximal contraction. Although the results from several studies support this theory, others do not.[49] It is also possible that inhibitory impulses are reduced, allowing more motor units to be activated, which is discussed further in the next section.

However, not all scientists agree that you can increase motor unit recruitment.[8] Some believe that the scientific evidence indicates that all motor units are recruited even in the untrained state. Unfortunately, the research techniques available to study this issue are indirect, and thus not precise, which could explain some of the confusion about motor unit recruitment.

Autogenic Inhibition

Inhibitory mechanisms in the neuromuscular system, such as the Golgi tendon organs, might be necessary to prevent the muscles from exerting more force than the bones and connective tissues can tolerate. This control is referred to as **autogenic inhibition**. During superhuman feats of strength, major damage often occurs to these structures, suggesting that the protective inhibitory mechanisms are overridden.

We discussed the function of Golgi tendon organs in chapter 2. When the tension on a muscle's tendons and internal connective tissue structures exceeds the threshold of the imbedded Golgi tendon organs, motor neurons to that muscle are inhibited. This reflex is called autogenic inhibition. Both the reticular formation in the brain stem and the cerebral cortex can also initiate and propagate inhibitory impulses.

Training can gradually reduce or counteract these inhibitory impulses, allowing the muscle to reach greater levels of strength.[1] Thus, strength gains may be achieved by reduced neurological inhibition. This theory is attractive because it can explain superhuman strength and strength gains in the absence of hypertrophy. Like any other theory, though, it must undergo the rigors of scientific testing before it can be accepted as fact.

> Autogenic inhibition may be attenuated with resistance training, allowing a greater force production from trained muscles independent of increases in muscle mass.

Other Neural Factors

In addition to increasing motor unit recruitment or decreasing neurological inhibition, other neural factors might contribute to strength gains with resistance training. One of these is referred to as coactivation of agonist and antagonist muscles (the agonist muscles are the primary movers, and the antagonist muscles act to impede the agonists). If we use forearm concentric action as an example, the biceps is the primary agonist and the triceps is the antagonist. If both were contracting with equal force development, no movement would occur. Thus, to maximize the force generated by an agonist, it is necessary to minimize the amount of coactivation.[17] Reduction in coactivation could explain a portion of strength gains attributed to neural factors, but its contribution likely would be small.

Rate coding is another potential factor that could increase force production of a muscle fol-

lowing training. Rate coding is a term used to describe the firing frequency or discharge rates of motor units. Although this has not been well studied, there is some evidence to support this as a possible factor.[17] Changes also have been noted in the morphology of the neuromuscular junction, with both increased and decreased activity levels that might be directly related to the muscle's force-producing capacity.[11]

Now that we have looked at potential neural factors that contribute to increased strength with resistance training, let's turn our attention to muscle hypertrophy. We will look at how a muscle increases in size and how this can contribute to an increase in strength or the force-producing capacity of the muscle.

Muscle Hypertrophy

How does a muscle's size increase? Two types of hypertrophy can occur: transient and chronic. **Transient hypertrophy** is the pumping-up of the muscle that happens during a single exercise bout. This results mainly from fluid accumulation (edema) in the interstitial and intracellular spaces of the muscle. This fluid is lost from the blood plasma.[42] Transient hypertrophy, as its name implies, lasts only for a short time. The fluid returns to the blood within hours after exercise.

Chronic hypertrophy refers to the increase in muscle size that occurs with long-term resistance training. This reflects actual structural changes in the muscle that can result from an increase in either or both the number of muscle fibers (**fiber hyperplasia**) and the size of existing individual muscle fibers (**fiber hypertrophy**). Controversy surrounds the theories that attempt to explain the underlying cause of this phenomenon, which has led to heated debate at scientific meetings. Of importance, however, is the recent finding that the eccentric component of training is important in maximizing increases in muscle fiber cross-sectional area.[14, 30] In one study, after 36 training sessions with subjects using only concentric actions or only eccentric actions, eccentric training resulted in an increase in fast-twitch (FT) fiber area approximately 10 times greater than that of concentric training, as well as substantially greater increases in strength.[32] Thus, training with only concentric actions could limit muscle hypertrophy and increases in muscle strength. Let's now look at the two postulated mechanisms for increasing muscle size with resistance training: fiber hypertrophy and fiber hyperplasia.

Fiber Hypertrophy

Early research suggested that the number of muscle fibers in each of your muscles is established by birth or shortly thereafter and that this number remains fixed throughout your life. If this is true, then whole muscle hypertrophy could result only from individual muscle fiber hypertrophy. This could be explained by

- more myofibrils,
- more actin and myosin filaments,
- more sarcoplasm,
- more connective tissue, or
- any combination of these.

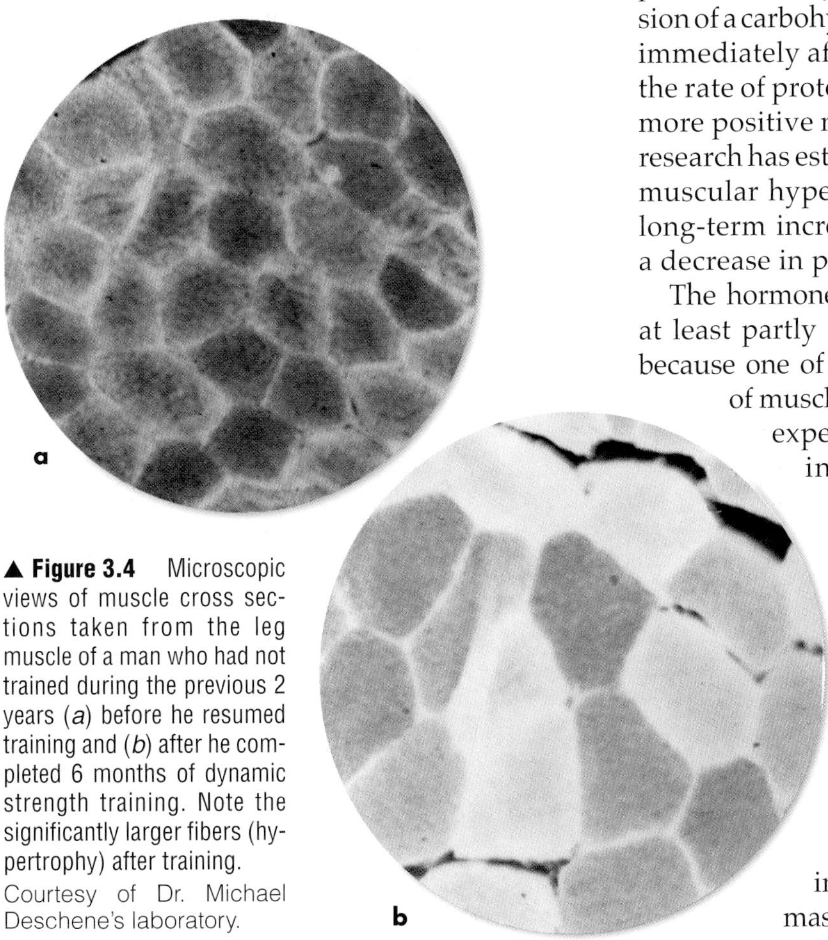

▲ **Figure 3.4**   Microscopic views of muscle cross sections taken from the leg muscle of a man who had not trained during the previous 2 years (*a*) before he resumed training and (*b*) after he completed 6 months of dynamic strength training. Note the significantly larger fibers (hypertrophy) after training.
Courtesy of Dr. Michael Deschene's laboratory.

As seen in the micrographs in figure 3.4, intense resistance training can significantly increase the cross-sectional area of muscle fibers. In this example, fiber hypertrophy is probably caused by increased numbers of myofibrils and actin and myosin filaments, which would provide more cross-bridges for force production during maximal contraction. Such dramatic enlargement of muscle fibers does not occur, however, in all cases of muscle hypertrophy.

Individual muscle fiber hypertrophy from resistance training appears to result from a net increase in muscle protein synthesis. The muscle's protein content is in a continual state of flux. Protein is always being synthesized and degraded. But the rates of these processes vary with the demands placed on the body. During exercise, protein synthesis decreases, while protein degradation apparently increases.[24] This pattern reverses during the postexercise recovery period, even to the point of a net synthesis of protein. The provision of a carbohydrate and protein supplement immediately after a training bout can reduce the rate of protein degradation, resulting in a more positive nitrogen balance.[56] In animals, research has established that exercise-induced muscular hypertrophy is accompanied by a long-term increase in protein synthesis and a decrease in protein degradation.

The hormone testosterone is thought to be at least partly responsible for these changes because one of its functions is the promotion of muscle growth (see chapter 5). Males experience a significantly greater increase in muscle size than females for the same resistance training program and even for the same relative increase in strength. Testosterone is an androgen, a substance that produces masculine characteristics. Anabolic steroids are also androgens, and it is well known that massive doses of anabolic steroids coupled with resistance training markedly increase muscle mass and strength (see chapter 15).

Fiber Hyperplasia

Recent research on animals suggests that hyperplasia may also be a factor in the hypertrophy of whole muscles. Studies on cats provide fairly clear evidence that fiber splitting occurs with extremely heavy weight training.[22] Cats were trained to move a heavy weight with a forepaw to get their food (figure 3.5). They learned to generate considerable force. With this intense strength training, selected muscle fibers appeared to actually split in half, and each half then increases to the size of the parent fiber. This is seen in the cross-sectional cuts through the muscle fibers shown in figure 3.6.

Subsequent studies, however, demonstrated that hypertrophy of selected muscles in chickens, rats, and mice that resulted from chronic exercise overload was attributable solely to hypertrophy of existing fibers, not hyperplasia.[20, 21, 59] In these studies, each fiber in the whole muscle was actually counted. These direct fiber counts revealed no change in fiber number.

This finding led the scientists who conducted the initial cat experiments to conduct an additional resistance training study with cats. This time they used actual fiber counts to determine if total muscle hypertrophy resulted from hyperplasia or fiber hypertrophy.[23] Fol-

lowing a resistance training program of 101 weeks, the cats were able to perform one-leg lifts of an average of 57% of their body weight, resulting in an 11% increase in muscle weight. Most important, the researchers found a 9% increase in the total number of muscle fibers, confirming that muscle fiber hyperplasia did occur.

The difference in results between the cat studies and those with other animals most likely is attributable to differences in the manner in which the animals were trained. The cats were trained with a pure form of resistance training: high resistance and low repetitions.

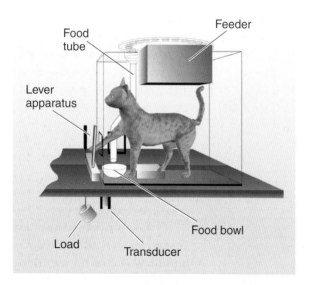

▲ **Figure 3.5**   Heavy resistance training in cats.

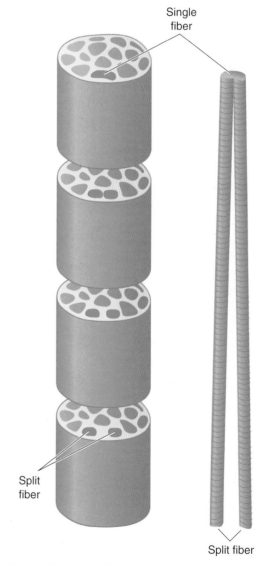

▲ **Figure 3.6**   Muscle fiber splitting. The drawn models were extrapolated from a series of microscopic slides.

The other animals were trained with more endurance-type activity: low resistance and high repetitions.

One additional animal model has been used to stimulate muscle hypertrophy associated with hyperplasia. Scientists have placed the anterior *latissimus dorsi* muscle of chickens in a state of chronic stretch by attaching weights to it, with the other wing serving as the normal control condition. In many of the studies that have used this model, the chronic stretch resulted in substantial hypertrophy and hyperplasia, although other studies using the same model were unable to replicate the finding of hyperplasia.[3]

Researchers are still uncertain about the roles played by hyperplasia and individual fiber hypertrophy in increasing human muscle size with resistance training. Most evidence indicates that individual fiber hypertrophy accounts for most whole-muscle hypertrophy. However, results of two studies on bodybuilders indicate that hyperplasia is possible in humans.

One study reported that the mean muscle fiber areas of the *vastus lateralis* and the deltoid muscles were smaller in a group of high-caliber bodybuilders than in a reference group of competitive powerlifters or weightlifters and were nearly identical to those in physical education students and non-strength-trained people. This suggests that individual fiber hypertrophy was not critical to the bodybuilders' gains in muscle mass.[57]

Similar results were found in a subsequent study comparing highly trained bodybuilders and active but untrained controls. Muscle fiber areas of the bodybuilders were similar to those of the control subjects, despite the fact that the bodybuilders had much greater limb circumferences.[38] The researchers also found more muscle fibers per motor unit in the bodybuilders than in the nonathlete control group. Because these bodybuilders had substantially greater muscle girth but normal muscle fiber cross-sectional area, these findings suggest that their number of muscle fibers increased. An alternate explanation is that these athletes had more muscle fibers at birth.

> Muscle fiber hyperplasia has been clearly shown to occur in animal models by using resistance training to induce muscle hypertrophy. Only a few studies, on the other hand, suggest evidence of hyperplasia in humans.

In contrast, at least one study has found large differences in muscle fiber area when bodybuilders were compared with male and female physical education students.[44] Mean fiber areas of the vastus lateralis for each of the three groups were as follows:

- Bodybuilders: 8,400 μm^2
- Male physical education students: 6,200 μm^2
- Female physical education students: 4,400 μm^2

The differences among these studies might be explained by the nature of the training load or stimulus. Training at high intensities or high resistances is thought to cause greater fiber hypertrophy, particularly of the FT fibers, than training at lower intensities or resistances.[35]

Only one longitudinal study has demonstrated the possibility of hyperplasia in male subjects who had previous recreational resistance training experience.[39] Following 12 weeks of intensified resistance training, the muscle fiber number in the *biceps brachii* of several of the 12 subjects appeared to increase significantly. It appears from this study that hyperplasia can occur in humans, but possibly only in certain subjects or under certain training conditions.

From the preceding information, it appears that fiber hyperplasia can occur in animals and possibly in humans. How are these new cells formed? As shown in figure 3.6, it is postulated that individual muscle fibers have the capacity to divide and split into two daughter cells, each of which can then develop into a functional muscle fiber. It has more recently been established that satellite cells, which are the myogenic stem cells involved in skeletal muscle regeneration, are likely involved in the generation of new muscle fibers. These cells are typically activated by muscle injury, and, as we see later in this chapter, muscle injury results from intense training, particularly eccentric-action training.[3] Muscle injury can lead to a cascade of responses, where satellite cells become activated and proliferate, migrate to the damaged region, and fuse to existing myofibers or combine and fuse to produce new myofibers.[31] This is illustrated in figure 3.7.

Integration of Neural Activation and Fiber Hypertrophy

Research conducted on resistance training adaptations indicates that early increases in voluntary strength, or maximal force production, are associated primarily with neural adaptations resulting in increased voluntary activation of muscle. This was clearly dem-

onstrated in a study of both men and women who participated in an 8-week, high-intensity resistance training program, training twice per week.[51] Muscle biopsies were obtained at the beginning of the study and every 2 weeks during the training period. Strength, measured according to the 1RM, increased substantially over the 8 weeks of training, with the greatest gains coming after the second week. Muscle biopsies, however, revealed only small, statistically insignificant increases in muscle fiber cross-sectional area by the end of the 8 weeks of training. Thus, the strength gains were largely the result of increased neural activation.

Long-term increases in strength generally are associated with hypertrophy of the trained muscle.[43] It takes time to build protein through a decrease in protein degradation, an increase

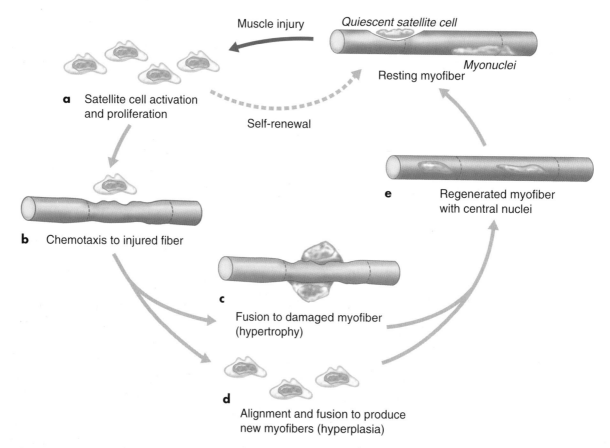

▲ **Figure 3.7** The satellite cell response to muscle injury. Muscle injury leads to (*a*) satellite cell activation and proliferation. The satellite cells migrate to the damaged region (*b*) and fuse to the damaged myofiber (*c*) or align and fuse to produce a new myofiber (*d*), both leading to a regenerated or new myofiber (*e*).

Reprinted, by permission, from T.J. Hawke and D.J. Garry, 2001, "Myogenic satellite cells, physiology to molecular biology," *Journal of Applied Physiology* 91: 534-551.

in protein synthesis, or both. Notable exceptions to this generalization have been found. A 6-month study of strength-trained athletes found that neural activation explained most of the strength gains during the most intensive training months and that hypertrophy was not a major factor.[29] It appears that neural factors make their greatest contribution during the first 8 to 10 weeks of training. Hypertrophy contributes little during the initial weeks of training but progressively increases its contribution, becoming the major contributor after 10 weeks of training.

> Early gains in strength appear to be more influenced by neural factors, but later long-term gains are largely the result of hypertrophy.

Muscle Atrophy and Decreased Strength With Inactivity

When a normally active or highly trained person reduces her level of activity or ceases training altogether, major changes occur in both muscle structure and function. This is illustrated by the results of two types of studies: studies in which entire limbs have been immobilized and studies where highly trained people stop training.

Immobilization

When a trained muscle suddenly becomes inactive through immobilization, major changes are initiated within that muscle in a matter of hours (chapter 12). During the first 6 h of immobilization, the rate of protein synthesis starts to decrease. This decrease likely initiates the start of muscular atrophy, which is the wasting away or decrease in the size of muscle tissue. Atrophy results from lack of muscle use and the consequent loss of muscle protein that accompanies the inactivity. Strength decreases are most dramatic during the first week of immobilization, averaging 3% to 4% per day.[4] This is associated with the atrophy but also with decreased neuromuscular activity of the immobilized muscle.

Immobilization appears to affect primarily the slow-twitch (ST) fibers. From various stud-

ies, researchers have observed disintegrated myofibrils, streaming Z disks (discontinuity of Z disks and fusion of the myofibrils), and mitochondrial damage in ST fibers. When muscle atrophies, both the cross-sectional fiber area and the percentage of ST fibers decrease. Whether ST fibers decrease from fiber necrosis (death) or conversion into FT fibers is unclear.[4]

Muscles can and often do recover from immobilization when activity is resumed. The recovery period is substantially longer than the period of immobilization but shorter than the original training period.

Cessation of Training

Similarly, significant muscle alterations can occur when you stop training. In one study, women resistance-trained for 20 weeks and then stopped training for 30 to 32 weeks. Finally they retrained for 6 weeks.[52] The training program focused on the lower extremity, using a full squat, leg press, and leg extension. Strength increases were dramatic, as seen in figure 3.8. Compare the women's strength after their initial training period (post-20) with their strength after detraining (pre-6). This represents the strength loss they experienced with cessation of training.

During the two training periods, increases in strength were accompanied by increases in the cross-sectional area of all fiber types and a decrease in the percentage of FT_b fibers. Detraining had relatively little effect on fiber cross-sectional area, although the FT fiber areas tended to decrease (figure 3.9).

To prevent losses in the strength gained through resistance training, basic maintenance programs must be established once the desired goals for strength development have been achieved. Maintenance programs are designed to provide sufficient stress to the muscles to maintain existing levels of strength while allowing a reduction in intensity, duration, or frequency of training.

In one study, men and women resistance-trained with knee extensions for either 10 or 18 weeks and then spent an additional 12 weeks with either no training or reduced training.[25] Knee extension strength increased 21.4% following the training period. Subjects who then

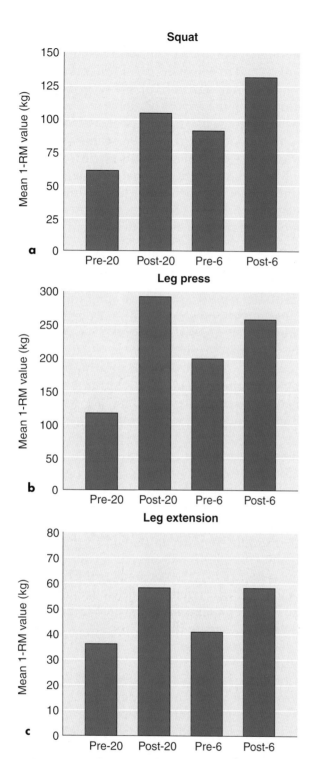

▲ Figure 3.8 Changes in muscle strength with resistance training in women. Pre-20 values indicate strength before starting training; post-20 values indicate the changes following 20 weeks of training; pre-6 values indicate the changes following detraining; and post-6 values indicate the changes following 6 weeks of retraining.

Adapted, by permission, from R.S. Staron et al., 1991, "Strength and skeletal muscle adaptations in heavy-resistance-trained women after detraining and retraining," *Journal of Applied Physiology* 70: 631-640.

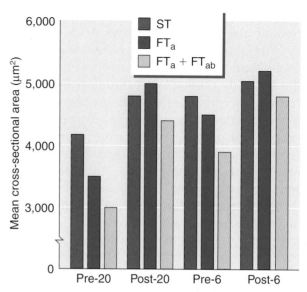

▲ Figure 3.9 Changes in mean cross-sectional areas for the major fiber types with resistance training in women over periods of training (post-20), detraining (pre-6), and retraining (post-6). FT$_{ab}$ is an intermediate fiber type. See figure 3.8 caption for more details.

stopped training lost 68% of their strength gains during the weeks they didn't train. But subjects who reduced their training week (from 3 days per week to 2 or from 2 to 1) did not lose strength. Thus, it appears that strength can be maintained for at least up to 12 weeks with reduced training frequency.

Fiber-Type Alterations

Can muscle fibers change from one type to another through resistance training? The earliest research concluded that neither speed (anaerobic) nor endurance (aerobic) training could alter the basic fiber type.[10, 19] These early studies did show, however, that fibers begin to take on certain characteristics of the opposite fiber type if the training is of the opposite kind (e.g., FT fibers might become more oxidative with aerobic training).

More recent research with animals has shown that fiber-type conversion is indeed possible under conditions of cross-innervation, in which an FT motor unit is artificially innervated by an ST motor neuron or vice versa. Also, chronic, low-frequency nerve stimulation transforms FT motor units into ST motor units within a matter of weeks.[41] Muscle fiber types in rats have changed in response to 15 weeks of high-intensity treadmill training, resulting in an increase in ST and FT_a fibers and a decrease in FT_b fibers.[26] The transition of fibers from FT_b to FT_a and from FT_a to ST was confirmed by several different histochemical techniques.

Staron and coworkers found evidence of fiber-type transformation in women as a result of heavy resistance training.[53] Substantial increases in static strength and in the cross-sectional area of all fiber types were noted following a 20-week heavy resistance training program for the lower extremity. The mean percentage of FT_b fibers decreased significantly, but the mean percentage of FT_a fibers increased. The reduction of FT_b fibers and the increase in FT_a fibers with resistance training has been consistently reported in a number of subsequent studies. The functional significance of this change in fiber type has not been clearly established.[36] More recent studies have shown that a combination of high-intensity resistance training and short-interval speed work can lead to a conversion of ST to FT_a fibers.[2]

Muscle Soreness

Muscle soreness can be present

- during the latter stages of an exercise bout and the immediate recovery period,
- between 12 and 48 h after a strenuous bout of exercise, or
- at both times.

Let's examine soreness during both periods.

Acute Muscle Soreness

Pain felt during and immediately after exercise can result from accumulation of the end products of exercise, such as H^+, and from tissue edema, mentioned earlier, which is caused by fluid shifting from the blood plasma into the tissues. Edema is the pumped-up sensation that you feel after heavy endurance or strength training. The pain and soreness usually disappear within a few minutes to several hours after the exercise. Thus, this soreness is often referred to as **acute muscle soreness.**

▶ Neural adaptations always accompany the strength gains that result from resistance training, but hypertrophy might or might not be present.

▶ Neural mechanisms leading to strength gains can include recruitment of more motor units, motor units recruited more synchronously, and decreases in autogenic inhibition from the Golgi tendon organs.

▶ Transient muscle hypertrophy is the pumped-up feeling you get immediately after an exercise bout. It results from edema and is short-lived.

▶ Chronic muscle hypertrophy occurs from repeated resistance training and reflects actual structural changes in the muscle.

▶ Although most muscle hypertrophy probably results from an increase in the size of individual muscle fibers (fiber hypertrophy), some evidence suggests that an increase in the number of muscle fibers (hyperplasia) also might be involved.

▶ Muscles atrophy (decrease in size and strength) when they become inactive, as with injury, immobilization, or cessation of training.

▶ Atrophy begins very quickly if training is stopped, but training can be reduced, as in a maintenance program, without resulting in atrophy or loss of strength.

▶ One fiber type can take on characteristics of the opposite type in response to training. Evidence indicates that one fiber type might actually be converted to the other type as a result of cross-innervation or chronic stimulation.

Delayed-Onset Muscle Soreness and Injury

Muscle soreness felt a day or two after a heavy bout of exercise is not totally understood, yet research over the past few years has given us greater insight into this phenomenon.

Because this pain does not occur immediately, it is referred to as **delayed-onset muscle soreness (DOMS).** In the following sections, we discuss some theories that attempt to explain this form of muscle soreness. As we consider these theories, realize that none of them yet has universal support. More research must be conducted before the exact mechanisms can be determined.

Almost all current theories acknowledge that eccentric action is the primary initiator of DOMS. This was initially observed in a study that investigated the relationship of muscle soreness to eccentric, concentric, and static actions.[54] This study found that a group who trained solely with eccentric actions experienced extreme muscle soreness, whereas the static- and concentric-action groups experienced little soreness. This idea was further explored in studies in which subjects ran on a treadmill for 45 min on two separate days, one day on a level grade and the other day on a 10% downhill grade.[45, 46] No muscle soreness was associated with the level running. But the downhill running, which required extensive eccentric action, resulted in considerable soreness within 24 to 48 h, even though blood lactate levels, previously thought to cause muscle soreness, were much higher with level running.

> Delayed-onset muscle soreness results primarily from eccentric action and is associated with actual muscle damage.

Let's examine some of the proposed explanations for exercise-induced DOMS.

Structural Damage

The presence of increased concentrations of several specific muscle enzymes in blood after intense exercise suggests that some structural damage may occur in the muscle membranes. These enzyme levels increase from 2 to 10 times their normal levels following bouts of heavy training. Recent studies support the idea that these changes might indicate various degrees of muscle tissue breakdown. Examination of tissue from the leg muscles of marathon runners has revealed remarkable damage to the muscle fibers after both training and marathon competition. The onset and timing of these muscle changes parallel the degree of muscle soreness experienced by the runners.

The electron micrograph in figure 3.10 shows muscle fiber damage as a result of marathon running.[27] In this case, the sarcolemma (cell membrane) was totally ruptured, allowing the cell's contents to float freely between the other normal fibers. Fortunately, not all damage to muscle cells is as severe.

Figure 3.11 shows changes in the contractile filaments and Z disks before and after a marathon race. Recall that Z disks are the points of contact for the contractile proteins. They provide structural support for the transmission of force when the muscle fibers are activated to shorten. Figure 3.11*b*, after the marathon, shows the Z disks pulled apart as a result of the force

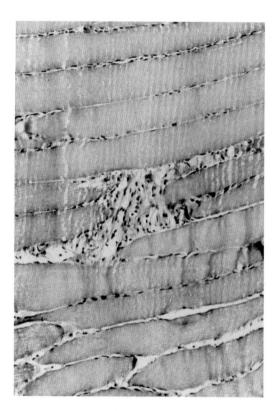

▲ **Figure 3.10**   An electron micrograph of a muscle sample taken immediately after a marathon, showing the disruption of the cell membrane in one muscle fiber.

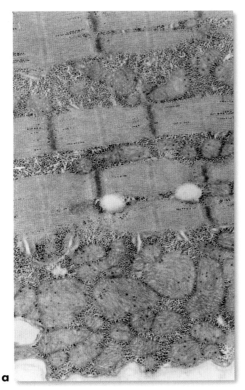 

▲ **Figure 3.11**   (*a*) An electron micrograph showing the normal arrangement of the actin and myosin filaments and Z disk configuration in the muscle of a runner before a marathon race. (*b*) A muscle sample taken immediately after a marathon race shows Z disk streaming caused by the eccentric actions of running.

Adapted from F.C. Hagerman et al., 1984.

of eccentric actions or stretching of the tightened muscle fibers.

Although the effects of muscle damage on performance are not fully understood, experts generally agree that this damage is responsible in part for the localized muscle pain, tenderness, and swelling associated with DOMS. However, blood enzyme levels might increase and muscle fibers might be damaged frequently during daily exercise that produces no muscle soreness.

Inflammatory Reaction

White blood cells serve as a defense against foreign materials that enter your body and against conditions that threaten the normal function of tissues. The white blood cell count tends to increase following activities that induce muscle soreness. This observation led some investigators to suggest that soreness results from inflammatory reactions in the muscle. But the link between these reactions and muscle soreness has been difficult to establish.

Researchers have tried to use drugs to block the inflammatory reaction, but these efforts have been unsuccessful in reducing either the amount of muscle soreness or the degree of inflammation.[37] Because both effects remain, conclusions about the role of inflammation in muscle soreness cannot be drawn from this research. More recent studies, however, are beginning to establish a link between muscle soreness and inflammation. For example, it is now recognized that substances released from injured muscle can act as "wound hormones," initiating the inflammatory process. Mononucleated cells in muscle are activated by the injury, providing the chemical signal to circulating inflammatory cells. Neutrophils (a type of white blood cell) invade the injury site and release cytokines (immunoregulatory substances), which then attract and activate additional inflammatory cells. Neutrophils possibly also release oxygen free radicals that can damage cell membranes. Macrophages (another type of cell of the immune system) then invade the damaged muscle fibers, removing debris through a process known as phagocytosis. Last, a second phase of macrophage invasion occurs, which is associated with muscle regeneration.[59]

Sequence of Events in DOMS

In 1984, Armstrong[5] conducted a review of possible mechanisms for exercise-induced DOMS. He concluded that DOMS is associated with

- elevations in plasma enzymes,
- myoglobinemia (presence of myoglobin in the blood), and
- abnormal muscle histology and ultrastructure.

He developed a model of DOMS that proposed the following sequence of events:

1. High tension in the contractile-elastic system of muscle results in structural damage to the muscle and its cell membrane.

2. The cell membrane damage disturbs calcium homeostasis in the injured fiber, resulting in necrosis (cell death) that peaks about 48 h after exercise.

3. The products of macrophage activity and intracellular contents (such as histamine, kinins, and K^+) accumulate outside the cells. These substances then stimulate the free nerve endings in the muscle. This process appears to be accentuated in eccentric exercise, in which large forces are distributed over relatively small cross-sectional areas of the muscle.

Recent comprehensive reviews have provided much greater insight into the cause of muscle soreness. We now are confident that muscle soreness results from injury or damage to the muscle itself, generally the muscle fiber and possibly the sarcolemma.[6] This damage sets up a chain of events that includes the release of intracellular proteins and an increase in muscle protein turnover. The damage and repair process involves calcium ions, lysosomes, connective tissue, free radicals, energy sources, inflammatory reactions, and intracellular and myofibrillar proteins. But the precise cause of skeletal muscle damage and the mechanisms of repair are not well understood. Some evidence suggests that this process is an important step in muscle hypertrophy.

Up to this point, our discussion of DOMS has focused on muscle injury. Edema, or the accumulation of fluids in the muscular compartment, also can lead to DOMS. This edema is likely the result of muscle injury but could occur independently of muscle injury. An accumulation of interstitial or intracellular fluid increases the tissue fluid pressure within the muscle compartment, which in turn activates pain receptors within the muscle.

DOMS and Performance

With DOMS comes a reduction in the force-generating capacity of the affected muscles. Whether the DOMS is the result of injury to the muscle or edema independent of muscle injury, the affected muscles are not able to exert as much force when the person is asked to apply maximal force, such as in the performance of a 1RM strength test. Maximal force-generating capacity gradually returns over days or weeks. It has been proposed that the loss in strength is the result of three factors:[61]

1. the physical disruption of the muscle as illustrated in figures 3.10 and 3.11;

2. failure within the excitation–contraction coupling process; and

3. loss of contractile protein.

Failure in excitation–contraction coupling appears to be the most important, particularly during the first 5 days. This is illustrated in figure 3.12.

Muscle glycogen resynthesis also is impaired when a muscle is damaged. It is generally normal for the first 6 to 12 h after exercise, but it slows or stops completely as the muscle undergoes repair, thus limiting the fuel-storage capacity of the injured muscle. Figure 3.13 illustrates the time sequence of the various factors associated with intense eccentric exercise, including pain, edema, plasma creatine

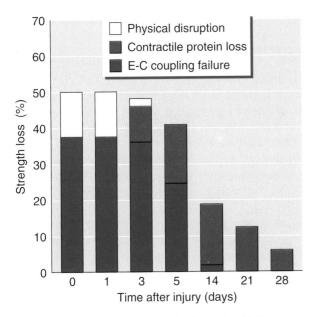

▲ **Figure 3.12** Estimated contributions of excitation–contraction (EC) coupling failure, decreased contractile protein content, and physical disruption to the decrease in strength following muscle injury.

Reprinted, by permission, from G.L. Warren et al., 2001, "Excitation-contraction uncoupling: Major role in contraction-induced muscle injury," *Exercise and Sport Sciences Reviews* 29: 82-87.[61]

kinase (a plasma enzyme marker of muscle fiber damage), glycogen depletion, ultrastructural damage in the muscle, and muscular weakness.

Reducing the Negative Effects of DOMS

Reducing the negative effects of DOMS is important for maximizing training gains. The eccentric component of muscle action could be minimized during early training, but this is not possible for athletes in most sports. An alternative approach is to start training at a very low intensity and progress slowly through the first few weeks. Yet another approach is to initiate the training program with a high-intensity, exhaustive training bout. Muscle soreness would be great for the first few days, but evidence suggests that subsequent training bouts would cause considerably less muscle soreness.[15] Because the factors associated with DOMS are also potentially important in stimulating muscle hypertrophy, DOMS is most likely necessary to maximize the training response.

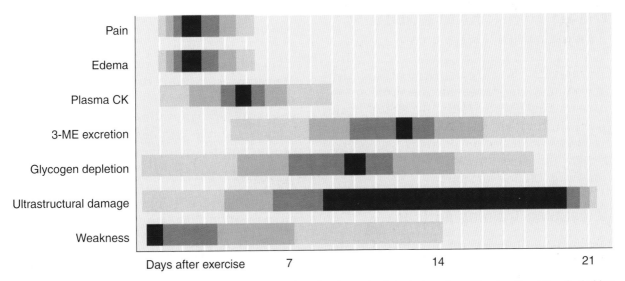

▲ **Figure 3.13**   The delayed response to eccentric exercise of various physiological markers. The density of the shaded bar corresponds to the intensity of the response at the indicated time. CK = creatine kinase.

Adapted, by permission, from W.J. Evans and J.G. Cannon, 1991, "The metabolic effects of exercise induced muscle damage," *Exercise and Sport Sciences Reviews* 19: 99-125.

Exercise-Induced Muscle Cramps

Few things frustrate athletes more than muscle cramps. Skeletal muscle cramps can come during the height of competition, immediately after competition, or at night in the deep of sleep. Muscle cramps are equally frustrating to scientists, because they have not been able to totally determine the cause of muscle cramping or how to treat or prevent cramps. Although muscle cramps can be the result of rare medical conditions, most exercise-induced or exercise-associated muscle cramps are unrelated to disease or medical disorder. Exercise-associated muscle cramps (EAMCs) have been defined as painful, spasmodic, involuntary contractions of skeletal muscles that occur during or immediately after exercise.[47] Nocturnal muscle cramps may or may not be associated with exercise.

Early research suggested that muscle cramps were caused by disturbances in fluid and electrolyte balance, associated with high rates of sweating. Although this might be the case for some EAMCs, more recent research proposes that EAMCs result from sustained α-motor neuron activity, which is caused by aberrant control at the spinal level.[47] Muscle fatigue appears to cause this lack of control through an effect on both Golgi tendon organs and muscle spindles. Spindle activity increases and tendon organ activity decreases.

Treatment includes rest, passive stretching of the affected muscle or muscle groups, and holding the muscle in a stretched position until muscle activation is relieved. Fluids also should be taken in if dehydration and electrolyte loss are suspected. To prevent EAMCs, the athlete should

- be well conditioned, to reduce the likelihood of muscle fatigue,
- regularly stretch the muscle groups prone to EAMCs,
- maintain fluid and electrolyte balance and carbohydrate stores, and
- reduce exercise intensity and duration if necessary.[47]

Designing Resistance Training Programs

Over the past 50 years, research has provided a substantial knowledge base concerning resistance training and its application to health and sport. The health aspects of resistance training are discussed in chapter 19. In this section, we concern ourselves primarily with the use of resistance training for sport.

Resistance Training Actions

Resistance training actions are related to the types of muscle actions. Resistance training can use static actions (also called isometric actions), dynamic actions, or both. Dynamic actions include the use of free weights, variable resistance, isokinetic actions, and plyometrics.

With free weights, such as barbells and dumbbells, the resistance or weight lifted remains constant throughout the dynamic range of movement. If you lift a 10-kg (22-lb) weight, it will always weigh 10 kg. In contrast, a variable-resistance action involves varying the resistance to try to match it to your strength curve. Figure 3.14 illustrates how your strength varies throughout the range of motion in a two-arm curl. Maximal strength production by the elbow flexors occurs at approximately 100° in the range of movement. These muscles are weakest at 60° (elbows fully flexed) and at 180° (elbows fully extended). In these positions, you are able to generate only 67% and 71%, respectively, of your maximal force-producing capabilities at the optimal angle of 100°.

When one is using free weights, the resistance or weight used to train the muscle is limited by the weakest point in the range of motion. If the person in figure 3.14 had the capacity to lift 45 kg (100 lb) at the optimal angle of 100°, then he would be able to lift

▶ Acute muscle soreness occurs late in an exercise bout and during the immediate recovery period.

▶ DOMS occurs a day or two after the exercise bout. Eccentric action seems to be the primary instigator of this type of soreness.

▶ Proposed causes of DOMS include structural damage to muscle cells and inflammatory reactions within the muscles.

▶ Armstrong's proposed model of the sequence of events that cause DOMS includes

1. structural damage,
2. impaired calcium homeostasis leading to necrosis,
3. accumulation of irritants, and
4. increased macrophage activity.

▶ Muscle soreness can be prevented or minimized by

1. reducing the eccentric component of muscle action during early training;
2. starting training at a low intensity and gradually increasing it; or
3. beginning with a high-intensity, exhaustive bout of eccentric-action exercise, which will cause much soreness initially but will decrease future pain.

▶ Muscle strength is reduced during the period of muscle soreness and is likely the result of three factors: physical disruption of the muscle, failure of the excitation–contraction coupling process, and loss of contractile protein.

▶ Muscle soreness may be an important part of maximizing the resistance training response.

▶ EAMCs are attributable to either or both fluid and electrolyte imbalances and muscle fatigue-associated sustained α-motor neuron activity, with increased muscle spindle activity and decreased Golgi tendon organ activity. Rest, passive stretching, and holding the muscle in the stretched position can be effective in treating EAMCs, and proper conditioning, stretching, and nutrition are possible prevention strategies.

only 32 kg (71 lb) at the fully extended position of 180°. Therefore, if he is starting with a barbell loaded with 31.5 kg, he can just barely move it from the fully extended position to start his lift. However, by the time he gets to an angle of 100° in his full range of motion, he is lifting only 70% of what he could maximally lift at that angle. Thus, with free weights you maximally tax only the weakest points in your range of motion and provide a relatively light resistance at the midrange (90-140°). This is why individuals performing the two-arm curl

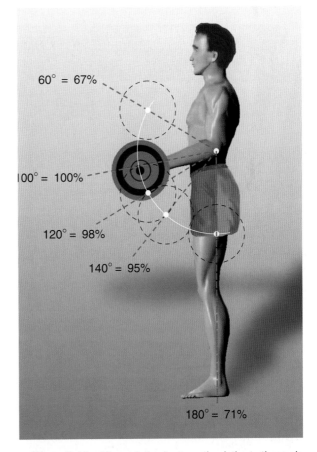

▲ **Figure 3.14**   The variation in strength relative to the angle of the elbow flexors during the two-arm curl. Strength is optimized at an angle of 100°. The maximal force-development capacity of the muscle group at a given angle is given as a percentage of the capacity at the optimal angle of 100°.

tend to greatly reduce their range of motion as they start to fatigue (referred to as "cheating"). They are simply trying to stay out of the weakest portion of their range of motion. The bottom line is that with free weights you can lift the maximal weight only for the weakest portion of your range of motion, which means that the strongest portion of your range of motion is never maximally taxed! However, as we shall see later, this does not mean that free weights are not the best way to train.

With a variable-resistance device, the resistance is decreased at the weakest points in the range of movement and increased at the strongest points. Variable resistance is the basis for several popular resistance training devices. The underlying theory is that the muscle can be more fully trained if it is forced to act at higher constant percentages of its capacity throughout each point in its range of movement. Figure 3.15 illustrates a variable-resistance device in which a cam alters the resistance through the range of motion.

> The ability of a muscle or muscle group to generate force varies throughout the full range of movement.

An isokinetic device holds movement speed constant. Whether you apply very light force or an all-out maximal muscle action, the speed of movement does not vary. Using either electronics, air, or hydraulics, the device can be preset to control the speed of movement (angular velocity) from $0°/s$ (static action) up to $300°/s$ or higher. An isokinetic device was illustrated in figure 3.2 (page 87). Theoretically, if properly motivated, the individual can contract the muscles at maximal force at all points in the range of motion.

Plyometrics, or jump training, uses the stretch reflex to facilitate recruitment of additional motor units. Plyometrics is discussed in more detail later in this chapter.

Training Needs Analysis

Fleck and Kraemer[18] suggested that a **needs analysis** be the first step in designing and prescribing a resistance training program for athletes. The needs analysis should include the following assessment:

- What major muscle groups need to be trained?
- What method of training should be used?
- What energy system should be stressed?
- What are the primary sites of concern for injury prevention?

Once you have completed this needs analysis, you can design and prescribe the resistance training program. You can now logically select

- the exercises that will be performed,
- the order in which they will be performed,
- the number of sets for each exercise,
- the rest periods between sets and between exercises, and

▲ **Figure 3.15** A variable-resistance training device that uses a cam to alter the resistance through the range of motion.

- the load to be used (amount of resistance).

The last point is quite important. The role of load or resistance in training for strength, power, endurance, and muscle size has been an area of great confusion.

Selecting the Appropriate Resistance

The actual resistance (weight) to be lifted generally is expressed as a percentage of your maximal capacity. Recall that a 1RM load is a maximum load: the highest resistance that can be moved only once. In contrast, a 25RM load (i.e., the peak resistance that can be lifted only 25 times before reaching fatigue) is very light. Strength development is optimized by few repetitions and high resistance (6RM or less), whereas muscular endurance is optimized by many repetitions and low resistance (20RM or more).

Less clear, at least from solid research, is how to best develop power. Fleck and Kraemer[18] suggested high-intensity training with resistances greater than that lifted at 10RM, varying the intensity over time (e.g., 1RM-5RM and 6RM-10RM) yet completing no more than five repetitions per set and emphasizing speed of movement. There should be a moderate to high number of sets for each exercise and moderate to long rest periods between sets and exercises.

When the training objective is to increase muscle size, an important objective for bodybuilders, the load should be set within a range of 6RM to 12RM, but the number of sets should be greater than three. Also, the rest interval should be very short, usually not more than 90 s.[18]

Selecting the Appropriate Number of Sets

From the early development of resistance training protocols in the 1940s through the 1960s, it generally has been assumed that you need to perform a minimum of three sets of each exercise to provide the greatest gains in muscle strength and size. A number of exercise scientists have recently challenged this assumption. On the basis of their research, it appears that a single set is just as effective as multiple sets for increasing muscle size and strength. In fact, of the studies that used appropriate controls, only one study demonstrated an advantage of multiple sets over a single set, and the magnitude of the difference in strength gains between three sets and one set was small.[50] This finding has substantial implications for designing resistance training programs: You can substantially reduce the total workout time, or you can include a greater variety of exercises instead of multiple sets within the same time period. A word of caution must be introduced here for dealing with individuals, particularly athletes, experienced in resistance training. Kraemer[33] showed a distinct advantage of multiple sets over a single set for training football players.

Periodization

Periodization refers to changes or variations in the resistance training program that are implemented over the course of a specific period of time, such as a year. Periodization varies the exercise stimulus to keep the individual from overtraining or becoming stale.

According to Fleck and Kraemer,[18] periodization consists of five phases in each training cycle. The first phase is characterized by high volume (many repetitions and sets) and low intensity. During the next three phases, volume is decreased and intensity is increased. Generally, these four phases are followed by an active recovery phase, in which either light resistance training or some unrelated activity is used to allow the body time to completely recover from the training cycle both physically and mentally. Once the active recovery phase is completed, the entire periodization cycle is repeated.

Periodization cycles can vary in duration from one cycle per year to two or three per year. An example of a periodization program is illustrated in table 3.2 on page 108. The number of repetitions and sets can be varied to accommodate the sport. The main idea is to gradually decrease volume while gradually increasing intensity. Each of the first four

Table 3.2

Periodization Training for Strength and Power Sports Using Two Cycles per Year

Variable	Phase I, hypertrophy	Phase II, strength	Phase III, power	Phase IV, peaking	Phase V, Active recovery
Sets	3-5	3-5	3-5	1-3	General activity or light resistance training
Repetitions	8-20	2-6	2-3	1-3	
Intensity	Low	High	High	Very high	
Duration (weeks)	6	6	6	6	2

Adapted, by permission, from M.H. Stone, H. O'Bryant, & J. Garhammer, 1991, A hypothetical model for strength training," *Sports Medicine* 21: 344.

phases emphasizes a different component of muscular fitness:

- Phase I: muscular hypertrophy (muscle size)
- Phase II: strength
- Phase III: power
- Phase IV: peak strength
- Phase V: active recovery

Forms of Resistance Training

Over the years, many claims have been made touting the advantages of specific forms of resistance training. Let's briefly evaluate some of these.

Static-Action Resistance Training

Static-action resistance training, also called **isometric training,** evolved in the early 20th century but gained great popularity and support in the mid-1950s as a result of research by several German scientists. These studies indicated that static resistance training caused tremendous strength gains and that those gains exceeded those resulting from dynamic-action procedures. Subsequent studies have not been able to reproduce the original studies' results, yet static actions remain an important form of training, particularly for postsurgical rehabilitation when the limb is immobilized and thus incapable of dynamic actions. Static actions

facilitate recovery and reduce muscle atrophy and strength loss.

Plyometrics

Plyometrics, or jump training, became popular during the late 1970s and early 1980s primarily for improving jumping ability. Proposed to bridge the gap between speed and strength training, plyometrics uses the stretch reflex to facilitate recruitment of additional motor units. It also loads both the elastic and contractile components of muscle. As an example, to develop knee extensor muscle strength, you jump up as high as you can, bringing the knees to the chest and grasping the shins with the hands before returning to the ground (see figure 3.16). A number of variations can be performed, including repetitive jumping on and off a box and jumping while wearing weight belts.

Bobbert[9] conducted an extensive review of both the scientific and coaching literature on plyometrics and concluded that the research is not sufficient to establish the superiority of plyometrics over more traditional resistance training techniques.

Eccentric Training

Another form of dynamic-action resistance training, called **eccentric training,** emphasizes the eccentric phase. With eccentric actions, the muscle's ability to resist force is approximately

▲ **Figure 3.16**   A plyometric tuck jump.

30% greater than with concentric actions. Subjecting the muscle to this greater training stimulus theoretically produces greater strength gains.

Early research was not able to show a clear advantage of eccentric training over either concentric- or static-action training.[7, 18] Most recently, however, several well-controlled studies showed the importance of including the eccentric phase of muscle action along with the concentric phase to maximize gains in strength and size.[14, 30, 32]

Free Weights

Many athletes are going back to simple free weights for resistance training instead of using the more complex resistance training equipment that has flooded the market over the past 30 years. Athletes and strength training coaches alike believe that free weights offer advantages that resistance machines do not provide. The athlete must control the weight being lifted. An athlete must recruit more motor units—not only in the muscles being trained, but also in additional muscles—to gain control of the bar, to stabilize the weight lifted, and to maintain body balance. When an athlete is training for a sport such as football, the experience with free weights more closely resembles competition.

Electrical Stimulation Training

A muscle can be stimulated by passing an electric current directly across it or its motor nerve. This technique, called **electrical stimulation training,** has been proven effective in a clinical setting. It is used to reduce the loss of strength and muscle size during periods of immobilization and to restore strength and size during rehabilitation. It also has been used experimentally in training healthy subjects (including athletes) because it can increase muscular strength.

However, the gains reported are no greater than those achieved with more conventional training. Athletes have used this technique to supplement their regular training programs, but no evidence shows any additional gains in strength, power, or performance from this supplementation.[28] Dudley and Harris[13] developed a strong theoretical basis for the use of electrical stimulation in both strength and power training of athletes, yet evidence from well-designed research studies to support this theory is lacking.

Specificity of Training Procedures

As discussed in the Introduction, training results are highly specific to the type of training program used. Long-distance running does little, if anything, to improve a powerlifter's maximal weight-lifting capacity. High-resistance weight training does little to improve a distance runner's marathon time. A resistance training program to develop strength and power should be carefully designed to match the requirements of the athlete's specific sport.

Strength gains from resistance training are highly specific to the speed of training. If a person trains at high velocities, testing at high velocities also shows maximal strength gains compared with testing at lower velocities. Most athletes should do at least part of their strength training at high velocities because most sport movements require high speed. High-velocity ballistic training is now known to induce specific neuromuscular adaptations. It is unclear if these adaptations are neural, within the central nervous system, or muscular (including the motor unit).[63]

Evidence also indicates that strength gains are highly specific to movement patterns. This suggests that the more closely the movement pattern mimics the actual sport performance, the greater the benefit from training.

At this time, how specific the resistance training program must be to provide maximal benefits is unclear. Also, resistance training may not improve performance. Tanaka and coworkers[56] found major increases in swimmers' strength following resistance training, but swimming performance did not improve any more than with swim training alone.

Resistance training should be as sport-specific as possible. At least part of the training should involve movements that closely mimic, in both pattern and speed, those needed for the athlete's sport or activity.

How do we train specifically to optimize our gains in strength, hypertrophy, power, and endurance? A group of scientists, renowned for their research on resistance training techniques, have agreed that specific training programs allow us to achieve the greatest results in each of these areas. However, for each area, the program is highly specific to that area and differs according to the training level of the individual. This group's recommendations were published as a Position Stand by the American College of Sports Medicine in 2002 (see table 3.3).[34]

▶ Resistance training can use static or dynamic actions. Dynamic actions include the use of free weights, variable resistance, isokinetic actions, and plyometrics.

▶ A needs analysis should be completed before a training program is designed to tailor the program to the athlete's specific needs.

▶ Low-repetition, high-resistance training enhances strength development, whereas high-repetition, low-intensity training optimizes muscular endurance.

▶ Periodization, through which various aspects of the training program are varied, is important to prevent overtraining or burnout. Typically the goal is to gradually decrease volume while increasing intensity. A typical cycle has four active phases, each emphasizing a different muscular fitness component, plus an active recovery phase.

▶ Strength gains are highly specific to the speed of training and the movement patterns used in training. For maximum benefit, a resistance training program must include activities quite similar to those done by the athlete in actual performance.

Resistance Training for Special Populations

Until recently, resistance training was widely regarded as appropriate only for young, healthy, male athletes. This narrow concept led many people to overlook the benefits of resistance training when planning their own activities. In this section, we first consider sex and age stereotypes, and then we summarize

Table 3.3

American College of Sports Medicine's Recommendations for Resistance Training Programs[a]

Emphasis	Training Level	Loading	Volume	Velocity	Frequency (times per week)
Strength	Novice	60-70% 1RM	1-3 sets, 8-12 reps	Slow, moderate	2-3
	Intermediate	70-80% 1RM	Multiple sets, 6-12 reps	Moderate	2-4
	Advanced	1RM[b]	Multiple sets, 1-12 reps[b]	Unintentionally slow to fast	4-6
Hypertrophy	Novice	60-70% 1RM	1-3 sets, 8-12 reps	Slow, moderate	2-3
	Intermediate	70-80% 1RM	Multiple sets, 6-12 reps	Slow, moderate	2-4
	Advanced	70-100% 1RM; emphasis on 70-85%[b]	Multiple sets, 1-12 reps with emphasis on 6-12 reps[b]	Slow, moderate, fast	4-6
Power	Novice	>80% 1RM—strength; 30-60% 1RM—velocity[b]	Train for strength	Moderate	2-3
	Intermediate	>80% 1RM—strength; 30-60% 1RM—velocity[b]	1-3 sets, 3-6 reps	Fast	2-4
	Advanced	>80% 1RM—strength; 30-60% 1RM—velocity[b]	3-6 sets, 1-6 reps[b]	Fast	4-6
Endurance	Novice	50-70% 1RM	1-3 sets, 10-15 reps	Slow—moderate reps / Moderate—high reps	2-3
	Intermediate	50-70% 1RM	Multiple sets, 10-15 reps or more	Slow—moderate reps / Moderate—high reps	2-4
	Advanced	30-80% 1RM[b]	Multiple sets, 10-25 reps or more[b]	Slow—moderate reps / Moderate—high reps	4-6

[a]These recommendations also include the type of muscle action (eccentric and concentric), single joint vs. multiple joint exercises, order or sequence of exercises, and the rest intervals.

[b]Periodized—see text for explanation of periodization.

Adapted, by permission, from W.L. Kraemer et al., 2002, "Progression models in resistance training for healthy adults," *Medicine and Science in Sports and Exercise* 34: 364-380.

the importance of this form of training for all athletes, regardless of their sex, age, or sport.

Sex and Age Differences

In recent years, considerable interest has focused on training for women, children, and the elderly. As mentioned earlier in this chapter, the widespread use of resistance training by women, either for sport or for health-related benefits, is rather recent. Substantial knowledge has developed since the early 1970s which reveals that women and men have the same ability to develop strength but that, on average, women may not be able to achieve peak values as high as those attained by men. This difference in strength is attributable primarily to muscle size differences related to sex differences in anabolic hormones. Resistance training techniques developed for and applied to men's training seem equally appropriate for women's training. Issues of strength and resistance training for women are covered in more detail in chapter 18.

> fyi   In 1984, the University of Arizona was the first NCAA Division I school to hire a woman as head strength coach for both the men's and women's athletic programs. The position went to Meg Ritchie, former Scottish discus thrower and shot-putter for Great Britain's Olympic team.

The wisdom of resistance training for children and adolescents has long been debated. The potential for injury, particularly growth-plate injuries from the use of free weights, has caused much concern. Many people once believed that children would not benefit from resistance training, based on the assumption that the hormonal changes associated with puberty are necessary for gaining muscle strength and mass. We now know that children and adolescents can train safely with minimal risk of injury if appropriate safeguards are followed. Furthermore, they can indeed gain both muscular strength and muscle mass (chapter 16).

Interest in resistance training procedures for the elderly has increased. A substantial loss of fat-free body mass accompanies aging. This loss reflects mainly the loss of muscle mass, largely because most people become less active as they age. When a muscle isn't used regularly, it loses function, with predictable atrophy and loss of strength.

Can resistance training in the elderly reverse this process? The elderly can indeed gain strength and muscle mass in response to resistance training. This fact has important implications for both their health and the quality of their lives (chapter 17). With maintained or improved strength, elderly people are less likely to fall. This is a significant benefit because falls are a major source of injury and debilitation for elderly people and often lead to death.

Resistance Training for Sport

Gaining strength, power, or muscular endurance simply for the sake of being stronger, being more powerful, or having greater muscular endurance is of relatively little importance to athletes unless it also improves their athletic performance. Resistance training by field-event athletes and competitive weightlifters makes intuitive sense. The need for resistance training by the gymnast, distance runner, baseball player, high jumper, or ballet dancer is less obvious.

We do not have extensive research to document the specific benefits of resistance training for every sport or for every event within a sport. But clearly each has basic strength requirements that must be met to achieve optimal performance. Training beyond these requirements may be unnecessary.

Training is costly in terms of time, and athletes can't afford to waste time on activities that won't result in better athletic performances. Thus, some performance measurement is imperative to evaluate any resistance training program's efficacy. To resistance train solely to become stronger, with no associated improvement in performance, is of questionable value. However, it should also be recognized that resistance training can reduce the risk of injury for most sports, because fatigued individuals are at an increased risk of injury.

► Resistance training can benefit almost everyone, regardless of his or her sex, age, or athletic involvement.

► Most athletes in most sports can benefit from resistance training if an appropriate program is designed for them. But to ensure that the program is working, performance should be assessed periodically and the training regime adjusted as needed.

In Closing . . .

In this chapter we have carefully considered the role of resistance training in increasing muscular strength and improving performance. We have examined how muscle strength is gained through both muscular and neural adaptations, what factors can lead to muscle soreness, and how to design an appropriate resistance training program that will meet the specific needs of the individual athlete. In the next chapter, we turn our attention away from the neuromuscular aspects of physical activity as we begin exploring how physical activity is fueled.

► Key Terms

1-repetition maximum (1RM)
acute muscle soreness
atrophy
autogenic inhibition
chronic hypertrophy
delayed-onset muscle soreness (DOMS)
eccentric training
electrical stimulation training
fiber hyperplasia
fiber hypertrophy
hypertrophy
isometric training
muscular endurance
needs analysis
periodization
plyometrics
power
resistance training

static-action resistance training
strength
transient hypertrophy

► Study Questions

1. Define and differentiate the terms *strength, power,* and *muscular endurance.* How does each component relate to athletic performance?

2. Discuss possible mechanisms that might account for superhuman feats of strength.

3. Discuss the different theories that have attempted to explain how muscles gain strength with training.

4. What is autogenic inhibition? How might it be important to resistance training?

5. Differentiate transient and chronic muscle hypertrophy.

6. What is fiber hyperplasia? How might it be related to gains in size and muscle strength with resistance training?

7. What is the definition of and the physiological basis for hypertrophy? For atrophy?

8. What is the physiological basis for muscle soreness?

9. Define and differentiate static, free weight, isokinetic, and variable-resistance training.

10. Describe five important principles that need to be considered when one is designing resistance training programs.

► References

1. Aagaard, P., Simonsen, E.B., Andersen, J.L., Magnusson, S.P., Halkjaer-Kristensen, J., & Dyhre-Poulsen, P. (2000). Neural inhibition during maximal eccentric and concentric quadriceps contraction: Effects of resistance training. *Journal of Applied Physiology,* **89,** 2249-2257.

2. Andersen, J.L., Schjerling, P., & Saltin, B. (2000). Muscle, genes and athletic performance. *Scientific American,* **283,** 48-55.

3. Antonio, A., & Gonyea, W.J. (1993). Skeletal muscle fiber hyperplasia. *Medicine and Science in Sports and Exercise,* **25**(12), 1333-1345.

4. Appell, H.-J. (1990). Muscular atrophy following immobilisation: A review. *Sports Medicine, 10,* 42-58.

5. Armstrong, R.B. (1984). Mechanisms of exercise-induced delayed-onset muscular soreness: A brief review. *Medicine and Science in Sports and Exercise, 16,* 529-538.

6. Armstrong, R.B., Warren, G.L., & Warren, J.A. (1991). Mechanisms of exercise-induced muscle fibre injury. *Sports Medicine, 12,* 184-207.

7. Atha, J. (1982). Strengthening muscle. *Exercise and Sport Sciences Reviews, 9,* 1-73.

8. Behm, D.G. (1995). Neuromuscular implications and applications of resistance training. *Journal of Strength and Conditioning Research, 9,* 264-274.

9. Bobbert, M.F. (1990). Drop jumping as a training method for jumping ability. *Sports Medicine, 9,* 7-22.

10. Costill, D.L., Coyle, E.F., Fink, W.F., Lesmes, G.R., & Witzmann, F.A. (1979). Adaptations in skeletal muscle following strength training. *Journal of Applied Physiology, 46,* 96-99.

11. Deschenes, M.R., Covault, J., Kraemer, W.J., & Maresh, C.M. (1994). The neuromuscular junction: Muscle fibre type differences, plasticity and adaptability of increased and decreased activity. *Sports Medicine, 17,* 358-372.

12. Duchateau, J., & Enoka, R.M. (2002). Neural adaptations with chronic activity patterns in able-bodied humans. *American Journal of Physical Medicine and Rehabilitation* **81** (11 supplement) 517-527.

13. Dudley, G.A., & Harris, R.T. (1992). Use of electrical stimulation in strength and power training. In P.V. Komi (Ed.), *Strength and power in sport* (pp. 329-337). Boston: Blackwell Scientific.

14. Dudley, G.A., Tesch, P.A., Miller, B.J., & Buchanan, P. (1991). Importance of eccentric actions in performance adaptations to resistance training. *Aviation, Space, and Environmental Medicine, 62,* 543-550.

15. Ebbeling, C.B., & Clarkson, P.M. (1989). Exercise-induced muscle damage and adaptation. *Sports Medicine, 7,* 207-234.

16. Enoka, R.M. (1988). Muscle strength and its development: New perspectives. *Sports Medicine, 6,* 146-168.

17. Enoka, R.M. (1997). Neural adaptations with chronic physical activity. *Journal of Biomechanics, 30,* 447-455.

18. Fleck, S.J., & Kraemer, W.J. (1997). *Designing resistance training programs* (2nd ed.). Champaign, IL: Human Kinetics.

19. Gollnick, P.D., Armstrong, R.B., Saltin, B., Saubert, C.W., IV, Sembrowich, W.L., & Shepherd, R.E. (1973). Effect of training on enzyme activity and fiber composition of human skeletal muscle. *Journal of Applied Physiology, 34,* 107-111.

20. Gollnick, P.D., Parsons, D., Riedy, M., & Moore, R.L. (1983). Fiber number and size in overloaded chicken anterior latissimus dorsi muscle. *Journal of Applied Physiology, 54,* 1292-1297.

21. Gollnick, P.D., Timson, B.F., Moore, R.L., & Riedy, M. (1981). Muscular enlargement and number of fibers in skeletal muscles of rats. *Journal of Applied Physiology, 50,* 936-943.

22. Gonyea, W.J. (1980). Role of exercise in inducing increases in skeletal muscle fiber number. *Journal of Applied Physiology, 48,* 421-426.

23. Gonyea, W.J., Sale, D.G., Gonyea, F.B., & Mikesky, A. (1986). Exercise induced increases in muscle fiber number. *European Journal of Applied Physiology, 55,* 137-141.

24. Goodman, M.N. (1988). Amino acid and protein metabolism. In E.S. Horton & R.L. Terjung (Eds.), *Exercise, nutrition, and energy metabolism* (pp. 89-99). New York: Macmillan.

25. Graves, J.E., Pollock, M.L., Leggett, S.H., Braith, R.W., Carpenter, D.M., & Bishop, L.E. (1988). Effect of reduced training frequency on muscular strength. *International Journal of Sports Medicine, 9,* 316-319.

26. Green, H.J., Klug, G.A., Reichmann, H., Seedorf, U., Wiehrer, W., & Pette, D. (1984). Exercise-induced fibre type transitions with regard to myosin, parvalbumin, and sarcoplasmic reticulum in muscles of the rat. *Pflugers Archiv: European Journal of Physiology, 400,* 432-438.

27. Hagerman, F.C., Hikida, R.S., Staron, R.S., Sherman, W.M., & Costill, D.L. (1984). Muscle damage in marathon runners. *Physician and Sportsmedicine, 12,* 39-48.

28. Hainaut, K., & Duchateau, J. (1992). Neuromuscular electrical stimulation and voluntary exercise. *Sports Medicine, 14,* 100-113.

29. Hakkinen, K., Alen, M., & Komi, P.V. (1985). Changes in isometric force and relaxation-time, electromyographic and muscle fibre characteristics of human skeletal muscle during strength training and detraining. *Acta Physiologica Scandinavica, 125,* 573-585.

30. Hather, B.M., Tesch, P.A., Buchanan, P., & Dudley, G.A. (1991). Influence of eccentric actions on skeletal muscle adaptations to resistance training. *Acta Physiologica Scandinavica, 143,* 177-185.

31. Hawke, T.J., & Garry, D.J. (2001). Myogenic satellite cells: Physiology to molecular biology. *Journal of Applied Physiology, 91,* 534-551.

32. Hortobagyi, T., Hill, J.P., Houmard, J.A., Fraser, D.D., Lambert, N.J., & Israel, R.G. (1996). Adaptive

responses to muscle lengthening and shortening in humans. *Journal of Applied Physiology, 80,* 765-772.

33. Kraemer, W.J. (1997). A series of studies—The physiological basis for strength training in American football: Fact over philosophy. *Journal of Strength and Conditioning Research, 11,* 131-142.

34. Kraemer, W.J., Adams, K., Cafarelli, E., Dudley, G.A., Dooly, C., Feigenbaum, M.S., Fleck, S.J., Franklin, B., Fry, A.C., Hoffman, J.R., Newton, R.U., Potteiger, J., Stone, M.H., Ratamess, N.A., & Triplett-McBride, T. (2002). ACSM Position Stand: Progression models in resistance training for healthy adults. *Medicine and Science in Sports and Exercise, 34,* 364-380.

35. Kraemer, W.J., Deschenes, M.R., & Fleck, S.J. (1988). Physiological adaptations to resistance exercise: Implications for athletic conditioning. *Sports Medicine, 6,* 246-256.

36. Kraemer, W.J., Fleck, S.J., & Evans, W.J. (1996). Strength and power training: Physiological mechanisms of adaptation. *Exercise and Sport Sciences Reviews, 24,* 363-397.

37. Kuipers, H., Keizer, H.A., Verstappen, F.T.J., & Costill, D.L. (1985). Influence of a prostaglandin-inhibiting drug on muscle soreness after eccentric work. *International Journal of Sports Medicine, 6,* 336-339.

38. Larsson, L., & Tesch, P.A. (1986). Motor unit fibre density in extremely hypertrophied skeletal muscle in man: Electrophysiological signs of muscle fiber hyperplasia. *European Journal of Applied Physiology, 55,* 130-136.

39. McCall, G.E., Byrnes, W.C., Dickinson, A., Pattany, P.M., & Fleck, S.J. (1996). Muscle fiber hypertrophy, hyperplasia, and capillary density in college men after resistance training. *Journal of Applied Physiology, 81,* 2004-2012.

40. McDonagh, M.J.N., & Davies, C.T.M. (1984). Adaptive response of mammalian skeletal muscle to exercise with high loads. *European Journal of Applied Physiology, 52,* 139-155.

41. Pette, D., & Vrbova, G. (1985). Neural control of phenotypic expression in mammalian muscle fibers. *Muscle and Nerve, 8,* 676-689.

42. Ploutz-Snyder, L.L., Convertino, V.A., & Dudley, G.A. (1995). Resistance exercise–induced fluid shifts: Change in active muscle size and plasma volume. *American Journal of Physiology, 269,* R536-R543.

43. Sale, D.G. (1988). Neural adaptation to resistance training. *Medicine and Science in Sports and Exercise, 20,* S135-S145.

44. Schantz, P., Randall-Fox, E., Hutchison, W., Tyden, A., & Åstrand, P.-O. (1983). Muscle fibre type distribution, muscle cross-sectional area and maximal voluntary strength in humans. *Acta Physiologica Scandinavica, 117,* 219-226.

45. Schwane, J.A., Johnson, S.R., Vandenakker, C.B., & Armstrong, R.B. (1983). Delayed-onset muscular soreness and plasma CPK and LDH activities after downhill running. *Medicine and Science in Sports and Exercise, 15,* 51-56.

46. Schwane, J.A., Watrous, B.G., Johnson, S.R., & Armstrong, R.B. (1983). Is lactic acid related to delayed-onset muscle soreness? *Physician and Sportsmedicine, 11*(3), 124-131.

47. Schwellnus, M.P. (1999). Skeletal muscle cramps during exercise. *Physician and Sportsmedicine, 27*(12), 109-115.

48. Semmler, J.G. (2002). Motor-unit synchronization and neuromuscular performance. *Exercise and Sport Sciences Reviews, 30,* 8-14.

49. Semmler, J.G., & Enoka, R.M. (2000). Neural contributions to changes in muscle strength. In Zatsiorsky V.M. (Ed.), *Biomechanics in sport* (pp. 3-20). Oxford, UK: Blackwell Science.

50. Starkey, D.B., Pollock, M.L., Ishida, Y., Welsch, M.A., Brechue, W.F., Graves, J.E., & Feigenbaum, M.S. (1996). Effect of resistance training volume on strength and muscle thickness. *Medicine and Science in Sports and Exercise, 28,* 1311-1320.

51. Staron, R.S., Karapondo, D.L., Kraemer, W.J., Fry, A.C., Gordon, S.E., Falkel, J.E., Hagerman, F.C., & Hikida, R.S. (1994). Skeletal muscle adaptations during early phase of heavy resistance training in men and women. *Journal of Applied Physiology, 76,* 1247-1255.

52. Staron, R.S., Leonardi, M.J., Karapondo, D.L., Malicky, E.S., Falkel, J.E., Hagerman, F.C., & Hikida, R.S. (1991). Strength and skeletal muscle adaptations in heavy-resistance-trained women after detraining and retraining. *Journal of Applied Physiology, 70,* 631-640.

53. Staron, R.S., Malicky, E.S., Leonardi, M.J., Falkel, J.E., Hagerman, F.C., & Dudley, G.A. (1990). Muscle hypertrophy and fast fiber type conversions in heavy-resistance-trained women. *European Journal of Applied Physiology, 60,* 71-79.

54. Stone, M.H., O'Bryant, H.S., & Garhammer, J. (1981). A hypothetical model for strength training. *Journal of Sports Medicine and Physical Fitness, 21*(4): 336, 342-351.

55. Talag, T.S. (1973). Residual muscular soreness as influenced by concentric, eccentric and static contractions. *Research Quarterly, 44,* 458-469.

56. Tanaka, H., Costill, D.L., Thomas, R., Fink, W.J., & Widrick, J.J. (1993). Dry-land resistance training for competitive swimming. *Medicine and Science in Sports and Exercise, 25,* 952-959.

57. Tarnopolsky, M. (1999). Protein metabolism in strength and endurance activities. In D.R. Lamb & R. Murray (Eds.), *The metabolic basis of performance in*

exercise and sport: Perspectives in exercise science and sports medicine (Vol. 12, pp. 125-157). Carmel, IN: Cooper.

58. Tesch, P.A., & Karlsson, J. (1985). Muscle fiber types and size in trained and untrained muscles of elite athletes. *Journal of Applied Physiology, 59,* 1716-1720.

59. Tidball, J.G. (1995). Inflammatory cell response to acute muscle injury. *Medicine and Science in Sports and Exercise, 27,* 1022-1032.

60. Timson, B.F., Bowlin, B.K., Dudenhoeffer, G.A., & George, J.B. (1985). Fiber number, area, and composition of mouse soleus muscle following enlargement. *Journal of Applied Physiology, 58,* 619-624.

61. Warren, G.L., Ingalls, C.P., Lowe, D.A., & Armstrong, R.B. (2001). Excitation-contraction uncoupling: Major role in contraction-induced muscle injury. *Exercise and Sport Sciences Reviews, 29,* 82-87.

62. Yao, W., Fuglevand, A.J., & Enoka, R.M. (2000). Motor-unit synchronization increases EMG amplitude and decreases force steadiness of simulated contractions. *Journal of Neurophysiology, 83,* 441-452.

63. Zehr, E.P., & Sale, D.G. (1994). Ballistic movement: Muscle activation and neuromuscular adaptation. *Canadian Journal of Applied Physiology, 19,* 363-378.

▶ Selected Readings

Armstrong, R.B. (1986). Muscle damage and endurance events. *Sports Medicine, 3,* 370-381.

Baechle, T.R., & Earle, R.W. (Ed.). (2000). *Essentials of strength training and conditioning* (2nd ed.). Champaign, IL: Human Kinetics.

Behm, D.G., & Sale, D.G. (1993). Velocity specificity of resistance training. *Sports Medicine, 15*(6), 374-388.

Carroll, T.J., Riek, S., & Carson, R.G. (2001). Neural adaptations to resistance training. *Sports Medicine, 31,* 829-840.

Chu, D.A. (1998). *Jumping into plyometrics* (2nd ed.). Champaign, IL: Human Kinetics.

Clarke, D.H. (1973). Adaptations in strength and muscular endurance resulting from exercise. *Exercise and Sport Sciences Reviews, 1,* 73-102.

Clarkson, P.M., & Sayers, S.P. (1999). Etiology of exercise-induced muscle damage. *Canadian Journal of Applied Physiology, 24,* 234-248.

Deschenes, M.R., Kraemer, W.J., Maresh, C.M., & Crivello, J.F. (1991). Exercise-induced hormonal changes and their effects upon skeletal muscle tissue. *Sports Medicine, 12,* 80-93.

Evans, W.J., & Cannon, J.G. (1991). The metabolic effects of exercise induced muscle damage. *Exercise and Sport Sciences Reviews, 19,* 99-125.

Garfinkel, S., & Cafarelli, E. (1992). Relative changes in maximal force, EMG, and muscle cross-sectional area after isometric training. *Medicine and Science in Sports and Exercise, 24*(11), 1220-1227.

Giddings, C.J., & Gonyea, W.J. (1992). Morphological observations supporting muscle fiber hyperplasia following weight-lifting exercise in cats. *Anatomical Record, 233,* 178-195.

Gonyea, W.J., Ericson, G.C., & Bonde-Petersen, F. (1977). Skeletal muscle fiber splitting induced by weight-lifting exercise in cats. *Acta Physiologica Scandinavica, 19,* 105-109.

Kannus, P., Jozsa, L., Renstrom, P., Jarvinen, M., Kvist, M., Lehto, M., Oja, P., & Vuori, I. (1992). The effects of training, immobilization and remobilization on musculoskeletal tissue. *Scandinavian Journal of Medicine and Science in Sports, 2,* 100-118, 164-176.

Kendall, B., & Eston, R. (2002). Exercise-induced muscle damage and the potential protective role of estrogen. *Sports Medicine, 32,* 103-123.

Komi, P.V. (Ed.). (1992). *Strength and power in sport.* Boston: Blackwell Scientific.

Larsson, L., & Ansved, T. (1985). Effects of long-term physical training and detraining on enzyme histochemical and functional skeletal muscle characteristics in man. *Muscle and Nerve, 8,* 714-722.

Ross, A., Leveritt, M., & Riek, S. (2001). Neural influences on sprint running: Training, adaptations and acute response. *Sports Medicine, 31,* 829-840.

Sale, D.G. (1992). Neural adaptation to strength training. In P.V. Komi (Ed.), *Strength and power in sport* (pp. 249-265). Boston: Blackwell Scientific.

Simoneau, J.A., Lortie, G., Boulay, M.R., Marcotte, M., Thibault, M.C., & Bouchard, C. (1985). Human skeletal muscle fiber type alteration with high-intensity intermittent training. *European Journal of Applied Physiology, 54,* 250-253.

Smith, L.L. (1991). Acute inflammation: The underlying mechanism in delayed onset muscle soreness? *Medicine and Science in Sports and Exercise, 23,* 542-551.

Stauber, W.T. (1989). Eccentric action of muscles: Physiology, injury and adaptation. *Exercise and Sport Sciences Reviews, 17,* 157-185.

Taylor, N.A.S., & Wilkinson, J.G. (1986). Exercise-induced skeletal muscle growth: Hypertrophy or hyperplasia? *Sports Medicine, 3,* 190-200.

Tesch, P.A. (1988). Skeletal muscle adaptations consequent to long-term heavy resistance exercise. *Medicine and Science in Sports and Exercise, 20*(Suppl.), S132-S134.

Tesch, P.A., & Larsson, L. (1982). Muscle hypertrophy in bodybuilders. *European Journal of Applied Physiology, 49,* 301-306.

PART II

Energy for Movement

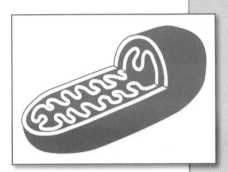

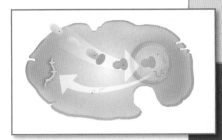

In part I, we studied the structure and function of the skeletal muscles, how muscles are controlled by the nervous system, and muscular and neural adaptations that occur in response to resistance training. We understand the basic mechanical aspects of muscles and how the muscles are controlled, but for muscles to produce movement they need energy. In part II, we turn our attention to how the body meets the energy needs of our skeletal muscles. In chapter 4, "Metabolism, Energy and the Basic Energy Systems," we examine our primary source of energy, adenosine triphosphate, and how it is provided through three energy systems. We discuss the body's energy expenditure, how it varies from rest to maximal exercise, and how fatigue can result if energy demands exceed energy supply. In chapter 5, "Hormonal Regulation of Exercise," we see how endocrine glands and their hormones help control energy metabolism and fluid and electrolyte balance. Finally, in chapter 6, "Metabolic Adaptations to Training," we study adaptations in the skeletal muscles and energy systems that occur in response to aerobic and anaerobic training and that can improve performance.

CHAPTER 4

METABOLISM, ENERGY, AND THE BASIC ENERGY SYSTEMS

All plants and animals depend on energy to sustain life. As humans, we derive this energy from food. Nearly all the foods we eat provide us with energy for normal cellular activity or energy that can be stored for use at a later time.

You cannot understand exercise physiology without understanding some key concepts about energy. In the previous chapters, we saw that movement does not occur without cost. We pay this cost with adenosine triphosphate (ATP), a form of chemical energy stored within our cells. We produce ATP by processes that are known collectively as metabolism, our focus for this chapter. In this chapter we review the biochemical processes that are basic to understanding how our muscles use food to create energy for movement. Then we discuss how measurement of energy production and consumption helps us understand the effects of acute and chronic exercise on performance and conditioning.

In 1978, Tom Osler, a world-ranked marathon and ultramarathon runner, came to the Ball State University Human Performance Laboratory to be studied during his attempt to run and walk continuously for 72 h. Laboratory measurements indicated that his muscles were using mostly carbohydrate for energy during the first hours of exercise. As the hours passed, more and more of the energy needed to continue this ordeal was obtained from fat. Ultimately, during the final 24 h of effort, nearly all his energy was provided by his body's fat stores, despite his continued intake of milk saturated with sugar and his consumption of an 18 × 21 in. (45.7 × 53.3 cm) birthday cake. Even with the intake of more than 9,000 kcal during the first 24 h, Tom was forced to end his effort at 70 h, exhausted and out of energy, having completed 200 miles (322 km).

All energy originates from the sun as light energy. Chemical reactions in plants (photosynthesis) convert light into stored chemical energy. In turn, we obtain energy by eating plants or animals that feed on plants. Energy is stored in food in the form of carbohydrates, fats, and proteins. These basic food components can be broken down in our cells to release the stored energy.

Because all energy eventually degrades to heat, the amount of energy released in a biological reaction is calculated from the amount of heat produced. Energy in biological systems is measured in **calories** (cal). By definition, 1 cal equals the amount of heat energy needed to raise 1 g of water 1° C, from 14.5° C to 15.5° C. In humans, energy is expressed in **kilocalories** (kcal), where 1 kcal is the equivalent of 1,000 cal. Some prefer to use the term *large calories,* or *Calories,* but this is technically not correct. When we speak of someone expending 3,000 cal per day, we really mean that person is expending 3,000 kcal per day.

Some free energy in the cells is used for growth and repair throughout the body. Such processes, as we saw earlier, build muscle mass during training and repair muscle damage after exercise or injury. Energy also is needed for active transport of many substances, such as glucose and Ca^{2+}, across cell membranes. Active transport is critical to the survival of cells and the maintenance of homeostasis. Myofibrils also use some of the energy released in our bodies to cause sliding of the actin and myosin filaments, resulting in muscle action and force generation, as we saw in chapter 1. This use is our main concern.

Energy Sources

Foods are composed primarily of carbon, hydrogen, oxygen, and—in the case of protein—nitrogen. Molecular bonds in foods are relatively weak and provide little energy when broken. Consequently, food is not used directly for cellular operations. Rather, the energy in food molecules' bonds is chemically released within our cells then stored in the form of a high-energy compound called **adenosine triphosphate** (ATP).

The formation of ATP provides the cells with a high-energy compound for storing and conserving energy.

At rest, the energy that your body needs is derived almost equally from the breakdown of carbohydrates and fats. Proteins are your body's building blocks, usually providing little energy for cellular function. During mild to severe muscular effort, more carbohydrate is used, with less reliance on fat. In maximal, short-duration exercise, ATP is generated almost exclusively from carbohydrate.

Carbohydrate

Your muscles' dependence on carbohydrate during exercise is related to **carbohydrate** availability and your muscles' well-developed system for carbohydrate metabolism. Carbohydrates are ultimately converted to glucose, a monosaccharide (one-unit sugar) that is transported via your blood to all body tissues. Under resting conditions, ingested carbohydrate is taken up by your muscles and liver and then converted into a more complex sugar molecule: **glycogen.** Glycogen is stored in the cytoplasm until your cells use it to form ATP. The glycogen stored in your liver is converted back to glucose as needed and then transported by your blood to active tissues, where it is metabolized.

Liver and muscle glycogen reserves are limited and can be depleted unless the diet contains a reasonable amount of carbohydrate. Thus, we rely heavily on dietary sources of starches and sugars to replenish our carbohydrate reserves. Without adequate carbohydrate intake, the muscles and liver can be deprived of their primary energy source.

Fat

Fat provides a sizable amount of energy during prolonged, less intense exercise. Body stores of potential energy in the form of fat are substantially larger than the reserves of carbohydrate. Table 4.1 provides an indication of the total body stores of these two energy sources in a very lean person (12% body fat). For the average middle-aged adult, the fat stores would be approximately twice as large, whereas the carbohydrate stores would be about the same. But fat is less accessible for cellular metabolism because it must first be reduced from its complex form, triglyceride, to its basic components, glycerol and **free fatty acids (FFAs).** Only FFAs are used to form ATP.

> Carbohydrate stores in the liver and skeletal muscle are limited to less than 2,500 kcal of energy, or the equivalent of the energy needed for about 40 km (25 mi) of running. Fat stores, however, generally exceed 70,000 kcal of energy.

Substantially more energy is derived from a given quantity of fat (9.4 kcal/g) than from the same amount of carbohydrate (4.1 kcal/g). Nonetheless, the rate of energy release from fat is too slow to meet all of the energy demands of intense muscular activity.

Table 4.1

Body Stores of Fuels and Energy

	g	kcal
Carbohydrates		
Liver glycogen	110	451
Muscle glycogen	500	2,050
Glucose in body fluids	15	62
Total	625	2,563
Fat		
Subcutaneous and visceral	7,800	73,320
Intramuscular	161	1,513
Total	7,961	74,833

Note. These estimates are based on an average body weight of 65 kg (143 lb) with 12% body fat.

Protein

Protein also can be used as an energy source, but it must first be converted into glucose. In the case of severe energy depletion or starvation, protein may even be used to generate FFAs for cellular energy. The process by which protein or fat is converted into glucose is called **gluconeogenesis.** The process of converting protein into fatty acids is termed **lipogenesis.**

Protein can supply up to 5% or 10% of the energy needed to sustain prolonged exercise. Only the most basic units of protein—the amino acids—can be used for energy. A gram of protein yields about 4.1 kcal.

Rate of Energy Release

To be useful, free energy must be released from chemical compounds at a controlled rate. This rate is partially determined by the choice of the primary fuel source. Large amounts of one particular fuel can cause cells to rely more on that source than on alternatives. This influence of energy availability is termed the *mass action effect.*

Specific protein molecules called enzymes control the rate of free-energy release. Many of these enzymes facilitate the breakdown (catabolism) of chemical compounds. The way these enzymes speed catabolism has been characterized as a "lock-and-key" mechanism, as illustrated in figure 4.1. However, many enzymes also become altered in structure after binding to the chemical compound. Thus, the structure and function of enzymes may be more complex than we have shown in figure 4.1, but the concept of the lock and key should give you the idea that the interactions between energy compounds (e.g., glucose) and enzymes are important to energy transfer within the cell. Although the enzyme names are quite complex, all end with the suffix -ase. For example, an important enzyme that acts on ATP is termed **adenosine triphosphatase (ATPase).**

Now that we know the energy sources, we can look at how that energy is formed. In the next section, we examine the production of the energy-storing compound ATP.

▶ Adenosine triphosphate (ATP) is the high-energy compound formed from the energy in food molecules' bonds, which is released and stored within our cells.

▶ We derive our energy from food sources: carbohydrates, fats, and proteins.

▶ The energy we derive from food is stored in the high-energy compound ATP.

▶ Carbohydrate and protein each provide about 4.1 kcal of energy per gram, compared with about 9.4 kcal/g for fat. But carbohydrate energy is more accessible than either protein or fat.

Bioenergetics: The Basic Energy Systems

An ATP molecule (figure 4.2a) consists of adenosine (a molecule of adenine joined to a molecule of ribose) combined with three inorganic phosphate (P_i) groups. Adenine is a nitrogenous base and ribose is a five-carbon

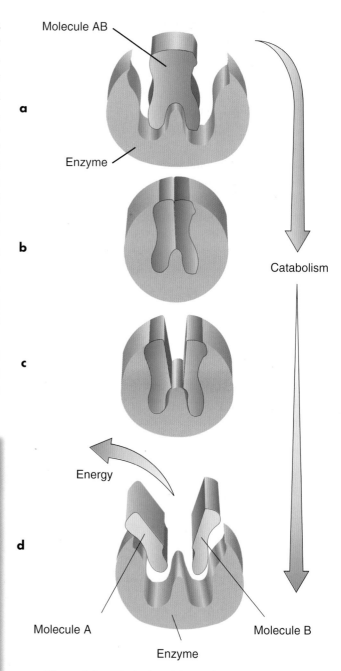

▲ **Figure 4.1**   The lock-and-key action of enzymes in the catabolism (breakdown) of compounds. The compound AB (*a*) is bound into the enzyme molecule, (*b*) undergoes modification while in combination with the enzyme, and (*c*) is broken down, thereby (*d*) releasing the energy that originally held the compound together.

sugar. When the ATP molecule is combined with water (hydrolysis) and acted on by the enzyme ATPase, the last phosphate group splits away, rapidly releasing a large amount of free energy (approximately 7.3 kcal per mole of ATP under standard conditions, but possibly up to 10 kcal per mole of ATP or greater within the cell). This reduces the ATP to **adenosine diphosphate (ADP)** and P_i (figure 4.2*b*). But how was that energy originally stored?

Through various chemical reactions, a phosphate group is added to a relatively low-energy compound, ADP, converting it to ATP, a process called *phosphorylation*. When these reactions occur without oxygen, the process is called **anaerobic metabolism.** When these reactions occur with the aid of oxygen, the overall process is called **aerobic metabolism,** and the aerobic conversion of ADP to ATP is *oxidative phosphorylation*.

Cells generate ATP through three different processes or systems:

1. The ATP–PCr system
2. The glycolytic system
3. The oxidative system

Let's now look at each of these systems.

ATP–PCr System

The simplest of the energy systems is the **ATP–PCr system,** shown in figure 4.3. In addition to ATP, your cells contain another high-energy

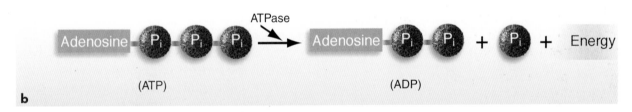

▲ **Figure 4.2** (*a*) The structure of an adenosine triphosphate (ATP) molecule, showing the high-energy phosphate bonds. (*b*) When the third phosphate on the ATP molecule is separated from adenosine by the action of adenosine triphosphatase (ATPase), energy is released.

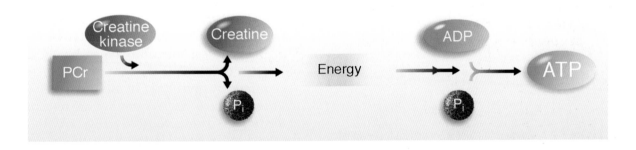

▲ **Figure 4.3** Adenosine triphosphate (ATP) can be recreated by binding an inorganic phosphate (P_i) to adenosine diphosphate (ADP, or adenosine plus two phosphates) with the energy derived from phosphocreatine (PCr).

phosphate molecule that stores energy. This molecule is called **phosphocreatine**, or **PCr** (also called creatine phosphate). Unlike ATP, energy released by the breakdown of PCr is not used directly to accomplish cellular work. Instead, it rebuilds ATP to maintain a relatively constant supply.

The release of energy from PCr is facilitated by the enzyme creatine kinase, which acts on PCr to separate P_i from creatine. The energy released can then be used to couple P_i to an ADP molecule, forming ATP. With this system, as energy is released from ATP by the splitting of a phosphate group, your cells can prevent ATP depletion by reducing PCr, providing energy to reform ATP from ADP and P_i.

This process is rapid and can be accomplished without any special structures within the cell. Although it can occur in the presence of oxygen, this process does not require oxygen, so the ATP–PCr system is said to be **anaerobic**.

During the first few seconds of intense muscular activity, such as sprinting, ATP is maintained at a relatively constant level, but the PCr level declines steadily as it is used to replenish the depleted ATP (see figure 4.4). At

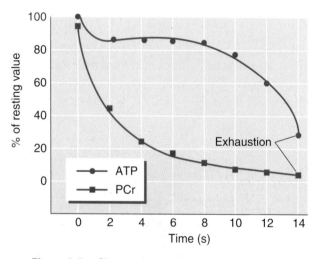

▲ **Figure 4.4** Changes in muscle adenosine triphosphate (ATP) and phosphocreatine (PCr) during 14 s of maximal muscular effort (sprinting). Although ATP is being used at a very high rate, the energy from PCr is used to synthesize ATP, preventing the ATP level from decreasing. However, at exhaustion, both ATP and PCr levels are low.

exhaustion, however, both ATP and PCr levels are quite low and are unable to provide the energy for further contractions and relaxations.

Thus, your capacity to maintain ATP levels with the energy from PCr is limited. Your ATP and PCr stores can sustain your muscles' energy needs for only 3 to 15 s during an all-out sprint. Beyond that point, the muscles must rely on other processes for ATP formation: the glycolytic and oxidative combustion of fuels.

Glycolytic System

Another method of ATP production involves the liberation of energy through the breakdown (lysis) of glucose. This system is called the **glycolytic system** because it involves **glycolysis**, which is the breakdown of glucose via special glycolytic enzymes. An overview of this process is depicted in figure 4.5.

Glucose accounts for about 99% of all sugars circulating in the blood. Blood glucose comes from the digestion of carbohydrate and the breakdown of liver glycogen. Glycogen is synthesized from glucose by a process called **glycogenesis.** Glycogen is stored in the liver or in muscle until needed. At that time, the glycogen is broken down to glucose-1-phosphate through the process of **glycogenolysis.**

Before either glucose or glycogen can be used to generate energy, it must be converted to a compound called glucose-6-phosphate. Conversion of a molecule of glucose requires one molecule of ATP. In the conversion of glycogen, glucose-6-phosphate is formed from glucose-1-phosphate without this energy expenditure. Glycolysis begins once the glucose-6-phosphate is formed.

Glycolysis ultimately produces pyruvic acid. This process doesn't require oxygen, but the use of oxygen determines the fate of the pyruvic acid formed by glycolysis. In this text, when we refer to the glycolytic system, we are referring to the process of **anaerobic glycolysis**, without the need for oxygen. In this case, the pyruvic acid is converted to lactic acid.

Glycolysis, which is far more complex than the ATP–PCr system, requires 12 enzymatic reactions for the breakdown of glycogen to lactic

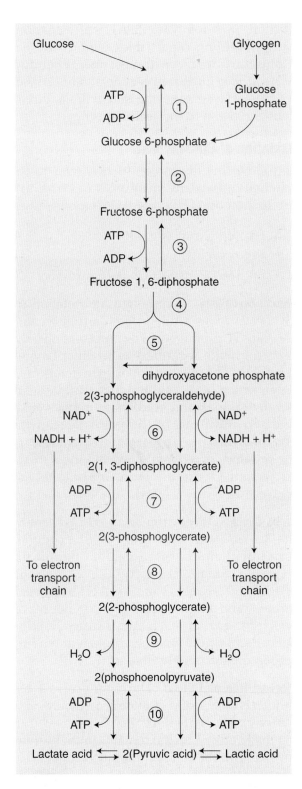

acid. All these enzymes operate within the cells' cytoplasm. The net gain from this process is 3 moles (mol) of ATP formed for each mole of glycogen broken down. If glucose is used instead of glycogen, the gain is only 2 mol of ATP because 1 mol is used for the conversion of glucose to glucose-6-phosphate.

This energy system does not produce large amounts of ATP. Despite this limitation, the combined actions of the ATP–PCr and glycolytic systems allow the muscles to generate force even when the oxygen supply is limited. These two systems predominate during the early minutes of high-intensity exercise.

Another major limitation of anaerobic glycolysis is that it causes an accumulation of lactic acid in the muscles and body fluids. In all-out sprint events lasting 1 or 2 min, the demands on the glycolytic system are high, and muscle lactic acid levels can increase from a resting value of about 1 mmol/kg of muscle to more than 25 mmol/kg. This acidification of muscle fibers inhibits further glycogen breakdown because it impairs glycolytic enzyme function. In addition, the acid decreases the fibers' calcium-binding capacity and thus may impede muscle contraction (see chapter 1).

A muscle fiber's rate of energy use during exercise can be 200 times greater than at rest. The ATP–PCr and glycolytic systems alone cannot supply all the needed energy. Furthermore, these two systems are not capable of supplying all of the energy needs for all-out activity lasting more than 2 min, nor are they efficient for generating ATP for long duration activity. Let's turn our attention to the third energy system, the oxidative system.

▲ **Figure 4.5**   The derivation of energy (adenosine triphosphate, ATP) via glycolysis. An overview of the breakdown of glucose (a 6-carbon molecule) and glycogen (a chain of glucose molecules) to two 3-carbon molecules of pyruvic acid. Note that there are roughly 10 separate steps in this anaerobic process.

> **fyi**   Lactic acid and **lactate** are not the same compound. Lactic acid is an acid with the chemical formula $C_3H_6O_3$. Lactate is any salt of lactic acid. When lactic acid releases hydrogen ions (H^+), the remaining compound joins with sodium ions (Na^+) or potassium ions (K^+) to form a salt. Anaerobic glycolysis produces lactic acid, but it quickly dissociates, and the salt—lactate—is formed. For this reason, the terms often are used interchangeably.

▶ ATP is generated through three energy systems:

1. The ATP–PCr system
2. The glycolytic system
3. The oxidative system

▶ In the ATP–PCr system, P_i is separated from PCr through the action of creatine kinase. The Pi can then combine with ADP to form ATP by using the energy released from the breakdown of PCr. This system is anaerobic, and its main function is to maintain ATP levels. The energy yield is 1 mol of ATP per 1 mol of PCr.

▶ The glycolytic system involves the process of glycolysis, through which glucose or glycogen is broken down to pyruvic acid via glycolytic enzymes. When glycolysis is conducted without oxygen, the pyruvic acid is converted to lactic acid. One mole of glucose yields 2 mol of ATP, but 1 mol of glycogen yields 3 mol of ATP.

▶ The ATP–PCr and glycolytic systems are major contributors of energy during the early minutes of high-intensity exercise.

Oxidative System

The final system of cellular energy production is the oxidative system. This is the most complex of the three energy systems, but we will avoid cumbersome details. The process by which the body disassembles fuels with the aid of oxygen to generate energy is called cellular respiration. Because oxygen is used, this is an aerobic process. This oxidative production of ATP occurs within special cell organelles: the mitochondria. In muscles, these are adjacent to the myofibrils and are also scattered throughout the sarcoplasm. (See figure 1.3, chapter 1.)

Muscles need a steady supply of energy to continuously produce the force needed during long-term activity. Unlike anaerobic ATP pro-

duction, the oxidative system has a tremendous energy-yielding capacity, so aerobic metabolism is the primary method of energy production during endurance events. This places considerable demands on the body's ability to deliver oxygen to the active muscles.

Oxidation of Carbohydrate

To follow this discussion, please refer to the schematic diagram in figure 4.6. Oxidative production of ATP involves three processes:

- Aerobic glycolysis (figure 4.6a)
- Krebs cycle (figure 4.6b)
- Electron transport chain (figure 4.6c)

Aerobic Glycolysis   In carbohydrate metabolism, glycolysis plays a role in both anaerobic and aerobic ATP production. The process of glycolysis is the same regardless of whether oxygen is present. The presence of oxygen determines only the fate of the end product, pyruvic acid. Recall that anaerobic glycolysis produces lactic acid and only 3 mol of ATP per mole of glycogen, or 2 mol of ATP per mole of glucose. In the presence of oxygen, however, the pyruvic acid is converted into a compound called **acetyl coenzyme A** (acetyl CoA).

Krebs Cycle   Once formed, acetyl CoA enters the **Krebs cycle** (citric acid cycle), a complex series of chemical reactions that permit the complete oxidation of acetyl CoA. At the end of the Krebs cycle, 2 mol of ATP have been formed, and the substrate (the compound on which the enzymes act, in this case, the original carbohydrate) has been broken down into carbon dioxide and hydrogen.

Electron Transport Chain   During glycolysis, hydrogen is released as glucose is metabolized to pyruvic acid. More hydrogen is released during the Krebs cycle. If it remains in the system, the inside of the cell becomes too acidic. What happens to this hydrogen?

The Krebs cycle is coupled to a series of reactions known as the **electron transport chain**. The hydrogen released during glycolysis and during the Krebs cycle combines with two coenzymes: nicotinamide adenine dinucleotide (NAD) and flavin adenine dinucleotide (FAD). These carry the hydrogen atoms to the electron transport chain, where they are split into protons and electrons. At the end of the chain, the H^+ combines with oxygen to form water, thus preventing acidification.

The electrons that were split from the hydrogen pass through a series of reactions (hence the name electron transport chain) and ultimately provide energy for the phosphorylation of ADP, thus forming ATP. Because this process relies on oxygen, it is referred to as oxidative phosphorylation. This is illustrated in figure 4.7.

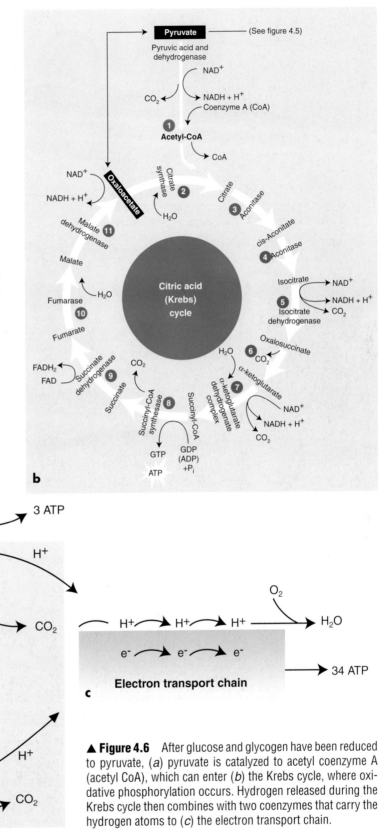

▲ **Figure 4.6** After glucose and glycogen have been reduced to pyruvate, (*a*) pyruvate is catalyzed to acetyl coenzyme A (acetyl CoA), which can enter (*b*) the Krebs cycle, where oxidative phosphorylation occurs. Hydrogen released during the Krebs cycle then combines with two coenzymes that carry the hydrogen atoms to (*c*) the electron transport chain.

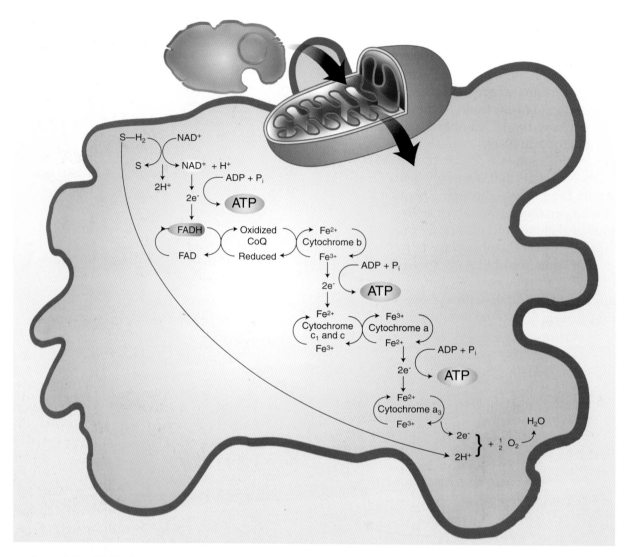

▲ **Figure 4.7**   Within the mitochondrion, adenosine triphosphate (ATP) is formed at three sites along the electron transport chain. This process is known as oxidative phosphorylation.

Energy Yield From Carbohydrate   The oxidative system of energy production can generate up to 39 molecules of ATP from one molecule of glycogen. If the process begins with glucose, the net gain is 38 ATP molecules (recall that one ATP molecule is used for conversion to glucose-6-phosphate before glycolysis begins). The energy gained is summarized in table 4.2.

It should be noted that the molecules of reduced NAD (termed NADH) formed in the cytoplasm cannot directly enter the mitochondria. They must donate their electrons to either NADH or reduced FAD (FADH) carrier molecules in the electron transport chain. Two cytoplasmic NADH molecules donating their electrons to mitochondrial NADH yield six ATP molecules, as opposed to only four ATP molecules when their electrons are donated to mitochondrial FADH. Thus, when FADH is the carrier, only up to 36 ATP molecules can be generated from glucose and 37 ATP molecules from glycogen.

Oxidation of Fat

As noted earlier, fat also contributes to muscles' energy needs. Muscle and liver glycogen stores may be able to provide only 1,500 to 2,500 kcal of energy, but the fat stored inside muscle fibers

Table 4.2

Energy Production From the Oxidation of Liver Glycogen

Stage of process	Direct	By oxidative phosphorylation[a]
Glycolysis (glucose to pyruvic acid)	3	4-6[b]
Pyruvic acid to acetyl coenzyme A	0	6
Krebs cycle	2	22
Subtotal	5	32-34
Total		37-39

[a]Refers to adenosine triphosphate (ATP) produced by transferring H^+ and electrons to the electron transport chain.

[b]The energy yield differs depending on whether reduced nicotinamide adenine dinucleotide (NADH) or reduced flavin adenine dinucleotide (FADH) is the carrier molecule to transport the electron through the mitochondrial membrane and the electron transport chain, with NADH yielding up to 39 molecules of ATP and FADH yielding 37 molecules of ATP.

and in fat cells can supply at least 70,000 to 75,000 kcal, even in a lean adult.

Although many chemical compounds (such as triglycerides, phospholipids, and cholesterol) are classified as fats, only triglycerides are major energy sources. Triglycerides are stored in fat cells and between and within skeletal muscle fibers. To be used for energy, a triglyceride must be broken down to its basic units: one molecule of glycerol and three molecules of FFAs. This process is called **lipolysis**, and it is carried out by enzymes known as lipases. The FFAs are the primary energy source, so they will be our focus.

Once freed from glycerol, FFAs can enter the blood and be transported throughout the body, entering muscle fibers by diffusion. Their rate of entry into the muscle fibers depends on the concentration gradient. Increasing the concentration of FFAs in the blood facilitates their transport into muscle fibers.

β-Oxidation   Although the various FFAs in the body differ structurally, their metabolism is essentially the same, as shown in figure 4.8. On entering the muscle fiber, FFAs are enzymatically activated with energy from ATP, preparing them for catabolism (breakdown) within the mitochondria. This enzymatic catabolism of fat by the mitochondria is termed **β-oxidation.**

In this process, the carbon chain of an FFA is cleaved into separate 2-carbon units of acetic acid. For example, if an FFA originally has a 16-carbon chain, β-oxidation yields eight molecules of acetyl CoA.

Krebs Cycle and the Electron Transport Chain   From this point on, fat metabolism follows the same path as carbohydrate metabolism. Acetyl CoA formed by β-oxidation enters the Krebs cycle. The Krebs cycle generates hydrogen, which is transported to the electron transport chain along with the hydrogen generated during β-oxidation to undergo oxidative phosphorylation (see figure 4.7). As in glucose metabolism, the by-products of FFA oxidation are ATP, H_2O, and carbon dioxide (CO_2). However, the complete combustion of an FFA molecule requires more oxygen because an FFA molecule contains considerably more carbon than a glucose molecule.

fyi   Although fat provides more kilocalories of energy per gram than carbohydrate, fat oxidation requires more oxygen than carbohydrate oxidation. The energy yield from fat is 5.6 ATP molecules per oxygen molecule used, compared with carbohydrate's yield of 6.3 ATP per oxygen molecule. Oxygen delivery is limited by the oxygen transport system, so carbohydrate is the preferred fuel during high-intensity exercise. Furthermore, the maximum rate of high-energy phosphate formation from lipid oxidation is too low to match the rate of utilization of high-energy phosphate during higher intensity exercise. This explains the reduction in an athlete's race pace when carbohydrate stores are depleted and fat, by default, becomes the predominant fuel source.

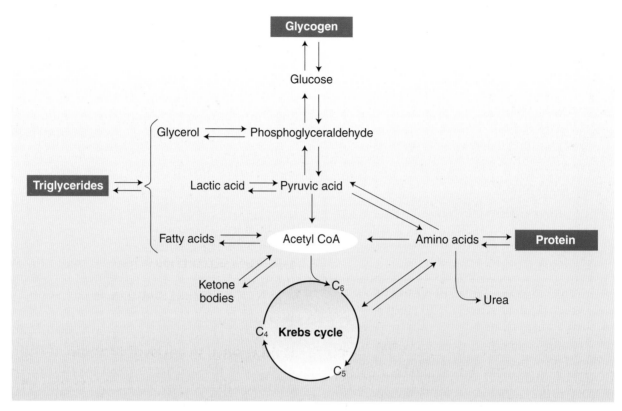

▲ **Figure 4.8**   The metabolism of fat, carbohydrate, and protein share some common pathways. Note that they are all reduced to acetyl coenzyme A (CoA) and enter the Krebs cycle.

The advantage of having more carbon in FFAs than in glucose is that more acetyl CoA is formed from the metabolism of a given amount of fat, so more acetyl CoA enters the Krebs cycle and more electrons are sent to the electron transport chain. This is why fat metabolism can generate so much more energy than glucose metabolism.

Consider the example of palmitic acid, a rather abundant 16-carbon FFA. The combined reactions of oxidation, the Krebs cycle, and the electron transport chain produce 129 molecules of ATP from one molecule of palmitic acid (as shown in table 4.3), compared with only 38 molecules of ATP from glucose or 39 from glycogen. Although this yield seems quite high, only about 40% of the energy released by the metabolism of either glucose or FFA molecules is captured to form ATP. The remaining 60% is given off as heat.

Table 4.3

Energy Production From the Oxidation of Palmitic Acid

	Adenosine triphosphate produced from 1 molecule of $C_{16}H_{32}O_2$	
Stage of process	**Direct**	**By oxidative phosphorylation**
Fatty acid activation	0	–2
β-oxidation	0	35
Krebs cycle	8	88
Subtotal	8	121
Total		129

Oxidation of Protein

As noted earlier, carbohydrates and fatty acids are our bodies' preferred fuels. But proteins, or rather the amino acids that form them, are also used. Some amino acids can be converted into glucose (by gluconeogenesis). Alternatively, some can be converted into various intermediates of oxidative metabolism (such as pyruvate or acetyl CoA) to enter the oxidative process.

Protein's energy yield is not as easily determined as that of carbohydrate or fat because protein also contains nitrogen. When amino acids are catabolized, some of the released nitrogen is used to form new amino acids, but the remaining nitrogen cannot be oxidized by the body. Instead it is converted into urea and then excreted, primarily in the urine. This conversion requires the use of ATP, so some energy is spent in this process.

When protein is broken down through combustion in the laboratory, the energy yield is 5.65 kcal/g. However, because of the energy expended in converting nitrogen to urea, when protein is metabolized in the body, the energy yield is only about 4.1 kcal/g, 27.4% less than the laboratory value.

To accurately assess the rate of protein metabolism, the amount of nitrogen being eliminated from the body must be determined. These measurements require urine collection for 12- to 24-h periods, clearly a time-consuming process. Because the healthy body uses little protein during rest and exercise (usually not more than 5% of total energy expended), estimates of energy expenditure generally ignore protein metabolism.

Interaction of the Three Energy Systems

The three energy systems do not work independently of one another. When a person is exercising at the highest intensity possible, from the shortest sprints (less than 10 s) to endurance events (greater than 30 min), each of the energy systems is contributing to the total energy needs of the body.[10] However, generally one energy system is predominant, except when there is a transition from the predominance of one energy system to another. As an example, in a 10-s, 100-m sprint, the ATP–PCr system is the predominant energy system, but both the anaerobic glycolytic and oxidative systems provide a small portion of the energy needed. At the other extreme, in a 30-min 10,000-m run, the oxidative system is predominant, but both the ATP–PCr and anaerobic glycolytic systems contribute. Figure 4.9 provides an estimate of the primary or predominant energy system for maximal intensity, all-out exercise of increasing duration.

Oxidative Capacity of Muscle

We have seen that the processes of oxidative metabolism have the highest energy yields. It would be ideal if these processes always functioned at peak capacity. But, as with all physiological systems, they operate within certain constraints. The oxidative capacity of muscle ($\dot{Q}O_2$) is a measure of its maximal capacity to use oxygen. This measurement is made in the laboratory where a small amount of muscle tissue can be tested to determine its capacity to consume oxygen when chemically stimulated to generate ATP. In this section, we'll look at the limitations of the muscles' oxidative capacity.

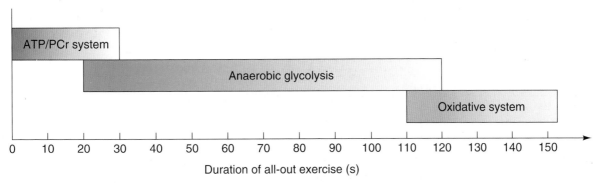

▲ **Figure 4.9**   Interaction of the energy systems with peak intensity exercise of increasing duration.

Enzyme Activity

Muscle fibers' capacity to oxidize carbohydrate and fat is difficult to determine. Numerous studies have shown a close relationship between a muscle's ability to perform prolonged aerobic exercise and the activity of its oxidative enzymes. Because many enzymes are required for oxidation, the enzyme activity of your muscle fibers provides a reasonable indication of their oxidative potential.

Measuring all the enzymes in muscles is impractical, so a few representatives have been selected to reflect the aerobic capacity of the fibers. The enzymes most frequently measured include succinate dehydrogenase and citrate synthase, mitochondrial enzymes involved in the Krebs cycle (see figure 4.6). Figure 4.10 illustrates the close relationship between succinate dehydrogenase activity in the vastus lateralis muscle and the muscle's oxidative capacity. Endurance athletes' muscles have oxidative enzyme activities nearly two to four times greater than those of untrained men and women.[6, 7, 11]

Fiber-Type Composition and Endurance Training

A muscle's fiber-type composition determines, in part, its oxidative capacity. As noted in chapter 1, slow-twitch (ST) fibers have a greater capacity for aerobic activity than the

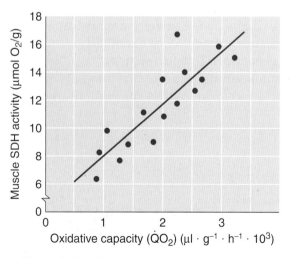

▲ **Figure 4.10** The relationship between muscle succinate dehydrogenase (SDH) activity and its oxidative capacity ($\dot{Q}O_2$), measured in a biopsy sample taken from the vastus lateralis.

fast-twitch (FT) fibers because ST fibers have more mitochondria and higher concentrations of oxidative enzymes. FT fibers are better suited for glycolytic energy production. Thus, in general, the more ST fibers in your muscles, the greater your muscles' oxidative capacity. Elite distance runners, for example, have been reported to possess more ST fibers, more mitochondria, and higher muscle oxidative enzyme activities than do untrained individuals.[11, 13]

Endurance training enhances the oxidative capacity of all fibers, especially FT fibers. Training that places demands on oxidative phosphorylation stimulates the muscle fibers to develop more mitochondria that also are larger and contain more oxidative enzymes. By increasing the fibers' enzymes for β-oxidation, this training also enables the muscle to rely more heavily on fat for ATP production.

Thus, with endurance training, even people with large percentages of FT fibers can increase their muscles' aerobic capacities. But it is generally agreed that an endurance-trained FT fiber will not develop the same high-endurance capacity as a similarly trained ST fiber.

Oxygen Needs

Although your muscles' oxidative capacity is determined by the number of mitochondria and the amount of oxidative enzymes present, oxidative metabolism ultimately depends on an adequate supply of oxygen. When you are at rest, your body's need for ATP is relatively small, requiring minimal oxygen delivery. As your exercise intensity increases, so do your energy demands. To meet them, your rate of oxidative ATP production also increases. In an effort to satisfy your muscles' need for oxygen, the rate and depth of your respiration increase, improving gas exchange in the lungs, and your heart beats faster, pumping more oxygenated blood to your muscles.

The human body stores little oxygen. Therefore, the amount of oxygen entering your blood as it passes through your lungs is directly proportional to the amount used by your tissues for oxidative metabolism. Consequently, a reasonably accurate estimate of aerobic energy production can be made by measuring the amount of oxygen consumed at the lungs.

▶ The oxidative system involves breakdown of fuels with the aid of oxygen. This system yields more energy than the ATP–PCr or glycolytic systems.

▶ Oxidation of carbohydrate involves glycolysis, the Krebs cycle, and the electron transport chain. The end result is H_2O, CO_2, and 38 or 39 ATP molecules per carbohydrate molecule.

▶ Fat oxidation begins with β-oxidation of FFAs and then follows the same path as carbohydrate oxidation: the Krebs cycle and the electron transport chain. The energy yield for fat oxidation is much higher than for carbohydrate oxidation, and it varies with the FFA being oxidized. However, the maximum rate of high-energy phosphate formation from lipid oxidation is too low to match the rate of utilization of high-energy phosphate during higher intensity exercise, and the energy yield of fat per oxygen molecule used is much less than that for carbohydrate.

▶ Protein oxidation is more complex because protein (amino acids) contains nitrogen, which cannot be oxidized. Protein contributes relatively little to energy production, generally less than 5%, so its metabolism is often overlooked.

▶ Your muscles' oxidative capacity depends on their oxidative enzyme levels, their fiber-type composition, and oxygen availability.

Measuring Energy Use During Exercise

The energy turnover in muscle fibers cannot be directly measured. But numerous indirect laboratory methods can be used to calculate the rate and quantity of energy expenditure when your body is at rest and during exercise. Several of these methods have been in use since the early 1900s. Others are new and only recently have been used in exercise physiology. In the following sections, we review some of these methods of measurement.

Direct Calorimetry

As noted earlier, only about 40% of the energy liberated during the metabolism of glucose and fats is used to produce ATP. The remaining 60% is converted to heat, so one way to gauge the rate and quantity of energy production is to measure your body's heat production. This technique is called direct calorimetry.

This approach was first described by Zuntz and Hagemann in the late 1800s.[21] They developed the calorimeter (illustrated in figure 4.11), which is an insulated, airtight chamber. The walls of the chamber contain copper tubing through which water is passed. When you're inside the chamber, the heat produced by your body radiates to the walls and warms the water. The water's temperature change is recorded, as are temperature changes in the air entering and leaving the chamber as you breathe. These changes are caused by the heat your body generates. Your metabolism can be calculated from the resulting values.

Calorimeters are expensive to construct and to use and are slow to generate results. Their only real advantage is that they measure heat directly. Although a calorimeter can provide an accurate measure of total-body energy expenditure, it cannot follow rapid changes in energy release. For that reason, energy metabolism during intense exercise cannot be studied with a calorimeter. Consequently, this method is seldom used today because it is easier and less expensive to measure energy expenditure by assessing the exchange of oxygen and carbon dioxide that occurs during oxidative phosphorylation.

Indirect Calorimetry

As noted earlier, glucose and fat metabolism depends on O_2 availability and produces CO_2 and water. The amount of O_2 and CO_2 exchanged in the lungs normally equals that used and released by body tissues. With this knowledge, you can estimate your caloric expenditure by measuring your respiratory gases. This method

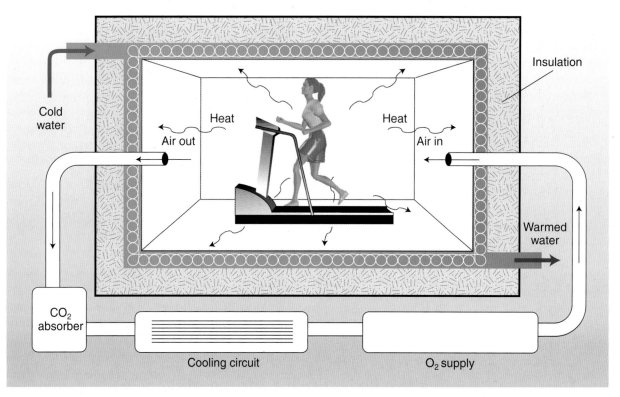

▲ **Figure 4.11**   A calorimetric chamber. The heat generated within the subject's body is transferred to the air and walls of the chamber (through conduction, convection, and evaporation). This heat produced by the subject is then measured by recording the temperature change in the air and water flowing through the chamber. This heat change is a measure of the subject's metabolic rate.

of estimating energy expenditure is called **indirect calorimetry** because heat production is not measured directly. Rather, it is calculated from the respiratory exchange of CO_2 and O_2.

Figure 4.12 shows some of the equipment used to measure CO_2 production and O_2 consumption. Although this equipment is cumbersome and limits movement, it has been adapted for use under a variety of conditions in the laboratory, on the playing field, and elsewhere.

Calculating Oxygen Consumption and Carbon Dioxide Production

By using the equipment illustrated in figure 4.12 to measure the volume and gas concentrations of both the air going into the lungs (inspired air) and the air coming out of the lungs (expired air), you can calculate the actual volume of oxygen consumed (VO_2) and volume of CO_2 produced (VCO_2). This can be

done breath-by-breath, or it can be done over discreet time periods. Generally, values are presented as oxygen consumed per minute ($\dot{V}O_2$) and CO_2 produced per minute ($\dot{V}CO_2$). The dot over the V ($\dot{V}$) is used to indicate a rate of O_2 consumption or CO_2 production. Calculation of $\dot{V}O_2$ and $\dot{V}CO_2$ require the following information.

- Volume of air inspired ($\dot{V}_I$)
- Volume of air expired ($\dot{V}_E$)
- Fraction of oxygen in the inspired air (F_IO_2)
- Fraction of CO_2 in the inspired air (F_ICO_2)
- Fraction of oxygen in the expired air (F_EO_2)
- Fraction of CO_2 in the expired air (F_ECO_2)

Indirect Calorimetry: Measuring Respiratory Gas Exchange

Respiratory gas exchange is determined by measuring the volume of O_2 and CO_2 that enters and leaves the lungs during a given period of time, usually at 1-min intervals. Because O_2 is removed from the inspired air in the alveoli and CO_2 is added to the alveolar air, the expired O_2 concentration is less than that inspired, whereas the CO_2 concentration is higher in expired air than inspired air. Consequently, the difference between inspired and expired air tells us how much O_2 is taken up and how much CO_2 is being produced. Because the body has only limited O_2 storage, the amount taken up at the lungs accurately reflects the body's use of O_2. Although a number of sophisticated and expensive methods are available for measuring the respiratory exchange of O_2 and CO_2, the simplest and oldest methods (i.e., Douglas bag and chemical gas analysis) are probably the most accurate, but they are relatively slow and permit only a few measurements during each session. Modern electronic computer systems for respiratory gas exchange measurements offer a large time savings and multiple measurements.

▲ **Figure 4.12** The equipment used to measure the respiratory exchange of oxygen and carbon dioxide.

Notice in figure 4.12 that the gas that is expired by the subject passes through a hose into a mixing chamber, where samples are pumped to electronic oxygen and carbon dioxide analyzers. The computer recording equipment uses the measurements of respired gas volume and expired oxygen and carbon dioxide concentrations to calculate O_2 uptake and CO_2 production.

The oxygen consumption, in liters of oxygen consumed per minute, is then calculated as follows:

$$\dot{V}O_2 = (\dot{V}_I \times F_IO_2) - (\dot{V}_E \times F_EO_2)$$

The CO_2 production is calculated like this:

$$\dot{V}CO_2 = (\dot{V}_E \times F_ECO_2) - (\dot{V}_I \times F_ICO_2)$$

An example of how to calculate oxygen consumption is illustrated in figure 4.13. This example uses data representative of values that would be expected at maximal rates of work.

Haldane Transformation

Over the years, scientists have attempted to simplify the actual calculation of oxygen consumption and CO_2 production. Several of the measurements needed in the preceding equations are known and do not change. The gas

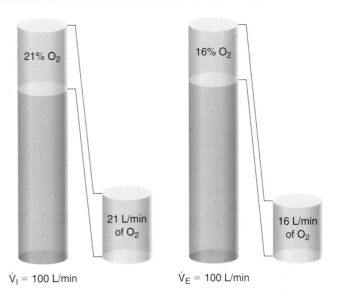

$\dot{V}O_2 = \dot{V}_I \times F_IO_2 - \dot{V}_E \times F_EO_2$
$\dot{V}O_2 = 100 \text{ L/min} \times 0.21 - 100 \text{ L/min} \times 0.16$
$\dot{V}O_2 = 21 \text{ L/min} - 16 \text{ L/min} = 5 \text{ L/min}$

▲ **Figure 4.13** Example of the calculation of oxygen consumption.

concentrations of the three gases that make up inspired air are known: oxygen accounts for 20.93%, CO_2 accounts for 0.04%, and nitrogen accounts for 79.03% of the inspired air. What about the volume of inspired and expired air? Aren't they the same such that you would only need to measure one of the two?

Inspired air volume equals expired air volume only when the volume of oxygen consumed equals the volume of CO_2 produced. When the volume of oxygen consumed is greater than the volume of CO_2 produced, $\dot{V}_I$ is greater than $\dot{V}_E$. Likewise, $\dot{V}_E$ is greater than $\dot{V}_I$ when the volume of CO_2 produced is greater than the volume of oxygen consumed. However, the one thing that is constant is that the volume of nitrogen inspired ($\dot{V}_I N_2$) is equal to the volume of nitrogen expired ($\dot{V}_E N_2$). Because $\dot{V}_I N_2 = \dot{V}_I \times F_I N_2$ and $\dot{V}_E N_2 = \dot{V}_E \times F_E N_2$, you can calculate $\dot{V}_I$ from $\dot{V}_E$ by using the following equation, which has been referred to as the **Haldane transformation**:

(1) $\dot{V}_I \times F_I N_2 = \dot{V}_E \times F_E N_2$,

which can be rewritten as

(2) $\dot{V}_I = (\dot{V}_E \times F_E N_2)/ F_I N_2$

Furthermore, because you are actually measuring the concentrations of O_2 and CO_2 in the expired gases, you can calculate $F_E N_2$ from the sum of $F_E O_2$ and $F_E CO_2$, or

(3) $F_E N_2 = 1 - (F_E O_2 + F_E CO_2)$

So, in pulling all of this information together, we can rewrite the equation for calculating $\dot{V}O_2$ as follows:

$\dot{V}O_2 = (\dot{V}_I \times F_I O_2) - (\dot{V}_E \times F_E O_2)$;

by substituting equation 2, you get the following:

$\dot{V}O_2 = [(\dot{V}_E \times F_E N_2)/(F_I N_2 \times F_I O_2)] - [(\dot{V}_E) \times (F_E O_2)]$;

by substituting known values for ($F_I O_2$) of 0.2093, and $F_I N_2$ of 0.7903, you get the following:

$\dot{V}O_2 = [(\dot{V}_E \times F_E N_2)/0.7903) \times 0.2093] - [(\dot{V}_E) \times (F_E O_2)]$;

by substituting equation 3, you get the following:

$\dot{V}O_2 = [(\dot{V}_E) \times (1 - (F_E O_2 + F_E CO_2)) \times (0.2093/0.7903)] - [(\dot{V}_E) \times (F_E O_2)]$;

or, simplified,

$\dot{V}O_2 = (\dot{V}_E) \times [(1 - (F_E O_2 + F_E CO_2)) \times (0.265)] - (\dot{V}_E) \times (F_E O_2)$,

or further simplified,

$\dot{V}O_2 = (\dot{V}_E) \times \{[(1 - (F_E O_2 + F_E CO_2)) \times (0.265)] - (F_E O_2)\}$

Respiratory Exchange Ratio

To estimate the amount of energy used by the body, it is necessary to know the type of food (carbohydrate, fat, or protein) being oxidized. The carbon and oxygen contents of glucose, FFAs, and amino acids differ dramatically. As a result, the amount of oxygen used during metabolism depends on the type of fuel being oxidized. Indirect calorimetry measures the amount of CO_2 released ($\dot{V}CO_2$) and oxygen consumed ($\dot{V}O_2$). The ratio between these two values is termed the **respiratory exchange ratio (RER)**.

$$RER = \dot{V}CO_2/\dot{V}O_2$$

In general, the amount of oxygen needed to completely oxidize a molecule of carbohydrate or fat is proportional to the amount of carbon in that fuel. For example, glucose ($C_6H_{12}O_6$) contains six carbon atoms. During glucose combustion, 6 molecules of oxygen are used to produce 6 CO_2 molecules, 6 H_2O molecules, and 38 ATP molecules:

$$6\ O_2 + C_6H_{12}O_6 \rightarrow 6\ CO_2 + 6\ H_2O + 38\ ATP$$

By evaluating how much CO_2 is released compared with the amount of O_2 consumed, we find that the respiratory exchange ratio is 1.0:

$$RER = \dot{V}CO_2/\dot{V}O_2 = 6\ CO_2/6\ O_2 = 1.0$$

As shown in table 4.4, the RER value varies with the type of fuels being used for energy. Free fatty acids have considerably more carbon and hydrogen but less oxygen than glucose. Consider palmitic acid, $C_{16}H_{32}O_2$. To completely oxidize this molecule to CO_2 and H_2O requires 23 molecules of oxygen:

16 C	+	16 O_2	→	16 CO_2
32 H	+	8 O_2	→	16 H_2O
Total	=	24 O_2 needed		
	−	1 O_2 provided by the palmitic acid		
		23 O_2 must be added		

Table 4.4

Caloric Equivalence of the Respiratory Exchange Ratio (RER) and % kcal From Carbohydrates and Fats

	Energy	% kcal	
RER	kcal/L O2	Carbohydrates	Fats
0.71	4.69	0.0	100.0
0.75	4.74	15.6	84.4
0.80	4.80	33.4	66.6
0.85	4.86	50.7	49.3
0.90	4.92	67.5	32.5
0.95	4.99	84.0	16.0
1.00	5.05	100.0	0.0

Ultimately, this oxidation results in 16 molecules of CO_2, 16 molecules of H_2O, and 129 molecules of ATP:

$$C_{16}H_{32}O_2 + 23\ O_2 \rightarrow 16\ CO_2 + 16\ H_2O + 129\ ATP$$

Combustion of this fat molecule requires significantly more oxygen than combustion of a carbohydrate molecule. During carbohydrate oxidation, approximately 6.3 molecules of ATP are produced for each molecule of O_2 used (38 ATP per 6 O_2), compared with 5.6 molecules of ATP per molecule of O_2 during palmitic acid metabolism (129 ATP per 23 O_2).

Although fat provides more energy than carbohydrate, more oxygen is needed to oxidize fat than carbohydrate. This means that the RER value for fat is substantially lower than for carbohydrate. For palmitic acid, the RER value is 0.70:

$$RER = \dot{V}CO_2/\dot{V}O_2 = 16/23 = 0.70$$

Once the RER value is determined from the calculated respiratory gas volumes, the value can be compared with a table (table 4.4) to determine the food mixture being oxidized. If, for example, the RER value is 1.0, the cells are using only glucose or glycogen, and each liter of oxygen consumed would generate 5.05 kcal. The oxidation of only fat would yield 4.69 kcal/L of O_2, and the oxidation of protein would yield 4.46 kcal/L of O_2 consumed. Thus, if the muscles were using only glucose and the body were consuming 2 L of O_2/min, then the rate of heat energy production would be 10.1 kcal/min (2 L/min · 5.05 kcal).

Limitations of Indirect Calorimetry

Even indirect calorimetry is far from perfect. Calculations of gas exchange assume that the body's O_2 content remains constant and that CO_2 exchange in the lung is proportional to its release from the cells. Arterial blood remains almost completely oxygen saturated (about 98%), even during intense effort. We can accurately assume that the oxygen being removed from the air we breathe is in proportion to its cellular uptake. Carbon dioxide exchange, however, is less constant. Body CO_2 pools are quite large and can be altered simply by deep breathing or by performing highly intense exercise. Under these conditions, the amount of CO_2 released in the lung may not represent that being produced in the tissues, so calculations of carbohydrate and fat used based on gas measurements appear to be valid only during rest or steady-state exercise.

Use of the RER also can lead to inaccuracies. Recall that protein is not completely oxidized in the body because nitrogen is not oxidizable. This makes it impossible to calculate the body's protein use from the RER. As a result, the RER is sometimes referred to as nonprotein RER because it simply ignores protein oxidation.

Traditionally, protein was thought to contribute little to the energy used during exercise, so exercise physiologists felt justified in using the nonprotein RER when making calculations. But more recent evidence suggests that in exercise lasting for several hours, protein may contribute up to 5% of the total energy.

The body normally uses a combination of fuels. RER values vary depending on the specific mixture being oxidized. At rest, the RER value is typically in the range of 0.78 to 0.80. During exercise, though, muscles rely increasingly on carbohydrate for energy, resulting in a higher RER. As exercise intensity increases, the muscles' carbohydrate demand also increases. As more carbohydrate is used, the RER value approaches 1.0.

This increase in the RER value to 1.0 reflects the demands on blood glucose and muscle glycogen, but it also may indicate that more CO_2 is being unloaded from the blood than is being produced by the muscles. At or near exhaustion, lactate accumulates in the blood. Your body tries to reverse this acidification by releasing more CO_2. Lactate accumulation increases CO_2 production because excess acid causes carbonic acid in the blood to be converted to CO_2. As a consequence, the excess CO_2 diffuses out of the blood and into the lungs for exhalation, increasing the amount of CO_2 released. For this reason, RER values approaching 1.0 may not accurately estimate the type of fuel being used by the muscles.

Another complication is that glucose production from the catabolism of amino acids and fats in the liver produces an RER below 0.70. Thus, calculations of carbohydrate oxidation from the RER value will be underestimated if energy is derived from this process.

Despite its shortcomings, indirect calorimetry still provides our best estimate of energy expenditure at rest and during submaximal exercise.

Isotopic Measurements of Energy Metabolism

In the past, determining an individual's total daily energy expenditure depended on recording food intake over several days and measuring body composition changes during that period. This method, although widely used, is limited by the individual's ability to keep accurate records and by the ability to match the individual's activities to accurate energy costs.

Fortunately, the use of isotopes has expanded our ability to investigate energy metabolism. Isotopes are elements with an atypical atomic weight. They can be either radioactive (radioisotopes) or nonradioactive (stable isotopes). As an example, carbon-12

(^{12}C) has a molecular weight of 12, is the most common natural form of carbon, and is nonradioactive. In contrast, carbon-14 (^{14}C) has two more neutrons than ^{12}C, giving it an atomic weight of 14. ^{14}C is created in the laboratory and is radioactive.

Carbon-13 (^{13}C) constitutes about 1% of the carbon in nature and is used frequently for studying energy metabolism. Because ^{13}C is nonradioactive, it is less easily traced within the body than ^{14}C. But although radioactive isotopes are easily detected in the body, they pose a hazard to body tissues and thus are used infrequently in human research.

^{13}C and other isotopes such as hydrogen 2 (deuterium, or ^2H) are used as tracers, meaning that they can be selectively followed in the body. Tracer techniques involve infusing isotopes into an individual and then following their distribution and movement.

Although the method was first described in the 1940s, studies that used doubly labeled water for monitoring energy expenditure during normal daily living in humans were not conducted until the 1980s. The subject ingests a known amount of water labeled with two isotopes ($^2H_2^{18}O$), hence the term doubly labeled water. The deuterium (^2H) diffuses throughout the body's water, and the oxygen-18 (^{18}O) diffuses throughout both the water and the bicarbonate stores (where much of the CO_2 derived from metabolism is stored). The rate at which the two isotopes leave the body can be determined by analyzing their presence in a series of urine, saliva, or blood samples. These turnover rates then can be used to calculate how much CO_2 is produced, and that value can be converted to energy expenditure by using calorimetric equations.

Because isotope turnover is relatively slow, energy metabolism must be measured for several weeks. Thus, this method is not well suited for measurements of acute exercise metabolism. However, its accuracy (more than 98%) and low risk make it well suited for determining day-to-day energy expenditure. Nutritionists have hailed the doubly labeled water method as the most significant technical advance of the past century in the field of energy metabolism.

▶ Direct calorimetry involves using a calorimeter to directly measure heat produced by the body.

▶ Indirect calorimetry involves measuring O_2 consumption and CO_2 production, calculating the RER value (the ratio of these two gas volumes), comparing the RER value with standard values to determine the foods being oxidized, and then calculating the energy expended per liter of oxygen consumed.

▶ The RER value at rest is usually 0.78 to 0.80. The RER value for the oxidation of fat is 0.70 and 1.00 for carbohydrates.

▶ Isotopes can be used to determine metabolic rate over long periods of time. They are injected or ingested into the body. The rates at which they are cleared can be used to calculate CO_2 production and then caloric expenditure.

Energy Expenditure at Rest and During Exercise

Now that we have discussed the energy-producing systems in our bodies, we can turn our attention to how this energy is used.

Resting Metabolic Rate

The rate at which your body uses energy is your metabolic rate. As noted earlier, estimates of energy expenditure during rest and exercise are based on measurement of whole-body oxygen consumption and its caloric equivalent. At rest, an average person consumes about 0.3 L of O_2/min. This equals 18 L/h or 432 L/day.

Now let's calculate this person's daily caloric expenditure. Recall that at rest, the body usually burns a mixture of carbohydrate and fat. An RER value of 0.80 when at rest is fairly common for most individuals on a mixed diet. The caloric equivalence of an RER value of 0.80 is 4.80 kcal per liter of O_2 consumed (from table 4.4). Using these common values, we can cal-

culate this individual's caloric expenditure as follows:

$$kcal/day = \text{liters of } O_2 \text{ consumed per day} \times \text{kcal used per liter of } O_2$$
$$= 432 \text{ L } O_2/day \times 4.80 \text{ kcal/L } O_2$$
$$= 2,074 \text{ kcal/day}$$

This value closely agrees with the average resting energy expenditure expected for a 70-kg (154-lb) man. Of course, it does not include the energy needed for normal daily activity.

One standardized measure of energy expenditure when at rest is the **basal metabolic rate** (BMR). The BMR is the rate of energy expenditure for an individual at rest in a supine position, measured immediately after at least 8 h of sleep and at least 12 h of fasting. This value reflects the minimum amount of energy required to carry on your body's essential physiological functions.

Your basal metabolic rate is directly related to your fat-free mass and generally is reported in kilocalories per kilogram of fat-free mass per minute ($kcal \cdot kg^{-1} \cdot min^{-1}$). The more fat-free mass, the more total calories expended in a day. Because women tend to have a lower fat-free mass and a greater fat mass than men, women tend to have lower BMRs than men of similar weight.

Your body's surface area is equally important. The more surface area you have, the more heat loss occurs across your skin, which raises your BMR because more energy is needed to maintain your body temperature. For this reason, the BMR is also often reported in kilocalories per square meter of body surface area per hour ($kcal \cdot m^{-2} \cdot h^{-1}$). Because we are discussing daily energy expenditure, we've opted for a simpler unit: kcal/day.

Many other factors affect your BMR. Among them are the following:

• Age: BMR gradually decreases with increasing age, generally because of a decrease in fat-free mass.

• Body temperature: BMR increases with increasing temperature.

• Stress: Stress increases activity of the sympathetic nervous system, which increases the BMR.

- Hormones: Thyroxine from the thyroid gland and epinephrine from the adrenal medulla both increase the BMR.

Instead of BMR, most researchers now use the term **resting metabolic rate (RMR)**, because most measurements follow the same conditions required for measuring BMR but don't require the individual to sleep over in a hospital or research laboratory. BMR and RMR values are essentially identical. The BMR and RMR ranges from 1,200 and 2,400 kcal/day. But the average total metabolic rate of an individual engaged in normal daily activity ranges from 1,800 to 3,000 kcal.

> **fyi** The energy expenditure for very large athletes engaged in intense daily training can exceed 10,000 kcal/day!

Metabolic Rate During Submaximal Exercise

As your body shifts from rest to exercise, your energy needs increase. Your metabolism increases in direct proportion to the increase in your rate of work, as shown in figure 4.14a. The subject exercised on a cycle ergometer for 5 min at 50 watts (W); oxygen consumption ($\dot{V}O_2$) increased from its resting value to a steady-state value within 1 to 2 min. The same subject then cycled the same or next day for 5 min at 100 W, and again a steady-state $\dot{V}O_2$ was reached in 1 or 2 min. In a similar manner, the subject cycled for 5 min at 150 W, 200 W, 250 W, and 300 W, respectively, and steady-state values were achieved at each power output. The steady-state $\dot{V}O_2$ values were plotted against their respective power outputs (right portion of figure 4.14a), showing clearly that there is a linear increase in the $\dot{V}O_2$ with increases in power output. The steady-state $\dot{V}O_2$ value represents the energy cost for that specific power output.

In more recent studies, it is clear that the $\dot{V}O_2$ response at higher rates of work does not follow the steady-state response pattern shown in figure 4.14a but rather follows that illustrated in figure 4.14b. It appears that at power outputs above the lactate threshold (the lactate response

is indicated by the dashed line in the right half of figure 4.14, *a* and *b*), the oxygen consumption continues to increase beyond the typical 1 to 2 min needed to reach a steady-state value. This increase has been called the slow component of oxygen uptake kinetics.[9, 18] Gaesser and Poole[9] speculated that this slow component is possibly associated with the additional oxygen cost associated with ventilation, shifting of metabolic substrate from fat to carbohydrate, or an increase in body temperature. The most likely mechanism, however, is an alteration in muscle fiber recruitment patterns, with the recruitment of more FT muscle fibers, which are less efficient (i.e., they require a higher $\dot{V}O_2$ to achieve the same power output).[9] A study by Barstow and colleagues[3] supported the speculation that the recruitment of a higher percentage of FT fibers is the predominant mechanism for this slow component.

There is a similar, but likely unrelated, phenomenon referred to as $\dot{V}O_2$ drift. $\dot{V}O_2$ **drift** is defined as a slow increase in $\dot{V}O_2$ during prolonged, submaximal, constant power output exercise. Unlike the slow component, $\dot{V}O_2$ drift is observed at power outputs well below lactate threshold, and the magnitude of the increase in $\dot{V}O_2$ drift is much less. Although not understood completely, $\dot{V}O_2$ drift is likely attributable to an increase in ventilation and in levels of circulating catecholamines.[15]

Maximal Capacity for Aerobic Exercise

In figure 4.14a, it is clear that when the subject cycles at 300 W, the $\dot{V}O_2$ response is not different from that achieved at 250 W. This indicates that the subject has reached a maximal limit of his ability to increase his $\dot{V}O_2$. This peak value is referred to as aerobic capacity, **maximal oxygen uptake**, or **$\dot{V}O_2$max**. $\dot{V}O_2$max is regarded by most as the best single measurement of cardiorespiratory endurance and aerobic fitness. This concept is further illustrated in figure 4.15, comparing the $\dot{V}O_2$max of a trained and an untrained man.

Although some sport scientists have suggested that $\dot{V}O_2$max is a good predictor of success in endurance events, the winner of a

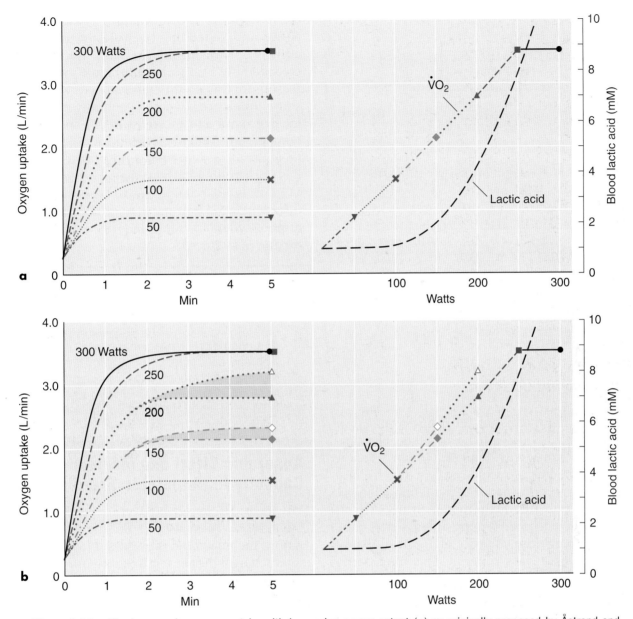

▲ **Figure 4.14** The increase in oxygen uptake with increasing power output (*a*) as originally proposed by Åstrand and Rodahl (1986, p. 300) and (*b*) as redrawn by Gaesser and Poole (1996, p. 36). See the text for a detailed explanation of the significance of this figure.

Reprinted, by permission, from G.A. Gaesser and D.C. Poole, 1996, "The slow component of oxygen uptake kinetics in humans," *Exercise and Sport Sciences Reviews* 24: 36.

marathon race cannot be predicted from the runner's laboratory-measured $\dot{V}O_2$max.[8] Likewise, an endurance-running performance test is only a modest predictor of one's $\dot{V}O_2$max. This suggests that a good performance requires more than a high $\dot{V}O_2$max.[4]

Also, research has documented that $\dot{V}O_2$max increases with physical training for only 8 to 12 weeks and then this value plateaus, despite continued higher intensity training. Although $\dot{V}O_2$max doesn't continue to increase, the participants continue to improve their endurance performance. People may develop a greater ability to perform at a higher percentage of their $\dot{V}O_2$max. Most runners, for example, can complete a 42-km (26.1-mi) race at an average pace that requires them to use approximately 75% to 80% of their $\dot{V}O_2$max.[4]

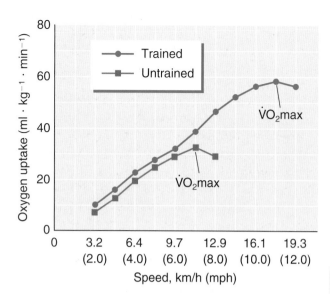

▲ **Figure 4.15** The relationship between exercise intensity (running speed) and oxygen uptake, illustrating $\dot{V}O_2$max in a trained and an untrained man.

Consider the case of Alberto Salazar, a near world-record holder in the marathon. His measured $\dot{V}O_2$max was 70 ml · kg⁻¹ · min⁻¹. That is below the $\dot{V}O_2$max expected based on his best marathon performance of 2 h 8 min. He was, however, able to run the marathon at 86% of his $\dot{V}O_2$max when performing at his racing pace, a percentage considerably higher than that of other world-class runners. This may partly explain his world-class running ability.

Because individuals' needs for energy vary with body size, $\dot{V}O_2$max generally is expressed relative to body weight, in milliliters of oxygen consumed per kilogram of body weight per minute (ml · kg⁻¹ · min⁻¹). This allows a more accurate comparison of different-sized individuals who exercise in weight-bearing events, such as running. In non-weight-bearing activities, such as swimming and cycling, endurance performance is more closely related to $\dot{V}O_2$max measured in liters per minute.

Normally active 18- to 22-year-old college students have average $\dot{V}O_2$max values of 38 to 42 ml · kg⁻¹ · min⁻¹ for women and 44 to 50 ml · kg⁻¹ · min⁻¹ for men. After the age of 25 to 30 years, inactive people's $\dot{V}O_2$max values decrease about 1% per year. This is probably attributable to a combination of biological aging and sedentary lifestyle. In addition, adult women generally have $\dot{V}O_2$max values considerably below those of adult men. Two reasons for this sex difference are body composition differences (women generally have less fat-free mass and more fat mass) and blood hemoglobin content (women have less, thus they have less oxygen-carrying capacity). But it is unclear how much of the sex difference in $\dot{V}O_2$max is attributable to actual physiological differences and how much might be caused by a more sedentary lifestyle. This is discussed further in chapter 18.

> **fyi** Aerobic capacities of 80 to 84 ml · kg⁻¹ · min⁻¹ have been observed among elite male long-distance runners and cross-country skiers. The highest $\dot{V}O_2$max value recorded for a man is that of a champion Norwegian cross-country skier who had a $\dot{V}O_2$max of 94 ml · kg⁻¹ · min⁻¹. The highest value recorded for a woman is 77 ml · kg⁻¹ · min⁻¹ for a Russian cross-country skier.[1] In contrast, poorly conditioned adults may have values below 20 ml · kg⁻¹ · min⁻¹.

Anaerobic Effort and Maximal Capacity for Anaerobic Exercise

We have discussed how your aerobic metabolism can be measured. But these methods ignore the anaerobic processes. How can the interaction of the aerobic (oxidative) processes and the anaerobic processes be evaluated? The most common methods for estimating anaerobic effort involve the examination of either the excess postexercise oxygen consumption or the lactate threshold. Let's consider both.

Postexercise Oxygen Consumption

Your body's ability to gauge your muscles' need for oxygen is not perfect. When you begin exercise, your oxygen transport system (respiration and circulation) does not immediately supply the needed quantity of oxygen to the active muscles. Your oxygen consumption requires several minutes to reach the required (steady-state) level at which the aerobic processes are fully functional, but your body's oxygen requirements increase immediately the moment exercise begins.

Because oxygen needs and oxygen supply differ during the transition from rest to exercise, your body incurs an oxygen deficit, as shown in figure 4.16, even with low levels of exercise. The oxygen deficit is calculated simply as the difference between the oxygen required for a given rate of work (steady state) and the oxygen actually consumed. Despite insufficient oxygen, your muscles still generate the ATP needed through the anaerobic pathways.

During the initial minutes of recovery, even though your muscles are no longer actively working, oxygen demand does not immediately decrease. Instead, oxygen consumption remains elevated temporarily (figure 4.16). This consumption, which exceeds that usually required when at rest, traditionally has been referred to as the oxygen debt. A more common term today is **excess postexercise oxygen consumption (EPOC).** The EPOC is the volume of oxygen consumed above that normally consumed at rest. Think of what happens when you finish an intense exercise bout: A quick run down the block to catch a departing bus or a fast climb up several flights of stairs leaves you with a rapid pulse and feeling out of breath. After several minutes of recovery, your pulse and breathing return to resting rates.

For many years, the EPOC curve was described as having two distinct components: an initial fast component and a secondary slow component. According to classical theory, the fast component of the curve represented the oxygen required to rebuild the ATP and PCr used during exercise, especially in its initial stages. Without sufficient oxygen, the high-energy phosphate bonds in these compounds were broken to supply the required energy. During recovery, these bonds would need to be reformed, via oxidative processes, to replenish the energy stores, or repay the debt. The slow component of the curve was thought to result from removal of accumulated lactate from the tissues, by either conversion to glycogen or oxidation to CO_2 and H_2O, thus providing the energy needed to restore glycogen stores.

According to this theory, both the fast and slow components of the curve were thought to reflect the anaerobic activity that had occurred during exercise. The belief was that by examining the postexercise oxygen consumption, one could estimate the amount of anaerobic activity that had occurred.

However, more recent studies concluded that the classical explanation of EPOC is too simplistic. For example, during the initial phase

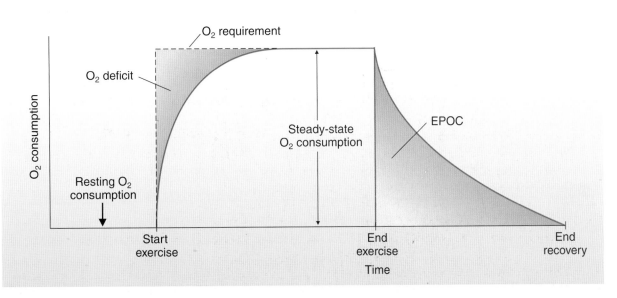

▲ **Figure 4.16**   Oxygen need during exercise and recovery, illustrating the oxygen deficit and excess postexercise oxygen consumption (EPOC).

of exercise, some oxygen is borrowed from the oxygen stores (hemoglobin and myoglobin). That oxygen must be replenished during recovery. Also, respiration remains temporarily elevated following exercise partly in an effort to clear CO_2 that has accumulated in the tissues as a by-product of metabolism. Body temperature also is elevated, which keeps the metabolic and respiratory rates high, thus requiring more oxygen, and elevated levels of norepinephrine and epinephrine during exercise have similar effects.

Thus, more is involved than is explained by the classical theory. The EPOC depends on many factors other than merely rebuilding ATP and PCr and clearing the lactate produced by anaerobic metabolism. The physiological mechanisms responsible for the EPOC still need to be clearly defined.

Lactate Threshold

Many investigators consider the lactate threshold to be a good indicator of an athlete's potential for endurance exercise. The **lactate threshold** is defined as the point at which blood lactate begins to accumulate above resting levels during exercise of increasing intensity. For example, a runner might be required to run on the treadmill at different speeds with a rest between each speed. After each run, a blood sample is taken from his or her fingertip, or from a catheter in one of the arm veins, from which we measure arterialized blood lactate. As illustrated in figure 4.17, the results of such testing can be used to plot the relationship between blood lactate and running velocity. At low running velocities, blood lactate levels remain at or near resting levels. But as running speed increases above about 13 km/h (10.0 mph), the blood lactate levels increase rapidly. The point at which blood lactate appears to increase above resting levels is termed the lactate threshold.

The lactate threshold has been thought to reflect the interaction of the aerobic and anaerobic energy systems. Some researchers have suggested that the lactate threshold represents a significant shift toward anaerobic glycolysis, which forms lactate. Consequently, the sudden increase in blood lactate with increasing effort also has been referred to as the anaerobic threshold. However, blood lactate concentration is determined not only by the production of lactate in skeletal muscle or other tissues but also by the clearance of lactate from the blood by the liver, skeletal muscle, cardiac muscle, and other tissues in the body. Thus, lactate threshold is best defined as that point in time

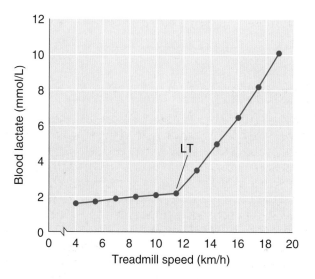

▲ **Figure 4.17**   The relationship between exercise intensity (running velocity) and blood lactate concentration. Blood samples were taken from a runner's arm vein and analyzed for lactate after the subject ran at each speed for 5 min. LT = lactate threshold.

during exercise of increasing intensity when the rate of lactate production exceeds the rate of lactate clearance or removal.

The lactate threshold usually is expressed in terms of the percentage of maximal oxygen uptake (%$\dot{V}O_2$max) at which it occurs. The ability to exercise at a high intensity without accumulating lactate is beneficial to the athlete because lactate accumulation contributes to fatigue. From the previous section, we learned that the major determinants of successful endurance performance are $\dot{V}O_2$max and the percentage of $\dot{V}O_2$max that an athlete can maintain for a prolonged period. The latter is probably related to the lactate threshold, because the lactate threshold is likely the major determinant of the pace that can be tolerated during a long-term endurance event. So the ability to perform at a higher percentage of $\dot{V}O_2$max probably reflects a higher lactate threshold. Consequently, a lactate threshold at 80% $\dot{V}O_2$max suggests a greater aerobic exercise tolerance than a threshold at 60% $\dot{V}O_2$max. Generally, in two individuals with the same maximal oxygen uptake, the person with the highest lactate threshold exhibits the best endurance performance.

Lactate threshold, when expressed as a percentage of $\dot{V}O_2$max, is one of the best determinants of an athlete's pace in endurance events such as distance running and cycling.

Economy of Effort

As you become more skillful at performing an exercise, your energy demands during that exercise are reduced. You become more efficient. This is illustrated in figure 4.18 by the data from two distance runners. At all running speeds faster than 11.3 km/h (7.0 mph), runner B used significantly less oxygen than runner A. These men had similar $\dot{V}O_2$max values (64-65 ml · kg^{-1} · min^{-1}), so runner B's lower submaximal energy use would be a decided advantage during competition.

These two runners competed on numerous occasions. During marathon races, they ran at paces requiring them to use 85% of their $\dot{V}O_2$max. On the average, runner B's running efficiency gave him a 13-min advantage in these competitions. Because their $\dot{V}O_2$max values are so similar but their energy needs so different during these events, much of runner B's competitive advantage can be attributed to his greater running efficiency. Unfortunately, we cannot explain the underlying causes of these efficiency differences.

fyi In untrained people, the lactate threshold typically occurs at around 50% to 60% of their $\dot{V}O_2$max. Elite endurance athletes may not reach lactate threshold until around 70% or 80% of $\dot{V}O_2$max.

Measuring Anaerobic Capacity

An acceptable method to determine one's anaerobic capacity is not available. Several methods have been described, but their validity has been challenged, and at best they offer only a rough estimate of one's anaerobic capacity. Early attempts to determine anaerobic capacity measured blood lactate after exhaustive exercise.[14,16] Although it is generally agreed that lactate in the blood indicates anaerobic glycolysis, it does not give a quantitative estimate of the anaerobic energy production. The maximal oxygen debt or maximal EPOC also has been proposed as an index of anaerobic capacity, but subsequent research has not supported this. In 1988, Medbø and coworkers[17] proposed the use of the maximal accumulated oxygen deficit as a measure of anaerobic capacity. Subsequently, a number of research studies confirmed that this test provides valid estimates of anaerobic capacity. Other tests that have shown great promise are the Wingate anaerobic test[2] and the critical power test.[12] Despite the limitations inherent in these methods, they remain our only indirect indicators of the metabolic potential of anaerobic capacity.

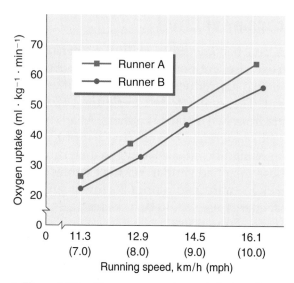

▲ **Figure 4.18**   The oxygen requirements for two distance runners while they ran at various speeds. Although they had similar $\dot{V}O_2$max values (64-65 ml · kg⁻¹ · min⁻¹), runner B was more efficient and therefore faster.

Various studies with sprint, middle-distance, and marathon runners have shown that marathon runners are generally the most efficient. In general, these ultra-long-distance runners use 5% to 10% less energy than middle-distance and sprint runners. However, this economy of effort has been studied at only relatively slow speeds (paces of 10-19 km/h, or 6-12 mph). We can reasonably assume that distance runners are less efficient at sprinting than runners who train specifically for short, faster races.

Variations in running form and the specificity of training for sprint and distance running may account for these differences in running economy. Film analyses reveal that middle-distance and sprint runners have significantly more vertical movement when running at 11 to 19 km/h (7-12 mph) than marathoners do. But such speeds are well below those required during middle-distance races and probably don't accurately reflect the running efficiency of competitors in shorter events of 1,500 m (1 mi) or less.

Performance in other athletic events might be even more affected by efficiency of movement than is running. Part of the energy expended during swimming, for example, is used to support the body on the surface of the water and to generate enough force to overcome the water's resistance to motion. Although the energy needed for swimming depends on body size and buoyancy, the efficient application of force against the water is the major determinant of swimming economy.

Characteristics of Successful Athletes in Aerobic Endurance Events

From our discussion of the metabolic characteristics of aerobic endurance athletes in this chapter and their muscle fiber type characteristics in chapter 1, it is clear that to be successful in aerobic endurance activities you need the following:

- High $\dot{V}O_2$max value
- High lactate threshold when expressed as a percentage of $\dot{V}O_2$max
- High economy of effort, or low $\dot{V}O_2$ value for the same rate of work
- High percentage of ST muscle fibers

From the limited data available, these four characteristics appear to be properly ranked in their order of importance. As an example, lactate threshold is the best predictor we have of actual race pace when you consider a group of elite distance runners. However, each of those runners already has a high $\dot{V}O_2$max. Although economy of effort is important, it does not vary that much between elite runners. Finally, having a high percentage of ST muscle fibers is helpful but not essential. The bronze medal winner in one of the Olympic marathon races had only 50% ST muscle fibers in his gastrocnemius muscle, one of the primary muscles used in running.

Energy Cost of Various Activities

The amount of energy expended for different activities varies with the intensity and type of exercise. The energy costs of many activities have been determined, usually by monitoring oxygen consumption during the activity to determine an average oxygen uptake per unit of time. Kilocalories of energy used per minute (kcal/min) then can be calculated from this value.

These values typically ignore the anaerobic aspects of exercise and the excess postexercise oxygen consumption. This omission is important because an activity that costs a total of 300 kcal during the actual exercise period may cost an additional 100 kcal during the recovery period. Thus, the total cost of that activity would be 400, not 300, kcal.

An average body requires 0.20 to 0.35 L of oxygen per minute to satisfy its resting energy requirements. This would amount to 1.0 to 1.8 kcal/min, 60 to 108 kcal/h, or 1,440 to 2,592 kcal/day. Obviously, any activity above resting levels will add to the projected daily expenditure. The range for total daily caloric expenditure is highly variable. It depends on many factors, including

- activity level,
- age,
- sex,
- size,
- weight, and
- body composition.

The energy costs of sport activities also differ. Some, such as archery or bowling, require only slightly more energy than when at rest. Others, such as sprinting, require so much energy that they can be maintained for only seconds. In addition to exercise intensity, the duration of the activity must be considered. For example, approximately 29 kcal/min is expended while a person is running at 25 km/h (15.5 mph), but this pace can be endured for only brief periods. Jogging at 11 km/h (7 mph), on the other hand, expends only 14.5 kcal/min, half that of running at 25 km/h (15.5 mph). But jogging can be maintained for considerably longer, resulting in a greater total energy expenditure.

Table 4.5 provides an estimate of energy expenditure for various activities for average mature men and women. These values are mere averages. Most activities involve moving the body mass, so these figures may vary considerably with individual differences such as those previously listed and with individual skill (efficiency of movement).

▶ The basal metabolic rate (BMR) is the minimum amount of energy required by your body to sustain basic cellular functions and is highly related to fat-free body mass and body surface area. It typically ranges from 1,200 to 2,400 kcal/day, but when daily activity is added, the typical daily caloric expenditure is 1,800 to 3,000 kcal/day.

▶ Your metabolism increases with increased exercise intensity, but your oxygen consumption is limited. Its peak value is $\dot{V}O_2$max. Successful aerobic performance is linked to a high $\dot{V}O_2$max and to the ability to perform for long periods at a high percentage of $\dot{V}O_2$max.

▶ The excess postexercise oxygen consumption (EPOC) is the elevation of metabolic rate above resting levels that occurs after exercise during the recovery period.

▶ Lactate threshold is that point at which blood lactate production begins to exceed the body's ability to clear or remove lactate, resulting in a rapid increase in blood lactate concentrations above resting levels during exercise of increasing intensity.

▶ Generally, individuals with higher lactate thresholds, expressed as a percentage of their $\dot{V}O_2$max, are capable of the best endurance performances.

▶ Aerobic endurance performance capacity is also associated with a high economy of effort, or low $\dot{V}O_2$ for the same rate of work.

Table 4.5

Energy Expenditure During Various Physical Activities

Activity	Men (kcal/min)	Women (kcal/min)	Relative to body mass (kcal · kg⁻¹ · min⁻¹)
Basketball	8.6	6.8	0.123
Cycling			
11.3 km/h (7.0 mph)	5.0	3.9	0.071
16.1 km/h (10.0 mph)	7.5	5.9	0.107
Handball	11.0	8.6	0.157
Running			
12.1 km/h (7.5 mph)	14.0	11.0	0.200
16.1 km/h (10.0 mph)	18.2	14.3	0.260
Sitting	1.7	1.3	0.024
Sleeping	1.2	0.9	0.017
Standing	1.8	1.4	0.026
Swimming (crawl), 4.8 km/h (3.0 mph)	20.0	15.7	0.285
Tennis	7.1	5.5	0.101
Walking, 5.6 km/h (3.5 mph)	5.0	3.9	0.071
Weightlifting	8.2	6.4	0.117
Wrestling	13.1	10.3	0.187

Note. Values presented are for a 70-kg (154-lb) man and a 55-kg (121-lb) woman. These values will vary depending on individual differences.

Fatigue and Its Causes

What exactly is the meaning of the term **fatigue** during exercise? Sensations of fatigue are markedly different when a person is exercising to exhaustion in events lasting 45 to 60 s, such as the 400-m run, than during prolonged exhaustive muscular effort, such as marathon running. We typically use the term fatigue to describe general sensations of tiredness and accompanying decrements in muscular performance.

Most efforts to describe underlying causes and sites of fatigue focus on

- the energy systems (ATP–PCr, anaerobic glycolysis, and oxidation),
- the accumulation of metabolic by-products, such as lactate,
- the nervous system, and
- the failure of the muscle fiber's contractile mechanism.

None of these alone can explain all aspects of fatigue. For example, although the lack of available energy can reduce the muscles' capacity to generate force, the energy systems are not wholly responsible for all forms of fatigue. The sensations of tiredness we often experience at the end of a workday have little to do with ATP availability. Fatigue also may result from the alteration of homeostasis by environmental stress. Many questions about fatigue remain unanswered.

Energy Systems and Fatigue

The energy systems are an obvious area to explore when considering possible causes of fatigue. When we feel fatigued, we often express it by saying, "I have no energy." But this use of the term *energy* is far removed from its physiological meaning. What role does energy, in this narrower physiological sense, play in fatigue during exercise?

PCr Depletion

Recall that PCr is used under anaerobic conditions to rebuild high-energy ATP as it is used and thus maintain your body's ATP stores. Biopsy studies of human thigh muscles have shown that during repeated maximal contractions, fatigue coincides with PCr depletion. Although ATP is directly responsible for the energy used during such activities, it is depleted less rapidly than PCr during muscular effort because ATP is being produced by other systems. But as PCr is depleted, your body's ability to quickly replace the spent ATP is seriously hindered. ATP use continues, but the ATP–PCr system is less able to replace it. Thus, ATP levels also decrease. At exhaustion, both ATP and PCr may be depleted (review figure 4.4). It now appears that P_i, which increases during intense short-term exercise because of the breakdown of PCr, is a potential cause of fatigue.[20]

To delay fatigue, the athlete must control the rate of effort through proper pacing to ensure that PCr and ATP are not prematurely exhausted. If the beginning pace is too rapid, the ATP and PCr available will quickly decrease, leading to early fatigue and inability to maintain the pace in the event's final stages. Training and experience allow the athlete to judge the optimal pace that permits the most efficient use of ATP and PCr for the entire event.

Glycogen Depletion

Muscle ATP levels are also maintained by the aerobic and anaerobic breakdown of muscle glycogen. In events lasting longer than a few seconds, muscle glycogen becomes the primary energy source for ATP synthesis. Unfortunately, glycogen reserves are limited and are depleted quickly.

As with PCr use, the rate of muscle glycogen depletion is controlled by the intensity of the activity. Increasing the work rate results in a disproportionate decrease in muscle glycogen. During sprint running, for example, muscle glycogen may be used 35 to 40 times faster than during walking. Muscle glycogen can be a limiting factor even during mild effort. The muscle depends on a constant supply of glycogen to meet the high energy demands of exercise.

Muscle glycogen is used more rapidly during the first few minutes of exercise than in the later stages, as seen in figure 4.19.[5] The

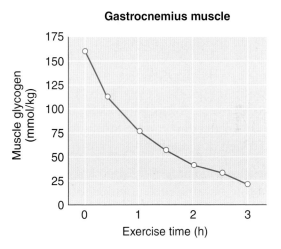

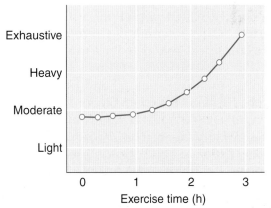

▲ **Figure 4.19**   The decline in gastrocnemius (calf) muscle glycogen (upper panel) during 3 h of treadmill running at 70% of $\dot{V}O_2$max. The subject's subjective rating of the effort was recorded in the lower panel. Note that the effort was rated as moderate for nearly 1.5 h of the run, although glycogen was decreasing steadily. Not until the muscle glycogen became quite low (less than 50 mmol/kg) did the rating of effort increase.

Adapted, by permission, from D.L. Costill, 1986, *Inside running: Basics of sports physiology* (Indianapolis: Benchmark Press). Copyright 1986 Cooper Publishing Group, Carmel, In.

illustration shows the change in muscle glycogen content in the subject's gastrocnemius (calf) muscle during the test. Although the subject ran the test at a steady pace, the rate of muscle glycogen metabolized from the gastrocnemius was greatest during the first 75 min.

The subject reported his perceived exertion (how difficult his effort seemed to be) at various times during the test. He felt only moderately stressed early in the run, when his glycogen stores were still high, even though he was using glycogen at a high rate. He didn't perceive severe fatigue until his muscle glycogen levels were nearly depleted. Thus, the sensation of fatigue in long-term exercise coincides with the decrease of muscle glycogen. Marathon runners commonly refer to the sudden onset of fatigue that they experience at 29 to 35 km (18-22 mi) as "hitting the wall." At least part of this sensation can be attributed to muscle glycogen depletion.

Depletion in Different Fiber Types Muscle fibers are recruited and deplete their energy reserves in selected patterns. The individual fibers most frequently recruited during exercise may become depleted of glycogen. This reduces the number of fibers capable of producing the muscular force needed for exercise.

This glycogen depletion is illustrated in figure 4.20, which shows a micrograph of muscle fibers taken from a runner after a 30-km (18.6-mi) run. Figure 4.20*a* has been stained to differentiate ST and FT fibers. One of the FT fibers is circled. Figure 4.20*b* shows a second sample from the same muscle, stained to show glycogen. The redder (darker) the stain, the more glycogen is present. Before the run, all fibers were full of glycogen and appeared red (not depicted). In figure 4.20*b* (after the run), the lighter ST fibers are almost completely depleted of glycogen. This suggests that ST fibers are used more heavily during endurance exercise that requires only moderate force development, such as the 30-km run.

The pattern of glycogen depletion from ST and FT fibers depends on the exercise inten-

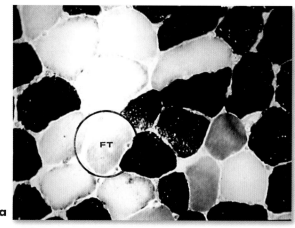

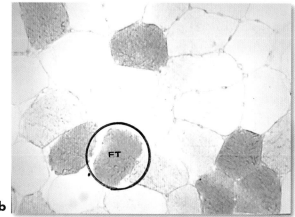

▲ **Figure 4.20**   Histochemical staining for (*a*) fiber type and (*b*) muscle glycogen after a 30-km run. A fast-twitch (FT) fiber is circled in figure 4.20*a*. Note that a number of FT fibers still have glycogen, as noted by their darker stain (figure 4.20*b*), whereas most of the slow-twitch (ST) fibers are empty of glycogen.

sity.[19] Recall that ST fibers are the first fibers to be recruited during light exercise. As muscle tension requirements increase, FT$_a$ fibers are added to the workforce. In exercise approaching maximal intensities, the FT$_b$ fibers are added to the pool of recruited fibers.

Depletion in Different Muscle Groups In addition to selectively depleting glycogen from ST or FT fibers, exercise may place unusually heavy demands on select muscle groups. In one study, subjects ran on a treadmill positioned for uphill, downhill, and level running for 2 h at 70% of $\dot{V}O_2$max. Figure 4.21 compares the resultant glycogen depletion in three muscles of the lower extremity:

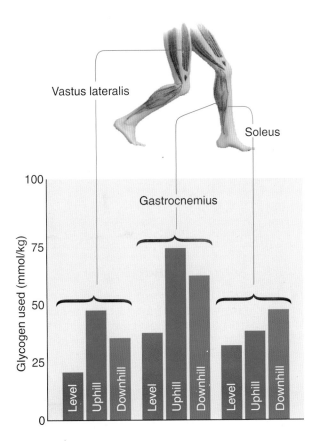

▲ **Figure 4.21**   Muscle glycogen use in the vastus lateralis, gastrocnemius, and soleus muscles during 2 h of level, uphill, and downhill running on a treadmill at 70% of $\dot{V}O_2$max. Note that the greatest glycogen use is in the gastrocnemius during uphill and downhill running.

- Vastus lateralis (knee extensor)
- Gastrocnemius (ankle extensor)
- Soleus (also an ankle extensor)

The results show that whether you run uphill, downhill, or on a level surface, the gastrocnemius uses more glycogen than the vastus lateralis or the soleus. This suggests that the ankle extensor muscles are more likely to become depleted during distance running than are the thigh muscles, isolating the site of fatigue to the lower leg muscles.

Depletion and Blood Glucose   Muscle glycogen alone can't provide enough carbohydrate for exercise lasting several hours. Glucose delivered by the blood to the muscles contributes a lot of energy during endurance exercise. The liver breaks down its stored glycogen to provide a constant supply of blood glucose. In the early stages of exercise, energy production requires relatively little blood glucose, but in the later stages of an endurance event, blood glucose may make a large contribution. To keep pace with the muscles' glucose uptake, the liver must break down increasingly more glycogen as exercise duration increases.

The liver's glycogen stores are limited, and it can't produce glucose rapidly from other substrates. Consequently, blood glucose levels can decrease when muscle uptake exceeds the liver's glucose output. Unable to obtain sufficient glucose from the blood, the muscles must rely more heavily on their glycogen reserves, accelerating muscle glycogen depletion and leading to earlier exhaustion.

Effects of Depletion on Performance   Not surprisingly, endurance performances improve when the muscle glycogen supply is elevated before the start of activity. The importance of muscle glycogen storage for endurance performance is discussed in chapter 14. For now, note that glycogen depletion and hypoglycemia (low blood sugar) limit performance in activities lasting 30 min or longer. Fatigue in shorter events more likely results from accumulation of metabolic by-products, such as lactate and H^+, within the muscles.

Metabolic By-Products and Fatigue

Recall that lactic acid is a by-product of anaerobic glycolysis. Although most people believe that lactic acid is responsible for fatigue and exhaustion in all types of exercise, lactic acid accumulates within the muscle fiber only during relatively brief, highly intense muscular effort. Marathon runners, for example, may have near-resting lactic acid levels at the end of the race, despite their exhaustion. As noted in the previous section, their fatigue is caused by inadequate energy supply, not excess lactic acid.

Sprints in running, cycling, and swimming all lead to large accumulations of lactic acid.

But the presence of lactic acid should not be blamed for the feeling of fatigue in itself. When not cleared, the lactic acid dissociates, converting to lactate and causing an accumulation of hydrogen ions. This H^+ accumulation causes muscle acidification, resulting in a condition known as acidosis.

Activities of short duration and high intensity, such as sprint running and sprint swimming, depend heavily on anaerobic glycolysis and produce large amounts of lactate and H^+ within the muscles. Fortunately, the cells and body fluids possess buffers, such as bicarbonate (HCO_3), that minimize the disrupting influence of the H^+. Without these buffers, H^+ would lower the pH to about 1.5, killing the cells. Because of the body's buffering capacity, the H^+ concentration remains low even during the most severe exercise, allowing muscle pH to decrease from a resting value of 7.1 to no lower than 6.6 to 6.4 at exhaustion.

However, pH changes of this magnitude adversely affect energy production and muscle contraction. An intracellular pH below 6.9 inhibits the action of phosphofructokinase, an important glycolytic enzyme, slowing the rate of glycolysis and ATP production. At a pH of 6.4, the influence of H^+ stops any further glycogen breakdown, causing a rapid decrease in ATP and ultimately exhaustion. In addition, H^+ may displace calcium within the fiber, interfering with the coupling of the actin–myosin cross-bridges and decreasing the muscle's contractile force. Most researchers agree that low muscle pH is the major limiter of performance and the primary cause of fatigue during maximal, all-out exercise lasting more than 20 to 30 s.

As seen in figure 4.22, reestablishing the preexercise muscle pH after an exhaustive sprint bout requires about 30 to 35 min of recovery. Even when normal pH is restored, blood and muscle lactate levels can remain quite elevated. However, experience has shown that an athlete can continue to exercise at relatively high intensities even with a muscle pH below 7.0 and a blood lactate level

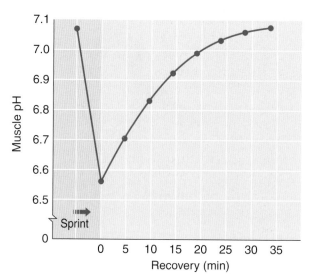

▲ **Figure 4.22** Changes in muscle pH during sprint exercise and recovery. Note the drastic decrease in muscle pH during the sprint and the gradual recovery to normal after the effort. Note that it took more than 30 min for pH to return to its pre-exercise level.

above 6 or 7 mmol/L, four to five times the resting value.

Currently, some coaches and sport physiologists are attempting to use blood lactate measurements to gauge the intensity and volume of training needed to produce an optimal training stimulus. Such measurements provide an index of training intensity, but they might not be related to the anaerobic processes or the state of acidosis in the muscles. Because lactate and H^+ are generated in the muscles, both diffuse out of the cells. They then are diluted in the body fluids and transported to other areas of the body to be metabolized. Consequently, blood lactate values depend on the rates of production, diffusion, and oxidation. A variety of factors can influence these processes, so lactate values are of questionable value when prescribing training.

Neuromuscular Fatigue

Thus far we have considered only factors within the muscle that might be responsible for fatigue. Evidence also suggests that under some circumstances, fatigue may result from an inability to activate the muscle fibers, a function of the nervous system. As noted in chapter 2, the nerve impulse is transmitted across the neuromuscular junction to activate the fiber's membrane, and it causes the fiber's sarcoplasmic reticulum to release calcium. The calcium, in turn, binds with troponin to initiate muscle contraction. Let's examine two possible neural mechanisms that could disrupt this process and possibly contribute to fatigue.

Neural Transmission

Fatigue may occur at the neuromuscular junction, preventing nerve impulse transmission to the muscle fiber membrane. Studies in the early 1900s clearly established such a failure of nerve impulse transmission in fatigued muscle. This failure may involve one or more of the following processes:

- The release or synthesis of acetylcholine (ACh), the neurotransmitter that relays the nerve impulse from the motor nerve to the muscle membrane, might be reduced.

- Cholinesterase, the enzyme that breaks down ACh once it has relayed the impulse, might become hyperactive, preventing sufficient concentration of ACh to initiate an action potential.

- Cholinesterase activity might become hypoactive (inhibited), allowing ACh to accumulate excessively, paralyzing the fiber.

- The muscle fiber membrane might develop a higher threshold.

- Some substance might compete with ACh for the receptors on the muscle membrane without activating the membrane.

- Potassium might leave the intracellular space of the contracting muscle, decreasing the membrane potential to half of its resting value.

Although most of these causes for a neuromuscular block have been associated with neuromuscular diseases (such as myasthenia gravis), they may also cause some forms of neuromuscular fatigue. Some evidence suggests that fatigue also may be attributable to calcium retention within the sarcoplasmic reticulum, which would decrease the calcium available for muscle contraction. In fact, depletion of PCr and lactate buildup might simply increase the rate of calcium accumulation within the sarcoplasmic reticulum. However, these theories of fatigue remain speculative.

Central Nervous System

The central nervous system (CNS) also might be a site of fatigue, although there is evidence both for and against this theory. Early studies showed that when a subject's muscles appeared to be nearly exhausted, verbal encouragement, shouting, or even direct electrical stimulation of the muscle could increase the strength of muscle contraction. These studies suggest that the limits of performance in exhaustive exercise may, to a great extent, be psychological. The precise mechanisms underlying such CNS fatigue are not fully understood. Whether this form of fatigue is isolated to the CNS or linked to peripheral nerve transmission is also difficult to determine.

The recruitment of muscle depends, in part, on conscious control. The psychological trauma of exhaustive exercise may consciously or subconsciously inhibit the athlete's willingness to tolerate further pain. The CNS may slow the exercise pace to a tolerable level to protect the athlete. Indeed, researchers generally agree that the perceived discomfort of fatigue precedes the onset of a physiological limitation within the muscles. Unless they are highly motivated, most individuals terminate exercise before their muscles are physiologically exhausted. To achieve peak performance, athletes train to learn proper pacing and tolerance for fatigue.

▶ Fatigue may result from depletion of PCr or glycogen; both situations impair ATP production.

▶ Lactic acid often has been blamed for fatigue, but it is actually the H^+ generated by lactic acid that lead to fatigue. The accumulation of H^+ decreases muscle pH, which impairs the cellular processes that produce energy and muscle contraction.

▶ Failure of neural transmission may be a cause of some fatigue. Many mechanisms can lead to such failure, and all need further research.

▶ The CNS also may cause fatigue, perhaps as a protective mechanism. Perceived fatigue usually precedes physiological fatigue, and athletes who feel exhausted often can be psychologically encouraged to continue.

In Closing . . .

In previous chapters, we discussed how muscles and the nervous system function together to produce movement. In this chapter we focused on metabolism. We considered the energy needed for movement. We saw how energy is stored in the form of ATP, examined the three systems that generate energy, and explored how energy production and availability can limit your performance. We also learned that your metabolic needs vary considerably. In the next chapter, we turn our attention to the regulation of metabolism as we focus on the endocrine system and hormonal regulation.

▶ Key Terms

acetyl coenzyme A (acetyl CoA)
adenosine diphosphate (ADP)
adenosine triphosphatase (ATPase)
adenosine triphosphate (ATP)
aerobic metabolism
anaerobic

anaerobic metabolism
anaerobic glycolysis
ATP–PCr system
basal metabolic rate (BMR)
β-oxidation
calorie (cal)
calorimeter
carbohydrate
direct calorimetry
electron transport chain
excess postexercise oxygen consumption (EPOC)
exhaustion
fatigue
free fatty acid (FFA)
gluconeogenesis
glycogen
glycogenesis
glycogenolysis
glycolysis
glycolytic system
Haldane transformation
indirect calorimetry
kilocalorie (kcal)
Krebs cycle
lactate
lactate threshold
lipogenesis
lipolysis
maximal oxygen uptake ($\dot{V}O_2$max)
oxidative capacity of muscle ($\dot{Q}O_2$)
oxidative system
phosphocreatine (PCr)
respiratory exchange ratio (RER)
resting metabolic rate (RMR)
$\dot{V}O_2$ drift

▶ Study Questions

1. What is the role of PCr in energy production?

2. Describe the relationship between muscle ATP and PCr during sprint exercise.

3. Why are the ATP–PCr and glycolytic energy systems considered anaerobic?

4. What role does oxygen play in the process of aerobic metabolism?

5. Describe the by-products of energy production from ATP–PCr, glycolysis, and oxidation.

6. What is the respiratory exchange ratio (RER)? Explain how it is used to determine the oxidation of carbohydrate and fat.

7. What is the relationship between oxygen consumption and energy production?

8. What is the lactate threshold?

9. How can we use measurements of oxygen consumption to estimate exercise efficiency?

10. Why do athletes with high $\dot{V}O_2$max values perform better in endurance events than those with lower values?

11. Why is oxygen consumption often expressed as milliliters of oxygen per kilogram of body weight per minute ($ml \cdot kg^{-1} \cdot min^{-1}$)?

12. Describe the possible causes of fatigue during exercise bouts lasting 15 to 30 s and 2 to 4 h.

▷ References

1. Åstrand, P.-O., & Rodahl, K. (1986). *Textbook of work physiology: Physiological bases of exercise* (3rd ed.). New York: McGraw-Hill.

2. Bar-Or, O. (1987). The Wingate Anaerobic Test: An update on methodology, reliability and validity. *Sports Medicine, 4,* 381-394.

3. Barstow, T.J., Jones, A.M., Nguyen, P.H., & Casaburi, R. (1996). Influence of muscle fiber type and pedal frequency on oxygen uptake kinetics of heavy exercise. *Journal of Applied Physiology, 81,* 1642-1650.

4. Costill, D.L. (1970). Metabolic responses during distance running. *Journal of Applied Physiology, 28,* 251-255.

5. Costill, D.L. (1986). *Inside running: Basics of sports physiology.* Indianapolis: Benchmark Press.

6. Costill, D.L., Daniels, J., Evans, W., Fink, W., Krahenbuhl, G., & Saltin, B. (1976). Skeletal muscle enzymes and fiber composition in male and female track athletes. *Journal of Applied Physiology, 40,* 149-154.

7. Costill, D.L., Fink, W.J., Flynn, M., & Kirwan, J. (1987). Muscle fiber composition and enzyme activities in elite female distance runners. *International Journal of Sports Medicine, 8,* 103-106.

8. Costill, D.L., & Fox, E.L. (1969). Energetics of marathon running. *Medicine and Science in Sports, 1*(2), 81-86.

9. Gaesser, G.A., & Poole, D.C. (1996). The slow component of oxygen uptake kinetics in humans. *Exercise and Sport Sciences Reviews, 24,* 35-70.

10. Gastin, P.B. (2001). Energy system interaction and relative contribution during maximal exercise. *Sports Medicine, 31,* 725-741.

11. Gollnick, P.D., Armstrong, R., Saubert, C., Piehl, K., & Saltin, B. (1972). Enzyme activity and fiber composition in skeletal muscle of untrained and trained men. *Journal of Applied Physiology, 33,* 312-319.

12. Hill, D.W. (1993). The critical power concept: A review. *Sports Medicine, 16,* 237-254.

13. Ivy, J.L., Withers, R.T., Van Handel, P.J., Elger, D.H., & Costill, D.L. (1980). Muscle respiratory capacity and fiber type as determinants of the lactate threshold. *Journal of Applied Physiology, 48,* 523-527.

14. Jacobs, I. (1986). Blood lactate implications for training and sports performance. *Sports Medicine, 3,* 10-25.

15. Kalis, J.K., Freund, B.J., Joyner, M.J., Jilka, S.M., Nittolo, J., & Wilmore, J.H. (1988). Effect of β-blockade on the drift in O_2 consumption during prolonged exercise. *Journal of Applied Physiology, 64,* 753-758.

16. Margaria, R., Cerretilli, P., Aghemo, P., & Sassi, G. (1963). Energy cost of running. *Journal of Applied Physiology, 18,* 367-370.

17. Medbø, J.I., Mohn, A.C., Tabata, I., Bahr, R., Vaage, O., & Sejersted, O.M. (1988). Anaerobic capacity determined by maximal accumulated O_2 deficit. *Journal of Applied Physiology, 64,* 50-60.

18. Poole, D.C., & Richardson, R.S. (1997). Determinants of oxygen uptake. *Sports Medicine, 24,* 308-320.

19. Vøllestad, N.K., & Blom, P.C.S. (1988). Effect of varying exercise intensity on glycogen depletion in human muscle fibers. *Acta Physiologica Scandinavica, 125,* 395-405.

20. Westerblad, H., Allen, D.G., & Lännergren, J. (2002). Muscle fatigue: Lactic acid or inorganic phosphate the major cause? *News in the Physiological Sciences, 17,* 17-21.

21.   Zuntz, N., & Hagemann, O. (1898). *Untersuchungen uber den Stroffwechsel des Pferdes bei Ruhe und Arbeit.* Berlin: Parey.

▷ Selected Readings

Bangsbo, J. (1998). Quantification of anaerobic energy production during intense exercise. *Medicine and Science in Sports and Exercise, 30,* 47-52.

Bergstrom, J. (1967). Local changes of ATP and phosphocreatine in human muscle tissue in connection with exercise. In *Physiology of muscular exercise* (Monograph No. 15, pp. 191-196). New York: American Heart Association.

Biolo, G., Tipton, K.D., Klein, S., & Wolfe, R.R. (1997). An abundant supply of amino acids enhances the metabolic effect of exercise on muscle protein. *American Journal of Physiology, 273,* E122-E129.

Blom, P., Vøllestad, N.K., & Costill, D.L. (1986). Factors affecting changes in muscle glycogen concentration during and after prolonged exercise. *Acta Physiologica Scandinavica, 128*(Suppl. 556), 67-74.

Bonen, A., Baker, S.K., & Hatta, H. (1997). Lactate transport and lactate transporters in skeletal muscle. *Canadian Journal of Applied Physiology, 22*(6), 531-552.

Brooks, G.A. (1987). Amino acid and protein metabolism during exercise and recovery. *Medicine and Science in Sports and Exercise, 19*(5), S150-S156.

Coggan, A.R., & Coyle, E.F. (1991). Carbohydrate ingestion during prolonged exercise: Effects on metabolism and performance. *Exercise and Sport Sciences Reviews, 19,* 1-40.

Costill, D.L., Coyle, E., Dalsky, G., Evans, W., Fink, W., & Hoopes, D. (1977). Effects of elevated plasma FFA and insulin on muscle glycogen usage during exercise. *Journal of Applied Physiology, 43,* 695-699.

Costill, D.L., Gollnick, P.D., Jansson, E.D., Saltin, B., & Stein, E.M. (1973). Glycogen depletion pattern in human muscle fibers during distance running. *Acta Physiologica Scandinavica, 89,* 374-383.

Costill, D.L., Jansson, E., Gollnick, P.D., & Saltin, B. (1974). Glycogen utilization in leg muscles of men during level and uphill running. *Acta Physiologica Scandinavica, 91,* 475-481.

Favero, T.G. (1999). Sarcoplasmic reticulum Ca^{2+} release and muscle fatigue. *Journal of Applied Physiology, 87,* 471-483.

Fitts, R.H. (1994). Cellular mechanisms of muscle fatigue. *Physiological Reviews, 74,* 49-84.

Gladden, L.B. (2000). Muscle as a consumer of lactate. *Medicine and Science in Sports and Exercise, 32,* 764-771.

Henriksson, J., & Reitman, J. (1976). Quantitative measures of enzyme activities in type I and type II muscle fibers of man after training. *Acta Physiologica Scandinavica, 97,* 392-397.

Holloszy, J.O., & Hansen, P.A. (1996). Regulation of glucose transport into skeletal muscle. *Reviews of Physiology, Biochemistry and Pharmacology, 129,* 99-193.

Holloszy, J.O., & Kohrt, W.M. (1996). Regulation of carbohydrate and fat metabolism during and after exercise. *Annual Review of Nutrition, 16,* 121-138.

Hultman, E. (1995). Fuel selection, muscle fibre. *Proceedings of the Nutrition Society, 54,* 107-121.

Ingjer, F. (1991). Maximal oxygen uptake as a predictor of performance ability in women and men elite cross-country skiers. *Scandinavian Journal of Medicine and Science in Sports, 1,* 25-30.

Ivy, J.L., Costill, D.L., & Maxwell, B.D. (1980). Skeletal muscle determinants of maximum aerobic power in man. *European Journal of Applied Physiology, 44*(1), 1-8.

Katz, A., & Sahlin, K. (1990). Role of oxygen in regulation of glycolysis and lactate production in human skeletal muscle. *Exercise and Sport Sciences Reviews, 18,* 1-28.

MacIntosh, B.R., & Rassier, D.E. (2002). What is fatigue? *Canadian Journal of Applied Physiology, 27,* 42-55.

Murgatroyd, P.R., Shetty, P.S., & Prentice, A.M. (1993). Techniques for the measurement of human energy expenditure: A practical guide. *International Journal of Obesity, 17,* 549-568.

O'Brien, M.J., Viguie, C.A., Mazzeo, R.S., & Brooks, G.A. (1993). Carbohydrate dependence during marathon running. *Medicine and Science in Sports and Exercise, 25,* 1009-1017.

Phillips, S.M., Tipton, K.D., Aarsland, A., Wolf, S.E., & Wolfe, R.R. (1997). Mixed muscle protein synthesis and breakdown after resistance exercise in humans. *American Journal of Physiology, 273,* E99-E107.

Richardson, R.S. (1998). Oxygen transport: Air to muscle cell. *Medicine and Science in Sports and Exercise, 30,* 53-59.

Romijn, J.A., Coyle, E.F., Sidossis, L.S., Zhang, X.-J., & Wolfe, R.R. (1995). Relationship between fatty acid delivery and fatty acid oxidation during strenuous exercise. *Journal of Applied Physiology, 79,* 1939-1945.

Scott, C.B., Roby, F.B., Lohman, T.B., & Bunt, J.C. (1991). The maximally accumulated oxygen deficit as an indicator of anaerobic capacity. *Medicine and Science in Sports and Exercise, 23*(5), 618-624.

Sjøgaard, G. (1991). Role of exercise-induced potassium fluxes underlying muscle fatigue: A brief review. *Canadian Journal of Physiology and Pharmacology, 69*, 238-245.

St. Clair Gibson, A., Lambert, M.I., & Noakes, T.D. (2001). Neural control of force output during maximal and submaximal exercise. *Sports Medicine, 31*, 637-650.

Tipton, K.D., Ferrando, A.A., Williams, B.D., & Wolfe, R.R. (1996). Muscle protein metabolism in female swimmers after a combination of resistance and endurance exercise. *Journal of Applied Physiology, 81*, 2034-2038.

Vandewalle, G.P., & Monod, H. (1987). Standard anaerobic exercise tests. *Sports Medicine, 4*, 268-289.

CHAPTER 5

HORMONAL REGULATION OF EXERCISE

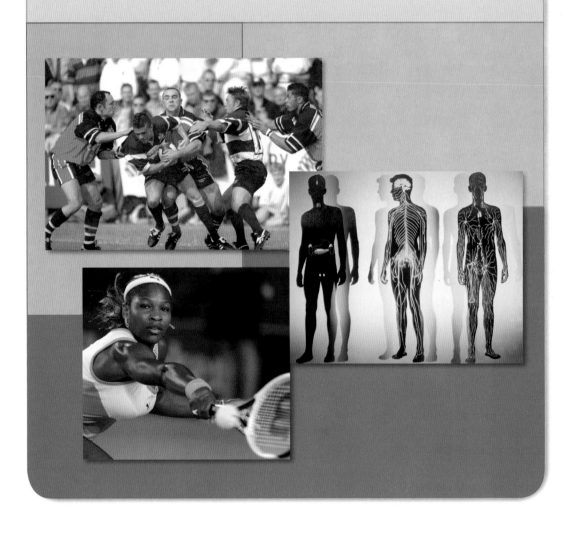

overview

During exercise, your body is faced with tremendous demands that lead to many physiological changes. The rate of energy use increases. Metabolic by-products that must be cleared often begin to accumulate. Water shifts between the fluid compartments and is lost through sweating. Even at rest, your body's internal environment is in a constant state of flux. But during exercise, it faces major changes that occur rapidly and frequently.

Yet we know that homeostasis must be maintained for you to survive. The more rigorous the exercise, the more difficult this maintenance becomes. Much of the regulation required during exercise is accomplished by the nervous system. But another system is in contact with virtually every cell in your body. It constantly monitors your body's internal milieu, noting all changes that occur and responding quickly to ensure that homeostasis is not drastically disrupted. It is your endocrine system, which exerts its control through the hormones it releases. In this chapter, we focus on the importance of your endocrine system in maintaining homeostasis even amidst all the internal chaos created during physical activity.

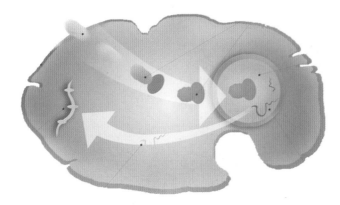

outline

In 1972, I (DLC) convinced my coauthor (JHW) to run 16 km (10 mi) per day for five successive days. To add to the physical stress, these runs were performed in the bright summer sun and heat (approximately 30-35° C, or 86-95° F) of Davis, California. We each lost 3 to 4 kg (6.6-8.8 lb) of sweat daily, leaving us somewhat dehydrated and overheated. Blood samples taken each morning during this period revealed that our hemoglobin levels and hematocrits (percentage of blood composed of red blood cells) were declining. Nevertheless, isotope studies of our blood revealed that we were not losing hemoglobin or blood cells. Rather, our plasma, the fluid part of our blood, was increasing day by day. Thus, it appeared that our bodies were trying to offset the detrimental effects of daily dehydration by retaining water to minimize the loss of plasma volume that accompanied such heavy sweating. But how did our bodies know that we needed to expand our plasma? What was responsible for the water retention? We now know that at least three hormones—aldosterone, renin, and antidiuretic hormone—all function to maintain appropriate plasma volume and minimize the risk of dehydration.

Muscular activity requires coordinated integration of many physiological and biochemical systems. Such integration is possible only if your body's various tissues and systems can communicate with each other. Although your nervous system is responsible for much of this communication, fine-tuning your body's physiological responses to any disturbance of its equilibrium is primarily the responsibility of your endocrine system. The endocrine and nervous systems work in concert to initiate and control movement and all physiological processes that movement involves. The nervous system functions quickly, having short-lived, localized effects, whereas the endocrine system functions much more slowly, having longer lasting and more general effects.

The endocrine system includes all tissues or glands that secrete **hormones.** The major endocrine glands are illustrated in figure 5.1. Endocrine glands secrete their hormones directly into the blood. Hormones act as chemical signals throughout the body. When secreted by the specialized endocrine cells, hormones are transported via the blood to specific **target cells**—cells that possess specific hormone receptors. On reaching their destinations, hormones can control the activity of the target tissue. A unique feature of hormones is that they travel away from the cells that secrete them and specifically affect the activities of other cells and organs. Some hormones affect many body tissues, whereas others target specific cells of the body.

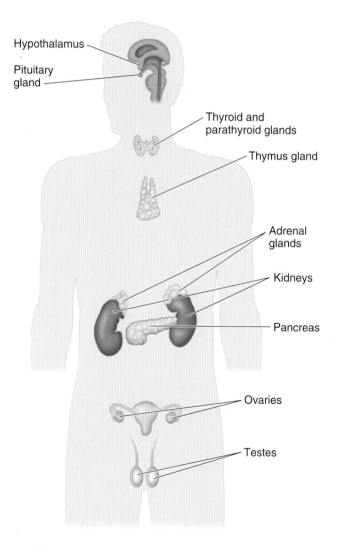

Hypothalamus

Pituitary gland

Thyroid and parathyroid glands

Thymus gland

Adrenal glands

Kidneys

Pancreas

Ovaries

Testes

▲ **Figure 5.1**   Locations of the major endocrine organs.

Nature of Hormones

Hormones are involved in most physiological processes, so their actions are relevant to many aspects of exercise and sport performance. Before examining the roles played by specific hormones, we need a better understanding of the nature of these substances. In the following sections, we examine the chemical nature of hormones and their general mechanisms of action.

Chemical Classification of Hormones

Hormones can be categorized as two basic types: steroid hormones and nonsteroid hormones. **Steroid hormones** have a chemical structure similar to cholesterol, and most are derived from cholesterol. For this reason, they are lipid soluble and diffuse rather easily through cell membranes. This group includes the hormones secreted by

- the adrenal cortex (such as cortisol and aldosterone),
- the ovaries (estrogen and progesterone),
- the testes (testosterone), and
- the placenta (estrogen and progesterone).

Nonsteroid hormones are not lipid soluble, so they cannot easily cross cell membranes. The nonsteroid hormone group can be subdivided into two groups: protein or peptide hormones and amino acid–derivative hormones. The two hormones from the thyroid gland (thyroxine and triiodothyronine) and the two from the adrenal medulla (epinephrine and norepinephrine) are amino acid hormones. All other nonsteroid hormones are protein or peptide hormones.

Hormone Actions

Because hormones travel in the blood, they contact virtually all body tissues. How, then, can they limit their effects to specific targets? This ability is attributable to the specific hormone receptors possessed by the target tissues.

The interaction between the hormone and its specific receptor has been compared with a lock (receptor) and key (hormone) arrangement, in which only the correct key can unlock a given action within the cells (see figures 5.2 and 5.3). The combination of a hormone bound to its receptor is referred to as a hormone–receptor complex.

Each cell typically has from 2,000 to 10,000 receptors. Receptors for nonsteroid hormones are located on the cell membrane, whereas those for steroid hormones are found either in the cell's cytoplasm or in its nucleus. Each hormone is usually highly specific for a single type of receptor and binds only with its specific receptors, thus affecting only tissues that contain those specific receptors.

Numerous mechanisms allow hormones to control the actions of cells. Let's examine the primary modes of action for both the steroid and the nonsteroid hormones.

Steroid Hormones

As mentioned earlier, steroid hormones are lipid soluble and thus pass easily through the cell membrane. Their mechanism of action is illustrated in figure 5.2. Once inside the cell, a steroid hormone binds to its specific receptors. The hormone–receptor complex then enters the nucleus, binds to part of the cell's DNA, and activates certain genes. This process is referred to as **direct gene activation**. In response to this activation, mRNA is synthesized within the nucleus. The mRNA then enters the cytoplasm and promotes protein synthesis. These proteins may be

- enzymes that can have numerous effects on cellular processes,
- structural proteins to be used for tissue growth and repair, or
- regulatory proteins that can alter enzyme function.

Nonsteroid Hormones

Because nonsteroid hormones cannot cross the cell membrane, they react with specific receptors outside the cell, on the cell membrane. A nonsteroid hormone molecule binds

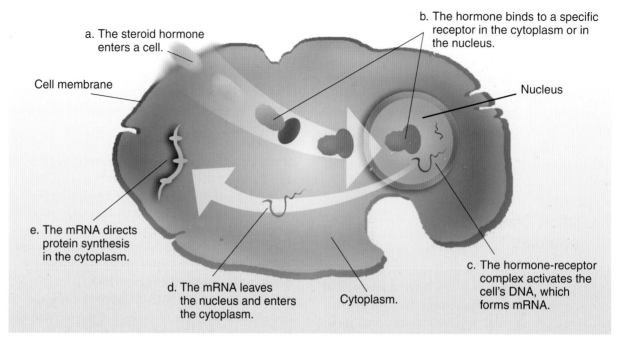

a. The steroid hormone enters a cell.

b. The hormone binds to a specific receptor in the cytoplasm or in the nucleus.

Cell membrane

Nucleus

e. The mRNA directs protein synthesis in the cytoplasm.

d. The mRNA leaves the nucleus and enters the cytoplasm.

Cytoplasm.

c. The hormone-receptor complex activates the cell's DNA, which forms mRNA.

▲ **Figure 5.2**   The mechanism of action of a steroid hormone, leading to direct gene activation.

to its receptor and triggers a series of enzymatic reactions that lead to the formation of an intracellular **second messenger**. The most studied and most widely distributed second messenger is **cyclic adenosine monophosphate** (cyclic AMP, or **cAMP**). This mechanism of action is depicted in figure 5.3. In this case, attachment of the hormone to the appropriate membrane receptor activates an enzyme, adenylate cyclase, situated within the cell membrane. This enzyme catalyzes the formation of cAMP from cellular adenosine triphosphate (ATP). cAMP then can produce specific physiological responses, which can include

- activation of cellular enzymes,
- change in membrane permeability,
- promotion of protein synthesis,
- change in cellular metabolism, or
- stimulation of cellular secretions.

Thus, nonsteroid hormones typically activate the cAMP system of the cell, which then alters intracellular functions.

> Hormones influence specific target tissues or cells through the unique interaction between the hormone and the specific receptors for that hormone on the cell membrane or within the cell.

Control of Hormone Release

Hormones appear to be released in relatively brief bursts, so plasma levels of specific hormones fluctuate over short periods such as an hour or less. But these levels also fluctuate over longer periods of time, showing daily or even monthly cycles (such as monthly menstrual cycles). How do endocrine glands know when to release their hormones?

Negative Feedback

Most hormone secretion is regulated by a **negative feedback system**. Secretion of a hormone causes some change in the body, and this change in turn inhibits further hormone secretion. Consider how a home thermostat works. When the room temperature decreases below some preset level, the thermostat signals

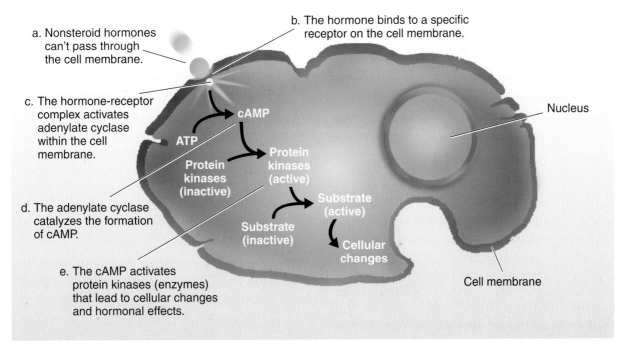

a. Nonsteroid hormones can't pass through the cell membrane.

b. The hormone binds to a specific receptor on the cell membrane.

c. The hormone-receptor complex activates adenylate cyclase within the cell membrane.

d. The adenylate cyclase catalyzes the formation of cAMP.

e. The cAMP activates protein kinases (enzymes) that lead to cellular changes and hormonal effects.

ATP

cAMP

Protein kinases (inactive)

Protein kinases (active)

Substrate (inactive)

Substrate (active)

Cellular changes

Nucleus

Cell membrane

▲ **Figure 5.3**   The mechanism of action of a nonsteroid hormone, using a second messenger (cyclic adenosine monophosphate, or cAMP) within the cell.

Prostaglandins

Prostaglandins, although technically not hormones, are often considered to be a third class of hormones. These substances are derived from a fatty acid, arachidonic acid, and they are associated with the plasma membranes of almost all body cells. Prostaglandins typically act as local hormones, exerting their effects in the immediate area where they are produced. But some also survive long enough to circulate through the blood to affect distant tissues. Prostaglandin release can be triggered by many stimuli, such as other hormones or a local injury. Their functions are quite numerous because there are several different types of prostaglandins. They often mediate the effects of other hormones. They are also known to act directly on blood vessels, increasing vascular permeability (which promotes swelling) and vasodilation. In this capacity, they are important mediators of the inflammatory response. They also sensitize the nerve endings of pain fibers; thus, they promote both inflammation and pain.

the furnace to produce heat. When the room temperature increases to the preset level, the thermostat's signal ends, and the furnace stops producing heat. When the temperature again falls below the preset level, the cycle begins anew. In the body, secretion of a specific hormone is similarly turned on or off by specific physiological changes.

Negative feedback is the primary mechanism through which your endocrine system maintains homeostasis. Let's consider the case of plasma glucose levels and the hormone insulin. When the plasma glucose concentration is high, the pancreas releases insulin. Insulin increases cellular uptake of glucose, lowering plasma concentration of glucose. When plasma glucose concentration returns to normal, insulin release is inhibited until the plasma glucose level increases again.

Number of Receptors

The plasma levels of specific hormones are not always the best indicators of actual hormone activity because the number of receptors on a cell can be altered to increase or decrease that cell's sensitivity to a certain hormone. Most commonly, an increased amount of a specific hormone decreases the number of cell receptors available to it. When this happens, the cell becomes less sensitive to that hormone, because with fewer receptors, less hormone can bind. This is referred to as **down-regulation,** or desensitization. In some people with obesity, for example, the number of insulin receptors on their cells appears to be reduced. Their bodies respond by increasing insulin secretion from the pancreas, so their plasma insulin levels increase. To obtain the same degree of plasma glucose control as normal, healthy people, these individuals must release much more insulin.

In a few instances, a cell may respond to the prolonged presence of large amounts of a hormone by increasing its number of available receptors. When this happens, the cell becomes more sensitive to that hormone because more can be bound at one time. This is referred to as **up-regulation.** In addition, one hormone occasionally can regulate the receptors for another hormone.

Endocrine Glands and Their Hormones

Now that we have discussed the general nature of hormones, we are prepared to look at specific hormones and their functions. The endocrine glands and their respective hormones are listed in table 5.1 on pages 166-167. This table also lists each hormone's target and actions. Keep in mind that the endocrine system is extremely complex. The presentation here has been greatly simplified to focus on those hormones of greatest importance to sport and physical activity. Realize also that research on hormone actions during exercise is limited, and

- ▶ Hormones can be classified as either steroid or nonsteroid. Steroid hormones are lipid soluble, and most are formed from cholesterol. Nonsteroid hormones are formed from proteins, peptides, or amino acids.

- ▶ Hormones generally are secreted into the blood and then circulate through the body to affect only their target cells. They act by binding in a lock-and-key manner with specific receptors found only in the target tissues.

- ▶ Steroid hormones pass through cell membranes and bind to receptors inside the cell. They use a mechanism called direct gene activation to cause protein synthesis.

- ▶ Nonsteroid hormones cannot enter the cells easily, so they bind to receptors on the cell membrane. This activates a second messenger within the cell, which in turn can trigger numerous cellular processes.

- ▶ A negative feedback system regulates secretion of most hormones.

- ▶ The number of receptors for a specific hormone can be altered to meet the body's demands. Up-regulation refers to an increase in receptors, and down-regulation is a decrease. These two processes change cell sensitivity to hormones.

available information is often conflicting. Much remains to be learned in this area.

Pituitary Gland

The pituitary gland, also referred to as the hypophysis, is a marble-sized gland at the base of the brain. This gland was at one time considered the human body's master gland because it secretes a number of hormones that affect a wide variety of other glands and organs. However, the secretory action of the pituitary itself is controlled by either neural mechanisms or other hormones secreted by the hypothalamus. Therefore, the pituitary gland is perhaps more appropriately thought of as the relay between central nervous system control centers and peripheral endocrine glands.

> The pituitary gland was once thought to be the master endocrine gland controlling many other glands and organs. It is now recognized that the pituitary gland is largely controlled by the hypothalamus.

The pituitary gland is composed of three lobes: anterior, intermediate, and posterior (figure 5.4). The intermediate lobe is very small and is thought to play little or no role in humans, but both the posterior and anterior lobes have major endocrine functions.

Posterior Lobe of the Pituitary Gland

The pituitary's posterior lobe is an outgrowth of neural tissue from the hypothalamus. For this reason, it is also referred to as the neurohypophysis. It secretes two hormones: antidiuretic hormone (ADH, or vasopressin) and oxytocin. But these hormones are actually produced in the hypothalamus. They travel through the neural tissue and are stored in vesicles within nerve endings in the posterior pituitary. These hormones are released into capillaries as needed in response to neural impulses from the hypothalamus.

Of the two posterior pituitary hormones, only ADH is known to play an important role in exercise. ADH promotes water conservation by increasing the water permeability of the kidneys' collecting ducts. As a result, less water is excreted in the urine.

Little is known about the effects of exercise on the secretions from the posterior lobe of the pituitary gland. We do know, however, that injecting a concentrated electrolyte solution into the blood causes the release of large quantities of ADH from this gland. This hormone's role in conserving body water minimizes the risk of dehydration during periods of heavy sweating and hard exercise.

Muscular activity and sweating cause electrolytes to become concentrated in the blood plasma. This is called hemoconcentration, and it increases the plasma osmolarity (the ionic concentration of dissolved substances in the plasma). This is the primary physiological stimulus for ADH release. The increased osmolarity is sensed by osmoreceptors in the hypothalamus. In response, the hypothalamus sends neural impulses to the posterior pituitary, stimulating ADH release. The ADH enters the blood, travels to the kidneys, and promotes wa-

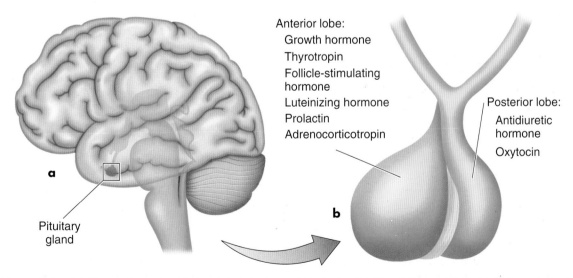

Anterior lobe:
Growth hormone
Thyrotropin
Follicle-stimulating hormone
Luteinizing hormone
Prolactin
Adrenocorticotropin

Posterior lobe:
Antidiuretic hormone
Oxytocin

Pituitary gland

▲ **Figure 5.4**   (*a*) The pituitary gland is located at the base of the brain. (*b*) The pituitary is divided into the anterior lobe, which secretes six different hormones; the intermediate lobe, which seems to have little function in humans; and the posterior lobe, which secretes antidiuretic hormone and oxytocin.

Table 5.1

The Endocrine Glands, Their Hormones, Target Organs, Controlling Factors, and Functions

Endocrine gland	Hormone	Target organ	Controlling factor	Major functions
Anterior pituitary	Growth hormone (GH)	All cells in the body	Hypothalamic GH-releasing hormone; GH-inhibiting hormone (somatostatin)	Promotes development and enlargement of all body tissues until maturation; increases rate of protein synthesis; increases mobilization of fats and use of fat as an energy source; decreases rate of carbohydrate use
	Thyrotropin (TSH)	Thyroid gland	Hypothalamic TSH-releasing hormone	Controls the amount of thyroxin and triiodothyronine produced and released by the thyroid gland
	Adrenocorticotropin (ACTH)	Adrenal cortex	Hypothalamic ACTH-releasing hormone	Controls the secretion of hormones from the adrenal cortex
	Prolactin	Breasts	Prolactin-releasing and inhibiting hormones	Stimulates milk production by the breasts
	Follicle-stimulating hormone (FSH)	Ovaries, testes	Hypothalamic FSH-releasing hormone	Initiates growth of follicles in the ovaries and promotes secretion of estrogen from the ovaries; promotes development of the sperm in the testes
	Luteinizing hormone (LH)	Ovaries, testes	Hypothalamic FSH-releasing hormone	Promotes secretion of estrogen and progesterone and causes the follicle to rupture, releasing the ovum; causes testes to secrete testosterone
Posterior pituitary	Antidiuretic hormone (ADH or vasopressin)	Kidneys	Hypothalamic secretory neurons	Assists in controlling water excretion by the kidneys; elevates blood pressure by constricting blood vessels
	Oxytocin	Uterus, breasts	Hypothalamic secretory neurons	Controls contraction of uterus; milk secretion
Thyroid	Thyroxine (T_4) and triiodothyronine (T_3)	All cells in the body	TSH and T_3 and T_4 concentrations	Increase the rate of cellular metabolism; increase rate and contractility of the heart
	Calcitonin	Bones	Plasma calcium concentrations	Controls calcium ion concentration in the blood
Parathyroid	Parathyroid hormone (PTH or parathormone)	Bones, intestines, and kidneys	Plasma calcium concentrations	Controls calcium ion concentration in the extracellular fluid through its influence on bones, intestines, and kidneys

Endocrine gland	Hormone	Target organ	Controlling factor	Major functions
Adrenal medulla	Epinephrine	Most cells in the body	Baroreceptors, glucose receptors, brain and spinal centers	Stimulates breakdown of glycogen in liver and muscle and lipolysis in adipose tissue and muscle; increases skeletal muscle blood flow; increases heart rate and contractility; increases oxygen consumption
	Norepinephrine	Most cells in the body	Baroreceptors, glucose receptors, brain and spinal centers	Stimulates lipolysis in adipose tissue and in muscle to a lesser extent; constricts arterioles and venules, thereby elevating blood pressure
Adrenal cortex	Mineralocorticoids (aldosterone)	Kidneys	Angiotensin and plasma potassium concentrations; renin	Increase sodium retention and potassium excretion through the kidneys
	Glucocorticoids (cortisol)	Most cells in the body	ACTH	Control metabolism of carbohydrates, fats, and proteins; exert an anti-inflammatory action
	Androgens and estrogens	Ovaries, breasts, and testes	ACTH (?)	Assist in the development of female and male sex characteristics
Pancreas	Insulin	All cells in the body	Plasma glucose and amino acid concentrations	Controls blood glucose levels by lowering glucose levels; increases use of glucose and synthesis of fat
	Glucagon	All cells in the body	Plasma glucose and amino acid concentrations	Increases blood glucose; stimulates the breakdown of protein and fat
	Somatostatin	Islets of Langerhans and intestines	Plasma glucose, insulin, and glucagon concentrations	Depresses the secretion of both insulin and glucagon
Kidney	Renin	Adrenal cortex	Plasma sodium concentrations	Assists in blood pressure control
	Erythropoietin (EPO)	Bone marrow	Low tissue oxygen concentrations	Stimulates erythrocyte production
Testes	Testosterone	Sex organs, muscle	FSH and LH	Promotes development of male sex characteristics, including growth of testes, scrotum, and penis, facial hair, and change in voice; promotes muscle growth
Ovaries	Estrogens and progesterone	Sex organs and adipose tissue	FSH and LH	Promote development of female sex organs and characteristics; increase storage of fat; assist in regulating the menstrual cycle

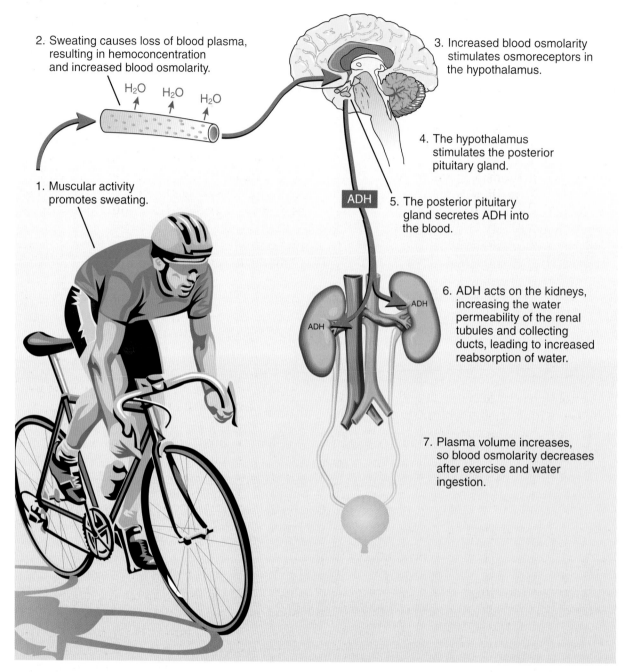

2. Sweating causes loss of blood plasma, resulting in hemoconcentration and increased blood osmolarity.

3. Increased blood osmolarity stimulates osmoreceptors in the hypothalamus.

H_2O H_2O H_2O

4. The hypothalamus stimulates the posterior pituitary gland.

1. Muscular activity promotes sweating.

ADH

5. The posterior pituitary gland secretes ADH into the blood.

ADH

ADH

6. ADH acts on the kidneys, increasing the water permeability of the renal tubules and collecting ducts, leading to increased reabsorption of water.

7. Plasma volume increases, so blood osmolarity decreases after exercise and water ingestion.

▲ **Figure 5.5**   The mechanism by which antidiuretic hormone (ADH) conserves body water.

ter retention in an effort to dilute the plasma electrolyte concentration back to normal levels. Figure 5.5 illustrates this process.

Anterior Lobe of the Pituitary Gland

The anterior pituitary, also called the adenohypophysis, secretes six hormones in response to **releasing factors** and **inhibiting factors** (hormones) secreted by the hypothalamus. Communication between the hypothalamus and the anterior lobe of the pituitary occurs through a specialized circulatory system that transports the releasing and inhibiting hormones from the hypothalamus to the anterior pituitary. The major functions of each of the anterior pituitary hormones, along with their releasing and inhibiting factors, are listed in table 5.1. Exercise appears to be a strong stimulant to the hypothalamus because exercise increases the release rate of all anterior pituitary hormones.

Of the six anterior pituitary hormones, four are tropic hormones, meaning they affect the functioning of other endocrine glands. The exceptions to this are growth hormone and prolactin. **Growth hormone** is a potent anabolic agent (a substance that promotes constructive metabolism). It promotes muscle growth and hypertrophy by facilitating amino acid transport into the cells. In addition, growth hormone directly stimulates fat metabolism (lipolysis) by increasing the synthesis of enzymes involved in this process. Growth hormone levels are elevated during aerobic exercise, apparently in proportion to the exercise intensity, and typically remain elevated for some time after exercise.

Thyroid Gland

The thyroid gland is located along the midline of the neck, immediately below the larynx. It secretes two important nonsteroid hormones, triiodothyronine (T_3) and thyroxine (T_4), which regulate metabolism in general, and an additional hormone, calcitonin, which assists in regulating calcium metabolism.

Triiodothyronine and Thyroxine

The two metabolic thyroid hormones share similar functions. **Triiodothyronine** and **thyroxine** increase the metabolic rate of almost all tissues and can increase the body's basal metabolic rate by as much as 60% to 100%. These hormones also

- increase protein synthesis (thus also enzyme synthesis),
- increase the size and number of mitochondria in most cells,
- promote rapid cellular uptake of glucose,
- enhance glycolysis and gluconeogenesis, and
- enhance lipid mobilization, increasing free fatty acid (FFA) availability for oxidation.

Release of **thyrotropin** (thyroid-stimulating hormone, or **TSH**) from the anterior pituitary increases during exercise. TSH controls the release of triiodothyronine and thyroxine, so the exercise-induced increase in TSH would be expected to stimulate the thyroid gland. Exercise indeed increases plasma thyroxine levels, but a delay occurs between the increase in TSH levels during exercise and the increase in plasma thyroxine levels. Furthermore, during prolonged submaximal exercise, thyroxine levels remain relatively constant after a sharp initial increase as exercise begins, and triiodothyronine levels tend to decrease.

Calcitonin

Calcitonin decreases the plasma calcium concentration. It acts primarily on two targets: the bones and the kidneys. In bone, calcitonin inhibits the activity of the osteoclasts (the bone-resorbing cells), thus inhibiting bone resorption. The osteoclasts may be calcitonin's only target in the bone. In the kidney, calcitonin increases urinary excretion of calcium by decreasing calcium reabsorption from the renal tubules.

Calcitonin is important primarily in children while their bones are rapidly growing and developing strength. This hormone is not a major regulator of calcium homeostasis in adults. But it does appear to offer some protection against excessive bone resorption.

Parathyroid Glands

The parathyroid glands are located on the back of the thyroid gland. They secrete **parathyroid hormone (PTH**, or parathormone), which is the principal regulator of plasma calcium concentration and also regulates plasma phosphate. A decrease in plasma calcium levels stimulates its release.

PTH exerts its effects on three targets: the bones, the intestines, and the kidneys. In the bone, PTH stimulates osteoclast activity. This increases bone resorption, which releases both calcium and phosphate into the blood. In the intestines, PTH increases calcium absorption indirectly by stimulating an enzyme that is required for the process. Increased intestinal absorption of calcium is accompanied by increased absorption of phosphate. Because PTH increases plasma levels of phosphate ions, the excess phosphate must be removed. This is accomplished by PTH's action in the kidneys, where it increases calcium reabsorption but decreases phosphate reabsorption, which promotes urinary excretion of phosphate.

> **fyi** Over an extended period, exercise increases bone formation. This results primarily from increased intestinal absorption of calcium ions (Ca^{2+}), decreased urinary excretion of Ca^{2+}, and increased PTH levels. Conversely, immobilization or complete bed rest promotes bone resorption. During such a period, PTH levels decrease.

Adrenal Glands

The adrenal glands are situated directly atop each kidney and are composed of the inner adrenal medulla and the outer adrenal cortex. The hormones secreted by these two parts are quite different, so we consider them separately.

Adrenal Medulla

The adrenal medulla produces and releases two hormones: **epinephrine** and **norepinephrine**, which are referred to as **catecholamines**. When the adrenal medulla is stimulated by the sympathetic nervous system, approximately 80% of its secretion is epinephrine and 20% is norepinephrine, although these percentages vary with different physiological conditions. The catecholamines have powerful effects similar to those of the sympathetic nervous system, but the hormones' effects last longer because these substances are removed from the blood relatively slowly. These two hormones prepare you for immediate action, eliciting the fight-or-flight response.

Epinephrine and norepinephrine help you face a real or perceived crisis. Although some of the specific actions of these two hormones differ, the two work together. Their combined effects include

- increased rate and force of heart contraction,
- increased metabolic rate,

- increased glycogenolysis (breakdown of glycogen to glucose) in the liver and muscle,
- increased release of glucose and FFAs into the blood,
- redistribution of blood to the skeletal muscles (through vasodilation of vessels supplying skeletal muscles and vasoconstriction of vessels to the skin and viscera),
- increased blood pressure, and
- increased respiration.

Release of epinephrine and norepinephrine is affected by a wide variety of factors, including changes in body position, psychological stress, and exercise. Figure 5.6a shows the changes in plasma levels of these hormones when individuals gradually increase their exercise intensity. Plasma norepinephrine levels increase markedly at work rates above 50% of $\dot{V}O_2$max. But the epinephrine level doesn't increase significantly until the exercise intensity exceeds 60% to 70% of $\dot{V}O_2$max. Figure 5.6b shows that during steady-state activity lasting more than 3 h at 60% of $\dot{V}O_2$max, blood levels of both hormones increase. When the exercise bout ends, epinephrine levels return to resting levels within only a few minutes of recovery, but norepinephrine can remain elevated for several hours.

Adrenal Cortex

The adrenal cortex secretes more than 30 different steroid hormones, referred to as corticosteroids. These generally are classified into three major types: mineralocorticoids, glucocorticoids, and gonadocorticoids (sex hormones).

Mineralocorticoids   The mineralocorticoids maintain electrolyte balance in the extracellular fluids, especially that of sodium (Na^+) and potassium (K^+). **Aldosterone** is the major mineralocorticoid, responsible for at least 95% of all mineralocorticoid activity. It works primarily by promoting renal reabsorption of sodium, thus causing the body to retain sodium. When sodium is retained, so is water; thus, aldosterone fights dehydration. Sodium retention also enhances potassium excretion,

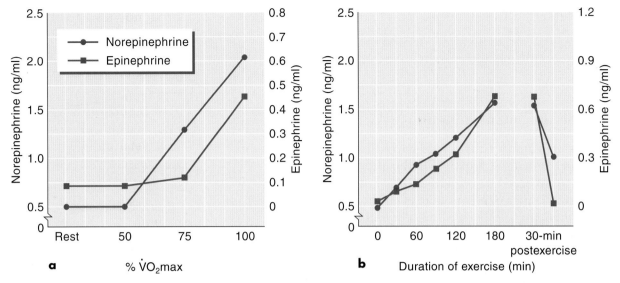

▲ **Figure 5.6**   Changes in blood concentrations of epinephrine and norepinephrine (*a*) at rest and at various intensities (%V̇O₂max) of treadmill running and (*b*) during 3 h of treadmill running at 60% V̇O₂max and during recovery.

so aldosterone plays a role in potassium balance as well. For these reasons, aldosterone secretion is stimulated by many factors, including decreased plasma sodium, decreased blood volume, decreased blood pressure, and increased plasma potassium concentration. We further discuss the actions of aldosterone later in this chapter.

Glucocorticoids   The glucocorticoids are essential to life. They enable us to adapt to external changes and stress. They also maintain fairly consistent plasma glucose levels even when we go for long periods without ingesting food. **Cortisol**, also known as hydrocortisone, is the major corticosteroid. It is responsible for about 95% of all glucocorticoid activity in the body. Cortisol is known to

- stimulate gluconeogenesis to ensure an adequate fuel supply;
- increase mobilization of FFAs, making them more available as an energy source;
- decrease glucose utilization, sparing it for the brain;
- stimulate protein catabolism to release amino acids for use in repair, enzyme synthesis, and energy production;
- act as an anti-inflammatory agent;

- depress immune reactions; and
- increase the vasoconstriction caused by epinephrine.

We discuss cortisol's important role in exercise later in this chapter, when we discuss the regulation of glucose and fat metabolism.

Gonadocorticoids   The adrenal cortex also synthesizes and releases the gonadocorticoids. These hormones are mostly androgens, although estrogens and progesterones are released in small amounts. These hormones are the same as those produced by the reproductive organs. Apparently these cortical secretions have little effect in adults; the amounts secreted are insignificant compared with the amounts released from the reproductive glands. Thus, the exact role played by gonadocorticoids is unclear.

Pancreas

The pancreas is located behind and slightly below the stomach. Its two major hormones are insulin and glucagon. These provide the major control of plasma glucose levels. When plasma glucose levels are elevated (**hyperglycemia**), such as after a meal, the pancreas receives signals to release insulin into the blood.

Among its actions, **insulin**

- facilitates glucose transport into the cells, especially those in muscle and connective tissue;
- promotes glycogenesis; and
- inhibits gluconeogenesis.

Insulin's main function is to reduce the amount of glucose circulating in the blood. But it is also involved in protein and fat metabolism, promoting cellular uptake of amino acids and enhancing synthesis of protein and fat.

The pancreas secretes **glucagon** when the plasma glucose concentration falls below normal levels **(hypoglycemia)**. Its effects generally oppose those of insulin. Glucagon promotes increased breakdown of liver glycogen to glucose (glycogenolysis) and increased gluconeogenesis. Both processes increase plasma glucose levels.

During exercise lasting 30 min or longer, the body attempts to maintain plasma glucose levels; however, insulin levels tend to decline, as shown in figure 5.7, *a* and *b*, respectively. Research has shown that the number or availability of insulin receptors increases during exercise, increasing the body's sensitivity to insulin. This reduces the need to maintain high plasma insulin levels for transporting glucose into the muscle cells. Plasma glucagon, on the other hand, shows a gradual increase throughout the period of exercise (see figure 5.7*c*). Glucagon primarily maintains plasma glucose concentrations by stimulating liver glycogenolysis. This increases glucose availability to the cells, maintaining adequate plasma glucose levels to meet increased metabolic demands. As can be seen in figure 5.7, the hormone response usually is blunted in trained people, and trained people are better able to maintain plasma glucose levels.

Gonads

The gonads are the reproductive glands: the testes and the ovaries. The hormones they secrete are generally anabolic, meaning they promote anabolism, the constructive phase of

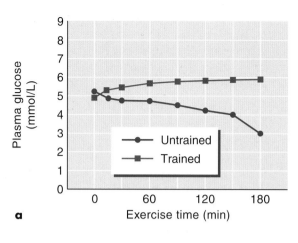

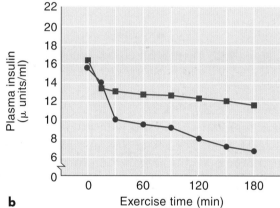

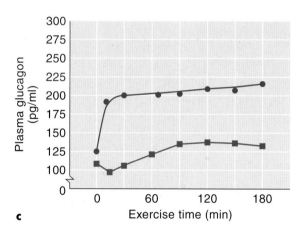

▲ **Figure 5.7**   Plasma glucose (*a*), insulin (*b*), and glucagon (*c*) levels in trained and untrained subjects during 3 h of cycling.

metabolism. The testes secrete androgens, of which testosterone is the most important. **Testosterone** is responsible for the development of male secondary sex characteristics and spermatogenesis. It is also essential for normal growth, development, and maturation of the male skeletal system. Another critical role of the androgens is the promotion of skeletal muscle growth, which is particularly important to our understanding of strength training and sex differences in muscle growth. The anabolic effects of testosterone are responsible in part for the muscle protein retention and muscle hypertrophy observed during strength training. This has led to some athletes' use of testosterone and other anabolic steroids to artificially promote muscle building beyond natural levels. We will discuss this illegal and dangerous use of testosterone and other anabolic steroids in more detail in chapter 15.

The ovaries secrete two types of hormones: estrogens and progesterone. **Estrogens** promote development of female secondary sex characteristics, the proliferative phase of the menstrual cycle, oogenesis, ovulation, and many changes during pregnancy. **Progesterone** promotes the secretory (luteal) phase of the menstrual cycle, prepares the uterus for pregnancy, and prepares the breasts for lactation.

Kidneys

Although the kidneys are not typically considered major endocrine organs, we discuss the kidneys here because they release a hormone called erythropoietin. **Erythropoietin** regulates red blood cell (erythrocyte) production by stimulating bone marrow cells. The red blood cells are essential for transporting oxygen to the tissues and removing carbon dioxide, so this hormone is extremely important in our adaptation to training and altitude. Research has shown that part of the adaptation when a person trains at altitude is an increased release of erythropoietin, which in turn stimulates more red blood cell production, increasing the blood's oxygen-carrying capacity. Because of this, some athletes have used injections of this hormone to build up their red blood cell count, hoping to gain an edge over their competitors. This is discussed further in chapter 15. The kidneys also release renin, a hormone and enzyme involved in blood pressure control and fluid and electrolyte balance. Renin is discussed in greater detail later in this chapter.

Hormonal Response to Exercise

The hormonal responses to an acute bout of exercise and exercise training are summarized in table 5.2 on pages 174-175. This table is limited to those hormones suspected of playing major roles in sport and physical activity.

Hormonal Effects on Metabolism and Energy

As noted in chapters 4 and 13, carbohydrate and fat metabolism is responsible for maintaining muscle ATP levels during prolonged exercise. Various hormones work to ensure glucose and FFA availability for muscle energy metabolism. In this section, we examine how the metabolism of glucose and fat is affected by these hormones during exercise. Because carbohydrate is the primary fuel used during both brief and prolonged exhaustive exercise, we first consider the hormones that regulate its availability.

Regulation of Glucose Metabolism During Exercise

As we saw in the previous chapter, for your body to meet the heightened energy demands of exercise, more glucose must be available to the muscles. Recall that glucose is stored in the body as glycogen, located primarily in the muscles and the liver. Glucose must be freed from storage, so glycogenolysis must increase. Glucose freed from the liver enters the blood to circulate throughout the body, allowing it access to the active tissues. Plasma glucose levels also can be increased through gluconeogenesis. Let's examine the hormones involved in both glycogenolysis and gluconeogenesis.

Plasma Glucose Levels

Four hormones work to increase the amount of circulating plasma glucose:

Table 5.2

Hormone Response to Acute Exercise and Change in Response With Exercise Training

Endocrine gland	Hormone	Response to acute exercise (untrained)	Effect of exercise training
Anterior pituitary	Growth hormone (GH)	Increases with increasing rates of work	Attenuated response at same rate of work
	Thyrotropin (TSH)	Increases with increasing rates of work	No known effect
	Adrenocorticotropin (ACTH)	Increases with increasing rates of work and duration	Attenuated response at same rate of work
	Prolactin	Increases with exercise	No known effect
	Follicle-stimulating hormone (FSH)	Small or no change	No known effect
	Luteinizing hormone (LH)	Small or no change	No known effect
Posterior pituitary	Antidiuretic hormone (ADH or vasopressin)	Increases with increasing rates of work	Attenuated response at same rate of work
	Oxytocin	Unknown	Unknown
Thyroid	Thyroxine (T_4) and triiodothyronine (T_3)	Free T_3 and T_4 increase with increasing rates of work	Increased turnover of T_3 and T_4 at same rate of work
	Calcitonin	Unknown	Unknown
Parathyroid	Parathyroid hormone (PTH or parathormone)	Increases with prolonged exercise	Unknown
Adrenal medulla	Epinephrine	Increases with increasing rates of work, starting at about 75% of $\dot{V}O_2$max	Attenuated response at same rate of work
	Norepinephrine	Increases with increasing rates of work, starting at about 50% of $\dot{V}O_2$max	Attenuated response at same rate of work
Adrenal cortex	Aldosterone	Increases with increasing rates of work	Unchanged
	Cortisol	Increases only at high rates of work	Slightly higher values
Pancreas	Insulin	Decreases with increasing rates of work	Attenuated response at same rate of work
	Glucagon	Increases with increasing rates of work	Attenuated response at same rate of work

Endocrine gland	Hormone	Response to acute exercise (untrained)	Effect of exercise training
Kidney	Renin	Increases with increasing rates of work	Unchanged
	Erythropoietin (EPO)	Unknown	Unchanged
Testes	Testosterone	Small increases with exercise	Resting levels decreased in male runners
Ovaries	Estrogens and progesterone	Small increases with exercise	Resting levels might be decreased in highly trained women

- Glucagon
- Epinephrine
- Norepinephrine
- Cortisol

The plasma glucose concentration during exercise depends on a balance between glucose uptake by the muscles and its release by the liver. At rest, glucose release from the liver is facilitated by glucagon, which promotes liver glycogen breakdown and glucose formation from amino acids. During exercise, glucagon secretion increases. Muscular activity also increases the rate of catecholamine release from the adrenal medulla, and these hormones (epinephrine and norepinephrine) work with glucagon to further increase glycogenolysis. Evidence suggests that cortisol levels also increase during exercise. Cortisol increases protein catabolism, freeing amino acids to be used within the liver for gluconeogenesis. Thus, all four of these hormones can increase the amount of plasma glucose by enhancing the processes of glycogenolysis and gluconeogenesis. In addition, growth hormone increases mobilization of FFAs and decreases cellular uptake of glucose, so less glucose is used by the cells (more remains in circulation), and the thyroid hormones promote glucose catabolism and fat metabolism.

The amount of glucose released by the liver depends on exercise intensity and the duration. As intensity increases, so does the rate of catecholamine release. This can cause the liver to release more glucose than is being taken up by the active muscles. Consequently, during or shortly after an explosive, short-term sprint

bout, blood glucose levels may be 40% to 50% above the resting level, illustrating that the glucose release by the liver is greater than the uptake by the muscles (see figure 5.8).

The greater the exercise intensity, the greater the catecholamine release, and thus the glycogenolysis rate is significantly increased. This process occurs not only in the liver but also in the muscle. The glucose released from the liver enters the blood to become available to the muscle. But the muscle has a more readily available source of glucose: its own glycogen. The muscle uses its own glycogen stores before using the plasma glucose during explosive, short-term exercise. Glucose released from the liver is not used as read-

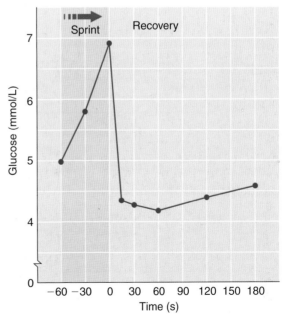

▲ **Figure 5.8** Changes in blood glucose during and after 60 s of sprint cycling.

ily, so it remains in circulation, elevating the plasma glucose. Following exercise, plasma glucose levels decrease as the glucose enters the muscle to replenish the depleted muscle glycogen stores.

During exercise bouts that last for several hours, however, the rate of liver glucose release more closely matches the muscles' needs, keeping plasma glucose at or only slightly above the resting levels. As muscle uptake of glucose increases, the liver's rate of glucose release also increases. In most cases, plasma glucose levels don't begin to decline until late in the activity as liver glycogen stores become depleted, at which time glucagon levels increase significantly. Glucagon and cortisol together enhance gluconeogenesis, providing more fuel.

Figure 5.9 illustrates the changes in plasma levels of epinephrine, norepinephrine, glucagon, cortisol, and glucose during 3 h of cycling. Although the hormonal regulation of glucose remains intact throughout such long-term activities, the liver's glycogen supply may become critically low. As a result, the liver's rate of glucose release may be unable to keep pace with the muscles' rate of glucose uptake. Under this condition, the plasma glucose level may decline, despite strong hormonal stimulation. Glucose ingestion during the activity can play a major role in maintaining plasma glucose levels.

> Plasma glucose levels are increased by glucagon, epinephrine, norepinephrine, and cortisol. This is important during exercise, particularly long-duration or high-intensity exercise, during which blood glucose levels might decline. Glucose ingestion during exercise also helps maintain plasma glucose levels.

Glucose Uptake by the Muscles

Merely releasing sufficient amounts of glucose into the blood does not ensure that the muscle cells will have enough glucose to meet their energy demands. The glucose not only must be released and delivered to these cells, but it also must be taken up by them. That job relies on insulin. Once glucose is delivered to the muscle, insulin facilitates its transport into the fibers.

Surprisingly, as seen in figure 5.10, plasma insulin levels tend to decrease during prolonged submaximal exercise, despite a slight increase in plasma glucose concentration and glucose uptake by muscle. This apparent contradiction between the plasma insulin concentrations and the muscles' need for glucose reminds us that a hormone's activity is determined not only by its concentration in blood but also by a cell's sensitivity to a given hormone. In this case, the cell's sensitivity to insulin may be as important as the amount of circulating hormone. Exercise may enhance insulin's binding to receptors on

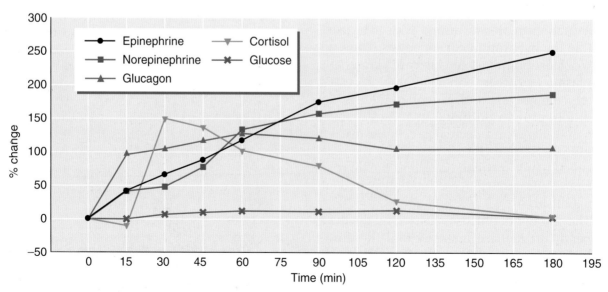

▲ **Figure 5.9**   Changes (as a percentage of preexercise values) in plasma levels of epinephrine, norepinephrine, glucagon, cortisol, and glucose during 3 h of cycling at 65% V̇O₂max.

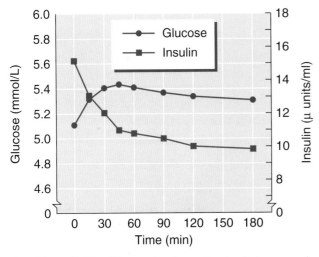

▲ **Figure 5.10**   Changes in plasma levels of glucose and insulin during prolonged cycling at 65% to 70% of V̇O₂max. Note the gradual decline in insulin throughout the exercise, suggesting an increased sensitivity to insulin during prolonged effort.

the muscle fiber.[5, 6] Muscle action, for reasons not completely understood, appears to have an insulin-like effect in recruiting receptors: More receptors appear on the cells, and their activity may be increased, thereby reducing the need for high levels of plasma insulin to transport glucose across the muscle cell membrane into the cell. This is important, because during exercise four hormones are trying to release glucose from its storage sites and create new glucose. High insulin levels would oppose their action, preventing this needed increase in plasma glucose supply.

Regulation of Fat Metabolism During Exercise

Although fat generally contributes less than carbohydrate to muscles' energy needs during exercise, mobilization and oxidation of FFAs are critical to performance in endurance exercise bouts. During such activity, carbohydrate reserves become depleted, and your body must rely more heavily on the oxidation of fat for energy production. When carbohydrate reserves are low (low plasma glucose and low muscle glycogen), the endocrine system can accelerate the oxidation of fats (lipolysis), thus ensuring that your muscles' energy needs can be met. Lipolysis is also enhanced

through the elevation of epinephrine and norepinephrine.

FFAs are stored as triglycerides in fat cells and inside muscle fibers. Adipose tissue triglycerides, however, must be broken down to release the FFAs, which are then transported to the muscle fibers. The rate of FFA uptake by active muscle correlates highly with the plasma FFA concentration. Increasing this concentration would increase cellular uptake of the FFA. We can assume that increased plasma FFA concentration increases FFA oxidation, because increased cellular uptake of FFA provides more FFA for oxidation.[1] Thus, the rate of triglyceride breakdown may determine, in part, the rate at which muscles use fat as a fuel source during exercise.

Triglycerides are reduced to FFAs and glycerol by a special enzyme called lipase, which is activated by at least four hormones:

- Cortisol
- Epinephrine
- Norepinephrine
- Growth hormone

In addition to having a role in gluconeogenesis, cortisol also accelerates the mobilization and use of FFAs for energy during exercise. Figure 5.11a illustrates the changes in the plasma concentrations of FFA and cortisol during prolonged exercise. Plasma cortisol levels peak after 30 to 45 min of exercise and then decrease to near normal levels. But the plasma FFA concentration continues to increase throughout the activity, meaning that lipase must continue to be activated by other hormones. The hormones that continue this process are the catecholamines and growth hormone. As shown in figure 5.11b, the plasma levels of these hormones continue to increase throughout exercise, progressively increasing FFA release and fat oxidation. The thyroid hormones also contribute to the mobilization and metabolism of FFAs but to a smaller degree than the previously mentioned hormones.

FFAs are a primary source of energy at rest and during exercise. They are derived from triglycerides through the action of the enzyme lipase, which breaks down triglycerides into FFA and glycerol.

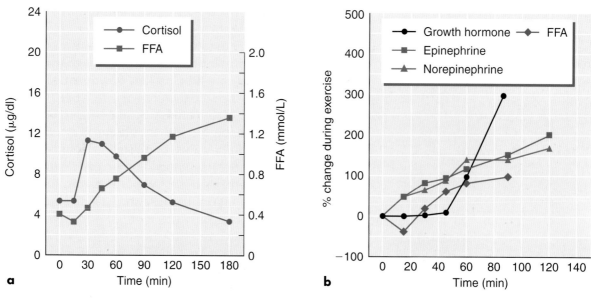

▲ **Figure 5.11**   Changes in plasma levels of (*a*) free fatty acid (FFA) and cortisol and (*b*) epinephrine, norepinephrine, growth hormone, and FFA during prolonged exercise.

▶ Plasma glucose is increased by the combined actions of glucagon, epinephrine, norepinephrine, and cortisol. These hormones promote glycogenolysis and gluconeogenesis, thus increasing the amount of glucose available for use as a fuel source.

▶ Insulin helps the released glucose enter the cells, where it can be used for energy production. But insulin levels decline during prolonged exercise, indicating that exercise facilitates the action of insulin so that less of the hormone is required during exercise than at rest.

▶ When carbohydrate reserves are low, the body turns more to fat oxidation for energy, and this process is facilitated by cortisol, epinephrine, norepinephrine, and growth hormone.

▶ Cortisol accelerates lipolysis, releasing FFAs into the blood so they can be taken up by the cells and used for energy production. But cortisol levels peak and then return to near normal levels during prolonged exercise. When this happens, the catecholamines and growth hormone take over cortisol's role.

Thus, the endocrine system plays a critical role in regulating ATP production during exercise and may be responsible for controlling the balance between carbohydrate and fat metabolism.

Hormonal Effects on Fluid and Electrolyte Balance During Exercise

Fluid balance during exercise is critical for optimal cardiovascular and thermoregulatory function. At the onset of exercise, water is shifted from the plasma volume to the interstitial and intracellular spaces.[2, 3] This water shift is specific to the muscle mass that is active and the intensity of effort. Metabolic by-products begin to accumulate in and around the muscle fibers, increasing the osmotic pressure there. Water is drawn into these areas. Also, increased muscular activity increases your blood pressure, which in turn drives water out of your blood. In addition, sweating increases during exercise. The combined effect of these actions is that your muscles and sweat glands gain water at the expense of your plasma volume. For example, running at approximately 75% of $\dot{V}O_2$max decreases plasma volume by 5% to 10%. Reduced plasma volume decreases your blood pressure and the amount of blood flow to your skin and muscles. Both of these effects can seriously impede athletic performance.

The endocrine system plays a major role in monitoring fluid levels and correcting imbalances, along with regulating the electrolyte balance, especially that of sodium. The two major hormones involved in this regulation are aldosterone and ADH, and the kidneys are their primary targets. Let's examine these hormones' effects.

Aldosterone and the Renin–Angiotensin Mechanism

The kidneys have a strong regulatory influence on blood pressure that also allows them to regulate fluid balance. Plasma volume is a major determinant of blood pressure: When plasma volume decreases, so does blood pressure. Your blood pressure is monitored by specialized cells within your kidneys. During exercise, these cells can be stimulated by decreased blood pressure, decreased blood flow to the kidneys through increased sympathetic nervous activity accompanying exercise, or direct stimulation from the sympathetic nerves.

Figure 5.12 shows the mechanism involved in renal control of blood pressure,

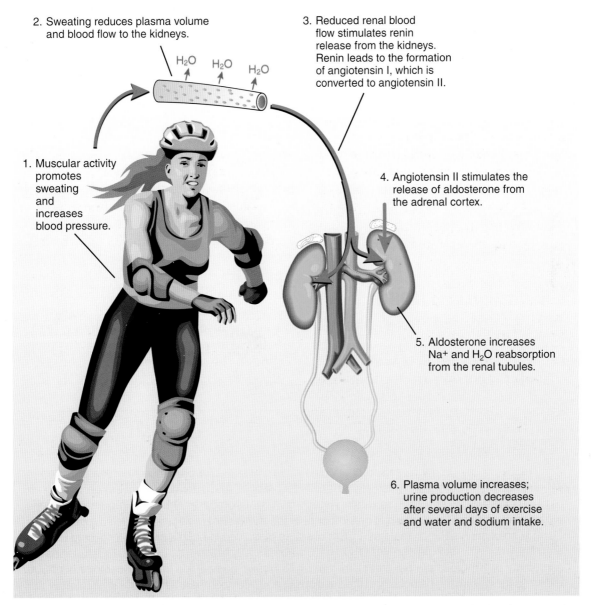

2. Sweating reduces plasma volume and blood flow to the kidneys.

H_2O H_2O H_2O

3. Reduced renal blood flow stimulates renin release from the kidneys. Renin leads to the formation of angiotensin I, which is converted to angiotensin II.

1. Muscular activity promotes sweating and increases blood pressure.

4. Angiotensin II stimulates the release of aldosterone from the adrenal cortex.

5. Aldosterone increases Na^+ and H_2O reabsorption from the renal tubules.

6. Plasma volume increases; urine production decreases after several days of exercise and water and sodium intake.

▲ **Figure 5.12**   The influence of water loss from plasma during exercise leads to a sequence of events that promotes sodium (Na^+) and water reabsorption from the renal tubules, thereby reducing urine production. In the hours after exercise when fluids are consumed, the elevated aldosterone levels cause an increase in the extracellular volume and an expansion of plasma volume.

the **renin–angiotensin mechanism**. The kidneys respond to decreased blood pressure or blood flow by forming an enzyme and hormone called **renin**. Renin, in turn, converts a plasma protein called angiotensinogen into an active form called angiotensin I. In the blood, angiotensin I is converted to angiotensin II when it encounters the enzyme **angiotensin converting enzyme (ACE)**. ACE inhibitors are a class of drugs used in the treatment of high blood pressure. They lower blood pressure by blocking, or inhibiting, the conversion of angiotensin I to angiotensin II. Angiotensin II acts in two ways. First, it is a potent arteriole constrictor. Through this action, your peripheral resistance increases, which raises your blood pressure. The second job of angiotensin II is to trigger aldosterone release from the adrenal cortex.

Recall that aldosterone's primary action is to promote sodium reabsorption in the kidneys. Because water follows sodium, this renal conservation of sodium causes your kidneys to retain water. The net effect is an effort to maintain your body's fluid content, thereby minimizing the loss of plasma volume while keeping blood pressure near normal. Figure 5.13 illustrates the changes in plasma volume and aldosterone concentrations during 2 h of exercise.

Antidiuretic Hormone (ADH)

The other major hormone involved in fluid balance is ADH. ADH is released in response to the increasing concentration (lower water content) of the blood. During exercise, the shifting of

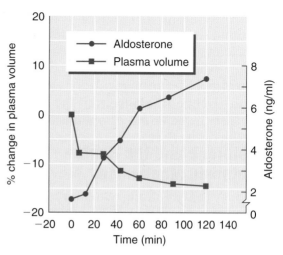

▲ **Figure 5.13**   Changes in plasma volume and aldosterone concentrations during 2 h of cycling exercise. Note that plasma volume declines rapidly during the first few minutes of exercise and then shows a smaller rate of decline despite large sweat losses. Plasma aldosterone concentration, on the other hand, increases rather steadily throughout the exercise.

water out of the plasma leaves the blood more concentrated, and sweating promotes dehydration, which also increases plasma concentration. This elevates the number of particles or osmoles per unit of plasma (see sidebar on osmolarity). The concentrated plasma circulates and reaches the hypothalamus, which houses osmoreceptors that constantly monitor blood osmolarity. When the osmolarity increases, the hypothalamus triggers ADH release from the posterior pituitary. Recall that ADH promotes water reabsorption in the kidneys, leading to water conservation. As with aldosterone, although the stimuli and mechanisms of action are different, the net effect of ADH secretion

Osmolarity

Body fluids contain many dissolved molecules and minerals. The presence of these particles in various body fluid compartments (i.e., intracellular, plasma, and interstitial spaces) generates an osmotic pressure or attraction to retain water within that compartment. The amount of osmotic pressure exerted by a body fluid is proportional to the number of molecular particles (osmoles, or Osm) in solution. A solution that has 1 Osm of solute dissolved in each liter of water is said to have an osmolarity of 1 osmole per liter (1 Osm/L), whereas a solution that has 0.001 Osm/L has an osmolarity of 1 milliosmole per liter (1 mOsm/L). Normally, body fluids have an osmolarity of 300 mOsm/L. Increasing the osmolarity of the solutions in one body compartment generally causes water to be drawn away from adjacent compartments that have a lower osmolarity (i.e., more water).

is also to increase the body's fluid content, restoring normal plasma volume and blood pressure.

Following the initial decrease, plasma volume remains relatively constant throughout exercise. In addition to the actions of aldosterone and ADH, some evidence suggests that, despite continued sweat losses during exercise, water returns from the exercising muscles back into the blood and protects plasma volume from further reduction. Also, as exercise progresses, the amount of metabolic water production via oxidation increases.

> Loss of fluid (plasma) from the blood results in a concentration of the constituents of the blood, a phenomenon referred to as hemoconcentration. Conversely, a gain of fluid into the blood results in a dilution of the constituents of the blood, which is referred to as hemodilution.

> ▶ The two primary hormones involved in the regulation of fluid balance are aldosterone and antidiuretic hormone.

> ▶ When plasma volume or blood pressure decreases, the kidneys form an enzyme called renin that converts angiotensinogen into angiotensin I, which later becomes angiotensin II. Angiotensin II increases peripheral arterial resistance, increasing the blood pressure.

> ▶ Angiotensin II also triggers the release of aldosterone from the adrenal cortex. Aldosterone promotes sodium reabsorption in the kidneys, which in turn causes water retention, thus increasing the plasma volume.

> ▶ ADH is released in response to increased plasma osmolarity. When osmoreceptors in the hypothalamus sense this increase, the hypothalamus triggers ADH release from the posterior pituitary.

> ▶ ADH acts on the kidneys, promoting water conservation. Through this mechanism, the plasma volume is increased, which dilutes the plasma solutes. Blood osmolarity decreases.

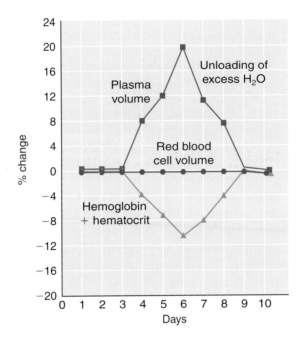

▲ **Figure 5.14**   Changes in plasma volume during 3 days of repeated exercise and dehydration. Subjects exercised in heat on days 3 to 6. Note the sudden decline in plasma volume when the subjects stop training (6th day). The changes in hemoglobin and hematocrit reflect the expansion and contraction of plasma volume during and after the 3-day training period.

Postexercise Hormone Activity and Fluid Balance

The hormonal influences of aldosterone and ADH persist for 12 to 48 h after exercise, reducing urine production and protecting the body from further dehydration.[3, 4] In fact, aldosterone's prolonged enhancement of Na^+ reabsorption causes the body's Na^+ concentration to increase above normal following an exercise bout. In an effort to compensate for this elevation in Na^+ levels, more of the water you ingest shifts into the extracellular compartment.

As shown in figure 5.14, individuals who are subjected to 3 repeated days of exercise and dehydration show a significant increase in plasma volume that continues to increase throughout the period of activity. This increase in plasma volume appears to parallel the body's retention of dietary Na^+. When the daily bouts of activity are terminated, the excess Na^+ and water are excreted in urine.

> **fyi** Hemoglobin is one of the substances diluted by plasma expansion. For this reason, some athletes who actually have normal hemoglobin levels may appear to be anemic as a consequence of Na^+-induced hemodilution. This condition, not to be confused with true anemia, can be remedied with a few days of rest, allowing time for aldosterone levels to return to normal and for the kidneys to unload the extra Na^+ and water.

Most athletes involved in heavy training have an expanded plasma volume, which dilutes various blood constituents. The actual amount of substances within the blood remains unaltered, but the substances are dispersed throughout a greater volume of water (plasma), so they are diluted. This phenomenon is called **hemodilution**.

In Closing . . .

In this chapter, we focused on the role of the endocrine system in regulating physiological processes that accompany exercise. We concentrated on the role of hormones in regulating the metabolism of glucose and fat and maintaining fluid balance. Our primary interest in this chapter was what happens during acute exercise. In the next chapter, we examine adaptations in the body's metabolic processes that occur as a result of training.

▷ Key Terms

aldosterone
angiotensin converting enzyme (ACE)
antidiuretic hormone (ADH)
calcitonin
catecholamines
cortisol
cyclic adenosine monophosphate (cAMP)
direct gene activation
down-regulation
epinephrine
erythropoietin
estrogen
glucagon
growth hormone

hemoconcentration
hemodilution
hormone
hyperglycemia
hypoglycemia
inhibiting factors
insulin
negative feedback system
nonsteroid hormones
norepinephrine
osmolarity
parathyroid hormone (PTH)
progesterone
prostaglandins
releasing factors
renin
renin–angiotensin mechanism
second messenger
steroid hormones
target cells
testosterone
thyrotropin (TSH)
thyroxine (T_4)
triiodothyronine (T_3)
up-regulation

▷ Study Questions

1. What is an endocrine gland, and what are the functions of hormones?
2. Explain the difference between steroid hormones and nonsteroid hormones.
3. How can hormones have very specific functions when they reach nearly all parts of the body through the blood?
4. How are plasma levels of specific hormones controlled?
5. Explain the relationship between the hypothalamus and the pituitary gland.
6. Briefly outline the major endocrine glands, their hormones, and the specific action of these hormones.
7. Which of the hormones outlined in question 6 are of major significance during exercise?

8. What is hemoconcentration, and how does the endocrine system relate to it?

9. Describe the hormonal regulation of metabolism during exercise. What hormones are involved, and how do they influence the availability of carbohydrates and fats for energy during exercise lasting for several hours?

10. Describe the hormonal regulation of fluid balance during exercise.

11. What is hemodilution, and how does the endocrine system relate to it?

▷ References

1. Costill, D.L., Coyle, E., Dalsky, G., Evans, W., Fink, W., & Hoopes, D. (1977). Effects of elevated plasma FFA and insulin on muscle glycogen usage during exercise. *Journal of Applied Physiology, 43,* 695-699.

2. Costill, D.L., & Fink, W.J. (1974). Plasma volume changes following exercise and thermal dehydration. *Journal of Applied Physiology, 37,* 521-525.

3. Dill, D.B., & Costill, D.L. (1974). Calculation of percentage changes in volumes of blood, plasma, and red cells in dehydration. *Journal of Applied Physiology, 37,* 247-248.

4. Edington, D.W., & Edgerton, V.R. (1976). *The biology of physical activity.* Boston: Houghton Mifflin.

5. Krotkiewshi, M., & Gorski, J. (1986). Effect of muscular exercise on plasma C-peptide and insulin in obese non-diabetics and diabetics, Type II. *Clinical Physiology, 6,* 499-506.

6. Sutton, J.R., & Farrell, P.A. (1988). Endocrine responses to prolonged exercise. In D. Lamb & R. Murray (Eds.), *Perspectives in exercise science and sports medicine: Prolonged exercise* (Vol. 1, pp. 153-208). Indianapolis: Benchmark Press.

▷ Selected Readings

Costill, D.L., Branam, G., Fink, W., & Nelson, R. (1976). Exercise induced sodium conservation: Changes in plasma renin and aldosterone. *Medicine and Science in Sports, 8,* 209-213.

Costill, D.L., Cote, R., Miller, E., Miller, T., & Wynder, S. (1975). Water and electrolyte replacement during repeated days of work in heat. *Aviation, Space, and Environmental Medicine, 46,* 795-800.

Farrell, P.A., Gates, W.K., Morgan, W.P., & Pert, C.B. (1983). Plasma leucine enkephalin-like radioreceptor activity and tension-anxiety before and after competitive running. In H.G. Knuttgen, J.A. Vogel, & J. Poortmans (Eds.), *Biochemistry of exercise* (pp. 637-644). Champaign, IL: Human Kinetics.

Galbo, H. (1983). *Hormonal and metabolic adaptation to exercise.* New York: Thieme-Stratton.

Ginzel, K.H. (1977). Interaction of somatic and autonomic functions in muscular exercise. *Exercise and Sport Sciences Reviews, 4,* 35-86.

Grossman, A., Bouloux, P., Price, P., Drury, P.L., Lam, K.S.L., Turner, T., Thomas, J., Besser, G.M., & Sutton, J. (1984). The role of opioid peptides in the hormonal responses to acute exercise in man. *Clinical Science, 67,* 483-491.

Kramer, W.J., Aguilear, B.A., Treada, M., Newton, R.U., Lynch, J.M., Rosendaal, G., McBride, J.M., Gordon, S.E., & Häkkinen, K. (1995). Responses of IGF-I to endogenous increases in growth hormone after heavy-resistance exercise. *Journal of Applied Physiology, 79,* 1310-1315.

Richter, E.A., & Sutton, J.R. (1994). Hormonal adaptation to physical activity. In C. Bouchard, R.J. Shephard, & T. Stephens (Eds.), *Physical activity, fitness, and health: International proceedings and consensus statement* (pp. 331-342). Champaign, IL: Human Kinetics.

Shangold, M.M. (1984). Exercise and the adult female: Hormonal and endocrine effects. *Exercise and Sport Sciences Reviews, 12,* 53-79.

Shephard, R.J., & Sidney, K.H. (1975). Effects of physical exercise on plasma growth hormone and cortisol levels in human subjects. *Exercise and Sport Sciences Reviews, 3,* 1-30.

Sutton, J.R., Farrell, P.A., & Harber, V.J. (1990). Hormonal adaptations to physical activity. In C. Bouchard, R. Shephard, T. Stephens, J. Sutton, & B. McPherson (Eds.), *Exercise fitness and health* (pp. 217-257). Champaign, IL: Human Kinetics.

Terjung, R. (1979). Endocrine response to exercise. *Exercise and Sport Sciences Reviews, 7,* 153-180.

Vander, A.J., Sherman, J.H., & Luciano, D.S. (1980). *Human physiology: The mechanisms of body function* (3rd ed.). New York: McGraw-Hill.

Wade, C.E. (1984). Response, regulation, and actions of vasopressin during exercise: A review. *Medicine and Science in Sports and Exercise, 16,* 506-511.

Weltman, A., Weltman, J.Y., Womack, C.J., Davis, S.E., Blumer, J.L., Gaesser, G.A., & Hartman, M.L. (1997). Exercise training decreases the growth hormone (GH) response to acute constant-load exercise. *Medicine and Science in Sports and Exercise, 29,* 669-676.

Winder, W.W. (1985). Regulation of hepatic glucose production during exercise. *Exercise and Sport Sciences Reviews, 13,* 1-31.

CHAPTER 6

METABOLIC ADAPTATIONS TO TRAINING

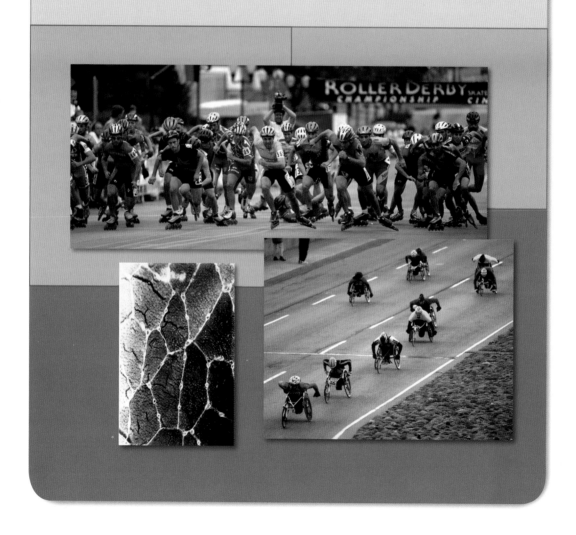

overview

We have discussed how our bodies use the foods we eat and the nutrients we store to produce the energy needed for physical activity, and we have examined the endocrine system's role in regulating the various metabolic processes that occur during exercise. But how can we maximize our potential to perform? How can we maximize our energy systems?

The answer is that we train. In this chapter, we turn our attention to the ways in which our bodies adapt to the repeated stimulus of training. We focus on the metabolic adaptations that enhance our capacity for physical activity. We discuss adaptations that occur within the muscles and in the energy systems that allow us to use our energy more efficiently. We examine the effects of both aerobic and anaerobic training and consider how we can maximize the improvements we gain from these types of training.

outline

Craig's wife thought he was crazy when, at age 37, he began jogging a few times per week with some friends. After all, he had never been interested in sports and had always shied away from any form of exercise. Despite his muscle soreness and inability to keep pace with the other runners, Craig increased his training to 6 days per week. After 8 weeks of this new exercise regimen, he suddenly realized that it was relatively easy to keep pace with the other runners. In fact, he had become the leader of the group and often felt as if he had to slow down to stay with the others. When the other runners decided to increase their training in preparation to run the Boston marathon, Craig considered it a challenge and joined the group in running 60 to 70 mi per week with Boston as his goal. During the race, he was able to run well ahead of his training partners, finishing more than an hour before them: His time of 2 h 45 min placed him in the top 5% of the more than 38,000 runners in the 1996 centennial race. Suddenly, it was apparent that Craig was gifted with an unusual capacity for endurance running, a talent that could be realized only after his body had been stimulated to adapt to the stresses of hard training.

Not all of the mechanisms that are responsible for improving muscular strength and anaerobic and aerobic power are fully understood. Nevertheless, we have been able to identify many of the metabolic and morphological changes that accompany days and weeks of exercise training. **Aerobic training**, or cardiorespiratory endurance training, for example, improves central and peripheral blood flow and enhances the capacity of the muscle fibers to generate greater amounts of adenosine triphosphate (ATP). **Anaerobic training**, on the other hand, increases muscular strength and promotes a greater tolerance for acid-base imbalances during highly intense effort. In the following pages, we discuss the importance of various metabolic changes that accompany physical training, and we discover how these changes can benefit physical performance.

Adaptations to Aerobic Training

Improvements in endurance that accompany daily aerobic training, such as jogging or swimming, result from many adaptations to the training stimulus. Some adaptations occur within the muscles, and many of these involve changes in the energy systems. Still other changes occur in the cardiovascular system, improving circulation to and within the muscles. In the following discussion, we focus on the muscular adaptations that occur with endurance training. (Cardiorespiratory adaptations to endurance training are discussed in detail in chapter 9.)

Changes in Aerobic Power

The most readily observable changes with aerobic training are an increased ability to perform prolonged submaximal exercise and an increase in one's maximal aerobic capacity ($\dot{V}O_2max$). Note, however, that there are wide individual variations in the degree of improvement in both **submaximal endurance** and $\dot{V}O_2max$ with any given training program. Whereas one individual may experience a 40% to 50% improvement in $\dot{V}O_2max$ as a consequence of endurance cycling, another person may show almost no change (less than 5%) as a result of the same training program (see sidebar).[14, 28] Of course, the status of one's physical fitness at the start of a training program will influence the magnitude of improvement observed during training. Individuals who already have a high level of fitness may show a smaller change in aerobic power than those who have led a sedentary life. Generally, the average increase in $\dot{V}O_2max$ from training studies that have used large numbers of subjects varies between 15% and 20%. As an example, Green and coworkers[14] observed a 15.6% increase in $\dot{V}O_2max$ in normally active men engaged in 2 h per day of cycling (at 62% $\dot{V}O_2max$) five to six times per week for 8 weeks. They noted that most of

Individual Differences in Response to Training—The HERITAGE Family Study

It has been clearly established that individuals differ considerably in their responses to a specific intervention, such as a drug or diet. The same is true for the response of $\dot{V}O_2$max to aerobic training, with studies showing improvements ranging from 0% to 50% or more in response to exactly the same training program.

The HERITAGE Family Study was funded by the National Institutes for Health in 1992 to study the possible genetic control of these variations in $\dot{V}O_2$max in response to aerobic training and the associated variations in the major risk factors for cardiovascular disease and type 2 diabetes. Funding has continued through the year 2004. Families were recruited for this study, including the natural mother and natural father and three or more of their children. Some families were included that did not meet these criteria.

Led by Dr. Claude Bouchard, executive director of the Pennington Biomedical Research Center, Louisiana State University, and pioneer in genetics research in the exercise sciences (see figure 6.1), a team of researchers from several universities (Arizona State University/Indiana University, Laval University in Canada, the University of Minnesota, and the University of Texas at Austin) recruited a total of 742 sedentary subjects from families of both white and black descent who completed the study. Another university, Washington University, was responsible for data quality control and analyses. These subjects completed a comprehensive battery of tests both before and after a 20-week program of aerobic training, with each training session supervised by an exercise physiologist. The test battery included physiological measures associated with aerobic fitness and clinical medicine markers associated with risk for cardiovascular disease and diabetes. The average increase in $\dot{V}O_2$max expressed per kilogram of body weight was 18%, but the range of increases varied from 0% to 53%. The increase in $\dot{V}O_2$max was influenced by genetics (maximal heritability = 47%; see figure 6.2), but was influenced very little by age, sex, and race.[3, 30] It is important to recognize from these data that each individual responds differently to the exact same exercise stimulus. We cannot expect the same improvement in all people! Just because a person has a low response to training does not mean that person did not follow the training program.

For more information on the HERITAGE Family Study, go to its Web site (http://www.pbrc.edu/heritage/home.htm).

▲ **Figure 6.1**   Dr. Claude Bouchard, Executive Director of the Pennington Biomedical Research Center, Louisiana State University, was the initiator of the HERITAGE Family Study.

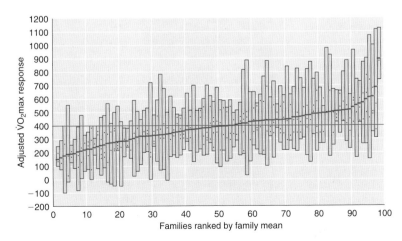

◀ **Figure 6.2**   Variations in the improvement in $\dot{V}O_2$max following 20 weeks of endurance training by family (The HERITAGE Family Study). Values represent the changes in $\dot{V}O_2$max in ml/min, with an average increase of 393 ml/min. Each family is enclosed within a bar, and each family member's value is represented within the bar.

Adapted, by permission, from C. Bouchard et al., 1999, "Familial aggregation of $\dot{V}O_2$max response to exercise training. Results from HERITAGE Family Study," *Journal of Applied Physiology* 87: 1003-1008.

this improvement occurred during the first 4 weeks of training.

There also appears to be an upper limit to the amount of improvement that can be achieved in one's aerobic power as a consequence of training. As training volume (e.g., distance run per training session) is increased, there appears to be a proportional gain in $\dot{V}O_2$max. Eventually, however, increasing the distance or time per training session will fail to improve aerobic power, despite longer and harder training sessions. The factors that regulate this "upper limit" are not fully understood but may be related to inherent factors that enable some individuals to achieve extremely high values (e.g., more than 80 ml $\cdot$ kg^{-1} $\cdot$ min^{-1}), whereas others are limited to aerobic capacities below 50 ml $\cdot$ kg^{-1} $\cdot$ min^{-1} despite equally rigorous training programs. Refer to chapter 9 for additional information on this topic.

Adaptations in Muscle

Repeated use of muscle fibers stimulates changes in their structure and function. Many of these changes were discussed in chapters 1 and 3 with reference to resistance training, but our main interest here is on aerobic training and the changes it produces in muscle fiber type, capillary supply, myoglobin content, mitochondrial function, and oxidative enzymes.

Muscle Fiber Type

As noted in chapter 1, aerobic activities such as jogging and low- to moderate-intensity cycling rely extensively on the slow-twitch (ST) fibers. In response to aerobic training, ST fibers become larger. That is, they have a larger cross-sectional area, although the magnitude of change depends on the intensity and duration of each training bout and the length of the training program. Increases of up to 25% have been reported. Fast-twitch (FT) fibers, because they are not being recruited to the same extent, generally do not increase cross-sectional area.

Most early studies found no change in the percentage of ST and FT fibers following aerobic training, but subtle changes were noted among FT fiber subtypes. FT$_b$ fibers are used less often than FT$_a$ fibers, and for that reason they have a lower aerobic capacity. Long-duration exercise may eventually recruit these fibers into action, demanding them to perform in a manner normally expected of the FT$_a$ fibers. This can cause some FT$_b$ fibers to take on the characteristics of the more oxidative FT$_a$ fibers. Recent evidence suggests that not only is there a transition of FT$_b$ to FT$_a$ fibers, but there can also be a transition of FT to ST fibers. The magnitude of change is generally small, not more than a few percentage points. As an example, in the HERITAGE Family Study,[27] a 20-week program of aerobic training increased ST fibers from 43.2% pretraining to 46.7% posttraining and decreased FT$_b$ fibers from 20.0% to 15.1%, with FT$_a$ remaining essentially unchanged. These more recent studies have included larger numbers of subjects and have taken advantage of improved measurement technology, both of which might explain why changes are now being recognized.

Capillary Supply

One of the most important adaptations to aerobic training is an increase in the number of capillaries surrounding each muscle fiber. The micrographs in figure 6.3 illustrate that endurance-trained men can have considerably more capillaries in their leg muscles than sedentary individuals.[17, 20] With long periods of aerobic training, the number of capillaries has been shown to increase by more than 15%.[27] Having more capillaries allows greater exchange of gases, heat, wastes, and nutrients between the

▲ **Figure 6.3** Micrographs of the capillaries around the muscle fibers show that capillaries are less numerous in (*a*) untrained men than in (*b*) trained male distance runners.

blood and the working muscle fibers. In fact, the increase in capillary density (i.e., increase in capillaries per muscle fiber) is potentially one of the most important alterations in response to training that allows the increase in $\dot{V}O_2$max. It is now clear that the diffusion of oxygen from the capillary to the mitochondria is a major factor limiting the maximal rate of oxygen consumption.[25] Increasing capillary density facilitates this diffusion, thus maintaining an environment well suited to energy production and repeated muscle contractions. Substantial increases in muscle capillary number occur within the first few weeks or months of training. But little research has been performed to determine what capillary changes occur with longer training periods.

> Aerobic training increases both the number of capillaries per muscle fiber and the number of capillaries for a given cross-sectional area of muscle. Both of these changes improve blood perfusion through the muscles, thereby enhancing the exchange of gases, wastes, and nutrients between the blood and muscle fibers.

Myoglobin Content

When oxygen enters the muscle fiber, it binds to **myoglobin,** a compound similar to hemoglobin. This iron-containing compound shuttles the oxygen molecules from the cell membrane to the mitochondria. The ST fibers contain large quantities of myoglobin, which gives these fibers their red appearance (myoglobin is a pigment that turns red when bound to oxygen). The FT fibers, on the other hand, are highly glycolytic, so they require (and have) little myoglobin, giving them a whiter appearance. More important, their limited myoglobin supply limits their oxygen capacity, resulting in poor aerobic endurance for these fibers.

Myoglobin stores oxygen and releases it to the mitochondria when oxygen becomes limited during muscle action. This oxygen reserve is used during the transition from rest to exercise, providing oxygen to the mitochondria during the lag between the beginning of exercise and the increased cardiovascular delivery of oxygen.

Myoglobin's precise contributions to oxygen delivery are not yet fully understood. But aerobic training has been shown to increase muscle myoglobin content by 75% to 80%. This adaptation would be expected only if it enhances a muscle's capacity for oxidative metabolism.

Mitochondrial Function

As noted in chapter 4, aerobic energy production is conducted in the mitochondria. Not surprisingly, then, aerobic training also induces changes in mitochondrial function that improve the muscle fibers' capacity to produce ATP. The ability to use oxygen and produce ATP via oxidation depends on the number, size, and efficiency of the muscle's mitochondria. All three of these qualities improve with aerobic training.

During one study that involved endurance training in rats, the actual number of mitochondria increased approximately 15% during 27 weeks of exercise.[19] At the same time, the average mitochondrial size also increased by about 35% over the entire period. We now know that as the volume of aerobic training increases, so do the number and size of the mitochondria.

> Skeletal muscle mitochondria increase in both size and number with aerobic training, providing the muscle with much more efficient oxidative metabolism.

Oxidative Enzymes

Regular endurance exercise has been shown to induce major adaptations in skeletal muscle, including an increase in the number and size of the muscles' mitochondria, as we have just discussed. These changes are further enhanced by an increase in mitochondrial efficiency. Recall from chapter 4 that the oxidative breakdown of fuels and the ultimate production of ATP depend on the action of **mitochondrial oxidative enzymes,** the special proteins that catalyze (i.e., speed up) the breakdown of nutrients to form ATP. Aerobic training increases these enzymes' activities. As a consequence of such training, exercise at a given intensity results in a smaller disturbance in homeostasis.

Figure 6.4 illustrates the changes in the activity of **succinate dehydrogenase (SDH)**, one of the muscles' key oxidative enzymes, during 7 months of gradually increased swimming training. Interestingly, although this enzyme activity continued to increase throughout the period of training, little change in the body's maximal oxygen uptake occurred during the final 6 weeks of training. This suggests that $\dot{V}O_2max$ might be more influenced by the circulatory system's limitations for transporting oxygen than by the muscles' oxidative potential.

The activities of muscle enzymes such as succinate dehydrogenase and citrate synthase are dramatically influenced by aerobic training. This is seen in figure 6.5, which compares the activities of these enzymes in untrained, moderately trained, and highly trained people.[9] Even moderate amounts of daily exercise increase these enzyme activities and thus the muscles' aerobic capacity. For example, jogging or cycling for as little as 20 min per day has been shown to increase the leg muscles' SDH activity by more than 25% that of sedentary individuals. Training more vigorously, such as 60 to 90 min per day, produces a 2.6-fold increase in this activity.

The training-induced increase in the activities of these oxidative enzymes reflects both the increase in the number and size of a muscle's mitochondria and an improved capacity for ATP production. Initially, the enzyme activity increase coincides with improvements in the individual's $\dot{V}O_2max$. But only a weak relationship, at best, exists between increased muscle oxidative enzyme activities and enhancement of $\dot{V}O_2max$.[13, 28] Holloszy and Coyle[18] suggested that the major metabolic consequence of mitochondrial changes induced by aerobic training is a slower use of muscle glycogen and a reduced production of lactate during exercise of a given intensity. Either effect can enhance endurance performance. This increase in the oxidative enzymes with aerobic training most likely improves your ability to sustain a higher exercise intensity, such as maintaining a faster race pace in a 10-km run. This improvement is related to the increase in lactate threshold discussed later in this chapter.[18]

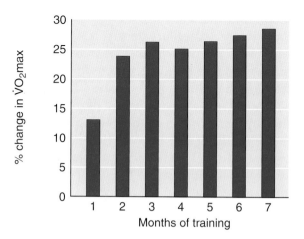

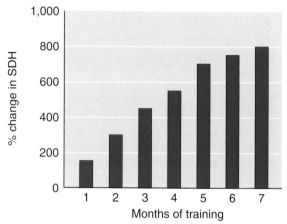

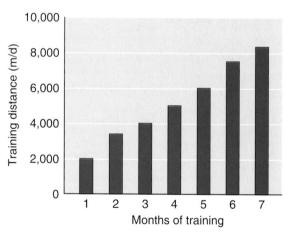

▲ **Figure 6.4**   The percentage of change in the activity of succinate dehydrogenase (SDH), one of the muscles' key oxidative enzymes, and maximal oxygen uptake during 7 months of swimming training. Interestingly, although this enzyme activity continues to increase with increasing levels of training, the swimmers' maximal oxygen uptake appears to level off after the first 8 to 10 weeks of training. It appears that the mitochondrial enzyme activities are not a direct indication of one's aerobic capacity and might not be the best indicators of whole-body endurance capacity.

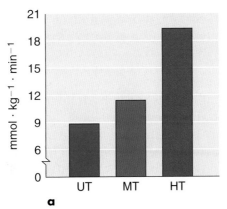
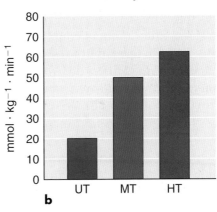

▲ **Figure 6.5**   Leg muscle (gastrocnemius) enzyme activities of untrained (UT) subjects, moderately trained (MT) joggers, and highly trained (HT) marathon runners. Enzyme levels shown are for (*a*) succinate dehydrogenase and (*b*) citrate synthase, two of many enzymes that participate in the oxidative production of adenosine triphosphate.

Adapted, by permission, from D.L. Costill et al., 1979, "Lipid metabolism in skeletal muscle of endurance-trained males and females." *Journal of Applied Physiology* 28: 251-255 and from D.L. Costill et al., 1979, "Adaptations in skeletal muscle following strength training," *Journal of Applied Physiology* 46: 96-99.

▶ Aerobic training stresses ST muscle fibers more than FT fibers. Consequently, the ST fibers tend to enlarge with training. There appears to be a small increase in the percentage of ST fibers in addition to a transition of FT_b to FT_a fibers.

▶ The number of capillaries supplying each muscle fiber increases with training.

▶ Aerobic training increases muscle myoglobin content by about 75% to 80%. Myoglobin stores oxygen.

▶ Aerobic training increases both the number and the size of the mitochondria.

▶ Activities of many oxidative enzymes are increased with aerobic training.

▶ All these changes that occur in the muscles, combined with adaptations in the oxygen transport system, enhance functioning of the oxidative system and improve endurance.

Adaptations Affecting Energy Sources

Aerobic training places repeated demands on muscles' stores of both glycogen and fat. Not surprisingly, our bodies adapt to this repeated stimulus to make energy production more efficient and to reduce our risk of fatigue. Let's examine the adaptations in the way a trained body metabolizes both carbohydrate and fat for energy.

Carbohydrate for Energy

Muscle glycogen is used extensively during each training bout, so the mechanisms responsible for its resynthesis are stimulated after each session, allowing the depleted glycogen stores to be replenished. With adequate rest and sufficient dietary carbohydrate intake, trained muscle stores considerably more glycogen than untrained muscle does. In one study, endurance exercise training doubled the rate of muscle glycogen accumulation or storage following a glycogen-depleting exercise bout.[15] Furthermore, when distance runners stop training for several days and eat a diet rich in carbohydrates (400-550 g/day), their muscle glycogen levels increase to nearly twice the levels of sedentary people who follow the same dietary regimen. More glycogen storage allows the athlete to better tolerate the subsequent demands of training because more fuel is available for use. Additional information regarding the dietary needs for training appears in chapter 13.

Fat for Energy

In addition to their greater glycogen content, endurance-trained muscle fibers contain substantially more fat (also termed lipid) stored as triglyceride than untrained fibers do. Although information about the mechanisms responsible for this improvement in fuel storage with aerobic training is limited, a 1.8-fold increase in muscle triglyceride content has been reported after only 8 weeks of endurance running.[11] In general, as shown in figure 6.6, the vacuoles that contain triglyceride are distributed throughout the muscle fiber but are generally close to the mitochondria. Thus, access to them for use as fuel during exercise is easy.

In addition, activities of many muscle enzymes responsible for the β-oxidation of lipids increase with endurance training. This adaptation enables endurance-trained muscle to burn or oxidize lipids more efficiently, reducing the demands placed on a muscle's glycogen supply. Muscle samples taken from men's thighs before and after cycle training have shown a 30% increase in the ability to oxidize free fatty acids (FFAs). Aerobic training also increases the rate at which FFAs are released from storage during prolonged exercise, making them readily available for the muscles' use.

Issekutz and coworkers[21] observed that this elevation of blood levels of FFAs enables the muscle to burn more fat and less carbohydrate. Subsequent studies have shown that elevated blood FFA levels can spare muscle glycogen, postponing exhaustion.[7, 26] For any given rate of work, trained individuals tend to use more

> With aerobic training, you become much more efficient at using fat as an energy source for exercise. This allows muscle and liver glycogen to be used at a slower rate.

fat and less carbohydrate for energy than untrained people do.

To summarize, improvements in muscles' aerobic energy system increase the capacity to produce energy, with a shift toward greater reliance on fat for ATP production. The improved capacity of endurance-trained muscles to use fat is caused by the following:

- Increased storage of fat within the muscle fiber
- Enhanced ability to mobilize FFAs
- Improved capacity to oxidize fat

In activities lasting several hours, these adaptations prevent early muscle glycogen depletion and thus ensure a continued supply of ATP. Thus, endurance performance is enhanced.

Balancing the Use of Carbohydrate and Fat

In recent years, efforts have been made to understand the factors that regulate the balance between carbohydrate and fat use during exercise and after aerobic training. As noted in chapter 4, carbohydrate is the predominant energy source with increasing levels of effort, whereas aerobic training results in a shift toward greater fat use at a given submaximal exercise intensity. In a review of the literature on this topic, Brooks and Mercier[4] proposed a "crossover" concept to explain the interaction of exercise intensity and training on the balance of carbohydrate and lipid use. They pointed out that at exercise levels below 45% of $\dot{V}O_2$max, lipid is the main substrate used for energy, whereas at more intense levels of effort (greater than 70% of $\dot{V}O_2$max) carbohydrate is the predominant substrate of choice. The "crossover point" is the power output at which energy from carbohydrate-derived fuels predominates over energy from lipids (figure 6.7). With endurance training, sympathetic nervous system activity decreases, thereby reducing carbohydrate use during submaximal exercise.

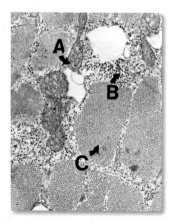

▲ **Figure 6.6** An electron micrograph showing (A) mitochondria, (B) muscle glycogen granules, and (C) triglyceride vacuoles.

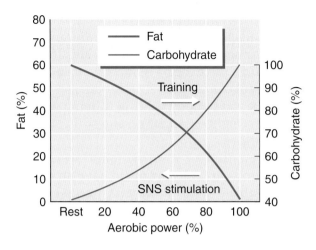

◀ **Figure 6.7**   The reliance on carbohydrate for energy increases with the intensity (% V̇O₂max) of exercise, whereas the use of fat tends to decline as the exercise intensity increases. Training tends to shift both curves to the right, while the stimulation from the sympathetic nervous system (SNS) declines. As a result, training tends to decrease the need for carbohydrate as the primary energy source but increases the use of fat.

Adapted, by permission, from G.A. Brooks and J. Mercier, 1994, "Balance of carbohydrate and lipid utilization during exercise: The 'crossover' concept," *Journal of Applied Physiology* 76: 2253-2261.

Muscle's Maximal Oxygen Uptake

In the laboratory, researchers can measure the aerobic capacity of a specimen of muscle obtained by a needle biopsy. When researchers grind the muscle sample in a solution containing other essential items, the mitochondria are stimulated to use oxygen and to produce ATP. As a result, the maximal rate at which the muscle's mitochondria use oxygen to generate ATP can be measured. This procedure therefore measures the muscle's maximal respiratory capacity, or $\dot{Q}O_2$. In effect, $\dot{Q}O_2$ is a measure of the muscle's maximal oxygen uptake (in contrast to $\dot{V}O_2$max, which is a measure of the body's maximal oxygen uptake).

In figure 6.8, samples from the gastrocnemius muscles of three groups of people show that untrained muscles have $\dot{Q}O_2$ values of about 1.5 microliters (μL) of oxygen per hour for each gram of muscle ($\mu L \cdot h^{-1} \cdot g^{-1}$). In contrast, muscles from individuals who expend 1,500 to 2,500 kcal/week during training (such as jogging 25-40 km/week, or 15.5-25 mi/week) have $\dot{Q}O_2$ values around 2.7 $\mu L \cdot h^{-1} \cdot g^{-1}$, which is 1.8 times that of untrained people. Highly trained marathon runners, such as those who expend more than 5,000 kcal/week in training (about 80 km/week, or 50 mi/week), have displayed $\dot{Q}O_2$ values in excess of 4.0 $\mu L \cdot h^{-1} \cdot g^{-1}$, nearly 2.7 times that of untrained people.

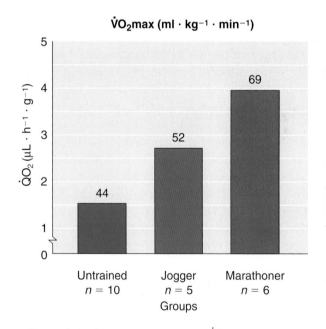

▲ **Figure 6.8**   Respiratory capacities ($\dot{Q}O_2$) in the gastrocnemius muscles and maximal oxygen uptake ($\dot{V}O_2$max) of untrained subjects, moderately trained joggers (25-40 km/week), and highly trained marathon runners (80 km/week). Note that $\dot{Q}O_2$ is somewhat proportional to training volume. Numbers above each column represent each group's whole body maximal oxygen uptake (ml $\cdot$ kg⁻¹ $\cdot$ min⁻¹).

► Endurance-trained muscle stores considerably more glycogen than does untrained muscle.

► Endurance-trained muscle also stores more fat (triglyceride) than does untrained muscle.

► The activities of many enzymes involved in β-oxidation of fat increase with training; thus, FFA levels increase. This increases the use of fat as an energy source, sparing glycogen.

Training the Aerobic System

Clearly, endurance-trained individuals have an advantage over those who are less trained. But how do they gain this advantage? In the following sections, we examine various aspects of aerobic training in an effort to understand how we can maximize aerobic benefits from training.

Volume of Training

Training adaptations are best achieved when an optimal amount of work is performed in each training session and over a given period of time. Although this optimal load may differ from one individual to another, observations with distance runners suggest that, on the average, the ideal training regimen may be equivalent to an energy expenditure of between 5,000 and 6,000 kcal/week (approximately 715-860 kcal/day). This translates to 80 to 95 km (50-60 mi) of running per week.[6] To achieve the same aerobic benefits, swimmers would need to swim roughly 4,000 to 6,000 m (4,375-6,560 yd) per day.[10] Of course, these are only estimates of the stimulus required for muscular conditioning. Some individuals may show greater improvement with less training, whereas others may require more.

How much your aerobic capacity improves is determined, in part, by how many calories you expend during each training bout and how much work you accomplish over a period of

weeks. Many athletes and coaches believe that this means that gains in aerobic endurance are proportional to the volume of training. If training volume were the most important stimulus for muscular adaptations, individuals who expend the most energy during training should have the highest $\dot{V}O_2$max values. But this is not the case.

The amount of improvement attainable with aerobic training seems to have an upper limit. Athletes who train with progressively greater workloads will eventually reach a maximal level of improvement beyond which additional increases in training volume won't improve endurance or $\dot{V}O_2$max. This is illustrated in figure 6.9, which shows the changes in $\dot{V}O_2$max for two distance runners before and during various levels of training. The runners' $\dot{V}O_2$max increased dramatically with the initiation of training (40 km/week, or 25 mi/week) and continued to improve when they increased their training to 80 km/week (50 mi/week). Beyond that level of training, however, these individuals had no additional gains in endurance. During a 1-month period, they increased their training to more than 350 km/week (217 mi/week) but still showed no further improvement in endurance.[6]

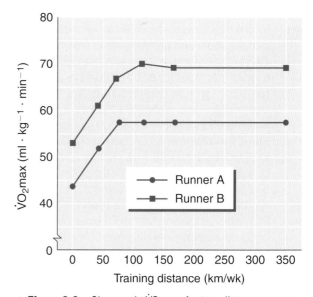

▲ **Figure 6.9** Changes in $\dot{V}O_2$max for two distance runners while they trained at various distances. Training 40 to 80 km/week (25-50 mi/week) improved aerobic capacity, whereas greater training distances produced no additional increase.

fyi As you increase your central circulatory capacity with endurance training, your whole body's aerobic capacity increases. There is also an increase in the maximal oxidative or respiratory capacity ($\dot{Q}O_2$) of your muscle. The highest average $\dot{Q}O_2$ value reported in human muscle (5.2 μL · h⁻¹ · g⁻¹) was from the deltoid muscles of swimmers who had expended more than 10,000 kcal/week during training.[5]

Some athletes try to perform more physical work by training twice each day, doubling their normal training volume. If an athlete is already expending 1,000 kcal/day in training, a second training session is unlikely to produce additional benefits. (This is discussed further in chapter 12.)

Intensity of Training

The degree of adaptation to aerobic training depends not only on training volume but also on training intensity. Let's briefly consider the importance of training intensity for performance in prolonged events. (This topic is examined more closely in chapter 12.)

Muscular adaptations are specific to both the speed and duration of effort performed during training. Runners, cyclists, and swimmers who incorporate intermittent high-intensity bouts of exercise into their training regimens show more improvement in performance than those who perform only long, slow, low-intensity training bouts. Long-distance, low-intensity training doesn't develop the neu-

rological patterns of muscle fiber recruitment and the high rate of energy production required for maximal endurance performance.

High-intensity speed training can include either intermittent exercise (intervals) or continuous exercise at near-competition pace. Let's look at the benefits from each type of training.

Interval Training

Although interval training has been used for many years, most athletes use it mainly to improve their anaerobic capacity. Consequently, most repeated exercise bouts are performed at speeds that produce large amounts of lactate. But this training format also can be used to develop the aerobic system. Repeated, fast-paced, brief exercise bouts with short rest intervals between bouts achieve the same aerobic benefits as long, high-intensity, continuous exercise.

This form of aerobic interval training has become the framework for aerobic conditioning, particularly in competitive swimming. It involves repeated, short efforts lasting from 30 s to 5 min (50-400 m of swimming) performed just under race pace but with very brief rest intervals (5-15 s). Such short rest intervals force the participant to exercise at an aerobic level and to rely little on the glycolytic lactate-producing system.

Table 6.1 offers an example of an aerobic interval training program for runners. Because volume is the key to successful aerobic training, the runner must perform a large number of these repeated runs. In this example, 20

Table 6.1

Example of Aerobic Intervals for Runners Training for a 10-km Race

Best 10 km (min:s)	Reps	Interval distance (m)	Rest (s)	Pace (min:s)
46:00	20	400	10–15	2:00
43:00	20	400	10–15	1:52
40:00	20	400	10–15	1:45
37:00	20	400	10–15	1:37
34:00	20	400	10–15	1:30

repetitions of 400 m are performed, resulting in a total run of 8,000 m (approximately 5 mi). The pace maintained is slightly slower than that used during the 10-km race, in this case slower by 8 to 10 s per 400 m. Yet this pace is generally faster than could be sustained easily during a continuous 8,000-m run. The difficulty of this interval set is that the prescribed rest between repetitions must be relatively brief—only about 10 to 15 s. Such short rest intervals allow little time for muscles to recover, yet they provide a brief respite from muscular stress.

Continuous Training

It can be argued that a single, continuous bout of high-intensity exercise can give the same aerobic benefits as a set of aerobic intervals. But some athletes find continuous endurance exercise boring. At present, no direct evidence shows that aerobic interval training produces greater muscular adaptations than continuous training bouts. Whether the training is performed as one long bout of continuous exercise or in a series of shorter intervals, the aerobic muscular benefits seem to be about the same. Personal preference, therefore, may be the deciding factor in choosing continuous or interval aerobic training.

▶ The ideal training regimen should have a caloric expenditure of about 5,000 to 6,000 kcal/week. There seems to be little benefit for the aerobic systems beyond this level.

▶ Intensity is also a critical factor in improving performance. Adaptations are specific to the speed and duration of training bouts, so those who perform at higher intensities must train at higher intensities.

▶ Aerobic interval training involves repeated bouts of high-intensity performance separated by brief rest periods. This training, although traditionally considered only anaerobic, generates aerobic benefits because the rest period is so brief that full recovery can't occur; thus, the aerobic system is stressed.

▶ Continuous training is done as one prolonged bout of exercise, but many athletes find it boring.

▶ The aerobic benefits from both interval training and continuous high-intensity training seem to be about the same.

Adaptations to Anaerobic Training

In muscular activities that require near-maximal force production, such as sprint running, cycling, and swimming, much of the energy needs are met by the ATP–phosphocreatine (PCr) system and the anaerobic breakdown of muscle glycogen (glycolysis). In the following discussion, we focus on the trainability of these two systems.

Changes in Anaerobic Power and Anaerobic Capacity

Unfortunately, scientists have had difficulty agreeing on an appropriate laboratory or field test of anaerobic power. Unlike aerobic power, where $\dot{V}O_2max$ is generally agreed to be the best marker, no single test adequately represents anaerobic power. Most research has been conducted by using three different tests of either or both anaerobic power and anaerobic capacity: the Wingate anaerobic test, the critical power test, and the maximal accumulated oxygen deficit test (see chapter 4). Of these three, the Wingate test has been the most widely used.

With the Wingate anaerobic test, the subject pedals a cycle ergometer at maximal speed for 30 s against a high braking force. The braking force is determined by the person's weight, sex, age, and level of training. Power output can be determined instantaneously throughout the 30-s test but is generally averaged over 3- to 5-s intervals. The peak power output is the highest mechanical power achieved at any stage in the test; it is generally achieved during the first 5 to 10 s and is considered an index of anaerobic power. The mean power output is computed as the average power output over the total 30-s period, and total work is simply obtained by multiplying the mean power output by 30 s. Mean power output and total work have been used as indexes of anaerobic capacity.[1]

With anaerobic training, such as sprint training on the track or on a cycle ergometer, many studies have found increases in both peak anaerobic power and anaerobic capacity. However, results have ranged widely across studies, from those that found no significant increases to those that reported increases of up to 25%.

Adaptations in Muscle

With anaerobic training, which includes sprint training and resistance training, there are changes in skeletal muscle that specifically reflect muscle fiber recruitment for these types of activities. As discussed in chapter 1, as you work at higher intensities, you recruit FT muscle fibers to a greater extent, but not exclusively, because ST fibers continue to be recruited. Overall, sprint and resistance types of activities use the FT muscle fibers significantly more than do aerobic activities. Consequently, both FT_a and FT_b muscle fibers undergo an increase in their cross-sectional areas. The cross-sectional area of ST fibers also is increased but usually to a lesser extent. Furthermore, with sprint training there appears to be a reduction in the percentage of ST fibers and an increase in the percentage of FT fibers, with the greatest change in FT_a fibers. In two of these studies, where subjects performed 15-s or 15-s and 30-s all-out sprints, the ST percentage decreased from 57% to 48% and FT_a increased from 32% to 38%.[22, 23] This shift of ST to FT fibers usually is not seen with just resistance training.

Adaptations in the Energy Systems

Just as aerobic training produces changes in the aerobic energy system, anaerobic training alters the ATP–PCr and anaerobic glycolytic energy systems. These changes are not exactly what we would predict, but they do improve performance of an anaerobic nature.

Adaptations in the ATP–PCr System

Activities that emphasize maximal muscle force production, such as sprinting and weightlift-

ing events, rely most heavily on the ATP–PCr system for energy. Maximal efforts lasting less than about 6 s place the greatest demands on the breakdown and resynthesis of ATP and PCr. Few studies have examined the training adaptations to short, maximal exercise bouts that are aimed specifically at developing the ATP–PCr system. In 1979, however, Costill and coworkers reported their findings from one such study.[8] The participants performed maximal knee extensions for training. One leg was trained by using 6-s maximal work bouts that were repeated 10 times. This type of training preferentially stresses the ATP–PCr energy system. The other leg was trained with repeated 30-s maximal bouts, which instead preferentially stress the glycolytic system.

Both forms of training produced the same muscular strength gains (about 14%) and the same resistance to fatigue. As seen in figure 6.10, the activities of the anaerobic muscle enzymes creatine kinase and myokinase increased as a result of the 30-s training bouts but were almost unchanged in the leg trained with repeated 6-s maximal efforts. This finding leads us to conclude that maximal sprint bouts (6-s) might improve muscular strength but contribute little to the mechanisms responsible for ATP breakdown. Such training

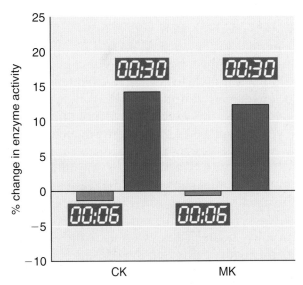

▲ **Figure 6.10**   Changes in creatine kinase (CK) and muscle myokinase (MK) activities as a result of 6-s and 30-s bouts of maximal anaerobic training.

enhances performance by improving strength but results in little or no improvement in the energy release from ATP and PCr.

Another study, however, showed improvements in these ATP–PCr enzyme activities with training bouts lasting only 5 s.[31] Regardless of the conflicting results, these studies suggest that the major value of training bouts that last only a few seconds (sprints) is the development of muscular strength. Such strength gains enable the individual to perform a given task with less effort, which reduces the risk of fatigue. Whether these changes allow the muscle to perform more anaerobic work remains unanswered, although a 60-s sprint-fatigue test suggests that short sprint-type anaerobic training does not enhance anaerobic endurance.[8]

Adaptations in the Glycolytic System

Anaerobic training (30-s bouts) increases the activities of several key **glycolytic enzymes.** The most frequently studied glycolytic enzymes are **phosphorylase, phosphofructokinase (PFK),** and **lactate dehydrogenase.** In one study, the activities of these three enzymes increased 10% to 25% with repeated 30-s training bouts but changed little with short (6-s) bouts that stress primarily the ATP–PCr system.[8] In a more recent study, 30-s maximal all-out sprints significantly increased hexokinase (56%) and PFK (49%) but not total phosphorylase activity or lactate dehydrogenase.[24]

Because both PFK and phosphorylase are essential to the anaerobic yield of ATP, such training might enhance glycolytic capacity and allow the muscle to develop greater tension for a longer period of time. However, as seen in figure 6.11, this conclusion is not supported by results of a 60-s sprint performance test, in which the subjects performed maximal knee extension and flexion. Power output and the rate of fatigue (shown by a decrease in power production) were affected to the same degree after sprint training with either 6- or 30-s training bouts. Thus, we must conclude that performance gains with these forms of training result from improvements in strength rather than improvements in the anaerobic yield of ATP.

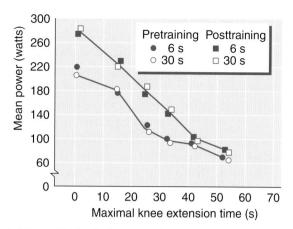

▲ **Figure 6.11** Performance in a 60-s sprint bout after training with 6-s and 30-s anaerobic bouts. Subjects are the same as in figure 6.8.

Anaerobic training increases the ATP–PCr and glycolytic enzymes but has no effect on the oxidative enzymes. Conversely, aerobic training increases the oxidative enzymes but has little effect on the ATP–PCr or glycolytic enzymes. This fact reinforces a recurring theme: Physiological alterations that result from training are highly specific to the type of training.

Adaptations in Lactate Threshold

Lactate threshold is a physiological marker that is closely associated with aerobic endurance performance—the higher the lactate threshold, the better the aerobic performance. However, because lactate is a product of anaerobic metabolism, we discuss lactate threshold in this section. Figure 6.12a illustrates the difference in lactate threshold between an endurance-trained individual and an untrained individual. This figure also accurately represents the changes in lactate threshold following a 6- to 12-month program of endurance training. In either case, in the trained state, you can exercise at a higher percentage of your $\dot{V}O_2$max before lactate begins to accumulate in the blood. Because race pace in aerobic endurance events is closely associated with lactate threshold,[12] this translates into a much faster race pace (see figure 6.12b). The reduction in lactate values at the same rate

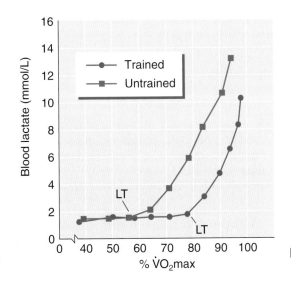

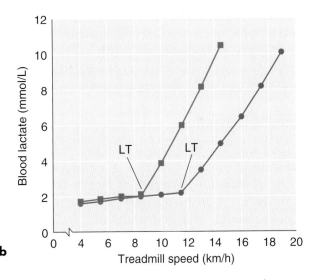

▲ **Figure 6.12** Changes in lactate threshold with training expressed as (*a*) a percentage of maximal oxygen uptake (% $\dot{V}O_2$max) and (*b*) an increase in speed on the treadmill. Lactate threshold (LT) occurs at a speed of 8.4 km/h in the untrained state and 11.6 km/h in the trained state.

of work is likely attributable to a combination of reduced lactate production and increased lactate clearance.[2]

The concentration of lactate in the blood following a fixed-pace swim or run provides an excellent means of monitoring the physiological changes that occur with training. As you become better trained, your blood lactate concentration is lower for the same rate of work. The results of a study using this technique are shown in figure 6.13. Blood lactate concentration following a standardized 200-m swim declined steadily in swimmers over a 7-month training period. This suggests that they were developing a greater aerobic capacity, a reduced reliance on the glycolytic system for energy, or perhaps both.

Other Adaptations to Anaerobic Training

How else might anaerobic training improve performance? In addition to strength gains, at least three other changes may enhance performance and delay fatigue in highly anaerobic events. These three changes are improvements in efficiency of movement, aerobic energetics, and buffering capacity.

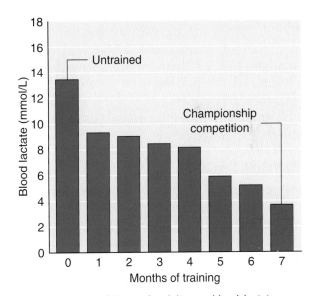

▲ **Figure 6.13** Effects of training on blood lactate concentration after a 200-m swim at a predetermined speed. The lowest lactate values were recorded during the period when the swimmers produced their best performances (i.e., in championship competition).

Efficiency of Movement

Training at high speeds improves your skill and coordination for performing at higher intensities. From our discussion of selective muscle fiber recruitment in chapter 1, we can assume that anaerobic training optimizes fiber recruitment to allow more efficient movement. Training at fast speeds and with heavy loads improves your efficiency, economizing your muscles' use of their energy supply.

Aerobic Energetics

Anaerobic training stresses more than the anaerobic energy systems. Part of the energy needed for sprints that last at least 30 s is derived from oxidative metabolism. Consequently, repeated bouts of sprint-type exercise (such as 30-s, maximal-effort bouts) also increase the muscles' aerobic capacity.[8, 28] Although this change is often small, we can reasonably expect that this enhancement of the oxidative potential of muscles helps the anaerobic energy systems meet muscle energy needs during highly anaerobic effort. Possibly related to this, sprint training (30-s bouts) also has been reported to increase the oxidative enzymes malate dehydrogenase (29%), succinate dehydrogenase (65%), and citrate synthase (36%).[24] Furthermore, sprint training (30-s all out sprints) has been shown to reduce anaerobic ATP generation, possibly the result of enhanced aerobic metabolism, thus increasing the time to fatigue.[16]

Buffering Capacity

Anaerobic training improves the muscles' capacity to tolerate the acid that accumulates within them during anaerobic glycolysis. As discussed in chapter 4, lactic acid accumulation is considered a major cause of fatigue during sprint-type exercise because the H^+ that dissociates from lactic acid is thought to interfere with both metabolism and the contractile process. **Buffers** (such as bicarbonate and muscle phosphates) combine with hydrogen to reduce the fiber's acidity; thus, buffers can delay the onset of fatigue during anaerobic exercise.

Muscle buffering capacity has been shown to increase by 12% to 50% with 8 weeks of anaerobic training.[29] Aerobic training, on the other hand, has no effect on buffer capacity. As with other training adaptations, changes in muscle buffering capacity are specific to the intensity of exercise performed during training. It should

be noted, however, that a recent study using 30-s all-out sprint bouts as the training stimulus did not find changes in muscle buffering capacity.[16]

> Anaerobic training improves muscle buffering capacity, but aerobic training does little to enhance the muscles' capacity to tolerate sprint-type activities.

With increased buffering capacity, sprint-trained subjects can accumulate more lactate in their blood and muscles during and following an all-out sprint to exhaustion than untrained individuals can. This is because the H^+ that dissociates from the lactic acid, not the lactate that accumulates, leads to fatigue. With enhanced buffering capacity, muscles can generate energy for longer periods before a critically high concentration of H^+ inhibits the contractile process. As discussed in chapter 4, once intracellular pH decreases below 6.9, critical glycolytic enzymes start to be inhibited. At a pH of 6.4, glycogen breakdown is no longer possible.

Interestingly, under similar circumstances (such as sprinting to exhaustion), endurance-trained subjects neither accumulate as much muscle lactate as sprint-trained individuals nor experience the unusually low pH values seen in sprint-trained individuals. This difference is difficult to explain, but muscle pH doesn't appear to limit sprint performance for endurance-trained people.

Specificity of the Training Response

Table 6.2 shows the activities of selected muscle enzymes from the three energy systems for untrained, anaerobically trained, and aerobically trained men. This table shows that aerobically trained muscles have significantly lower glycolytic enzyme activities. Thus, they might have less capacity for anaerobic metabolism or they might rely less on energy from glycolysis. More research is needed to explain the implications of the muscular changes accompanying both anaerobic and aerobic training, but this table clearly shows the high degree of specificity to a given training stimulus.

Table 6.2

Selected Muscle Enzyme Activities (mmol · g⁻¹ · min⁻¹) for Untrained, Anaerobically Trained, and Aerobically Trained Men

	Untrained	Anaerobically trained	Aerobically trained
Aerobic enzymes			
Oxidative system			
Succinate dehydrogenase	8.1	8.0	20.8[a]
Malate dehydrogenase	45.5	46.0	65.5[a]
Carnitine palmityl transferase	1.5	1.5	2.3[a]
Anaerobic enzymes			
ATP–PCr system			
Creatine kinase	609.0	702.0[a]	589.0
Myokinase	309.0	350.0[a]	297.0
Glycolytic system			
Phosphorylase	5.3	5.8	3.7[a]
Phosphofructokinase	19.9	29.2[a]	18.9
Lactate dehydrogenase	766.0	811.0	621.0

[a]Significant difference from the untrained value.

▶ Anaerobic training bouts improve anaerobic performance, but the improvement appears to result more from strength gains than from improvements in the functioning of the anaerobic energy systems.

▶ Anaerobic training also improves the efficiency of movement, and more efficient movement requires less energy expenditure.

▶ Although sprint-type exercise is anaerobic by nature, part of the energy used during longer sprint bouts comes from oxidation, so muscle aerobic capacity also can be increased with this type of training.

▶ Muscle buffering capacity is increased by anaerobic training, permitting higher muscle and blood lactate levels.

Improved buffering capacity allows the H⁺ that dissociates from lactic acid to be neutralized, thus delaying fatigue.

▶ Changes in muscle enzyme activity are highly specific to the type of training.

In Closing . . .

In this part, we have focused on the energy needed for physical performance. In chapter 4, we reviewed the systems responsible for the production of energy. In chapter 5, we discussed the regulatory role of the endocrine system. Finally, in this chapter, we looked at the metabolic adaptations within the muscles and in the energy systems themselves that accompany aerobic and anaerobic training, and we saw how these types of training can improve performance.

In part III, we turn our attention to the oxygen transport system. We begin in chapter 7 with a discussion of the cardiovascular system and its role in transporting oxygen, fuels, nutrients, and hormones throughout our bodies.

▶ Key Terms

aerobic interval training

aerobic training

anaerobic training

buffers

continuous training

creatine kinase

glycolytic enzymes

interval training

lactate dehydrogenase

mitochondrial oxidative enzymes

muscle buffering capacity

myoglobin

myokinase

phosphofructokinase (PFK)

phosphorylase

sprint training

submaximal endurance

succinate dehydrogenase (SDH)

▶ Study Questions

1. What are the effects of aerobic and anaerobic training on muscle fibers?

2. How does aerobic training improve oxygen delivery to the muscle fibers?

3. What effect does aerobic training have on the type of fuels used during exercise?

4. Describe the factors responsible for the improvements in muscle respiratory capacity ($\dot{Q}O_2$) that occur with aerobic training.

5. Give examples of interval training sessions that might be used to develop the ATP–PCr, glycolytic, and oxidative systems for a runner.

6. What changes occur in muscle during anaerobic training that might reduce its fatigability during highly glycolytic exercise?

7. What changes might be expected in the lactate threshold as a result of aerobic training?

8. Describe the relationship between running speed and blood lactate accumulation.

9. Describe the changes in muscle buffering capacity resulting from aerobic and anaerobic training. How might these changes improve performance?

▶ References

1. Bar-Or, O. (1987). The Wingate anaerobic test: An update on methodology, reliability and validity. *Sports Medicine,* **4,** 381-394.

2. Bergman, B.C., Wolfel, E.E., Butterfield, G.E., Lopaschuk, G.D., Casazza, G.A., Horning, M.A., & Brooks, G.A. (1999). Active muscle and whole body lactate kinetics after endurance training in men. *Journal of Applied Physiology,* **87,** 1684-1696.

3. Bouchard, C., An, P., Rice, T., Skinner, J.S., Wilmore, J.H., Gagnon, J., Pérusse, L., Leon, A.S., & Rao, D.C. (1999). Familial aggregation of $\dot{V}O_2$max response to exercise training: Results from the HERITAGE Family Study. *Journal of Applied Physiology,* **87,** 1003-1008.

4. Brooks, G.A., & Mercier, J. (1994). Balance of carbohydrate and lipid utilization during exercise: The "crossover" concept. *Journal of Applied Physiology,* **76,** 2253-2261.

5. Costill, D.L. (1985). Practical problems in exercise physiology research. *Research Quarterly,* **56,** 379-384.

6. Costill, D.L. (1986). *Inside running: Basics of sports physiology.* Indianapolis: Benchmark Press.

7. Costill, D.L., Coyle, E., Dalsky, G., Evans, W., Fink, W.W., & Hoopes, D. (1977). Effects of elevated plasma FFA and insulin on muscle glycogen usage during exercise. *Journal of Applied Physiology,* **43,** 695-699.

8. Costill, D.L., Coyle, E.F., Fink, W.F., Lesmes, G.R., & Witzmann, F.A. (1979). Adaptations in skeletal muscle following strength training. *Journal of Applied Physiology: Respiratory Environmental Exercise Physiology,* **46,** 96-99.

9. Costill, D.L., Fink, W.J., Ivy, J.L., Getchell, L.H., & Witzmann, F.A. (1979). Lipid metabolism in skeletal muscle of endurance-trained males and females. *Journal of Applied Physiology,* **28,** 251-255.

10. Costill, D.L., Thomas, R., Robergs, R.A., Pascoe, D.D., Lambert, C.P., Barr, S.I., & Fink, W.J. (1991). Adaptations to swimming training: Influence of training volume. *Medicine and Science in Sports and Exercise,* **23,** 371-377.

11. Essen, B., Hagenfeldt, L., & Kaijser, L. (1977). Utilization of blood-borne and intramuscular substrates during continuous and intermittent exercise in man. *Journal of Physiology, 265,* 480-506.

12. Farrell, P.A., Wilmore, J.H., Coyle, E.F., Billing, J.E., & Costill, D.L. (1979). Plasma lactate accumulation and distance running performance. *Medicine and Science in Sports and Exercise, 11,* 338-344.

13. Gollnick, P.D., Armstrong, R.B., Saubert, C.W., IV, Piehl, K., & Saltin, B. (1972). Enzyme activity and fiber composition in skeletal muscle of untrained and trained men. *Journal of Applied Physiology, 33,* 312-319.

14. Green, H.J., Jones, S., Ball-Burnett, M., Farrance, B., & Ranney, D. (1995). Adaptations in muscle metabolism to prolonged voluntary exercise and training. *Journal of Applied Physiology, 78,* 138-145.

15. Greiwe, J.S., Hickner, R.C., Hansen, P.A., Racette, S.B., Chen, M.M., & Holloszy, J.O. (1999). Effects of endurance exercise training on muscle glycogen accumulation in humans. *Journal of Applied Physiology, 87,* 222-226.

16. Harmer, A.R., McKenna, M.J., Sutton, J.R., Snow, R.J., Ruell, P.A., Booth, J., Thompson, M.W., MacKay, N.A., Stathis, C.G., Crameri, R.M., Carey, M.F., & Eager, D.M. (2000). Skeletal muscle metabolic and ionic adaptations during intense exercise following sprint training in humans. *Journal of Applied Physiology, 89,* 1793-1803.

17. Hermansen, L., & Wachtlova, M. (1971). Capillary density of skeletal muscle in well-trained and untrained men. *Journal of Applied Physiology, 30,* 860-863.

18. Holloszy, J.O., & Coyle, E.F. (1984). Adaptations of skeletal muscle to endurance exercise and their metabolic consequences. *Journal of Applied Physiology, 56,* 831-838.

19. Holloszy, J.O., Oscai, L.B., Mole, P.A., & Don, I.J. (1971). Biochemical adaptations to endurance exercise in skeletal muscle. In B. Pernow & B. Saltin (Eds.), *Muscle metabolism during exercise* (51-61). New York: Plenum Press.

20. Ingjer, F. (1979). Capillary supply and mitochondrial content of different skeletal muscle fiber types in untrained and endurance trained men: A histochemical and ultra structural study. *European Journal of Applied Physiology, 40,* 197-209.

21. Issekutz, B., Miller, H.I., Paul, P., & Rodahl, K. (1965). Aerobic work capacity and plasma FFA turnover. *Journal of Applied Physiology, 20,* 293-296.

22. Jacobs, I., Esbjörnsson, M., Sylvén, C., Holm. I., & Jansson, E. (1987). Sprint training effects on muscle myoglobin, enzymes, fiber types, and blood lactate. *Medicine and Science in Sports and Exercise, 19,* 368-374.

23. Jansson, E., Esbjörnsson, M., Holm. I., & Jacobs, I. (1990). Increase in the proportion of fast-twitch muscle fibres by sprint training in males. *Acta Physiologica Scandinavica, 140,* 359-363.

24. MacDougall, J.D., Hicks, A.L., MacDonald, J.R., McKelvie, R.S., Green, H.J., & Smith, K.M. (1998). Muscle performance and enzymatic adaptations to sprint interval training. *Journal of Applied Physiology, 84,* 2138-2142.

25. McGuire, B.J, & Secomb, T.W. (2001). A theoretical model for oxygen transport in skeletal muscle under conditions of high oxygen demand. *Journal of Applied Physiology, 91,* 2255-2265.

26. Rennie, M.J., Winder, W.W., & Holloszy, J.O. (1976). A sparing effect of increased plasma fatty acids on muscle and liver glycogen content in exercising rat. *Biochemistry Journal, 156,* 647-655.

27. Rico-Sanz, J., Rankinen, T., Joanisse, D.R., Leon, A.S., Skinner, J.S., Wilmore, J.H., Rao, D.C., & Bouchard, C. 2003. Familial resemblance for muscle phenotypes in The Heritage Family Study. *Medicine and Science in Sports and Exercise,* 35 (8): 1360-1366.

28. Saltin, B., Nazar, K., Costill, D.L., Stein, E., Jansson, E., Essen, B., & Gollnick, P.D. (1976). The nature of the training response: Peripheral and central adaptations to one-legged exercise. *Acta Physiologica Scandinavica, 96,* 289-305.

29. Sharp, R.L., Costill, D.L., Fink, W.J., & King, D.S. (1986). Effects of eight weeks of bicycle ergometer sprint training on human muscle buffer capacity. *International Journal of Sports Medicine, 7,* 13-17.

30. Skinner, J.S., Jaskólski, A., Jaskólska, A., Krasnoff, J., Gagnon, J., Leon, A.S., Rao, D.C, Wilmore, J.H., & Bouchard, C. (2001). Age, sex, race, initial fitness, and response to training: The HERITAGE Family Study. *Journal of Applied Physiology, 90,* 1770-1776.

31. Thorstensson, A. (1975). Enzyme activities and muscle strength after sprint training in man. *Acta Physiologica Scandinavica, 94,* 313-318.

▷ Selected Readings

Beltz, J.D., Costill, D.L., Thomas, R., Fink, W.J., & Kirwan, J.P. (1988). Energy demands of interval training for competitive swimming. *Journal of Swimming Research,* 4(3), 5-9.

Brooks, G.A. (1985). Response to Davis' manuscript. *Medicine and Science in Sports and Exercise, 17,* 19-21.

Costill, D.L., Fink, W.J., Hargreaves, M., King, D.S., Thomas, R., & Fielding, R. (1985). Metabolic characteristics of skeletal muscle during detraining from competitive swimming. *Medicine and Science in Sports and Exercise, 17,* 339-343.

Davis, J.A. (1985). Anaerobic threshold: Review of the concept and directions for future research. *Medicine and Science in Sports and Exercise,* **17,** 6-18.

Davis, J.A., Frank, M.H., Whipp, B.J., & Wasserman, K. (1979). Anaerobic threshold alterations caused by endurance training in middle-aged men. *Journal of Applied Physiology,* **46,** 1039-1046.

Hargreaves, M. (Ed.). (1995). *Exercise metabolism.* Champaign, IL: Human Kinetics.

Henriksson, J., & Reitman, J.S. (1977). Time course of changes in human skeletal muscle succinate dehydrogenase and cytochrome oxidase activities and maximal oxygen uptake with physical activity and inactivity. *Acta Physiologica Scandinavica,* **99,** 91-97.

Hickson, R.C. (1980). Interference of strength development by simultaneously training for strength and endurance. *European Journal of Applied Physiology,* **45,** 255-263.

Houmard, J.A., Costill, D.L., Mitchell, J.B., Park, S.H., & Chenier, T.C. (1991). The role of anaerobic ability in middle distance running performance. *European Journal of Applied Physiology,* **62,** 40-43.

Houston, M.E. (1995). *Biochemistry primer for exercise science.* Champaign, IL: Human Kinetics.

Kiens, B., Essen-Gustavsson, B., Christensen, N.J., & Saltin, B. (1993). Skeletal muscle substrate utilization during submaximal exercise in man: Effect of endurance training. *Journal of Physiology,* **469,** 459-478.

Lesmes, G.R., Costill, D.L., Coyle, E.F., & Fink, W.J. (1978). Muscle strength and power changes during maximal isokinetic training. *Medicine and Science in Sports,* **10,** 266-269.

Pette, D., & Staron, R.S. (2000). Myosin isoforms, muscle fiber types, and transitions. *Microscopy Research and Techniques,* **50,** 500-509.

Prud'homme, D., Bouchard, G., Leblanc, C., Landry, F., & Fontaine, E. (1984). Sensitivity of maximal aerobic power to training is genotype-dependent. *Medicine and Science in Sports and Exercise,* **16,** 489-493.

Ross, A., & Leveritt, M. (2001). Long-term metabolic and skeletal muscle adaptations to short-sprint training. *Sports Medicine,* **31,** 1063-1082.

Saltin, B., Blomqvist, G., Mitchell, J.H., Johnson, R.L., Jr., Wildenthal, K., & Chapman, C.B. (1968). Response to submaximal and maximal exercise after bed rest and training. *Circulation,* **38**(Suppl. 7), 1-78.

Saltin, B., & Karlsson, J. (1971). Muscle ATP, CP, and lactate during exercise after physical conditioning. In B. Pernow & B. Saltin (Eds.), *Muscle metabolism during exercise* (pp. 395-399). New York: Plenum Press.

Saltin, B., & Rowell, L.B. (1980). Functional adaptations to physical activity and inactivity. *Federation Proceedings,* **39,** 1506-1513.

Wasserman, K. (1984). The anaerobic threshold measurement to evaluate exercise performance. *American Review of Respiratory Diseases,* **129**(Suppl.), S35-S40.

P A R T   I I I

Cardiovascular and Respiratory Function and Performance

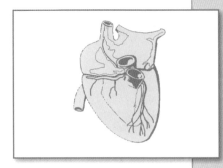

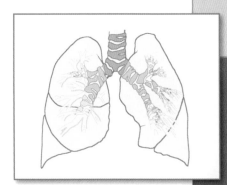

In the previous part, we learned how the body produces energy through metabolism to fuel its movement. In part III, we focus on how the cardiovascular and respiratory systems provide oxygen and fuel to the active muscles, how they rid the body of carbon dioxide and metabolic wastes, and how these systems adapt to aerobic training. In chapter 7, "Cardiovascular Control During Exercise," we look at the structure and function of the cardiovascular system: the heart, the blood vessels, and the blood. Our primary focus is on how this system provides active muscles with an adequate blood supply to meet their demands during increasing rates of work. In chapter 8, "Respiratory Regulation During Exercise," we examine the mechanics and regulation of breathing, the process of gas exchange in the lungs and at the muscles, and the transportation of oxygen and carbon dioxide in the blood. We also see how this system regulates the body's pH within a very narrow range. In chapter 9, "Cardiovascular and Respiratory Adaptations to Training," we study the concept of endurance capacity. We consider how endurance capacity is evaluated in the laboratory and how it improves with aerobic training. We concentrate on cardiorespiratory adaptations that occur in response to aerobic training and that improve performance.

CHAPTER 7

CARDIOVASCULAR CONTROL DURING EXERCISE

overview

Your cardiovascular system—which includes your heart, blood vessels, and blood—has many roles, including nurturer, protector, and even garbage hauler. This system must reach every cell in your body, and it must be able to respond immediately to any change in your internal environment to keep all body systems functioning at peak efficiency. Even when you are at rest, your cardiovascular system constantly works to meet the demands of your body's tissues. But during exercise, you place numerous and far more urgent demands on this system.

In this chapter, we explore the amazing role that the cardiovascular system plays in physical activity. In the first section of the chapter, we review the structure and function of the cardiovascular system, emphasizing some of its complexities. In the second section, we focus on how the cardiovascular system responds to increased demands during exercise. You will learn how each component of this system adapts to changes in the body's internal environment that result from increased rates of physical activity and how the system controls our ability to perform.

outline

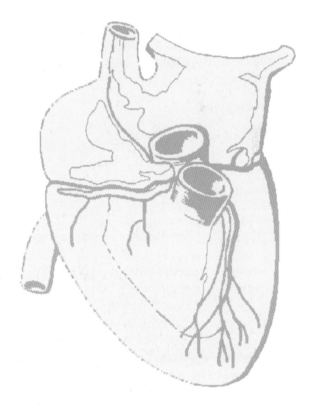

On January 5, 1988, the sports world lost one of its greatest athletes. "Pistol Pete" Maravich, former National Basketball Association star, collapsed and died of cardiac arrest at 40 years of age during a pickup basketball game. His death came as a shock, and the cause of his death surprised the medical experts. Maravich's heart was abnormally enlarged, primarily because he was born with only a single coronary artery on the right side of his heart—he was missing the two coronary arteries that supply the left side of the heart! The medical community was amazed that this single right coronary artery had taken over supplying the left side of Maravich's heart and that this adaptation had allowed him to compete for many years as one of the top players in the history of basketball. Although Maravich's death was a tragedy and shocked the sports world, he was able to perform at the highest level of competition for 10 years in one of the most physically demanding sports. More recently, a number of highly promising high school and college athletes have had their lives cut short by sudden cardiac death. Most of these deaths are attributable to hypertrophic cardiomyopathy, a disease characterized by an abnormally enlarged heart muscle mass, generally involving the left ventricle. In about half of the cases, the disease is inherited. Although this remains the major cause of sudden cardiac death in adolescent and young adult athletes (~36%), it is relatively rare with an estimated occurrence of between 1 and 2 cases per 1 million athletes annually.

The cardiovascular system serves a number of important functions in the body, most of which support other physiological systems. The major cardiovascular functions fall into five categories:

- Delivery
- Removal
- Transport
- Maintenance
- Prevention

Consider some examples. The cardiovascular system delivers oxygen and nutrients to, and removes carbon dioxide and metabolic waste products from, every cell in the body. It transports hormones from endocrine glands to their target receptors. The system maintains body temperature, and the blood's buffering capabilities help control the body's pH. The cardiovascular system maintains appropriate fluid levels to prevent dehydration and helps prevent infection from invading organisms.

Although this is an abbreviated list, the cardiovascular functions listed here are important for understanding the physiological bases of physical activity. But before examining specific cardiovascular responses to activity, we need to review the components of the cardiovascular system and how they work together.

Structure and Function of the Cardiovascular System

The cardiovascular system is impressive in its ability to respond immediately to your body's many and ever-changing needs. All bodily functions and virtually every cell in your body depend in some way on this system.

Any system of circulation requires three components:

- A pump (the heart)
- A system of channels (the blood vessels)
- A fluid medium (the blood)

Let's examine each of these separately.

Heart

The heart, shown in figure 7.1, has two atria that act as receiving chambers and two ventricles acting as sending units. About the size of your fist and located in the center of the thoracic cavity, the heart is the primary pump that circulates blood through the entire vascular system. It is enclosed in a tough membranous sac called the **pericardium.** The thin cavity between the pericardium and the heart is filled with pericardial fluid, which is necessary to

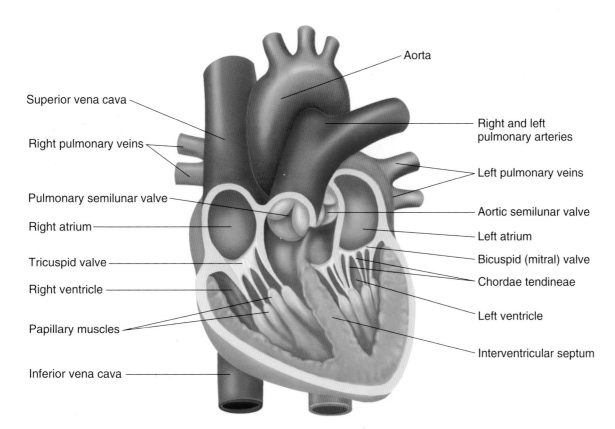

▲ **Figure 7.1**   The anatomy of the human heart.

Blood Flow Through the Heart

Blood that has coursed its way between the cells of the body, delivering oxygen and nutrients and picking up waste products, returns through the great veins—the superior vena cava and inferior vena cava—to the right atrium. This chamber receives all the body's deoxygenated blood.

From the right atrium, blood passes through the tricuspid valve into the right ventricle. This chamber pumps the blood through the pulmonary semilunar valve into the pulmonary artery, which carries the blood to the right and left lungs. Thus, the right side of the heart is known as the pulmonary side, sending the blood that has circulated throughout the body into the lungs for reoxygenation.

After receiving a fresh supply of oxygen, the blood exits the lungs through the pulmonary veins, which carry it back to the heart and into the left atrium. All freshly oxygenated blood is received by this chamber. From the left atrium, the blood passes through the bicuspid (mitral) valve into the left ventricle. Blood leaves the left ventricle by passing through the aortic semilunar valve into the aorta, which ultimately sends it out to all body parts and systems. The left side of the heart is known as the systemic side. It receives the oxygenated blood from the lungs and then sends it out to supply all body tissues.

Myocardium and Its Blood Supply

Cardiac muscle is collectively called the **myocardium.** Myocardial thickness varies directly with the stress placed on the walls of the heart chambers. The left ventricle is the most powerful of the four heart chambers. Through its contractions, this chamber must pump blood through the entire body. When the body is sitting or standing, the left ventricle must contract with enough force to overcome the effect of gravity, which tends to pool blood in the lower extremities.

The left ventricle's tremendous power is reflected by the greater size (hypertrophy) of its

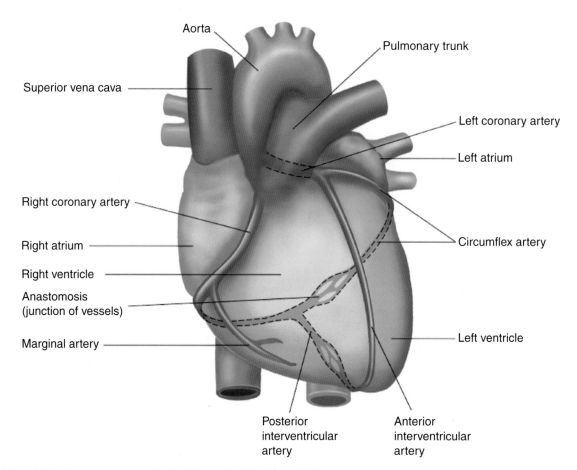

Aorta

Pulmonary trunk

Superior vena cava

Left coronary artery

Left atrium

Right coronary artery

Circumflex artery

Right atrium

Right ventricle

Anastomosis
(junction of vessels)

Marginal artery

Left ventricle

Posterior
interventricular
artery

Anterior
interventricular
artery

▲ **Figure 7.2**   The coronary circulation, illustrating the right and left coronary arteries and their major branches.

muscular wall compared with the other heart chambers. This hypertrophy is simply the result of the demands placed on the left ventricle at rest or under normal conditions of moderate activity. With more vigorous exercise—particularly intense aerobic activity, during which the working muscles' need for blood increases considerably—the demands on the left ventricle are high. Over time it responds by increasing its size, like skeletal muscle.

Although striated in appearance, the myocardium differs from skeletal muscle in one important way. Cardiac muscle fibers are anatomically interconnected end to end by dark-staining regions called intercalated disks. These disks have desmosomes, which are structures that anchor the individual cells together so that they don't pull apart during contraction, and gap junctions, which allow rapid transmission of the impulse that signals contraction.

The myocardium, just like skeletal muscle, must have a blood supply to deliver oxygen and nutrients and remove waste products. Although blood courses through each chamber of the heart, little nourishment comes from this blood supply. The primary blood supply to the heart is provided by the right and left coronary arteries, which arise from the base of the aorta and encircle the outside of the myocardium (figure 7.2). The right coronary artery supplies the right side of the heart, dividing into two primary branches, the marginal artery and the posterior interventricular artery. The left coronary artery, also referred to as the left main coronary artery, also divides into two major branches, the circumflex artery and the anterior interventricular artery, also called the anterior descending coronary artery. The posterior interventricular artery and the anterior interventricular artery merge, or anastomose, in the lower posterior area of the heart, as does the circumflex. These arteries are very susceptible to atherosclerosis, or narrowing, which can lead to coronary artery disease. This

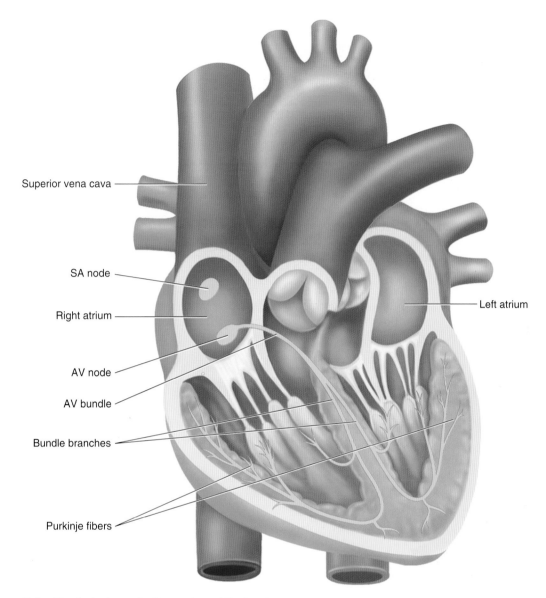

Superior vena cava

SA node

Right atrium

AV node

AV bundle

Bundle branches

Purkinje fibers

Left atrium

▲ **Figure 7.3**   The intrinsic conduction system of the heart.

disease is discussed in much greater detail in chapter 20. Anomalies also sometimes occur in the coronary arteries, as we saw in our chapter opening.

To understand how cardiac contractions are coordinated, we must understand how the signal for contraction originates and travels through the heart. These functions are accomplished by the cardiac conduction system.

Cardiac Conduction System

Cardiac muscle has the unique ability to generate its own electrical signal, called autoconduc-

tion, which allows it to contract rhythmically without any external stimulation. With neither neural nor hormonal stimulation, the intrinsic heart rate (HR) averages 70 to 80 beats (contractions) per minute but can drop well below this rate in endurance-trained people.

Figure 7.3 illustrates the four components of the cardiac conduction system:

- Sinoatrial (SA) node
- Atrioventricular (AV) node
- AV bundle (bundle of His)
- Purkinje fibers

Heart Murmur

The four heart valves prevent back-flow of blood, ensuring one-way flow through the heart. These valves maximize the amount of blood pumped out of the heart during contraction. A heart murmur is a condition in which abnormal heart sounds are detected with the aid of a stethoscope. Normally, a heart valve makes a distinct clicking sound when it snaps shut. With a murmur, the click is replaced by a sound similar to blowing. This abnormal sound can indicate the turbulent flow of blood through a narrowed or leaky valve. It could also indicate errant blood flow through a hole in the wall (septum) separating the right and left sides of the heart (septal defect).

Heart murmurs are quite common in growing children and adolescents. During growth periods, valve development doesn't always keep up with enlargement of the heart openings. Valves can leak even in an adult. With mitral valve prolapse, the mitral (bicuspid) valve allows some blood to flow back into the left atrium during ventricular contraction. This disorder, common in adults (6-17% of the population), including athletes, usually has little clinical significance unless there is significant back-flow.

Most murmurs in athletes are benign, affecting neither the heart's pumping nor the athlete's performance. But heart murmurs can indicate diseased valves, such as those with stenosis, in which the valve is narrowed and often thickened and rigid. This condition can require surgical replacement of the valve.

The impulse for heart contraction is initiated in the **sinoatrial (SA) node,** a group of specialized cardiac muscle fibers located in the posterior wall of the right atrium. Because this tissue generates the impulse, typically at a frequency of about 60 to 80 beats/min, the SA node is known as the heart's pacemaker, and the rhythm it establishes is called the sinus rhythm. The electrical impulse generated by the SA node spreads through both atria and reaches the **atrioventricular (AV) node,** located in the right atrial wall near the center of the heart. As the impulse spreads through the atria, they are signaled to contract, which they do almost immediately.

The AV node conducts the impulse from the atria into the ventricles. The impulse is delayed by about 0.13 s as it passes through the AV node, and then it enters the AV bundle. This delay allows the atria to fully contract before the ventricles do, maximizing ventricular filling. The AV bundle travels along the ventricular septum and then sends right and left bundle branches into both ventricles. These branches send the impulse toward the apex of the heart and then outward. Each bundle branch subdivides into many smaller ones that spread throughout the entire ventricular wall.

These terminal branches of the AV bundle are the Purkinje fibers. They transmit the impulse through the ventricles approximately six times faster than through the rest of the cardiac conduction system. This rapid conduction allows all parts of the ventricle to contract at about the same time.

fyi   Occasionally, chronic problems develop within the cardiac conduction system, hampering its ability to maintain appropriate sinus rhythm throughout the heart. In such cases, an artificial pacemaker can be surgically installed. This is a small, battery-operated electrical stimulator, usually implanted under the skin, with electrodes attached to the right ventricle. An electrical stimulator is useful, for example, to treat a condition called AV block. With this disorder, the SA node fires its impulse, but the impulse is blocked at the AV node and can't reach the ventricles. The artificial pacemaker takes over the role of the disabled AV node, supplying the needed impulse and thus controlling ventricular contraction.

Extrinsic Control of Heart Activity

Although the heart initiates its own electrical impulses (intrinsic control), their timing and effects can be altered. Under normal conditions, this is accomplished primarily through three extrinsic systems:

- The parasympathetic nervous system
- The sympathetic nervous system
- The endocrine system (hormones)

Although an overview of these systems' effects is offered here, they are discussed in more detail in chapters 2 and 5.

The parasympathetic system, a branch of the autonomic nervous system, acts on the heart through the vagus nerve (cranial nerve X). At rest, parasympathetic system activity predominates in a state referred to as vagal tone. The vagus nerve has a depressant effect on the heart: It slows impulse conduction and thus decreases the heart rate. Maximal vagal stimulation can lower the heart rate to between 20 and 30 beats/min. The vagus nerve also decreases the force of cardiac contraction.

The sympathetic nervous system, the other branch of the autonomic system, has opposite effects. Sympathetic stimulation increases impulse conduction speed and thus heart rate. Maximal sympathetic stimulation allows the heart rate to soar up to 250 beats/min. Sympathetic input also increases the contraction force. The sympathetic system predominates during times of physical or emotional stress, when the body's demands are higher. After the stress subsides, the parasympathetic system again predominates.

The endocrine system exerts its effect through the hormones released by the adrenal medulla: norepinephrine and epinephrine (see chapter 5). These hormones are also known as catecholamines. Like the sympathetic nervous system, these hormones stimulate the heart, increasing its rate. In fact, release of these hormones is triggered by sympathetic stimulation during times of stress, and their actions prolong the sympathetic response.

Normal resting heart rate (RHR) typically varies between 60 and 85 beats/min. With extended periods of endurance training (months to years), the RHR can decrease to 35 beats/min or less. We have observed an RHR of 28 beats/min in a world-class, long-distance run-

ner. These lower RHRs are postulated to result from a reduced intrinsic heart rate and from increased parasympathetic stimulation (vagal tone), with reduced sympathetic activity probably playing a lesser role.

> Heart rate is established by the SA node, the pacemaker, but can be altered by the sympathetic and parasympathetic nervous systems and the endocrine system.

Cardiac Arrhythmias

Occasionally, disturbances in the normal sequence of cardiac events can lead to an irregular heart rhythm, called an arrhythmia. These disturbances vary in degree of seriousness. Bradycardia and tachycardia are two types of arrhythmias. Bradycardia means "slow heart" and indicates an RHR lower than 60 beats/min, whereas tachycardia means "fast heart" and indicates a resting rate greater than 100 beats/min. With these arrhythmias, the sinus rhythm itself usually is altered. The heart's function may be normal, but its timing is abnormal, which can affect circulation. Symptoms of both arrhythmias include fatigue, dizziness, lightheadedness, and fainting. Tachycardia also can cause palpitations.

Other arrhythmias also occur. For example, premature ventricular contractions, which result in the feeling of skipped or extra beats, are relatively common and result from impulses originating outside the SA node. Atrial flutter, in which the atria contract at rates of 200 to 400 beats/min, and atrial fibrillation, in which the atria contract in a rapid and uncoordinated manner, are more serious arrhythmias that cause the atria to pump little or no blood. Ventricular tachycardia, defined as three or more consecutive premature ventricular contractions, is a very serious arrhythmia that can lead to ventricular fibrillation, in which contraction of the ventricular tissue is uncoordinated. When this happens, the heart cannot pump blood. Most cardiac deaths result from ventricular fibrillation. Use of a defibrillator to shock the heart back into a normal sinus rhythm must occur within minutes if the victim is to survive. Cardiopulmonary resuscitation (CPR) imposes a normal rhythm on the heart and can maintain life for several hours, but chances of survival are greater if emergency treatment, including defibrillation, is provided quickly.

Interestingly, most highly trained endurance athletes develop low RHRs, an advantageous adaptation, as a result of training. Also, the heart rate naturally accelerates during physical activity to meet the increased demands of exertion. These adaptations should not be confused with pathological bradycardia or tachycardia, which are abnormal alterations in the RHR that usually indicate a pathological disturbance.

Electrocardiogram (ECG)

The electrical activity of the heart can be recorded to monitor cardiac changes or diagnose potential cardiac problems. The principle involved is simple: Body fluids are good electrical conductors. Electrical impulses generated in the heart are conducted through body fluids to the skin, where they can be detected and printed out by a sensitive machine called an **electrocardiograph** (figure 7.4a). This printout is called an **electrocardiogram,** or ECG (figure 7.4b). Three components of the ECG represent important aspects of cardiac function:

- The P wave
- The QRS complex
- The T wave

The P wave represents atrial depolarization and occurs when the electrical impulse travels from the SA node through the atria to the AV node. The QRS complex represents ventricular depolarization and occurs as the impulse spreads from the AV bundle to the **Purkinje fibers** and through the ventricles. The T wave represents ventricular repolarization. Atrial repolarization cannot be seen, because it occurs during ventricular depolarization (QRS complex).

ECGs are often obtained during exercise, and these are valuable diagnostic tests. As exercise intensity increases, the heart must beat faster and work harder to deliver more blood to active muscles. If the heart is diseased, an indication may show up on the ECG as the heart increases

its rate of work. Exercise ECGs are also invaluable tools for research in exercise physiology because they provide a convenient method for tracking cardiac changes during acute and chronic exercise.

> An ECG provides a graphic record of the electrical activity of the heart and can be used to diagnose that someone has had a heart attack in the past or is at risk for a heart attack in the future.

> ▶ The atria receive blood into the heart; the ventricles eject blood from the heart.
>
> ▶ Because the left ventricle must produce more power than other chambers, its myocardium is thicker due to hypertrophy.
>
> ▶ Cardiac tissue is capable of autoconduction and has its own conduction system. It initiates its own pulse without neural or hormonal control.
>
> ▶ The SA node is the heart's pacemaker, establishing its rate and coordinating activity throughout the heart.
>
> ▶ Heart rate and contraction strength can be altered by the autonomic nervous system and the endocrine system.
>
> ▶ The ECG is a recording of the heart's electrical activity. An exercise ECG can be used to detect underlying cardiac disorders.

Terminology of Cardiac Function

The following terms are essential to understanding the work done by the heart and to our later discussions of cardiac response during activity: cardiac cycle, stroke volume, ejection fraction, and cardiac output ($\dot{Q}$).

Cardiac Cycle The **cardiac cycle** includes all events that occur between two consecutive heartbeats. In mechanical terms, it consists of all heart chambers undergoing a relaxation phase (diastole) and a contraction phase (systole). During diastole, the chambers fill

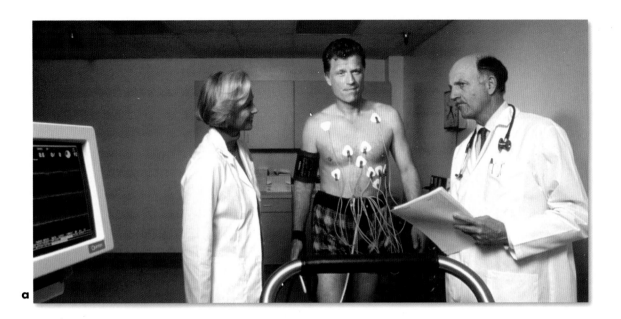

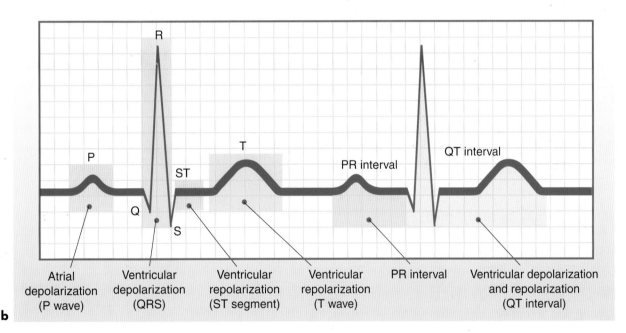

▲ **Figure 7.4**   (*a*) Taking an exercise electrocardiogram. The subject is advised to let go of the handrail during testing (after he steadies himself). (*b*) A graphic illustration of the various phases of the resting electrocardiogram.

with blood. During systole, the chambers contract and expel their contents. The diastolic phase is longer than the systolic phase. Consider an individual with a heart rate of 74 beats/min. At this HR, the entire cardiac cycle takes 0.81 s to complete (60 s/74 beats). Of the total cardiac cycle at this rate, diastole accounts for 0.50 s, or 62% of the cycle, and systole accounts for 0.31 s, or 38%. As the

HR increases, these time intervals shorten proportionately.

Refer to the normal ECG in figure 7.4. One cardiac cycle spans the time between one systole and the next. Ventricular contraction (systole) begins during the QRS complex and ends in the T wave. Ventricular relaxation (diastole) occurs during the T wave and continues until the next contraction. You can see from this illustration that although the heart seems to always be at work, it actually spends slightly more time in the resting phase than in the working phase.

The interactions of the various events of the heart are illustrated in figure 7.5. This is called a Wiggers diagram, named after the physiologist who created it. The diagram integrates information from the electrical conduction signals (ECG), heart sounds primarily from the heart valves, pressure changes within the heart chambers, and left ventricular volume.

Stroke Volume   During systole, a certain volume of blood is ejected from the left ventricle. This amount is the **stroke volume (SV)** of the heart, or the volume of blood pumped per stroke (contraction). This is depicted in figure 7.6a. To understand stroke volume, consider the amount of blood in the ventricle before and after contraction. At the end of diastole, just before contraction, the ventricle has completed filling. The volume of blood it now contains is called the **end-diastolic volume (EDV)**. At the end of systole, just after contraction, the ventricle has completed its ejection phase. The volume of blood remaining in the ventricle is called the **end-systolic volume (ESV)**. Stroke volume is the volume of blood that was ejected and is merely the difference between the amount originally there and the amount remaining in the ventricle after contraction. So stroke volume is simply the difference between EDV and ESV, or SV = EDV − ESV.

Ejection Fraction   The proportion of the blood pumped out of the left ventricle each beat is the **ejection fraction** (EF). This value, as seen in figure 7.6b, is determined by dividing the

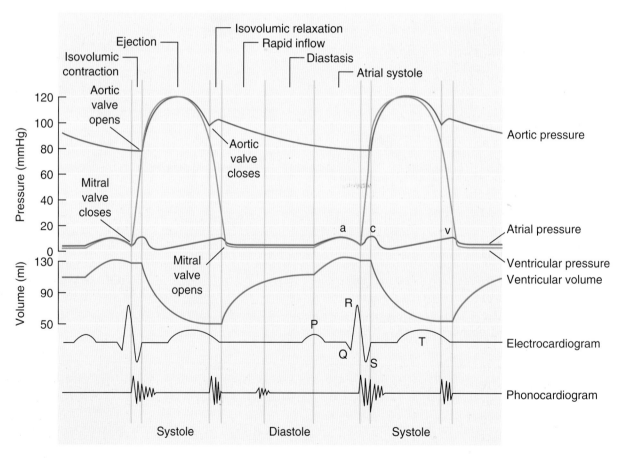

▲ **Figure 7.5**   The Wiggers diagram, illustrating the events of the cardiac cycle for left ventricular function. Integrated into this diagram are the changes in left atrial and ventricular pressure, aortic pressure, ventricular volume, electrical activity (electrocardiogram), and heart sounds.

stroke volume by EDV. It reveals how much of the blood entering the ventricle is actually ejected during contraction. The EF, generally expressed as a percentage, averages 60% at rest. Thus, 60% of the blood in the ventricle at the end of diastole is ejected with the next contraction and 40% remains.

Cardiac Output   Cardiac output ($\dot{Q}$), as shown in figure 7.6c, is the total volume of blood pumped by the ventricle per minute, or simply the product of HR and SV. The SV at rest in the standing position averages between 60 and 80 ml of blood in most adults. Thus, at an RHR of 70 beats/min, the resting cardiac output will vary between 4.2 and 5.6 L/min. The average adult body contains about 5 L of blood, so this means that the equivalent of our total blood volume is pumped through our hearts about once every minute.

$$SV = EDV - ESV$$
$$EF = (SV/EDV) \times 100$$
$$\dot{Q} = HR \times SV$$

Understanding the mechanical activity of the heart provides a basis for understanding the work of the cardiovascular system, but the heart is only one part of this system. Let's next turn our attention to the vast system of vessels that carry the blood to all body tissues.

Vascular System

The vascular system contains a series of vessels that transport blood from the heart to the tissues and back: the arteries, arterioles, capillaries, venules, and veins.

   Arteries are typically the largest, most muscular, and most elastic vessels, and they always carry blood away from the heart to the arterioles. The aorta is the great artery transporting blood from the left ventricle to all regions of the body, and the vena cava is the great vein transporting blood back to the right atrium from all regions of the body above (superior vena cava) and below (inferior vena cava) the heart.

   From the **arterioles**, blood enters the **capillaries**, the narrowest vessels, often with walls

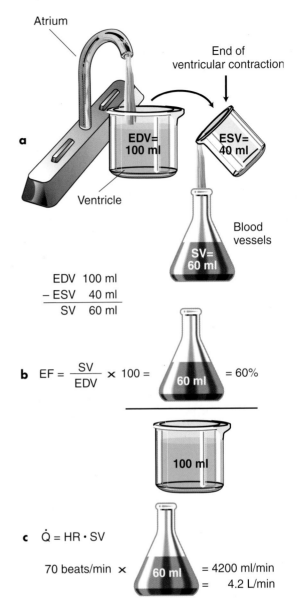

▲ **Figure 7.6**   Calculations of (*a*) stroke volume (SV), which is the difference between end-diastolic volume (EDV) and end-systolic volume (ESV), (*b*) ejection fraction (EF), and (*c*) cardiac output ($\dot{Q}$).

only one cell thick. Virtually all exchange between the blood and the tissues occurs at the capillaries. Blood leaves the capillaries to begin the return trip to the heart in the **venules,** and the venules form larger vessels—the **veins**—that complete the circuit.

   During contraction, when blood is forced out of the left ventricle under high pressure, the aortic semilunar valve is forced open. When this valve is open, its flaps block the entrances to the coronary arteries. As the pressure in the aorta decreases, the semilunar valve closes, and these entrances are exposed so that blood can then enter the coronary arteries. This design ensures that the coronary arteries are spared the very high blood pressure created by contraction of the left ventricle, thus protecting these vessels from damage.

Return of Blood to the Heart

Because we spend so much time in an upright position, the cardiovascular system requires assistance to overcome the force of gravity when blood returns from the lower parts of the body to the heart. Three basic mechanisms assist in this process:

- Breathing
- The muscle pump
- Valves

Each time you inhale and exhale, changes in pressure in the abdominal and thoracic cavities assist blood return to the heart. With exercise, the skeletal muscles in the legs and abdomen also contract. During breathing and skeletal muscle contraction, the veins in those areas where muscles are contracting and in the abdominal and thoracic cavities are compressed, and blood is pushed upward toward the heart. These actions are aided by a series of valves in the veins that allow blood to flow in only one direction, thus preventing back-flow and pooling of the blood in the lower body (see figure 7.7).

Distribution of Blood

Distribution of blood to the various body tissues varies tremendously depending on the immediate needs of a specific tissue and of the whole body. At rest under normal conditions, the most metabolically active tissues receive the greatest blood supply. The liver and kidneys combined receive almost half the blood being circulated (27% and 22%, respectively), and resting skeletal muscles receive only about 15%.

During exercise, blood is redirected to the areas where it is needed most. During heavy endurance exercise, for example, this redistri-

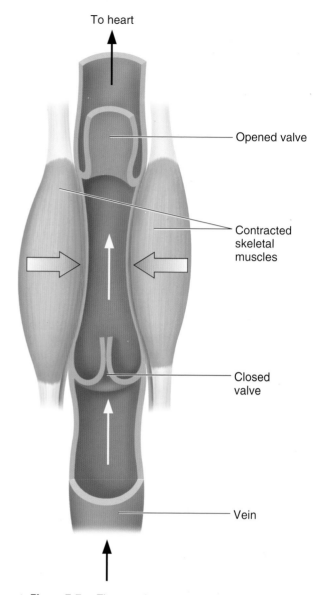

▲ **Figure 7.7**   The muscle pump. As the skeletal muscles contract, they squeeze the veins in the legs and assist in the return of blood to the heart. Valves within the veins ensure the unidirectional flow of blood back to the heart.

bution is rather remarkable: Muscles receive up to 80% or more of the available blood. This, along with increases in cardiac output (to be discussed later), allows up to 25 times more blood flow to active muscles.

Similarly, after you eat a big meal, your digestive system receives more blood than when your digestive system is empty. During increasing environmental heat stress, the skin's blood supply increases as the body attempts to maintain normal temperature. When we consider that the needs of the various body tissues are constantly changing, it is indeed amazing that the cardiovascular system can respond so efficiently, guaranteeing an adequate blood supply to the areas where it is most needed.

> The body has tremendous capacity to redistribute its blood away from areas where the need is low to areas where there is increased need. Skeletal muscle normally receives about 15% of the total blood flow, or cardiac output, at rest. This can increase to 80% or more during heavy endurance exercise.

Distribution of blood to various areas is controlled primarily by the arterioles. These vessels have two important characteristics. They have a strong muscular wall that can significantly alter vessel diameter, and they also respond to the mechanisms that control blood flow: autoregulation and extrinsic neural control. Let's examine these mechanisms.

Autoregulation   Local control of blood distribution is called autoregulation because the arterioles in specific areas control themselves. Autoregulation refers to the vessels' ability to self-regulate their own blood flow depending on the immediate needs of the tissues they supply. The arterioles can undergo vasodilation, opening up to allow more blood to enter an area in need.

This increased blood flow is a direct response to changes in the tissue's local chemical environment. Oxygen demand appears to be the strongest stimulus. As the tissue's oxygen use increases, available oxygen is diminished. Local arterioles dilate to allow more blood, and thus more oxygen, to perfuse that area. Other chemical changes that can stimulate increased blood flow are decreases in other nutrients and increases in by-products (carbon dioxide, K^+, H^+, lactic acid) or inflammatory chemicals. Nitric oxide, produced in the endothelium (inner lining) of arterioles, is a potent vasodilator through its action on smooth muscle cell relaxation. Its role during exercise in humans has not been clearly established, but it likely is important.[11] Acetylcholine and adenosine also have been proposed as active vasodilators. Increased blood supply can either bring in needed substances or clear out harmful ones.

Extrinsic Neural Control   Although the concept of autoregulation explains local redistribution of blood within an organ or tissue mass, it can't explain how the cardiovascular system as a whole knows to send less blood to one part of the body when more is needed elsewhere. Redistribution at the system or body level is controlled by neural mechanisms. This is known as extrinsic neural control of blood flow, because the control comes from outside the specific area (extrinsic) instead of from inside the tissues (intrinsic) as in autoregulation.

Blood flow to all body parts is regulated largely by the sympathetic nervous system. The muscle within the walls of all vessels of systemic circulation is supplied by sympathetic nerves. In most vessels, stimulation by these nerves causes the muscle cells to contract, constricting that vessel so that less blood can pass through it.

Under normal conditions, the sympathetic nerves transmit impulses continuously to the blood vessels, keeping the vessels moderately constricted to maintain adequate blood pressure. This state of partial constriction is referred to as vasomotor tone. When sympathetic stimulation increases, further constriction of the blood vessels in a specific area decreases blood flow into that area and allows more blood to shift elsewhere. But if sympathetic stimulation decreases below that needed to maintain tone, constriction of vessels in that area is lessened, so the vessels dilate, increasing blood flow into that area. Therefore, sympathetic stimulation will cause vasoconstriction in most vessels, but blood flow is altered by either increasing or decreasing the amount of vasoconstriction relative to normal vasomotor tone.

The sympathetic system also can cause vasodilation directly through some of its fibers. A different type of sympathetic fiber supplies some blood vessels in skeletal muscles and in the heart. Stimulation of these fibers causes vasodilation, which increases blood flow into the muscles and heart. This system functions during the classic fight-or-flight response and increases blood flow into skeletal muscles and the heart in times of crisis. This response is also active during exercise, when the skeletal muscles and the heart are working harder, requiring much more blood than when at rest.

> Blood flow can be controlled at the local tissue level by both intrinsic autoregulation and extrinsic neural control. The sympathetic nervous system plays a major role in redirecting blood flow from areas of low need to areas of high need.

Redistribution of Venous Blood   We have discussed the mechanisms that control the redistribution of blood from one area of the body to another. But we must also consider how the blood normally is distributed in the vascular system. At rest, the blood volume is distributed among the vasculature as shown in figure 7.8. The majority of blood is located in the venous return channels (veins, venules, venous sinuses). Thus, the venous system provides a large reservoir of blood readily available to meet increased need. When this need arises, sympathetic stimulation of the venules and veins constricts these vessels. This rapidly redistributes blood from peripheral venous circulation back to the heart and then out to those areas that have greater needs. Not only is blood diverted away from other tissues, but more blood is sent into arterial circulation from the venous system, thereby ensuring a substantial increase of blood flow to a needy area.

Blood Pressure

Blood pressure is the pressure exerted by the blood on the vessel walls, and the term usually refers to arterial blood pressure. It is expressed by two numbers: the **systolic blood pressure (SBP)** and the **diastolic blood pressure (DBP)**. The higher number is the SBP; it represents the highest pressure in the artery and corresponds to ventricular systole of the heart. Ventricular contraction pushes the blood through the arteries with tremendous force, which exerts high pressure on the arterial walls. The lower num-

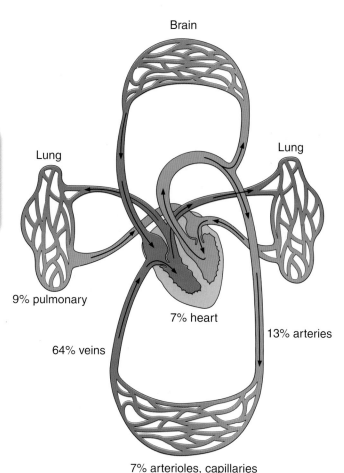

▲ Figure 7.8   Blood distribution within the vasculature when the body is at rest.

ber is the DBP and represents the lowest pressure in the artery, corresponding to ventricular diastole when the ventricle is filling.

Mean arterial pressure (MAP) represents the average pressure exerted by the blood as it travels through the arteries. Mean arterial pressure can be estimated from the DBP and SBP as follows:

$$MAP = DBP + [0.333 \times (SBP - DBP)]$$

To illustrate, with an SBP of 120 mmHg and a DBP of 80 mmHg, the MAP = 80 + [0.333 × (120 − 80)] = 93 mmHg. Note that this relationship is not a simple average of the systolic and diastolic values. The heart is in diastole longer than in systole, so the arteries experience diastolic pressure longer than systolic pressure; that is factored into the equation.

Alterations in blood pressure are controlled largely by the specific changes in the arteries, arterioles, and veins previously described. Generalized constriction of blood vessels increases blood pressure, and generalized dilation reduces it. Hypertension is the clinical term describing the condition in which blood pressure is chronically elevated above normal, healthy values. The cause of hypertension is unknown in approximately 90% of cases, but it usually can be controlled effectively by weight loss, proper diet, and exercise, although appropriate medication also may be required. Hypertension is discussed in more detail in chapter 20.

▶ Blood returns to the heart through the veins, assisted by breathing, the muscle pump, and valves within the vessels.

▶ Blood is distributed throughout the body based on the needs of the individual tissues. The most active tissues receive the most blood.

▶ Redistribution of blood is controlled locally by autoregulation. Autoregulation causes vasodilation in response to local chemical changes, thus increasing blood supply to the area.

▶ Extrinsic neural control of distribution is accomplished by the sympathetic nervous system, primarily through vasoconstriction (although some vasodilation occurs in vessels supplying active skeletal muscles and the heart).

▶ SBP is the highest pressure within the vascular system, whereas DBP is the lowest pressure there. MAP is the average pressure on the vessel walls.

Blood

The third component of any system of circulation is a circulating substance. In the human body, this is the blood and lymph. These fluids are responsible for transporting various materials between the different cells or tissues of the body.

Recall from basic physiology the relationship between blood and lymph: Some blood plasma filters out of the capillaries into the tissues, becoming interstitial (tissue) fluid. Much of the interstitial fluid returns to the capillaries after exchange occurs, but less is returned than was originally filtered out. The excess fluid enters the lymph capillaries, and it is then referred to as lymph, which ultimately returns to the blood.

Clearly, the lymphatic system plays a crucial role in maintaining appropriate fluid levels in the tissues as well as maintaining proper blood volume by ensuring that interstitial fluid is returned to circulation. This function becomes more important during exercise, when increased blood flow to the active muscles and increased blood pressure lead to the formation of more interstitial fluid. The lymphatic system minimizes swelling in the active areas and keeps the cardiovascular system working efficiently. This system is extremely important to coordinated physiological function and general health. But beyond its role in fluid return, the lymphatic system is not a major area of concern for exercise and sport physiology. We instead focus on the blood.

Blood serves many useful purposes in regulating normal body function. The three functions of primary importance to exercise and sport are

- transportation,
- temperature regulation, and
- acid-base (pH) balance.

We are most familiar with blood's transportation functions. In addition, blood is critical in temperature regulation during physical activity; it picks up heat from the body core or from areas of increased metabolic activity and dissipates that heat throughout the body during normal conditions and to the skin when the body is overheated (see chapter 10). Blood can buffer the acids produced by anaerobic metabolism, maintaining the proper pH for efficient activity of metabolic processes (see chapter 8).

Blood Volume and Composition

The total volume of blood in the body varies considerably with the individual's size and state of training. Larger blood volumes are associated with larger body size and high levels of endurance training. The blood volumes of people of average body size and normal physical activity (not training aerobically) generally range from 5 to 6 L in men and 4 to 5 L in women.

Blood is composed of plasma (primarily water) and formed elements (see figure 7.9). Plasma normally constitutes about 55% to 60% of total blood volume but can decrease by 10% of its normal amount or more with intense exercise in heat or can increase by 10% or more with endurance training or acclimatization to heat and humidity. Approximately 90% of the plasma volume is water, 7% consists of plasma proteins, and the remaining 3% includes cellular nutrients, electrolytes, enzymes, hormones, antibodies, and wastes.

The formed elements, which normally constitute about 40% to 45% of total blood volume, are the red blood cells (erythrocytes), white blood cells (leukocytes), and platelets (thrombocytes). Red blood cells constitute more than 99% of the formed-element volume; white blood cells and platelets together account for less than 1%. The percentage of the total blood volume composed of cells or formed elements is referred to as the **hematocrit.**

> The hematocrit is the ratio of the formed elements in the blood (red cells, white cells, and platelets) to the total blood volume.

White blood cells protect the body from disease organism invasion either by directly destroying the invading agents through phagocytosis (ingestion) or by forming antibodies to destroy them. Adults have about 7,000 white blood cells per cubic millimeter of blood.

The remaining formed elements are the blood platelets. These are not really cells at all but rather cell fragments. These small disks are required for blood coagulation (clotting), which prevents excessive blood loss. We are most concerned with red blood cells, however, so our discussion focuses on them.

Red Blood Cells

Mature red blood cells (erythrocytes) have no nucleus, so they can't reproduce. They must be replaced with new cells. The normal life span of a red blood cell is only about 4 months. Thus, these cells are continuously produced and destroyed at about equal rates. This balance is very important, because adequate oxygen delivery to body tissues depends on having a sufficient number of carriers: the red blood cells. Decreases in their count or function can hinder oxygen delivery and thus affect performance.

Red blood cells can be destroyed during exercise. The cell membrane appears to be disrupted by the wear and tear associated with increased circulation rate and by increased body temperature. Studies have even demonstrated that the constant pounding of the sole of the foot in the shoe during distance running can increase fragility and destruction of red blood cells.

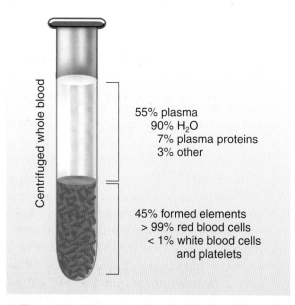

▲ **Figure 7.9** The composition of whole blood, illustrating the plasma volume (fluid portion) and the cellular volume (red cells, white cells, and platelets) after the blood sample has been centrifuged.

Red blood cells transport oxygen, which is primarily bound to their hemoglobin. **Hemoglobin** is composed of a protein (globin) and a pigment (heme). Heme contains iron, which binds oxygen. Each red blood cell contains approximately 250 million hemoglobin molecules, each able to bind four oxygen molecules, so each red blood cell can bind up to a billion molecules of oxygen! There is an average of 15 g of hemoglobin per 100 ml of whole blood. Each gram of hemoglobin can combine with 1.33 ml of oxygen, so as much as 20 ml of oxygen can be bound for each 100 ml of blood.

Blood Viscosity

Viscosity refers to the thickness or stickiness of the blood. The more viscous a fluid, the more resistant it is to flow. The viscosity of blood is normally about twice that of water. Blood viscosity, and thus resistance to flow, increases with higher hematocrits.

Because of oxygen transport by the red blood cells, an increase in their number would be expected to maximize oxygen transport. But if an increase in red blood cell count is not accompanied by a similar increase in plasma volume, blood viscosity will increase, which could restrict blood flow. This generally is not a problem unless the hematocrit reaches 60% or more.

Conversely, the combination of a low hematocrit with a high plasma volume, which decreases the blood's viscosity, appears to have certain benefits for the blood's transport function because the blood can flow more easily. Unfortunately, a low hematocrit frequently results from a reduced red blood cell count, as in diseases such as anemia. Under these circumstances, the blood can flow easily, but it contains fewer carriers, so oxygen transport is impeded. For physical activity, a low hematocrit with a normal or slightly elevated number of red blood cells is desirable. This combination should facilitate oxygen transport. Many endurance athletes achieve this condition as part of their cardiovascular system's normal adaptation to training. This adaptation is discussed in chapter 9.

fyi When we donate blood, the removal of one unit, or nearly 500 ml, represents approximately an 8% to 10% reduction in both the total blood volume and the number of circulating red blood cells. Donors are advised to drink plenty of fluids. Because plasma is primarily water, simple fluid replacement returns plasma volume to normal within 24 to 48 h. However, it takes at least 6 weeks to reconstitute the red blood cells because they must go through full development before they are functional. Blood loss greatly compromises the performance of endurance athletes by reducing oxygen delivery capacity.

▶ Blood and lymph transport materials to and from body tissues.

▶ Fluid from the plasma enters the tissues, becoming interstitial fluid. Most interstitial fluid returns to the capillaries, but some enters the lymphatic system as lymph, eventually returning to the blood.

▶ Blood is about 55% to 60% plasma and 40% to 45% formed elements.

▶ Oxygen is transported primarily by binding to the hemoglobin in red blood cells.

▶ As blood viscosity increases, so does resistance to flow.

Cardiovascular Response to Exercise

Now that we have reviewed the basic anatomy and physiology of the cardiovascular system, we can look specifically at how this system responds to the increased demands placed on the body during exercise. During exercise, oxygen demand in the active muscles increases sharply, and more nutrients are used. Metabolic processes speed up, so more waste is created. During prolonged exercise or exercise in a hot environment, body temperature increases. In intense exercise, H^+ concentration increases in the muscles and blood, lowering their pH.

Numerous cardiovascular changes occur during exercise. All share a common goal: they allow the system to meet the increased demands placed on it and carry out its functions with maximal efficiency. To better understand the changes that occur, we must look more closely at specific cardiovascular functions. We next examine changes in all components of the cardiovascular system, looking specifically at the following:

- Heart rate
- Stroke volume
- Cardiac output
- Blood flow
- Blood pressure
- Blood

We then see how these changes are integrated to provide for the body's needs.

Heart Rate

Heart rate (HR) is one of the simplest and most informative of the cardiovascular parameters. Measuring it involves simply taking the subject's pulse, usually at the radial or carotid artery sites. HR reflects the amount of work the heart must do to meet the increased demands of the body when engaged in activity. To understand this, we must compare the HR at rest and during exercise.

Resting Heart Rate

Resting heart rate (RHR) averages 60 to 80 beats/min. In middle-aged, unconditioned, sedentary individuals, the resting rate can exceed 100 beats/min. In highly conditioned, endurance-trained athletes, resting rates in the range of 28 to 40 beats/min have been reported. Your RHR typically decreases with age. It is also affected by environmental factors; for example, it increases with extremes in temperature and altitude.

Before the start of exercise, your preexercise heart rate usually increases well above normal resting values. This is called an anticipatory response. This response is mediated through release of the neurotransmitter norepinephrine from your sympathetic nervous system and

the hormone epinephrine from your adrenal glands. Vagal tone probably also decreases. Because the preexercise HR is elevated, reliable estimates of actual RHR should be made only under conditions of total relaxation, such as early in the morning before the subject rises from a restful night's sleep. Preexercise HRs should not be used as estimates of RHR.

Heart Rate During Exercise

When you begin to exercise, your HR increases directly in proportion to the increase in exercise intensity (see figure 7.10), until you are near the point of exhaustion. As you approach that point, your HR begins to level off. This indicates that you are approaching your maximum value. The maximum heart rate (HRmax) is the highest HR value you achieve in an all-out effort to the point of exhaustion. This is a highly reliable value that remains constant from day to day and changes only slightly from year to year.

HRmax can be estimated based on your age because HRmax shows a slight but steady decrease of about one beat per year beginning at 10 to 15 years of age. Subtracting your age from 220 provides an approximation of your average

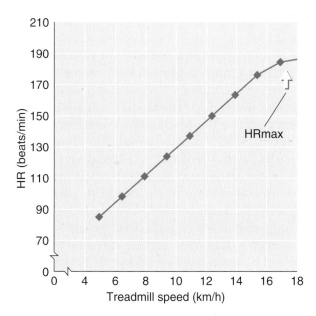

▲ Figure 7.10   Changes in heart rate (HR) while a subject walks, jogs, and runs with increasing speed on a treadmill. The heart rate increases in direct proportion to the increase in speed, eventually reaching a maximal rate (HRmax).

HRmax. However, this is only an estimate—individual values vary considerably from this average value. To illustrate, for a 40-year-old person, HRmax would be estimated to be 180 beats/min (HRmax = 220 – 40). However, 68% of all 40-year-olds have actual HRmax values between 168 and 192 beats/min (mean ± 1 standard deviation), and 95% fall between 156 and 204 beats/min (mean ± 2 standard deviations). This demonstrates the potential for error in estimating a person's HRmax. A more recent equation has been developed to estimate HRmax from age, which appears to be more accurate, particularly for those under 20 and over 50 years of age. In this equation, HRmax = 208 – (0.7 × age).[23]

> To estimate HRmax:
>
> HRmax = 220 – age in years, or
> HRmax = 208 – (0.7 × age in years)

When the rate of work is held constant at submaximal levels of exercise, HR increases fairly rapidly until it reaches a plateau. This plateau is the steady-state heart rate, and it is the optimal heart rate for meeting the circulatory demands at that specific rate of work. For each subsequent increase in intensity, HR will reach a new steady-state value within 1 to 2 min. However, the more intense the exercise, the longer it takes to achieve this steady-state value.

The concept of steady-state heart rate forms the basis for several tests that have been developed to estimate physical fitness. In one such test, individuals are placed on an exercise device, such as a cycle ergometer, and exercise at two or three standardized rates of work. Those in better physical condition (i.e., those with better cardiorespiratory endurance capacity) will have lower steady-state HRs at a given rate of work than those who are less fit. Thus, steady-state HR is a valid predictor of heart efficiency: A lower rate reflects a more efficient heart.

When exercise is performed at a constant rate over a prolonged period, particularly under conditions of heat stress, the HR tends to drift upward instead of maintaining its steady-state value. This response is part of a phenomenon called cardiovascular drift (discussed later in this chapter).

Stroke Volume

Stroke volume (SV) also changes during exercise to allow the heart to work more efficiently. It has become increasingly clear that for near-maximal and maximal rates of work, stroke volume is a major determinant of cardiorespiratory endurance capacity. Let's examine the basis for this.

Stroke volume is determined by four factors:

1. The volume of venous blood returned to the heart
2. Ventricular distensibility (the capacity to enlarge the ventricle)
3. Ventricular contractility (the capacity of the ventricle to contract)
4. Aortic or pulmonary artery pressure (the pressure against which the ventricles must contract)

The first two factors influence the filling capacity of the ventricle, determining how much blood is available for filling the ventricle and the ease with which the ventricle is filled at the available pressure. The last two factors influence the ventricle's ability to empty, determining the force with which blood is ejected and the pressure against which it must flow in the arteries. These four factors directly control the alterations in stroke volume in response to increasing exercise intensity.

Stroke Volume Increase With Exercise

Researchers agree that stroke volume increases above resting values during exercise. But there are conflicting reports about stroke volume changes as a person goes from very low rates of work to maximal work or exhaustion. Most researchers agree that stroke volume increases with increasing rates of work but only up to exercise intensities between 40% and 60% of maximal capacity. At that point, SV is thought to plateau, remaining essentially unchanged

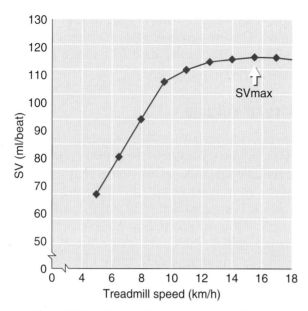

▲ **Figure 7.11** Changes in stroke volume (SV) while a subject walks, jogs, and runs with increasing speed on a treadmill. The stroke volume increases in direct proportion to the increase in speed up to about 40% to 60% of maximal exercise intensity.

up to and including the point of exhaustion as shown in figure 7.11. Others have reported that SV continues to increase up through maximal exercise intensities. This is discussed in more detail in the sidebar on page 228.

When the body is in an upright position, stroke volume can almost double from resting to maximal values. For example, in active but untrained individuals, stroke volume increases from about 60 to 70 ml at rest to 110 to 130 ml during maximal exercise. In highly trained endurance athletes, stroke volume can increase from 80 to 110 ml at rest to 160 to 200 ml during maximal exercise. During supine exercise, such as swimming, stroke volume also increases but usually by only about 20% to 40%—not nearly as much as in an upright position. Why does body position make such a difference?

When the body is in the supine position, blood does not pool in the lower extremities. Because of this, blood returns more easily to the heart, which means that resting stroke volume values are much higher in the supine position than in the upright position. Thus, the increase in stroke volume with maximal exercise is not as great in the supine position as in the up-

right position. Interestingly, the highest stroke volume attainable in upright exercise is only slightly greater than the resting value in the reclining position. The majority of the stroke volume increase during low to moderate intensities of exercise in the upright position appears to be compensating for the force of gravity.

Explanations of Stroke-Volume Increase

Although there is agreement that stroke volume increases from rest to exercise, just how this increase occurs has historically been difficult to document. One explanation is the Frank–Starling law, which states that the primary factor in controlling stroke volume is the extent to which the ventricle stretches. When the ventricle stretches more, it will contract with more force. For example, if a larger volume of blood enters the chamber when your ventricle fills during diastole, the ventricle's walls will be stretched more than when a smaller volume enters. To eject the greater amount of blood, your ventricle must react to this increased stretching by contracting more strongly. This is referred to as the **Frank–Starling mechanism**. Another explanation is that stroke volume can increase if the ventricle's contractility is greater, even without an increased end-diastolic volume.

Several newer cardiovascular diagnostic techniques have made it possible to determine exactly how stroke volume changes with exercise. Echocardiography (using sound waves) and radionuclide techniques (using electromagnetic radiation) have been used successfully to determine how the heart chambers respond to increasing oxygen demands during exercise. With both techniques, continuous pictures can be taken of the heart at rest and up to near-maximal intensities of exercise.

Figure 7.12 illustrates the results of one study of normal, active, but untrained subjects.[18] In this study, participants were tested during both supine and upright cycle ergometry under four conditions:

1. Rest
2. Low-intensity exercise
3. Intermediate-intensity exercise
4. Peak-intensity exercise

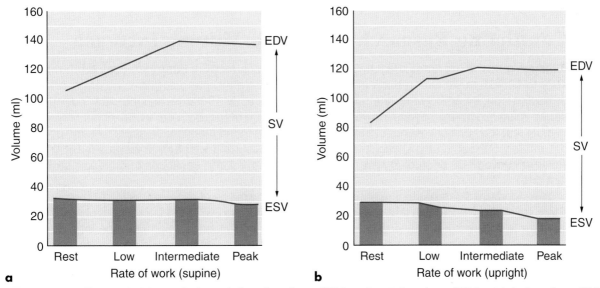

▲ Figure 7.12 Changes in left ventricular end-diastolic volume (EDV), end-systolic volume (ESV), and stroke volume (SV) at rest and during low-, intermediate-, and peak-intensity exercise when the subject is (*a*) in the supine position and (*b*) in the upright position.

Reprinted, by permission, from L. R. Poliner et al., 1980, "Left ventricular performance in normal subjects: A comparison of the responses to exercise in the upright and supine position," *Circulation* 62: 528-534.

As you go from rest to exercise of increasing intensities, an increase in left ventricular end-diastolic volume (greater filling) would indicate that the Frank–Starling mechanism is operating, and a decrease in the left ventricular end-systolic volume (greater emptying) would indicate an increased degree of contractility.

The results, illustrated in figure 7.12, indicate that both the Frank–Starling mechanism and increased contractility are important in increasing stroke volume. The Frank–Starling mechanism appears to have its greatest influence at the lower rates of work, and contractility has its greatest effects at the higher exercise intensities. Several other studies support this interpretation.

Recall that heart rate increases with exercise intensity. The plateau or small decrease in left ventricular EDV at higher exercise intensities could be caused by reduced ventricular filling time. One study observed that ventricular filling time decreased from about 500 to 700 ms at rest to about 150 ms at higher HRs (about 150-200 beats/min).[24] Therefore, with increasing work rates approaching HRmax, the diastolic filling time could be shortened enough to limit filling. As a result, EDV might plateau or start to decrease.

For the Frank–Starling mechanism to work, the amount of blood entering the ventricle must increase. For this to occur, venous blood return to the heart must increase. This can happen rapidly with redistribution of blood by sympathetic activation of arteries and arterioles in inactive areas of the body and general sympathetic activation of the venous system. Also, the muscles are more active during exercise, so their pumping action increases. In addition, respiration increases, so intrathoracic and intra-abdominal pressure increases. All these changes enhance venous return.

We have discussed two factors that can contribute to an increase in stroke volume with increasing intensity of exercise: increased venous return and increased ventricular contractility. A third factor that also contributes to the increase in stroke volume during exercise is a decrease in total peripheral resistance attributable to increased vasodilation of the blood vessels going to active skeletal muscle. This decrease in total peripheral resistance allows the left ventricle to contract against less resistance, facilitating emptying of the blood from this chamber.

Conflicting Research on Stroke-Volume Increase

Although researchers agree that stroke volume increases as work rates increase up to around 40% to 60% of maximum, reports about what happens after that point differ widely. A review of studies conducted between the mid-1960s and the early 1990s reveals no clear pattern of stroke volume increase beyond the 40% to 60% work rate range. Both early and recent research reveals that the conflict continues. Several studies have found a plateau in stroke volume at approximately 50% of $\dot{V}O_2$max, with little or no change occurring with further increases.[1, 5, 13, 17, 22] However, several other studies have shown that stroke volume continues to increase beyond that rate.[5, 7, 12, 20, 26]

This apparent disagreement might be the result of the mode of exercise testing or the participant's training level. Studies that show plateaus in the 40% to 60% $\dot{V}O_2$max range typically have used cycle ergometers. Previous studies have shown that blood is trapped in the legs during cycle ergometer exercise. Thus, the plateau in stroke volume might be unique to exercise on cycle ergometers, resulting from decreased venous return of blood from the legs.

In studies where stroke volume continued to increase up to maximal rates of exercise, subjects were generally highly trained athletes. Many highly trained athletes, including highly trained cyclists tested on a cycle ergometer,[10] can continue to increase their SV after exceeding the 40% to 60% $\dot{V}O_2$max level, perhaps because of adaptations to training. The increases in cardiac output and SV with increasing rates of work, as represented by increasing HR, in elite athletes, trained university distance runners, and untrained university students, are illustrated in figure 7.13.[26]

Finally, stroke volume is difficult to assess, particularly at higher work rates, so differences between studies could result from differences in the techniques used to measure cardiac output or stroke volume and the accuracy of these techniques at different exercise intensities.

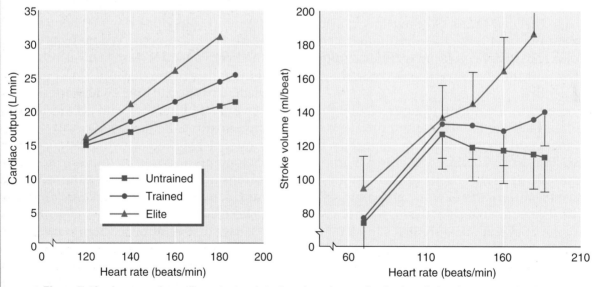

▲ **Figure 7.13** Increases in cardiac output and stroke volume in untrained university students, trained university distance runners, and elite athletes.

Adapted, by permission, from B. Zhou et al., 2001, "Stroke volume does not plateau during graded exercise in elite male distance runners," *Medicine and Science in Sports and Exercise* 33: 1849-1854.

Cardiac Output

Now that we have discussed both of the components of cardiac output—heart rate and stroke volume—we can put this information together to understand what happens to cardiac output during exercise. Changes in cardiac output, because it is the product of heart rate and stroke volume ($\dot{Q}$ = HR × SV), are predictable with increasing work rates, as seen in figure 7.14. The resting value for cardiac output is approximately 5.0 L/min, but this will vary in proportion to the size of the person. Cardiac output increases directly with increasing exercise intensity to between 20 and 40 L/min. The absolute value varies with body size and endurance conditioning. The linear relationship between cardiac output and work rate should not be surprising, though, because the major purpose of the increase in cardiac output is to meet the muscles' increased demand for oxygen.

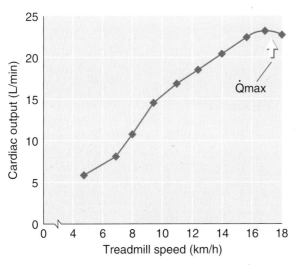

▲ **Figure 7.14**   Changes in cardiac output ($\dot{Q}$) while a subject walks, jogs, and runs with increasing speed on a treadmill. The cardiac output increases in direct proportion to the increase in speed, eventually reaching a maximal value ($\dot{Q}$max).

Overall Changes in Cardiac Function

To see how HR, SV, and $\dot{Q}$ vary under various conditions of rest and exercise, consider the following example. From a reclining position, you rise to a sitting position and then stand. You start walking. Your walk gives way to a jog, and finally you are running. How does your heart respond? As you progress from the reclined position to a full run, your cardiovascular system continuously makes adjustments that allow you to progressively increase your work rate.

If your heart rate when you were reclining was 50 beats/min, it will increase to about 55 beats/min when sitting and to about 60 beats/min when standing. Why does your heart rate increase? When your body shifts from a reclining to a sitting position and then to a standing position, stroke volume immediately decreases. This is primarily because gravity causes blood to pool in your legs, which reduces the volume of blood returning to your heart. At the same time, your heart rate increases. This increase in heart rate when you change to an upright posture is simply an adaptation to maintain the cardiac output, because $\dot{Q}$ = HR × SV.

> During exercise, cardiac output increases primarily to match the need for increased oxygen supply to the working muscles.

As you begin your activity, moving from standing to walking, your heart rate increases from about 60 to about 90 beats/min. Heart rate increases to 140 beats/min with moderate-paced jogging and can reach 180 beats/min or more with a fast-paced run. The increase in heart rate when you change positions at rest maintains your cardiac output, but the increase that comes with greater activity allows delivery of substantially more blood to your working muscles to meet the increasing oxygen requirements. Stroke volume also increases with exercise, further increasing cardiac output. These relationships are illustrated in figure 7.15.

The relationships between heart rate and stroke volume differ by activity. As an example, more blood is pooled in your legs when you are cycling on the cycle ergometer than when running or swimming, resulting in a lower maximal stroke volume. This is attributable to a shorter period of relaxation of the

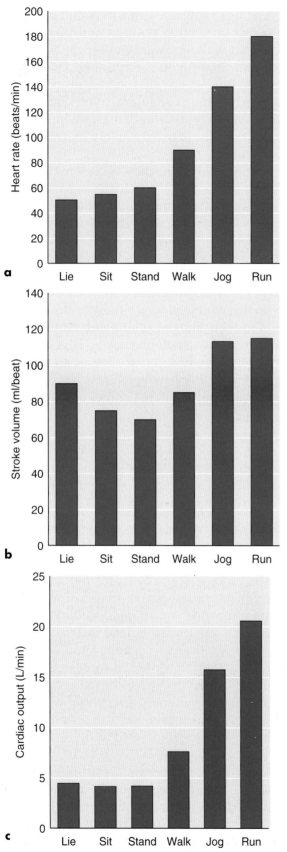

▲ Figure 7.15   Changes in heart rate, stroke volume, and cardiac output with changes in posture (lying supine, sitting, and standing upright) and with exercise (walking at 5 km/h, jogging at 11 km/h, and running at 16 km/h).

leg muscles when you cycle, reducing the time available for the veins to empty. Less blood is pooled in the legs when you swim than during running or cycling. The supine position of swimming allows more blood to return to the heart, resulting in a higher maximal stroke volume potential. Trained cyclists, however, appear to partially overcome this pooling in the legs and can achieve a stroke volume and cardiac output while cycling on a cycle ergometer similar to that obtained while running on a treadmill.

During the initial stages of exercise in untrained individuals, increased cardiac output is caused by an increase in both heart rate and stroke volume. When the level of exercise exceeds 40% to 60% of the individual's capacity, stroke volume has either plateaued or begun to increase at a much slower rate. Thus, further increases in cardiac output are largely the result of increases in heart rate. Stroke volume is likely to contribute more during the higher intensities of exercise in those people who are highly trained, as we discussed earlier.

▶ As exercise intensity increases, heart rate increases proportionately, up to maximal exercise intensities where you achieve HRmax.

▶ Stroke volume (the amount of blood ejected with each contraction) also increases proportionately with increasing exercise intensity but usually achieves its maximal value at about 40% to 60% of $\dot{V}O_2max$ in untrained individuals. Highly trained individuals likely can continue to increase SV up to maximal exercise intensities.

▶ Increases in HR and SV increase cardiac output. Thus, more blood is forced out of the heart during exercise than when at rest, and circulation speeds up. This ensures that adequate supplies of the needed materials—oxygen and nutrients—reach the tissues and that waste products, which build up much more rapidly during exercise, are quickly cleared away.

Blood Flow

We now understand those changes that increase cardiac output during exercise, but the cardiovascular system is even more efficient at getting blood to areas where it is needed than these cardiac adaptations suggest. Recall from our earlier discussion that the vascular system can redistribute blood so that areas with the greatest need receive more blood than areas with low demands. We now turn our attention to changes in blood flow during exercise.

Redistribution of Blood During Exercise

Blood flow patterns change markedly as you move from rest to exercise. Through the action of the sympathetic nervous system, blood is redirected away from areas where it is not essential to those areas that are active during exercise. Only 15% to 20% of the resting cardiac output goes to muscle, but during exhaustive exercise, the muscles receive 80% to 85% of the cardiac output. This shift in blood flow to the muscles is accomplished primarily by reducing blood flow to the kidneys, liver, stomach, and intestines. Figure 7.16 illustrates a typical distribution of blood throughout the body at rest and during heavy exercise. The values are expressed both as relative percentages of the total blood available and as absolute volumes.

As the body starts to overheat, either as a direct result of exercise or because of high environmental temperatures, more blood is redirected to the skin to conduct heat away from the body's core to its periphery, where heat is lost to the environment. The more blood that flows to the skin, the less that is available for muscles, which explains why most endurance athletic performances in the heat are well below average.

When considering blood redistribution within the body, we must think about each of the mechanisms responsible for these changes, but we must remember that they all work together. To illustrate this, we next examine what happens to blood flow during exercise, focusing on the needs of the skeletal muscles.

As exercise begins, the active skeletal muscles rapidly experience a need for increased blood

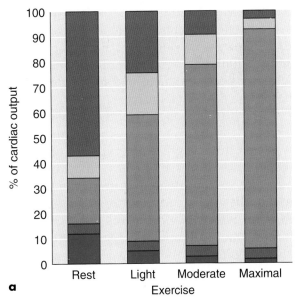

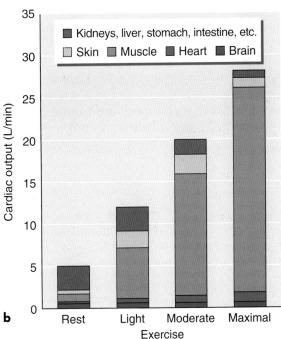

▲ **Figure 7.16**   The distribution of cardiac output at rest and during exercise, expressed (a) relative to total blood volume and (b) in absolute values.

Data from A.J. Vander, J.H. Sherman, & D.S. Luciano, 1985, *Human physiology: The mechanisms of body function*, 4th ed. (New York: McGraw-Hill).

supply. This need is met through a generalized sympathetic stimulation of vessels in those areas to which blood flow is to be reduced (e.g., the digestive system and the kidneys). This constricts vessels in those areas and thus diverts blood flow to the skeletal muscles, where it is needed. In contrast, in the skeletal muscles, sympathetic stimulation to the constrictor fibers in the vessel walls decreases, and sympathetic stimulation to the vasodilator fibers increases. Thus, these vessels dilate, and additional blood flows into the active muscles.

Also, the metabolic rate of the muscle tissue increases during exercise. As a result, metabolic waste products begin to accumulate. Increased metabolism causes an increase in acidity, carbon dioxide, and temperature in the muscle tissue. These local changes trigger vasodilation through autoregulation, increasing blood flow through the local capillaries. Autoregulation is also triggered by the low partial pressure of oxygen in the tissue or a reduction in oxygen bound to hemoglobin (increased oxygen demand), the act of muscle contraction, and possibly other vasoactive substances released as a result of contraction.

Regulation of body temperature is controlled in much the same way. During heavy exercise (or even at rest in a hot environment), heat builds up in the body and must be dissipated. To accomplish this, reduced sympathetic stimulation of the superficial vessels in

the skin causes them to dilate, redirecting or shunting blood there. This promotes heat loss, because heat from deep in the body can be released when blood moves close to the skin. This allows maintenance of body temperature, although body temperature does increase with exercise, as we show in chapter 10. Conversely, when exposed to a cold environment, the body conserves heat by increasing sympathetic stimulation to vessels in the skin, causing them to constrict to divert blood away from the cold skin.

Cardiovascular Drift

With prolonged aerobic exercise or aerobic exercise in a hot environment, at a constant rate of work, SV gradually decreases and HR increases. Systemic and pulmonary arterial pressure also decline. These alterations, illustrated in figure 7.17, have been referred to collectively as **cardiovascular drift**,[19] and they are generally associated with increasing body temperature. It has been hypothesized that cardiovascular drift is the result of a progressive increase in the fraction of cardiac output directed to the

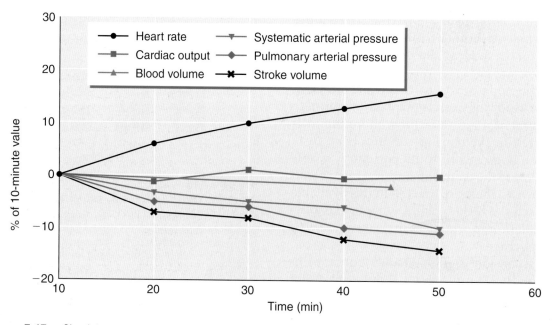

▲ **Figure 7.17**   Circulatory responses to prolonged, moderately intense exercise in the upright position in a neutral 20° C environment illustrating cardiovascular drift. Values are expressed as the percentage of change from the values measured at the 10-min point of the exercise.

vasodilated skin in an attempt to lose body heat and attenuate the increase in body core temperature.[19] With more blood in the skin for the purpose of cooling the body, less blood is available to return to the heart. There is also a small decrease in blood volume resulting from sweating and from a generalized shift of plasma across the capillary membrane into the surrounding tissues. These combine to decrease central venous filling pressure, which decreases venous return to the right side of the heart and reduces the EDV. With the reduction in EDV, SV is reduced (SV = EDV − ESV). The heart rate compensates for the decrease in stroke volume by increasing, in an effort to maintain cardiac output ($\dot{Q}$ = HR × SV).

Recent work at the University of Texas at Austin by Dr. Edward Coyle and colleagues[4, 9] has questioned the preceding explanation for cardiovascular drift. In their earlier studies, these researchers were able to greatly attenuate cardiovascular drift by maintaining subjects in a fully hydrated state. In their most recent research, they found that when a β-receptor-blocking drug was administered so as to have its effect after 10 to 15 min of exercise, it allowed the heart rate and stroke volume to remain constant (no drift) over 60 min of exercise. Furthermore, in the control trial—without β-blockade—the authors saw the expected decrease in stroke volume and increase in heart rate, but skin blood flow remained relatively stable after the first 15 min throughout the 60 min of exercise, suggesting that skin blood flow, reflecting the volume of blood in the skin, is unrelated to cardiovascular drift. β-blocking drugs act directly on the β-receptors, and in the heart they reduce both resting and exercise HR. Evidently, the reduction in exercise HR with β-blockade allowed stroke volume to be maintained.

Competition for Blood Supply

When the demands of exercise are added to all the other demands of the body, competition for the limited volume of blood available can occur. Consider the following research, in which researchers examined the relationship between the timing of feeding and competition for blood supply. Working with miniature pigs,

McKirnan and coworkers[16] studied the effects of feeding versus fasting on the distribution of blood flow during exercise. The pigs were divided into two groups. One group fasted for 14 to 17 h before exercise. The other group ate their morning ration in two feedings: Half the ration was fed 90 to 120 min before exercise and the other half 30 to 45 min before exercise. Both groups of pigs ran at approximately 65% of their $\dot{V}O_2$max.

Blood flow to the hindlimb muscles during exercise was 18% lower in the fed group than in the fasted group. Gastrointestinal blood flow increased 23% in the fed group. Waaler and colleagues[25] reported similar results in humans, concluding that the redistribution of gastrointestinal blood flow to the working muscles is less marked after a meal than before a meal. This finding suggests that athletes should be very careful in timing their meals before competition to maximize blood flow to the active muscles during exercise.

Blood is redistributed in the body primarily to meet the demands of active tissues. We next turn our attention to the driving force behind blood flow: blood pressure.

Blood Pressure

When examining differences in blood pressure during exercise, we must distinguish between systolic and diastolic pressure, because they show different changes. With whole-body endurance activity, systolic blood pressure increases in direct proportion to increased exercise intensity. A systolic pressure that starts out at 120 mmHg at rest can exceed 200 mmHg at exhaustion. Systolic pressures of 240 to 250 mmHg have been reported in normal, healthy, highly trained athletes at maximal levels of aerobic-type exercise.

Increased systolic blood pressure results from the increased cardiac output ($\dot{Q}$) that accompanies increasing rates of work. It helps drive the blood quickly through the vasculature. Also, blood pressure determines how much fluid leaves the capillaries, entering the tissues and carrying needed supplies. Thus increased systolic pressure facilitates the delivery process.

Diastolic blood pressure changes little if any during endurance exercise, regardless of the intensity. Remember that diastolic pressure reflects the pressure in the arteries when the heart is at rest. None of the changes we have discussed alter this pressure significantly, so there is no reason to expect it to increase. Increases in diastolic pressure of 15 mmHg or more are considered abnormal responses to exercise and are one of several indications for immediately stopping a diagnostic exercise test. Figure 7.18 illustrates a typical blood pressure response to both leg and arm cycling exercise with increasing rates of work.

As seen in figure 7.18, when you exercise the upper and lower body musculature at the same absolute rate of energy expenditure, the use of upper body musculature causes a greater blood pressure response. This is most likely attributable to the smaller muscle mass and vasculature of the upper body compared with the lower body. This size difference results in more resistance to blood flow and thus an increase in blood pressure to overcome this resistance.

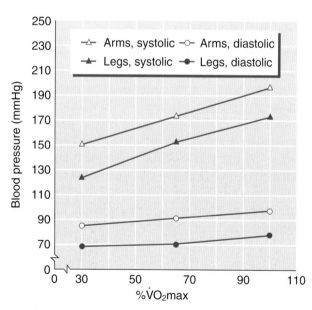

▲ **Figure 7.18**   Blood pressure response to both leg and arm cycling at the same relative rates of oxygen consumption (% V̇O₂max).

Adapted, by permission, from P.-O. Åstrand et al., 1965, "Intraarterial blood pressure during exercise with different muscle groups," *Journal of Applied Physiology* 20: 253-256.

This difference in the systolic blood pressure response to upper and lower body exercise has important implications for the heart. Myocardial oxygen uptake and myocardial blood flow are directly related to the product of heart rate and systolic blood pressure. This value is referred to as the double product (DP = HR × SBP). With static or dynamic resistance exercise or upper body work, the double product is elevated, indicating a much higher cost to the heart.

Blood pressure reaches a steady state during submaximal steady-state endurance exercise. As work intensity increases, so does systolic blood pressure. If steady-state exercise is prolonged, the systolic pressure might start to decrease gradually, but diastolic pressure remains constant. The decrease in systolic blood pressure, if it occurs, is a normal response and simply reflects increased arteriole dilation in the active muscles, which decreases the **total peripheral resistance** (recall from basic physiology that blood pressure = cardiac output × total peripheral resistance).

Blood pressure responses to resistance exercise, such as weightlifting, are exaggerated. With high-intensity resistance training, blood pressure can exceed 480/350 mmHg.[15] In such exercise, use of the **Valsalva maneuver** is quite common. This maneuver occurs when a person tries to exhale while the mouth, nose, and glottis are closed. This action causes an enormous increase in intrathoracic pressure. Much of the subsequent blood pressure increase results from the body's effort to overcome the high internal pressures created during the Valsalva maneuver.

Blood

We have now examined how the heart and blood vessels respond to exercise. The remaining component of the cardiovascular system is the blood: the fluid that carries needed substances to the tissues and clears away harmful substances. As metabolism increases during exercise, the functions of the blood become more vital for efficient performance. We now examine changes that occur within the blood to meet these increased demands.

Oxygen Content

At rest, the blood's oxygen content varies from 20 ml of oxygen per 100 ml of arterial blood to 14 ml of oxygen per 100 ml of venous blood returning to the right atrium. The difference between these two values (20 ml – 14 ml = 6 ml) is referred to as the **arterial–mixed venous oxygen difference, or a-$\bar{v}O_2$ difference**. This value represents the extent to which oxygen is extracted, or removed, from the blood as it passes through the body.

With increasing rates of exercise, the a-$\bar{v}O_2$ difference increases progressively. The a-$\bar{v}O_2$ difference can increase approximately threefold from rest to maximal levels of exercise (see figure 7.19). This reflects a decreasing venous oxygen content, because arterial oxygen content changes little from rest up to maximal exertion. We should note, however, that there have been reports of decreased arterial oxygen content in highly trained athletes at maximal levels of exercise.[6] This is discussed in greater detail in chapter 8. More oxygen is required by the active muscles, so more oxygen is extracted from the blood. The venous oxygen content decreases, approaching zero in the active muscles, but the mixed venous blood in the right atrium of the

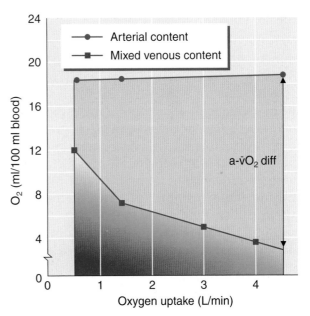

▲ **Figure 7.19** Changes in arterial-mixed venous oxygen difference, a-$\bar{v}O_2$ difference, from low levels to maximal levels of exercise.

heart rarely decreases below 4 ml of oxygen per 100 ml of blood. This is because the blood returning from the active tissues is mixed with blood from inactive areas as it returns to the heart. Oxygen use in the inactive tissues is far lower than in the active muscles.

The Fick Principle and the Fick Equation

In the 1870s, a cardiovascular physiologist by the name of A. Fick developed a principle critical to our understanding of the basic relationships between heart rate, stroke volume, a-$\bar{v}O_2$ difference, and oxygen consumption. The Fick principle states that the amount of a substance removed from or taken up by an organ per unit of time is equal to the arterial concentration minus the venous concentration of that substance multiplied by the blood flow through the organ. If the organ is represented by the total human body, you can express the relationship between the body's oxygen consumption ($\dot{V}O_2$, the amount of substance removed from the blood), the a-$\bar{v}O_2$ difference (arterial and venous concentrations of oxygen), and cardiac output ($\dot{Q}$, blood flow through the body) as an equation, referred to as the Fick equation:

$$\dot{V}O_2 = \dot{Q} \times \text{a-}\bar{v}O_2 \text{ diff}$$

which can be rewritten as

$$\dot{V}O_2 = HR \times SV \times \text{a-}\bar{v}O_2 \text{ diff}$$

This basic relationship comes up frequently throughout the remainder of this book.

Plasma Volume

With the onset of exercise, there is an almost immediate loss of blood plasma volume to the interstitial fluid space. This probably results from two factors. As blood pressure increases, the hydrostatic pressure within the capillaries increases. Thus, the increase in blood pressure forces water from the vascular compartment to the interstitial compartment. Also, as metabolic waste products build up in the active muscle, intramuscular osmotic pressure increases, which attracts fluid to the muscle.

A 10% to 20% or greater reduction in plasma volume can occur with prolonged work. Similarly, 15% to 20% decreases in plasma volume have been observed in 1-min bouts of exhaustive exercise.[21] With resistance training, the plasma volume loss is proportional to the intensity of the effort, with losses ranging from 7.7% when exercising at 40% of the 1-repetition maximum up to 13.9% when training at 70%.[2]

If exercise intensity or environmental conditions cause sweating, additional plasma loss can be expected. Although the major source of fluid for sweating is the interstitial fluid, this fluid will be diminished as sweating continues. This increases the osmotic pressure in the interstitial space, which causes even more plasma to move into the tissues. Intracellular fluid volume is impossible to measure directly and accurately, but research suggests that fluid is also lost from the intracellular compartment and even from the red blood cells, which may shrink.

A reduction of plasma volume likely will impair performance. For long-duration activities in which heat loss is a problem, the total flow of blood to active tissues must be reduced to allow increasingly more blood to be diverted to the skin in an attempt to lose body heat. Reduced plasma volume also increases blood viscosity, which can impede blood flow and thus limit oxygen transport, especially if the hematocrit exceeds 60%.

In activities that last several minutes or less, body fluid changes and temperature regulation are of little practical importance. As exercise duration increases, however, body fluid changes and temperature regulation become important to efficient performance. For the football player, the Tour de France cyclist, or the marathon runner, these processes are crucial, not only for competition but also for survival. Deaths have occurred from dehydration and hyperthermia during or as a result of various sport activities. These issues are discussed in detail in chapter 10.

Hemoconcentration

When plasma volume is reduced, hemoconcentration occurs. This means that the fluid portion of the blood is reduced, and the cellular and protein portions represent a larger fraction of the total blood volume, as seen in figure 7.20. Thus, they become more concentrated in the blood. This hemoconcentration increases red blood cell concentration substantially—by up to 20% or 25%. Hematocrits can increase from 40% to 50%. However, the total number or content of red blood cells is unlikely to change substantially.

As the hematocrit increases, the net effect, even without an increase in the total number of red blood cells, is to increase the amount of red blood cells per unit of blood because the cells are more concentrated. As the red blood cell concentration increases, so does the blood's per-unit hemoglobin content. This substantially

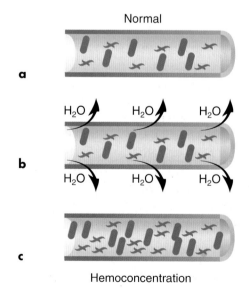

▲ **Figure 7.20** A blood vessel with normal blood concentration (*a*). Hemoconcentration occurs when H_2O leaves the blood vessels (*b*), increasing the concentration of the substances that remain in the blood (*c*).

increases the blood's oxygen-carrying capacity, which is advantageous during exercise and provides a distinct advantage at altitude at rest and during submaximal exercise, as we see in chapter 11.

At one time, the hemoconcentration response was believed to reflect an addition of red blood cells from the spleen to the blood to facilitate oxygen transport. This view assumed that because many animals, such as the horse, can increase the number of circulating red blood cells by releasing cells stored in the spleen, humans can also. This explanation was largely discounted in the 1950s and 1960s because limited evidence suggested that the spleen does not serve that function in humans. We now know that the spleen has a storage capacity of about 50 ml of concentrated red blood cells. Flamm and coworkers,[8] using radionuclide techniques, reported a progressive decrease in the spleen's total blood volume with increasing work rates. The increase in hematocrit paralleled a decrease in the spleen's blood volume. This was confirmed in a subsequent study.[14] The importance of the spleen's role during exercise has yet to be fully determined.

Blood pH

Finally, blood pH can change considerably with moderate- to high-intensity exercise. Recall that neutral pH is 7.0; a pH greater than 7.0 is alkaline, or basic; and less than 7.0 is acidic (see chapter 8). At rest, arterial blood pH remains constant at about 7.4—slightly alkaline.

Little change occurs from rest up to an exercise intensity of about 50% of maximal aerobic capacity. As intensity increases above 50%, pH starts to decrease as the blood becomes more acidic. This decrease is gradual at first but becomes more rapid as the body approaches exhaustion. Blood pH values of 7.0 or lower have been reported following maximal sprint-type exercise. The pH in active muscle decreases even more, to 6.5 or lower.

The decrease in blood pH results primarily from an increased reliance on anaerobic metabolism and corresponds to increases in blood lactate observed with increasing exercise intensity. Blood and muscle pH are discussed in much more detail in chapter 8.

Integration of the Exercise Response

As you can see from our discussion of all the changes in cardiovascular function that take place when you start to exercise, the cardiovascular system is extremely complex. You may be overwhelmed by all this information! Fortunately, Dr. Edward Coyle[3] developed an excellent flow diagram that illustrates how the body is able to integrate all these cardiovascular

▶ The changes that occur in the blood during exercise demonstrate that the blood is carrying out its necessary tasks. Here are the major changes seen:

1. The a-$\bar{v}O_2$ difference increases. This happens because the venous oxygen concentration decreases during exercise, reflecting increased extraction of oxygen from the blood for use by the active tissues.

2. Plasma volume decreases during exercise. The fluid (water) is pushed out of the capillaries by increases in hydrostatic pressure as blood pressure increases, and fluid is drawn into the muscles by the increased osmotic pressure that results from waste accumulation. However, with prolonged exercise or exercise in hot environments, increasingly more plasma volume is lost through sweating in an attempt to maintain body temperature, placing the person at risk of dehydration.

3. Hemoconcentration occurs as plasma volume (water) decreases. Although the actual number of red blood cells might not increase, the relative number of red blood cells per unit of blood increases, which increases oxygen-carrying capacity.

4. Blood pH can change significantly during exercise, becoming more acidic as it decreases from the slightly alkaline resting value of 7.4 down to 7.0 or lower. The muscle pH decreases even further. The decrease in pH results primarily from increased blood lactate accumulation with increasing exercise intensity.

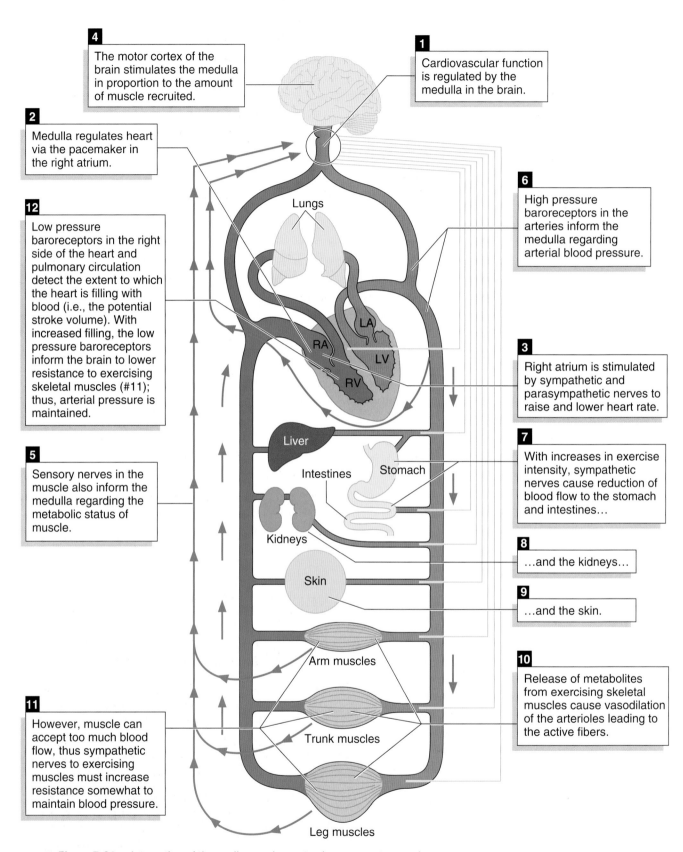

4
The motor cortex of the brain stimulates the medulla in proportion to the amount of muscle recruited.

1
Cardiovascular function is regulated by the medulla in the brain.

2
Medulla regulates heart via the pacemaker in the right atrium.

12
Low pressure baroreceptors in the right side of the heart and pulmonary circulation detect the extent to which the heart is filling with blood (i.e., the potential stroke volume). With increased filling, the low pressure baroreceptors inform the brain to lower resistance to exercising skeletal muscles (#11); thus, arterial pressure is maintained.

6
High pressure baroreceptors in the arteries inform the medulla regarding arterial blood pressure.

3
Right atrium is stimulated by sympathetic and parasympathetic nerves to raise and lower heart rate.

5
Sensory nerves in the muscle also inform the medulla regarding the metabolic status of muscle.

7
With increases in exercise intensity, sympathetic nerves cause reduction of blood flow to the stomach and intestines…

8
…and the kidneys…

9
…and the skin.

10
Release of metabolites from exercising skeletal muscles cause vasodilation of the arterioles leading to the active fibers.

11
However, muscle can accept too much blood flow, thus sympathetic nerves to exercising muscles must increase resistance somewhat to maintain blood pressure.

Lungs
LA
RA
LV
RV
Liver
Intestines
Stomach
Kidneys
Skin
Arm muscles
Trunk muscles
Leg muscles

▲ **Figure 7.21** Integration of the cardiovascular system's response to exercise.

Adapted, by permission, from E.F. Coyle, 1991, "Cardiovascular function during exercise: Neural control factors," *Sports Science Exchange* 4(34): 1-6. Copyright 1991 by Gatorade Sports Science Institute.

responses to provide for its needs during exercise. This diagram is presented in figure 7.21. Key areas and responses are labeled and summarized to help you see how all this activity is coordinated. It is important to note that although the body attempts to meet the blood flow needs of the muscle, it will do so only if blood pressure is not compromised. Maintenance of arterial blood pressure appears to be the highest priority of the cardiovascular system, irrespective of the environment and the body's other needs.

In Closing . . .

In this chapter, we reviewed the structure and function of the cardiovascular system and examined how this system responds during exercise to meet the increased needs of active muscles. We explored its role in transporting and delivering oxygen and nutrients to the active tissues while clearing away metabolic wastes, including carbon dioxide. Knowing how substances are moved within the body, we can now look more closely at the movement of oxygen and carbon dioxide. In the next chapter, we explore the respiratory system, considering how oxygen is moved into and throughout the body, how oxygen is delivered to the active tissues, and how carbon dioxide is cleared away from them. Then we examine how respiratory function changes to meet the demands of an active body.

▶ Key Terms

arterial–venous oxygen difference, a-$\bar{v}O_2$ difference
arteries
arterioles
atrioventricular (AV) node
autoregulation
bradycardia
capillaries
cardiac cycle
cardiac output ($\dot{Q}$)
cardiovascular drift
diastolic blood pressure (DBP)

ejection fraction (EF)
electrocardiogram (ECG)
electrocardiograph
end-diastolic volume (EDV)
end-systolic volume (ESV)
extrinsic neural control
Frank–Starling mechanism
hematocrit
hemoconcentration
hemoglobin
hypertension
maximum heart rate (HRmax)
mean arterial pressure (MAP)
myocardium
pericardium
premature ventricular contraction
Purkinje fibers
resting heart rate (RHR)
sinoatrial (SA) node
steady-state heart rate
stroke volume (SV)
systolic blood pressure (SBP)
tachycardia
total peripheral resistance
Valsalva maneuver
vasodilation
veins
ventricular fibrillation
ventricular tachycardia
venules

▶ Study Questions

1. Describe the structure of the heart, the pattern of blood flow through the valves and chambers of the heart, how the heart as a muscle is supplied with blood, and what happens when the resting heart must suddenly supply an exercising body.

2. What events take place that allow the heart to contract, and how is heart rate controlled?

3. What is the difference between systole and diastole, and how do they relate to SBP and DBP?

4. How is blood flow to the various regions of the body controlled? How does blood flow vary with exercise?

5. Describe how heart rate, stroke volume, and cardiac output respond to increasing rates of work.

6. How do you determine HRmax? What are alternative methods using indirect estimates? What are the major limitations to these indirect estimates?

7. Describe two important mechanisms for returning blood back to the heart when you are exercising in an upright position.

8. What are the major cardiovascular adjustments that your body makes when you are overheated during exercise?

9. What is cardiovascular drift? Why might this be a problem with prolonged exercise?

10. Describe the primary functions of blood.

11. What changes occur in the plasma volume and red blood cells with increasing levels of exercise? With prolonged exercise in the heat?

▷ References

1. Åstrand, P.-O., Cuddy, T.E., Saltin, B., & Stenberg, J. (1964). Cardiac output during submaximal and maximal work. *Journal of Applied Physiology,* **19,** 268-274.

2. Collins, M.A., Cureton, K.J., Hill, D.W., & Ray, C.A. (1989). Relation of plasma volume change to intensity of weight lifting. *Medicine and Science in Sports and Exercise,* **21,** 178-185.

3. Coyle, E.F. (1991). Cardiovascular function during exercise: Neural control factors. *Sports Science Exchange,* **4**(34), 1-6.

4. Coyle, E.F., & González-Alonso, J. (2001). Cardiovascular drift during prolonged exercise: New perspectives. *Exercise and Sport Sciences Reviews,* **29,** 88-92.

5. Crawford, M.H., Petru, M.A., & Rabinowitz, C. (1985). Effect of isotonic exercise training on left ventricular volume during upright exercise. *Circulation,* **72,** 1237-1243.

6. Dempsey, J.A. (1986). Is the lung built for exercise? *Medicine and Science in Sports and Exercise,* **18,** 143-155.

7. Ekblom, B., & Hermansen, L. (1968). Cardiac output in athletes. *Journal of Applied Physiology,* **25,** 619-625.

8. Flamm, S.D., Taki, J., Moore, R., Lewis, S.F., Keech, F., Maltais, F., Ahmad, M., Callahan, R., Dragotakes, S., Alpert, N., & Strauss, H.W. (1990). Redistribution of regional and organ blood volume and effect on cardiac function in relation to upright exercise intensity in healthy human subjects. *Circulation,* **81,** 1550-1559.

9. Fritzsche, R.G., Switzer, T.W., Hodgkinson, B.J., & Coyle, E.F. (1999). Stroke volume decline during prolonged exercise is influenced by the increase in heart rate. *Journal of Applied Physiology,* **86,** 799-805.

10. Gledhill, N., Cox, D., & Jamnik, R. (1994). Endurance athletes' stroke volume does not plateau: Major advantage is diastolic function. *Medicine and Science in Sports and Exercise,* **26,** 1116-1121.

11. Green, D.J., O'Driscoll, G., Blanksby, B.A., & Taylor, R.R. (1996). Control of skeletal muscle blood flow during dynamic exercise: Contribution of endothelium-derived nitric oxide. *Sports Medicine,* **21,** 119-146.

12. Hermansen, L., Ekblom, B., & Saltin, B. (1970). Cardiac output during submaximal and maximal treadmill and bicycle exercise. *Journal of Applied Physiology,* **29,** 82-86.

13. Higginbotham, M.B., Morris, K.G., Williams, R.S., McHale, P.A., Coleman, R.E., & Cobb, F.R. (1986). Regulation of stroke volume during submaximal and maximal upright exercise in normal man. *Circulation Research,* **58,** 281-291.

14. Laub, M., Hvid-Jacobsen, K., Hovind, P., Kanstrup, I.-L., Christensen, N.J., & Nielsen, S.L. (1993). Spleen emptying and venous hematocrit in humans during exercise. *Journal of Applied Physiology,* **74,** 1024-1026.

15. MacDougall, J.D., Tuxen, D., Sale, D.G., Moroz, J.R., & Sutton, J.R. (1985). Arterial blood pressure response to heavy resistance exercise. *Journal of Applied Physiology,* **58,** 785-790.

16. McKirnan, M.D., Gray, C.G., & White, F.C. (1991). Effects of feeding on muscle blood flow during prolonged exercise in miniature swine. *Journal of Applied Physiology,* **70,** 1097-1104.

17. Plotnick, G.D., Becker, L.C., Fisher, M.L., Gerstenblith, G., Renlund, D.G., Fleg, J.L., Weisfeldt, M.L., & Lakatta, E.G. (1986). Use of the Frank–Starling mechanism during submaximal versus maximal upright exercise. *American Journal of Physiology (Heart and Circulation Physiology),* **251,** H1101-H1105.

18. Poliner, L.R., Dehmer, G.J., Lewis, S.E., Parkey, R.W., Blomqvist, C.G., & Willerson, J.T. (1980). Left ventricular performance in normal subjects: A comparison

of the responses to exercise in the upright and supine position. *Circulation, 62,* 528-534.

19. Rowell, L.B. (1993). *Human cardiovascular control.* New York: Oxford University Press.

20. Scruggs, K.D., Martin, N.B., Broeder, C.E., Hofman, Z., Thomas, E.L., Wambsgans, K.C., & Wilmore, J.H. (1991). Stroke volume during submaximal exercise in endurance-trained normotensive subjects and in untrained hypertensive subjects with beta-blockade (propranolol and pindolol). *American Journal of Cardiology, 67,* 416-421.

21. Sejersted, O.M., Vøllestad, N.K., & Medbø, J.I. (1986). Muscle fluid and electrolyte balance during and following exercise. *Acta Physiologica Scandinavica,* **128**(Suppl. 556), 119-127.

22. Stenberg, J., Åstrand, P.-O., Ekblom, B., Royce, J., & Saltin, B. (1967). Hemodynamic response to work with different muscle groups, sitting and supine. *Journal of Applied Physiology, 22,* 61-70.

23. Tanaka, H., Monahan, D.K., & Seals, D.R. (2001). Age-predicted maximal heart rate revisited. *Journal of the American College of Cardiology, 37,* 153-156.

24. Turkevich, D., Micco, A., & Reeves, J.T. (1988). Noninvasive measurement of the decrease in left ventricular filling time during maximal exercise in normal subjects. *American Journal of Cardiology, 62,* 650-652.

25. Waaler, B.A., Eriksen, M., & Janbu, T. (1990). The effect of a meal on cardiac output in man at rest and during moderate exercise. *Acta Physiologica Scandinavica, 140,* 167-173.

26. Zhou, B., Conlee, R.K., Jensen, R., Fellingham, G.W., George, J.D., & Fisher, A.G. (2001). Stroke volume does not plateau during graded exercise in elite male distance runners. *Medicine and Science in Sports and Exercise, 33,* 1849-1854.

▶ Selected Readings

Åstrand, P.-O., Ekblom, B., Messin, R., Saltin, B., & Stenberg, J. (1965). Intraarterial blood pressure during exercise with different muscle groups. *Journal of Applied Physiology, 20,* 253-256.

Buckwalter, J.B., & Clifford, P.S. (2001). The paradox of sympathetic vasoconstriction in exercising skeletal muscle. *Exercise and Sport Sciences Reviews, 29,* 159-163.

Carlsten, A., & Grimby, G. (1966). *The circulatory response to muscular exercise in man.* Springfield, IL: Charles C Thomas.

Clausen, J.P. (1977). Effect of physical training on cardiovascular adjustments to exercise in man. *Physiological Reviews, 57,* 779-815.

Hughson, R.L., & Tschakovsky, M.E. (1999). Cardiovascular dynamics at the onset of exercise. *Medicine and Science in Sports and Exercise, 31,* 1005-1010.

Laughlin, M.H., & Armstrong, R.B. (1985). Muscle blood flow during locomotor exercise. *Exercise and Sport Sciences Reviews, 13,* 95-136.

Raven, P.B., Potts, J.T., & Shi, X. (1997). Baroreflex regulation of blood pressure during dynamic exercise. *Exercise and Sport Sciences Reviews, 25,* 368-389.

Rowell, L.B. (1974). Human cardiovascular adjustments to exercise and thermal stress. *Physiological Reviews, 54,* 75-159.

Rowell, L.B. (1986). *Human circulation: Regulation during physical stress.* New York: Oxford University Press.

Saltin, B. (1985). Hemodynamic adaptations to exercise. *American Journal of Cardiology, 55,* 42D-47D.

Saltin, B., Bouschel, R., Secher, N., & Mitchell, J. (Eds.). (2000). *Exercise and circulation in health and disease.* Champaign, IL: Human Kinetics.

Saltin, B., & Rowell, L.B. (1980). Functional adaptations to physical activity and inactivity. *Federation Proceedings, 39,* 1506-1513.

Senay, L.C., Jr., & Pivarnik, J.M. (1985). Fluid shifts during exercise. *Exercise and Sport Sciences Reviews, 13,* 335-387.

Steingart, R.M., Wexler, J., Slagle, S., & Scheuer, J. (1984). Radionuclide ventriculographic responses to graded supine and upright exercise: Critical role of the Frank–Starling mechanism at submaximal exercise. *American Journal of Cardiology, 53,* 1671-1677.

Sullivan, M.J., Cobb, F.R., & Higginbotham, M.B. (1991). Stroke volume increase by similar mechanisms during upright exercise in normal men and women. *American Journal of Cardiology, 67,* 1405-1412.

Vanoverschelde, J.-L., Essamri, B., Vanbutsele, R., D'Hondt, A.-M., Cosyns, J., Detry, J.-L.R., & Melin, J.A. (1993). Contribution of left ventricular diastolic function to exercise capacity in normal subjects. *Journal of Applied Physiology,* **74**(5), 2225-2233.

RESPIRATORY REGULATION DURING EXERCISE

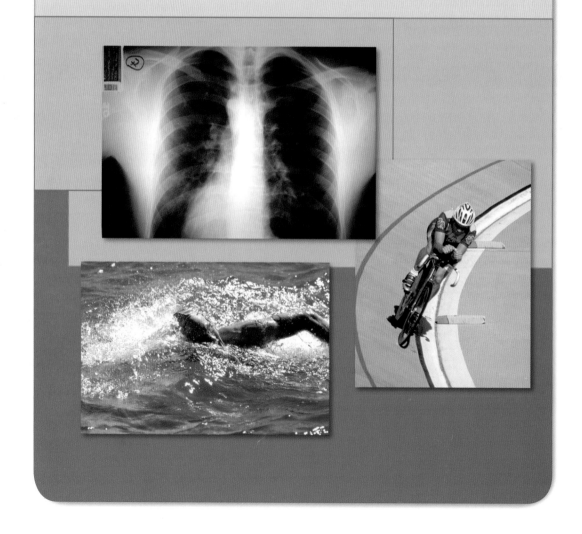

overview

We cannot live without oxygen. As we learned in chapter 4, oxygen is essential for the energy production that fuels our bodies' activities. Endurance performance depends on the delivery of sufficient oxygen to our muscles and adequate cellular uptake of this gas once it arrives there. But at the same time, the metabolic processes occurring in our active muscles are generating another gas—carbon dioxide—which, unlike oxygen, is toxic. Normal cellular activity requires oxygen but is impaired when carbon dioxide levels increase.

Our muscles' crucial needs for adequate oxygen supply and carbon dioxide clearance are met by the respiratory system. As we saw in chapter 7, the cardiovascular system transports these gases. But the respiratory system—our focus in this chapter—brings oxygen into our cells and rids us of excess carbon dioxide. We begin with an overview of the steps involved in respiration and gas exchange, and then we examine how these processes are regulated. We consider how the respiratory system functions when we are exercising and how it can limit performance. Finally, we consider the respiratory system's unique role in maintaining acid–base balance throughout our body and the significance of this balance during physical activity.

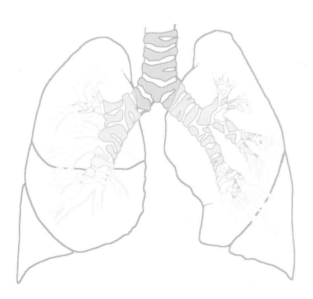

outline

Because swimming requires controlled breathing, many competitive swimmers try to improve their breath-holding ability by attempting to swim several pool lengths underwater without breathing. While I (DLC) was coaching a high school team, a group of my swimmers tried to see who could swim the farthest underwater. They took repeated deep breaths before diving into the pool and swam until the need to breathe became intolerable. Most were able to make two lengths (total of 50 yd or 46 m) of the pool. But one swimmer continued on for a third length, made the turn, and headed into a fourth length. Halfway back, and at a depth of 3 or 4 ft (about 1 m), he suddenly stopped swimming and became motionless. One of the other swimmers quickly dove into the pool and recovered his unconscious body. Fortunately, when he reached the surface, he began breathing and recovered consciousness within 15 to 20 s. Needless to say, we never used that training ploy again.

The respiratory and cardiovascular systems combine to provide an efficient delivery system that carries oxygen to and removes carbon dioxide from our body tissues. This transportation involves four separate processes:

- Pulmonary ventilation (breathing), which is the movement of air into and out of the lungs
- Pulmonary diffusion, which is the exchange of oxygen and carbon dioxide between the lungs and the blood
- Transport of oxygen and carbon dioxide via the blood
- Capillary gas exchange, which is the exchange of oxygen and carbon dioxide between the capillary blood and the metabolically active tissues

The first two processes are referred to as **external respiration** because they involve moving gases from outside the body into the lungs and then the blood. Once the gases are in the blood, they must be transported to the tissues. When blood arrives at the tissues, the fourth step of respiration occurs. This gas exchange between the blood and the tissues is called **internal respiration**. Thus, external and internal respiration are linked by the circulatory system. In the following sections, we examine all four components of respiration.

Pulmonary Ventilation

Pulmonary ventilation, commonly referred to as breathing, is the process by which we move air into and out of our lungs. The anatomy of the respiratory system is illustrated in figure 8.1. Air typically is drawn into the lungs through the nose, although the mouth also must be used when the demand for air exceeds the amount that can comfortably be brought in through the nose. Bringing air in through the nose has certain advantages over mouth breathing. The air is warmed and humidified as it swirls through the bony irregular surfaces (turbinates or conchae) inside the nose. Of equal importance, the turbinates churn the inhaled air, causing dust and other particles to contact and adhere to the nasal mucosa. This filters out all but the tiniest particles, minimizing irritation and the threat of respiratory infections. From the nose and mouth, the air travels through the pharynx, larynx, trachea, bronchi, and bronchioles, until it finally reaches the smallest respiratory units: the alveoli. The alveoli are the sites of gas exchange in the lungs.

Breathing through the nose helps humidify and warm the air during inhalation and filters out foreign particles from the air.

The lungs are not directly attached to the ribs. Rather, they are suspended by the pleural sacs. The pleural sacs have a double wall: the parietal pleura, which lines the thoracic wall, and the visceral or pulmonary pleura, which lines the outer aspects of the lung. These pleural walls envelop the lungs and have a thin film of fluid between them that reduces friction during respiratory movements. In addition, these

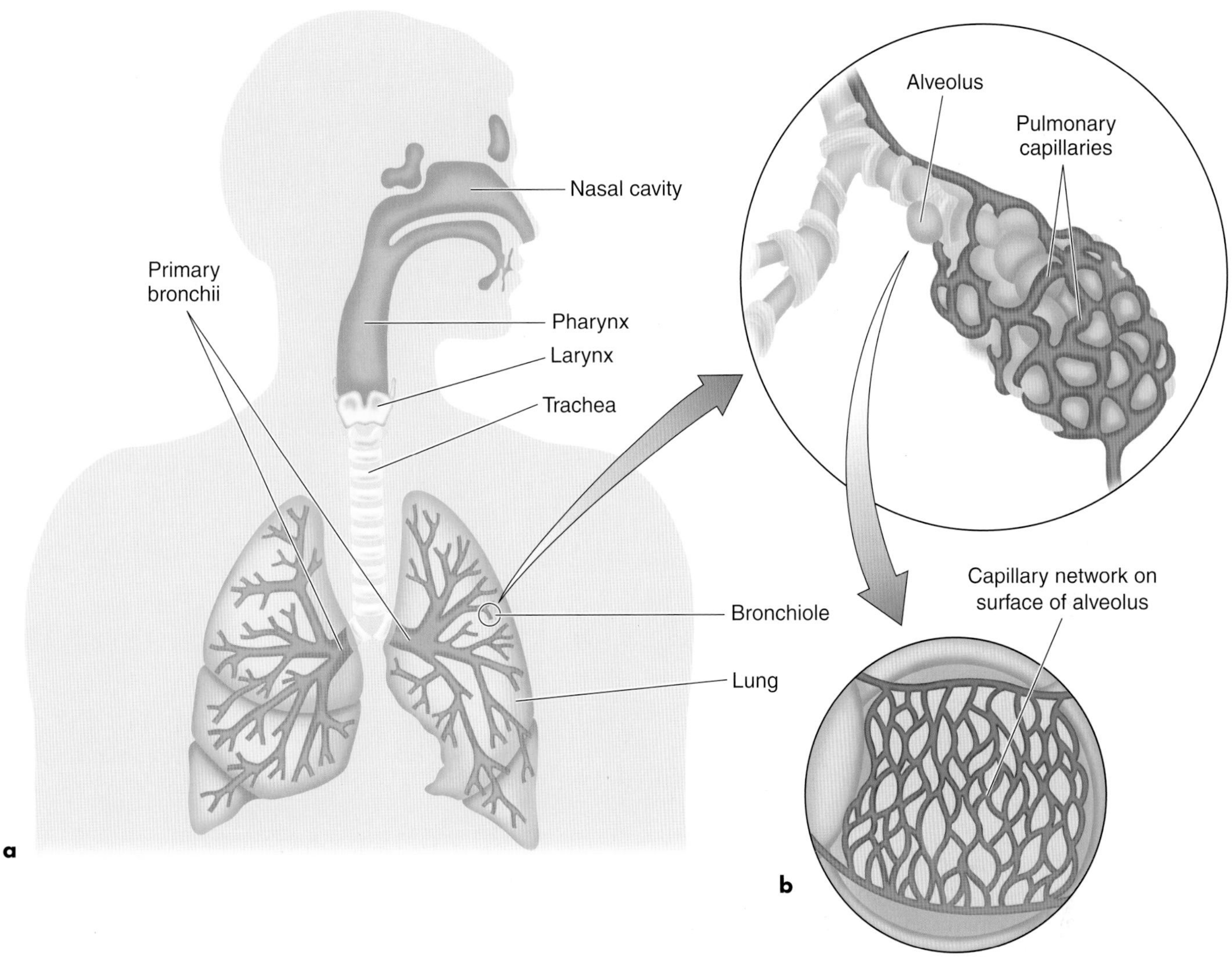

▲ **Figure 8.1**   (*a*) The anatomy of the respiratory system, illustrating the respiratory tract (i.e., nasal cavity, pharynx, trachea, and bronchii). (*b*) The enlarged view of the alveolus shows the regions of gas exchange between the alveolus and pulmonary blood in the capillaries.

sacs are connected to the lungs and to the inner surface of the thoracic cage, causing the lungs to take the shape and size of the rib or thoracic cage as the chest expands and contracts.

These relationships between the lungs, the pleural sacs, and the thoracic cage determine airflow into and out of the lungs. Let's examine the two phases involved: inspiration and expiration.

Inspiration

Inspiration is an active process involving the diaphragm and the external intercostal muscles. Figure 8.2*a* shows the resting positions of the diaphragm and the thoracic cage, or thorax. With inspiration, the ribs and sternum are moved by the external intercostal muscles. The ribs swing up and out, much like the movement of a bucket handle. The sternum swings up and forward, much like the movement of a pump handle. At the same time, the diaphragm contracts, flattening down toward the abdomen. These actions are illustrated in figure 8.2*b*.

These actions expand all three dimensions of the thoracic cage, in turn expanding and stretching the lungs. When the lungs are expanded, the air within them has more space to fill, so the pressure within the lungs decreases. As a result, the pressure in the lungs (intrapulmonary pressure) is less than the air pressure outside the body. Because the respiratory tract is open to the outside, air rushes into the lungs to reduce this pressure difference. This is how air is brought into the lungs during inspiration.

During forced or labored breathing, such as during heavy exercise, inspiration is further assisted by the action of other muscles, such as the scaleni (anterior, middle, and posterior) and sternocleidomastoid in the neck and the pectorals in the chest. These muscles help raise the ribs even more than during regular breathing.

> **fyi** The pressure changes required for adequate ventilation at rest are really quite small. For example, at standard atmospheric pressure (760 mmHg), inspiration may decrease the pressure in the lungs (intrapulmonary pressure) by only about 3 mmHg. However, during maximal respiratory effort, such as during exhaustive exercise, the intrapulmonary pressure can decrease by 80 to 100 mmHg!

Expiration

At rest, **expiration** is usually a passive process involving relaxation of the inspiratory muscles and elastic recoil of the lung tissue. As the diaphragm relaxes, it returns to its normal upward, arched position. As the external

intercostal muscles relax, the ribs and sternum lower back into their resting positions (figure 8.2c). While this happens, the elastic nature of the lung tissue causes it to recoil to its resting size. This increases the pressure in the thorax, so air is forced out of the lungs. Thus, expiration is accomplished.

During forced breathing, expiration becomes a more active process. The internal intercostal muscles actively pull the ribs down. This action can be assisted by the latissimus dorsi and quadratus lumborum muscles. Contracting the abdominal muscles increases the intra-abdominal pressure, forcing the abdominal viscera upward against the diaphragm and accelerating its return to the domed position. These muscles also pull the rib cage down and inward.

The changes in intra-abdominal and intra-thoracic pressure that accompany forced breathing also help return venous blood back to the heart. As these pressures increase, they are transmitted to the great veins—the pulmonary veins and venae cavae—that transport blood back to the heart. When the pressures decrease, the veins return to their original size and fill with blood. The changing pressures within the abdomen and thorax squeeze the blood in the veins, assisting its return through a milking action. This is an essential part of venous return. Similarly, muscle contractions during exercise also produce this type of milking action to assist venous return.

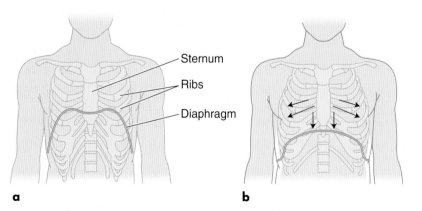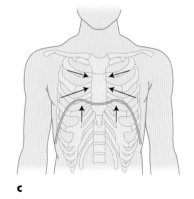

a   b   c

▲ **Figure 8.2** The process of inspiration and expiration, showing how movement of the ribs and diaphragm can increase and decrease the size of the thorax. Note the size of the ribcage at rest (a). The dimensions of the lungs and the thoracic cage increase during inspiration (b), forming a negative pressure that draws air into the lungs. During expiration (c), the lung volume decreases, thereby forcing air out of the lungs.

▶ Pulmonary ventilation (breathing) is the process by which air is moved into and out of the lungs. It has two phases: inspiration and expiration.

▶ Inspiration is an active process in which the diaphragm and the external intercostal muscles contract, increasing the dimensions, and thus the volume, of the thoracic cage. This decreases the pressure in the lungs, allowing air to flow in.

▶ Normal expiration is a passive process. The inspiratory muscles and diaphragm relax and the elastic tissue of the lungs recoils, returning the thoracic cage to its smaller, normal dimensions. This increases the pressure in the lungs and forces air out.

▶ Forced or labored inspiration and expiration are active processes, dependent on muscle actions.

Pulmonary Diffusion

Gas exchange in the lungs, called **pulmonary diffusion**, serves two major functions:

1. It replenishes the blood's oxygen supply, which is depleted at the tissue level, where it is used for oxidative energy production.

2. It removes carbon dioxide from returning venous blood.

Pulmonary diffusion has two requirements: air that brings oxygen into and takes carbon dioxide out of the lungs and blood to receive the oxygen and give up carbon dioxide. Air is brought into the lungs during pulmonary ventilation; now gas exchange must occur between this air and the blood.

Blood from most of the body returns through the venae cavae to the pulmonary (right) side of the heart. From the right ventricle, this blood is pumped through the pulmonary artery to the lungs, ultimately working its way into the pulmonary capillaries. These capillaries form a dense network around the alveolar sacs. These vessels are so small that the red blood cells must pass through them in single file, exposing each cell to the surrounding lung tissue. This is where pulmonary diffusion occurs.

Respiratory Membrane

Gas exchange between the air in the alveoli and the blood in the pulmonary capillaries occurs across the **respiratory membrane** (also called the alveolar–capillary membrane). This membrane, depicted in figure 8.3, is composed of

- the alveolar wall,
- the capillary wall, and
- their basement membranes.

The respiratory membrane is very thin, measuring only 0.5 to 4.0 μm. As a result, the gases in the nearly 300 million alveoli are in close proximity to the blood circulating through the capillaries. Nevertheless, this membrane presents a potential barrier for gas exchange. Let's now examine how this exchange occurs.

Partial Pressures of Gases

The air we breathe is a mixture of gases. Each exerts a pressure in proportion to its concentration in the gas mixture. The individual pressures from each gas in a mixture are referred to as **partial pressures.** According to Dalton's law, the total pressure of a mixture of gases equals the sum of the partial pressures of the individual gases in that mixture.

> The total pressure of a mixture of gases equals the sum of the partial pressures of the individual gases in that mixture.

Consider the air we breathe. It is composed of 79.04% nitrogen (N_2), 20.93% oxygen (O_2), and 0.03% carbon dioxide (CO_2). At sea level, the atmospheric (or barometric) pressure is approximately 760 mmHg, which is also referred to as standard atmospheric pressure. This is considered the total pressure, or 100%. Thus, if the total atmospheric pressure is 760 mmHg, then the partial pressure of nitrogen (PN_2) in air is 600.7 mmHg (79.04% of the total 760 mmHg pressure). Oxygen's partial pressure

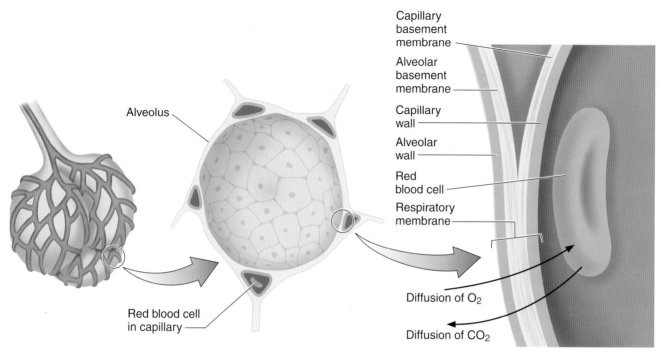

▲ Figure 8.3   The anatomy of the respiratory membrane, showing the exchange of oxygen and carbon dioxide between an alveolus and pulmonary capillary blood.

(PO_2) is 159.1 mmHg (20.93% of 760 mmHg), and carbon dioxide's partial pressure (PCO_2) is 0.2 mmHg (0.03% of 760 mmHg).

Gases in our bodies are dissolved in fluids, such as blood plasma. According to **Henry's law,** gases dissolve in liquids in proportion to their partial pressures, depending also on their solubilities in the specific fluids and on the temperature. A gas's solubility in blood is a constant, and blood temperature also remains relatively constant. Thus, the most critical factor for gas exchange between the alveoli and the blood is the pressure gradient between the gases in the two areas.

Gas Exchange in the Alveoli

Differences in the partial pressures of the gases in the alveoli and the gases in the blood create a pressure gradient across the respiratory membrane. This forms the basis of gas exchange during pulmonary diffusion. If the pressures on each side of the membrane were equal, the gases would be at equilibrium and unlikely to move. But the pressures are not equal. Let's first consider oxygen and its partial pressure gradients.

Oxygen Exchange

The PO_2 of air outside the body at standard atmospheric pressure is 159 mmHg. But this pressure decreases to about 105 mmHg when air is inhaled and enters the alveoli. The inhaled air mixes with the air in the alveoli, and the alveolar air contains a lot of water vapor and carbon dioxide that contribute to the total pressure there. Fresh air that ventilates the lungs is constantly mixed with the air in the alveoli while some of the alveolar gases are exhaled to the environment. As a result, alveolar gas concentrations remain relatively stable.

The blood, stripped of much of its oxygen by the tissues, typically enters the pulmonary capillaries with a PO_2 of about 40 mmHg (see figure 8.4). This is about 60 to 65 mmHg less than the PO_2 in the alveoli. In other words, the pressure gradient for oxygen across the respiratory membrane is typically about 65 mmHg. As noted earlier, this pressure gradient drives the oxygen from the alveoli into the blood to balance the pressure of the oxygen on each side of the membrane.

The PO_2 in the alveoli stays relatively steady at about 105 mmHg. As the deoxygenated

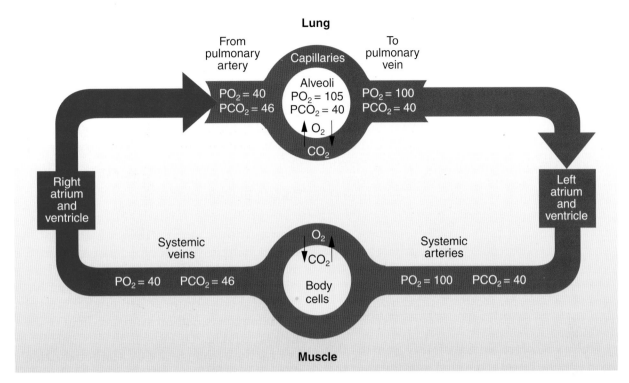

▲ Figure 8.4   Partial pressure of oxygen (PO_2) and carbon dioxide (PCO_2) in blood as a result of gas exchange in the lungs and gas exchange between the capillary blood and tissues.

blood enters the pulmonary artery, the PO_2 in the blood is only about 40 mmHg. But as the blood moves along the pulmonary capillaries, more exchange occurs. By the time the pulmonary blood reaches the venous end of these capillaries, the PO_2 in the blood equals that in the alveoli (approximately 105 mmHg), and the blood is now considered to be fully oxygenated (see figure 8.5). The blood leaving the lungs through the pulmonary veins and subsequently returning to the systemic (left) side of the heart has a rich supply of oxygen to deliver to the tissues. Notice, however, that the PO_2 in the pulmonary vein is 100 mmHg, not the 105 mmHg found in the alveolar air and pulmonary capillaries. This difference is attributable to the fact that about 2% of the blood is shunted from the aorta directly to the lung to meet the oxygen needs of the lung. It joins the blood coming out of the gas exchange area, without having taken part in gas exchange, thus decreasing the PO_2 of the blood returning to the heart.

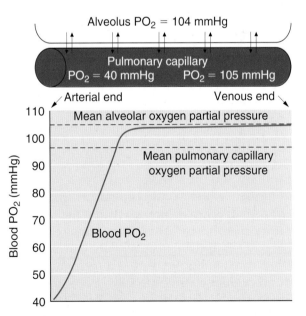

▲ Figure 8.5   Uptake of oxygen from the alveolar air into the pulmonary capillary.

Adapted from *Textbook of Medical Physiology*, 9th ed, A.C. Guyton and J.E. Hall, pg. 514, Copyright 1996, with permission from Elsevier.

It is important to remember that venous blood, once in the heart, leaves the right side of the heart through the pulmonary arteries on its way to the lungs; thus, these arteries transport deoxygenated blood. Conversely, once the deoxygenated blood travels through the lungs and becomes oxygenated, it returns to the heart via the pulmonary veins; thus, these veins transport oxygenated blood. We tend to think of arteries as transporters of oxygenated blood and veins as transporters of deoxygenated blood, but this is true only of systemic circulation (see chapter 7).

> The greater the pressure gradient across the respiratory membrane, the more rapidly oxygen diffuses across it.

The rate at which oxygen diffuses from the alveoli into the blood is referred to as the **oxygen diffusion capacity** and is expressed as the volume of oxygen that diffuses through the membrane each minute for a pressure difference of 1 mmHg. At rest, the oxygen diffusion capacity is about 21 ml of oxygen per minute per 1 mmHg of pressure difference between the alveoli and the pulmonary capillary blood. Although the partial pressure gradient between venous blood coming into the lung and the alveolar air is about 65 mmHg (105 mmHg – 40 mmHg), the oxygen diffusion capacity is calculated on the basis of the mean pressure in the pulmonary capillary, which has a substantially higher PO_2, as illustrated in figure 8.5. The gradient between the mean partial pressure of the pulmonary capillary and the alveolar air is approximately 11 mmHg, which would provide a diffusion of 231 ml of oxygen per minute through the respiratory membrane.[8] During maximal exercise, the oxygen diffusion capacity may increase by up to three times the resting rate. In fact, rates of more than 80 ml/min have been observed among highly trained rowers.

The increase in oxygen diffusion capacity from rest to exercise is caused by a relatively inefficient, sluggish circulation through the lungs at rest, which results primarily from limited perfusion of the upper regions of the lungs attributable to gravity. During maximal exercise, however, blood flow through the lungs is greater, primarily because of elevated blood pressure, thereby increasing lung perfusion. Athletes with large aerobic capacities often also have greater oxygen diffusion capacities. This is likely the combined result of increased cardiac output, increased alveolar surface area, and reduced resistance to diffusion across respiratory membranes.

Carbon Dioxide Exchange

Carbon dioxide exchange, like oxygen exchange, moves along a pressure gradient. As shown in figure 8.4, the blood passing from the right side of the heart through the alveoli has a PCO_2 of about 46 mmHg. In the alveoli, air has a PCO_2 of about 40 mmHg. Although this results in a relatively small pressure gradient of only about 6 mmHg, it is more than adequate to allow for the exchange of CO_2. Carbon dioxide's diffusion coefficient is 20 times greater than that of oxygen, so CO_2 can diffuse across the respiratory membrane much more rapidly.

The partial pressures of gases involved in pulmonary diffusion are summarized in table 8.1. Note that the total pressure in the venous blood is only 705 mmHg, 55 mmHg lower than the total pressure in dry air and alveolar air. This is the result of a much greater decrease in PO_2 compared with the increase in PCO_2 as the blood goes through the body's tissues.

Transport of Oxygen and Carbon Dioxide

We have considered how we bring air into our lungs via pulmonary ventilation and how gas exchange occurs via pulmonary diffusion. Next we must consider how gases are transported in our blood to deliver oxygen to the tissues and to remove the carbon dioxide that the tissues produce. We consider separately the transport of each gas.

Oxygen Transport

Oxygen is transported by the blood either combined with hemoglobin in the red blood

Table 8.1

Partial Pressures of Respiratory Gases at Sea Level

Gas	% in dry air	Partial pressure (mmHg)				
		Dry air	Alveolar air	Arterial blood	Venous blood	Diffusion gradient
Total	100.00	760	760	760	706[a]	0
H_2O	0.00	0	47	47	47	0
O_2	20.93	159.1	105	100	40	60
CO_2	0.03	0.2	40	40	46	6
N_2	79.04	600.7	568	573	573	0

[a]See text for an explanation of the decrease in total pressure.

▶ Pulmonary diffusion is the process by which gases are exchanged across the respiratory membrane in the alveoli.

▶ The amount of gas exchange that occurs across the membrane depends primarily on the partial pressure of each gas, although gas solubility and temperature are also important. Gases diffuse along a pressure gradient, moving from an area of higher pressure to one of lower pressure. Thus, oxygen enters the blood and carbon dioxide leaves it.

▶ Oxygen diffusion capacity increases as you move from rest to exercise. When your body needs more oxygen, oxygen exchange is facilitated.

▶ The pressure gradient for carbon dioxide exchange is less than for oxygen exchange, but carbon dioxide's diffusion coefficient is 20 times greater than that of oxygen, so carbon dioxide crosses the membrane easily, even without a large pressure gradient.

cells (greater than 98%) or dissolved in the blood plasma (less than 2%). Only about 3 ml of oxygen is dissolved in each liter of plasma. Assuming a total plasma volume of 3 to 5 L, only about 9 to 15 ml of oxygen can be carried in the dissolved state. This limited amount of oxygen cannot adequately meet the needs of even resting body tissues, which generally require more than 250 ml of oxygen per minute (depending on body size). Fortunately, hemoglobin, contained in the body's 4 to 6 billion red blood cells, allows the blood to transport nearly 70 times more oxygen than can be dissolved in plasma.

Hemoglobin Saturation

Each molecule of hemoglobin can carry four molecules of oxygen. When oxygen binds to hemoglobin, it forms oxyhemoglobin; hemoglobin that is not bound to oxygen is referred to as deoxyhemoglobin. The binding of oxygen to hemoglobin depends on the PO_2 in the blood and the bonding strength, or affinity, between hemoglobin and oxygen. The middle curve in figure 8.6a shows an oxygen–hemoglobin dissociation curve, which reveals the amount of **hemoglobin saturation** at different PO_2 values. A high blood PO_2 results in almost complete hemoglobin saturation, which means that the maximal amount of oxygen is bound. But as the PO_2 decreases, so does hemoglobin saturation.

Many factors can influence hemoglobin saturation. If, for example, the blood becomes more acidic, the dissociation curve shifts to the right. This indicates that more oxygen is being unloaded from the hemoglobin at the tissue level. This rightward shift of the curve (see figure 8.6*a*), attributable to a decline in pH, is referred to as the Bohr effect.[8] The pH in the lungs is generally high, so hemoglobin passing through the lungs has a strong affinity for oxygen, encouraging high saturation. At the tissue level, however, the pH is lower, causing oxygen to dissociate from hemoglobin, thereby supplying oxygen to the tissues. With exercise, the ability to unload oxygen to the muscles increases as the muscle pH decreases.

Blood temperature also affects oxygen dissociation. As shown in figure 8.6*b*, increased blood temperature shifts the dissociation curve to the right, indicating that oxygen is unloaded more efficiently at higher temperatures. Because of this, the hemoglobin unloads more oxygen when blood circulates through the metabolically heated active muscles. In the lungs, where the blood might be a bit cooler, hemoglobin's affinity for oxygen is increased, which encourages oxygen binding.

> Increased temperature and hydrogen ion (H^+) concentration (lowered pH) in an exercising muscle affect the oxygen dissociation curve, allowing more oxygen to be unloaded to supply the active muscle.

Blood Oxygen-Carrying Capacity

The oxygen-carrying capacity of blood is the maximal amount of oxygen the blood can transport. It depends primarily on the blood hemoglobin content. Each 100 ml of blood contains an average of 14 to 18 g of hemoglobin in men and 12 to 16 g in women. Each gram of hemoglobin can combine with about 1.34 ml of oxygen, so the oxygen-carrying capacity of blood is approximately 16 to 24 ml per 100 ml of blood when blood is fully saturated with oxygen. At rest, as the blood passes through the lungs, it is in contact with the alveolar air for approximately 0.75 s. This is sufficient time for hemoglobin to bind nearly all the oxygen it can hold, resulting in 98% saturation. At high intensities of exercise, the contact time is greatly reduced, which can reduce the binding of hemoglobin to oxygen and decrease the saturation.

People with low hemoglobin contents, such as those with anemia, have reduced oxygen-carrying capacities. Depending on the severity of the condition, these people might feel few effects of anemia while they are at rest, because their cardiovascular system can compensate for reduced blood oxygen content by increasing cardiac output. However, during activities in which oxygen delivery can become a limitation, such as highly intense aerobic effort, reduced blood oxygen content limits their energy production and performance.

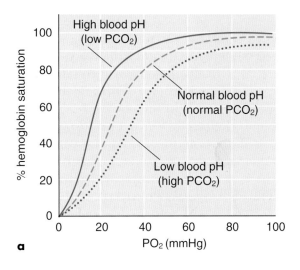

a

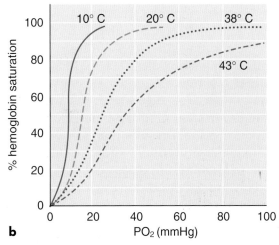

b

▲ **Figure 8.6** The effects of (*a*) blood pH and (*b*) blood temperature on the shape of the normal oxygen–hemoglobin dissociation curve.

Carbon Dioxide Transport

Carbon dioxide also relies on the blood for transportation. Once carbon dioxide is released from the cells, it is carried in the blood primarily in three forms:

- Dissolved in plasma
- As bicarbonate ions resulting from the dissociation of carbonic acid
- Bound to hemoglobin

Let's examine each method of transportation.

Dissolved Carbon Dioxide

Part of the carbon dioxide released from the tissues is dissolved in plasma. But only a small amount, typically just 7% to 10%, is transported this way. This dissolved carbon dioxide comes out of solution where the PCO_2 is low, such as in the lungs. There it diffuses out of the capillaries into the alveoli to be exhaled.

Bicarbonate Ion

The vast majority of carbon dioxide is carried in the form of bicarbonate ion. This form accounts for the transport of 60% to 70% of the carbon dioxide in the blood. Carbon dioxide and water molecules combine to form carbonic acid (H_2CO_3). This acid is unstable and quickly dissociates, freeing a hydrogen ion (H^+) and forming a bicarbonate ion (HCO_3^-):

$$CO_2 + H_2O \rightarrow H_2CO_3 \rightarrow H^+ + HCO_3^-$$
$$\text{(carbonic} \qquad \text{(bicarbonate}$$
$$\text{acid)} \qquad \text{ion)}$$

The H^+ subsequently binds to hemoglobin, and this binding triggers the Bohr effect, mentioned previously, which shifts the oxygen–hemoglobin dissociation curve to the right. Thus, the formation of bicarbonate ion enhances oxygen unloading. Through this mechanism, hemoglobin acts as a buffer, binding and neutralizing the H^+ and thus preventing any significant acidification of the blood. We discuss the issue of acid–base balance in more detail later in this chapter.

When the blood enters the lungs, where the PCO_2 is lower, the H^+ and bicarbonate ions rejoin to form carbonic acid, which then splits into carbon dioxide and water:

$$H^+ + HCO_3^- \rightarrow H_2CO_3 \rightarrow CO_2 + H_2O$$

The carbon dioxide that is thus re-formed can enter the alveoli and be exhaled.

> The majority of carbon dioxide produced by the active muscle is transported back to the lungs in the form of bicarbonate ions.

Carbaminohemoglobin

Carbon dioxide transport also can occur when the gas binds with hemoglobin, forming a compound called carbaminohemoglobin. The compound is so named because carbon dioxide binds with amino acids in the globin part of the hemoglobin molecule, rather than with the heme group as oxygen does. Because carbon dioxide binding occurs on a different part of the hemoglobin molecule than does oxygen binding, the two processes do not compete. However, carbon dioxide binding varies with the oxygenation of the hemoglobin (deoxyhemoglobin binds carbon dioxide more easily than oxyhemoglobin) and the partial pressure of CO_2 (carbon dioxide is released from hemoglobin when PCO_2 is low). Thus, in the lungs, where the PCO_2 is low, carbon dioxide is readily released from the hemoglobin, allowing it to enter the alveoli to be exhaled.

Gas Exchange at the Muscles

We have considered how our respiratory and cardiovascular systems bring air into our lungs, exchange oxygen and carbon dioxide in the alveoli, and transport oxygen to the muscles (and carbon dioxide away from them). We can now consider the delivery of oxygen from the capillary blood to the muscle tissue and the removal of metabolically produced carbon dioxide. This gas exchange between the tissues and the blood in the capillaries is the fourth and final step in gas transportation–internal respiration.

▶ Oxygen is transported in the blood primarily bound to hemoglobin (as oxyhemoglobin), although a small part of it is dissolved in blood plasma.

▶ Hemoglobin oxygen saturation decreases when

1. PO_2 decreases,
2. pH decreases, or
3. temperature increases.

Each of these conditions can reflect increased local oxygen demand. They increase oxygen unloading in the needy area.

▶ Hemoglobin is usually about 98% saturated with oxygen. This is a much higher oxygen content than our bodies require, so the blood's oxygen-carrying capacity seldom limits performance.

▶ Carbon dioxide is transported in the blood primarily as bicarbonate ion. This prevents the formation of carbonic acid, which can cause H^+ to accumulate, decreasing the pH. Smaller amounts of carbon dioxide are carried either dissolved in the plasma or bound to hemoglobin.

Arterial-Venous Oxygen Difference

At rest, the oxygen content of arterial blood is about 20 ml of oxygen per 100 ml of blood. As shown in figure 8.7a, this value decreases to 15 or 16 ml of oxygen per 100 ml as the blood passes through the capillaries into the venous system and returns to the right side of the heart. This difference in oxygen content between arterial and venous blood is referred to as the **arterial-mixed venous oxygen difference, or a-v̄O₂ difference** (see chapter 7). The term *mixed venous* (v̄) refers to the oxygen content of blood in the right atrium, which comes from all parts of the body, active and inactive. It reflects the 4 to 5 ml of oxygen per 100 ml of blood taken up by the tissues. The amount of oxygen taken up is proportional to its use for oxidative energy production. Thus, as the rate of oxygen use increases, the a-v̄O₂ difference also increases. It can increase to 15 to 16 ml per 100 ml of blood during maximal levels of

endurance exercise (figure 8.7b). However, at the level of the contracting muscle, the **a-vO₂ difference** during intense exercise can increase to 17 to 18 ml per 100 ml of blood. Note that there is not a bar over the v in this instance because we are now looking at local muscle venous blood, not mixed venous blood in the right atrium. During such an effort, the blood unloads more oxygen to the active muscles because the PO_2 in the muscles is substantially lower than in arterial blood.

The a-v̄O₂ difference increases from a resting value of about 4 to 5 ml per 100 ml of blood up to values of 15 ml per 100 ml of blood during intense exercise. This increase reflects an increased extraction of oxygen from arterial blood by active muscle, thus decreasing the oxygen content of the venous blood. It is important to understand that we are talking about the *whole body* a-v̄O₂ difference, which is comparing the oxygen content of the arterial blood coming out of the left ventricle with the oxygen content of the venous blood returning to the right atrium. Remember, the blood returning to the right atrium is coming from all parts of the body, active and inactive. Therefore, the venous oxygen content will not decrease to values much lower than 4 to 5 ml of oxygen per 100 ml of venous blood.

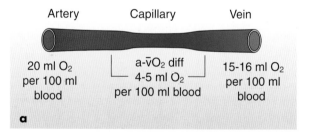

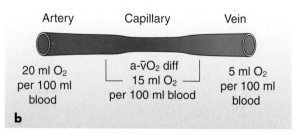

▲ **Figure 8.7** The arterial-mixed venous oxygen difference, or a-v̄O₂ difference, across the lung (*a*) at rest and (*b*) during intense aerobic exercise.

Factors Influencing Oxygen Delivery and Uptake

The rates of oxygen delivery and uptake depend on three major variables:

- Oxygen content of blood
- Amount of blood flow
- Local conditions

As we begin to exercise, each of these variables must be adjusted to ensure increased oxygen delivery to our active muscles. We have discussed the oxygen content of the blood and know that under normal circumstances, hemoglobin is 98% saturated with oxygen. Any reduction in the blood's normal oxygen-carrying capacity would hinder oxygen delivery and reduce cellular uptake of oxygen. Likewise, a reduction in the PO_2 of the arterial blood would lower the partial pressure gradient, limiting the unloading of oxygen at the tissue level. This is what happens at altitude (see sidebar).

As discussed in chapter 7, exercise increases blood flow through the muscles. As more blood carries oxygen through the muscles, less oxygen must be removed from each 100 ml of blood (assuming the demand is unchanged). Thus, increased blood flow improves oxygen delivery and uptake.

Many local changes in the muscle during exercise affect oxygen delivery and uptake. For example, muscle activity increases muscle

Oxygen Delivery at Altitude

You have likely had the experience of trying to exercise at altitude, particularly at elevations above 1,500 to 2,000 m (4,920-6,560 ft), whether it be skiing, jogging, bicycling, or any other type of aerobic activity. Trying to maintain your normal pace of exercise at such elevations, you find that your breathing is labored, your heart rate is considerably higher than at sea level, and you feel like you are totally out of shape. Why does altitude limit our performance?

Most people, when asked this question, answer, "There is less oxygen in the air at altitude." This seems like a logical answer, but it isn't technically correct. The percentage of oxygen in the air at altitude is exactly the same as at sea level—20.93%. What is different, however, is that the partial pressure of oxygen is reduced at altitude, and this is the direct result of the reduction in total atmospheric pressure. As an example, as you go from sea level to an altitude of 2,500 m (8,200 ft), the changes shown in the following table occur.

Altitude	Sea level	2,500 m
Total atmospheric pressure (mmHg)	760	560
PO_2 in dry atmospheric air (mmHg)	159	117
PO_2 in alveolar air (mmHg)	105	77
PO_2 in arterial blood	100	72
PO_2 in venous blood	40	40
Partial pressure gradient—driving force	60	32

Thus, the limiting factor at altitude is the large reduction in the oxygen partial pressure gradient, which greatly reduces the driving force to get oxygen from the alveoli into the capillaries and from the capillaries into the tissue. This is discussed in detail in chapter 11.

acidity because of lactate production. Also, muscle temperature and carbon dioxide concentration both increase because of increased metabolism. All these changes increase oxygen unloading from the hemoglobin molecule, facilitating oxygen delivery and uptake by the muscles.

During maximal exercise, however, when we push our bodies to the limit, changes in any of these areas can impair oxygen delivery and restrict our abilities to meet oxidative demands. We discuss these potential limitations later in this chapter.

Carbon Dioxide Removal

Carbon dioxide exits the cells by simple diffusion in response to the partial pressure gradient between the tissue and the capillary blood. For example, muscles generate carbon dioxide through oxidative metabolism, so the PCO_2 in muscles is relatively high compared with that in the capillary blood. Consequently, CO_2 diffuses out of the muscles and into the blood to be transported to the lungs.

We have completed our discussion of how oxygen is brought into the body and delivered to the tissues and how carbon dioxide is removed from the tissues and delivered to the lungs to be cleared. These processes are summarized in figure 8.8.

> ▶ The a-$\bar{v}O_2$ difference is the difference in the oxygen content of arterial and venous blood throughout the body. This measure reflects the amount of oxygen taken up by the tissues.
>
> ▶ Oxygen delivery to the tissues depends on the oxygen content of the blood, the amount of blood flow to the tissues, and local conditions (e.g., tissue temperature and PO_2).
>
> ▶ Carbon dioxide exchange at the tissues is similar to oxygen exchange, except that carbon dioxide leaves the muscles, where it is formed, and enters the blood to be transported to the lungs for clearance.

Regulation of Pulmonary Ventilation

Maintaining homeostatic balance in blood PO_2, PCO_2, and pH requires a high degree of coordination between the respiratory and circulatory systems. Much of this coordination is accomplished by involuntary regulation of pulmonary ventilation. This control is not yet fully understood, although many of the intricate neural controls have been identified. Let's examine some of them.

Mechanisms of Regulation

The respiratory muscles are under the direct control of motor neurons, which are in turn regulated by **respiratory centers** (inspiratory and expiratory) located within the brain stem (in the medulla oblongata and pons, see chapter 2, page 69). These centers establish the rate and depth of breathing by sending out periodic impulses to the respiratory muscles.

However, the respiratory centers don't act alone in controlling breathing. Breathing also is regulated by a changing chemical environment in the body. For example, sensitive areas in the brain respond to changes in carbon dioxide and H^+ levels. When these levels increase, signals are sent to the inspiratory center activating the neural circuitry to increase the rate and depth of respiration, which increases the removal of carbon dioxide and H^+. In addition, chemoreceptors in the aortic arch (the aortic bodies) and in the bifurcation of the common carotid artery (the carotid bodies) are sensitive primarily to blood changes in PO_2 but also respond to changes in H^+ concentration and PCO_2. Overall, of the various stimuli, PCO_2 appears to be the strongest stimulus for the regulation of breathing. When carbon dioxide levels become too high, recall that carbonic acid forms, then quickly dissociates, giving off H^+. If H^+ accumulates, the blood becomes too acidic (pH decreases). Thus, an increased PCO_2 stimulates the inspiratory center to increase respiration, not to bring in more oxygen but to rid the body of excess carbon dioxide and minimize pH changes.

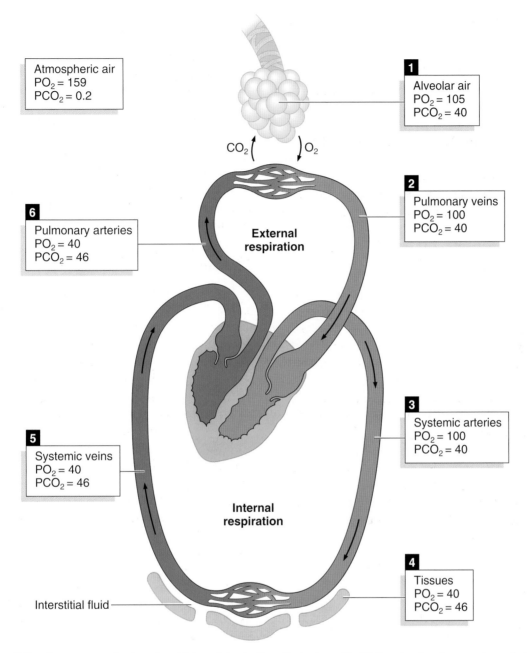

Atmospheric air
$PO_2 = 159$
$PCO_2 = 0.2$

1
Alveolar air
$PO_2 = 105$
$PCO_2 = 40$

CO_2 O_2

6
Pulmonary arteries
$PO_2 = 40$
$PCO_2 = 46$

External respiration

2
Pulmonary veins
$PO_2 = 100$
$PCO_2 = 40$

3
Systemic arteries
$PO_2 = 100$
$PCO_2 = 40$

5
Systemic veins
$PO_2 = 40$
$PCO_2 = 46$

Internal respiration

4
Tissues
$PO_2 = 40$
$PCO_2 = 46$

Interstitial fluid

▲ **Figure 8.8**   A summary of external and internal respiration. Beginning with (1) the alveolar air, we can see that oxygen moves into (2) the pulmonary veins; the blood remains saturated as it flows through (3) the systemic arteries. As blood passes through (4) the tissue capillaries, it gives up some of its oxygen and gains carbon dioxide. The deoxygenated blood then returns through (5) the venous system to (6) the pulmonary artery and capillaries, where it begins the process of oxygenation/deoxygenation all over again.

In addition to the chemoreceptors, other neural mechanisms influence breathing. The pleurae, bronchioles, and alveoli in the lungs contain stretch receptors. When these areas are excessively stretched, that information is relayed to the expiratory center. The expiratory center responds by shortening the duration of an inspiration, which decreases the risk of overinflating the respiratory structures. This response is known as the Hering–Breuer reflex.

We can exert some voluntary control over our breathing through the cerebral motor cortex. However, this voluntary control can

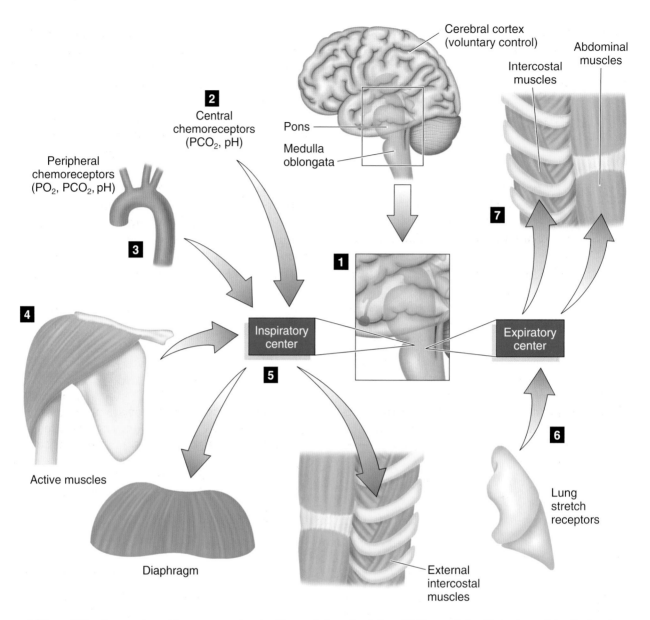

▲ **Figure 8.9** An overview of the processes involved in respiratory regulation. (1) The medulla oblongata contains the inspiratory and expiratory centers. When (2) central chemoreceptors, (3) peripheral chemoreceptors, and (4) active muscles stimulate the inspiratory center, the inspiratory center stimulates (5) the external intercostal and diaphragm muscles to contract to increase the volume of the thorax, thereby drawing air into the lungs. (6) This stretching of the lungs triggers the expiratory center to contract (7) the intercostal and abdominal muscles, causing the thoracic volume to decrease and force air out of the lungs.

be overridden by the involuntary control of the respiratory center. Try to hold your breath for 5 min. At some point, regardless of your conscious decision to suppress breathing, your carbon dioxide and H+ levels become quite high, your oxygen level decreases, and your inspiratory center decides that breathing is imperative and forces you to inhale.

We see that many control mechanisms are involved in the regulation of breathing, as shown in figure 8.9. Such simple stimuli as emotional distress or an abrupt change in the temperature of your surroundings can affect breathing. But all these control mechanisms are essential. The goal of respiration is to maintain appropriate levels of the blood and tissue gases and to maintain proper pH for normal cellular function. Even relatively small changes in any of these, if they are not carefully controlled, can impair physical activity and jeopardize health.

Pulmonary Ventilation During Exercise

The onset of physical activity is accompanied by a two-phase increase in ventilation. An almost immediate, marked increase occurs, followed by a continued, more gradual increase in the depth and rate of breathing. This is shown in figure 8.10 for light, moderate, and heavy exercise. This two-phase adjustment suggests that the initial increase in ventilation is produced by the mechanics of body movement. As exercise begins, but before any chemical stimulation occurs, the motor cortex becomes more active and transmits stimulatory impulses to the inspiratory center, which responds by increasing respiration. Also, proprioceptive feedback from the active skeletal muscles and joints provides additional input about the movement, and the respiratory center can adjust its activity accordingly.

The more gradual second phase of the respiratory increase is produced by changes in the temperature and chemical status of the arterial blood. As exercise progresses, increased metabolism in the muscles generates more heat, carbon dioxide, and H^+. All these factors enhance oxygen unloading in the muscles, which increases the a-$\bar{v}O_2$ difference. Also, more carbon dioxide enters the blood, increasing blood levels of both carbon dioxide and H^+. This is sensed by the chemoreceptors, which in turn stimulate the inspiratory center, increasing rate and depth of respiration. Some researchers have suggested that chemoreceptors in the muscles might also be involved. In addition, data suggest that receptors in the right ventricle of the heart send information to the inspiratory center so that increases in cardiac output can stimulate breathing during the early minutes of exercise.

> **fyi** Pulmonary ventilation increases during exercise, up to near-maximal rates of work, in direct proportion to the body's metabolic needs. At lower exercise intensities, this is accomplished by increases in **tidal volume** (the amount of air moved in and out of the lungs during regular breathing). At higher intensities, the rate of respiration also increases. Maximal rates of pulmonary ventilation depend on body size. Maximal ventilation rates of approximately 100 L/min are common for smaller individuals, but rates exceeding 200 L/min are found in larger individuals.

At the end of exercise, the muscles' energy demands decrease almost immediately to resting levels. But pulmonary ventilation returns to normal at a relatively slow rate. If the rate of breathing perfectly matched the metabolic demands of the tissues, respiration would decrease to the resting level within seconds after exercise. But respiratory recovery takes several minutes, which suggests that postexercise breathing is regulated primarily by acid–base balance, PCO_2, and blood temperature.

Breathing Irregularities During Exercise

Ideally, during exercise our breathing is regulated in a way that maximizes our ability to

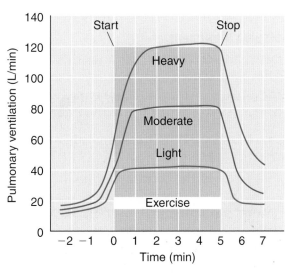

▲ **Figure 8.10**   The ventilatory response to light, moderate, and heavy exercise. The subject exercised at each of the three intensities for 5 min. The ventilation volume tended to plateau at a steady-state value at the light and moderate intensities but continued to increase at the heavy intensity.

perform. Unfortunately, this doesn't always occur. Respiratory irregularities can accompany exercise and hinder performance. Let's examine a few.

Dyspnea

The sensation of **dyspnea** (shortness of breath) during exercise is most common among individuals in poor physical condition who attempt to exercise at levels that significantly elevate their arterial carbon dioxide and H^+ concentrations. As noted earlier, both stimuli send strong signals to the inspiratory center to increase the rate and depth of ventilation. Although exercise-induced dyspnea is sensed as an inability to breathe, the underlying cause is an inability to readjust the blood PCO_2 and H^+. Failure to reduce these stimuli during exercise appears to be related to poor conditioning of respiratory muscles. Despite a strong neural drive to ventilate the lungs, the respiratory muscles fatigue easily and are unable to reestablish normal homeostasis.

Hyperventilation

The anticipation of or anxiety concerning exercise, as well as some respiratory disorders, can cause a sudden increase in ventilation that exceeds the metabolic need for oxygen. Such overbreathing is termed **hyperventilation.** At rest, voluntary hyperventilation decreases the normal PCO_2 of 40 mmHg in the alveoli and arterial blood to about 15 mmHg. As carbon dioxide levels decrease, so do H^+ levels; thus, blood pH increases. These effects reduce the ventilatory drive. Because the blood leaving the lungs is nearly always about 98% saturated with oxygen, an increase in the alveolar PO_2 does not increase the oxygen content of the blood. Consequently, the reduced desire to breathe and the improved ability to hold one's breath after hyperventilating result from carbon dioxide unloading rather than increased blood oxygen. When performed for only a few seconds, such deep, rapid breathing can lead to lightheadedness and even loss of consciousness. This phenomenon reveals the sensitivity of the respiratory system's regulation of carbon dioxide and pH.

In hope of reducing respiratory distress, swimmers often hyperventilate before com-

petition. Breath holding during competitive swimming offers some advantages to stroke mechanics, so most swimmers hyperventilate immediately before starting sprint events. Although they might feel little desire to breathe during the first 8 to 10 s of the race, their alveolar and arterial oxygen contents can decline to critically low levels because oxygen is being used but not replaced. This can impair muscle oxidation and oxygen delivery to the central nervous system. The benefits of preexercise hyperventilation aren't clear, but it may impair performance rather than improve it. Perhaps future studies will provide some insight into the effects of this practice.

Preexercise hyperventilation is also common among people who attempt to perform long underwater swims, as noted in the opening story at the start of this chapter. Hyperventilating before a dive decreases the drive to breathe but doesn't increase the body's oxygen stores. Breath holding usually becomes intolerable when the PCO_2 in arterial blood reaches 50 mmHg. Unfortunately, during a dive that is preceded by hyperventilation, the oxygen content of the blood can reach critically low levels long before carbon dioxide accumulation signals the swimmer to surface and breathe. These events can cause a diver to lose consciousness before experiencing a desire to breathe.

Valsalva Maneuver

The **Valsalva maneuver** is a respiratory procedure that is frequently performed in certain types of exercise that can be potentially dangerous. This occurs when the individual

- closes the glottis (the opening between the vocal cords),
- increases the intra-abdominal pressure by forcibly contracting the diaphragm and the abdominal muscles, and
- increases the intrathoracic pressure by forcibly contracting the respiratory muscles.

As a result of these actions, air is trapped and pressurized in the lungs. This maneuver is often performed during the lifting of heavy objects as the person attempts to stabilize the back and chest wall.

The high intra-abdominal and intrathoracic pressures restrict venous return by collapsing the great veins. This maneuver, if held for an extended period of time, can greatly reduce the volume of blood returning to the heart, decreasing cardiac output. Although the Valsalva maneuver can be helpful in certain circumstances, this maneuver can be dangerous and should be avoided by people who have hypertension or known cardiovascular limitations.

▶ The respiratory centers in the brain stem set the rate and depth of breathing.

▶ Central chemoreceptors in the brain respond to changes in concentrations of carbon dioxide and H^+. When either of these increase, the inspiratory center increases respiration.

▶ Peripheral receptors in the arch of the aorta and the bifurcation of the common carotid artery respond primarily to changes in blood oxygen levels but also to changes in carbon dioxide and H^+ levels. If oxygen levels drop too low, or if CO_2 and H^+ levels increase, these chemoreceptors relay their information to the inspiratory center, which in turn increases respiration.

▶ Stretch receptors in the air passages and lungs can cause the expiratory center to shorten the period of inspiration to prevent overinflation of the lungs. In addition, we can exert some voluntary control over respiration.

▶ During exercise, ventilation shows an almost immediate increase, resulting from increased inspiratory center stimulation caused by the muscle activity itself. This is followed by a more gradual increase that results from the increase in temperature and chemical changes in the arterial blood that are caused by the muscular activity.

▶ Breathing irregularities associated with exercise include dyspnea, hyperventilation, and performance of the Valsalva maneuver.

Ventilation and Energy Metabolism

During long periods of mild steady-state activity, ventilation appears to match the rate of energy metabolism. It tends to vary in proportion to the volume of oxygen consumed and the volume of carbon dioxide produced by the body. Let's examine how closely breathing matches the consumption of oxygen.

Ventilatory Equivalent for Oxygen

The ratio between the volume of air ventilated ($\dot{V}_E$) and the amount of oxygen consumed by the tissues ($\dot{V}O_2$) in a given amount of time indicates breathing economy. This ratio is referred to as the **ventilatory equivalent for oxygen**, or $\dot{V}_E/\dot{V}O_2$. It is typically measured in liters of air breathed per liter of oxygen consumed per minute.

At rest, the $\dot{V}_E/\dot{V}O_2$ can range from 23 to 28 L of air per liter of oxygen consumed. This value changes very little during mild exercise, such as walking. But when work intensity increases to near maximum, the $\dot{V}_E/\dot{V}O_2$ can be greater than 30 L of air per liter of oxygen consumed. In general, however, the $\dot{V}_E/\dot{V}O_2$ remains relatively constant over a wide range of exercise levels. This indicates that the control systems for breathing are properly matched to the body's need for oxygen. Even in activities such as swimming, where breathing must be synchronized with the arm-stroke cycle, the $\dot{V}_E/\dot{V}O_2$ does not differ from that of other free-breathing activities.

Ventilatory Breakpoint

As exercise intensity increases toward maximum, at some point ventilation increases disproportionately compared with oxygen consumption. This point is called the **ventilatory breakpoint**, illustrated in figure 8.11. When your work rate exceeds 55% to 70% of your $\dot{V}O_2$max, oxygen delivery to the muscles can no longer support the oxygen requirements of oxidation. To compensate, more energy is derived from glycolysis. This results in increased lactic acid production and accumulation. This lactic acid combines with sodium bicarbonate (which buffers acid) and forms sodium lactate, water, and carbon dioxide. As we know, the increase in carbon dioxide stimulates chemoreceptors that signal the inspiratory center to increase ventilation. Thus, the ventilatory breakpoint reflects the respiratory response to increased carbon dioxide levels. Ventilation increases dramatically beyond the ventilatory breakpoint, as seen in figure 8.11.

> Ventilation increases during exercise in direct proportion to the rate of work being performed, up to the ventilatory breakpoint. Beyond this point, ventilation increases disproportionately as the body tries to clear excess CO_2.

Running speed (km/h)	$\dot{V}_E/\dot{V}O_2$
9.6	21.5
10.8	20.0
12.0	20.4
13.2	20.3
14.4	24.9
15.6	33.3

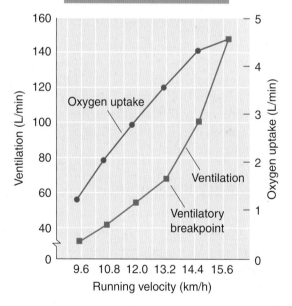

▲ **Figure 8.11** Changes in pulmonary ventilation ($\dot{V}_E$) and oxygen uptake ($\dot{V}O_2$) during exercise, illustrating the ventilatory breakpoint (graph). From these data it is possible to calculate the ratio between ventilation and oxygen uptake (table).

Anaerobic Threshold

The disproportionate increase in ventilation without an increase in oxygen consumption led to early speculation that the ventilatory breakpoint might be related to the lactate threshold (that point at which blood lactate production exceeds lactate clearance during a graded exercise test). Ventilatory breakpoint reflects a disproportionate increase in the volume of carbon dioxide produced per minute ($\dot{V}CO_2$) relative to the oxygen consumption. Recall from chapter 4 that the respiratory exchange ratio (RER) is the ratio of carbon dioxide production to oxygen consumption. Thus, the disproportionate increase in carbon dioxide production also causes RER to increase.

The increased $\dot{V}CO_2$ was thought to result from excess carbon dioxide being released from bicarbonate buffering of lactic acid. Wasserman and McIlroy[13] coined the term **anaerobic threshold** to describe this phenomenon because they assumed the sudden increase in CO_2 reflected a shift toward more anaerobic metabolism. They used the increase in RER as the marker of anaerobic threshold and believed that this was a good noninvasive alternative to blood sampling for detecting the onset of anaerobic metabolism.

Over the years, this concept has been refined considerably. The most accurate technique for identifying anaerobic threshold now appears to involve monitoring both the ventilatory equivalent for oxygen ($\dot{V}_E/\dot{V}O_2$) and the **ventilatory equivalent for carbon dioxide ($\dot{V}_E/\dot{V}CO_2$)**, which is the ratio of the volume of air expired ($\dot{V}_E$) to the volume of carbon dioxide produced ($\dot{V}CO_2$). The most specific criterion for estimating anaerobic threshold[2] is a systematic increase in $\dot{V}_E/\dot{V}O_2$ without a concomitant increase in $\dot{V}_E/\dot{V}CO_2$. This is illustrated in figure 8.12. The $\dot{V}_E/\dot{V}CO_2$ remains relatively constant, indicating that ventilation matches the body's need to remove CO_2. The increase in $\dot{V}_E/\dot{V}O_2$ indicates that the increase in ventilation to remove CO_2 is disproportionate to the body's need to provide O_2.

Anaerobic threshold has been used as a noninvasive estimate of lactate threshold, and under most conditions the two occur at about the same point in time during an incremental

Anaerobic threshold reflects the lactate threshold under most conditions, although the relationship is not always exact.

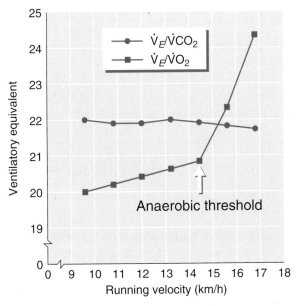

▲ **Figure 8.12** Changes in the ventilatory equivalent for carbon dioxide ($\dot{V}_E/\dot{V}CO_2$) and the ventilatory equivalent for oxygen ($\dot{V}_E/\dot{V}O_2$) during increasing intensities of running. Note that the breakpoint of the anaerobic threshold at a running velocity of 14.4 km/h is evident only in the $\dot{V}_E/\dot{V}O_2$ ratio.

▶ During mild, steady-state exercise, ventilation accurately reflects the rate of energy metabolism. Ventilation parallels oxygen uptake. The ratio of air ventilated to oxygen consumed is the ventilatory equivalent of oxygen ($\dot{V}_E/\dot{V}O_2$).

▶ The ventilatory breakpoint is the point at which ventilation abruptly increases disproportionately to the increase in oxygen consumption. This increase in $\dot{V}_E$ reflects the need to remove excess carbon dioxide.

▶ The anaerobic threshold can be determined by identifying the point at which $\dot{V}_E/\dot{V}O_2$ shows a sudden increase while $\dot{V}_E/\dot{V}CO_2$ stays relatively stable. Anaerobic threshold has been used as a noninvasive estimate of lactate threshold.

exercise bout or at about the same percentage of maximal oxygen uptake. However, there are exceptions.[1] For example, people with McArdle's disease are incapable of increasing blood lactate and H^+ levels during exercise because of a lack of muscle phosphorylase. They demonstrate a clear anaerobic threshold during exercise of increasing intensity, even though blood lactate concentration remains at resting levels. Depleting the glycogen stores before exercise also alters the relationship between anaerobic threshold and lactate threshold.

Respiratory Limitations to Performance

Like all tissue activity, lung ventilation and gas transport in the body require energy. Most of this energy is used by the respiratory muscles during pulmonary ventilation. At rest, the respiratory muscles account for only about 2% of the total oxygen uptake. As the rate and depth of ventilation increase, so do their energy costs. The diaphragm, the intercostal muscles, and the abdominal muscles can account for up to 11% of the total oxygen consumed during heavy exercise and up to 15% of the cardiac output. During recovery, breathing continues to demand much energy, accounting for 9% to 12% of the total oxygen consumed.

Although the muscles of respiration are heavily taxed during exercise, ventilation is sufficient to prevent an increase in alveolar carbon dioxide or a decline in alveolar PO_2 during activities lasting only a few minutes. Even during maximal effort, ventilation usually is not pushed to a person's maximal capacity to voluntarily move air in and out of the lungs. This capacity is called the **maximal voluntary ventilation**. However, considerable evidence suggests that pulmonary ventilation might be a limiting factor in highly trained subjects during maximal exhaustive exercise.[3, 4, 11]

Some researchers have suggested that heavy breathing for several hours (such as during marathon running) can cause glycogen depletion and fatigue of the respiratory muscles. However, untrained rats during exercise experienced a substantial sparing of their respiratory muscle glycogen compared with muscle glycogen in their hindlimbs. Unfortunately, similar data are not available for humans, but our respiratory muscles apparently are better designed for long-term activity than are the muscles in our extremities. The diaphragm, for example, has two to three times more oxidative capacity (oxidative enzymes and mitochondria) and capillary density than other skeletal muscle. Consequently, the diaphragm can obtain more energy from fat oxidation than can other muscles.

Airway resistance and gas diffusion in the lungs do not limit exercise in a normal, healthy individual. The volume of air inspired can increase 20- to 40-fold with exercise—from ~5 L/min at rest up to 100 to 200 L/min with maximal exertion. Airway resistance, however, is maintained at near-resting levels by airway dilation (through an increase in the laryngeal aperture and bronchodilation). During submaximal efforts, blood leaving the lungs remains nearly saturated with oxygen (~98%). Maximal exercise, on the other hand, may impose too large a demand on lung gas exchange, resulting in a decline in arterial PO_2 and arterial oxygen saturation (i.e., **exercise-induced arterial hypoxemia**). Although this does not appear to be a problem in untrained or moderately trained individuals, approximately 40% to 50% of elite endurance athletes experience a significant reduction in arterial oxygenation during exercise approaching exhaustion.[10] Thus, the respiratory system is well designed to accommodate the demands of heavy breathing during short- and long-term physical effort. However, individuals who consume unusually large amounts of oxygen during exhaustive exercise can face some respiratory limitations.

In some cases, highly trained distance runners' performances can be limited by their respiratory systems. They cannot sufficiently ventilate their lungs to prevent a decrease in arterial blood PO_2, leading to reduced hemoglobin saturation.

The respiratory system also can limit performance in people with abnormally restricted or obstructed airways. For example, asthma causes constriction of the bronchial tubes and swelling of their mucous membranes. These effects cause considerable resistance to ventilation and shortness of breath. Exercise is known to have an adverse effect on some people with asthma. The mechanism or mechanisms through which exercise induces airway obstruction in individuals with asthma are unknown, despite extensive study. Thorough reviews of this topic have been presented by Sly[12] and Eggleston.[5]

▶ Up to 11% of the body's total oxygen consumption and 15% of the cardiac output during heavy exercise can occur in the respiratory muscles.

▶ Pulmonary ventilation is usually not a limiting factor for performance, even during maximal effort, although it can limit performance in up to 50% of elite endurance athletes.

▶ The respiratory muscles seem to be better designed for avoiding fatigue during long-term activity than muscles of the extremities.

▶ Airway resistance and gas diffusion usually do not limit performance in normal, healthy individuals.

▶ The respiratory system can limit performance in people with restrictive or obstructive respiratory disorders.

Respiratory Regulation of Acid–Base Balance

As noted earlier, intense muscular activity often results in the production and accumulation of lactate and H^+. These can impair energy production and reduce muscle contractile force. Although the body's regulation of acid–base balance involves more than control of respiration, we include it here because the respiratory system plays a crucial role in rapid adjustment of the body's acid–base status during and immediately after exercise.

Acids, such as lactic acid and carbonic acid, release H^+. As noted in chapter 4, the metabolism of carbohydrate, fat, or protein produces inorganic acids that dissociate, increasing the H^+ concentration in body fluids. To minimize the effects of free H^+, the blood and muscles contain base substances that combine with, and thus buffer or neutralize, the H^+:

$$H^+ + buffer \rightarrow H\text{-}buffer$$

Under resting conditions, body fluids have more bases (such as bicarbonate, phosphate, and proteins) than acids, resulting in a tissue pH that ranges from 7.1 in muscle to 7.4 in arterial blood. The tolerable limits for arterial blood pH extend from 6.9 to 7.5, although the extremes of this range can be tolerated only for a few minutes (see figure 8.13). An H^+ concentration elevated above normal (low pH) is referred to as acidosis, whereas a decrease in H^+ below the normal concentration (high pH) is termed alkalosis.

The pH of intra- and extracellular body fluids is kept within a relatively narrow range by

- chemical buffers,
- pulmonary ventilation, and
- kidney function.

The three major chemical buffers in the body are bicarbonate (HCO_3^-), inorganic phosphates (P_i), and proteins. In addition to these, the hemoglobin in the red blood cells, as we mentioned earlier, also is a major buffer. Table 8.2 illustrates the relative contributions of these buffers in handling acids in the blood. Recall that bicarbonate combines with H^+ to form carbonic acid, thereby eliminating its acidifying influence. The carbonic acid in turn forms carbon dioxide and water in the lungs. The CO_2 is then exhaled and only water remains.

The amount of bicarbonate that combines with H^+ equals the amount of acid buffered. When lactic acid decreases the pH from 7.4 to 7.0, more than 60% of the bicarbonate initially present in the blood has been used. Even under resting conditions, the acid produced by the end products of metabolism would eliminate a major portion of the bicarbonate from the blood if there were no other way of removing H^+ from the body. Fortunately, the blood and

these buffers are required only to transport metabolic acids from their sites of production (the muscles) to the lungs or kidneys, where they can be removed. Once transportation is completed, the buffer molecules can be used again.

In the muscle fibers and the kidney tubules, H^+ is primarily buffered by phosphates, such as phosphoric acid and sodium phosphate. Unfortunately, less is known about the capacity of the buffers housed in the cells, although we know that the cells contain more protein

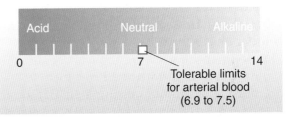

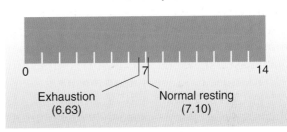

▲ **Figure 8.13**   Tolerable limits for arterial blood pH and muscle pH at rest and at exhaustion. Note the small range of physiological tolerance for both muscle and blood pH.

Table 8.2	
Buffering Capacity of Blood Components	
Buffer	**Slykes[a]**
Bicarbonate	18.0
Hemoglobin	8.0
Proteins	1.7
Phosphates	0.3
Total	28.0

[a]Milliequivalents of hydrogen ions taken up by each liter of blood from pH 7.4 to 7.0.

and phosphates and less bicarbonate than do the extracellular fluids.

As noted earlier, any increase in free H^+ in the blood stimulates the respiratory center to increase ventilation. This facilitates the binding of H^+ and bicarbonate and the removal of carbon dioxide. The end result is a decrease in free H^+ and an increase in blood pH. Thus, both the chemical buffers and the respiratory system provide temporary means of neutralizing the acute effects of exercise acidosis. To maintain a constant buffer reserve, the accumulated H^+ is removed from the body via the kidneys and urinary system. The kidneys filter H^+ from the blood along with other waste products. This provides a way to eliminate H^+ from the body while maintaining the concentration of extracellular bicarbonate.

During sprint exercise, muscles generate a large amount of lactate and H^+, which lowers the muscle pH from a resting level of 7.08 to less than 6.70. As shown in table 8.3, an all-out sprint for 400 m decreases leg-muscle pH to 6.63 and increases muscle lactate from a resting value of 1.2 mmol/kg to 19.7 mmol/kg of muscle. As noted earlier, such disturbances in acid–base balance impair muscle contractility and its capacity to generate adenosine triphosphate (ATP). Lactate and H^+ accumulate in the muscle, in part because they do not freely diffuse across fiber membranes. Despite the great production of lactate and H^+ during the 60 s required to run 400 m, these by-products diffuse throughout the body fluids and reach equilibrium after only about 5 to 10 min of recovery. Five minutes after the exercise, the runners evaluated for table 8.3 had blood pH values of 7.10 and blood lactate values of 12.3 mmol/L, compared with a resting pH of 7.40 and a resting lactate level of 1.5 mmol/L.

Reestablishing normal resting levels of blood and muscle lactate after such an exhaustive exercise bout is a relatively slow process, often requiring 1 to 2 h. As shown in figure 8.14, recovery of blood lactate to the resting level is facilitated by continued, lower intensity exercise, called active recovery.[9] After a series of

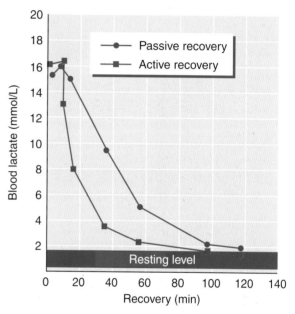

▲ **Figure 8.14**   Effects of active and passive recovery on blood lactate levels after a series of exhaustive sprint bouts. Note that the blood lactate removal rate is faster when the subjects perform exercise during recovery than when they rest during recovery.

Table 8.3

Blood and Muscle pH and Lactate Concentration After a 400-m Run

		Muscle		Blood	
Runner	**Time (s)**	**pH**	**Lactate (mmol/kg)**	**pH**	**Lactate (mmol/kg)**
1	61.0	6.68	19.7	7.12	12.6
2	57.1	6.59	20.5	7.14	13.4
3	65.0	6.59	20.2	7.02	13.1
4	58.5	6.68	18.2	7.10	10.1
Average	60.4	6.63	19.7	7.10	12.3

Air Pollution

During the past 30 years, concern has increased about possible problems associated with exercising in polluted air. The air in many cities is contaminated with small quantities of gases and particles not naturally found in the air we breathe. When air becomes stagnant or when a temperature inversion occurs, some of these pollutants reach concentrations that significantly impair athletic performance. The major contaminants of concern are carbon monoxide, ozone, and sulfur oxides.

Carbon monoxide (CO) is an odorless gas, derived from the burning of various fuels and tobacco smoke, that rapidly enters the blood when breathed and can be lethal. Hemoglobin's affinity for carbon monoxide is approximately 250 times greater than its affinity for oxygen, so hemoglobin preferentially binds the carbon monoxide. Blood carbon monoxide levels are directly related to the carbon monoxide levels in the inspired air. Several studies have reported linear

decreases in $\dot{V}O_2$max with increases in blood levels of carbon monoxide. Literature reviews on this topic have concluded that the reduction in $\dot{V}O_2$max is not statistically significant until blood carbon monoxide levels exceed 4.3%, although performance time on the treadmill has been reduced at carbon monoxide levels as low as 2.7%.[6, 7] Submaximal exercise at less than 60% of $\dot{V}O_2$max doesn't appear to be affected until blood carbon monoxide levels exceed 15%.

Ozone (O_3) is the most common photochemical oxidant and is the result of the reaction between ultraviolet light and internal combustion engine emissions. It produces many subjective complaints when its concentration in inspired air is high. Eye irritation, chest tightness, breathlessness, coughing, and nausea are common complaints. Ozone especially affects the respiratory tract. Decrements in lung function occur with increased ozone concentrations as well as with increased exposure and ventilation. $\dot{V}O_2$max has been found to be significantly decreased following 2 h of intermittent exercise with exposure to 0.75 parts per million (ppm) of ozone. This decrease in $\dot{V}O_2$max is likely associated with reduced oxygen transfer at the lung, resulting from reduced alveolar air exchange.

Sulfur dioxide (SO_2) generated from fossil fuels is another contaminant of concern. Research on sulfur dioxide and exercise is limited, but we know that air levels of this gas above 1.0 ppm cause significant discomfort and are detrimental to performance.[6] Sulfur dioxide is primarily an upper airway and bronchial irritant.

Certain cities have initiated air pollution or smog alerts. These are usually color coded, with colors indicating pollution severity. Standards need to be established nationally and air monitored accordingly. Increasing evidence points to the wisdom of canceling all games and practices when pollution levels put athletes' health at risk.

We hope that current and future research will provide a clearer understanding of the limitations imposed by air pollutants.

exhaustive sprint bouts, the participants in this study either sat quietly (passive recovery) or exercised at an intensity of 50% $\dot{V}O_2$max. Blood lactate is removed more quickly during active recovery because the activity maintains elevated blood flow through the active muscles, which in turn enhances both lactate diffusion out of the muscles and lactate oxidation.

Although blood lactate remains elevated for 1 to 2 h after highly anaerobic exercise, blood and muscle H^+ concentrations return to normal within 30 to 40 min of recovery. Chemical buffering, principally by bicarbonate, and respiratory removal of excess carbon dioxide are responsible for this relatively rapid return to normal acid–base homeostasis.

▶ Excess H^+ (decreased pH) impairs muscle contractility and ATP formation.

▶ The respiratory system plays an integral role in maintaining acid–base balance.

▶ Whenever H^+ levels start to increase, the inspiratory center responds by increasing respiration. Removing carbon dioxide is an essential means for reducing H^+ concentrations. Carbon dioxide is transported primarily bound to bicarbonate. Once it reaches the lungs, carbon dioxide is formed again and exhaled.

▶ Whenever H^+ levels begin to increase, whether from carbon dioxide or lactate accumulation, bicarbonate ion can buffer the H^+ to prevent acidosis.

In Closing . . .

In chapter 7, we discussed the role of the cardiovascular system during exercise. In this chapter we looked at the role played by the respiratory system. We also considered the limitations that these systems can impose on our abilities to perform. In the next chapter, we examine the physiological adaptations that occur in both the cardiovascular and the respiratory systems when they are exposed to the repeated stimulus of training. We will see how these adaptations improve these systems' abilities to meet our bodies' demands and how they can improve performance.

▶ Key Terms

anaerobic threshold

arterial-mixed venous oxygen difference, or a-$\bar{v}O_2$ diff

arterial-venous oxygen difference, or a-vO_2 diff

dyspnea

exercise-induced arterial hypoxemia

expiration

external respiration

hemoglobin saturation

Henry's law

hyperventilation

inspiration

internal respiration

maximal voluntary ventilation

oxygen diffusion capacity

partial pressure

pulmonary diffusion

pulmonary ventilation

respiratory centers

respiratory membrane

tidal volume

Valsalva maneuver

ventilatory breakpoint

ventilatory equivalent for carbon dioxide ($\dot{V}_E/\dot{V}CO_2$)

ventilatory equivalent for oxygen ($\dot{V}_E/\dot{V}O_2$)

▶ Study Questions

1. Describe the anatomical structures involved in pulmonary ventilation.

2. Identify the muscles associated with breathing and their function in pulmonary ventilation.

3. What are the partial pressures of oxygen and carbon dioxide in inspired air, alveolar air, and arterial and mixed venous blood?

4. In what forms are oxygen and carbon dioxide transported in the blood?

5. What are the chemical stimuli that control the depth and rate of breathing? How do they control respiration during exercise? How are they affected during voluntary hyperventilation?

6. What other stimuli control ventilation during exercise?

7. What is the ventilatory equivalent for oxygen? What is the ventilatory equivalent for carbon dioxide?

8. Define ventilatory breakpoint and anaerobic threshold.

9. Define lactate threshold and anaerobic threshold. How are these terms related?

10. What role does the respiratory system play in acid–base balance?

11. What is the normal resting pH for arterial blood? For muscle? How are these values

changed as a result of exhaustive sprint exercise?

12. What are the primary buffers in the blood? In muscles?

13. How long does it take blood pH and lactate levels to return to normal after an all-out sprint?

▷ References

1. Anderson, G.S., & Rhodes, E.C. (1989). A review of blood lactate and ventilatory methods of detecting transition thresholds. *Sports Medicine, 8,* 43-55.

2. Davis, J.A. (1985). Anaerobic threshold: Review of the concept and directions for future research. *Medicine and Science in Sports and Exercise, 17,* 6-18.

3. Dempsey, J.A. (1986). Is the lung built for exercise? *Medicine and Science in Sports and Exercise, 18,* 143-155.

4. Dempsey, J.A., & Wagner, P.D. (1999). Exercise-induced arterial hypoxemia. *Journal of Applied Physiology, 87,* 1997-2006.

5. Eggleston, P.A. (1986). Pathophysiology of exercise-induced asthma. *Medicine and Science in Sports and Exercise, 18,* 318-321.

6. Folinsbee, L.J. (1995). Heat and air pollution. In M.L. Pollock & D.H. Schmidt (Eds.), *Heart disease and rehabilitation* (3rd ed., pp. 327-342). Champaign, IL: Human Kinetics.

7. Folinsbee, L.J., & Raven, P.B. (1984). Exercise and air pollution. *Journal of Sports Sciences, 2,* 57-75.

8. Guyton, A.C., & Hall, J.E. (1996). *Textbook of medical physiology* (9th ed.). Philadelphia: Saunders.

9. Hermansen, L. (1981). Effect of metabolic changes on force generation in skeletal muscle during maximal exercise. In R. Porter & J. Whelan (Eds.), *Human muscle fatigue: Physiological mechanisms* (pp. 75-88). London: Pitman Medical.

10. Powers, S.K., Martin, D., & Dodd, S. (1993). Exercise-induced hypoxaemia in elite endurance athletes: Incidence, causes and impact on $\dot{V}O_2$max. *Sports Medicine, 16,* 14-22.

11. Prefaut, C., Durand, F., Mucci, P., & Caillaud, C. (2000). Exercise-induced arterial hypoxaemia in athletes. *Sports Medicine, 30,* 47-61.

12. Sly, R.M. (1986). History of exercise-induced asthma. *Medicine and Science in Sports and Exercise, 18,* 314-317.

13. Wasserman, K., & McIlroy, M.B. (1964). Detecting the threshold of anaerobic metabolism in cardiac patients during exercise. *American Journal of Cardiology, 14,* 844-852.

▷ Selected Readings

Babcock, M.A., & Dempsey, J.G. (1994). Pulmonary system adaptations: Limitations to exercise. In C. Bouchard, R.J. Shephard, & T. Stephens (Eds.), *Physical activity, fitness, and health* (pp. 320-330). Champaign, IL: Human Kinetics.

Brooks, G.A., Fahey, T.D., White, T.P. & Baldwin, K.M. (2000). *Exercise physiology: Human bioenergetics and its applications, 3rd Ed.* Mountain View, CA: Mayfield.

Costill, D.L. (1970). Metabolic responses during distance running. *Journal of Applied Physiology, 28,* 251-255.

Costill, D.L., Barnett, A., Sharp, R., Fink, W.J., & Katz, A. (1983). Leg muscle pH following sprint running. *Medicine and Science in Sports and Exercise, 15,* 325-329.

Costill, D.L., Verstappen, F., Kuipers, H., Janssen, E., & Fink, W. (1984). Acid–base balance during repeated bouts of exercise: Influence of HCO_3^-. *International Journal of Sports Medicine, 5,* 228-231.

Dempsey, J.G. (1985). Exercise and chemoreception. *American Review of Respiratory Disease, 129,* 31-34.

Forster, H.V. (2000). Exercise hyperpnea: Where do we go from here? *Exercise and Sport Sciences Reviews, 28,* 133-137.

Gavin, T.P., Derchak, A., & Stager, J.M. (1998). Ventilation's role in the decline in $\dot{V}O_2$max and SaO_2 in acute hypoxic exercise. *Medicine and Science in Sports and Exercise, 30,* 195-198.

Hardarson, T., Skarphedinsson, J.O., & Sveinsson, T. (1998). Importance of the lactate anion in control of breathing. *Journal of Applied Physiology, 84,* 411-416.

Harms, C.A., & Stager, J.M. (1995). Low chemoresponsiveness and inadequate hyperventilation contribute to exercise-induced hypoxemia. *Journal of Applied Physiology, 79,* 575-580.

Mateika, J.H., & Duffin, J. (1995). A review of the control of breathing during exercise. *European Journal of Applied Physiology, 71,* 1-27.

Paterson, D.J. (1997). Potassium and breathing in exercise. *Sports Medicine, 23,* 149-163.

Powers, S.K., & Howley, E.T. (2001). *Exercise physiology: Theory and application to fitness and performance, 4th Ed.* Dubuque, IA: McGraw Hill.

Sharp, R.L., Costill, D.L., Fink, W.J., & King, D.S. (1986). Effects of eight weeks of bicycle ergometer sprint training on human muscle buffer capacity. *International Journal of Sports Medicine, 7,* 13-17.

Sutton, J.R., Jones, N.L., & Toews, C.J. (1981). Effect of pH on muscle glycolysis during exercise. *Clinical Science, 61,* 331-338.

CHAPTER 9

CARDIOVASCULAR AND RESPIRATORY ADAPTATIONS TO TRAINING

During a single bout of exercise, the human body is quite adept at adjusting its cardiovascular and respiratory functioning to adequately meet the heightened demands of active muscles. When these systems are faced with these demands repeatedly, such as with daily training, they adapt in ways that allow the body to improve its endurance performance. For example, the miler can run a faster mile. The physiological and metabolic processes that bring oxygen into the body, distribute it, and allow it to be used by active tissues become and remain highly efficient at these tasks. In this chapter, we examine adaptations in cardiovascular and respiratory function in response to training and how such adaptations affect an athlete's endurance capacity and performance.

Lance Armstrong, 31-year-old cyclist from the United States, won his fifth consecutive Tour de France in 2003. The Tour is considered the world's most prestigious bicycle race and the most grueling endurance event ever. Armstrong completed the 21-day, 3,434-km (2,129-mi) ordeal in a total time of 83 h 41 min at an average speed of about 40 km/h (~25 mph). The cyclists who compete have to climb multiple mountains and then ride at breakneck speeds down mountainsides during some stages and compete in individual and team time trials on other stages. How are these athletes able to compete in this race? They must train specifically to develop their cardiorespiratory endurance capacities.

Cardiovascular and respiratory **endurance**, or what has been termed cardiorespiratory or aerobic endurance, is probably the one component of training that is least understood by coaches and athletes. Training programs for many nonendurance athletes completely ignore the aerobic endurance factor. This is understandable, because for maximum improvement in performance, training should be highly specific to the particular sport or activity in which the athlete participates, and endurance is frequently not recognized as important to nonendurance activities. The reasoning then is, Why waste valuable training time if the result is not improved performance?

The problem with this reasoning is that often, although it might not be obvious, a nonendurance sport does indeed have an endurance, or aerobic, component. For example, if you play football, you and your coach might fail to recognize the importance of cardiorespiratory endurance as part of your total training program. From all outward appearances, football is an anaerobic, or burst-type, activity consisting of repeated bouts of high-intensity work of short duration. Seldom does a run exceed 40 to 60 yd (37-55 m), and even this is followed by a substantial rest interval. The need for endurance is not readily apparent.

What you and your coach might fail to consider is that this burst-type activity must be repeated many times during the game. With a high aerobic endurance level, you could maintain the quality of your burst activity throughout the game, and you would still be relatively fresh during the fourth quarter.

Similar questions have been asked concerning the importance of including resistance training as a part of the total training program for sports that do not demand high levels of strength. Yet, athletes in almost all endurance sports are doing some resistance training to increase, or at least maintain, basic strength levels.

Sport scientists have now demonstrated the importance of endurance training for almost all types of sports or activities, including burst-type, such as football, soccer, and basketball; moderate-intensity and skilled, such as baseball and golf; endurance-type, such as running, cycling, and swimming.

Many authorities now believe that football teams that fall apart in the final quarter have ignored the endurance component in their training programs. A similar case can be made for athletes in most sports. But before we look specifically at how endurance can improve performance, we must review what endurance is.

Endurance: Muscular and Cardiorespiratory

Endurance is a term that describes two separate but related concepts: muscular endurance and cardiorespiratory endurance. Each makes a unique contribution to athletic performance, so each differs in importance to different athletes.

For sprinters, endurance is the quality that allows them to sustain a high speed over the full distance of, for example, a 100- or 200-m race. This quality is muscular endurance, the ability of a single muscle or muscle group to sustain high-intensity, repetitive, or static exercise. This type of endurance is also exemplified by the weightlifter, boxer, and wrestler. The exercise or activity can be rhythmic and repetitive in nature, such as performing multiple repetitions of the bench press for the weightlifter and jabbing for the boxer. Or the activity can be more static, such as a sustained muscle action when a wrestler attempts to pin an opponent to the mat. In either case, the resulting fatigue is confined to a specific muscle group, and the activity's duration is usually no more than 1 or 2 min. Muscular endurance is highly related to muscular strength and anaerobic development.

Whereas muscular endurance is specific to individual muscles, cardiorespiratory endurance relates to the body as a whole. Specifically, it refers to your body's ability to sustain prolonged, rhythmic exercise. This type of endurance is used by the cyclist, distance runner, or endurance swimmer who completes long distances at a fairly fast pace. Cardiorespiratory endurance is related to the development of cardiovascular and respiratory systems and thus aerobic development. This is why the term aerobic endurance is used to represent cardiorespiratory endurance.

> Cardiorespiratory endurance, or aerobic endurance, is the ability of the whole body to sustain prolonged exercise.

In chapter 6 we reviewed training adaptations for muscular endurance. In this chapter we focus on cardiorespiratory endurance. Now that we understand the concept of cardiorespiratory endurance, we are ready to examine the physiological bases for it. We focus primarily on endurance-type training but occasionally refer to burst- and resistance-type training specifically by name.

Evaluating Cardiorespiratory Endurance Capacity

To study training effects on endurance, we need a means for evaluating an individual's cardiorespiratory endurance capacity with which we can easily monitor his or her improvement during the training program.

Maximal Endurance Capacity: $\dot{V}O_2$max or Aerobic Power

Most sport scientists regard $\dot{V}O_2$max, representing aerobic power, as the best objective laboratory measure of maximal cardiorespiratory endurance capacity. Recall from chapter 4 that $\dot{V}O_2$max is defined as the highest rate of oxygen consumption attainable during maximal or exhaustive exercise. As you increase exercise intensity, oxygen consumption will eventually either plateau or decrease slightly, even with further increases in intensity, indicating you have reached your $\dot{V}O_2$max.

With endurance training, more oxygen can be delivered and consumed than in an untrained state, as we discussed in chapter 6. Previously untrained subjects have average increases in $\dot{V}O_2$max of 15% to 20% after a 6-month training program. These improvements allow you to perform endurance activities at a higher work rate or faster pace, improving your performance potential. Figure 9.1 on page 274 illustrates the increase in $\dot{V}O_2$max with 12 months of endurance training in a previously untrained individual. In this example, $\dot{V}O_2$max increased by 28%. Some improvement in cardiorespiratory function can occur with burst-type anaerobic training and with resistance-type training, but these improvements in $\dot{V}O_2$max are generally small.[27]

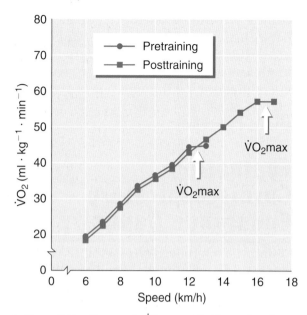

▲ **Figure 9.1** Changes in $\dot{V}O_2$max with 12 months of endurance training. $\dot{V}O_2$max increased from 39 to 57 ml · kg⁻¹ · min⁻¹, a 46% increase. Peak speed during the treadmill test increased from 12 km/h to 17 km/h.

Submaximal Endurance Capacity

In addition to increasing maximal endurance capacity, endurance training also increases **submaximal endurance capacity**, which is much more difficult to evaluate because there is no one physiological variable that we can measure before and after training to objectively quantify its change with training. Most scientists have used performance measures to quantify submaximal endurance capacity. As an example, several tests have been developed to determine the average absolute power output a person can maintain during a fixed period of time on a cycle ergometer. For running and swimming, the average speed or velocity a person can maintain during a fixed period of time would be a similar type of test. Generally, these tests will last at least 30 min but usually not more than 90 min.

Submaximal endurance capacity is more closely related to actual competitive endurance performance and likely is determined by both the person's $\dot{V}O_2$max and his or her lactate threshold. With endurance training, submaximal endurance capacity also increases. This would be expected, however, because we have already learned that both $\dot{V}O_2$max and lactate threshold increase with endurance training.

Cardiovascular Adaptations to Training

Numerous cardiovascular adaptations occur in response to training. We will look at changes in the following cardiovascular parameters:

- Heart size
- Stroke volume
- Heart rate
- Cardiac output
- Blood flow
- Blood pressure
- Blood volume

But first, let's review how these components relate to oxygen transport.

Oxygen Transport System

Cardiorespiratory endurance is intimately related to your body's ability to deliver sufficient oxygen to meet your active tissues' needs. Oxygen transportation and delivery are major functions shared by both your cardiovascular and respiratory systems. All components of these two systems that are related to the transport of oxygen are collectively referred to as the **oxygen transport system.**

The functioning of the oxygen transport system is defined by the interaction of the cardiac output and the arterial-venous oxygen difference, or a-$\overline{v}O_2$ difference, as discussed in chapter 8. Cardiac output (stroke volume × heart rate) tells us how much oxygen-carrying blood leaves the heart in 1 min. The a-$\overline{v}O_2$ difference, which is the difference between the oxygen content of the arterial blood and the oxygen content of the venous blood, tells us how much oxygen is extracted by the tissues. The product of these values, the **Fick equation,** tells us the rate at which oxygen is being consumed by the body tissues:

$$\dot{V}O_2 = \text{stroke volume} \times \text{heart rate} \times \text{a-}\overline{v}O_2 \text{ diff}$$

The active tissues' oxygen demand naturally increases during exercise. Your endurance depends on your oxygen transport system's ability to deliver sufficient oxygen to these active tissues to meet their heightened demands. As you reach maximal levels of exercise, heart size, blood flow, blood pressure, and blood volume can limit your maximal ability to transport oxygen. Endurance training elicits numerous changes in the components of the oxygen transport system that enable it to function more effectively. Now we can examine some of these components and their adaptations to training.

Heart Size

In response to increased work demand, the heart's mass and volume increase with training. Cardiac muscle, like skeletal muscle, undergoes hypertrophy as a result of chronic endurance training. At one time, **cardiac hypertrophy** induced by exercise—**"athlete's heart,"** as it was called—caused great concern because experts generally believed that enlargement of the heart always reflected a pathological state. Fortunately, cardiac hypertrophy is now recognized as a normal adaptation to chronic exercise training.

The left ventricle, as we discussed in chapter 7, is the hardest working heart chamber and undergoes the greatest change. It is now recognized that the extent and location of changes in heart size depend on the type of exercise performed. For example, during resistance-type training, the left ventricle must contract against high blood pressure in the systemic circulation, a situation referred to as a high afterload. It was postulated that to overcome this high afterload, the heart muscle compensated by increasing its size (wall thickness), thereby increasing its contractility. From chapter 7 we learned that blood pressure during resistance training can exceed 480 to 350 mmHg. This presents a considerable resistance that must be overcome by the left ventricle. Thus, the increase in its muscle mass is in direct response to this resistance training.

With endurance-type training, left ventricular filling increases. This is largely attributable to a training-induced increase in plasma volume (discussed later in this chapter) that increases left ventricular end-diastolic volume (an increased preload). Related to this, a decrease in heart rate at rest and during exercise at the same rate of work allows a longer dia-

stolic filling period. The increases in plasma volume and diastolic filling period increase left ventricular chamber size at the end of diastole.

It was originally believed that this increase in chamber size was the only change in the left ventricle caused by endurance training. In fact, studies have verified that this increase does indeed occur.[15] But more recent research has revealed that myocardial wall thickness also increases with endurance training, rather than just with resistance training.[14] Using magnetic resonance imaging, Milliken and colleagues[32] found that highly trained, competitive cross-country skiers, endurance cyclists, and long-distance runners had greater left ventricular masses than did nonathletic control subjects. These authors also found that left ventricular mass was highly correlated with $\dot{V}O_2$max, or aerobic power.

Fagard[14] conducted the most extensive review of the existing research literature in 1996, focusing on long-distance runners (135 athletes and 173 controls), cyclists (69 athletes and 65 controls), and strength athletes (178 athletes, including weight- and powerlifters, bodybuilders, wrestlers, throwers, and bobsledders, and 105 controls). For each group, the athletes were matched by age and body size with a group of sedentary control subjects. For each group of runners, cyclists, and strength athletes, the internal diameter of the left ventricle (LVID), the interventricular septal thickness (IVST), the posterior wall thickness (PWT), and the left ventricular mass (LVM) were greater in the athletes compared with their age- and sized-matched controls. The LVID is an index of chamber size, whereas the other three are indexes of myocardial mass and thickness. These comparisons are shown in figure 9.2*a* (LVID), *b* (IVST), *c* (PWT), and *d* (LVM). Figure 9.2*e* illustrates the percentage difference between the athletic group and the control group for LVID, mean wall thickness (MWT—average of IVST and PWT), and LVM. In all cases, the values for the athletic groups were significantly greater than for their age- and size-matched controls.

Most studies conducted on heart size changes with training have been cross-sectional, where trained individuals are compared with

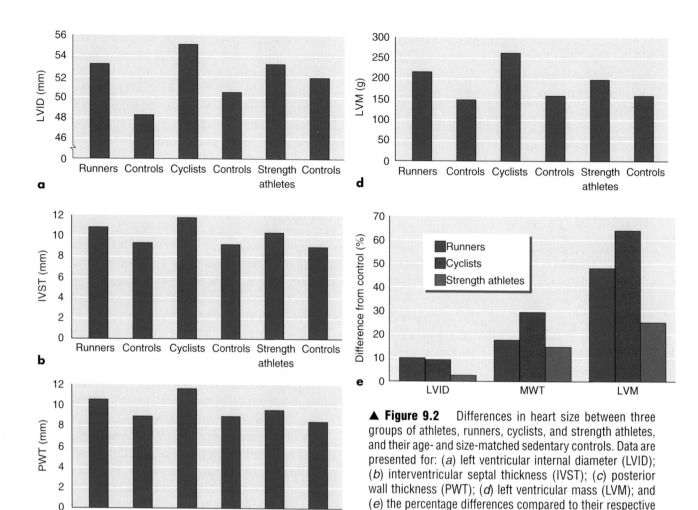

▲ **Figure 9.2**   Differences in heart size between three groups of athletes, runners, cyclists, and strength athletes, and their age- and size-matched sedentary controls. Data are presented for: (*a*) left ventricular internal diameter (LVID); (*b*) interventricular septal thickness (IVST); (*c*) posterior wall thickness (PWT); (*d*) left ventricular mass (LVM); and (*e*) the percentage differences compared to their respective control groups. Mean wall thickness (MWT) is the average of the IVST and PWT measurements.

sedentary, untrained individuals. We can learn much from cross-sectional studies, but they do not provide us with the same information that we could get from studying a group of untrained people who train for months or years, focusing on their changes from the untrained to the trained state. Certainly a portion of the differences that we see in figure 9.2 can be attributed to genetics, not training. However, a number of studies have actually followed individuals from an untrained state to a trained state, and others have followed individuals from a trained state to an untrained state. These studies have reported increases in heart size with training and decreases with detraining. So, training does bring about changes, but they are likely not as large as the differences we see in figure 9.2.

Stroke Volume

As a result of endurance training, stroke volume shows an overall increase. Stroke volume at rest is substantially higher after an endurance training program than it is before training. This training-induced increase is also seen during both standardized submaximal exercise and maximal exercise. This increase is illustrated in figure 9.3, which shows the changes in stroke volume of a subject who exercised at increasing intensities up to maximal intensity before and after a 6-month aerobic endurance training program. Typical values for stroke volume at rest and during maximal exercise in untrained, trained, and highly trained athletes are listed in table 9.1. The wide variation in stroke volume values for any given cell within this table is largely attributable to differences in body

▶ Cardiorespiratory endurance refers to your body's ability to sustain prolonged, rhythmic exercise. It is highly related to your aerobic development.

▶ Most sport scientists regard $\dot{V}O_2$max—the highest rate of oxygen consumption obtainable during maximal or exhaustive exercise—to be the best indicator of cardiorespiratory endurance.

▶ Cardiac output represents how much blood leaves the heart each minute, whereas $a\text{-}\overline{v}O_2$ difference indicates how much oxygen is extracted from the blood by the tissues. The product of these values tells us the rate of oxygen consumption: $\dot{V}O_2$ = stroke volume × heart rate × $a\text{-}\overline{v}O_2$ difference.

▶ Of the chambers of the heart, the left ventricle changes the most in response to endurance training.

▶ The internal dimensions of the left ventricle increase, mostly in response to an increase in ventricular filling.

▶ Left ventricular wall thickness and mass also increase, increasing the strength potential of that chamber's contractions.

size. Absolute stroke volume values at rest and during exercise are not merely a function of a person's state of training. They also reflect differences in body size. Larger people typically have greater stroke volumes. This is very important to remember when you are comparing stroke volumes of different people. But what causes the increase in stroke volume with training?

After training, the left ventricle fills more completely during diastole than it does in an untrained heart. As we discuss later, blood plasma volume increases with training, which means that more blood is available to enter the ventricle, increasing end-diastolic volume. The rate of a trained heart also is lower at rest and at the same absolute rate of work than an untrained heart, allowing an increase in the diastolic filling time. More blood entering the ventricle increases the stretching of the ventricular walls; by the Frank–Starling mechanism (see chapter 7), this results in a stronger elastic recoil.

The thickness of the posterior and septal walls of the left ventricle increase slightly with endurance training. Increased ventricular muscle mass can cause a more forceful contraction. This increased contractility would cause the end-systolic volume to decrease because more blood would be forced out of the heart as a result of a more powerful contraction, leaving less blood in the left ventricle after systole. This is augmented by the decrease in blood pressure that occurs with training, thus reducing the systemic peripheral resistance.

Increased contractility, coupled with the reduction in systemic peripheral resistance and

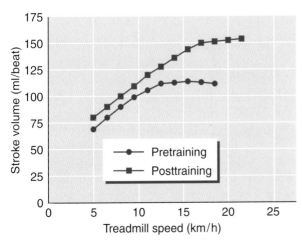

▲ **Figure 9.3** Changes in stroke volume with endurance training while walking, jogging, and running on a treadmill at increasing velocities.

Table 9.1		
Stroke Volumes (SV) for Different States of Training		
Subjects	**SVrest (ml)**	**SVmax (ml)**
Untrained	50-70	80-110
Trained	70-90	110-150
Highly trained	90-110	150->220

Measuring Heart Size

The measurement of heart size has been of interest to cardiologists for years because a hypertrophied or enlarged heart is typically a pathological condition indicating the presence of cardiovascular disease. More recently, exercise scientists have been interested in heart size as it relates to the training state and performance of the athlete or exercising individual. How is heart size measured? Initial studies used x-rays to estimate the size of the heart, but the techniques were not very accurate, and the information obtained was limited to only a few crude measurements. Since the 1970s, studies of athletes and people who participate in endurance training have used echocardiography to more accurately measure the size of the heart and its chambers. Echocardiography uses the technique of ultrasound, in which beams of high-frequency sound waves are directed through the chest wall to the heart. These sound waves are emitted from a transducer placed on the chest, and once they contact the various structures of the heart, they rebound back to a sensor, which is able to capture a moving picture of the heart. You can visualize the size of the heart's chambers, thicknesses of its walls, and heart valve action. There are several forms of echocardiography: M-mode echocardiography, which provides a one-dimensional, or "ice-pick," view of the heart; two-dimensional echocardiography, which provides a two-dimensional picture of the heart structures; and Doppler echocardiography, which is used more often to measure blood flow. Figure 9.4 illustrates two-dimensional echocardiography being conducted and the resulting echocardiogram.

 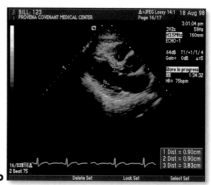

▲ **Figure 9.4** Two-dimensional echocardiography, illustrating (*a*) the procedure and (*b*) the resulting echocardiogram.

the increased elastic recoil that results from greater diastolic filling, increases the ejection fraction in the trained heart. More blood enters the left ventricle, and a greater percentage of what enters is forced out with each contraction, so stroke volume increases.

These stroke volume changes are well illustrated by a study in which older men endurance trained for 1 year.[12] Their cardiovascular function was evaluated before and after training. The subjects performed running, treadmill, and cycle ergometer exercise for 1 h each day, 4 days per week. They exercised at intensities of 60% to 80% of $\dot{V}O_2$max, with brief bouts of exercise exceeding 90% of $\dot{V}O_2$max.

The results are shown in figure 9.5. End-diastolic volume increased at rest and throughout submaximal exercise. The ejection fraction increased, and this increase was associated with a decreased end-systolic volume, both suggesting increased contractility of the left ventricle. $\dot{V}O_2$max increased by 23%, indicating a substantial improvement in endurance.

A stronger heart, a reduced systemic peripheral resistance, and the availability of a greater blood volume appear to account for the increases in resting, submaximal, and maximal stroke volumes after an endurance training program.

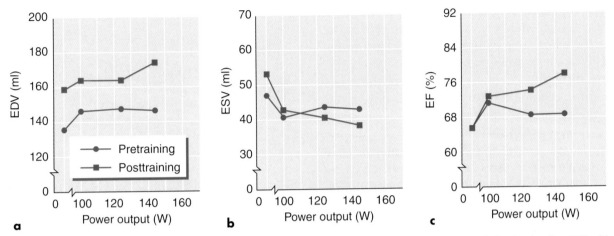

▲ **Figure 9.5**   Differences in (*a*) end-diastolic volume (EDV), (*b*) end-systolic volume (ESV), and (*c*) ejection fraction (EF) with increasing power output in individuals before and after endurance training. See the text for an explanation of these findings.
Adapted, by permission, from A.A. Ehsani et al., 1991, "Exercise training improves left ventricular systolic function in older men," *Circulation* 83: 96-103.

▶ After endurance training, stroke volume increases during rest, submaximal exercise, and maximal exertion.

▶ A major factor leading to the stroke volume increase is an increased end-diastolic volume, probably caused by an increase in blood plasma and greater diastolic filling time.

▶ Another major factor is increased left ventricular contractility. This is caused by hypertrophy of the cardiac muscle and increased elastic recoil, which results from increased stretching of the chamber with more diastolic filling. Reduced systemic blood pressure lowers the resistance to the flow of blood pumped from the left ventricle.

Heart Rate

Now that we have examined one aspect of cardiac output, we can turn our attention to the other factor in the equation: heart rate. Studies that have directly monitored the oxygen consumption of the heart have shown that the heart rate, both at rest and during exercise, is a good index of how hard the heart is working. Because active muscle requires more oxygen than resting muscle, it is not surprising that the heart's oxygen consumption and thus the

amount of work it performs are directly related to the heart's contraction rate. Let's examine how training affects heart rate.

Resting Heart Rate

The heart rate at rest can decrease markedly as a result of endurance training. If you are a sedentary individual with an initial resting heart rate of 80 beats/min, your heart rate can decrease by approximately 1 beat/min each week for the first few weeks of training. So, after 10 weeks of moderate endurance training, your resting heart rate could decrease from 80 to 70 beats/min. The actual mechanisms responsible for this decrease are not entirely known, but training appears to increase parasympathetic activity in the heart while decreasing sympathetic activity.

Not all research studies have found resting heart rate to decrease dramatically with aerobic endurance training. In a highly controlled study, the HERITAGE Family Study, which involved more than 700 subjects, 20 weeks of intense endurance training in previously sedentary individuals resulted in only a small decrease in resting heart rate, from 65.4 to 62.8 beats/min.[48] Several other studies with fewer subjects found essentially the same results: little or no change in resting heart rate. A good argument can be made for why resting heart rate should decrease after aerobic endurance training, because plasma volume is usually increased, which allows a greater return of blood to the heart, thereby increasing stroke volume. This is why studies showing little or no change are difficult to explain. Somewhere there is a logical explanation for this, but to date it has eluded exercise scientists.

> **fyi** Highly conditioned endurance athletes often have resting heart rates lower than 40 beats/min, and some have values lower than 30 beats/min! Lance Armstrong, whom we detailed at the beginning of this chapter, has a resting heart rate of 32 to 34 beats/min.

Recall from chapter 7 that bradycardia is a clinical term indicating a heart rate of less than 60 beats/min. In untrained individuals, bradycardia is usually the result of abnormal cardiac function or a diseased heart. Therefore, it is necessary to differentiate between training-induced bradycardia, which is a natural response to endurance training, and pathological bradycardia, which can be a serious cause for concern.

Submaximal Heart Rate

During submaximal exercise, greater aerobic conditioning results in proportionally lower heart rates at a specified rate of work. This is illustrated in figure 9.6, which shows the heart rate of an individual exercising on a treadmill both before and after training. At each specified work rate, indicated by the speed at which the subject is walking or running, the posttraining heart rate is lower than the heart rate before training. After a 6-month endurance training program of moderate intensity, decreases in heart rate of 10 to 30 beats/min are common at the same standardized submaximal rate of work, the decrease being greater at higher rates of work.

These decreases indicate that the heart becomes more efficient through training. In carrying out its necessary functions, a conditioned heart performs less work than an unconditioned heart.

Maximum Heart Rate

A person's maximum heart rate (HRmax) tends to be stable and usually remains relatively un-

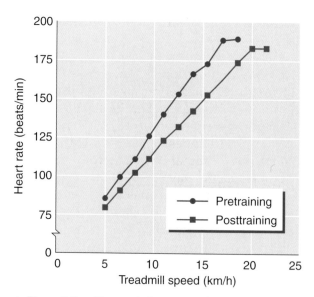

▲ **Figure 9.6** Changes in heart rate with endurance training while walking, jogging, and running on a treadmill at increasing velocities.

changed after endurance training. However, several studies have suggested that for people whose untrained HRmax values exceed 180 beats/min, HRmax might be slightly reduced after training. Also, highly conditioned endurance athletes tend to have lower HRmax values than untrained individuals of the same age, although this is not always the case. Athletes over 60 years old sometimes have higher HRmax values than untrained people of the same age. Why might HRmax decrease with training?

Heart Rate and Stroke Volume Interactions

During exercise, your heart rate combines with your stroke volume to provide an appropriate cardiac output for the rate of work you are performing. At maximal or near-maximal rates of work, your body might adjust your heart rate to provide the optimal combination of heart rate and stroke volume to maximize your cardiac output. If your heart rate is too fast, diastole, the period of ventricular filling, is reduced, and your stroke volume might be compromised.[45] For example, if your HRmax is 180 beats/min, your heart beats three times per second. Each cardiac cycle thus lasts for only 0.33 s. Diastole is as short as 0.150 s or less. This allows very little time for your ventricles to fill. As a consequence, your stroke volume could decrease at high heart rates where filling time is compromised.

However, if your heart rate slows, your ventricles would have longer to fill. Perhaps this is why highly trained endurance athletes tend to have lower HRmax values: Their hearts have adapted to training by drastically increasing their stroke volumes, so lower HRmax values can provide optimal cardiac output.

All disciplines have their dilemmas, and this is one for exercise physiology: Which comes first—does increased stroke volume decrease heart rate, or does decreased heart rate increase stroke volume? This question remains unanswered. In any event, the combination of increased stroke volume and decreased heart rate is a very efficient way for the heart to meet the body's demands. The heart expends less energy by contracting less often but more forcefully than it would if contraction frequency increased. Changes in heart rate and stroke volume in response to training go hand in hand and share a common goal: to allow the heart to expel the maximal amount of oxygenated blood at the lowest energy cost.

Heart Rate Recovery

During exercise, as discussed in chapter 7, your heart rate must increase to meet the demands of your active muscles. When the exercise bout is finished, your heart rate does not instantly return to its resting level. Instead, it remains elevated for a while, slowly returning to its resting rate. The time it takes for your heart rate to return to its resting rate is called the **heart rate recovery period.**

Following a period of training, as shown in figure 9.7, heart rate returns to its resting level much more quickly after exercise than it does before training. This is true after standardized submaximal exercise as well as after maximal exercise.

Because the heart rate recovery period is shortened by endurance training, this measurement has been used as an index of cardiorespiratory fitness. In general, a more fit person recovers faster after a standardized rate of work than a less fit person. However, factors other than training level can affect heart

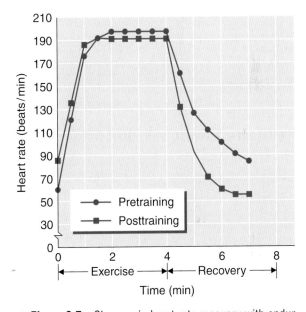

▲ **Figure 9.7** Changes in heart rate recovery with endurance training following completion of an all-out maximal work bout of 4 min duration.

Optimizing the Heart Rate–Stroke Volume Relationship

To better understand the concept of optimizing the heart rate–stroke volume relationship, consider the maximal breathing capacity test. In this test, the objective is to inspire and expire the largest volume of air possible in a fixed period of time, usually 15 s. During the test, you are encouraged to breathe as deeply as possible. But you are also encouraged to breathe in and out as rapidly as possible. Try it! It is impossible to breathe both deeply and rapidly simultaneously. Obviously, an optimal combination of respiratory rate and depth will produce the greatest ventilation volume in a given period of time. The same principle is thought to apply to heart rate and stroke volume. At an optimal heart rate, your stroke volume is maximized to provide the greatest cardiac output.

rate recovery time. For example, exercise in hot environments or at high altitudes can prolong heart rate elevation. Some people undergo a stronger sympathetic nervous system response during exercise than others, and this also could prolong heart rate elevation.

The heart rate recovery curve is an excellent tool for tracking a person's progress during a training program. But because of the potential influence of other factors, it should not be used to compare one individual with another.

> Resting heart rate usually is decreased substantially (more than 10 beats/min) following aerobic endurance training, although several studies have shown little or no change. After endurance training, submaximal heart rates always decrease when one is exercising at the same rate of work, generally by 10 to 20 beats/min or more. Maximal heart rates generally do not change.

Resistance Training and Heart Rate

Our discussion about heart rate so far has focused on endurance training. Some research has shown that resistance training can reduce the heart rate at rest and at standardized rates of submaximal exercise. But not all studies have confirmed these reductions.

The reductions in heart rate that have been reported are much less than the typical reductions obtained from endurance training. These changes appear to depend on the following characteristics of the resistance training program:[43]

- Training volume
- Training intensity
- Training duration
- Length of rest periods between sets
- Amount of muscle mass used

The mechanisms responsible for this decrease in heart rate with resistance training have not been determined but could be related to changes in heart size and contractility resulting from training, as discussed earlier in this chapter.

▶ Resting heart rate can decrease considerably as a result of endurance training. In a sedentary person, the decrease is typically about 1 beat/min per week during initial training. Some studies have reported much smaller decreases. Highly trained endurance athletes often have resting rates of 40 beats/min or less.

▶ Heart rate during submaximal exercise also decreases, often by about 10 to 30 beats/min following 6 months of moderate training. A person's submaximal heart rate decreases proportionately with the amount of training completed, and the magnitude of decrease is greater at higher rates of work.

▶ Maximal heart rate either remains unchanged or decreases slightly with training. When a decrease occurs, it probably allows for optimal stroke volume to maximize cardiac output.

▶ The heart rate recovery period decreases with increased endurance, making this value well suited for tracking an individual's progress with training. However, this value is not useful for comparing fitness levels of different people.

▶ Resistance training can also reduce heart rates; however, these decreases are not as reliable or as large as those seen with endurance training.

Cardiac Output

We have looked at the effects of training on the two components of cardiac output: stroke volume and heart rate. We have seen that stroke volume increases, but heart rate at rest and for the same rate of work generally decreases. How does this affect cardiac output?

Cardiac output at rest and during submaximal exercise at a given rate of work doesn't change much following endurance training. For exercise at the same submaximal work rate, cardiac output might decrease slightly. This could be the result of an increase in the a-$\bar{v}O_2$ difference, reflecting greater oxygen extraction by the tissues, or a decrease in the rate of oxygen consumption, reflecting an increased efficiency. Generally, cardiac output changes to match the oxygen consumption required for any given rate of work or intensity of exercise.

However, cardiac output increases considerably at maximal rates of work, as seen in figure 9.8. This increase results primarily from the increase in maximal stroke volume, because HRmax changes little, if any. Maximal cardiac output ranges from 14 to 20 L/min in untrained people, 25 to 35 L/min in trained people, and 40 L/min or more in large, highly conditioned endurance athletes. These absolute values, however, are greatly influenced by body size.

> ▶ Cardiac output at rest or during submaximal levels of exercise remains unchanged or decreases slightly after training.
>
> ▶ Cardiac output at maximal levels of exercise increases considerably. This is largely the result of the substantial increase in maximal stroke volume.

Blood Flow

Now that we have discussed training-induced changes in cardiac structure and function, we next turn our attention to changes in the vasculature, beginning with blood flow.

We know that active muscles need considerably more oxygen and nutrients. To meet these needs, more blood must be brought into these muscles during exercise. As the muscles become better trained, the cardiovascular system adapts to increase blood flow to them. Four factors account for this enhanced blood supply to muscle following training:

- Increased capillarization of trained muscles
- Greater opening of existing capillaries in trained muscles
- More effective blood redistribution
- Increased blood volume

To permit increased blood flow, new capillaries develop in trained muscles. This allows the blood to more fully perfuse the tissues. This increase in capillaries usually is expressed as an increase in the number of capillaries per muscle fiber, or the **capillary-to-fiber ratio**. Table 9.2 illustrates the differences in capillary-to-fiber ratios between well-trained and untrained men, both before and after exercise.[21]

The existing capillaries in trained muscles can open up more, which increases blood flow through the capillaries and into the muscles. This adaptation is easily accomplished because endurance training also increases blood volume; because more blood is present in the system to begin with, shifting more into the capillaries will not severely compromise venous return.

A more effective redistribution of the cardiac output also can increase blood flow to the active muscles. Blood flow is directed to the active musculature and shunted away from areas that don't need high flow. Venous compliance also can decrease with endurance training as a result of increased venous tone. That means that the veins are not as easily distended by the blood, so less blood pools in the venous system, thereby increasing the amount of arterial blood available for working muscles.

Blood flow can increase to the more active areas in even a specific muscle group. Armstrong and Laughlin[1] demonstrated that endurance-trained rats could redistribute blood flow to

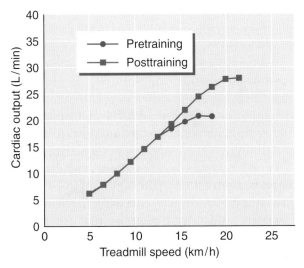

▲ **Figure 9.8** Changes in cardiac output with endurance training while walking, jogging, and running on a treadmill at increasing velocities.

Table 9.2

Capillaries and Muscle Fibers per mm², Capillary-to-Fiber Ratio, and Diffusion Distance in Well-Trained and Untrained Men

Groups	Capillaries per mm²	Muscle fibers per mm²	Capillary-to-fiber ratio	Diffusion distance[a]
Well-trained				
Preexercise	640	440	1.5	20.1
Postexercise	611	414	1.6	20.3
Untrained				
Preexercise	600	557	1.1	20.3
Postexercise	599	576	1.0	20.5

Note. This table illustrates the larger size of the muscle fibers in the well-trained men in that they had fewer fibers for a given area (fibers per mm²). They also had an approximately 50% higher capillary-to-fiber ratio than the untrained men.

[a]Diffusion distance is expressed as the average half-distance between capillaries on the cross-sectional view expressed in micrometers.

Adapted, by permission, from L. Hermansen and M. Wachtlova, 1971, "Capillary density of skeletal muscle in well-trained and untrained men," *Journal of Applied Physiology* 30: 860-863.

their most active tissues during exercise better than untrained rats could. The researchers used radiolabeled microspheres, radioactive particles that are injected into the bloodstream. By using a counter, the scientists could trace the distribution of these microspheres throughout the body. The total blood flow to the hind limbs did not differ between the trained and untrained rats during exercise. However, the trained rats distributed more of their blood to the most active muscle fibers.

Finally, the body's total blood volume increases, providing more blood to meet the body's many needs during endurance activity. The mechanisms responsible for this are discussed later in this chapter.

> The increase in blood flow to muscle is one of the most important factors for increased aerobic endurance capacity and performance. This increase is attributable to better capillary supply (both new capillaries and greater opening of existing capillaries), more blood being diverted to the active muscles, and increased blood volume.

Blood Pressure

Following endurance training, arterial blood pressure is reduced at the same submaximal exercise work rate, but at maximal work rates, systolic blood pressure is increased and diastolic pressure decreased.[6, 48] Resting blood pressure generally is lowered in people who are borderline or moderately hypertensive before training. This reduction occurs in both systolic and diastolic blood pressure. Decreases average approximately 10 mmHg for systolic pressure and 8 mmHg for diastolic pressure in hypertensive subjects and slightly less in borderline hypertensives.[16] The mechanisms underlying this reduction are unknown.

Although resistance-type exercise can cause large increases in both systolic and diastolic blood pressure during lifting of heavy weights (see chapter 7), chronic exposure to these high pressures does not elevate resting blood pressure.[43] Hypertension is not common in competitive weightlifters or in strength and power athletes. In fact, the cardiovascular system can respond to resistance training by lowering resting blood pressure. Hagberg and co-workers[20] followed a group of borderline-hypertensive adolescents through 5 months of weight training. The subjects' resting systolic blood pressures decreased significantly. These reductions were somewhat greater than those resulting from endurance training.

> ▶ Blood flow to muscles is increased by endurance training.
>
> ▶ Increased blood flow results from four factors:
>   1. Increased capillarization
>   2. Greater opening of existing capillaries
>   3. More effective blood redistribution
>   4. Increased blood volume
>
> ▶ Resting blood pressure generally is reduced by endurance training in those with borderline or moderate hypertension.
>
> ▶ Endurance training results in a reduction in blood pressure at the same submaximal work rate.
>
> ▶ At maximal rates of work, systolic blood pressure is increased and diastolic blood pressure is decreased.

Blood Volume

As has been mentioned several times in this chapter, endurance training increases blood volume. This effect is greater with more intense training. Furthermore, this effect occurs rapidly. This increased blood volume results primarily from an increase in plasma volume, but there is also an increase in red blood cells. The time course for the increase of each of these is quite different. Let's now look at how each specifically contributes to the increase in blood volume.

Plasma Volume

The increase in plasma volume is thought to be caused by two mechanisms. First, exercise increases the release of antidiuretic hormone and aldosterone. Recall from chapter 5 that these hormones cause the kidneys to retain water, which increases blood plasma. Second, exercise increases the amount of plasma proteins, particularly albumin.[50] Recall from basic physiology that plasma proteins are the major basis for the blood's osmotic pressure. As plasma protein concentration increases, so does osmotic pressure, and more fluid is retained in the blood. Thus, both mechanisms work together to increase the fluid portion of the blood, the blood plasma. It is likely that the initial phase of rapid plasma volume increase is the result of the increased plasma albumin, which is noted within the first hour of recovery from the first training bout.[33] Nearly all of the increase in blood volume during the first 2 weeks of training can be explained by the increase in plasma volume, as it takes about 2 to 3 weeks of endurance training before you start to see increases in red blood cell volume. This is illustrated in figure 9.9.

Red Blood Cells

An increase in red blood cell volume also contributes to the overall increase in blood volume, but this increase has not been found

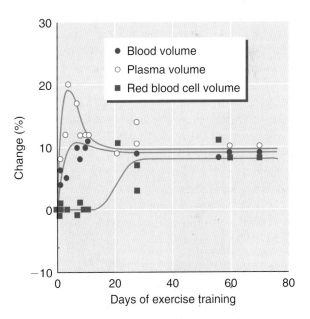

▲ **Figure 9.9** Estimated time course of percent changes in total blood volume, plasma volume, and red blood cell volume.

Data published by Sawka et al.[41] representing a total of 18 different studies where all three volumes were reported.

consistently.[19] When red blood cell volume has been shown to increase, plasma volume usually has increased much more.[19] Because of this, although the actual number of red blood cells increases, the hematocrit—the ratio of the red blood cell volume to the total blood volume—actually can decrease. Figure 9.10 illustrates this apparent paradox. Notice that the hematocrit is reduced even though there has been a slight increase in red blood cells. A trained athlete's hematocrit can decrease to a level where the athlete appears to be anemic on the basis of a relatively low concentration of red cells and hemoglobin (pseudoanemia). It is interesting that some highly trained endurance athletes, particularly adult female endurance athletes, tend to have hematocrit and hemoglobin values similar to untrained individuals' values. It is postulated that a gradual increase in red cell mass over years of training returns the ratio of cell mass to total volume to more normal levels.

The change in the ratio of plasma to cells resulting from a greater increase in the fluid por-

tion reduces the blood's viscosity. Reduced viscosity may facilitate blood movement through the blood vessels, particularly through the smallest vessels, such as the capillaries. Research has shown that low blood viscosity enhances oxygen delivery to the active muscle mass.

Both the total amount (absolute values) of hemoglobin and the total number of red blood cells are typically above normal in highly trained athletes, although these values relative to total blood volume are below normal. This ensures that the blood has more than ample oxygen-carrying capacity to meet the body's needs at all times. The turnover rate of red blood cells also may be higher with intense training. This might be an advantage, provided that production equals or exceeds destruction, because younger cells are more efficient in transporting oxygen.[42]

Plasma Volume, Stroke Volume, and $\dot{V}O_2$max

An increase in plasma volume is one of the most significant changes that occur with endurance training. Recall from our previous discussion that the increase in plasma volume is a major factor in the increase in stroke volume that results from training. Stroke volume, in turn, affects oxygen delivery.

As plasma volume increases, so does blood volume. Consequently, more blood enters the heart. As more blood enters the heart, stroke volume increases. At maximal rates of work, HRmax generally remains relatively stable, so an increased stroke volume allows maximal cardiac output to increase. Increasing maximal cardiac output makes more oxygen available to working muscles, thus allowing $\dot{V}O_2$max to increase.

> The increase in blood volume following aerobic endurance training is attributable to increases in both plasma volume and red blood cell volume; both changes facilitate the delivery of oxygen to active muscles.

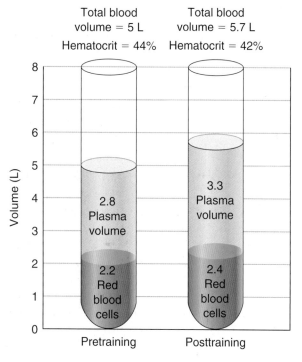

▲ **Figure 9.10** Increases in total blood volume and plasma volume with endurance training. Note that, although the hematocrit (percentage of red blood cells) decreases from 44% to 42%, the total volume of red blood cells increases by 10%.

This sequence of events—an increase in plasma volume leading to an increase in stroke volume, which in turn increases $\dot{V}O_2$max—also has been demonstrated in untrained individuals whose plasma volumes were expanded by an infusion of 6% dextran (a carbohydrate similar to starch or glycogen).[24] The opposite is also true: A plasma volume reduction (produced by detraining highly trained athletes or by placing healthy men in a position with head tilted downward −5° for 20 h) results in significant decreases in stroke volume and equivalent decreases in $\dot{V}O_2$max.[8, 18]

> ▶ Blood volume increases as a result of endurance training.
>
> ▶ The increase in blood volume initially is caused by an increase in blood plasma volume.
>
> ▶ Red blood cell volume also increases, but the increase in plasma volume is typically higher, resulting in a relatively greater fluid portion of the blood (lower hematocrit).
>
> ▶ Increased plasma volume decreases blood viscosity, which can improve circulation and oxygen availability.
>
> ▶ Research has shown that changes in plasma volume are highly correlated with changes in stroke volume and $\dot{V}O_2$max, making the training-induced increase in plasma volume one of the most significant training effects.

Respiratory Adaptations to Training

No matter how efficient the cardiovascular system is at supplying adequate amounts of blood to tissues, endurance would be hindered if the respiratory system didn't bring in enough oxygen to meet oxygen demands. Respiratory system functioning usually does not limit performance because ventilation can be increased to a greater extent than cardiovascular function. But, as with the cardiovascular system, the respiratory system undergoes specific adaptations to endurance training to maximize its efficiency. Let's consider some of these adaptations.

Pulmonary Ventilation

After training, pulmonary ventilation is essentially unchanged or slightly reduced at rest and slightly reduced at the same submaximal work rates. But maximal pulmonary ventilation is substantially increased. In untrained subjects, maximal pulmonary ventilation typically increases from a beginning rate of about 100 to 120 L/min to about 130 to 150 L/min or more following training. Pulmonary ventilation rates typically increase to about 180 L/min in highly trained athletes and can exceed 200 L/min in very large, highly trained endurance athletes. Two factors can account for the increase in maximal pulmonary ventilation following training: increased tidal volume and increased respiratory rate at maximal exercise. Research recently revealed that training the inspiratory muscles can significantly improve performance. In a study of 14 competitive female rowers, 11 weeks of inspiratory muscle training substantially improved the rowing distance covered in a 6-min all-out effort (3.5%) and the time to complete a 5000-m time trial (3.1%).[46]

> **fyi** Large, highly trained endurance athletes, such as rowers, can have maximal pulmonary ventilation rates in excess of 240 L/min, fully twice the rate typical of untrained individuals!

Ventilation is usually not considered a limiting factor for endurance exercise performance. However, some evidence suggests that at some point in a highly trained person's adaptation, the pulmonary system's capacity for oxygen transport won't be able to meet the demands of the limbs and the cardiovascular system.[9] This results in what has been termed exercise-induced arterial hypoxemia, where arterial oxygen saturation decreases below 96%. Mild exercise-induced arterial hypoxemia is defined as an arterial saturation of 93% to 95%, moderate as 88% to 93%, and severe as less than 88%.[10]

Pulmonary Diffusion

Pulmonary diffusion, which is the gas exchange occurring in the alveoli, is unaltered at rest and during standardized submaximal exercise following training. However, it increases during maximal exercise. Pulmonary blood flow (blood coming from the heart to the lungs) appears to increase following training, particularly the flow to the upper regions of the lungs when a person is sitting or standing. This increases lung perfusion. More blood is brought into the lungs for gas exchange, and at the same time ventilation increases so that more air is brought into the lungs. This means that more alveoli will be actively involved in pulmonary diffusion. The net result is that pulmonary diffusion increases.

Arterial-Venous Oxygen Difference

The oxygen content of arterial blood changes very little with training. Even though total hemoglobin is increased, the amount of hemoglobin per unit of blood is the same or even slightly reduced. The arterial-venous oxygen difference, or a-$\bar{v}O_2$ difference, however, does increase with training, particularly at maximal levels of exercise. This increase results from a lower mixed venous oxygen content, which means that the blood returning to the heart, which is a mixture of venous blood from all body parts, not just the active tissues, contains less oxygen than it would in an untrained person. This reflects both greater oxygen extraction at the tissue level and a more effective distribution of total blood volume (more goes to the active tissues).

In summary, the respiratory system is quite adept at bringing adequate amounts of oxygen into the body. For this reason, the respiratory system seldom limits endurance performance.

Although the largest part of the increase in $\dot{V}O_2$max results from the increases in cardiac output and muscle blood flow, the increase in a-$\bar{v}O_2$ difference also plays a key role. This increase in a-$\bar{v}O_2$ difference is attributable to a more effective distribution of arterial blood away from inactive tissue to the active tissue, so that more of the blood coming back to the right atrium has gone through active muscle.

Not surprisingly, the major training adaptations noted in the respiratory system are apparent during maximal exercise, when all systems are being maximally stressed.

▶ The combined effect of increased tidal volume and respiration rate is an increase in pulmonary ventilation during maximal effort following training.

▶ Pulmonary diffusion at maximal work rates increases, probably because of increased ventilation and increased lung perfusion.

▶ The a-$\bar{v}O_2$ difference increases with training, most notably at maximal levels of work, reflecting increased oxygen extraction by the tissues and more effective blood distribution to the active tissues.

Metabolic Adaptations to Training

Now that we have discussed training changes in both the cardiovascular and respiratory systems, we are ready to look at how these systems integrate with metabolism in the active tissues. Because metabolic adaptations to training were discussed in chapter 6, our discussion focuses briefly on

- lactate threshold,
- respiratory exchange ratio, and
- oxygen consumption.

Lactate Threshold

Endurance training increases lactate threshold. In other words, after training you can perform at a higher rate of work, such as an increased running velocity (see figure 6.12b on page 199), and at a higher absolute rate of oxygen consumption without raising your blood lactate above resting levels. Even though $\dot{V}O_2$max also increases, the lactate threshold occurs at a higher percentage of $\dot{V}O_2$max after training. Thus, blood lactate concentrations at each level of a standardized, graded exercise test above lactate threshold are lower following endurance training.

This increase in lactate threshold appears to be the result of several factors, including a greater ability to clear lactate produced in the muscle and a reduction in lactate production for the same work rate. The latter change is the net result of an increase in skeletal muscle enzymes coupled with a shift in metabolic substrate as a result of training.

> The increase in lactate threshold is a major factor in the improved performance of aerobically trained endurance athletes. This increase allows them to increase their race pace; it has been clearly shown that lactate threshold is highly related to race pace.

Maximal blood lactate concentration at the point of exhaustion increases slightly following endurance training. This increase is relatively small, particularly when compared with the magnitude of increase seen with sprint-type training.

Respiratory Exchange Ratio

Recall from chapter 4 that the **respiratory exchange ratio (RER)** is the ratio of carbon dioxide released to oxygen consumed during nutrient metabolism. RER reflects the type of substrates being used as an energy source.

After training, the RER decreases at both absolute and relative submaximal rates of work. These changes are attributable to a greater utilization of free fatty acids instead of carbohydrate at these work rates following training. This shift in substrate utilization was discussed in chapter 6.

However, at maximal levels of work, the RER increases in trained individuals. This increase reflects a sustained hyperventilation with excessive carbon dioxide and results from the ability to perform at maximal levels for longer periods of time than possible before training, which most likely reflects an increased psychological drive.

Resting and Submaximal Oxygen Consumption

Oxygen consumption ($\dot{V}O_2$) at rest generally is unchanged following endurance training. Several studies have suggested that resting metabolic rates are elevated in highly trained endurance athletes.[35] However, the results of longitudinal training studies, in which resting metabolic rate is measured in the same individuals before and after training, are not conclusive. Although several studies have found increases, most have found no change. Where increases have been found, the subjects have usually been 50 years of age or older. In the HERITAGE Family Study, with a large number of subjects and with duplicate measures of resting metabolic rate both before and after 20 weeks of training, there was no evidence of an increased metabolic rate after training.[49]

At submaximal levels of exercise, $\dot{V}O_2$ is either unchanged or slightly reduced following training. In the large HERITAGE Family Study, with more than 700 participants, training reduced submaximal $\dot{V}O_2$ by 3.5% at a work rate of 50 W. There was a corresponding reduction in cardiac output at 50 W, reinforcing the strong interrelationship between $\dot{V}O_2$ and cardiac output.[47] A decrease in $\dot{V}O_2$ during submaximal exercise could result from an increase in metabolic efficiency, an increase in mechanical efficiency (performing the same physical work with less extraneous movement), or a combination of both.

It is possible that the reduced $\dot{V}O_2$ for a standardized submaximal work rate reflects a practice effect. Subjects who use an ergometer (e.g., a treadmill or cycle ergometer) with which they have no prior experience probably feel awkward on the device the first time they use it. As a result, they might expend more energy during the first trial but then perform at a lower energy cost the second or third time they exercise on that device simply because they have become more familiar with using it. In the HERITAGE Family Study, where a reduction in submaximal $\dot{V}O_2$ was observed, subjects had many practice trials on the cycle ergometer before actual testing, so it is unlikely that the results of this study reflect a practice effect.

Another potential problem could arise if the exercise device is weight dependent (i.e., the work performed depends on your body weight, such as on a treadmill). In this case, any weight loss from training would decrease your $\dot{V}O_2$ because you are doing less work; thus the decreased $\dot{V}O_2$ doesn't necessarily reflect changes in efficiency but rather the loss in body weight. Therefore, any observed change in submaximal $\dot{V}O_2$ with endurance training might not be the result of cardiovascular or metabolic adaptations to training.

Maximal Oxygen Consumption

As we pointed out early in the chapter, most researchers regard $\dot{V}O_2$max as the best indicator of cardiorespiratory endurance capacity. Now that we have covered the various physiological adaptations that occur with endurance training, it is not surprising to find that $\dot{V}O_2$max increases substantially in response to endurance training. Increases of 4% to 93% have been reported.[36] An increase of 15% to 20% is typical for an average person who was sedentary before training and who trains at 75% of his or her capacity three times per week, 30 min per day, for 6 months.[36] The $\dot{V}O_2$max of a sedentary individual can increase from an initial value of 35 ml · kg^{-1} · min^{-1} to 42 ml · kg^{-1} · min^{-1} as a result of such a program. This is far below the values we see in world-class endurance athletes, whose values generally range from 70 to 94 ml · kg^{-1} · min^{-1}.

The factors responsible for increased $\dot{V}O_2$max have been identified, and we have discussed many in this chapter. But at one time, much controversy surrounded their importance. Two theories were proposed to explain these increases with training.

Limitation of Oxidative Enzymes

One theory held that endurance performance usually is limited by the lack of sufficient amounts of oxidative enzymes in the mitochondria. Proponents of this theory provided impressive evidence that endurance training programs substantially increase the amount of these oxidative enzymes. This would allow active tissue to use more of the available oxygen, resulting in a higher $\dot{V}O_2$max. In addition, proponents supported their case by pointing out that endurance training increases both the size and number of muscle mitochondria. Thus, this theory argues, the main limitation of maximal oxygen consumption is the inability of the existing mitochondria to use the available oxygen beyond a certain rate. This theory is referred to as the utilization theory.

Limitation of Oxygen Delivery

The second theory proposed that central and peripheral circulatory factors limit endurance capacity. These circulatory factors would preclude delivery of sufficient amounts of oxygen to the active tissues. According to this theory, improvement in $\dot{V}O_2$max following endurance training results from increased blood volume, increased cardiac output (via stroke volume), and a better perfusion of active muscle with blood. This theory is referred to as the presentation theory.

Again, impressive research evidence strongly supports this theory. In one study, subjects breathed a mixture of carbon monoxide and air during exercise to exhaustion.[34] $\dot{V}O_2$max decreased in direct proportion to the percentage of carbon monoxide breathed. The carbon monoxide molecules bonded to approximately 15% of the total hemoglobin; this percentage agreed with the percentage reduction in $\dot{V}O_2$max. In another study, approximately 15% to 20% of each subject's total blood volume was removed.[13] $\dot{V}O_2$max decreased by approximately the same relative amount. Reinfusion of the subjects' packed red blood cells approximately 4 weeks later increased $\dot{V}O_2$max above baseline or control conditions. In both studies, the reduction in the oxygen-carrying capacity of the blood—by either blocking hemoglobin or removing whole blood—resulted in the delivery of less oxygen to the active tissues and a corresponding reduction in $\dot{V}O_2$max. Similarly, studies have shown that breathing oxygen-enriched mixtures, in which the partial pressure of oxygen in the inspired air is substantially increased, increases endurance capacity.

These and subsequent studies indicate that the available oxygen supply is the major limiter of endurance performance. Saltin and Rowell,[40] in an excellent review article, concluded that oxygen transport to the working muscles, not the available mitochondria and oxidative enzymes, limits $\dot{V}O_2$max. They argued that increases in $\dot{V}O_2$max with training are largely attributable to increased maximal blood flow and increased muscle capillary density in the active tissues. The major skeletal muscle adaptations (including increased mitochondrial content and respiratory capacity of the muscle fibers) appear more closely related to the ability to perform prolonged, high-intensity, submaximal exercise, or what we termed submaximal endurance capacity at the beginning of this chapter.[23]

Table 9.3 summarizes the expected physiological changes that occur with endurance training. The changes pre- to posttraining in a previously inactive man are compared with values for a world-class male endurance runner.

▶ Lactate threshold increases with endurance training, which allows you to perform at higher rates of work and levels of oxygen consumption without increasing your blood lactate above resting levels. Maximal blood lactate levels can increase slightly.

▶ With endurance training, the RER decreases at submaximal work rates, indicating greater use of free fatty acids, but increases at maximal effort.

▶ Oxygen consumption generally remains unchanged at rest and decreases slightly or remains unaltered during submaximal exercise following endurance training.

▶ $\dot{V}O_2$max increases substantially following endurance training, but the amount of increase possible is limited in each individual. The major limiting factor appears to be oxygen delivery to the active muscles.

Long-Term Improvement in Cardiorespiratory Endurance

Although an individual's highest attainable $\dot{V}O_2$max usually is reached within 18 months of intense endurance conditioning, endurance performance continues to improve with continued training for many additional years. Improvement in endurance performance without improvement in $\dot{V}O_2$max is probably attributable to the body's ability to perform at increasingly higher percentages of $\dot{V}O_2$max for extended periods.

Consider, for example, a young male runner who starts training with an initial $\dot{V}O_2$max of 52.0 ml · kg^{-1} · min^{-1}. He reaches his genetically determined peak $\dot{V}O_2$max of 71.0 ml · kg^{-1} · min^{-1} after 2 years and is unable to increase it further, even with more intensive workouts. At this point, as shown in figure 9.11, the young runner is able to run at 75% of his $\dot{V}O_2$max (0.75 × 71.0 = 53.3 ml · kg^{-1} · min^{-1}) in a 10 km (6.2-mi) race. After an additional 2 years of intensive training, his $\dot{V}O_2$max is unchanged, but he is now able to compete at 88% of his $\dot{V}O_2$max (0.88 × 71.0 = 62.5 ml · kg^{-1} · min^{-1}). Obviously, by being able to sustain an oxygen uptake of 62.5 ml · kg^{-1} · min^{-1}, he is able to run at a much faster pace.

This increase in performance without an increase in $\dot{V}O_2$max is the result of an increase in lactate threshold, because race pace is directly related to the $\dot{V}O_2$ value at lactate threshold, as we have seen in previous chapters.

Factors Affecting the Response to Aerobic Training

We have discussed general trends in adaptations that occur in response to endurance training. However, we must always remember that we are talking about adaptations in individuals and that everyone does not respond in the same manner. Several factors that can affect individual response to aerobic training

Table 9.3

Expected Physiological Alterations Resulting From Endurance Training in a Previously Inactive Man, Compared With Values for a Male World-Class Endurance Athlete

Variables	Sedentary male		World-class endurance runner
	Pretraining	**Posttraining**	
Cardiovascular			
HR at rest (beats/min)	75	65	45
HRmax (beats/min	185	183	174
SV at rest (ml/beat)	60	70	100
SVmax (ml/beat)	120	140	200
$\dot{Q}$ at rest (L/min)	4.5	4.5	4.5
$\dot{Q}$max (L/min	22.2	25.6	34.8
Heart volume (ml)	750	820	1,200
Blood volume (L)	4.7	5.1	6.0
Systolic BP at rest (mmHg)	135	130	120
Systolic BPmax (mmHg)	200	210	220
Diastolic BP at rest (mmHg)	78	76	65
Diastolic BPmax (mmHg)	82	80	65
Respiratory			
$\dot{V}_E$ at rest (L/min)	7	6	6
$\dot{V}_E$max (L/min)	110	135	195
TV at rest (L)	0.5	0.5	0.5
TVmax (L)	2.75	3.0	3.9
VC (L)	5.8	6.0	6.2
RV (L)	1.4	1.2	1.2
Metabolic			
a-$\bar{v}$O$_2$ diff at rest (ml/100ml)	6.0	6.0	6.0
a-$\bar{v}$O$_2$ diff max (ml/100 ml)	14.5	15.0	16.0
$\dot{V}$O$_2$ at rest (ml · kg^{-1} · min^{-1})	3.5	3.5	3.5
$\dot{V}$O$_2$max (ml · kg^{-1} · min^{-1})	40.7	49.9	81.9
Blood lactate at rest (mmol/L)	1.0	1.0	1.0
Blood lactate max (mmol/L)	7.5	8.5	9.0
Body composition			
Weight (kg)	79	77	68
Fat weight (kg)	12.6	9.6	5.1
Fat-free weight (kg)	66.4	67.4	62.9
Fat (%)	16.0	12.5	7.5

Note. HR = heart rate; SV = stroke volume; $\dot{Q}$ = cardiac output; BP = blood pressure; $\dot{V}_E$ = ventilation; TV = tidal volume; VC = vital capacity; RV = residual volume; a-$\bar{v}$O$_2$ diff = arterial-mixed venous oxygen difference; $\dot{V}$O$_2$ = oxygen consumption.

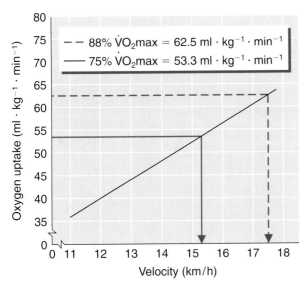

▲ Figure 9.11 Change in race pace with continued training after maximal oxygen uptake stops increasing beyond 71.0 ml · kg⁻¹ · min⁻¹.

must be considered. Let's turn our attention to them now.

Level of Conditioning and $\dot{V}O_2$max

The higher the initial state of conditioning, the smaller is the relative improvement for the same program of training. In other words, if two people, one sedentary and the other partially trained, undergo the same endurance training program, the sedentary person will show the greatest relative improvement. You can look at this in terms of how well trained the people are when they begin the program—those who are less trained have the most room for improvement.

It appears that in fully mature athletes, the highest attainable $\dot{V}O_2$max is reached within 8 to 18 months of heavy endurance training, indicating that each athlete has a finite attainable level of oxygen consumption. This finite range potentially is influenced by training in early childhood,[17] but this observation is currently conjecture; it needs to be substantiated by experimental research.

Heredity

Maximal oxygen consumption levels depend on genetic limits. You should not take this to mean that each individual has an exact $\dot{V}O_2$max that cannot be exceeded. Rather, a range of $\dot{V}O_2$max values seems to be predetermined by an individual's genetic makeup, and that individual's highest attainable $\dot{V}O_2$max should fall in that range.

Klissouras[25] studied the genetic basis of $\dot{V}O_2$max in a series of studies conducted in the late 1960s and early 1970s, and more recently Bouchard and his colleagues[5] examined this topic. Research has found that identical (monozygous) twins have similar $\dot{V}O_2$max values, whereas the variability for dizygous (fraternal) twins is much greater. Figure 9.12 illustrates this. Each symbol represents a pair of brothers. Brother A's $\dot{V}O_2$max value is indicated by the symbol's position on the x axis and brother B's $\dot{V}O_2$max value on the y axis. Similarity in the siblings' $\dot{V}O_2$max values is noted by comparing the x and y coordinates of the symbol (i.e., how close it falls to the diagonal line $x = y$ on the graph). Similar results were found for endurance capacity, determined by the maximal amount of work performed in an all-out, 90-min ride on a cycle ergometer.

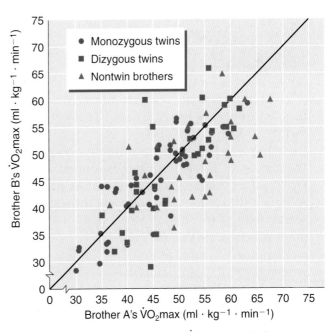

▲ Figure 9.12 Comparisons of $\dot{V}O_2$max in twin (monozygous and dizygous) and nontwin brothers.

Adapted, by permission, from C. Bouchard et al., 1986, "Aerobic performance in brothers, dizygotic and monozygotic twins," *Medicine and Science in Sports and Exercise* 18: 639-646.[5]

Bouchard and colleagues[4] concluded that heredity accounts for between 25% and 50% of the variance in $\dot{V}O_2$max values. This means that of all factors influencing $\dot{V}O_2$max, heredity alone is responsible for one quarter to one half of the total influence. World-class athletes who have stopped endurance training continue for many years to have high $\dot{V}O_2$max values in their sedentary, deconditioned state. Their $\dot{V}O_2$max values may decrease from 85 to 65 ml · kg⁻¹ · min⁻¹, but this deconditioned value is still very high.

Heredity also potentially explains the fact that some people have relatively high $\dot{V}O_2$max values yet have no history of endurance training. In a study that compared untrained men who had $\dot{V}O_2$max values below 49 ml · kg⁻¹ · min⁻¹ with untrained men who had $\dot{V}O_2$max values above 62.5 ml · kg⁻¹ · min⁻¹, those with high values were distinguished by having higher blood volume values, leading to higher stroke volume and cardiac output values at maximal rates of exercise. The higher blood volumes in the high $\dot{V}O_2$max group were possibly genetically determined.[30]

> Heredity is a major determinant of aerobic capacity, accounting for as much as half of the variation in $\dot{V}O_2$max values.

Thus, both genetic and environmental factors influence $\dot{V}O_2$max values. The genetic factors probably establish the boundaries for the athlete, but endurance training can push $\dot{V}O_2$max to the upper limit of these boundaries. Dr. Per-Olof Åstrand, one of the most highly recognized exercise physiologists during the second half of the 20th century, stated on numerous occasions that the best way to become a champion Olympic athlete is to be selective when choosing your parents!

Age

Age also can influence $\dot{V}O_2$max. However, $\dot{V}O_2$max values that have been reported in the research literature could lead to an improper interpretation of true age differences if compared across ages. Figure 9.13 illustrates

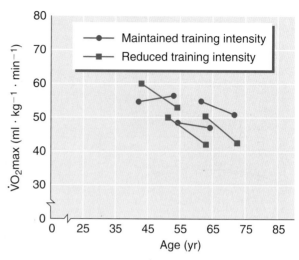

▲ **Figure 9.13**   Changes in $\dot{V}O_2$max with age in a group of athletes who maintained training intensity and in a group who reduced training intensity. Individual data points represent longitudinal data from men followed for 10 years.

Data from M.L. Pollock et al., 1987, "Effect of age and training on aerobic capacity and body composition of master athletes," *Journal of Applied Physiology* 62: 725-731.[37]

the $\dot{V}O_2$max values of two groups of older endurance athletes: those who maintained their training intensity and those who reduced it.[37] For those men who continued to train at the same intensity, the rate of decline in $\dot{V}O_2$max was attenuated. This indicates that age-related decreases might result partly from an age-related decrease in activity levels. This decrease is not an absolute trend: Endurance training of untrained elderly subjects substantially increases $\dot{V}O_2$max.[26] This is illustrated in figure 9.14 and is discussed in greater detail in chapter 17.

Sex

Healthy untrained girls and women have much lower $\dot{V}O_2$max values (20-25% lower) than healthy untrained boys and men. However, highly conditioned female endurance athletes have values much closer to those of highly trained male endurance athletes (about 10% lower). This is discussed in greater detail in chapter 18. Representative ranges of $\dot{V}O_2$max values for athletes and nonathletes are presented in table 9.4 by age, sex, and sport.

Table 9.4

Maximal Oxygen Uptake Values (ml · kg^{-1} ·min^{-1}) for Nonathletes and Athletes

Group or sport	Age	Males	Females
Nonathletes	10-19	47-56	38-46
	20-29	43-52	33-42
	30-39	39-48	30-38
	40-49	36-44	26-35
	50-59	34-41	24-33
	60-69	31-38	22-30
	70-79	28-35	20-27
Baseball/softball	18-32	48-56	52-57
Basketball	18-30	40-60	43-60
Bicycling	18-26	62-74	47-57
Canoeing	22-28	55-67	48-52
Football	20-36	42-60	—
Gymnastics	18-22	52-58	36-50
Ice hockey	10-30	50-63	—
Jockey	20-40	50-60	—
Orienteering	20-60	47-53	46-60
Racquetball	20-35	55-62	50-60
Rowing	20-35	60-72	58-65
Skiing, alpine	18-30	57-68	50-55
Skiing, nordic	20-28	65-94	60-75
Ski jumping	18-24	58-63	—
Soccer	22-28	54-64	50-60
Speed skating	18-24	56-73	44-55
Swimming	10-25	50-70	40-60
Track and field, discus	22-30	42-55	—
Track and field, running	18-39	60-85	50-75
	40-75	40-60	35-60
Track and field, shot put	22-30	40-46	—
Volleyball	18-22	—	40-56
Weightlifting	20-30	38-52	—
Wrestling	20-30	52-65	—

Responders and Nonresponders

For years, researchers have found wide variations in improvement of V̇O₂max with aerobic training. This variation is illustrated in figure 9.14, showing the improvements in V̇O₂max of a group of older men and women who endurance trained for 9 to 12 months.[26] Improvement in V̇O₂max ranged from 0% to 43%, even though all the subjects completed exactly the same training program.

In the past, scientists have assumed that these variations result from differing degrees of compliance with the training program. Good compliers should have the highest percentage of improvement, and poor compliers should show little or no improvement. The idea of comparing compliers with noncompliers has now been replaced with the concept of comparing responders with nonresponders. Given the same training stimulus, implying full compliance with the program, substantial variations

occur in the percentage improvements in V̇O₂max values of different people.

Bouchard[2] has clearly established that the response to a training program is also genetically determined. This is illustrated in figure 9.15. Ten pairs of identical twins completed a 20-week endurance training program; the improvements in V̇O₂max, expressed as percentages, are plotted for each twin pair: Twin A on the x axis and Twin B on the y axis. Notice the similarity in response for each twin pair. Yet across twin pairs, improvement varied from 0% to 40%. These results, and those from other studies, indicate that there will be **responders** (large improvement) and **nonresponders** (little or no improvement) among groups of people who experience identical training programs.

Results from the HERITAGE Family Study also support a strong genetic component affecting the magnitude of increase in V̇O₂max with endurance training. Families, including the natural mother and father and three or more of their children, trained 3 days/week for 20 weeks, initially exercising at heart rate equal to 55% of their V̇O₂max for 35 min/day

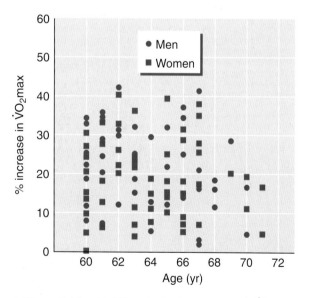

▲ **Figure 9.14**   Variations in the improvement in V̇O₂max with 12 months of training in 60- to 71-year old men and women. The graph shows that individuals respond differently to training even when following the same program. Age partly influences these variations.

Data from W.M. Kohrt et al., 1991, "Effects of gender, age and fitness level on response of V̇O₂max to training in 60-71 year olds," *Journal of Applied Physiology* 71: 2004-2011.

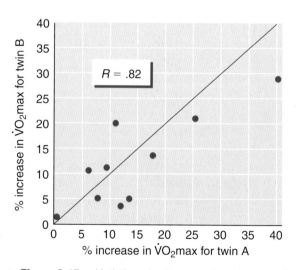

▲ **Figure 9.15**   Variations in the percentage increase in V̇O₂max for identical twins undergoing the same 20-week training program.

From D. Prud'homme et al., 1984, "Sensitivity of maximal aerobic power to training is genotype-dependent," *Medicine and Science in Sports and Exercise* 16(5): 489-493. Copyright 1984 by American College of Sports Medicine. Adapted by permission.

and progressing to a heart rate equal to 75% of their $\dot{V}O_2$max for 50 min/day by the end of the 14th week, which they maintained for the last 6 weeks.[3] The average increase in $\dot{V}O_2$max was about 17% but varied from 0% to more than 50%. Figure 6.2 (page 187) illustrates the improvement in $\dot{V}O_2$max for each subject in each family. Maximal heritability was estimated at 47%, and this is obvious from the figure where those who are high responders are clustered in the same families and those who are low responders are clustered in the same families.

It is now clear that this is a genetic phenomenon, not a result of compliance or noncompliance. This important point must be considered when one is conducting training studies and designing training programs. You must allow for individual differences.

> When examining training effects, you must always remember that individual differences cause variation in subjects' responses to the training program. Even with identical programs, differences will be apparent. Genetics accounts for much of this variation in response.

Specificity of Training

Physiological adaptations in response to physical training are highly specific to the nature of the training activity. Furthermore, the more specific the training program is to a given sport or activity, the greater the improvement in performance in that sport or activity. The concept of **specificity of training** is very important for cardiorespiratory adaptations. This concept, as mentioned in previous chapters, is also important when testing athletes.

To accurately measure endurance improvements, athletes should be tested while they are engaged in an activity similar to the sport or activity in which they usually participate. Consider one study of highly trained rowers, cyclists, and cross-country skiers. Their $\dot{V}O_2$max values were tested while they performed two types of work: uphill running on a treadmill and maximal performance of their specific sport activity.[44] The important finding,

shown in figure 9.16, was that the $\dot{V}O_2$max values attained by all the athletes during their sport-specific activity were as high as or higher than the values obtained on the treadmill. For many of these athletes, $\dot{V}O_2$max values were substantially higher during their sport-specific activity.

In nonathletes, uphill running previously had been shown to consistently produce the highest $\dot{V}O_2$max values. The widespread assumption was that this result would hold true for athletes, but that assumption is not correct.

The concept of training specificity is further illustrated in a study by Magel and his associates.[29] They studied $\dot{V}O_2$max improvements with swim training (1 h/day, 3 days/week, for 10 weeks). Subjects performed maximal treadmill running and tethered swimming tests both before and after training. The swimming $\dot{V}O_2$max increased by 11.2% following the

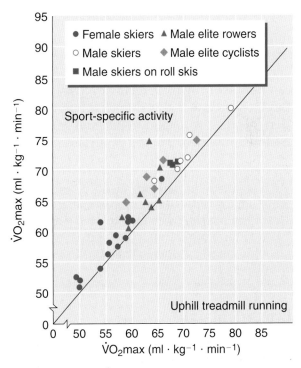

▲ **Figure 9.16** $\dot{V}O_2$max values during uphill treadmill running versus sport-specific activities in selected groups of athletes.

Adapted, by permission, from S.B. Strømme, F. Ingjer, and H.D. Meen, 1977, "Assessment of maximal aerobic power in specifically trained athletes," *Journal of Applied Physiology* 42: 833-837.

10-week swim training period. However, the running $\dot{V}O_2$max increased by only 1.5%, not a statistically significant change from the pretraining value. If the treadmill alone had been used for testing, the researchers would have concluded that swim training had no influence on cardiorespiratory endurance capacity!

One of the most elegant designs to study the concept of specificity of training involves one-legged exercise training, in which the untrained opposite leg is used as the control. In one study, subjects were placed in three groups: One group sprint trained one leg and endurance trained the other leg; one group sprint trained one leg and let the other leg remain untrained; and the last group endurance trained one leg and let the other leg remain untrained.[39] Improvement in $\dot{V}O_2$max and lowered heart rate and blood lactate response at submaximal work rates were found only when exercise was performed with the endurance-trained leg.

Much of the training response occurs in the specific muscles that have been trained, possibly even in individual motor units in a specific muscle. Studies conducted in this area indicate that this observation applies to both metabolic and cardiorespiratory responses to training. We can now see why road-racing cyclists, such as those referred to in this chapter's opening section, need to incorporate hill and mountain training into their training programs.

> Close attention must be given to selecting the appropriate training program. It must be carefully matched with the athlete's individual needs to maximize the physiological adaptations to training, thereby optimizing the athlete's performance.

Cross-Training

Cross-training refers to training for more than one sport at the same time or training for several different fitness components (such as endurance, strength, and flexibility) at one time. The athlete who trains by swimming, running, and cycling in preparation for competing in a triathlon is an example of the former, and the athlete involved in heavy resistance training and high-intensity cardiorespiratory training at the same time is an example of the latter.

Very little research data are available concerning multisport training. In any cross-training program of this nature, it is important to determine how best to partition the available training time to optimize performance in each sport. Although certain aspects of training are the same for all endurance activities, most training is highly specific to a particular sport.

For the athlete training for cardiorespiratory endurance and strength at the same time, the studies conducted to date indicate that gains in strength, power, and endurance can result. However, the gains in muscular strength and power are less when strength training is combined with endurance training than when strength training alone is done.[11, 28] The opposite does not appear to be true: Improvement of aerobic capacity with endurance training does not appear to be attenuated by including a resistance training program. In fact, short-term endurance can be increased with resistance training.[22] Although most studies support the conclusion that concurrent strength and endurance training limits gains in strength and power, one well-controlled study did not find this. McCarthy and colleagues[31] reported similar gains in strength, muscle hypertrophy, and neural activation in a group of previously untrained subjects who underwent concurrent high-intensity strength training and cycle endurance training when compared with a group who performed only high-intensity strength training.

Cardiorespiratory Endurance and Performance

Many people regard cardiorespiratory endurance as the most important component of physical fitness. It is an athlete's major defense against fatigue. Low endurance capacity leads to fatigue, even in the more sedentary sports or activities. For any athlete, regardless of the sport or activity, fatigue represents a major deterrent to optimal performance. Even minor fatigue can hinder the athlete's total performance:

- Muscular strength is decreased.
- Reaction and movement times are prolonged.
- Agility and neuromuscular coordination are reduced.
- Whole-body movement speed is slowed.
- Concentration and alertness are reduced.

The decline in concentration and alertness associated with fatigue is particularly important. The athlete can become careless and more prone to serious injury, especially in contact sports. Even though these decrements in performance might be small, they can be just enough to cause an athlete to miss the critical free throw in basketball, the strike zone in baseball, or the 20-ft putt in golf.

All athletes can benefit from maximizing their endurance. Even golfers, whose sport is relatively sedentary, can improve. Improved endurance can allow golfers to complete a round of golf with less fatigue and to better withstand long periods of walking and standing.

For the sedentary, middle-aged adult, numerous health factors indicate that cardiovascular endurance should be the primary emphasis of training. Training for health and fitness is discussed at length in part VII of this book.

The extent of endurance training needed varies considerably from one athlete to the next. It depends on the athlete's current endurance capacity and the endurance demands of the chosen activity. The marathon runner uses endurance training almost exclusively, with limited attention to strength, flexibility, and speed. The baseball player, however, places very limited demands on endurance capacity, so endurance conditioning is not as highly emphasized. Nevertheless, baseball players could gain substantially from endurance running, even if only at a moderate pace for 5 km (3.1 mi) per day for 3 days per week. As a benefit

of training, baseball players would have little or no leg trouble (a frequent complaint), and they would be able to complete a doubleheader with little or no fatigue.

Adequate cardiovascular conditioning must be the foundation of any athlete's general conditioning program. Many athletes in nonendurance activities never incorporate even moderate endurance training into their training programs. Those who have done so are well aware of their improved physical condition and its impact on their athletic performance.

▶ Although $\dot{V}O_2$max has an upper limit, endurance performance can continue to improve for years with continued training.

▶ An individual's genetic makeup predetermines a range for his or her $\dot{V}O_2$max and accounts for 25% to 50% of the variance in $\dot{V}O_2$max values. Heredity also largely explains individual variations in response to identical training programs.

▶ Age-related decreases in aerobic capacity result partly from decreased activity.

▶ Highly conditioned female endurance athletes have $\dot{V}O_2$max values only about 10% lower than those of highly conditioned male endurance athletes.

▶ For athletes to maximize cardiorespiratory gains from training, the training should be specific to the type of activity that an athlete usually performs.

▶ Resistance training in combination with endurance training does not appear to restrict improvement in aerobic capacity and may increase short-term endurance, but it can limit improvement in strength and power when compared with gains from resistance training alone.

▶ All athletes can benefit from maximizing their endurance.

Modeling Endurance Performance

Physiologists commonly establish models to help explain how various physiological factors or variables work together to affect a specific outcome or component of performance. Dr. Edward Coyle,[7] a noted exercise physiologist at The University of Texas at Austin, has provided an excellent example of this in his attempt to model the factors that affect aerobic endurance performance, specifically performance velocity. This model (see figure 9.17) describes how various physiological factors interact to determine the average speed at which an athlete can run, cycle, or swim an endurance event.

Most of the variables included in this model have been discussed in this chapter and in previous chapters. Resistance to movement refers to the drag encountered by the athlete and equipment (e.g., bicycle) while moving through air or water. Performance power is the average power production that the individual is capable of producing over the period of time that it takes to complete an event and is usually expressed in watts (W). Performance $\dot{V}O_2$ is the highest rate of whole-body oxygen consumption that an individual can maintain for the duration of the event. The functional abilities and morphological components have already been discussed in this and previous chapters. It is beyond the scope of this book to go into further detail on this specific model. Our intent in introducing this model is to help you understand how scientists attempt to integrate scientific information into a framework that allows them to understand human performance. Once a model has been developed, it is tested and refined. Then the information is translated for the coach and athlete to be implemented in the design of training programs and in the planning of race strategies, among other practical applications.

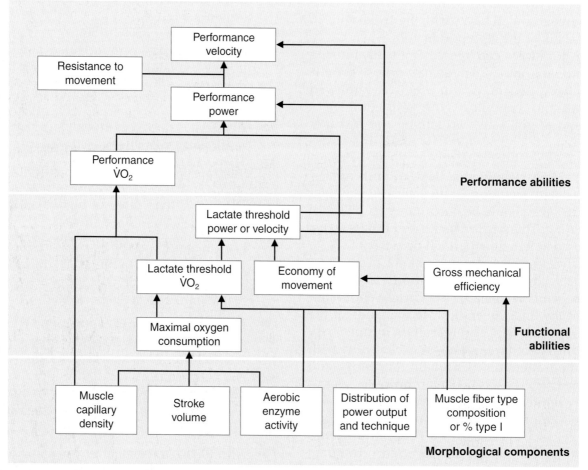

▲ **Figure 9.17** Modeling endurance performance.

Adapted, by permission, from E.F. Coyle, 1995, "Integration of the physiological factors determining endurance performance ability," *Exercise and Sport Sciences Reviews* 23: 25-63.

In Closing . . .

In this chapter, we examined how the cardiovascular and respiratory systems adapt to chronic exposure to the stress of training. We concentrated on how these adaptations can improve cardiorespiratory endurance. This chapter concludes our review of how body systems respond to both acute and chronic exercise. Now that we have completed our examination of how the body responds to internal changes, we can turn our attention to the external world. In the next part, we focus on the body's adaptations to varying environmental conditions, beginning in the next chapter by considering how external temperature can affect performance.

▶ Key Terms

athlete's heart

capillary-to-fiber ratio

cardiac hypertrophy

cardiorespiratory endurance

cross-training

endurance

Fick equation

heart rate recovery period

muscular endurance

nonresponders

oxygen transport system

respiratory exchange ratio (RER)

responders

specificity of training

submaximal endurance capacity

$\dot{V}O_2$max

▶ Study Questions

1. Differentiate between muscular endurance and cardiovascular endurance.

2. What is maximal oxygen uptake ($\dot{V}O_2$max)? How is it defined physiologically, and what determines its limits?

3. Of what importance is $\dot{V}O_2$max to endurance performance?

4. Describe the changes in the oxygen transport system that occur with endurance training.

5. What is possibly the most important adaptation that the body makes in response to endurance training, which allows for an increase in both $\dot{V}O_2$max and performance?

6. What metabolic adaptations occur in response to endurance training?

7. Explain the two theories that have been proposed to account for improvements in $\dot{V}O_2$max with endurance training. Which of these has the greatest validity today? Why?

8. How important is genetic potential in a developing young athlete?

9. Why would cardiovascular endurance conditioning be important for athletes in nonendurance sports?

▶ References

1. Armstrong, R.B., & Laughlin, M.H. (1984). Exercise blood flow patterns within and among rat muscles after training. *American Journal of Physiology,* **246,** H59-H68.

2. Bouchard, C. (1990). Discussion: Heredity, fitness, and health. In C. Bouchard, R.J. Shephard, T. Stephens, J.R. Sutton, & B.D. McPherson (Eds.), *Exercise, fitness, and health* (pp. 147-153). Champaign, IL: Human Kinetics.

3. Bouchard, C., An, P., Rice, T., Skinner, J.S., Wilmore, J.H., Gagnon, J., Pérusse, L., Leon, A.S., & Rao, D.C. (1999). Familial aggregation of $\dot{V}O_2$max response to exercise training: Results from the HERITAGE Family Study. *Journal of Applied Physiology,* **87,** 1003-1008.

4. Bouchard, C., Dionne, F.T., Simoneau, J.-A., & Boulay, M.R. (1992). Genetics of aerobic and anaerobic performances. *Exercise and Sport Sciences Reviews,* **20,** 27-58.

5. Bouchard, C., Lesage, R., Lortie, G., Simoneau, J.A., Hamel, P., Boulay, M.R., Perusse, L., Theriault, G., & Leblanc, C. (1986). Aerobic performance in brothers, dizygotic and monozygotic twins. *Medicine and Science in Sports and Exercise,* **18,** 639-646.

6. Clausen, J.P. (1977). Effect of physical training on cardiovascular adjustments to exercise in man. *Physiological Reviews,* **57,** 779-816.

7. Coyle, E.F. (1995). Integration of the physiological factors determining endurance performance ability. *Exercise and Sport Sciences Reviews, 23,* 25-63.

8. Coyle, E.F., Hemmert, M.K., & Coggan, A.R. (1986). Effects of detraining on cardiovascular responses to exercise: Role of blood volume. *Journal of Applied Physiology, 60,* 95-99.

9. Dempsey, J.A. (1986). Is the lung built for exercise? *Medicine and Science in Sports and Exercise, 18,* 143-155.

10. Dempsey, J.A., & Wagner, P.D. (1999). Exercise-induced arterial hypoxemia. *Journal of Applied Physiology, 87,* 1997-2006.

11. Dudley, G.A., & Fleck, S.J. (1987). Strength and endurance training: Are they mutually exclusive? *Sports Medicine, 4,* 79-85.

12. Ehsani, A.A., Ogawa, T., Miller, T.R., Spina, R.J., & Jilka, S.M. (1991). Exercise training improves left ventricular systolic function in older men. *Circulation, 83,* 96-103.

13. Ekblom, B., Goldbarg, A.M., & Gullbring, B. (1972). Response to exercise after blood loss and reinfusion. *Journal of Applied Physiology, 33,* 175-180.

14. Fagard, R.H. (1996). Athlete's heart: A meta-analysis of the echocardiographic experience. *International Journal of Sports Medicine, 17,* S140-S144.

15. Fagard, R.H. (1997). The athlete's heart and cardiovascular disease: Impact of different sports and training on cardiac structure and function. *Cardiology Clinics, 15,* 397-412.

16. Fagard, R.H., & Tipton, C.M. (1994). Physical activity, fitness, and hypertension. In C. Bouchard, R.J. Shephard, & T. Stephens (Eds.), *Physical activity, fitness, and health* (pp. 633-655). Champaign, IL: Human Kinetics.

17. Fahey, T.D., Del Valle-Zuris, A., Oehlsen, G., Trieb, M., & Seymour, J. (1979). Pubertal stage differences in hormonal and hematological responses to maximal exercise in males. *Journal of Applied Physiology, 46,* 823-827.

18. Gaffney, F.A., Nixon, J.V., Karlsson, E.S., Campbell, W., Dowdey, A.B.C., & Blomqvist, C.G. (1985). Cardiovascular deconditioning produced by 20 hours of bedrest with head-down tilts (–5 degrees) in middle-aged healthy men. *American Journal of Cardiology, 56,* 634-638.

19. Green, H.J., Sutton, J.R., Coates, G., Ali, M., & Jones, S. (1991). Response of red cell and plasma volume to prolonged training in humans. *Journal of Applied Physiology, 70,* 1810-1815.

20. Hagberg, J.M., Ehsani, A.A., Goldring, D., Hernandez, A., Sinacore, D.R., & Holloszy, J.O. (1984). Effect of weight training on blood pressure and hemodynamics in hypertensive adolescents. *Journal of Pediatrics, 104,* 147-151.

21. Hermansen, L., & Wachtlova, M. (1971). Capillary density of skeletal muscle in well-trained and untrained men. *Journal of Applied Physiology, 30,* 860-863.

22. Hickson, R.C., Dvorak, B.A., Gorostiaga, E.M., Kurowski, T.T., & Foster, C. (1988). Potential for strength and endurance training to amplify endurance performance. *Journal of Applied Physiology, 65,* 2285-2290.

23. Holloszy, J.O., & Coyle, E.F. (1984). Adaptations of skeletal muscle to endurance exercise and their metabolic consequences. *Journal of Applied Physiology, 56,* 831-838.

24. Hopper, M.K., Coggan, A.R., & Coyle, E.F. (1988). Exercise stroke volume relative to plasma-volume expansion. *Journal of Applied Physiology, 64,* 404-408.

25. Klissouras, V. (1971). Adaptability of genetic variation. *Journal of Applied Physiology, 31,* 338-344.

26. Kohrt, W.M., Malley, M.T., Coggan, A.R., Spina, R.J., Ogawa, T., Ehsani, A.A., Bourey, R.E., Martin, W.H., III, & Holloszy, J.O. (1991). Effects of gender, age and fitness level on response of $\dot{V}O_2$max to training in 60-71 yr olds. *Journal of Applied Physiology, 71,* 2004-2011.

27. Kraemer, W.J., Deschenes, M.R., & Fleck, S.J. (1988). Physiological adaptations to resistance exercise: Implications for athletic conditioning. *Sports Medicine, 6,* 246-256.

28. Leveritt, M., Abernethy, P.J., Barry, B.K., & Logan, P.A. (1999). Concurrent strength and endurance training: A review. *Sports Medicine, 28,* 413-427.

29. Magel, J.R., Foglia, G.F., McArdle, W.D., Gutin, B., Pechar, G.S., & Katch, F.I. (1975). Specificity of swim training on maximum oxygen uptake. *Journal of Applied Physiology, 38,* 151-155.

30. Martino, M., Gledhill, N., & Jamnik, V. (2002). High $\dot{V}O_2$max with no history of training is primarily due to high blood volume. *Medicine and Science in Sports and Exercise, 34,* 966-971.

31. McCarthy, J.P., Pozniak, M.A., & Agre, J.C. (2002). Neuromuscular adaptations to concurrent strength and endurance training. *Medicine and Science in Sports and Exercise, 34,* 511-519.

32. Milliken, M.C., Stray-Gundersen, J., Peshock, R.M., Katz, J., & Mitchell, J.H. (1988). Left ventricular mass as determined by magnetic resonance imaging in male endurance athletes. *American Journal of Cardiology, 62,* 301-305.

33. Nagashima, K., Mack, G.W., Haskell, A., Nishiyasu, T., & Nadel, E.R. (1999). Mechanism for the posture-specific plasma volume increase after a single

intense exercise protocol. *Journal of Applied Physiology,* **86,** 867-873.

34.   Pirnay, F., Dujardin, J., Deroanne, R., & Petit, J.M. (1971). Muscular exercise during intoxication by carbon monoxide. *Journal of Applied Physiology,* **31,** 573-575.

35.   Poehlman, E.T., Melby, C.L., & Goran, M.I. (1991). The impact of exercise and diet restriction on daily energy expenditure. *Sports Medicine,* **11,** 78-101.

36.   Pollock, M.L. (1973). Quantification of endurance training programs. *Exercise and Sport Sciences Reviews,* **1,** 155-188.

37.   Pollock, M.L., Foster, C., Knapp, D., Rod, J.L., & Schmidt, D.H. (1987). Effect of age and training on aerobic capacity and body composition of master athletes. *Journal of Applied Physiology,* **62,** 725-731.

38.   Prud'homme, D., Bouchard, C., LeBlanc, C., Landrey, F., & Fontaine, E. (1984). Sensitivity of maximal aerobic power to training is genotype-dependent. *Medicine and Science in Sports and Exercise,* **16,** 489-493.

39.   Saltin, B., Nazar, K., Costill, D.L., Stein, E., Jansson, E., Essen, B., & Gollnick, P.D. (1976). The nature of the training response: Peripheral and central adaptations to one-legged exercise. *Acta Physiologica Scandinavica,* **96,** 289-305.

40.   Saltin, B., & Rowell, L.B. (1980). Functional adaptations to physical activity and inactivity. *Federation Proceedings,* **39,** 1506-1513.

41.   Sawka, M.N., Convertino, V.A., Eichner, E.R., Schnieder, S.M., & Young, A.J. (2000). Blood volume: Importance and adaptations to exercise training, environmental stresses, and trauma/sickness. *Medicine and Science in Sports and Exercise,* **32,** 332-348.

42.   Smith, J.A. (1995). Exercise, training and red blood cell turnover. *Sports Medicine,* **19,** 9-31.

43.   Stone, M.H., Fleck, S.J., Triplett, N.T., & Kraemer, W.J. (1991). Health- and performance-related potential of resistance training. *Sports Medicine,* **11,** 210-231.

44.   Strømme, S.B., Ingjer, F., & Meen, H.D. (1977). Assessment of maximal aerobic power in specifically trained athletes. *Journal of Applied Physiology,* **42,** 833-837.

45.   Turkevich, D., Micco, A., & Reeves, J.T. (1988). Noninvasive measurement of the decrease in left ventricular filling time during maximal exercise in normal subjects. *American Journal of Cardiology,* **62,** 650-652.

46.   Volianitis, S., McConnell, A.K., Koutedakis, Y., McNaughton, L., Backx, K., & Jones, D.A. (2001). Inspiratory muscle training improves rowing performance. *Medicine and Science in Sports and Exercise,* **33,** 803-809.

47.   Wilmore, J.H., Stanforth, P.R., Gagnon, J., Rice, T., Mandel, S., Leon, A.S., Rao, D.C., Skinner, J.S., & Bouchard, C. (2001). Cardiac output and stroke volume changes with endurance training: The HERITAGE Family Study. *Medicine and Science in Sports and Exercise,* **33,** 99-106.

48.   Wilmore, J.H., Stanforth, P.R., Gagnon, J., Rice, T., Mandel, S., Leon, A.S., Rao, D.C., Skinner, J.S., & Bouchard, C. (2001). Heart rate and blood pressure changes with endurance training: The HERITAGE Family Study. *Medicine and Science in Sports and Exercise,* **33,** 107-116.

49.   Wilmore, J.H., Stanforth, P.R., Hudspeth, L.A., Gagnon, J., Daw, E.W., Leon, A.S., Rao, D.C., Skinner, J.S., & Bouchard, C. (1998). Alterations in resting metabolic rate as a consequence of 20 wk of endurance training: The HERITAGE Family Study. *American Journal of Clinical Nutrition,* **68,** 66-71.

50.   Yang, R.C., Mack, G.W., Wolfe, R.R., & Nadel, E.R. (1998). Albumin synthesis after intense intermittent exercise in human subjects. *Journal of Applied Physiology,* **84,** 584-592.

▶ Selected Readings

Convertino, V.A. (1991). Blood volume: Its adaptation to endurance training. *Medicine and Science in Sports and Exercise,* **23,** 1338-1348.

Dowell, R.T. (1983). Cardiac adaptations to exercise. *Exercise and Sport Sciences Reviews,* **11,** 99-117.

Hudlicka, O. (1977). Effect of training on macro- and microcirculatory changes in exercise. *Exercise and Sport Sciences Reviews,* **5,** 181-230.

Joyner, M.J., & Shastry, S. (2000). Vascular endothelial growth factor and capillary density in exercise training. *Exercise and Sport Sciences Reviews,* **28,** 97-98.

Krip, B., Gledhill, N., Jamnik, V., & Warburton, D. (1997). Effect of alterations in blood volume on cardiac function during maximal exercise. *Medicine and Science in Sports and Exercise,* **29,** 1469-1476.

Lash, J.M. (1998). Training-induced alterations in contractile function and excitation-contraction coupling in vascular smooth muscle. *Medicine and Science in Sports and Exercise,* **30,** 60-66.

Longhurst, J.C., & Stebbins, C.L. (1997). The athlete's heart and cardiovascular disease: The power athlete. *Cardiology Clinics,* **15,** 413-429.

MacRae, H.S.-H., Dennis, S.C., Bosch, A.N., & Noakes, T.D. (1992). Effects of training on lactate production and removal during progressive exercise in humans. *Journal of Applied Physiology,* **72,** 1649-1656.

Mier, C.M., Turner, M.J., Ehsani, A.A., & Spina, R.J. (1997). Cardiovascular adaptations to 10 days of cycle exercise. *Journal of Applied Physiology,* **83,** 1900-1906.

Moore, R.L., & Korzick, D.H. (1995). Cellular adaptations of the myocardium to chronic exercise. *Progress in Cardiovascular Diseases*, **37**, 371-396.

Perrault, H., & Turcotte, R.A. (1994). Exercise-induced cardiac hypertrophy: Fact or fallacy? *Sports Medicine*, **17**, 288-308.

Rowell, L.B. (1986). *Human circulation regulation during physical stress.* New York: Oxford University Press.

Saltin, B. (1990). Cardiovascular and pulmonary adaptation to physical activity. In C. Bouchard, R.J. Shephard, T. Stephens, J.R. Sutton, & B.D. McPherson (Eds.), *Exercise, fitness, and health* (pp. 187-203). Champaign, IL: Human Kinetics.

Sexton, W.L., Korthuis, R.J., & Laughlin, M.H. (1988). High-intensity exercise training increases vascular transport capacity of rat hindquarters. *American Journal of Physiology*, **254**, H274-H278.

Tanaka, H. (1994). Effects of cross-training: Transfer of training effects on VO$_2$max between cycling, running and swimming. *Sports Medicine, 18*, 330-339.

PART IV

Environmental Influences on Performance

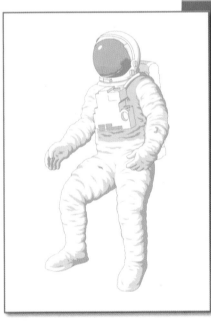

In previous sections, we discussed how the various body systems coordinate their activities to allow us to perform physical activity. We also saw how these systems adapt when exposed to the stress of various types of training. In part IV, we turn our attention to how the body responds and adapts when challenged to exercise under unusual environmental conditions. In chapter 10, "Exercise in Hot and Cold Environments: Thermoregulation," we examine the mechanism by which the body can regulate its internal temperature both at rest and during exercise. Then we consider how the body responds and adapts to exercise in the heat and cold, along with health risks that are associated with physical activity in each environment. In chapter 11, "Exercise in Hypobaric, Hyperbaric, and Microgravity Environments," we discuss the unique challenges that the body faces when performing physical activity under conditions of reduced atmospheric pressure (altitude), increased atmospheric pressure (diving), and reduced gravity (space travel).

CHAPTER 10

EXERCISE IN HOT AND COLD ENVIRONMENTS: THERMOREGULATION

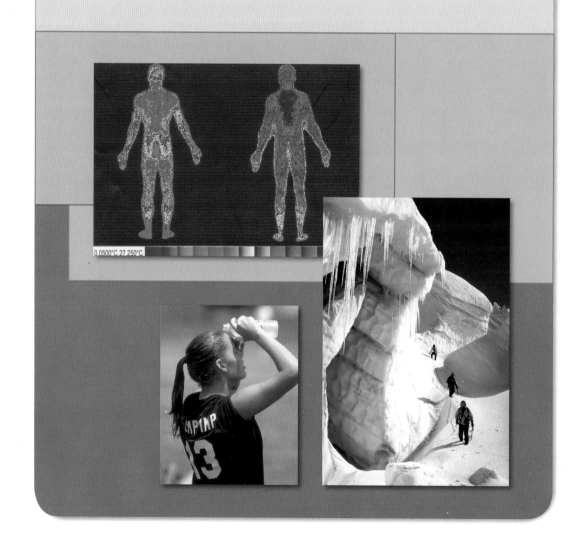

Our body's physiology enables us to tolerate the demands of physical stress. As we have seen in the preceding chapters, our bodies attempt to maintain a constant internal environment (i.e., homeostasis) despite the chaos created by muscular activity. We have seen how different parts of the body communicate and function together, and we have examined the physiological responses to acute exercise and the body's adaptations to training. Yet our discussion thus far has focused only on our internal environment, without considering the demands imposed by our surroundings.

Beginning with this chapter, we change our focus. Now that we know how the body meets the demands of exercise, we are ready to see how the body meets these demands when they are coupled with demands imposed by the external environment. In this chapter, we examine the impact of extreme temperatures on performance.

Korey Stringer, a 335-lb right tackle for the Minnesota Vikings professional football team, was one of the National Football League's top offensive players. Stringer collapsed on the practice field on July 31, 2001, during an intense practice at the Vikings' training facility in Mankato, Minnesota. He had been training on a cloudless day in the heat, with the heat index reaching 110° F. On the previous day, the first day of training camp, he had failed to complete the full team workout. Having lost consciousness, he was taken to an air-conditioned trailer, and eventually was taken by ambulance to a local hospital, where they recorded a core body temperature of 108.8° F. He died approximately 13 h later—a victim of heat stroke. Athletes are highly susceptible to heat stroke, particularly football players who have to start training for their sport during the hottest months of the year. Furthermore, the padding and uniforms worn by football players limit their ability to lose heat, and many do not drink sufficient fluids.
Source: Sports Illustrated; July 29, 2002; 97 (4): 56-60.

The stresses of physical exertion often are complicated by environmental thermal conditions. Performing in extreme heat or cold places a heavy burden on the mechanisms that regulate body temperature. Although these mechanisms are amazingly effective in regulating body heat under normal conditions, mechanisms of **thermoregulation** can be inadequate when we are subjected to extreme heat or cold. Fortunately, our bodies are able to adapt to such environmental stresses with continued exposure over time.

In the following discussion, we focus on the physiological responses to acute and chronic exercise in both hot and cold environments. Specific health risks are associated with exercise in both temperature extremes, so we also discuss the prevention of temperature-related illness and injuries during exercise. The online study guide (www.humankinetics.com/physiologyofsportandexercise/osg) covers some ways of assessing and calculating mean body temperature.

Mechanisms of Body Temperature Regulation

Humans are homeothermic, which means that internal body temperature is kept nearly constant throughout life. Although your temperature varies from day to day, and even from hour to hour, these fluctuations are usually no more than about 1.0° C (1.8° F). Only during prolonged heavy exercise, illness, or extreme conditions of heat or cold do body temperatures deviate from the normal range of 36.1 to 37.8° C (97.0-100.0° F).

Body temperature reflects a careful balance between heat production and heat loss. Whenever this balance is disturbed, your body temperature changes. Recall from chapter 4 that a large part of the energy your body generates degrades to heat, the lowest form of energy. All metabolically active tissues produce heat that can be used to maintain the internal temperature of your body. But if your body's heat

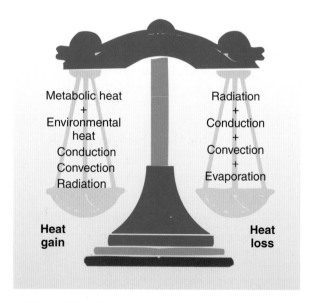

▲ **Figure 10.1** The balance between heat gained from internal metabolic processes and from external environmental factors and heat lost by radiation, conduction, convection, and evaporation.

production exceeds its heat loss, your internal temperature increases. Your ability to maintain a constant internal temperature depends on your ability to balance the heat you gain from metabolism and from the environment with the heat that your body loses. This balance is depicted in figure 10.1. Now let's examine the mechanisms by which heat is transferred between you and your surroundings.

Transfer of Body Heat

For your body to transfer heat to the environment, the heat produced in your body must have access to the outside world. The heat from deep in your body (the core) is moved by the blood to your skin (the shell). Once heat nears your skin, it can be transferred to the environment by any of four mechanisms: conduction, convection, radiation, and evaporation. These are illustrated in figures 10.1 and 10.2.

Conduction and Convection

Heat **conduction** involves the transfer of heat from one material to another through direct molecular contact. As an example, heat generated deep in your body can be conducted through adjacent tissue until it reaches your body's surface. It can then be conducted to your clothing or to the air that is in direct contact with your skin. Conversely, if a hot object is pressed against your skin, heat from the object will be conducted to your skin, warming it.

Convection, on the other hand, involves moving heat from one place to another by the motion

of a gas or a liquid across the heated surface. Although we're not always aware of it, the air around us is in constant motion. As it circulates around us, passing over the skin, it sweeps away the air molecules that have been warmed by their contact with the skin. The greater the movement of the air (or liquid, such as water), the greater the rate of heat removal by convection. Thus, in an environment that is cooler than your skin temperature, conduction permits the transfer of heat from your skin to materials that contact it (e.g., water or air), whereas convection involves moving the heated material away from the skin. When combined with conduction, convection can also cause the body to gain heat in a very hot environment, when the surroundings are hotter than the skin.

Although conduction and convection constantly remove body heat when the air temperature is lower than your skin temperature, their

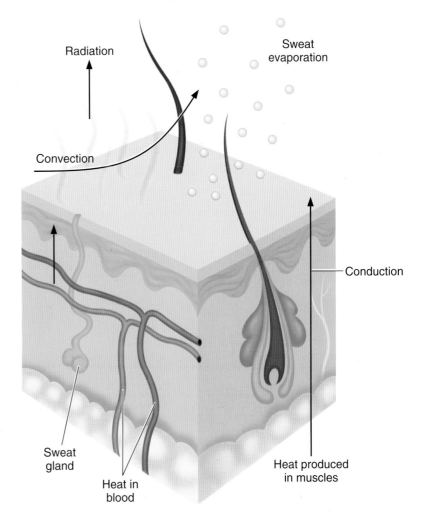

▶ **Figure 10.2** The removal of heat from the skin. Heat is delivered to the body surface via the arterial blood and also by conduction through the subcutaneous tissue. When the temperature of the skin is greater than that of the environment, heat is removed by conduction, convection, radiation, and sweat evaporation; when the environmental temperature exceeds skin temperature heat is removed only by evaporation.

contribution to your body's total heat loss in air is relatively small—only about 10% to 20%. However, if you are submerged in cold water, the amount of heat dissipated from your body to the water is nearly 26 times greater than when you are exposed to a similar air temperature.

When air or water temperature is greater than your skin temperature, your body will gain heat through conduction and convection. We often consider these processes to be mechanisms of heat loss, forgetting that when the environmental temperature exceeds skin temperature, the gradient works in the opposite direction.

Radiation

In a body at rest, radiation is the primary method for discharging the body's excess heat. At normal room temperature (typically 21-25° C, or 69.8-77° F), the nude body loses about 60% of its excess heat by radiation. The heat is given off in the form of infrared rays, which are a type of electromagnetic wave. Figure 10.3 shows two infrared thermograms of an individual.

Your skin constantly radiates heat in all directions to the objects around it, such as clothing, furniture, and walls, but it also can receive radiational heat from surrounding objects that are warmer. If the temperature of the surrounding objects is greater than that of your skin, your body will experience a net heat gain via radiation. A tremendous amount of radiational heat is received from exposure to the sun.

Evaporation

Evaporation is the primary avenue for heat dissipation during exercise. As a fluid evaporates, heat is lost. Evaporation accounts for about 80% of the total heat loss when you are physically active but for only about 20% of body heat loss at rest. Some evaporation occurs without our awareness. This is referred to as insensible water loss and happens wherever body fluid is brought into contact with the external environment, such as in your lungs, at the mucosa (such as that lining your mouth), and at your skin.

Insensible water loss removes about 10% of the total metabolic heat produced by the body. But this mechanism for heat loss is relatively constant, so when your body needs to lose more heat, this mechanism cannot help. Instead, as body temperature increases, sweat production increases. As sweat reaches the skin, it is converted from a liquid to a vapor by heat from the skin. Thus, sweat evaporation becomes increasingly important as body temperature increases. Evaporation of 1 L of sweat results in the loss of 580 kcal (2,428 kJ).

The relative contributions of each of the four heat loss mechanisms are summarized in table 10.1. The data presented are taken both at rest,

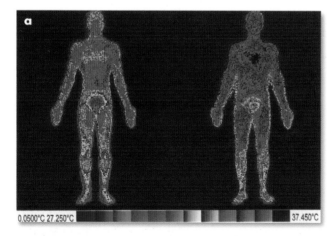
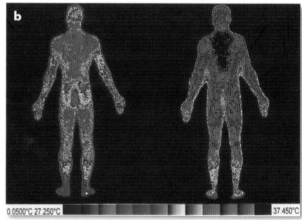

▲ **Figure 10.3** Thermograms of the body showing the variations in radiant (infrared) heat leaving the front (*a*) and rear (*b*) surfaces of the body both before (left) and after (right) running outside at a temperature of 30° C and 75% humidity. The color scale at the bottom of each photo provides the temperature variation for the changes in color. These thermograms were provided by Dr. David D. Pascoe, Department of Health and Human Performance, Auburn University, Alabama.

Courtesy of JohnEric Smith, Joe Molloy, and David D. Pascoe.

when the body produces about 1.5 kcal of heat per minute, and during prolonged exercise at 70% of $\dot{V}O_2$max, when heat production is 10 times higher, or about 15 kcal/min. These values are simple averages, because individual metabolic heat production varies with body size, composition, and temperature. Environmental conditions, such as air velocity, humidity, and sun exposure, will also affect these values. Figure 10.4 shows the complex interaction between the mechanisms of body heat balance (production and loss) and environmental conditions.[7]

Table 10.1

Estimated Caloric Heat Loss at Rest and During Prolonged Exercise

Mechanism of heat loss	Rest		Exercise	
	% total	kcal/min	% total	kcal/min
Conduction and convection	20	0.3	15	2.2
Radiation	60	0.9	5	0.8
Evaporation	20	0.3	80	12.0
Total	100	1.5	100	15.0

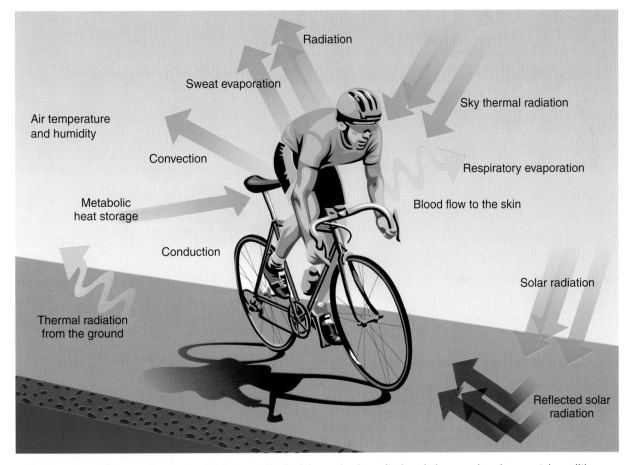

▲ **Figure 10.4**   The complex interaction between the body's mechanisms for heat balance and environmental conditions.
Adapted, by permission, from C.V. Gisolfi and C.B. Wenger, 1984, "Temperature regulation during exercise: Old concepts, new ideas," *Exercise and Sport Sciences Reviews* 12: 339-372.

Humidity and Heat Loss

The water vapor content, or humidity, of the air plays a major role in heat loss, especially by evaporation. When humidity is high, the air already contains many water molecules. This decreases its capacity to accept more water because the concentration gradient is decreased. Thus, high humidity limits sweat evaporation and heat loss. Low humidity, on the other hand, offers an ideal opportunity for sweat evaporation and heat loss. But this too can pose problems. If water evaporates from the skin more rapidly than sweat is produced, the skin can become too dry.

Humidity affects our perception of **thermal stress.** Consider two situations: exposure to dry desert air at 32.2° C (90.0° F) with 10% relative humidity, compared with exposure to air at the same temperature with 90% relative humidity. You sweat profusely in the dry desert, but evaporation occurs so rapidly that you are not aware that you are sweating. But in air that is already 90% saturated with water, little sweat can evaporate. The result is a continuous bath of sweat dripping from your skin. Very little heat is removed, and you feel very uncomfortable.

Even at moderate environmental temperatures, humidity is a primary concern because evaporation is the major method of heat loss during exercise. If the air is saturated with water, almost no evaporation can occur. Your body might not be able to shed all its excess heat when

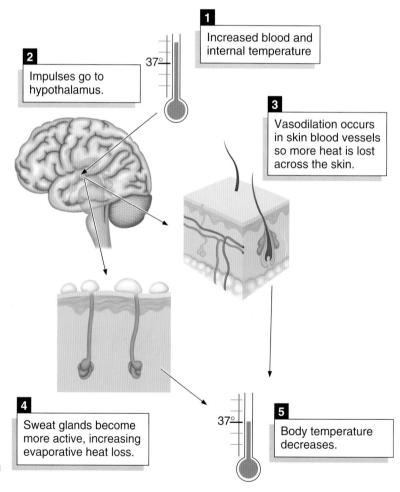

▲ **Figure 10.5a, Hyperthermia**   An overview of the role of the hypothalamus in controlling body temperature. *(a)* When the body becomes overheated (hyperthermia), the hypothalamus is stimulated to activate increased blood flow to the skin and sweat production to eliminate excess body heat.

faced with high temperature, high humidity, and prolonged, intense exercise. Consequently, your body temperature can increase to critical levels, seriously jeopardizing your health.

Fortunately, the mechanisms for heat transfer to the skin and for sweat production are well developed. Generally, except under extreme conditions, the removal of heat from the body depends on the gradient between the skin temperature and the environment.

Control of Heat Exchange

Internal body temperature (measured rectally) when at rest is kept at approximately 37° C (98.6° F). During exercise, because the body is often unable to dissipate heat as rapidly as it is produced, a person can develop an internal temperature exceeding 40° C (104° F), with a temperature above 42° C (107.6° F) in active muscles. The muscles' energy systems become more chemically efficient with a small increase in muscle temperature, but internal body temperatures above 40° C (104° F) can adversely affect the nervous system and reduce further efforts to unload excess heat. How does your body regulate its internal temperature? The hypothalamus plays a central role (see figure 10.5).

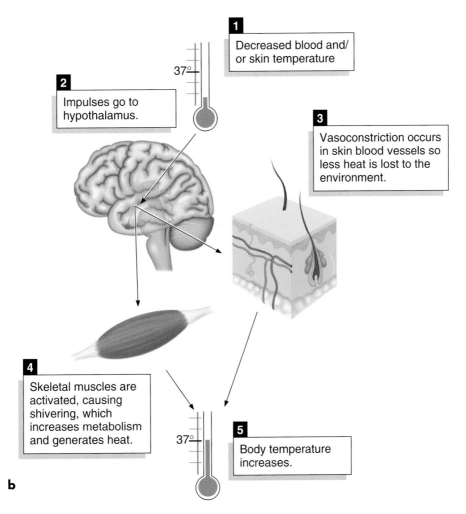

1
Decreased blood and/or skin temperature

2
Impulses go to hypothalamus.

3
Vasoconstriction occurs in skin blood vessels so less heat is lost to the environment.

4
Skeletal muscles are activated, causing shivering, which increases metabolism and generates heat.

5
Body temperature increases.

b

▲ **Figure 10.5*b*, Hypothermia**   An overview of the role of the hypothalamus in controlling body temperature. (*b*) When the blood and body tissues are below the normal set point of 37° C (hypothermia), the hypothalamus relays signals to the skin to prevent body heat loss and activates shivering to produce more heat.

Sweat must evaporate to provide cooling. Sweat that drips off the skin provides little or no cooling.

The Hypothalamus: The Body's Thermostat

The mechanisms that control body temperature are analogous to the thermostat that controls the air temperature in your home, although your body's mechanisms function in a more complex manner and generally with greater precision than your home heating and cooling system. Sensory receptors, called **thermoreceptors,** detect changes in your body temperature and relay this information to your body's thermostat: the hypothalamus. In response, the hypothalamus activates mechanisms that regulate the heating or cooling of your body. Like your home thermostat, the hypothalamus has a predetermined temperature, or set point, that it tries to maintain. This is your normal body temperature. The smallest deviation from this set point signals your **thermoregulatory center,** located within your hypothalamus, to readjust your body temperature.

Changes in your body temperature are sensed by two sets of thermoreceptors: central receptors and peripheral receptors. Central receptors are located in your hypothalamus and monitor the temperature of your blood as it circulates through your brain. These central receptors are sensitive to blood temperature changes as small as 0.01° C (0.018° F). Changes in the temperature of the blood passing through the hypothalamus trigger reflexes that help you conserve or eliminate body heat as needed.

Peripheral receptors, located in your skin, monitor the temperature around you. They provide information to the hypothalamus and cerebral cortex, which allow you to consciously perceive temperature so that you can voluntarily control your exposure to heat or cold. You might decide to go to a more temperate environment or to select appropriate clothing. However, during sweat evaporation, your skin can feel cold while the interior of your body is hyperthermic (overheated). In this case, your skin receptors would incorrectly notify your hypothalamus and your cerebral cortex that you are chilled when, in fact, you might be nearing a critically high temperature.

Effectors That Alter Body Temperature

When your body temperature fluctuates, your normal temperature usually can be restored by the actions of four effectors:

Sweat glands. When either your skin or your blood is heated, your hypothalamus sends impulses to your sweat glands, commanding them to actively secrete sweat that moistens the skin. The hotter you are, the more sweat you produce. The evaporation of this moisture, as discussed earlier, removes heat from your skin's surface.

Smooth muscle around arterioles. When your skin and blood are heated, your hypothalamus sends signals to the smooth muscle in the walls of the arterioles that supply your skin, causing them to dilate. This increases blood flow to your skin. The blood carries heat from the deeper parts of your body to your skin, where the heat dissipates to the environment through conduction, convection, radiation, or evaporation.

Skeletal muscle. Skeletal muscle is called into action when you need to generate more body heat. In a cold environment, the thermoreceptors in your skin relay signals to your hypothalamus. Similarly, whenever your blood temperature decreases, the change is noted by the central receptors in the hypothalamus. In response to this neural input, your hypothalamus activates the brain centers that control muscle tone. These centers stimulate shivering, which is a rapid, involuntary cycle of contraction and relaxation of skeletal muscles. This increased muscle activity generates heat to either maintain or increase your body temperature.

Endocrine glands. The effects of several hormones cause your cells to increase their metabolic rates. Increased metabolism affects heat balance because it increases heat production.

Cooling the body stimulates the release of thyroxine from the thyroid gland. Thyroxine can elevate the metabolic rate throughout the body by more than 100%. Also, recall that epinephrine and norepinephrine (the catecholamines) mimic and enhance the activity of the sympathetic nervous system. Thus, they directly affect the metabolic rate of virtually all body cells.

Physiological Responses to Exercise in the Heat

Heat production is beneficial when you exercise in a cold environment, because it helps maintain normal body temperature. However, even when you exercise in a thermally neutral environment, such as 21 to 26° C (70-79° F), the metabolic heat load places a considerable burden on the mechanisms that control body temperature. In this section, we examine some physiological changes that occur in response to exercise while the body is exposed to heat stress and the impact that these changes can have on performance. For this discussion, heat stress means any environmental condition that increases body temperature and jeopardizes homeostasis.

Cardiovascular Function

As we learned in chapter 7, exercise increases the demands on the cardiovascular system. When the need to regulate body temperature is added, the cardiovascular system can become burdened during exercise in the heat. The circulatory system transports the heat generated in the muscles to the surface of the body, where the heat can be transferred to the environment. To accomplish this during exercise in the heat, a large part of the cardiac output must be shared by the skin and the working muscles. Because blood volume is limited, exercise poses a complex problem:

▶ Humans are homeothermic, meaning that they maintain a constant internal body temperature, usually in the range of 36.1 to 37.8° C (97.0-100.0° F).

▶ Body heat is transferred by conduction, convection, radiation, and evaporation. At rest, most heat is lost via radiation, but during exercise, evaporation becomes the most important avenue of heat loss.

▶ Higher humidity decreases the capacity to lose heat by evaporation.

▶ The hypothalamus houses your thermoregulatory center. It acts as a thermostat, monitoring your temperature and accelerating heat loss or heat production as needed.

▶ Two sets of thermoreceptors provide temperature information to your thermoregulatory center. The peripheral receptors in the skin relay information about the temperature of your skin and the environment around it. Central receptors in your hypothalamus transmit information about your internal body temperature.

▶ Effectors stimulated by the hypothalamus can alter your body temperature. Increased skeletal muscle activity increases your temperature by increasing metabolic heat production. Increased sweat gland activity decreases your temperature by increasing evaporative heat loss. Smooth muscle in the arterioles can cause these vessels to dilate to direct blood to the skin for heat transfer or to constrict to retain heat deep in the body. Metabolic heat production can be increased by the actions of hormones such as thyroxine and the catecholamines.

Increased blood flow to one of these areas automatically decreases blood flow to the others.

Consider what happens when you are running at a fast pace on a hot day. The exercise increases the demand for blood flow and oxygen delivery to your muscles. It also increases metabolic heat production. This excess heat can be dissipated only if blood flow increases to your skin, transferring the heat to your body's surface. However, blood flow to your skin cannot increase as much as necessary if your muscles are to receive all the blood flow that they need. So the demands of muscles impair heat transfer to the skin.

At the same time, your thermoregulatory center instructs your cardiovascular system to direct more blood flow to your skin. The superficial blood vessels dilate to bring more of the warm blood to your body's surface. This restricts the amount of blood available to your active muscles, limiting their endurance capacity. Thus, the cardiovascular demands of exercise and those of thermoregulation compete for the limited blood supply.

> Exercising in hot environments sets up a competition between the active muscles and the skin for the limited blood supply. The muscles need blood and the oxygen it delivers to sustain activity; the skin needs blood to facilitate heat loss to keep the body cool.

To maintain a constant cardiac output while shunting blood to the periphery, the cardiovascular system must make some noticeable adjustments. The redistribution of blood reduces the volume of blood that returns to the heart, which reduces the end-diastolic volume. This in turn reduces the stroke volume. Cardiac output remains reasonably constant throughout a 30-min exercise bout in a hot (36° C, or 96.8° F) or temperate (20° C, or 68° F) environment, despite a steady decrease in stroke volume. A gradual upward drift in heart rate compensates for the decrease in stroke volume throughout exercise. This phenomenon, known as cardiovascular drift, was discussed in chapter 7.

At some point, though, your body can no longer compensate for the increasing demands of exercise: Neither your muscles nor your skin can receive adequate blood flow. Consequently, any factor that tends to overload the cardiovascular system or to interfere with heat dissipation can drastically impair your performance and increase your risk of overheating. Not surprisingly, this means that the best endurance performances are achieved in cool conditions. Seldom are records set in endurance events, such as distance running, when the environmental heat stress is great.

Energy Production

Fink and colleagues[6] demonstrated that, in addition to increasing body temperature and heart rate, exercise in the heat also increases oxygen uptake, causing the working muscles to use more glycogen and to produce more lactate compared with exercise in the cold. As shown in figure 10.6a, repeated bouts (15 min) of exercise in the heat (40° C, 104° F) increased the subjects' heart rates and oxygen uptakes significantly compared with exercise in a cool environment (9° C, 48° F). As described earlier, a warmer environment places greater stress on the cardiovascular system, which increases the heart rate. Also, increased sweat production and respiration demand more energy, which requires a higher oxygen uptake. As seen in figure 10.6b, the compromised blood flow to the muscles during exercise in the heat leads to a greater use of muscle glycogen and production of more lactic acid. Thus, exercise in the heat can hasten glycogen depletion and increase muscle lactate, both of which are known to contribute to the sensations of fatigue and exhaustion. Febbraio[5] hypothesized that increased muscle temperature impairs skeletal muscle function and metabolism, which can lead to fatigue. Increased carbohydrate utilization appears to be related to the increased secretion of epinephrine with elevated body temperature, or hyperthermia.

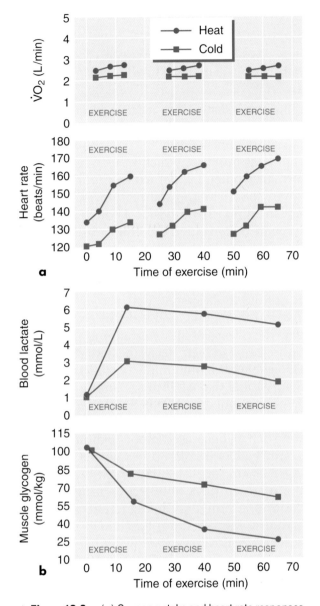

▲ **Figure 10.6**   (*a*) Oxygen uptake and heart rate responses during exercise in hot (40° C, 15% humidity) and cold (9° C, 55% humidity) conditions; (*b*) changes in blood lactate and muscle glycogen during cycling in the same conditions.

Adapted, by permission, from W. Fink, D.L. Costill, P. Van Handel, and L. Getchell, 1975, "Leg muscle metabolism during exercise in the heat and cold," *European Journal of Applied Physiology* 34: 183-190.

Body Fluid Balance: Sweating

Under some conditions, the temperature of the environment approaches and can exceed both the skin and deep body temperatures. As mentioned earlier, this makes evaporation far

more important for heat loss, because radiation, convection, and conduction are less effective as environmental temperature increases. In fact, these mechanisms lead to heat gain when environmental temperature exceeds skin temperature. Increased dependence on evaporation means an increased demand for sweating.

The sweat glands are controlled by stimulation of the hypothalamus. Elevated blood temperature causes the hypothalamus to transmit impulses through the sympathetic nerve fibers to the millions of sweat glands distributed over the body's surface. The sweat glands are tubular structures extending through the dermis and epidermis, opening onto the skin, as illustrated in figure 10.7.

Sweat is formed by the secretory portion of the sweat gland. As the filtrate sweat passes through the duct of the gland, sodium and chloride gradually are reabsorbed back into the

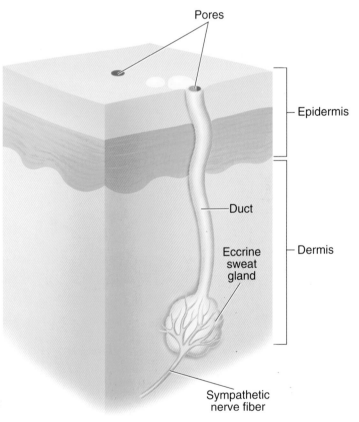

▲ **Figure 10.7**   Anatomy of an eccrine sweat gland that is innervated by a sympathetic nerve.

surrounding tissues and then into the blood. During light sweating, the filtrate sweat travels slowly through the tubules, allowing time for almost complete reabsorption of sodium and chloride. Thus, the sweat that forms during light sweating contains very little of these minerals by the time it reaches the skin. However, when the sweating rate increases during exercise, the filtrate moves more quickly through the tubules, allowing less time for reabsorption. As a result, the sodium and chloride content of sweat can be considerably higher.

As seen in table 10.2, the mineral content of trained and untrained subjects' sweat is significantly different. With training and repeated heat exposure, aldosterone can strongly stimulate the sweat glands, causing them to reabsorb more sodium and chloride. Unfortunately, the sweat glands apparently do not have a similar mechanism for conserving other electrolytes. Potassium, calcium, and magnesium are not reabsorbed by the sweat glands and are therefore found in the same concentrations in both sweat and plasma.

While performing heavy exercise in hot conditions, the body can lose more than 1 L of sweat per hour per square meter of body surface. This means that during intense effort on a hot and humid day (high level of heat stress), an average-sized individual (50-75 kg, or 110-165 lb) might lose 1.6 to 2.0 L of sweat, or about 2.5% to 3.2% of body weight each hour. A person can lose a critical amount of body water in only a few hours of exercise in these conditions.

fyi Sweat rates as high as 3 to 4 L/h have been observed, but these rates cannot be sustained for more than several hours. Maximal daily sweat rates are in the range of 10 to 15 L.

A high rate of sweating reduces blood volume. This limits the volume of blood available to supply the needs of the muscles and to prevent heat buildup, which in turn reduces performance potential, particularly for endurance activities. In long-distance runners, sweat losses can approach 6% to 10% of body weight. Such severe dehydration can limit subsequent sweating and make the individual susceptible to heat-related illnesses. Chapter 13 provides a detailed discussion of dehydration and the value of fluid replacement.

Loss of both minerals and water by sweating triggers the release of aldosterone and antidiuretic hormone (ADH). Recall that aldosterone is responsible for maintaining appropriate sodium levels and that ADH maintains fluid balance. As mentioned in chapter 5, aldosterone is released from the adrenal cortex in response to stimuli such as decreased blood sodium content, reduced blood volume, or reduced blood pressure. During acute exercise in the heat and during repeated days of exercise in the heat, this hormone limits sodium excretion from the kidneys. More sodium is retained by the body, which in turn promotes water retention. Because of this, plasma and interstitial fluid volumes can increase 10% to 20%. This allows

Table 10.2

Sodium, Chloride, and Potassium Concentrations in the Sweat of Trained and Untrained Subjects During Exercise

Subjects	Sweat Na⁺ (mmol/L)	Sweat Cl⁻ (mmol/L)	Sweat K⁺ (mmol/L)
Untrained males	90	60	4
Trained males	35	30	4
Untrained females	105	98	4
Trained females	62	47	4

Data from the Human Performance Laboratory, Ball State University.

the body to retain water and sodium in preparation for additional exposure to the heat and subsequent sweat losses.

Similarly, exercise and body water loss stimulate the posterior pituitary gland to release ADH. This hormone stimulates water reabsorption from the kidneys, which further promotes fluid retention in the body. Thus, the body attempts to compensate for loss of minerals and water during periods of heat stress and heavy sweating by reducing their loss in urine.

> ▶ During exercise in the heat, the heat loss mechanisms compete with the active muscles for more of the limited blood volume. Thus, neither area is adequately supplied under extreme conditions.
>
> ▶ Although cardiac output may remain reasonably constant, stroke volume may decline, resulting in a gradual upward drift in heart rate.
>
> ▶ Oxygen uptake also increases during constant-rate exercise in the heat.
>
> ▶ Sweating increases during exercise in the heat, and this can quickly lead to dehydration and excessive electrolyte loss. To compensate, the release of aldosterone and ADH increases, causing sodium and water retention, which can expand the plasma volume.

Health Risks During Exercise in the Heat

Despite the body's defenses against overheating, excessive heat production by active muscles, heat gained from the environment, and conditions that prevent the dissipation of excess body heat may elevate the internal body temperature to levels that impair normal cellular functions. Under such conditions, excessive heat gains pose a risk to one's health, as we highlighted in the opening anecdote of this chapter. Air temperature alone is not an accurate index of the total physiological stress imposed on the body in a hot environment.

At least four variables must be taken into account:

- Air temperature
- Humidity
- Air velocity
- Amount of thermal radiation

All these factors influence the degree of heat stress that a person experiences. The contributions of each factor to the total heat stress are not clearly understood because their contributions vary with changing environmental conditions.

An individual exercising on a bright, sunny day with an air temperature of 23° C (73.4° F) and no measurable wind experiences considerably more heat stress than someone exercising in the same air temperature but under cloud cover and with a slight breeze. At temperatures above skin temperature, which is normally 30 to 32° C (86-89.6° F), radiation, conduction, and convection substantially add to the body's heat load rather than acting as avenues for heat loss. How, then, can we judge the amount of heat stress to which an individual may be exposed?

> Heat stress is not accurately reflected by air temperature alone. Humidity, air velocity (or wind), and thermal radiation also contribute to the total heat stress that you experience when exercising in the heat.

Measuring Heat Stress

Through the years, efforts have been made to quantify atmospheric variables into a single index. In the 1970s, the **wet bulb globe temperature (WBGT)** was devised to simultaneously account for conduction, convection, evaporation, and radiation (see figure 10.8 on page 320). It provides a single temperature reading to estimate the cooling capacity of the surrounding environment.

The dry bulb measures the actual air temperature (T_{DB}). The wet bulb is kept moist. As

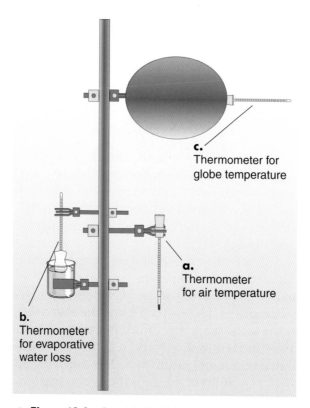

▲ Figure 10.8 A wet bulb globe temperature apparatus showing the thermometers for (*a*) air temperature, (*b*) evaporative water loss, and (*c*) globe temperature.

water evaporates from this bulb, its temperature (T_{WB}) will be lower than the dry bulb's, simulating the effect of sweat evaporating from the skin. The difference between the wet and dry bulb temperatures indicates the environment's capacity for cooling by evaporation. In still air with 100% humidity, these two bulb temperatures are the same because evaporation is impossible. Lower humidity and moving air promote evaporation, increasing the difference between these two bulb temperatures. The black globe absorbs radiated heat. Thus, its temperature (T_G) is a good indicator of the environment's capacity for transmitting radiated heat.

The temperatures from these three bulbs can be put together by using the following equation to estimate the overall atmospheric challenge to body temperature in a specific environment:

$$WBGT = 0.1\ T_{DB} + 0.7\ T_{WB} + 0.2\ T_G$$

This measurement of thermal stress has received considerable attention in recent years and is now used by coaches and athletic trainers to anticipate the health risks associated with athletic competitions in thermally stressful environments.

$$WBGT = 0.1\ T_{DB} + 0.7\ T_{WB} + 0.2\ T_G$$

Heat-Related Disorders

Exposure to the combination of external heat stress and the inability to dissipate metabolically generated heat can lead to three heat-related disorders (see figure 10.9): heat cramps, heat exhaustion, and heat stroke

Heat Cramps

Heat cramps, the least serious of the three heat disorders, are characterized by severe cramping of the skeletal muscles. They involve primarily the muscles that are most heavily used during exercise. This disorder is probably brought on by the mineral losses and dehydration that accompany high rates of sweating, but a cause-and-effect relationship has not been fully established. Heat cramps are treated by moving the stricken individual to a cooler location and administering fluids or a saline solution.

Heat Exhaustion

Heat exhaustion typically is accompanied by such symptoms as extreme fatigue, breathlessness, dizziness, vomiting, fainting, cold and clammy or hot and dry skin, hypotension (low blood pressure), and a weak, rapid pulse. It is caused by the cardiovascular system's inability to adequately meet the body's needs. Recall that during exercise in heat, your active muscles and your skin, through which excess heat is lost, compete for a share of your total blood volume. Heat exhaustion results when these simultaneous demands are not met. Heat exhaustion typically occurs when your blood volume decreases, by either excessive fluid loss or mineral loss from sweating.

With heat exhaustion, the thermoregulatory mechanisms are functioning but cannot dissi-

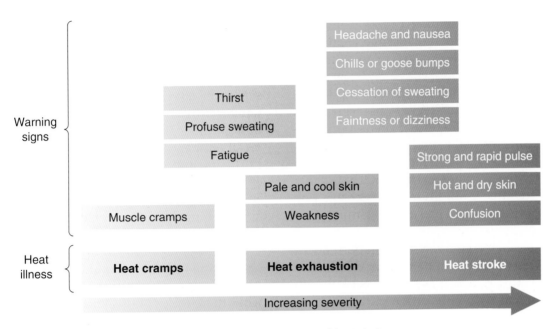

▲ **Figure 10.9** The warning signs of heat cramps, heat exhaustion, and heat stroke.
© PepsiCo 1995. Reprinted with permission.

pate heat quickly enough because insufficient blood volume is available to allow adequate distribution to the skin. Although the condition often occurs during mild to moderate exercise in the heat, it is not generally accompanied by a high rectal temperature. Some people who collapse from heat stress exhibit symptoms of heat exhaustion but have internal temperatures below 39° C (102.2° F). People who are poorly conditioned or not acclimatized to the heat are more susceptible to heat exhaustion.

Treatment for victims of heat exhaustion involves rest in a cooler environment with their feet elevated to avoid shock. If the person is conscious, administration of salt water is usually recommended. If the person is unconscious, medically supervised intravenous administration of saline solution is recommended. If allowed to progress, heat exhaustion can deteriorate to heat stroke.

Heat Stroke

As we saw in our opening story, **heat stroke** is a life-threatening heat disorder that requires immediate medical attention. It is characterized by

- an increase in internal body temperature to a value exceeding 40° C (104° F),
- cessation of sweating,
- hot and dry skin,
- rapid pulse and respiration,
- usually hypertension (high blood pressure), and
- confusion or unconsciousness.

If left untreated, heat stroke progresses to coma, and death quickly follows. Treatment involves rapidly cooling the person's body in a bath of cold water or ice or wrapping the body in wet sheets and fanning the victim.

Heat stroke is caused by failure of the body's thermoregulatory mechanisms. Body heat production during exercise depends on exercise intensity and body weight, so heavier athletes run a higher risk of overheating than lighter athletes when exercising at the same rate and when both are about equally acclimatized to the heat.

For the athlete, heat stroke is a problem associated not only with extreme conditions. Studies have reported rectal temperatures above 40.5° C (104.9° F) in marathon runners who successfully completed races conducted under relatively moderate thermal conditions (e.g., 21.1° C, or 70° F, and 30% relative humidity).[3, 22] Even in shorter events, the body's core temperature can reach life-threatening levels. As early as 1949, Robinson[18] observed rectal temperatures of 41° C (105.8° F) in runners competing in events lasting only about 14 min, such as the 5-km race. Following a 10-km race conducted with an air temperature of 29.5° C (85.1° F), 80% relative humidity, and bright sun, one runner who collapsed had a rectal temperature of 43° C (109.4° F)![2] Without proper medical attention, such fevers can result in permanent central nervous system damage or death. Fortunately, this runner was rapidly cooled with ice and recovered without complications.

> When exercising in the heat, if you suddenly feel chilled and goose bumps form on your skin, stop exercising, get into a cool environment, and drink plenty of cool fluids. The body's thermoregulatory system has become confused and thinks that the body temperature needs to increase even more! Left untreated, this condition can lead to heat stroke and death.

Preventing Hyperthermia

We can do little about environmental conditions. Thus, in threatening conditions, athletes must decrease their effort in order to reduce their heat production and their risk of developing hyperthermia (high body temperature). All athletes, coaches, and sports organizers should be able to recognize the symptoms of hyperthermia. Fortunately, our subjective sensations are well correlated with our body temperatures, as indicated in table 10.3. Although there is generally little concern when rectal temperature remains below 40° C (104° F) during prolonged exercise, athletes who experience throbbing pressure in their heads and chills should realize that they are rapidly approaching a dangerous situation that could prove fatal if they continue to exercise.

To prevent heat disorders, several simple precautions should be taken. Competition and practice should not be held outdoors when the WBGT is more than 28° C (82.4° F). As mentioned earlier, because the WBGT reflects humidity as well as absolute temperature, it reflects the true physiological heat stress more accurately than does standard air temperature. Scheduling practices and contests either in the early morning or at night avoids the severe heat stress of midday. Fluids should be readily available, and athletes should be required to drink as much as they can, stopping every 10 to 20 min for a fluid break in warm temperatures.

Clothing is another important consideration. Obviously, the more clothing that is worn, the less body area exposed to the environment to allow heat exchange. The foolish practice of exercising in a rubberized suit to promote weight loss is an excellent illustration of how a dangerous microenvironment (the isolated

Table 10.3

Subjective Symptoms Associated With Overheating

Rectal temperature	Symptoms
40-40.5° C (104-105° F)	Cold sensation over stomach and back with piloerection (goose bumps)
40.5-41.1° C (105-106° F)	Muscular weakness, disorientation, and loss of postural equilibrium
41.1-41.7° C (106-107° F)	Diminished sweating, loss of consciousness and hypothalamic control
>42.2° C (>108° F)	Death

environment inside the suit) can be created in which temperature and humidity can reach a sufficiently high level to block all heat loss from the body. This can rapidly lead to heat exhaustion or heat stroke. Football uniforms are another example. Areas that are covered by sweat-soaked clothing and padding are exposed to 100% humidity and higher temperatures, reducing the gradient between the body surface and the environment.

Athletes should wear as little clothing as possible when heat stress is a potential limitation to thermoregulation. They should always underdress because the metabolic heat load will soon make extra clothing an unnecessary burden. Clothing should be loosely woven to allow the skin to unload as much heat as possible and light colored to reflect heat back to the environment.

It is also important to maintain hydration of the body, since the body loses considerable water through sweating. This is discussed in considerable detail in chapter 13. Briefly, it has been clearly shown that drinking fluid both before and during exercise can greatly reduce the negative effects of exercising in the heat. Adequate fluid intake will attenuate the increase in core body temperature and heart rate normally seen when a person exercises in the heat. This is illustrated in figure 10.10.

The American College of Sports Medicine has provided guidelines to help distance runners prevent these heat-related injuries.[1] A modified list of these recommendations appears in table 10.4.

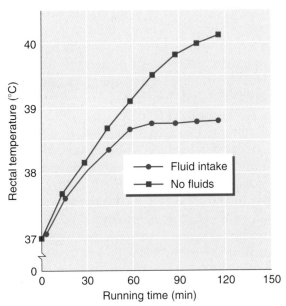

▲ **Figure 10.10**   Effects of fluid intake on core (rectal) body temperature during a 2-h run. Subjects were given fluid during one trial and completed a second trial without fluid on a separate day. Note that fluid intake did not have much influence until about 45 min, after which time body heat storage was reduced compared with the no-fluid trial.

Adapted, by permission, from D.L. Costill et al., 1970, "Fluid ingestion during distance running," *Archives of Environmental Health* 21: 520-525.

▶ Heat stress involves more than just the air temperature. Perhaps the most accurate means for measuring heat stress is the WBGT, which measures air temperature and accounts for the heat exchange potential through conduction, convection, evaporation, and radiation in a specific environment.

▶ Heat cramps are probably caused by losses of fluids and minerals that result from excessive sweating.

▶ Heat exhaustion results from the inability of the cardiovascular system to adequately meet the needs of the active muscles and the skin. It is brought on by reduced blood volume, typically caused by excessive loss of fluids and minerals through prolonged heavy sweating, and by competition for the existing blood volume between the skin and active muscles. Although it is not in itself life threatening, heat exhaustion can deteriorate to heat stroke if untreated.

▶ Heat stroke is caused by failure of the body's thermoregulatory mechanisms. If untreated, it progresses and can be fatal.

▶ Several precautions must be taken when one is planning to exercise in the heat. These include canceling the event if the environmental heat stress is too high (WBGT above 28° C, or 82.4° F), wearing proper clothing, being alert to the signs of hyperthermia, and ensuring adequate fluid intake, both before and during exercise.

Table 10.4

Guidelines for Distance Runners Competing Under Conditions of Heat Stress

1. Distance races should be scheduled to avoid extremely hot and humid conditions. If the WBGT is above 28° C (82° F), consider canceling the race.

2. Summer events should be scheduled in the early morning or evening to minimize solar radiation and unusually high air temperatures.

3. An adequate supply of fluid must be available before the start of the race, along the racecourse, and at the end of the event. Runners should be encouraged to replace their sweat losses or consume 150 to 300 ml (5.3 to 10.5 oz) every 15 min during the race.

4. Cool or cold (ice) water immersion is the most effective means of cooling a collapsed hyperthermic runner.

5. Runners should be aware of the early symptoms of hyperthermia, including dizziness, chilling, headache or throbbing pressure in the temporal region, and loss of coordination.

6. Race officials should be aware of the warning signs of an impending collapse in hot environments and should warn runners to slow down or stop if they appear to be in difficulty.

7. Organizational personnel should reserve the right to stop runners who exhibit clear signs of heat stroke or heat exhaustion.

These recommendations are based on the position stands published by the American College of Sports Medicine in 1987 and 1995.[1]

Acclimatization to Exercise in the Heat

How can we prepare for prolonged activity in the heat? Does training in the heat make us more tolerant of thermal stress? Many studies have investigated these questions and have concluded that repeated exercise in the heat causes a relatively fast adjustment that enables us to perform better in hot conditions.

Effects of Heat Acclimatization

Repeated, prolonged exercise bouts in the heat cause a relatively rapid improvement in your ability to eliminate excess body heat, which reduces your risk of heat exhaustion and heat stroke. This process, termed **heat acclimatization,** results in many adjustments in sweating and blood flow. The rate of sweating during exercise in the heat increases with heat acclimatization, and the amount of sweat produced often increases in the most exposed body areas and in the areas that are most effective at dissipating body heat. At the beginning of exercise, sweating starts earlier in an acclimatized person, which improves heat tolerance. As a result, skin temperatures are lower. This increases the temperature gradient from deep in the body to the skin and the environment. Because heat loss is facilitated, less blood must flow to the skin for body heat transfer, so more blood is available for the active muscles. In addition, the sweat produced is more dilute following training in the heat, so the body's mineral stores are conserved.

Because the body's heat-loss capacity at a given rate of work is enhanced by training, body temperatures during exercise increase less after training in the heat than before training (figure 10.11a).[10] Also, the heart rate increases less in response to standardized submaximal exercise after training in the heat (figure 10.11b). This adaptation results from increased blood volume, reduced blood flow to the skin, or both. Either of these changes increases the stroke volume. Investigators have found that an increase in blood volume accompanies heat acclimatization. However, this change is temporary, and blood volume usually returns to

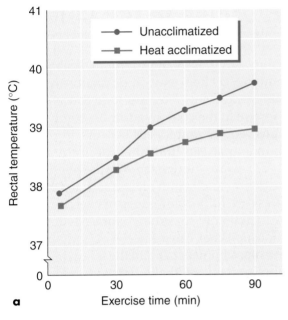

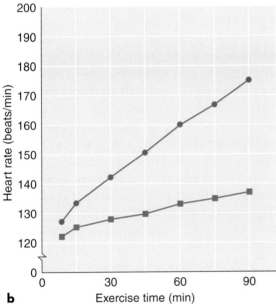

▲ **Figure 10.11**   Differences in (*a*) rectal temperature and (*b*) heart rate before training in the heat (unacclimatized) and after training in the heat (heat acclimatized).

Adapted, by permission, from D.S. King et al., 1984, "Muscle metabolism during exercise in the heat in unacclimatized and acclimatized humans," *Journal of Applied Physiology* 59: 1350-1354.[10]

original levels within 10 days; this probably relates to the body's efforts to retain sodium, thereby expanding the plasma volume.

In addition, after heat acclimatization, more work can be done before the onset of fatigue or exhaustion. Recall that exercise at a given intensity in the heat requires the use of more

muscle glycogen than the same effort done in cooler air. As a result, repeated days of training in the heat can rapidly deplete muscle glycogen and cause chronic fatigue in unacclimatized people. Heat acclimatization reduces the rate of muscle glycogen use by as much as 50% to 60%, reducing this risk.

Achieving Heat Acclimatization

Heat acclimatization requires more than mere exposure to a hot environment. It depends on

- the environmental conditions during each exercise session,
- the duration of heat exposure, and
- the rate of internal heat production (exercise intensity).

Although the research literature is not in total agreement on this point, apparently an athlete must exercise in a hot environment to attain acclimatization that carries over to exercise in the heat. Simply sitting in a hot environment, such as a sauna, for long periods each day will not prepare the individual for physical exertion in the heat, at least not to the same extent as will exercising in the heat.

How can the athlete maximize heat acclimatization? Although most individuals must be exposed to the heat to gain full adjustment, they can gain partial heat tolerance simply by training, even in a cooler environment. It is interesting that when athletes become acclimatized to a given level of heat stress, they can also perform better in cooler environments. But to gain maximal benefits, athletes who train in environments cooler than those in which they will compete must achieve heat acclimatization before the contest or event. Heat acclimatization will improve their performance and reduce the associated physiological stress and risk of heat injury.

If athletes must compete in hot weather, at least part of their training should be conducted in the warmest part of the day. Early morning and evening training will not fully prepare an athlete to tolerate the midday heat. Normal workouts in the heat for 5 to 10 days should provide nearly total heat acclimatization. Workout intensity should be reduced to 60%

or 70% during the first few days to prevent excessive heat stress. Of course, care must be taken to guard against heat disorders such as heat exhaustion and heat stroke. Those in training should be alert to symptoms of heat disorders and should consume as much fluid as possible.

You can adapt to heat (undergo heat acclimatization) by exercising in the heat for 1 or more hours each day for 5 to 10 days. Cardiovascular changes generally occur in the first 3 to 5 days, but changes in the sweating mechanisms generally take up to 10 days.

▶ Repeated exposure to heat stress gradually improves your ability to lose excess heat. This process of adaptation is called heat acclimatization.

▶ With heat acclimatization, you start to sweat earlier and the rate of sweating increases, particularly in areas that are well exposed and are the most efficient at promoting heat loss. This reduces skin temperature, which increases the thermal gradient from the internal to external body and promotes heat loss.

▶ Core temperature and heart rate during exercise are reduced, whereas stroke volume increases with heat acclimatization. Plasma volume increases, contributing to an increase in stroke volume that aids the delivery of more blood to the active muscles and skin when necessary.

▶ Heat acclimatization reduces the rate of muscle glycogen use, delaying the onset of fatigue.

▶ Heat acclimatization requires exercise in a hot environment, not merely exposure to heat.

▶ The amount of heat acclimatization attained depends on the conditions to which you are exposed during each session, the duration of the exposure, and your rate of internal heat production.

Exercise in the Cold

Increasing year-round participation in such sport activities as the triathlon, scuba diving, running, cycling, and long-distance swimming has sparked new interest in and concerns about exercise in the cold. In addition, some occupations require employees to work in cold conditions that can limit their performance. For these reasons, understanding the physiological responses and health risks associated with cold stress are important issues in exercise science. We define cold stress here as any environmental condition that causes a loss of body heat that threatens homeostasis. In the following discussion we focus on the two major cold stressors: air and water.

The hypothalamus has a temperature set point of about 37° C (98.6° F), but daily fluctuations in the body temperature can be as much as 1° C. A decrease in either skin or blood temperature provides feedback to the thermoregulatory center (hypothalamus) to activate the mechanisms that will conserve body heat and increase heat production. The primary means by which our bodies avoid excessive cooling are shivering, nonshivering thermogenesis, and peripheral vasoconstriction.

Because these mechanisms or effectors of heat production and conservation are often inadequate, we also must rely on clothing, muscles, and subcutaneous fat to help insulate our deep body tissues from the environment.

Shivering—a rapid, involuntary cycle of contraction and relaxation of skeletal muscles—can cause a four- to fivefold increase in the body's resting rate of heat production. **Nonshivering thermogenesis** involves stimulation of metabolism by the sympathetic nervous system. Increasing the metabolic rate increases the amount of internal heat production.

Peripheral vasoconstriction occurs as a result of sympathetic stimulation to the smooth muscle surrounding the arterioles in the skin. This stimulation causes the smooth muscles to contract, which constricts the arterioles, reduces the blood flow to the shell of the body, and prevents unnecessary heat loss. The metabolic rate of the skin cells also decreases as the skin's temperature decreases, so the skin requires less oxygen.

Factors Affecting Body Heat Loss

As in heat stress, the body's ability to meet the demands of thermoregulation is limited when exposed to extreme cold. Too much heat loss can occur. The mechanisms of conduction, convection, radiation, and evaporation, which usually perform so effectively in dissipating metabolically produced heat during exercise in warm conditions, can dissipate heat faster than the body produces it in a cold environment.

Pinpointing the exact conditions that permit excessive body heat loss and eventual **hypothermia** (low body temperature) is difficult. Thermal balance depends on a wide variety of factors that affect the gradient between body heat production and heat loss. Generally speaking, the larger the difference between the temperature of the skin and the cold environment, the greater the heat loss. However, a number of anatomical and environmental factors can influence the rate of heat loss. Let's consider a few.

> **fyi** When exercising in the cold, do not overdress. When you overdress, your body can become hot and initiate sweating. As the sweat soaks through the clothing, evaporation rapidly removes the heat, and you become chilled.

Body Size and Composition

Insulating the body against the cold is the most obvious protection against hypothermia. Subcutaneous fat is an excellent source of insulation.[8] Skinfold measurements of subcutaneous fat thickness are a good indicator of an individual's tolerance for cold exposure. The thermal conductivity of fat (its capacity for transferring heat) is relatively low, so it impedes heat transfer from the deep tissues to the body surface. People who have more fat mass conserve heat more efficiently in the cold.

The rate of heat loss also is affected by the ratio of body surface area to body mass. Tall, heavy individuals have a small surface area to body mass ratio, which makes them less susceptible to hypothermia. As shown in table 10.5, small children tend to have a large area to mass ratio compared with adults. This makes

Table 10.5

Body Weight, Height, Surface Area, and Surface Area/Mass Ratios for an Average-Sized Adult and Child

Person	Weight (kg)	Height (cm)	Surface area (cm²)	Area/ mass ratio
Adult	85	183	210	2.47
Child	25	100	79	3.16

it more difficult for them to maintain normal body temperature in the cold.

Women tend to have more body fat than men, but true sex differences in cold tolerance are minimal. Some studies have shown that the added subcutaneous fat in women might give them an advantage during cold-water immersion.[9] When men and women of similar body fat mass, size, and fitness are compared, little difference is noted in body temperature regulation with exposure to the cold.

> **fyi** The body's insulating shell consists of two regions: the superficial skin together with subcutaneous fat, and the underlying muscle. When skin temperatures drop below normal, constriction of the blood vessels supplying the skin and contraction of the skeletal muscles increase the shell's insulating properties. But it is estimated that vasoconstricted inactive muscle can provide as much as 85% of the body's total insulation during exposure to extreme cold. This represents a resistance to heat loss that is two to three times greater than that of the overlying fat and skin.[15, 17]

Windchill

As with heat, the air temperature alone is not a valid index of the amount of thermal stress from cold experienced by the individual. Wind creates a chill factor, known as the **windchill**, by increasing the rate of heat loss via convection and conduction. Also, the more humid the air, the greater the physiological stress. A dry, still day at 10° C (50° F) in the direct sun can be comfortable. Yet on a moist, windy day with complete cloud cover, the cold at this same temperature can be quite penetrating. Table 10.6 lists equivalent temperatures for various ambient air temperatures and wind velocities.

Heat Loss in Cold Water

More research has been conducted on cold exposure in water than air, so we will focus on the effects of immersion in cold water. Whereas radiation and sweat evaporation are the primary mechanisms for heat loss in air, conduction allows the greatest heat transfer during immersion in water. As mentioned earlier, water has a thermal conductivity about 26

times greater than air. This means that heat loss by conduction is 26 times faster in water than in air. When all heat-transfer mechanisms are considered (radiation, conduction, convection, and evaporation), the body generally loses heat four times faster in water than it does in air of the same temperature.

Humans generally maintain a constant internal temperature when they remain inactive in water at temperatures down to about 32° C (89.6° F). But when the water temperature decreases lower, they become hypothermic at a rate proportional to either the duration of their exposure or the thermal gradient.[12, 13] Because of the large drain of heat from a body immersed in cold water, prolonged exposure or unusually cold conditions can lead to extreme hypother-

Table 10.6

Windchill Factor Chart

Estimated wind speed (km/h)	Actual thermometer reading (°F)											
	50	40	30	20	10	0	−10	−20	−30	−40	−50	−60
	Equivalent temperature (°F)											
Calm	50	40	30	20	10	0	−10	−20	−30	−40	−50	−60
8	48	37	27	16	6	−5	−15	−26	−36	−47	−57	−68
16.1	40	28	16	4	−9	−24	−33	−46	−58	−70	−83	−95
24.1	36	22	9	−5	−18	−32	−45	−58	−72	−85	−99	−112
32.2	32	18	4	−10	−25	−39	−53	−67	−82	−96	−110	−124
40.2	30	16	0	−15	−29	−44	−59	−74	−88	−104	−118	−133
48.3	28	13	−2	−18	−33	−48	−63	−79	−94	−109	−125	−140
56.3	27	11	−4	−20	−35	−51	−67	−82	−98	−113	−129	−145
64.4	26	10	−6	−21	−37	−53	−69	−85	−100	−116	−132	−148
	LITTLE DANGER				INCREASING DANGER				GREAT DANGER			

Note: Wind speeds >64.4 km/h have little additional effect

Adapted, by permission, from American College of Sports Medicine, 1987, "Prevention of thermal injuries during distance running," Medicine and Science in Sports and Exercise 19: 529-533, and "Heat and cold illnesses during distance running: ACSM position stand," *Medicine and Science in Sports and Exercise* 28(12): 1-x.

mia and death. Individuals immersed in water at 15° C (59° F) experience a decrease in rectal temperature of about 2.1° C (3.8° F) per hour. If the water temperature were lowered to 4° C (39.2° F), rectal temperature would decrease at a rate of 3.2° C (5.8° F) per hour.[14] The rate of heat loss is further accelerated if the cold water is moving around the individual because heat loss by convection increases. As a result, survival time in cold water under these conditions is quite brief. Victims can become weak and lose consciousness within minutes.

In a 3-year study of factors limiting long-distance swimmers, Pugh and Edholm[16] observed a variety of responses to cold-water immersion (water temperature below 21° C, or 69.8° F). Subcutaneous fat appeared to play an important role in thermal insulation against the cold water, because subjects with obesity (about 30% body fat) could swim for 6 h 50 min in water at 11.8° C (53.2° F) with virtually no change in rectal temperature. But swimmers with relatively low body fatness (about 10%) experienced severe discomfort, and their rectal temperatures dropped to 33.7° C (92.7° F) after only 30 min of swimming in the same water temperature.

If the metabolic rate is low, such as when at rest, then even moderately cool water can cause hypothermia. But exercise increases the metabolic rate and offsets some of the heat loss. For example, although heat loss increases when one is swimming at high speeds (because of convection), the swimmer's accelerated rate of metabolic heat production more than compensates for the greater heat transfer. As shown in figure 10.12, subjects with low body fat (about 8%) maintained constant internal tem-

perature during exercise in water as cold as 17.4° C (63.3° F) when their metabolic rate was increased to about 15 kcal/min.[20] This was true even though their skin temperatures averaged about 17° C (31° F) below their internal temperatures! When the exercise ended, however, the skin warmed rapidly, and the body core began to cool. For competition and training, water temperatures between 23.9 to 27.8° C (75-82° F) seem appropriate.

Physiological Responses to Exercise in the Cold

We have seen how the body must struggle to maintain its internal temperature when exposed to a cold environment. Now we can consider what happens when you add the demands of physical performance to that struggle. How does the body respond to exercise when it is also dealing with exposure to the cold?

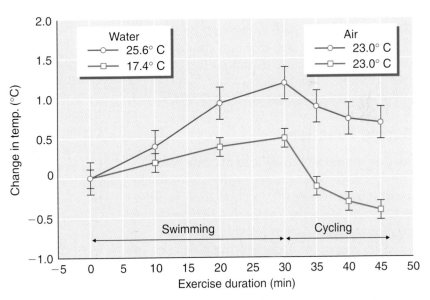

▲ **Figure 10.12**   Changes in rectal temperature during 30 min of swimming in normal water temperature (25.6° C, 78.1° F) and in relatively cool water (17.4° C, 63.3° F), followed by 15 min of cycling in air at 23° C.

From T.A. Trappe et al., 1995, "Thermal responses to swimming in three water temperatures: Influence of a wet suit," *Medicine and Science in Sports and Exercise* 27: 1214-1221. Copyright 1995 by Williams & Wilkins. Adapted by permission.

- ▶ Shivering (involuntary muscle contractions) increases metabolic heat production to help maintain or increase body temperature in the cold.

- ▶ Nonshivering thermogenesis accomplishes the same goal but through stimulation of the sympathetic nervous system and by the action of hormones such as thyroxine and the catecholamines.

- ▶ Peripheral vasoconstriction decreases the transfer of core heat to the skin, thus decreasing heat loss to the environment.

- ▶ Body size is an important consideration for heat loss. Both increased surface area and reduced subcutaneous fat facilitate the loss of body heat to the environment. So those who have a small surface area / body mass ratio and those with more fat are less susceptible to hypothermia.

- ▶ Wind increases heat loss by convection and conduction, so this effect, known as windchill, must be considered along with air temperature during cold exposure.

- ▶ Immersion in cold water tremendously increases heat loss through conduction. Exercise generates metabolic heat to offset some of this loss.

Muscle Function

Cooling a muscle causes it to become weaker. The nervous system responds to muscle cooling by altering the normal muscle fiber recruitment patterns.[4] Some researchers have suggested that this change in fiber selection for force development decreases the efficiency of the muscle's actions. Both muscle shortening velocity and power decrease significantly when temperature is lowered. If people attempt to work at the same velocity and power output when the muscle is at 25° C (77° F) as when it is at 35° C (95° F), they experience fatigue earlier. Thus, they have the choice of either performing an activity at a decreased velocity or expending more energy.

If clothing insulation and exercise metabolism are sufficient to maintain the athlete's body temperature in the cold, exercise performance may be unimpaired. However, as fatigue sets in and muscle activity slows, body heat production gradually decreases. Long-distance running, swimming, and skiing in the cold can expose the participant to such conditions. At the beginning of these activities, the athlete can exercise at a rate that generates sufficient internal heat to maintain body temperature. However, late in the activity, when the energy reserves have diminished, exercise intensity declines, and this reduces metabolic heat production. Subsequent hypothermia causes the individual to become even more fatigued and less capable of generating heat. In these conditions, the athlete is confronted with a potentially dangerous situation.

Metabolic Responses

As we learned earlier, prolonged exercise increases the mobilization and oxidation of free fatty acids (FFA). The primary stimulus for this increased lipid metabolism is the release of catecholamines (epinephrine and norepinephrine) into the vascular system. Exposure to cold markedly increases epinephrine and norepinephrine secretion, but FFA levels increase substantially less than during prolonged exercise in warmer conditions. Cold exposure triggers vasoconstriction in the vessels supplying the skin and subcutaneous tissues. The subcutaneous tissue is the major storage site for lipids (adipose tissue), so this vasoconstriction reduces the blood flow to the area from which the FFA would be mobilized. Thus, FFA levels do not increase as much as the elevated levels of epinephrine and norepinephrine would indicate.

Blood glucose plays an important role in cold tolerance and exercise endurance. Hypoglycemia (low blood sugar), for example, suppresses shivering and significantly reduces rectal temperature. The reasons for these changes are unknown. Fortunately, the blood glucose level is maintained reasonably well during cold exposure. Muscle glycogen, on the other hand, is used at a somewhat higher rate in cold water than in warmer conditions.[23] However, studies on exercise metabolism in the cold are limited, and our knowledge regarding hormonal regulation of metabolism in the cold is too limited to support any definitive conclusions.

> ▶ When muscle is cooled, it is weakened, and fatigue occurs more rapidly.
>
> ▶ During prolonged exercise in the cold, as energy supplies diminish and exercise intensity declines, a person becomes increasingly susceptible to hypothermia.
>
> ▶ Exercise triggers release of the catecholamines, which increase the mobilization and use of free fatty acids for fuel. But in the cold, vasoconstriction impairs circulation to the subcutaneous fat tissue, so this process is attenuated.

Health Risks During Exercise in the Cold

If humans had retained the ability of lower animals, such as reptiles, to tolerate low body temperatures, we could survive extreme hypothermia. Unfortunately, the evolution of thermoregulation in humans has been accompanied by a loss in vital tissues' ability to function when they are cooled by more than a few degrees. Let's briefly examine what happens during hypothermia and frostbite.

Hypothermia

It has been shown that individuals who are immersed in near-freezing water will die within a few minutes when their rectal temperature decreases from a normal level of 37° C (98.6° F) to 24 or 25° C (75.2 or 77° F). Cases of accidental hypothermia and data obtained from surgical patients who are intentionally made hypothermic reveal that the lethal lower limit of body temperature is usually between 23 and 25° C (73.4-77° F), although patients have recovered after having rectal temperatures below 18° C (64.4° F).[13] In 1958, a woman deliberately cooled to a rectal temperature of 9° C (48.2° F) under

anesthesia was satisfactorily revived despite a cardiac arrest lasting for more than 60 min.[14]

Once the body temperature falls below 34.5° C (94.1° F), the hypothalamus begins to lose its ability to regulate body temperature. This ability is completely lost when the internal temperature decreases to about 29.5° C (85.1° F). This loss of function is associated with slowing of metabolic reactions to one half their normal rates for each 10° C (18° F) decline in cellular temperature. As a result, cooling the body can cause drowsiness and even coma.

Cardiorespiratory Effects

The hazards of excessive cold exposure include potential injury to both peripheral tissues and the life-supporting cardiovascular and respiratory systems. The most important effect of hypothermia is on the heart. Death from hypothermia has resulted from cardiac arrest while respiration was still functional. Cooling primarily influences the sinoatrial node, the heart's pacemaker. As early as 1912, Knowlton and Starling[11] demonstrated by cooling heart–lung samples from dogs that cooling the heart tissue leads to a progressive decline of heart rate followed by cardiac arrest. The combined decrements in core temperature and heart rate result in a rapid decline in cardiac output.

Many people have questioned whether rapid, deep breathing of cold air can damage or freeze the respiratory tract. In fact, the cold air that passes into the mouth and trachea is rapidly warmed, even when the temperature of inhaled air is less than –25° C (–13° F).[21] Even at this temperature, when a person is at rest and breathing primarily through his or her nose, the air is warmed to about 15° C (59° F) by the time it has traveled about 5 cm (2 in.) into the nasal passage. As shown in figure 10.13 (on page 332), extremely cold air entering the nose is quite warm by the time it reaches the back of the nasal passage, thereby posing no threat of damage to the throat, trachea, or lungs. Mouth breathing, which often occurs during exercise, may result in cold irritation to the mouth, pharynx, trachea, and even bronchi when the air temperature is below –12° C (10° F). Excessive cold exposure also affects respiratory function by decreasing respiratory rate and volume.

Air temperature

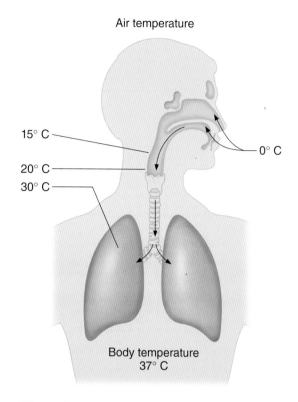

15° C

20° C

30° C

0° C

Body temperature
37° C

▲ **Figure 10.13**   The warming of inspired air as it moves through the respiratory tract.

Treatment for Hypothermia

Mild hypothermia can be treated by protecting the affected person from the cold and providing dry clothing and warm beverages. Moderate to severe cases of hypothermia require gentle handling to avoid initiating a cardiac arrhythmia. This requires slowly rewarming the victim. Severe cases of hypothermia require hospital facilities and medical care. Recommendations to prevent cold exposure–related injuries were outlined by the American College of Sports Medicine in its 1995 position stand on cold illnesses during distance running.[1]

Frostbite

Exposed skin can freeze when its temperature is lowered just a few degrees below the freezing point (0° C, 32° F). Because of the warming influence of circulation and metabolic heat production, the environmental air temperature (including windchill; see table 10.6) required to freeze exposed fingers, nose, and ears is about –

29° C (–20° F). Recall from our earlier discussion that peripheral vasoconstriction helps the body retain heat. Unfortunately, during exposure to extreme cold, the circulation in the skin can decrease to the point that the tissue dies from lack of oxygen and nutrients. This is commonly called **frostbite.** If not treated early, frostbite injuries can be serious, leading to gangrene and loss of tissue. Frostbitten parts should be left untreated until they can be thawed, preferably in a hospital, without risk of refreezing.

> ▶ The hypothalamus begins to lose its ability to regulate body temperature if body temperature drops below 34.5° C (94.1° F).
>
> ▶ Hypothermia primarily affects the heart's sinoatrial node, decreasing the heart rate, which in turn reduces cardiac output.
>
> ▶ Breathing cold air does not freeze the respiratory passages or the lungs when ventilation is low.
>
> ▶ Exposure to extreme cold decreases respiratory rate and volume.
>
> ▶ Frostbite occurs as a consequence of the body's attempts to prevent heat loss. Vasoconstriction in the skin reduces blood flow, so the skin rapidly cools. This reduced blood flow, combined with the lack of oxygen and nutrients, causes the skin tissue to die.

Acclimatization to Exercise in the Cold

Information about **cold acclimatization** is limited, but some data suggest that chronic daily exposure to cold water increases subcutaneous body fat.[9] Much of our information regarding habituation to the cold has been obtained from observations of the aborigines, natives of the Australian outback, who are normally exposed to low temperatures at night and high temperatures during the day.[19] Compared with unacclimatized Europeans, the aborigines were able to sleep more comfortably in the cold with little protection, and they experienced only minor changes in their metabolism and rectal temperatures. The Europeans, on the other hand, experienced significant distress and considerable difficulty in maintaining normal body temperature.

Although some data suggest that repeated exposure to the cold alters peripheral blood flow and skin temperatures, these changes are small, and the findings are inconclusive. Field studies have shown that chronic exposure of some areas of the skin, such as the hands, can provide greater cold tolerance. For example, fishermen who must work with their hands in cold water for many hours develop increased vasodilation and local warming of this exposed skin. The rate and degree of adjustment to these conditions have not been fully explained. Thus, acclimatization to cold is not as thoroughly understood as acclimatization to environmental heat stress.

> ▶ Cold acclimatization has not been well studied, so our knowledge is limited.
>
> ▶ Repeated exposure to the cold may alter peripheral blood flow and skin temperatures, allowing greater cold tolerance.

In Closing . . .

In this chapter, we began our examination of how the external environment affects the body's ability to perform physical work. We looked at the effects of extreme heat and cold stress and the body's responses to them. We considered the health risks associated with these temperature extremes and how the body can try to adapt to these conditions through acclimatization.

In the next chapter, we examine additional environmental extremes: hypobaric, hyperbaric, and microgravity environments.

▶ Key Terms

aldosterone
antidiuretic hormone (ADH)
cold acclimatization

conduction
convection
evaporation
frostbite
heat acclimatization
heat cramps
heat exhaustion
heat stroke
hyperthermia
hypothermia
nonshivering thermogenesis
peripheral vasoconstriction
radiation
shivering
thermal stress
thermoreceptors
thermoregulation
thermoregulatory center
wet bulb globe temperature (WBGT)
windchill

▶ Study Questions

1. What are the four major avenues for loss of body heat?

2. Which of these four pathways is most important for controlling body temperature when at rest? During exercise?

3. What happens to the body temperature during exercise, and why?

4. Why is humidity an important factor when performing in the heat? Why are wind and cloud cover important?

5. What is the purpose of the wet bulb globe temperature (WBGT)? What does it measure?

6. Differentiate between heat cramps, heat exhaustion, and heat stroke.

7. What physiological adaptations occur that allow a person to acclimatize to exercise in the heat?

8. How does the body minimize excessive heat loss during cold exposure?

9. What dangers are associated with cold-water immersion?

10. What factors should be considered to provide maximal protection when exercising in the cold?

▷ References

1. American College of Sports Medicine. (1995). Heat and cold illnesses during distance running: ACSM position stand. *Medicine and Science in Sports and Exercise, 28*(12), i-x.

2. Costill, D.L. (1986). *Inside running: Basics of sports physiology.* Indianapolis: Benchmark Press.

3. Costill, D.L., Kammer, W.F., & Fisher, A. (1970). Fluid ingestion during distance running. *Archives of Environmental Health, 21,* 520-525.

4. Faulkner, J.A., Claflin, D.R., & McCully, K.K. (1987). Muscle function in the cold. In J.R. Sutton, C.S. Houston, & G. Coates (Eds.), *Hypoxia and cold* (pp. 429-437). New York: Praeger.

5. Febbraio, M.A. (2000). Does muscle function and metabolism affect exercise performance in the heat? *Exercise and Sport Sciences Reviews, 28,* 171-176.

6. Fink, W., Costill, D.L., Van Handel, P., & Getchell, L. (1975). Leg muscle metabolism during exercise in the heat and cold. *European Journal of Applied Physiology, 34,* 183-190.

7. Gisolfi, C.V., & Wenger, C.B. (1984). Temperature regulation during exercise: Old concepts, new ideas. *Exercise and Sport Sciences Reviews, 12,* 339-372.

8. Hayward, M.G., & Keatinge, W.R. (1981). Roles of subcutaneous fat and thermoregulatory reflexes in determining ability to stabilize body temperature in water. *Journal of Physiology* (London), *320,* 229-251.

9. Kang, B.S., Song, S.H., Suh, C.S., & Hong, S.K. (1963). Changes in body temperature and basal metabolic rate of the ama. *Journal of Applied Physiology, 18,* 483-488.

10. King, D.S., Costill, D.L., Fink, W.J., Hargreaves, M., & Fielding, R.A. (1985). Muscle metabolism during exercise in the heat in unacclimatized and acclimatized humans. *Journal of Applied Physiology, 59,* 1350-1354.

11. Knowlton, F.P., & Starling, E.H. (1912). The influence of variations in temperature and blood-pressure on the performance of the isolated mammalian heart. *Journal of Physiology, 44,* 206-219.

12. Molnar, G.W. (1946). Survival of hypothermia by man immersed in the ocean. *Journal of the American Medical Association, 131,* 1046-1050.

13. Newburgh, L.H. (1949). *Physiology of heat regulation.* Philadelphia: Saunders.

14. Niazi, S.A., & Lewis, F.J. (1958). Profound hypothermia in man. *Annals of Surgery, 147,* 264-266.

15. Pendergast, D.R. (1988). The effect of body cooling on oxygen transport during exercise. *Medicine and Science in Sports and Exercise, 20*(Suppl.), S171-S176.

16. Pugh, L.G., & Edholm, D.G. (1955). The physiology of channel swimmers. *Lancet, 2,* 761-767.

17. Rennie, D.W. (1988). Tissue heat transfer in water: Lessons from the Korean divers. *Medicine and Science in Sports and Exercise, 20,* S177.

18. Robinson, S. (1949). Physiological adjustments to heat. In L.H. Newburgh (Ed.), *Physiology of heat regulation and the science of clothing* (pp. 193-231). Philadelphia: Saunders.

19. Scholander, P.F., Hammel, H.T., Hart, J.S., Lemessurier, D.H., & Steen, J. (1958). Cold adaptation in Australian aborigines. *Journal of Applied Physiology, 13,* 211-218.

20. Trappe, T.A., Starling, R.D., Jozsi, A.C., Goodpaster, B.H., Trappe, S.W., Nomura, T., Obara, S., & Costill, D.L. (1995). Thermal responses to swimming in three water temperatures: Influence of a wet suit. *Medicine and Science in Sports and Exercise, 27,* 1214-1221.

21. Webb, P. (1951). Air temperature in respiratory tracts of resting subjects in the cold. *Journal of Applied Physiology, 4,* 378-382.

22. Wyndham, C.H. (1973). The physiology of exercise under heat stress. *Annual Review of Physiology, 35,* 193-220.

23. Young, A.J., Sawka, M.N., Neufer, P.D., Muza, S.R., Askew, E.W., & Pandolf, K.B. (1989). Thermoregulation during cold water immersion is unimpaired by low muscle glycogen levels. *Journal of Applied Physiology, 66,* 1809-1816.

▷ Selected Readings

Alexander, L. (1946). *Treatment of shock from prolonged exposure to cold especially in water* (Item No. 24, File No. 26-37). Washington, DC: Combined Intelligence Objectives Subcommittee.

Aoyagi, Y., McLellan, T.M., & Shephard, R.J. (1997). Interactions of physical training and heat acclimation. *Sports Medicine, 23,* 173-210.

Armstrong, L.E. (2000). *Performing in extreme environments.* Champaign, IL: Human Kinetics.

Armstrong, L.E., & Maresh, C.M. (1991). The induction and decay of heat acclimatisation in trained athletes. *Sports Medicine, 12,* 302-312.

Armstrong, L.E., & Maresh, C.M. (1995). Exercise-heat tolerance of children and adolescents. *Pediatric Exercise Science, 7,* 239-252.

Cheuvront, S.N., & Haymes, E.M. (2001) Thermoregulation and marathon running. *Sports Medicine, 31,* 743-762.

Doubt, T.J. (1991). Physiology of exercise in the cold. *Sports Medicine, 11,* 367-381.

Edwards, R.H.T., Harris, R.C., Hultman, E., Kaijser, L., Koh, D., & Nordesjo, L.O. (1972). Effect of temperature on muscle energy metabolism and endurance during successive isometric contractions, sustained to fatigue, of the quadriceps muscle in man. *Journal of Physiology, 220,* 335-352.

Ferretti, G., Veicsteinas, A., & Rennie, D.W. (1988). Regional heat flows of resting and exercising men immersed in cool water. *Journal of Applied Physiology, 64,* 1239-1248.

González-Alonso, J., Calbet, J.A.L., & Nielsen, B. (1999). Metabolic and thermodynamic responses to dehydration-induced reductions in muscle blood flow in exercising humans. *Journal of Physiology, 520,* 577-589.

Hughson, R.L., Standi, L.A., & Mackie, J.M. (1983). Monitoring road racing in the heat. *Physician and Sportsmedicine, 11,* 94-105.

Jentjens, R.L.P.G., Wagenmakers, A.J.M., & Jeukendrup, A.E. (2002). Heat stress increases muscle glycogen use but reduces the oxidation of ingested carbohydrates during exercise. *Journal of Applied Physiology, 92,* 1562-1572.

Keatinge, W.R. (1969). *Survival in cold water.* Oxford, UK: Blackwell Scientific.

Kenney, W.L. (1997). Thermoregulation at rest and during exercise in healthy older adults. *Exercise and Sport Sciences Reviews, 25,* 41-76.

Kenney, W.L., & Anderson, R.K. (1988). Response of older and younger women in dry and humid heat without fluid replacement. *Medicine and Science in Sports and Exercise, 20,* 155.

Layden, J.D., Patterson, M.J., & Nimmo, M.A. (2002). Effects of reduced ambient temperature on fat utilization during submaximal exercise. *Medicine and Science in Sports and Exercise, 34,* 774-779.

Moran, D.S. (2001). Potential applications of heat and cold stress indices to sporting events. *Sports Medicine, 31,* 909-917.

Nadel, E.R. (Ed.). (1977). *Problems with temperature regulation during exercise.* New York: Academic Press.

Sallis, R., & Chassay, C.M. (1999). Recognizing and treating common cold-induced injury in outdoor sports. *Medicine and Science in Sports and Exercise, 31,* 1367-1373.

Sawka, M.N., & Wegner, C.B. (1988). Physiological responses to acute-exercise heat stress. In K.B. Pandolf, M.N. Sawka, & R.R. Gonzalez (Eds.), *Human performance physiology and environmental medicine at terrestrial extremes* (97-151). Indianapolis: Benchmark Press.

Shephard, R.J. (1993). Metabolic adaptations to exercise in the cold. *Sports Medicine, 16,* 266-289.

Siple, P.A., & Passel, C.F. (1945). Measurement of dry atmospheric cooling in subfreezing temperatures. *Proceedings of the American Physiological Society, 89,* 177-199.

Sutton, J.R. (1986). Thermal problems in the master athlete. In J.R. Sutton & R.M. Brock (Eds.), *Sports medicine for the mature athlete* (pp. 125-132). Indianapolis: Benchmark Press.

Toner, M.M., & McArdle, W.D. (1988). Physiological adjustments of a man to cold. In K.B. Pandolf, M.N. Sawka, & R.R. Gonzalez (Eds.), *Human performance physiology and environmental medicine at terrestrial extremes* (361-399). Indianapolis: Benchmark Press.

Wegner, C.B. (1988). Human heat acclimatization. In K.B. Pandolf, M.N. Sawka, & R.R. Gonzalez (Eds.), *Human performance physiology and environmental medicine at terrestrial extremes* (153-197). Indianapolis: Benchmark Press.

Young, A.J. (1988). Human adaptation to cold. In K.B. Pandolf, M.N. Sawka, & R.R. Gonzalez (Eds.), *Human performance physiology and environmental medicine at terrestrial extremes* (401-434). Indianapolis: Benchmark Press.

Young, A.J. (1990). Energy substrate utilization during exercise in extreme environments. *Exercise and Sport Sciences Reviews, 18,* 65-117.

CHAPTER 11

EXERCISE IN HYPOBARIC, HYPERBARIC, AND MICROGRAVITY ENVIRONMENTS

overview

We have heard accounts of grueling attempts to scale Mount Everest. We have heard of the bitter cold, the avalanches, and the failures. Some of us, from our armchairs, have joined in the undersea adventures of the late Jacques Cousteau, exploring the magnificent beauty under the ocean along with the team from his ship, the Calypso. And we have followed the journeys of our astronauts as they have ventured into space—and walked on the moon.

The physical conditions of each of these environments are so different from those to which we are accustomed that exposure to these extreme environments alters the body's functioning. The body must deal with low pressure at altitude, high pressure underwater, and reduced gravity in space.

In this chapter we examine the conditions experienced in these specialized environments, how these conditions affect our bodies, and how they affect performance. We also consider health risks associated with each environment and how we can adapt to these extreme conditions.

outline

Sports competitions at altitude traditionally have been associated with performance impairment. As a result, there were many complaints when it was announced that the 1968 Olympic Games would be held in Mexico City, at an altitude of 2,240 m (7,350 ft) above sea level. At least two athletes who participated in those games were glad to perform in the rarified air at that moderate altitude. Bob Beamon soared almost 0.6 m (2 ft) farther than the previous world record in the long jump, and Lee Evans beat the world record in the 400-m run by nearly 0.24 s. These records stood for nearly 20 years, indicating that conditions at the altitude of Mexico City likely contributed to the stellar performances in these relatively short-duration, explosive events.

Our previous discussions of the physiological responses to exercise were based on the conditions that exist at or near sea level, where the barometric pressure averages about 760 mmHg, the **partial pressure of oxygen** (PO_2) is about 159 mmHg, and we experience normal gravitational force. Although the human body tolerates reasonable fluctuations in these conditions, large variations from these values pose special problems. This is evident when mountain climbers ascend to higher altitudes, when divers are exposed to pressurized conditions underwater, or when astronauts are in space. Any of these situations can seriously impair physical performance and can even jeopardize life.

Barometric pressure is reduced at altitude. This situation is referred to as a **hypobaric environment** (low atmospheric pressure). The lower atmospheric pressure also means a lower PO_2, which limits pulmonary diffusion and oxygen transport to the tissues. This reduces oxygen delivery to the body tissues, resulting in **hypoxia** (oxygen deficiency). On the other hand, when immersed in water, the body is exposed to greater pressure. Thus, the underwater world is a **hyperbaric environment** (high atmospheric pressure). Gases breathed under these conditions must be pressurized to equal the force of water against the chest wall. This means that the lung and body tissues are presented with gas pressures well in excess of those experienced at sea level. Breathing pressurized gases has little effect on the transport of oxygen and carbon dioxide in the body, but the increased partial pressure of several gases can lead to life-threatening complications.

Our third environment of concern is **microgravity**, in which the body experiences a reduced gravitational force. Although athletes certainly do not compete in space, space exploration has revealed some physiological challenges that are relevant to exercise and sport physiology.

In the following discussion, we examine the special characteristics of hypobaric, hyperbaric, and microgravity environments and how these conditions alter our physiological responses to physical activity. We focus on the impact of these environments on oxygen transport and examine health risks associated with each.

Hypobaric Environments: Exercising at Altitude

Clinical problems associated with altitude were reported as early as 400 B.C.[47] However, most of the early concerns about ascent to high altitudes focused on the cold conditions at altitude rather than the limitations imposed by rarefied air. The initial discoveries that led to our current understanding of the reduced oxygen pressure at altitude can be credited to three scientists. Torricelli (ca. 1644) developed the mercury barometer, an instrument that permits accurate measurement of atmospheric gas pressures. A few years later (1648), Pascal demonstrated a reduction in barometric pressure at high altitudes.[47] Nearly 130 years later (1777), Lavoisier described oxygen and the other gases that contribute to the total barometric pressure.[47]

The deleterious effects of high altitude on humans that are caused by low oxygen tension (hypoxia) were subsequently recognized by Bert in the late 1800s.[2] More recently, the selection of Mexico City, with an elevation of 2,240 m (7,350 ft) above sea level, for the 1968 Olympic Games drew considerable attention to the effects of altitude on physical performance. For our discussion, the term *altitude* refers to elevations above 1,500 m (4,921 ft), because few physiological effects on performance are reported below that level.

Conditions at Altitude

Before examining how altitude affects performance, we must consider what special conditions exist in such hypobaric environments. Let's look at how the atmosphere at altitude differs from that near sea level.

Atmospheric Pressure at Altitude

Air has weight. The barometric pressure at any place on Earth is related to the weight of the air in the atmosphere above that point. At sea level, for example, the air extending to the outermost reaches of the Earth's atmosphere (approximately 38.6 km, or 24 mi) exerts a pressure equal to 760 mmHg. At the summit of Mount Everest, the highest point on Earth (8,848 m, or 29,028 ft), the pressure exerted by the air above is only about 250 mmHg. These and other differences are depicted in figure 11.1.

The barometric pressure on Earth does not remain constant. Rather, it varies with changes in climatic conditions, time of year, and the specific site at which the measurement is taken. On Mount Everest, for example, the mean barometric pressure varies from 243 mmHg in January to nearly 255 mmHg in June and July. Also, the Earth's atmosphere bulges outward slightly at the equator, which increases the barometric pressure there by a few millimeters of mercury above standard pressure. These points, of little interest to people living near sea level, are of considerable physiological importance for anyone attempting to climb Mount Everest without supplemental oxygen.

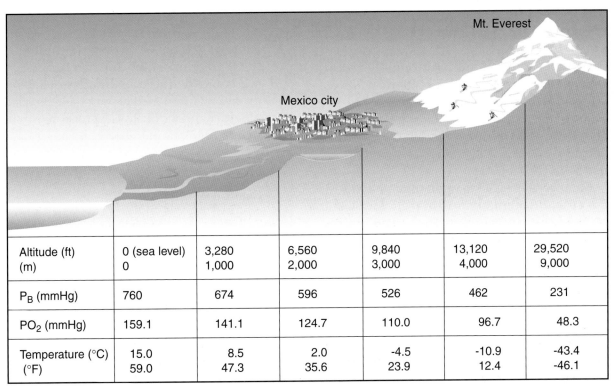

Altitude (ft) (m)	0 (sea level) 0	3,280 1,000	6,560 2,000	9,840 3,000	13,120 4,000	29,520 9,000
P$_B$ (mmHg)	760	674	596	526	462	231
PO$_2$ (mmHg)	159.1	141.1	124.7	110.0	96.7	48.3
Temperature (°C) (°F)	15.0 59.0	8.5 47.3	2.0 35.6	-4.5 23.9	-10.9 12.4	-43.4 -46.1

▲ **Figure 11.1** Differences in atmospheric conditions at sea level up through an altitude of 9,000 m (29,520 ft). Note that the partial pressure of oxygen decreases from 159.1 mmHg at sea level to 48.3 mmHg at an altitude of 9,000 m.

Although atmospheric pressure varies, the percentages of gases in the air that we breathe remain unchanged from sea level to high altitude. At any elevation, the air always contains 20.93% oxygen, 0.03% carbon dioxide, and 79.04% nitrogen. Only the partial pressures change. As shown in figure 11.1, the pressure that oxygen molecules exert at various altitudes is directly affected by the barometric pressure, and a change in the partial pressure of oxygen has a significant effect on the partial pressure gradient between the blood and the tissues. This point is discussed later in this chapter.

> The mixture of gases in the air we breathe at altitude is identical to that at sea level—oxygen (O_2) is 20.93%, carbon dioxide (CO_2) is 0.03%, and nitrogen (N_2) is 79.04% of the total air. The partial pressure of each gas, however, is reduced in direct proportion to the increase in altitude. The reduced PO_2 decreases performance at altitude, attributable to a reduced pressure gradient that hinders oxygen transport to the tissues.

Air Temperature at Altitude

Air temperature decreases at a rate of about 1° C (1.8° F) for every 150 m (about 490 ft) of ascent, as shown in table 11.1. The average temperature near the summit of Mount Everest is estimated to be about –40° C (–40° F), whereas at sea level the temperature would be about 15° C (59° F). The combination of low temperatures and high winds at altitude poses a serious risk of cold-related disorders, such as hypothermia and windchill injuries.

Because of the cold temperatures at altitude, the absolute humidity is extremely low. Cold air holds very little water. Thus, even if air is fully saturated with water (100% relative humidity), the actual amount of water contained in the air is small. The partial pressure of water at 20° C (68° F) is about 17 mmHg. But at –20° C (–4° F), this pressure decreases to only about 1 mmHg. The very low humidity at high altitude promotes dehydration. In fact, a large volume of body water is lost through respiratory evaporation attributable to the dry air and increased respiration rate (discussed later) experienced at altitude. The dry air also increases evaporative water loss through sweating during exercise at altitude.

> Air temperature decreases as altitude increases. This decrease in temperature is accompanied by a decrease in the amount of water vapor in the air. As a result, this drier air can lead to dehydration through increased insensible water loss.

Solar Radiation at Altitude

The intensity of solar radiation increases at high altitude for two reasons. First, because you are positioned higher in the atmosphere, light travels through less of the atmosphere before reaching you. For this reason, less of the sun's radiation, especially the ultraviolet rays, is absorbed by the atmosphere at altitude. Second, because atmospheric water normally absorbs a substantial amount of the sun's radiation, the limited water vapor found at altitude also increases your exposure. Solar radiation is further amplified if you are also exposed to reflective light from snow, which is usually found at higher elevations.

> ▶ Altitude presents a hypobaric environment (one in which the atmospheric pressure is reduced). Altitudes of 1,500 m (4,921 ft) or more have a notable physiological impact on the exercise performance of the human body.
>
> ▶ Although the percentages of the gases in the air we breathe remain constant regardless of altitude, the partial pressures of each of these gases vary with atmospheric pressure.
>
> ▶ Air temperature decreases as altitude increases. Cold air can hold little water, so the air at altitude is dry. These two factors increase your susceptibility to cold-related disorders and dehydration when at altitude.
>
> ▶ Because the atmosphere is thinner and drier at altitude, solar radiation is more intense at higher elevations.

Physiological Responses to Altitude

We have looked at some of the unique environmental conditions associated with altitude. Now we can examine how exposure to these conditions affects you. In this section we examine how your body responds to altitude, emphasizing responses that can affect performance. Our main concerns are respiratory, cardiovascular, and metabolic responses.

Most of this discussion deals with the physiological responses of unacclimatized men at altitude. Unfortunately, few studies on the effects of altitude have included women or children, populations whose sensitivity to the conditions of altitude might differ from that described here.

Respiratory Responses to Altitude

Adequate oxygen supply to the muscles is essential to physical performance and, as we saw in chapter 8, depends on an adequate supply of oxygen being brought into the body, transport of that oxygen to the muscles, and adequate oxygen uptake by the muscles. Any deficiency in these steps can impair performance. Let's see how altitude affects these processes.

Pulmonary Ventilation   Pulmonary ventilation (breathing) increases at higher altitudes, both at rest and during exercise. Because the number of oxygen molecules in a given volume of air is less at higher altitudes, more air must be inspired to supply as much oxygen as during normal breathing at sea level. Ventilation increases to bring in a larger volume of air.

> You ventilate greater volumes of air at altitude because air is less dense.

Increased ventilation acts much the same as hyperventilation at sea level. The amount of carbon dioxide in the alveoli is reduced. Carbon dioxide follows the pressure gradient, so more diffuses out of the blood, where its pressure is relatively high, and into the lungs to be exhaled. This increased carbon dioxide clearance allows blood pH to increase, a condition known as **respiratory alkalosis.** In an effort

to prevent this condition, the kidneys excrete more bicarbonate ion. Recall that bicarbonate ions buffer the carbonic acid formed from carbon dioxide. Thus, a reduction in bicarbonate ion concentration reduces the blood's buffering capacity. More acid remains in the blood, and the alkalosis can be reversed.

Pulmonary Diffusion and Oxygen Transport   Under resting conditions, pulmonary diffusion does not limit the exchange of gases between the alveoli and the blood. If gas exchange were limited or impaired, less oxygen would enter the blood, so the arterial PO_2 would be lower than the alveolar PO_2. Instead, these two values are about equal. For a resting person at sea level, the amount of oxygen entering the blood is determined by the alveolar PO_2 and the rate of blood flow through the pulmonary capillaries.

Recall that the PO_2 at sea level is 159 mmHg; however, it decreases to 118 mmHg at an elevation of 2,439 m (8,000 ft). As a result, the PO_2 within the alveoli and the pulmonary capillaries also decreases. Consequently, hemoglobin saturation decreases from about 98% at sea level to approximately 90% to 92% at an elevation of 2,439 m (8,000 ft). This small decrease in the hemoglobin saturation was once believed to reduce $\dot{V}O_2$max by approximately 15% and thus restrict performance at this altitude. However, as we will soon see, this $\dot{V}O_2$max reduction is really the result of the low arterial PO_2 that accompanies the decrease in barometric pressure at altitude.

Gas Exchange at the Muscles   Arterial PO_2 at sea level is about 100 mmHg, and the PO_2 in body tissues is consistently about 40 mmHg at rest, so the difference, or the pressure gradient, between the arterial PO_2 and the tissue PO_2 at sea level is about 60 mmHg. However, when you move to an elevation of 2,439 m (8,000 ft), your arterial PO_2 decreases to about 60 mmHg, whereas your tissue PO_2 remains at 40 mmHg. Thus, the pressure gradient decreases from 60 mmHg at sea level to only 20 mmHg at the higher altitude as shown in table 11.1 on page 342. This is an almost 70% reduction in the diffusion gradient! Because the diffusion gradient is responsible for driving the oxygen from your

Table 11.1

Changes in Partial Pressure Diffusion Gradient for Oxygen at 8,000 ft (2,439 m) altitude

Partial pressure (mmHg)	Sea level	8,000 ft (2,439 m)
Arterial PO_2	100	60
Muscle PO_2	40	40
Diffusion gradient	60	20

blood into your tissues, this change in arterial PO_2 at altitude is an even greater consideration than the small 6% to 8% reduction in hemoglobin saturation that occurs.

Maximal Oxygen Uptake   Maximal oxygen uptake decreases as altitude increases (see figure 11.2). $\dot{V}O_2$max decreases little until the atmospheric PO_2 drops below 131 mmHg. This generally occurs at an altitude of 1,600 m (5,249 ft)—about the elevation of Denver, Colorado.

As shown in figure 11.3, men climbing Mount Everest in a 1981 expedition experienced a change in $\dot{V}O_2$max from about 62 ml · kg⁻¹ · min⁻¹ at sea level to only about 15 ml · kg⁻¹ · min⁻¹ near the mountain's peak. Normal resting oxygen requirements are about 3.5 ml · kg⁻¹ · min⁻¹, so, without supplemental oxygen, these men had little capacity for physical effort at this elevation. A study by Pugh and coworkers[30] showed that men with $\dot{V}O_2$max values of 50 ml · kg⁻¹ · min⁻¹ at sea level would be unable to exercise, or even to move, near the peak of Mount Everest because their $\dot{V}O_2$max values at that altitude would decrease to 5 ml · kg⁻¹ · min⁻¹. Thus, most normal people with sea-level $\dot{V}O_2$max values below 50 ml · kg⁻¹ · min⁻¹ would not be able to survive without supplemental oxygen at the summit of Mount Everest because their $\dot{V}O_2$max values at such an altitude would be too low to sustain their body tissues. Enough oxygen would be consumed only to meet their resting requirements.

Below an altitude of 1,600 m (5,249 ft), altitude appears to have little effect on $\dot{V}O_2$max and endurance performance. Above 1,600 m, however, $\dot{V}O_2$max decreases approximately 8% to 11% for every 1,000-m (3,281-ft) increase.

Cardiovascular Responses to Altitude

As the respiratory system becomes increasingly stressed as altitude increases, so does the cardiovascular system, which undergoes substantial changes to compensate for the decrease in PO_2 that accompanies increased altitude. Let's examine a few of these changes.

Blood Volume   Within the first few hours of arriving at altitude, a person's plasma volume begins to progressively decrease, and it plateaus by the end of the first few weeks. This decrease in plasma volume is the result of both respiratory water loss and the body's conscious attempt to reduce plasma volume. The dehydration experienced during ascent to high altitude increases respiratory water loss. The body also appears to dump plasma volume during the first 24 to 48 h at altitude. The combination of respiratory water loss and the dumping of plasma volume reduces total plasma volume by up to 25%. Initially, the result of this plasma loss is an increase in the number of red blood cells per unit of blood, allowing more oxygen to be delivered to the muscles for a given cardiac output. Because this reduction in plasma volume occurs with little or no change in the total red blood cell count, it results in a higher hematocrit but a smaller total blood volume than at lower altitudes. The diminished plasma volume eventually returns to normal levels. In addition, continued exposure to high altitude triggers release of **erythropoietin**, the hormone responsible for stimulating erythrocyte (red blood cell) production. This increases the total number of red blood cells. These adaptations ultimately result in a greater total blood volume, which allows the person to partially

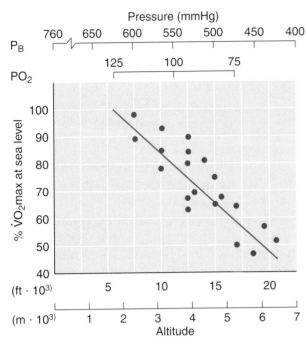

▲ Figure 11.2 Changes in maximal oxygen uptake ($\dot{V}O_2$max) with decrements in barometric pressure (P_B) and partial pressure of oxygen (PO_2). Values for $\dot{V}O_2$max are recorded as percentages of $\dot{V}O_2$max attained at sea level (P_B = 760 mmHg). Note that at the altitudes of Mexico City (2,240 m); Leadville, Colorado (3,180 m); and Nuñoa, Peru (4,000 m), your $\dot{V}O_2$max would be significantly below your capacity at sea level or in Denver (1,600 m).

Data from E.R. Buskirk et al., 1967, "Maximal performance at altitude and on return from altitude in conditioned runners," *Journal of Applied Physiology* 23: 259-266.

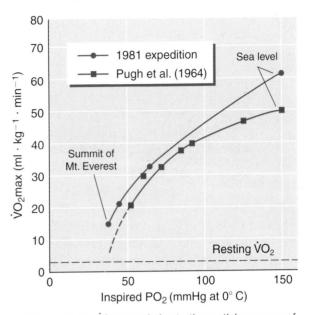

▲ Figure 11.3 $\dot{V}O_2$max relative to the partial pressure of oxygen (PO_2) of the inspired air.

Adapted, by permission, from J.B. West et al., 1983, "Maximal exercise at extreme altitudes on Mount Everest," *Journal of Applied Physiology* 55: 688-698.

compensate for the lower PO_2 experienced at altitude.

Cardiac Output   As we have seen, the amount of oxygen delivered to the muscles by a given volume of blood is limited at altitude because the reduced PO_2 causes a reduced diffusion gradient. A logical means to compensate for this is to increase the volume of blood delivered to the active muscles. At rest and during submaximal exercise, this is accomplished by increasing cardiac output. Recall that cardiac output is the product of stroke volume and heart rate, so increasing either of these will increase cardiac output.

During standardized submaximal exercise performed during the first few hours at altitude, stroke volume is decreased when compared to sea level values (attributable to the re-

duced plasma volume). Fortunately, heart rate is increased to compensate for the decrease in stroke volume and to slightly increase cardiac output. However, making the heart take on this extra workload for prolonged periods is not an efficient way to ensure sufficient oxygen delivery to the body's active tissues. Consequently, after a few days at altitude, the muscles begin extracting more oxygen from the blood (increasing the arteriovenous oxygen difference), which reduces the demand for increased cardiac output, in turn reducing the need for an elevated heart rate. It has been shown that after 10 days at high altitude, cardiac output during a given exercise bout is lower than it was at sea level before these adaptations to altitude occurred.[17]

At maximal or exhaustive work levels at higher altitudes, both maximal stroke volume and maximal heart rate are decreased. The decrease in stroke volume is directly related to the decrease in plasma volume. Maximal heart rate may be somewhat lower at high altitude as a consequence of a decrease in the response to sympathetic nervous system activity, possibly attributable to a reduction in β-receptors

▶ The hypoxic conditions (diminished oxygen supply) at altitude alter many of the body's normal physiological responses. Pulmonary ventilation increases, resulting in a hyperventilated state in which too much carbon dioxide can be cleared, leading to respiratory alkalosis. In response, the kidneys excrete more bicarbonate ion, so less acid can be buffered.

▶ Pulmonary diffusion is not hindered by altitude, but oxygen transport is slightly impaired because hemoglobin saturation at altitude is reduced, although by only a small amount.

▶ The diffusion gradient that allows oxygen exchange between the blood and active tissue is substantially reduced at moderate and high altitudes; thus, oxygen uptake is impaired. A decrease in plasma volume of up to 25%, concentrating the red blood cells and allowing more oxygen to be transported per unit of blood, partially compensates for this impaired oxygen uptake.

▶ Maximal oxygen consumption decreases in proportion to the decrease in atmospheric pressure. During submaximal work upon the initial ascent to altitude, the body increases its cardiac output by increasing the heart rate to compensate for the decrease in the pressure gradient that drives oxygen exchange.

▶ During maximal work at altitude, stroke volume and heart rate are both lower, which reduces cardiac output. This reduced cardiac output, combined with the decreased pressure gradient, severely impairs oxygen delivery and uptake.

▶ Because oxygen delivery is restricted at altitude, oxidative capacity is decreased. More anaerobic energy production must occur, as evidenced by increased blood lactate levels for a given submaximal work rate above lactate threshold. However, at maximal work rates, lactate levels are lower, perhaps because maximal

exercise is limited by an inability to tolerate maximal exercise loads or because the muscle's glycolytic capacity is limited by an intolerance to H^+ accumulation (i.e., decreased buffering capacity) on initial ascent to altitude.

(receptors in the heart that respond to sympathetic nerve activation, thus increasing the heart rate). The subsequent effect of these changes in maximal heart rate and stroke volume is obvious: Maximal cardiac output decreases. With a decreased diffusion gradient to push oxygen from the blood into the muscles coupled with this reduction in maximal cardiac output, we can easily understand why both $\dot{V}O_2max$ and aerobic performance are hindered at altitude. Thus, hypobaric conditions significantly limit oxygen delivery to the muscles, reducing your capacity to perform high-intensity aerobic activities.

Pulmonary Hypertension Blood pressure in the pulmonary arteries is increased during exercise at altitude. These pressure changes are observed in both acclimatized and unacclimatized subjects.[18, 23] The cause for this altitude-induced pulmonary hypertension is not fully understood, but this condition is presumed to indicate some structural changes in the pulmonary arteries in addition to **hypoxic vasoconstriction**.[18] Pulmonary diffusion is not greatly affected at altitude. This is attributable to both the increased pulmonary artery blood pressure and increased ventilation, which allow greater blood flow to the upper regions of the lung. This increases the total area of blood and lung interface.

Metabolic Adaptations to Altitude

Given the hypoxic conditions at altitude, we would expect anaerobic metabolism to increase during exercise to meet the body's energy demands because oxidation would be limited. If this occurs, we would expect lactic acid production to increase at any given work rate above lactate threshold. This is in fact the case, except at maximal effort, during which lactate concentration in the muscles and blood is lower.[16, 39] Some researchers propose that this depression in lactic acid concentration at maximal effort is attributable to the body's inability to reach a work rate that fully taxes the energy systems. It has also been suggested that living at altitude reduces the blood bicarbonate concentration, thereby limiting the muscle's capacity to tolerate hydrogen ion (H^+) accumulation. Consequently, the production of energy and lactic acid via glycolysis may be limited. To date, there is no conclusive explanation why a lower lactate concentration occurs during maximal effort at altitude than at sea level.

Performance at Altitude

The difficulty of exercise at high altitude has been described by many climbers. In 1925, E.G. Norton[29] gave the following account of climbing without supplemental oxygen at 8,600 m (28,208 ft): "Our pace was wretched. My ambition was to do 20 consecutive paces uphill without a pause to rest and pant elbow on bent knee, yet I never remember achieving it—13 was nearer the mark." In this section we briefly consider how performance is affected by altitude.

Endurance Activity

Obviously, activities of long duration that place considerable demands on oxygen transport and the aerobic energy system are the most severely affected by the hypobaric conditions at altitude. At the summit of Mount Everest, $\dot{V}O_2max$ is reduced to 10% to 25% of its value at sea level. This severely limits the body's exercise capacity. Because $\dot{V}O_2max$ is reduced by a certain percentage, individuals with larger aerobic capacities can perform a standard work task with less perceived effort and with less cardiovascular stress at altitude than those with lower $\dot{V}O_2max$ values. This may explain

Endurance athletes can prepare themselves for competition at altitude through high-intensity endurance training at sea level, or at whatever elevation they live, for the purpose of increasing their $\dot{V}O_2max$. Then, on arrival at altitude, competition at any given rate of work can be performed at a lower percentage of their $\dot{V}O_2max$.

how Messner and Habeler were able to reach the summit of Everest without supplemental oxygen in 1978. They obviously possessed high sea-level $\dot{V}O_2max$ values.

Anaerobic Sprinting, Jumping, and Throwing Activities

Whereas endurance events are impaired at altitude, anaerobic sprint activities that last less than a minute (such as swimming sprints) are generally not impaired by moderate altitude and can be improved. Such activities place minimal demands on the oxygen transport system and aerobic metabolism. Instead, most of the energy is provided through the adenosine triphosphate-phosphocreatine and glycolytic systems.

In addition, the thinner air at altitude provides less aerodynamic resistance to athletes' movements. At the 1968 Olympic Games, for example, the thinner air of Mexico City clearly aided the performances of certain athletes, as we saw at the beginning of this chapter. At Mexico City, world or Olympic records were set or tied in the men's 100-m, 200-m, 400-m, 800-m, long jump, and triple jump events and in the women's 100-m, 200-m, 400-m, 800-m, 4 × 100 relay, and long jump events. Similar results occurred in both men's and women's swimming for the events up to 800 m.

Exhaustive Activity

Exhaustive exercise at high altitude produces lower lactate levels in the blood and muscle than exercise performed at sea level. As mentioned earlier, with limited oxygen uptake and increased reliance on anaerobic energy production, we would expect the muscles to produce more, rather than less, lactate for a given all-out effort. But studies in the 1930s showed remarkable changes in peak blood lactate values during exhaustive exercise at high altitude (7.9 mmol/L at sea level compared with 1.9 mmol/L at 5,340 m, or 17,520 ft). This phenomenon might be related to decreased muscle glycolytic enzyme activities, reduced buffering capacity, or the reduced power output that occurs in an all-out effort at altitude rather than at sea level. More recent studies by Saltin and colleagues,[33] however, have shown that muscle buffer capacity is increased with altitude training. One would therefore anticipate that altitude-trained individuals might be capable of generating more lactate during exhaustive exercise, but this is not the case. At this time, there does not appear to be an explanation for the paradox of lower lactate at altitude.

> ▶ Endurance activity suffers the most in hypobaric conditions because oxidative energy production is limited.
>
> ▶ Anaerobic sprint activities that last 2 min or less are generally not impaired at moderate altitude.
>
> ▶ The thinner air at altitude provides less resistance to movement; this is a major reason for the amazing performances of sprint runners and long and triple jumpers at the 1968 Olympic Games in Mexico City.

Acclimatization: Prolonged Exposure to Altitude

A number of investigations have examined human habituation to altitude. When people are exposed to altitude for days and weeks, their bodies gradually adjust to the lower oxygen tension in the air. But, however well they acclimatize to the conditions at high altitude, these people never fully compensate for hypoxia. Even endurance-trained athletes who live at altitude for years never attain the level of performance or the $\dot{V}O_2$max values that they might achieve at sea level.

In the following sections, we examine some of the adaptations that occur with prolonged altitude exposure. We focus on blood, muscle, and cardiorespiratory adaptations.

Blood Adaptations

During the first weeks at altitude, the number of circulating erythrocytes increases. The lack of oxygen at altitude stimulates the release of erythropoietin. Within the first 3 h after the athlete arrives at a high elevation, the blood's erythropoietin concentration increases, reaching a maximum within 24 to 48 h.[52] After a person lives at 4,000 m (13,120 ft) for about 6 months, his or her total blood volume increases by about 9% to 10% as a result not only of this altitude-induced stimulation of erythrocyte production but also of plasma volume expansion (discussed later).[30]

The percentage of total blood volume composed of erythrocytes is referred to as the hematocrit. Residents in the central Andes of Peru (4,540 m, or 14,895 ft) have an average hematocrit of 60% to 65%. This is considerably higher than the average hematocrit of sea-level residents, which is only 45% to 48%. However, during 6 weeks of exposure to the Peruvian altitude, sea-level residents have shown remarkable increases in their hematocrit levels, up to an average of 59%.

As the volume of erythrocytes increases, so does the blood's hemoglobin content. As noted in figure 11.4, blood hemoglobin concentration tends to increase proportionately with increases in elevation. These data are for men. Limited data exist for women, however the data show a similar trend but with a lower concentration than men at a given altitude. These adaptations improve the oxygen-carrying capacity of a fixed volume of blood.

At altitude, plasma volume initially decreases, and the number of red blood cells increases; both changes result in more blood cells and thus more hemoglobin for a given volume of blood. This increases the viscosity and oxygen-carrying capacity of blood, which are beneficial at rest and during low levels of exercise, but increased viscosity can impair blood flow during near-maximal levels of exercise.

The reduction in plasma volume reduces total blood volume, thus reducing maximal cardiac output. But with acclimatization, as plasma volume increases over several weeks at altitude and as red blood cells continue to increase, maximal cardiac output increases but does not return to sea-level values. Thus, maximal oxygen delivery capacity is increased with acclimatization but not to the extent needed to achieve sea-level $\dot{V}O_2$max values.

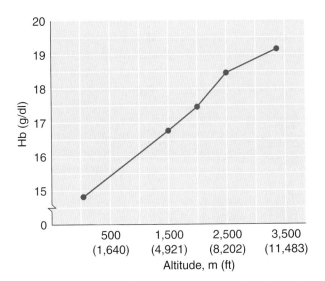

▲ **Figure 11.4** Hemoglobin (Hb) concentrations in male residents at various altitudes.

Muscle Adaptations

Although few attempts have been made to study muscular changes that occur during exposure to altitude, sufficient muscle biopsy data exist that indicate that muscles undergo significant structural and metabolic changes during ascent to altitude. Table 11.2 summarizes some of the muscle adaptations that accompanied 4 to 6 weeks of chronic hypoxia during expeditions to Mount Everest and Mount McKinley. Muscle fiber areas decreased, thus decreasing total muscle area. Capillary density in the muscles increased, which allowed more blood and oxygen to be delivered to the muscles. The cause for these changes is currently being debated. Muscles' inability to meet exercise demands at high altitude might be related to a decrease in their mass and their ability to generate adenosine triphosphate.

Prolonged exposure to high altitude frequently causes a loss of appetite and a noticeable weight loss. During a 1992 expedition to climb Mount McKinley, six men experienced an average weight loss of 6 kg (13.2 lb; D.L. Costill et al., unpublished data, 1992). Although part of this loss represents a general decrease in body weight and extracellular water, all the men experienced a noticeable decrease in muscle mass. As shown in table 11.2, this is associated with a significant decrease in the cross-sectional area of both the slow-twitch (ST) and the fast-twitch (FT) muscle fibers. It seems logical to assume that this decrease in muscle mass is associated with loss of appetite and a wasting of muscle protein. Perhaps future studies on nutrition and body composition in mountain climbers will provide a more thorough explanation of the incapacitating influences of high altitude on muscle.

Table 11.2

Changes in Muscle Structure and Metabolic Potential During 4 to 6 Weeks of Chronic Hypoxia

Parameter	Change	% change
Muscle area	Decreased	11-13
Slow-twitch fiber area	Decreased	21-25
Fast-twitch fiber area	Decreased	19
Capillary density (capillaries per mm²)	Increased	13
Succinate dehydrogenase	Decreased	25
Citrate synthase	Decreased	21
Phosphorylase	Decreased	32
Phosphofructokinase	Decreased	48

Several weeks at high altitude (above 2,500 m, or 8,200 ft) reduce the metabolic potential of muscle. Although this may not occur at lower elevations, at the higher elevations of Mount McKinley and Mount Everest, the mitochondria and glycolytic enzyme activities of the leg muscles (vastus lateralis and gastrocnemius) are significantly reduced after 3 to 4 weeks. This suggests that, in addition to receiving less oxygen, muscles lose some of their capacity to perform oxidative phosphorylation and to perform both aerobic and anaerobic exercise. Unfortunately, no muscle biopsy data have been obtained from residents at high altitudes to determine whether these individuals experience any muscular adaptations as a consequence of living at these elevations.

Cardiorespiratory Adaptations

One of the most important adaptations to altitude is an increase in pulmonary ventilation, both at rest and during exercise. Ventilation is stimulated by the decreased oxygen content of the inspired air at altitude. At an elevation of 4,000 m (13,123 ft), ventilation can increase by 50% at rest and during submaximal exercise. However, as noted earlier, such hyperventilation also promotes the unloading of CO_2 and the alkalization of blood. To prevent the blood from becoming abnormally alkaline, the amount of blood bicarbonate decreases rapidly during the first few days at altitude and remains depressed throughout the stay at high elevations.

Studies conducted on endurance-trained runners in the late 1960s reported that the decrement in $\dot{V}O_2$max when they first reached higher altitude improved little during their exposure to hypoxia. Aerobic capacity remained unchanged after 18 to 57 days at altitude.[1, 6, 17] Although the runners who previously had been exposed to altitude were more tolerant of hypoxia, their $\dot{V}O_2$max values and running performance were not significantly improved with acclimatization. Because of the blood changes that occur during these relatively long visits to altitude, this lack of improvement in aerobic endurance was somewhat unexpected. Perhaps these trained subjects had already attained maximal training adaptations and were

unable to further adapt in response to altitude exposure. Or perhaps the reduced PO_2 of altitude made it more difficult for them to train at the same intensity and with the same volume as at sea level.

▶ Hypoxic conditions stimulate the release of erythropoietin, which increases erythrocyte (red blood cell) production. More red blood cells mean more hemoglobin. Although plasma volume decreases initially, which also concentrates the hemoglobin, it eventually returns to normal. Normal plasma volume plus additional red blood cells increases total blood volume. All these changes increase the blood's oxygen-carrying capacity.

▶ Total muscle mass decreases when at altitude, as does total body weight. Part of this decrease is from dehydration and appetite suppression. There is also protein breakdown in the muscles.

▶ Other muscle adaptations include decreased fiber area, increased capillary supply, and decreased metabolic enzyme activities.

▶ The decrease in $\dot{V}O_2$max with initial exposure to altitude does not improve much during several weeks of exposure.

Physical Training and Performance

We have considered the major changes that occur as the human body becomes acclimatized to altitude and how these adaptations affect performance at altitude. But what happens when you train at altitude to improve performance at sea level, or when you train at sea level but must compete at altitude?

Altitude Training for Sea-Level Performance

Can altitude training improve sea-level performance? Most studies have shown no improvement in sea-level performance following altitude training. In the few studies where altitude training was found to influence postaltitude sea-level performance, the subjects were not well trained before going to altitude. This makes it difficult to determine how much of their postaltitude improvement was attributable solely to training, independent of altitude.

Studying athletes at altitude poses a difficult problem because they are often unable to train at the same volume and intensity of effort as when at sea level. This was demonstrated in a group of elite female cyclists who performed self-selected maximum power outputs during high-intensity interval training. They completed trials under the following conditions: breathing atmospheric air (normoxia) and breathing a hypoxic gas mixture simulating 2,100 m or 6,888 ft. The athletes' sustained (10-min) and short-term (15-s) power outputs at maximal intensity were reduced under hypoxic conditions.[4]

In addition, the conditions at moderate to high altitude often cause athletes to dehydrate and to lose fat-free mass. These and other conditions tend to diminish the athletes' fitness and their tolerance for intense training. As a result, studies are difficult to interpret, and the debate over the value of altitude training for optimal performance continues.

> Athletes have used altitude training in an attempt to improve endurance performance. This form of training, however, is expensive (because of transportation, housing, food, and other costs), and the existing research conducted on endurance athletes does not fully support its effectiveness.

A strong theoretical argument can be made for altitude training. First, altitude training evokes substantial tissue hypoxia (reduced oxygen supply). This is thought to be essential for initiating the conditioning response. Second, the altitude-induced increase in red blood cell mass and hemoglobin levels improves oxygen delivery on return to sea level. Although evidence suggests that these latter changes are transient, lasting only several days, this still should provide an advantage for the athlete. Recent studies, however, have shown that typical training while living at altitude does not improve sea-level performance more than good sea-level training does. In addition, living at sea level and training in a hypobaric chamber to simulate altitude do not appear to provide any advantage over sea-level training. How might we use altitude to better prepare endurance athletes for competition? This is discussed in the accompanying sidebar.

Training for Performance at Altitude

What can athletes who normally train at sea level but must compete at altitude do to prepare most effectively for competition? Although research thus far is not conclusive, it appears that athletes have two options. One option is to compete within 24 h of arrival at higher altitudes. This does not provide much acclimatization, but the altitude exposure is brief enough that the classic symptoms of altitude sickness are not yet totally manifest. After the first 24 h, the athlete's physical condition worsens because of physiological responses to altitude, such as dehydration and sleep disturbances.

Another option is to train at higher altitudes for at least 2 weeks before competing. But not even 2 weeks are sufficient for total acclimatization: That would require a minimum of 4 to 6 weeks. A third option, for team sports requiring considerable endurance (such as basketball, volleyball, soccer, or football), is several weeks of intense aerobic training at sea level to increase the athletes' $\dot{V}O_2$max, which will let them compete at altitude at a lower relative intensity than those who haven't prepared in a similar manner.

Training for optimal adaptations at altitude requires an elevation between 1,500 m (4,921 ft), which is considered the lowest level at which an effect will be noticed, and 3,000 m (9,843 ft), which is the highest level for efficient conditioning. Work capacity is reduced during the initial days at altitude. For this reason, when first reaching higher altitudes, athletes should reduce workout intensity to between 60% and 70% of sea-level intensity, gradually working up to full intensity within 10 to 14 days.

Adaptations to altitude are generally responses to the hypoxia experienced there, so we might anticipate that similar adaptations could be achieved simply by breathing gases

Living High, Training Low

Researchers at the Institute for Exercise and Environmental Medicine in Dallas, Texas, conducted a series of studies in the mid-1990s to investigate altitude training for enhancing endurance performance. When living and training at altitude, athletes are faced with the problem that the intensity of training is reduced because aerobic capacity and cardiorespiratory function are reduced at altitude. Thus, although athletes gain certain physiological benefits from being at altitude, they lose training adaptations associated with higher intensities of training. The researchers' studies investigated the potential of having subjects live at moderate altitude but train at low altitude, where training intensity is not compromised. In one study,[26] researchers divided 39 competitive runners into three equal groups: One (the high–low group) lived at moderate altitude (2,500 m, or 8,200 ft) and trained at low altitude (1,250 m, or 4,100 ft); one group (high–high) lived and trained at moderate altitude (2,500 m, or 8,200 ft); and one group (low–low) lived and trained at low altitude (150 m, or 490 ft). Using a 5,000-m time trial as the primary performance outcome measure, the researchers found that the high–low group was the only group to significantly improve their running performance, even though both the high–low and high–high groups increased their $\dot{V}O_2$max values by 5% in direct proportion to their increase in red cell mass. Thus, there appear to be performance benefits from living at moderate altitude but going to lower elevations to maximize training intensity.

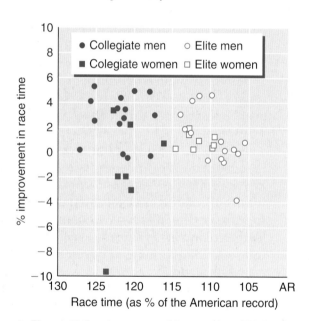

▲ **Figure 11.5** Improvement in race time (%) in elite men and women runners[38] and collegiate men and women runners[26] following 4 weeks of living at altitude but training at 1,250 m. See the text for details.

This was tested more recently by the same scientists working with a group of 14 elite men and 8 elite women runners, with all but two ranked in the United States top 50 for their event. These athletes lived at 2,500 m (8,200 ft) and trained at 1,250 m (4,100 ft) over a period of 27 days. Testing was conducted at sea level both the week before and the week after the 27 days of living at altitude. Sea-level 3,000-m time trial performance increased by 1.1% and $\dot{V}O_2$max increased by 3.2% as a result of this intervention.[37] Figure 11.5 illustrates the difference in race time performance for both studies, with values expressed as the percentage change before and after altitude exposure. These differences are plotted by the prealtitude race time, expressed as a percentage of the existing U.S. record in the event at the time of the time trial.

The favorable results of these studies on "living high and training low" have stimulated considerable interest in how this concept might be applied without having to send athletes to altitude to live. One approach has been to develop a hypoxic apartment where athletes sleep and live. The gas mixture inside the apartment is adjusted so that nitrogen represents a higher percentage of the inspired air, reducing the percentage of oxygen in the inspired air as well as its partial pressure. Pioneered by Finnish sport scientists, these apartments can simulate altitudes between 2,000 m (6,560 ft) and 3,000 m (9,840 ft) by adjusting the nitrogen and oxygen percentages in the inspired air to reduce the partial pressure of oxygen down to those levels associated with 2,000 to 3,000 m altitude. Hypoxic sleeping devices or tents also have been proposed. Unfortunately, at this time, few carefully controlled scientific data exist to confirm whether these apartments or sleeping devices actually improve performance and physiological function.[51]

with a low PO_2. But no evidence supports the idea that brief periods (1-2 h/day) of breathing hypobaric gases induces even a partial adaptation similar to that observed at altitude. On the other hand, Daniels and Oldridge[13] observed that alternate periods (lasting between 5 and 14 days) of training at 2,300 m (7,546 ft) and at sea level adequately stimulated altitude acclimatization. Staying at sea level for up to 11 days did not interfere with the usual adjustments to altitude, as long as training was maintained.

> ▶ Most studies show that training at altitude leads to no significant improvement in sea-level performance. The physiological changes that do occur, such as increased red blood cell production, are transient but could offer an advantage during the first few days after returning to sea level. This is still an area of debate. Living at high altitudes and training at low altitudes may be the best alternative.
>
> ▶ Athletes who must perform at altitude should do so within 24 h of arrival, before the detrimental changes that occur at altitude become too great.
>
> ▶ Alternatively, athletes who must perform at altitude could train at an altitude of 1,500 m (4,920 ft) to 3,000 m (9,840 ft) for at least 2 weeks prior to performing. This allows their bodies time to adapt to hypoxic and other environmental conditions at altitude.

Health Risks of Acute Exposure to Altitude

In addition to the cold, wind, and solar radiation that confront those who ascend to moderate and high altitudes, some people can experience symptoms of **acute altitude (mountain) sickness.** This disorder is characterized by symptoms such as headache, nausea, vomiting, dyspnea (difficult breathing), and insomnia. These symptoms typically begin 6 to 96 h after arriving at high altitude. Although not life threatening, acute altitude sickness can

be incapacitating for several days or longer. In some cases, the condition can worsen. The victim can develop the more lethal altitude illnesses of high-altitude pulmonary edema or high-altitude cerebral edema. Let's examine all three of these conditions, their causes, and precautions that can be taken to avoid them.

Acute Altitude Sickness

The incidence of acute altitude sickness varies with the altitude, the rate of ascent, and the individual's susceptibility.[19] Several studies have been conducted to determine the frequency of acute altitude sickness in groups of hikers. Reports vary widely, ranging from a frequency of 0.1% to 53% at altitudes of 3,000 to 5,500 m (9,843-18,045 ft). Forster,[15] however, reported that 80% of those who ascended to the top of Mauna Kea (4,205 m, or 13,796 ft) on the island of Hawaii experienced some symptoms of acute altitude sickness. The symptoms vary widely at 2,500 to 3,500 m (8,200-11,480 ft), altitudes commonly experienced by recreational skiers and hikers. At these elevations, the incidence of acute altitude sickness is about 6.5% for men and 22.2% for women. (The reason for this sex difference is unclear.[38])

Although the underlying cause of acute altitude sickness is not fully understood, several studies indicate that those people who experience the greatest distress also have a low ventilatory response to hypoxia.[20, 22, 27] Some people experience reduced breathing rate and depth during acute exposures to moderate and high altitude. This reduced ventilation allows carbon dioxide to accumulate in the tissues, and this may induce most of the symptoms associated with altitude sickness.

Another side effect of acute altitude sickness is an inability to sleep despite marked fatigue. Studies have shown that the inability to achieve satisfying sleep at altitude is associated with an interruption in the sleep stages.[48] In addition, some people suffer interrupted breathing, called Cheyne-Stokes breathing, which prevents them from relaxing and falling to sleep. Cheyne-Stokes breathing is characterized by alternate rapid breathing and slow, shallow breathing, usually including intermittent periods in which breathing completely stops.

The incidence of this irregular breathing pattern increases with altitude, occurring 24% of the time at 2,440 m (8,005 ft), 40% of the time at 4,270 m (14,009 ft), and 100% of the time at altitudes above 6,300 m (20,669 ft).[46, 49]

How can athletes avoid acute altitude sickness? We would like to think that well-trained individuals would be less susceptible to this disorder than poorly trained individuals, but no evidence indicates that physical conditioning prevents the symptoms of altitude sickness. Even athletes who are highly endurance trained before altitude exposure seem to have little protection against the effects of hypoxia.

Acute altitude sickness usually can be prevented by a gradual ascent to altitude and spending periods of a few days at lower elevations. A gradual ascent of no more than 300 m (984 ft) per day at elevations above 3,000 m (9,843 ft) has been suggested to minimize the risks of altitude sickness. Two drugs have been used to reduce the symptoms of those who develop acute altitude sickness: acetazolamide and dexamethasone. Both drugs must be used with medical supervision. Of course, the definitive treatment for severe acute mountain sickness is a retreat to lower altitude.

High-Altitude Pulmonary Edema

Unlike acute mountain sickness, **high-altitude pulmonary edema (HAPE)**, which is the accumulation of fluids in the lungs, is life threatening. The cause of HAPE is unknown. It seems to occur most frequently in people who rapidly ascend to altitudes above 2,700 m (8,858 ft). This disorder occurs in otherwise healthy people and has been reported more often in children and young adults. The fluid accumulation interferes with air movement into and out of the lungs, leading to shortness of breath and excessive fatigue. Disruption of normal breathing impairs blood oxygenation, causing blueness of the lips and fingernails, mental confusion, and loss of consciousness. HAPE is treated by administering supplemental oxygen and moving the victim to a lower altitude.

High-Altitude Cerebral Edema

Some rare cases of **high-altitude cerebral edema (HACE)**, which is fluid accumulation in the cranial cavity, have been reported. This condition is characterized by mental confusion, progressing to coma and death. Most cases have been reported at altitudes greater than 4,300 m (14,108 ft). The cause of HACE is unknown, but the treatment is administration of supplemental oxygen and descent to a lower altitude.

> ▶ Acute altitude sickness typically causes symptoms such as headaches, nausea, vomiting, dyspnea, and insomnia. These usually appear 6 to 96 h after arrival at altitude.
>
> ▶ The exact cause of acute altitude sickness is not known, but many researchers suspect that the symptoms may result from carbon dioxide accumulation in the tissues.
>
> ▶ Acute altitude sickness usually can be avoided by a gradual ascent to altitude, climbing no more than 300 m (984 ft) per day at elevations above 3,000 m (9,843 ft). Medications can also be used to reduce the symptoms.
>
> ▶ High-altitude pulmonary edema and cerebral edema—which involve accumulation of fluid in the lungs and cranial cavity, respectively—are life-threatening conditions. Both are treated by oxygen administration and descent.

Hyperbaric Environments: Exercising Underwater

Recreational scuba diving and snorkel diving present a unique challenge to human physiology. Aside from the thermal effects of water (chapter 10), the body must endure the effects of a hyperbaric environment—an environment in which the pressure is greater than at sea level. This environment increases the pressure of the gases contained in the paranasal sinuses, the respiratory tract, and the gastrointestinal tract and the gases dissolved in body fluids.

In the following discussion we consider the physiological effects a person experiences when submerged underwater.

Water Immersion and Gas Pressures

A balloon that is filled with air shrinks rapidly when it is pushed only a few feet below the water's surface. Even above water, the air we breathe is under the pressure of the weight of the atmosphere (1 atmosphere, or 760 mmHg at sea level). When you descend to a depth of 10 m (33 ft) underwater, the water exerts an additional 760 mmHg of pressure on you. Because water is denser than air, the pressure under 10 m of water equals the pressure experienced with a 6,000-m (19,685-ft) descent into a mine shaft.

Recall that volume and pressure are inversely related, so as pressure increases, volume decreases. As shown in figure 11.6, the air you breathe into your lungs at the surface will be compressed to one half its volume when you descend to a depth of 10 m (33 ft). If you continue to descend to greater depths, the gas volume becomes progressively smaller. At a depth of 30 m (98 ft), for example, your lung volume will be reduced to 25% of its volume at the surface.

Conversely, air taken into your lungs at a depth of 10 m (33 ft) will expand to twice its original volume by the time you reach the water's surface. Let's consider an example of diving while breathing from a **self-contained underwater breathing apparatus (scuba)**. It would be extremely dangerous for you to take in a deep breath at a depth of 10 m and then hold this breath as you ascend to the surface. As you ascend, the air would expand in your lungs. Before you reached the surface, your lungs would overdistend, rupturing the alveoli and causing pulmonary hemorrhage and lung collapse. (This condition, known as spontaneous pneumothorax, is discussed later.) If air bubbles enter the circulatory system as a result of this extensive damage, emboli develop and can block major vessels, leading to extensive tissue damage, if not death. Thus, it is important for divers to always exhale as they ascend to the surface.

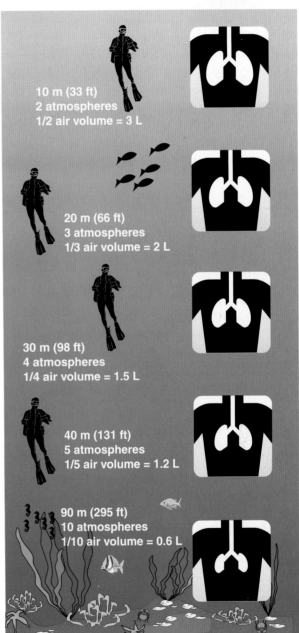

Sea level
1 atmosphere
Air volume = 6 L

10 m (33 ft)
2 atmospheres
1/2 air volume = 3 L

20 m (66 ft)
3 atmospheres
1/3 air volume = 2 L

30 m (98 ft)
4 atmospheres
1/4 air volume = 1.5 L

40 m (131 ft)
5 atmospheres
1/5 air volume = 1.2 L

90 m (295 ft)
10 atmospheres
1/10 air volume = 0.6 L

▲ **Figure 11.6**   The relationship between depth of water submersion and the volume of air in a diver's lungs.

Whereas the gases in the body are compressible, water and body fluids are noncompressible and thus are not measurably affected by either water depth or increased pressure. Nevertheless, we cannot ignore the pressure exerted by water on the gases (oxygen, nitrogen, and carbon dioxide) dissolved in body fluids. Breathing air at a depth of 10 m doubles the partial pressure of each of the gases (table 11.3). At depths approaching 30 m, the partial pressures of the dissolved gases are four times greater than on the surface. This increase in their partial pressures causes more molecules of these gases to dissolve in the body fluids. During too rapid an ascent, the partial pressures of the gases in the body fluids will exceed the water pressure. As a result, the tissue gases can come out of solution, forming bubbles. We discuss this condition, known as decompression sickness, later.

> ▶ Submersion in water exposes the human body to a hyperbaric environment, one in which the external pressure is greater than at sea level.
>
> ▶ Because air volume decreases as pressure increases, air that is in the body before it goes underwater is compressed when the body is submerged. Conversely, the air taken in when the diver is submerged expands during ascent.
>
> ▶ More molecules of gas are forced into solution when the body is submerged, but with a rapid ascent they come out of solution and can form bubbles.

Cardiovascular Response to Water Immersion

Water immersion reduces the stress on the cardiovascular system. Immersing the body to the neck applies pressure to the lower body, which tends to minimize blood pooling and facilitate blood return to the heart, reducing the cardiovascular system's work. In addition, plasma volume tends to increase, as seen by a reduction in hemoglobin and hematocrit. As a result, resting heart rate can drop by 5 to 8 beats/min with only partial body immersion. In addition, placing the face in the water lowers the heart rate even further. This is the result of a facial reflex common to many mammals.

In some diving animals (such as beavers, seals, and whales) heart rate can decrease (bradycardia) by 90% during diving. In humans, bradycardia is typically 60% of the predive heart rate.[36] For example, if your predive heart rate is 70 beats/min, it might decrease to only 40 to 45 beats/min during submersion. Cold water can exaggerate the decrease in both resting and exercise heart rate during immersion.[21] From a clinical point of view, the incidence of irregularities in cardiac conduction increases significantly when the water temperature is low. In other words, diving in cold water is associated with greater bradycardia and a higher incidence of cardiac arrhythmias. Of greatest importance is the fact that at a given exercise effort (i.e., oxygen uptake) in the water, a person's heart rate can be 10 to 12 beats/min lower than during similar effort on land, as seen in figure 11.7.

Table 11.3

Effects of Water Depth on the Partial Pressure of Inspired Oxygen (PO_2) and Nitrogen (PN_2)

Depth (m)	Total pressure (mmHg)	PO_2 (mmHg)	PN_2 (mmHg)
0	760	159	600
10	1,520	318	1,201
20	2,280	477	1,802
30	3,040	636	2,402

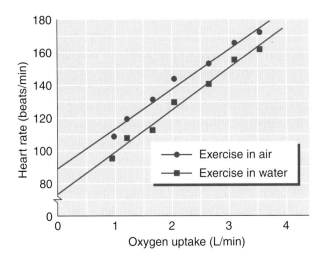

▲ **Figure 11.7**   The relationship between oxygen uptake and heart rate during exercise in air and water. On average, heart rates at various levels of exercise are about 10 to 12 beats/min lower in water than in air.

Breath-Hold Diving

Breath-hold diving is the oldest form of diving and is still used for recreation and work. The length of time that swimmers can hold their breath is determined by the breakpoint at which the stimulus to breathe becomes too strong to resist. The urge to breathe while underwater results from a buildup of arterial carbon dioxide, which is the greatest breathing stimulus, as you should recall from chapter 8. Voluntarily increasing either the rate or depth of breathing before a dive, as by hyperventilating, increases carbon dioxide removal from the body tissues. This can lengthen the time that you can hold your breath before you must breathe.[28] Remember, though, that hyperventilation does not increase the oxygen content of the blood. Thus, although hyperventilation may increase your breath-holding ability, it does not increase your oxygen reserves. Arterial oxygen can decrease to critically low levels in some individuals, causing them to lose consciousness before the carbon dioxide buildup in their blood forces them to the surface to breathe.[12]

Generally, swimming at the surface poses no pressure problems for the body's air compartments (lungs, respiratory passages, sinuses, middle ear, etc.), but breath-hold diving a meter or two below the surface can quickly pressurize these compartments. This can cause some discomfort to the ears and sinuses unless the gases trapped in these compartments are equalized to the pressure of the water by holding the nose closed and blowing air into the middle ear and sinuses.

As a breath-hold diver descends, the chest wall is squeezed and the volume of air in the lungs is reduced because of the increasing water pressure surrounding the diver. Eventually, lung volume can decrease to the residual volume of the lung, but no smaller. The residual volume is the amount of air that remains in the lungs after maximal exhalation, so it is the amount that cannot be exhaled. If the diver attempts to descend below that depth, the blood vessels in the lungs and respiratory tract can rupture because the blood pressure in the vessels exceeds the air pressure. Because of this, the depth limit for breath-hold diving is determined by the ratio between the diver's total lung volume (TLV) and residual volume (RV), called the **TLV/RV ratio**.

On average, adults have a TLV/RV ratio of 4:1 or 5:1. Water pressure at 20 to 30 m (66-98 ft) is usually sufficient to compress the volume of the chest and lungs to the residual volume. But people with large total lung volumes and small residual volumes can descend to greater depths before this point is reached. The ama (Japanese pearl divers) work daily at depths near the limits predicted by the TLV/RV ratio.

Gases trapped in the body are not the diver's only concern when facing changing water pressures with depth. The air trapped in the diver's goggles or mask also is compressed. Compression of this air can limit the maximum depth for unassisted diving because blood vessels in the eyes and face can rupture if the air is too compressed. To minimize this problem, pearl divers who regularly descend to depths of at least 5 m (16 ft) wear goggles that trap only a very small air volume or that can be pressure equalized with air from the nose or the mouth. Goggles that reduce the amount of air that can be compressed minimize the risk of injury to the blood vessels of the eye.

Scuba Diving

The air drawn into the lungs when the chest is submerged under only a few feet of water must be pressurized to equal the water pressure. Scuba is the most popular apparatus for accomplishing this task. This equipment, created by Jacques Cousteau in 1943, is illustrated in figure 11.8 and consists of the following four components:

- One or more tanks of air compressed to about 2,000-3,000 psi
- A first-stage regulator valve for reducing the pressure from the tank to a lower, breathable pressure (about 140 psi)
- A second-stage regulator valve that releases air on demand at a pressure equal to that of the water
- A one-way breathing valve that allows the pressurized air to be drawn into the lungs and expired into the water

Because expired air does not return to the tanks, this form of scuba is referred to as an open-circuit-demand type. How long the diver can remain underwater depends on the depth of the dive: Deeper dives demand greater airflow to compensate for the greater water pressure. Because the amount of air the diver needs varies with depth, the supply of air from a scuba tank is limited by the depth of the dive. The contents of a single air tank, for example, can be exhausted in only a few minutes at a depth of 60 to 70 m (197-230 ft) but can last for 30 to 40 min when the diver is at a depth of 6 to 7 m (20-23 ft).

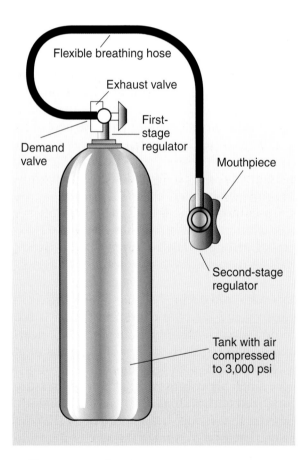

▲ **Figure 11.8** The open-circuit-demand type of scuba system. When the diver draws air from the tank, the regulator adjusts the pressure of the gas to match the surrounding water pressure, thereby enabling the diver to inhale.

▶ Water reduces the stress on the cardiovascular system, reducing its workload. When the body is submerged, the increased pressure on the lower body facilitates the return of blood to the heart. Plasma volume also increases. Because of these factors, resting and exercise heart rate decrease even when the body is only partially submerged. This effect is enhanced by cold water.

▶ Divers often practice hyperventilation before breath-hold diving to increase the time they can hold their breath. But this can lead to dangerously low oxygen levels, which can cause divers to lose consciousness underwater.

▶ During breath-hold diving, the gases in the body become pressurized even when swimming at a depth of only 1 to 2 m (3-6 ft) below the surface. At greater depths, the volume of air in the lungs can be reduced to the residual volume, but no smaller.

▶ The depth limit to breath-hold diving is determined by the ratio of total lung volume to residual volume. Those with large TLV/RV ratios can safely dive deeper than those with smaller ratios.

▶ Scuba diving can alleviate many of the problems faced during breath-hold diving because the diver breathes pressurized air while submerged.

Health Risks of Hyperbaric Conditions

The development of the underwater breathing apparatus enabled deeper and longer duration dives. However, this capability also presents divers with additional health risks. As a diver descends, the air in the breathing apparatus must be pressurized to equal the pressure exerted by the water. In turn, this increases the partial pressures of all gases in the mixture. This increases the pressure gradient that drives oxygen and nitrogen into the body tissues, and the increased alveolar partial pressure of carbon dioxide decreases the pressure gradient that allows it to be cleared by the lungs. Thus, breathing oxygen, carbon dioxide, and nitrogen under pressure can cause the body tissues to accumulate toxic levels of these gases.

Oxygen Poisoning

Exposure to a PO_2 ranging from 318 to 1,500 mmHg has been shown to have severe effects, particularly on the lungs and the central nervous system.[7, 35] A high PO_2 in inspired air can force enough oxygen into solution in the plasma that the dissolved oxygen can supply the metabolic needs of the diver, resulting in less oxygen unloading from hemoglobin at the tissues. Because of this, the hemoglobin in the venous blood remains highly saturated with oxygen.

Carbon dioxide does not bind as well to hemoglobin that is fully saturated with oxygen, so reduced oxygen unloading impairs carbon dioxide elimination via hemoglobin. In addition, when the diver breathes oxygen at a PO_2 greater than 318 mmHg (twice the normal atmospheric PO_2), cerebral blood vessels can become constricted, severely restricting the circulation to the central nervous system. This can result in such symptoms as visual distortion, rapid and shallow breathing, and convulsions. In some cases, this high PO_2 can irritate the respiratory tract, eventually leading to pneumonia. This condition, which results from excessive oxygen, is referred to as **oxygen poisoning.**

Decompression Sickness

The high partial pressure of nitrogen experienced during diving forces more nitrogen into the blood and tissues. If the diver attempts to ascend too rapidly, this additional nitrogen cannot be delivered to and released from the lungs quickly enough, and it becomes trapped as bubbles in the circulatory system and the tissues. This causes severe discomfort and pain and is referred to as **decompression sickness,** or the bends. Typically, this disorder is characterized by aching in the elbows, shoulders, and knees, where nitrogen bubbles accumulate. If bubbles become emboli in the blood, they can interfere with normal circulation, and this can be fatal.

Treatment involves placing the diver in a recompression chamber (see figure 11.9 on page 358). The air pressure is increased (recompressed) in the chamber to simulate that experienced while diving, and then it is gradually returned to the ambient pressure. **Recompression** forces the nitrogen back into solution, and then a gradual decrease in pressure allows the nitrogen to escape through the respiratory system.

To prevent decompression sickness, charts have been created that provide timetables for ascending from various depths. If, for example, a diver were to submerge to a depth of about 15 m (49 ft) for an hour, decompression would not be needed. But if the diver spent an hour at a depth of about 30 m (98 ft), slow decompression would be necessary. Strict adherence to the timetable for a specific diving depth allows a safe ascent without decompression sickness.

Nitrogen Narcosis

Although nitrogen is not metabolically active, meaning it doesn't participate in biological processes in our bodies, at high pressures—such as during deep dives—it can act much like an anesthetic gas. The resulting condition is referred to as **nitrogen narcosis,** or **rapture of the deep.** The diver develops symptoms similar to alcohol intoxication. The effect worsens as depth, and consequently pressure, increase. In fact, research suggests that for every 15-m (49-ft) increase in depth, this effect equals that of one martini ingested on an empty stomach.

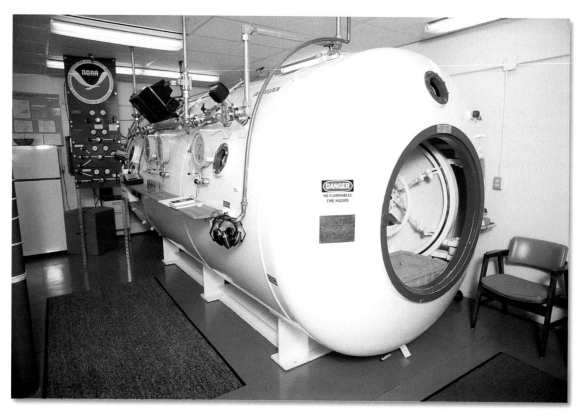

▲ **Figure 11.9** A hyperbaric chamber is used to increase the partial pressure of gases in and around the body and then slowly reduce them. Thus, individuals who have been exposed to high pressures can slowly return to lower atmospheric pressures and gradually release the excess gas (e.g., nitrogen) that has built up in their body tissues.

Divers at depths of 30 m (98 ft) or more have impaired judgment but might not recognize the problem. Poor judgment during diving can be life threatening, so most divers who descend below 30 m breathe a specialized gas mixture containing mostly helium.

Spontaneous Pneumothorax

Breathing pressurized gas at a depth of more than a meter or two below the surface can create a major problem if the gas is not expelled during ascent. A full breath taken at 2 m (6.6 ft) and held will expand enough during ascent to overdistend the lungs. This can rupture alveoli, allowing gas to enter the pleural space and in turn collapse the lung. This is known as a **spontaneous pneumothorax**. At the same time, small air bubbles can enter the pulmonary blood and form air emboli, which can become trapped in the vessels of other tissues, blocking circulation to those tissues. Severe blockage of the vessels supplying the lungs, myocardium, and central nervous system can cause death.

Fortunately, this condition can be prevented simply by keeping the mouth open and exhaling during ascent, allowing the compressed air in the respiratory passageways to escape.

Ruptured Eardrum

Failure to equalize the air pressure in the sinuses and middle ear during ascent and descent can rupture the small blood vessels and membranes that line these cavities. Pressure in the middle ear is normally allowed to equalize through the eustachian tube (which connects the middle ear to the throat). Inability to equalize the pressure in the middle ear creates unequal force against the eardrum, causing considerable pain. Under severe conditions, as during ascent or descent in deep water, an inability to equalize this pressure can rupture the eardrum.

During diving, the middle ear and the sinuses usually can be equalized by blowing with moderate pressure against the closed nostrils. But because upper respiratory infections and sinusitis can cause swelling of the membranes in the sinuses and eustachian tube, scuba and breath-hold diving are not recommended for people suffering from these conditions.

Some of the health risks associated with hyperbaric conditions are depicted in figure 11.10. The preceding sections do not cover all of the health risks associated with hyperbaric conditions. Rather, we have provided only an overview of some common and more serious risks. Diving can be dangerous for the inexperienced, and even the most experienced divers can get into trouble if they don't follow proper procedures or if they ignore the health risks associated with this sport.

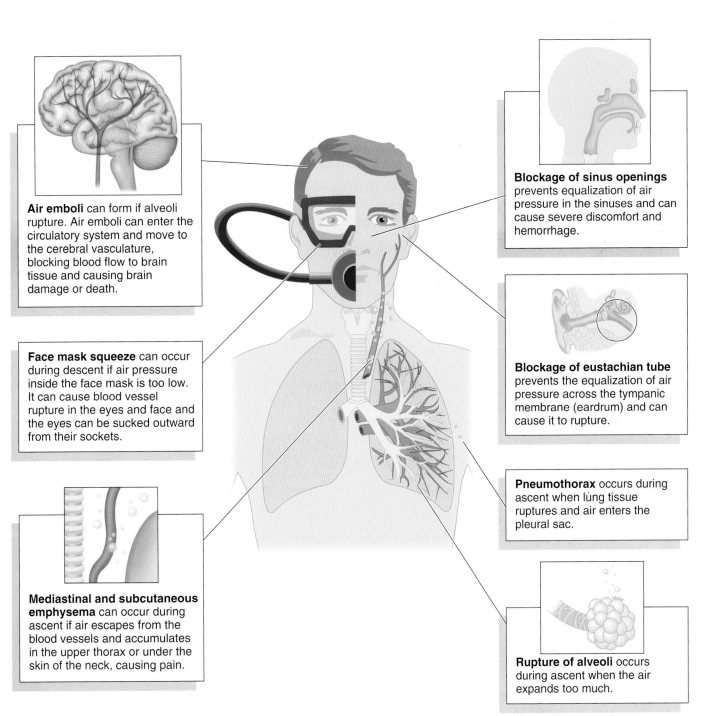

Air emboli can form if alveoli rupture. Air emboli can enter the circulatory system and move to the cerebral vasculature, blocking blood flow to brain tissue and causing brain damage or death.

Face mask squeeze can occur during descent if air pressure inside the face mask is too low. It can cause blood vessel rupture in the eyes and face and the eyes can be sucked outward from their sockets.

Mediastinal and subcutaneous emphysema can occur during ascent if air escapes from the blood vessels and accumulates in the upper thorax or under the skin of the neck, causing pain.

Blockage of sinus openings prevents equalization of air pressure in the sinuses and can cause severe discomfort and hemorrhage.

Blockage of eustachian tube prevents the equalization of air pressure across the tympanic membrane (eardrum) and can cause it to rupture.

Pneumothorax occurs during ascent when lung tissue ruptures and air enters the pleural sac.

Rupture of alveoli occurs during ascent when the air expands too much.

▲ **Figure 11.10**   Some health risks associated with hyperbaric conditions.

Staying Underwater

A series of SEALAB projects conducted by the U.S. Navy have enabled divers to live at depths of 60 to 260 m (197-853 ft) for up to 30 days. To allow these long stays, the Navy developed a technique called saturation diving. The principle behind this technique is that, at a given depth, the amount of metabolically inactive gases (such as nitrogen) that can dissolve in body tissues is limited. When a person lives in a pressurized housing for about 24 h, the body tissues become saturated with nitrogen gas. After saturation is reached, the tissues do not absorb significantly more nitrogen, no matter how long the diver stays at that depth. For time-consuming underwater work, it is far more efficient to stay at depth and complete the job than to repeatedly return to the surface, spending hours in decompression before each return.

Although brief ascents and descents to depths of nearly 100 m (328 ft) are possible with adequate precautions and decompression, the Navy's saturation diving program with SEALAB I, II, and III has brought to light some pathological problems associated with prolonged habitation in hyperbaric conditions. These problems generally relate to the narcotic effects of nitrogen. Substituting helium for nitrogen appears to cause far fewer effects, although it is difficult for divers to talk (helium causes the voice to sound like Donald Duck's).

▶ Breathing gases under pressure can cause gases to accumulate in the body in toxic levels, so precautions must be taken when one is diving with pressurized gases.

▶ Oxygen poisoning occurs when PO_2 values are above 318 mmHg. The tissues get their oxygen from that dissolved in the plasma; thus, less oxygen is removed from the hemoglobin. This impairs the binding of carbon dioxide to hemoglobin, so less carbon dioxide is removed by this route. Also, high PO_2 causes vasoconstriction in the cerebral vessels, which decreases blood flow to the brain.

▶ Decompression sickness (the bends) results from ascending too rapidly. The nitrogen dissolved in the body cannot be removed by the lungs quickly enough, so it forms bubbles. The bubbles can form emboli, which can be fatal. To treat this condition, the diver must undergo recompression to force the nitrogen back into solution and then undergo gradual decompression at a rate that allows the nitrogen to be removed during normal breathing. Tables have been formulated that specify how much time must be al-

lowed for ascent from various depths, and divers must adhere strictly to these timetables.

▶ Nitrogen narcosis (rapture of the deep) results from the narcotic effects of nitrogen when its partial pressure is high, such as during depth diving. The symptoms are similar to alcohol intoxication. Judgment is impaired, which can lead to fatal mistakes.

▶ Spontaneous pneumothorax and ruptured eardrum are other health risks associated with the changing pressures experienced when diving.

Microgravity Environments: Exercising in Space

The human body has a tremendous capacity to adapt to considerable environmental variations. In this and the previous chapter, we have discussed adaptations in response to heat, cold, humidity, and hypobaric (altitude) and hyperbaric (diving) conditions. Now we shift our focus to an unusual condition that most of us will never experience: prolonged microgravity.

Earth's gravity produces a standard acceleration of 1 g (g is the symbol used for acceleration due to gravity). Microgravity refers to reduced gravity, so the term is used to describe any condition where the gravitational force is less than on the Earth's surface (less than 1 g). For example, the moon's gravitational pull is only about 17% of that experienced on Earth, or 0.17 g. The term microgravity is often used to describe conditions in space because the body might not always be in a true weightless or 0-g state.

It is interesting that most physiological changes that accompany exposure to microgravity in many respects mimic detraining responses seen in athletes during periods of inactivity or immobilization or the changes associated with aging that likely result from reduced activity. This link is substantiated by evidence that exercise training during periods of exposure to microgravity is a successful countermeasure against the physiological deterioration that occurs in space. For these reasons, and because space exploration is ongoing, the effects of microgravity on physical activity is an area of increasing interest for exercise and sport physiologists.

> Most physiological changes that occur as a result of extended exposure to microgravity conditions during spaceflight are similar to those seen with detraining in athletes and with reduced activity in older people.

Physiological Alterations With Chronic Microgravity Exposure

Microgravity presents a challenge to normal body functions. An object's weight, which reflects the strength of the gravity pulling on it, decreases as the object moves away from the Earth's surface. At a distance of 12,875 km (8,000 mi), for example, a body's weight is only about 25% of its value on Earth. At 337,962 km (210,000 mi), the body becomes weightless because the gravitational acceleration is 0 g. If your body is weightless, your weight-bearing bones and antigravitational muscles (those that maintain your posture) are unloaded. Reducing the stress on both bone and muscle eventually leads to their deterioration and reduces their ability to function. Similar reductions are seen in cardiovascular function.

But what might be perceived as maladaptation might be in fact a needed adaptation to microgravity. In this section, we briefly review physiological changes that occur with chronic exposure to microgravity, focusing on muscle, bone, cardiovascular function, and body mass and composition. Whenever possible, we use actual data from spaceflight. Otherwise, the data represent those obtained in simulated microgravity.

Muscle

Studies have shown rapid changes in muscle structure and function when limbs are immobilized by a cast or when a leg is suspended. Muscle atrophy results primarily from decreased protein synthesis.[32] Experiments with rats have concluded that the protein synthesis rate decreases approximately 35% in the first several hours and up to 50% in the first several days of immobilization, leading to a net loss of muscle protein.[3, 40] The resulting muscle atrophy can be substantial over time. Similar studies with humans, however, have shown a slower loss in muscle mass and protein during muscle unloading (i.e., bed rest and leg suspension).[42,50] Also, there is a major difference between immobilization and the microgravity conditions of spaceflight. With immobilization, the immobilized muscle undergoes little or no activation. During spaceflight, however, the muscles are activated and shorten but exert far less force to perform a given task as a result of the loss of gravity. During a 17-day Space Shuttle mission in 1996 (STS-78), little or no change in calf muscle strength was found among four crew members, despite the fact that the muscle fibers decreased in cross-sectional area by about 8% to 11%.[14]

Bed-rest studies that simulate microgravity have found major decreases, both in strength and cross-sectional area, of both ST and FT muscle fibers; FT fibers tended to be most affected.[10] These relationships are illustrated in figure 11.11; note the close agreement for

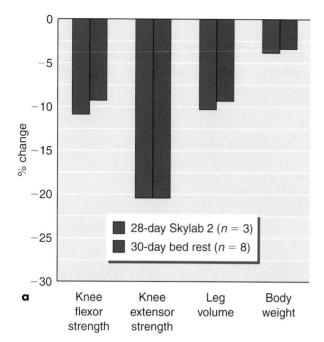

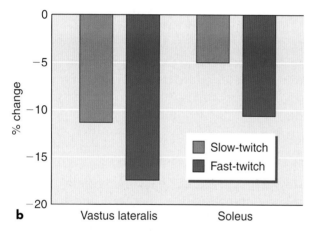

▲ **Figure 11.11** (*a*) Leg strength, leg volume, and body weight changes with 30 days of bed rest and a 28-day spaceflight. (*b*) Changes in cross-sectional area of slow-twitch and fast-twitch muscle fibers following 30 days of bed rest.

Adapted, by permission, from V.A. Convertino, 1991, "Neuromuscular aspects in development of exercise countermeasures," *The Physiologist* 34: S125-S128.

changes in leg strength, leg volume, and body weight between a 30-day bed-rest trial to simulate microgravity and the 28-day Skylab 2 spaceflight. Unlike the calf muscles in the 17-day Space Shuttle mission discussed in the previous paragraph, the upper leg muscles during this 28-day Skylab 2 mission did lose strength—up to 20% for knee extensor strength. The difference in results from these two flights

could be explained by the duration of the flight, the in-flight exercise program, the muscle group, or any combination of these three.

> Strength and the cross-sectional areas of ST and FT muscle fibers decrease with exposure to simulated and actual microgravity.

We must recognize the potential for muscle atrophy and loss of strength from exposure to microgravity. But results from the Skylab 4 mission and subsequent missions suggest that a well-designed exercise program can greatly attenuate the loss of muscle size and function.[41] More effective resistance training programs need to be developed to minimize the loss of muscle function during periods of microgravity. Astronauts could be faced with emergency situations in space or on their return to Earth that might require high levels of strength. Because the postural muscles are most affected, creative exercise devices must be devised to place an adequate stress on these muscles in the absence of gravity.

Bone

Most of the body's larger bones depend on the daily loading of gravitational forces. There has been great concern that extended space explorations of 18 months or longer could result in significant skeletal degeneration and calcium loss, increasing the risk of bone fracture on return to Earth. Calcium-balance studies from the Gemini, Apollo, and Skylab missions indicated a negative calcium balance, largely attributable to increases in both urinary and fecal calcium excretion. Urinary excretion of hydroxyproline (a constituent of the bone protein collagen) also increased, which indicates bone resorption.

Early studies of the Gemini astronauts produced remarkable estimates of bone mineral loss:

- 2% to 15% in the calcaneus (heel)
- 3% to 25% in the radius
- 3% to 16% in the ulna

These estimates were later scaled down when it was discovered that a technical error had caused an overestimation of bone mineral loss. Conversely, no bone mineral losses were reported in the same bones following the flights of Apollo 14 and Apollo 16, and only slight calcaneal losses of 5% to 6% were reported in two Apollo 15 crew members.[31] With the Skylab missions, no bone mineral loss was observed in the radius or ulna, but calcaneal losses were approximately 4%, similar to what occurs during bed rest. The calcaneus is a weight-bearing bone, but the radius and ulna are not.

Microgravity generally results in bone mineral losses of about 4% in the weight-bearing bones, but the magnitude of this response depends on the length of exposure to microgravity. Furthermore, there is considerable individual variability, as illustrated in figure 11.12.[45] These data also demonstrate considerably greater losses in weight-bearing (tibia; Figure 11.12b) compared with non-weight-bearing bone (radius; Figure 11.12a) and in cancellous compared with cortical weight-bearing bone.

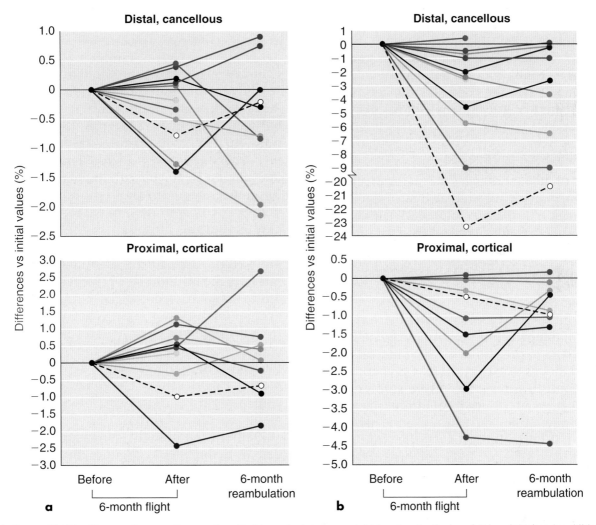

▲ **Figure 11.12** Changes in cancellous and cortical bone in (a) non-weight-bearing (radius) and (b) weight-bearing (tibia) bone in cosmonauts who were either controls (0 days of spaceflight) or experimental subjects (60-460 days of spaceflight). Data from Vico, et al. (2000). Effects of long-term microgravity exposure on cancellous and cortical weight-bearing bones of cosmonauts. *Lancet,* **355,** 1607-1611.[45]

Microgravity generally results in bone mineral losses of about 4% in the weight-bearing bones, but the magnitude of this response depends on the length of exposure to microgravity.

The mechanisms responsible for these bone changes have not been resolved. Bone formation could be retarded, bone resorption could be increased, or both. Furthermore, long-term consequences of these changes have not been established. It is not yet clear whether these bone mineral losses will be regained or whether repeated exposures to microgravity are cumulative so that astronauts will lose additional bone with each subsequent mission. It is reasonably clear, however, that the observed changes in bone with microgravity or bed rest result from the mechanical unloading of bone when the bone is no longer exposed to the normal gravitational or muscular forces experienced on Earth.

Cardiovascular Function

One of the first changes that occur in response to microgravity or simulated microgravity is a reduction in plasma volume. When the body is in microgravity, because there is a reduced hydrostatic pressure, blood no longer pools in the lower extremities as it does at $1.0\ g$. As a consequence, more blood returns to the heart, which leads to transient increases in cardiac output and arterial blood pressure. These increases are accompanied by an increase in the arterial pressure in the kidneys, which in turn causes the kidneys to excrete the excess volume. This response to increased blood pressure is referred to as pressure diuresis. Antidiuretic hormone, aldosterone, angiotensin, and atrial natriuretic factor (a hormone secreted by the heart's atria) also play roles in blood volume control, but pressure diuresis exerts the greatest control over blood volume in the microgravity state. These adaptations are essential to allow the body to regain its control over blood pressure.

The reduced blood volume serves astronauts well while they remain in microgravity. But it presents a serious problem on their return to a 1-g environment, where the body is again sub-jected to the hydrostatic pressure effect but now has a smaller blood volume. Astronauts have experienced postural (orthostatic) hypotension and fainting during their first few hours back in a normal 1-g environment because their blood volume was insufficient to meet all their circulatory needs.

Microgravity removes most of the effects of hydrostatic pressure experienced in a 1-g environment, causing the body to dump a large percentage of its plasma volume. Although this allows excellent regulation of cardiovascular function at rest and during exercise in space, it presents major orthostatic hypotension problems on return to Earth's atmosphere.

Soviet cosmonauts' preflight and in-flight resting cardiac functions and blood pressures were assessed during both the 23-day Salyut-1 mission and the 63-day Salyut-4 mission. The in-flight measurements were made between days 13 and 21 for Salyut-1 and at day 56 for Salyut-4. There were no differences between preflight and in-flight values for heart rate, stroke volume, and cardiac output, although in-flight systolic blood pressure was slightly elevated (see table 11.4). Also, during the longer Salyut-4 mission, the in-flight heart rate response to a 5-min standardized exercise bout on a cycle ergometer was not significantly different from the preflight, Earth-based value. Similar comparisons were made for the three Skylab missions. For exercise bouts at a constant submaximal work rate, the astronauts' preflight and in-flight heart rate and blood pressure responses did not differ.[9]

Preflight and in-flight submaximal exercise data from the 140-day flight of the orbital station Salyut-6 are also presented in table 11.4. The cosmonauts exercised at a constant work rate for 5 min. No change from preflight values was noted for any of the variables during the first month, but stroke volume, cardiac output, and systolic blood pressure all decreased and heart rate increased through 119 days. These variations were relatively small and could

Table 11.4

Cardiovascular Function at Rest and During Exercise During the Salyut-1, Salyut-4, and Salyut-6 Missions

	Heart rate (beats/min)	Stroke volume (ml/beat)	Cardiac output (L/min)	Systolic blood pressure (mmHg)	Diastolic blood pressure (mmHg)
Resting values Salyut-1 23-day mission					
Preflight	64 ± 5	94 ± 3	6.0 ± 0.5	113 ± 7	73 ± 3
In-flight, days 13-21	65 ± 5	96 ± 9	6.1 ± 0.1	122 ± 4[a]	80 ± 2[a]
Salyut-4 63-day mission					
Preflight	65 ± 3	84 ± 5	5.5 ± 0.3	120 ± 5	86 ± 4
In-flight, day 56	65 ± 3	90 ± 2	5.9 ± 0.3	130 ± 6[a]	86 ± 1
Exercise values Salyut-6 140-day mission (standardized rate of work, 750 kgm/min)					
Preflight	113 ± 5	136 ± 19	15.3 ± 1.5	156 ± 2	70 ± 4
In-flight, day 29	113 ± 4	133 ± 20	15.0 ± 1.7	157 ± 9	69 ± 2
In-flight, day 41	124 ± 12[a]	134 ± 30	16.2 ± 2.2[a]	156 ±1	70 ± 3
In-flight, day 62	116 ± 8	120 ± 26[a]	13.7 ± 2.1[a]	149 ± 2[a]	68 ±5
In-flight, day 97	122 ± 7[a]	131 ± 34[a]	15.8 ± 3.2	144 ± 4[a]	73 ± 2
In-flight, day 119	128 ± 13[a]	112 ± 20[a]	14.1 ± 1.2[a]	146 ± 3[a]	69 ± 1

Note. Values represent mean ± SE.

[a]All cosmonauts' in-flight values changed in the same direction compared with preflight values.

Adapted, by permission, V.A. Convertino, 1990, "Physiological adaptations to weightlessness: Effects on exercise and work performance," *Exercise and Sport Sciences Reviews* 18:119-166.

reflect an inadequate in-flight exercise program.

It could be argued that the results of these in-flight tests indicate that the adaptations to microgravity are precise and appropriate for the environment. This is confirmed by the studies of the Skylab 4 astronauts, whose $\dot{V}O_2$max values all increased from preflight through the completion of their 84-day mission (see figure 11.13 on page 366).[34] These increases in $\dot{V}O_2$max were at least partially attributable to the training that took place each day during the

mission (see figure 11.14). The important point, however, is that these astronauts did not experience cardiorespiratory deconditioning during this extended period of microgravity.

The crew's ability to adapt successfully and quickly on their return to Earth is a concern with any space mission. We already mentioned the possibility of postural hypotension on return to full gravity. Furthermore, echocardiograms from seven members of four Space Shuttle crews indicated increases in heart rate, mean arterial pressure, and systemic vascular

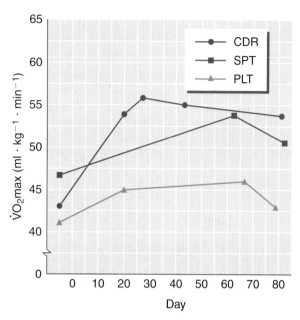

▲ Figure 11.13   Changes in V̇O₂ max from preflight through 84 days in space for the three astronauts aboard Skylab 4.
Data from D.F. Sawin, J.A. Rummel, and E.L. Bichel, 1975, "Instrumental personal exercise during long-duration space flights," *Aviation, Space, and Environmental Medicine* 46: 394-400.

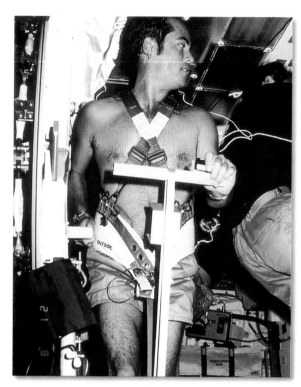

▲ Figure 11.14   Exercise training and testing aboard the Space Shuttle flight STS-78 in 1996.

resistance and decreases in end-diastolic volume and stroke volume within 1 h postflight.[5] End-diastolic volume remained depressed 7 to 14 days postflight. These changes can be explained at least partially by reduced plasma volume.

Convertino and colleagues,[11] however, suggested that other factors might be involved. They reported that increased leg venous compliance, because of which the veins of the legs can hold more blood, may contribute to the decreased end-diastolic volume. They found a decrease in calf muscle size with simulated microgravity, which allowed greater distension of the veins. The resulting blood pooling in the legs would decrease blood return to the heart, decreasing end-diastolic volume. Countermeasures during flight that are designed to minimize atrophy of the postural muscles might help prevent this.

Body Mass and Composition

Body mass and composition have been found to change substantially as a result of both bed rest and microgravity during space flight. The 33 Apollo crew members lost an average of 3.5 kg (7.7 lb), and the 9 Skylab crew members lost an average of 2.7 kg (6 lb). Individual weight change varied considerably, from a gain of 0.1 kg (0.2 lb) to a loss of 5.9 kg (13 lb).[24] The loss of mass in flights of 1 to 3 days appears to be largely caused by fluid loss. In flights of 12 days or longer, fluid loss accounts for approximately 50% of the mass loss, and the remaining loss is primarily from fat and protein. The Skylab missions provided an opportunity for comprehensive analysis of the composition of the mass loss.[24] The average 2.7-kg (6-lb) mass loss included

- 1.1 kg (2.4 lb) from total body water,
- 1.2 kg (2.6 lb) from fat,
- 0.3 kg (0.7 lb) from protein, and
- 0.1 kg (0.2 lb) from other sources.

The fat loss most likely resulted from an inadequate energy intake.

Exercise As a Countermeasure to Microgravity

Spaceflight and exposure to both short and long periods of microgravity are reasonably well tolerated. The astronaut has a remarkable ability to adapt to this unique environment, allowing normal or near-normal function in microgravity. But several of these adaptations pose potential problems when the astronaut returns to Earth, as we briefly described earlier. The primary areas of concern are the reductions in muscle, bone, and cardiovascular regulation of blood pressure. Possible decreases in strength and the increased risk of postural hypotension are of major concern because of the potential problems they present during landing: What if the astronauts couldn't get out of their space vehicle in the event of a crash or a fire? This has led space scientists to search for appropriate countermeasures to increase the crew's likelihood of successfully completing a mission.

In-flight exercise training is one of the proposed countermeasures that seems an obvious choice. Data from the Skylab missions have clearly shown that increasing exercise time and providing a variety of exercise equipment greatly attenuate muscle strength losses and even increase $\dot{V}O_2$max.[34, 41] In addition, in-flight bouts of maximal exercise might be important in preparing astronauts for return to a 1-g environment.[8] Research suggests that even a single maximal bout of exercise can cause a transient increase in plasma volume and in the sensitivity of the arterial receptors that monitor blood pressure, allowing $\dot{V}O_2$max to be maintained.

Finally, more attention must be focused on the design of equipment that will allow effective resistance training in microgravity as a means to preserve muscle function. Convertino[10] suggested that equipment that has an eccentric component might be important because the muscle and bone can be loaded with higher resistances by using eccentric actions. Training against higher resistances might also preserve the calcium content of weight-bearing bones.

Exercise may be one of the most effective countermeasures during spaceflight to prepare astronauts for successful adaptation on their return to Earth. Currently, the types and amounts of exercise required to maintain normal muscle strength and endurance are under debate. Although cosmonauts in the Russian space program are instructed to perform 2 h of endurance and resistance exercise each day while in space, recent evidence suggests that it may take considerably less activity to prevent strength loss and retain aerobic capacity.[25, 43, 44]

The importance of applying exercise physiology to space physiology is just now being realized. Unfortunately, research opportunities for exploring the physiological effects of microgravity are limited, and research on simulated microgravity is not as accurate as research conducted in space. Nonetheless, this promises to be an exciting area of future research for many exercise physiologists.

Health Risks of Microgravity

A number of health risks associated with microgravity are known, yet others are not yet fully recognized. We have discussed the changes in muscle, bone, and cardiovascular function that could compromise the health of astronauts either during space flight or on their return to earth's 1-g environment. It appears that the changes in muscle and cardiovascular function are short term and that these functions return to preflight levels within days or weeks. The recovery of bone loss, however, has not been followed as extensively. It is unclear if bone lost during spaceflight is fully recovered. This could present health problems later in life, if not in the short term.

Other health risks of microgravity include exposure to radiation, serious illness that cannot be treated because of the inability to abort the mission, neurovestibular dysfunction (disruptions of the sensory endorgans in the inner ear leading to vertigo, nausea, disequilibrium and disturbances in the control of posture and locomotion), psychological distress, and many others. Scientists are presently working on these issues with the hope of launching a mission to Mars by the year 2020. A mission to Mars would take a minimum of 3 years to complete, underlining the complexity of any potential health problem an astronaut could have.

In Closing . . .

Activities are seldom conducted under ideal environmental conditions. Heat, cold, humidity, altitude, and the underwater environment each present unique problems that affect the physiological demands of exercise. Space exploration has revealed that the body faces unique challenges during exposure to microgravity. This chapter and the preceding chapter summarized the nature of these various environmental stresses and how we can cope with them.

Much of our discussion thus far has dealt with how physiological variables and environmental stress can hinder our performance. In the next part, we examine various ways to optimize performance. We begin by looking at the importance of the amount of training, considering what happens when we train either too much or too little.

▶ Key Terms

acute altitude (mountain) sickness

decompression sickness (bends)

erythropoietin

high-altitude cerebral edema (HACE)

high-altitude pulmonary edema (HAPE)

hyperbaric environment

hypobaric environment

hypoxia

hypoxic vasoconstriction

microgravity

nitrogen narcosis (rapture of the deep)

oxygen poisoning

partial pressure of oxygen (PO_2)

recompression

respiratory alkalosis

self-contained underwater breathing apparatus (scuba)

spontaneous pneumothorax

TLV/RV ratio

▶ Study Questions

1. Describe the conditions at altitude that limit physical activity.

2. What types of activities are detrimentally influenced by exposure to high altitude?

3. Describe the physiological adjustments that accompany acclimatization to altitude.

4. Would an endurance athlete who trained at altitude be able to perform better during subsequent sea-level performance? Why or why not?

5. What environmental conditions are unique to underwater diving?

6. What effect does immersion have on heart rate? What causes this effect?

7. What health risks are associated with diving and using scuba equipment?

8. Describe the physiological and pathological problems that face scuba divers who descend to 30 m or more.

9. Under what conditions should a diver consider the need for decompression?

10. Define microgravity. What does 1 g represent?

11. What happens to muscle during the first few days of being in a cast, hindlimb suspension, or microgravity? Which muscles are most vulnerable to microgravity and why?

12. What happens to bone when exposed to simulated microgravity? Which bones are most vulnerable?

13. What physiological alterations occur with exposure to microgravity that reduce plasma volume?

14. How does $\dot{V}O_2$max change with prolonged exposure to microgravity? Consider preflight, in-flight, and postflight values.

15. What countermeasures to microgravity are likely to help the astronaut on his or her return to Earth?

▷ References

1. Adams, W.C., Bernauer, E.M., Dill, D.B., & Bomar, J.B. (1975). Effects of equivalent sea level and altitude training on $\dot{V}O_2$ and running performance. *Journal of Applied Physiology, 39,* 262-266.

2. Bert, P. (1943). *La pression barometrique* (M.A. Hitchcock & F.A. Hitchcock, Trans.). Columbus, OH: College Book Co.

3. Booth, F.W. (1982). Effect of limb immobilization on skeletal muscle. *Journal of Applied Physiology, 52,* 1113-1118.

4. Brosnan, M.J., Martin, D.T., Hahn, A.G., Gore, C.J., & Hawley, J.A. (2000). Impaired interval exercise responses in elite female cyclists at moderate simulated altitude. *Journal of Applied Physiology, 89,* 1819-1824.

5. Bungo, M.W., Goldwater, D.J., Popp, R.L., & Sandler, H. (1987). Echocardiographic evaluation of space shuttle crewmembers. *Journal of Applied Physiology, 62,* 278-283.

6. Buskirk, E.R., Kollias, J., Piconreatigue, E., Akers, R., Prokop, E., & Baker, P. (1967). Physiology and performance of track athletes at various altitudes in the United States and Peru. In R.F. Goddard (Ed.), *The effects of altitude on physical performance* (pp. 65-71). Chicago: Athletic Institute.

7. Clark, J.M., & Lambertsen, C.J. (1971). Pulmonary oxygen toxicity: A review. *Pharmacology Review, 23,* 37-133.

8. Convertino, V.A. (1987). Potential benefits of maximal exercise just prior to return from weightlessness. *Aviation, Space, and Environmental Medicine, 58,* 568-572.

9. Convertino, V.A. (1990). Physiological adaptations to weightlessness: Effects on exercise and work performance. *Exercise and Sport Sciences Reviews, 18,* 119-166.

10. Convertino, V.A. (1991). Neuromuscular aspects in development of exercise countermeasures. *Physiologist, 34,* S125-S128.

11. Convertino, V.A., Doerr, D.F., & Stein, S.L. (1989). Changes in size and compliance of the calf after 30 days of simulated microgravity. *Journal of Applied Physiology, 66,* 1509-1512.

12. Cotes, J.E. (1968). *Lung function: Assessment and application in medicine* (2nd ed.). Philadelphia: Davis.

13. Daniels, J., & Oldridge, N. (1970). Effects of alternate exposure to altitude and sea level on world-class middle-distance runners. *Medicine and Science in Sports, 2,* 107-112.

14. Fitts, R.H., Widrick, J.J., Knuth, S.T., Blaser, C.A., Karhanek, M., Trappe, S.W., Trappe, T.A., & Costill, D.L. (1997). Force-velocity and force-power properties of human muscle fiber after spaceflight [Abstract]. *Medicine and Science in Sports and Exercise, 29,* S190.

15. Forster, P.J.G. (1985). Effect of different ascent profiles on performance at 4200 m elevation. *Aviation, Space, and Environmental Medicine, 56,* 785-794.

16. Green, H.J., Sutton, J., Young, P., Cymerman, A., & Houston, C.S. (1989). Operation Everest II: Muscle energetics during maximal exhaustive exercise. *Journal of Applied Physiology, 66,* 142-150.

17. Grover, R., Reeves, J., Grover, E., & Leathers, J. (1967). Muscular exercise in young men native to 3,100 m altitude. *Journal of Applied Physiology, 22,* 555-564.

18. Groves, B.M., Reeves, J.T., Sutton, J.R., Wagner, P.D., Cymerman, A., Malconian, M.K., Rock, P.B., Yound, P.M., & Houston, C.S. (1987). Operation Everest II: Elevated high-altitude pulmonary resistance unresponsive to oxygen. *Journal of Applied Physiology, 63,* 521-530.

19. Hackett, P.H., Rennie, D., & Levine, H.D. (1976). The incidence, importance, and prophylaxis of acute mountain sickness. *Lancet, 2,* 1149-1154.

20. Hackett, P.H., Roach, R.C., Schoene, R.B., Harrison, G.L., & Mills, W.J., Jr. (1988). Abnormal control of ventilation in high-altitude pulmonary edema. *Journal of Applied Physiology, 64,* 1268-1272.

21. Hong, S.K., Song, S.H., Kim, P.K., & Suh, C.S. (1967). Seasonal observations on the cardiac rhythm during diving in the Korean ama. *Journal of Applied Physiology, 23,* 18-22.

22. King, A.B., & Robinson, S.M. (1972). Ventilation response to hypoxia and acute mountain sickness. *Aerospace Medicine, 43,* 419-421.

23. Kronenberg, R.S., Safar, P., Lee, J., Wright, F., Noble, W., Wahrenbrock, E., Hickey, R., Nemoto, E., & Severinghaus, J.W. (1971). Pulmonary artery pressure and alveolar gas exchange in man during acclimatization to 12,470 ft. *Journal of Clinical Investigation, 50,* 827-837.

24. Leonard, J.I., Leach, C.S., & Rambaut, P.C. (1983). Quantitation of tissue loss during prolonged space flight. *American Journal of Clinical Nutrition, 38,* 667-679.

25. Levine, B.D., Lane, L.D., Watenpaugh, D.E., Gaffney, F.A., Buckey, J.C., & Blomqvist, C.G. (1996). Maximal exercise performance after adaptation to microgravity. *Journal of Applied Physiology, 81*(2), 686-694.

26. Levine, B.D., & Stray-Gundersen, J. (1997) "Living high–training low": Effect of moderate-altitude acclimatization with low-altitude training on performance. *Journal of Applied Physiology, 83,* 102-112.

27. Mathews, L., Gopinathan, P.M., & Purkayastha, S.S. (1983). Chemoreceptor sensitivity and maladaptation to high altitude in man. *European Journal of Applied Physiology, 51,* 137-144.

28. Mithoefer, J.C. (1965). Breath-holding. In W.O. Fenn & H. Rahn (Eds.), *Handbook of physiology* (Section 3, Vol. 2, pp. 1011-1025). Washington, DC: American Physiological Society.

29. Norton, E.G. (1925). *The fight for Everest: 1924.* London: Arnold.

30. Pugh, L.C.G.E., Gill, M., Lahiri, J., Milledge, J., Ward, M., & West, J. (1964). Muscular exercise at great altitudes. *Journal of Applied Physiology, 19,* 431-440.

31. Rambaut, P.C., Smith, M.C., Mack, P.B., & Vogel, J.M. (1975). Skeletal response. In R.S. Johnston, L.F. Dietlein, & C.A. Berry (Eds.), *Biomedical results of Apollo* (NASA SP-368, pp. 303-322). Washington, DC: National Aeronautics and Space Administration.

32. Rennie, M.J., Edwards, R.H.T., Emery, P.W., Halliday, D., Lundholm, K., & Millward, D.J. (1983). Depressed protein synthesis is the dominant characteristic of muscle wasting and cachexia. *Clinical Physiology, 3,* 387-398.

33. Saltin, B., Kim, C.K., Terrados, N., Larsen, H., Svedenhag, J., & Rolf, C.J. (1995). Morphology, enzyme activities and buffer capacity in leg muscles of Kenyan and Scandinavian runners. *Scandinavian Journal of Medicine and Science in Sports, 5,* 222-230.

34. Sawin, C.F., Rummel, J.A., & Michel, E.L. (1975). Instrumented personal exercise during long-duration space flights. *Aviation, Space, and Environmental Medicine, 46,* 394-400.

35. Smith, J.L. (1899). The pathological effects due to increase of oxygen tension in the air breathed. *Journal of Physiology* (London), *24,* 19-35.

36. Song, S.H., Lee, W.K., Chung, Y.A., & Hong, S.K. (1969). Mechanism of apneic bradycardia in man. *Journal of Applied Physiology, 27,* 323-327.

37. Stray-Gundersen, J., Chapman, R.F., & Levine, B.D. (2001) "Living high–training low" altitude training improves sea level performance in male and female elite runners. *Journal of Applied Physiology, 91,* 1113-1120.

38. Sutton, J., & Lazarus, L. (1973). Mountain sickness in the Australian Alps. *Medical Journal of Australia, 1,* 545-546.

39. Sutton, J.R., Reeves, J.T., Wagner, P.D., Groves, B.M., Cymerman, A., Malconian, M.K., Rock, P.B.,

Young, P.M., Walter, S.D., & Houston, C.S. (1988). Operation Everest II: Oxygen transport during exercise at extreme simulated altitude. *Journal of Applied Physiology, 64,* 1309-1321.

40. Thomason, D.B., & Booth, F.W. (1990). Atrophy of the soleus muscle by hindlimb unweighting. *Journal of Applied Physiology, 68,* 1-12.

41. Thornton, W.E., & Rummel, J.A. (1977). Muscular deconditioning and its prevention in space flight. In R.S. Johnston & S.F. Dietlein (Eds.), *Biomedical results from Skylab* (NASA SP-377, pp. 191-197). Washington, DC: National Aeronautics and Space Administration.

42. Trappe, S.W., Trappe, T.A., Costill, D.L., & Fitts, R.H. (1996). Human calf muscle function in response to bed rest [Abstract]. *Medicine and Science in Sports and Exercise, 28,* S146.

43. Trappe, S.W., Trappe, T.A., Costill, D.L., Lee, G.A., Widrick J.J., & Fitts, R.H. (1997). Effects of spaceflight on human calf muscle morphology and function [Abstract]. *Medicine and Science in Sports and Exercise, 29,* S190.

44. Trappe, T.A., Trappe, S.W., Costill, D.L., Lee, G.A., Fitts, R.H., & Widrick, J.J. (1997). Oxygen uptake during submaximal and maximal cycling before, during, and after spaceflight [Abstract]. *Medicine and Science in Sports and Exercise, 29,* S190.

45. Vico, L., Collet, P., Guignandon, A., Lafage-Proust, M.-H., Thomas, T., Rehailia, M., & Alexandre, C. (2000). Effects of long-term microgravity exposure on cancellous and cortical weight-bearing bones of cosmonauts. *Lancet, 355,* 1607-1611.

46. Waggener, T.B., Brusil, P.J., Kronauer, R.E., Gabel, R.A., & Inbar, G.F. (1984). Strength and cycle time of high altitude ventilation patterns in unacclimatized humans. *Journal of Applied Physiology, 56,* 576-581.

47. Ward, M.P., Milledge, J.S., & West, J.B. (1989). *High altitude medicine and physiology.* Philadelphia: University of Pennsylvania Press.

48. West, J.B., Lahiri, S., Gill, M.B., Milledge, J.S., Pugh, L.C.G.E., & Ward, M.P. (1962). Arterial oxygen saturation during exercise at high altitude. *Journal of Applied Physiology, 17,* 617-621.

49. West, J.B., Peters, R.M., Aksnes, G., Maret, K.H., Milledge, J.S., & Schoene, R.B. (1986). Nocturnal periodic breathing at altitudes of 6300 and 8050 m. *Journal of Applied Physiology, 61,* 280-287.

50. Widrick, J.J., Romatowski, J.G., Bain, J.L.W., Trappe, S.W., Trappe, T.A., Thompson, J.L., Costill, D.L., Riley, D.A., & Fitts, R.H. (1997). Effect of 17 days bed rest on peak isometric force and unloaded shortening velocity of human soleus fibers. *American Physiological Society, 273*(42), C1690-C1699.

51. Wilber, R.L. (2001) Current trends in altitude training. *Sports Medicine*, **31**, 249-265.

52. Wolfel, E.E., Groves, B.M., Brooks, G.A., Butterfield, G.E., Mazzeo, R.S., Moore, L.G., Sutton, J.R., Bender, P.R., Dahms, T.E., McCullough, R.E., McCullough, R.G., Huang, S.-Y., Sun, S.-F., Grover, R.F., Hultgren, H.N., & Reeves, J.T. (1991). Oxygen transport during steady-state submaximal exercise in chronic hypoxia. *Journal of Applied Physiology*, **70**, 1129-1136.

▷ Selected Readings

Bender, P.R., Grove, B.M., McCullough, R.E., McCullough, R.G., Trad, L., Young, A.J., Cymerman, A., & Reeves, J.T. (1989). Decreased exercise muscle lactate release after high altitude acclimatization. *Journal of Applied Physiology*, **67**, 1456-1462.

Cerretelli, P., & Hoppeler, H. (1996). Morphologic and metabolic response to chronic hypoxia: The muscle system. In M.J. Fregly & C.M. Blatteis (Eds.), *Handbook of physiology, Section 4: Environmental physiology* (Vol. 2, pp. 1155-1181). New York: Oxford University Press.

Convertino, V.A. (1991). Carotid-cardiac baroreflex: Relation with orthostatic hypotension following simulated microgravity and implications for development of countermeasures. *Acta Astronautica*, **23**, 9-17.

Convertino, V.A. (1992). Effects of exercise and inactivity on intravascular volume and cardiovascular control mechanisms. *Acta Astronautica*, **24**, 1-7.

Convertino, V.A. (1996). Exercise as a countermeasure for physiological adaptation to prolonged spaceflight. *Medicine and Science in Sports and Exercise*, **28**, 999-1014.

Coote, J.H. (1995). Medicine and mechanisms in altitude sickness. *Sports Medicine*, **20**, 148-159.

Fitts, R.H., Riley, D.R., & Widrick, J.J. (2000). Microgravity and skeletal muscle. *Journal of Applied Physiology*, **89**, 823-839.

Fregly, M.J., & Blatteis, C.M. (1996). *Handbook of physiology, Section 4: Environmental physiology* (Vol. 1). New York: Oxford University Press.

Fulco, C.S., & Cymerman, A. (1990). Human performance and acute hypoxia. In K. Pandolf, M. Sawka, & R. Gonzalez (Eds.), *Human performance physiology and environmental medicine at terrestrial extremes* (pp. 467-495). Indianapolis: Benchmark Press.

Gaffney, F.A., Nixon, J.V., Karlsson, E.S., Campbell, W., Dowdey, A.B.C., & Blomqvist, C.G. (1985). Cardiovascular deconditioning produced by 20 hours of bedrest with head-down tilt (–5 degrees) in middle-aged healthy men. *American Journal of Cardiology*, **56**, 634-638.

Greenleaf, J.E., Bulbulian, R., Bernauer, E.M., Haskell, W.L., & Moore, T. (1989). Exercise-training protocols for astronauts in microgravity. *Journal of Applied Physiology*, **67**, 2191-2204.

Grover, R.F., Weil, J.V., & Reeves, J.T. (1986). Cardiovascular adaptation to exercise at high altitude. *Exercise and Sport Sciences Reviews*, **14**, 269-302.

Hargens, A.R., & Watenpaugh, D.E. (1996). Cardiovascular adaptation to spaceflight. *Medicine and Science in Sports and Exercise*, **28**, 977-982.

Harrison, M.H. (1986). Athletes, astronauts and orthostatic tolerance. *Sports Medicine*, **3**, 428-435.

Heath, D., & Williams, D.R. (1989). *High-altitude medicine and pathology*. London: Butterworths.

Johnston, R.S., Dietlein, L.F., & Berry, C.A. (Eds.). (1975). *Biomedical results of Apollo* (NASA SP-368). Washington, DC: National Aeronautics and Space Administration.

Kirby, C.R., Ryan, M.J., & Booth, F.W. (1992). Eccentric exercise training as a countermeasure to non-weight-bearing soleus muscle atrophy. *Journal of Applied Physiology*, **73**, 1894-1899.

Lenfant, C., & Sullivan, K. (1971). Adaptation to high altitude. *New England Journal of Medicine*, **284**, 1298-1309.

Margaria, R. (Ed.). (1967). *Exercise at altitude*. Amsterdam: Excerpta Medica Foundation.

Morey-Holton, E.R., Whalen, R.T., Arnaud, S.B., & Van Der Meulen, M.C. (1996). The skeleton and its adaptation to gravity. In M.J. Fregly & C.M. Blatteis (Eds.), *Handbook of physiology, Section 4: Environmental physiology* (Vol. 1, pp. 691-719). New York: Oxford University Press.

Muza, S.R. (1990). Hyperbaric physiology and human performance. In K. Pandolf, M. Sawka, & R. Gonzalez (Eds.), *Human performance physiology and environmental medicine at terrestrial extremes* (pp. 565-589). Indianapolis: Benchmark Press.

Nicogossian, A.E., Huntoon, C.L., & Pool, S.L. (Eds.). (1989). *Space physiology and medicine* (2nd ed.). Philadelphia: Lea & Febiger.

Pigman, E.C. (1991). Acute mountain sickness: Effects and implications for exercise at intermediate altitude. *Sports Medicine*, **12**, 71-79.

Shiraki, K., & Claybaugh, J.R. (1995). Effects of diving and hyperbaria on responses to exercise. *Exercise and Sport Sciences Reviews*, **23**, 459-485.

Stegemann, J., Essfeld, D., & Hoffmann, U. (1985). Effects of a 7-day headdown tilt (–6°) on the dynamics of oxygen uptake and heart rate adjustment in upright exercise. *Aviation, Space, and Environmental Medicine*, **56**, 410-414.

Thompson, C.A., Tatro, D.L., Ludwig, D.A., & Convertino, V.A. (1990). Baroreflex responses to acute changes in blood volume in humans. *American Journal of Physiology, 259,* R792-R798.

Turner, R.T. (2000). What do we know about the effects of spaceflight on bone? *Journal of Applied Physiology, 89,* 840-847.

West, J.B. (2000). Historical perspectives: Physiology in microgravity. *Journal of Applied Physiology, 89,* 379-384.

West, J.B., Boyer, S.J., Graber, D.J., Hackett, P.H., Maret, K.H., Milledge, J.S., Peters, R.M., Jr., Pizzo, C.J., Samaja, M., Sarnquist, F.H., Schoene, R.B., & Winslow, R.M. (1983). Maximal exercise at extreme altitudes on Mount Everest. *Journal of Applied Physiology, 55,* 688-698.

Westerterp, K.R. (2001). Energy and water balance at high altitude. *News in the Physiological Sciences, 16,* 134-137.

Wolski, L.A., McKenzie, D.C., & Wenger, H.A. (1996). Altitude training for improvements in sea level performance. *Sports Medicine, 22,* 251-263.

Young, A.J., & Young, P.M. (1990). Human acclimatization to high terrestrial altitude. In K. Pandolf, M. Sawka, & R. Gonzalez (Eds.), *Human performance physiology and environmental medicine at terrestrial extremes* (pp. 497-543). Indianapolis: Benchmark Press.

Zernicke, R.F., Vailas, A.C., & Salem, G.J. (1990). Biomechanical response of bone to weightlessness. *Exercise and Sport Sciences Reviews, 18,* 167-192.

Zuntz, N., Loewy, A., Muller, F., & Caspari, W. (1906). *Hohenklima und Bergwanderrungen in ihret Wirkung auf den Menschen.* Berlin: Bong.

P A R T   V

Optimizing Performance in Sport

We now understand how the body responds to an acute bout of exercise, how it adapts to chronic training, and how it adjusts to environmental extremes. We can now apply that knowledge to optimizing athletic performance. In part V, we focus on how athletes can best prepare for competition. In chapter 12, "Training for Sport," we discuss the effects of changing the amount of training stress and explore how too much or too little training can impair performances. In chapter 13, "Nutrition and Sport," we evaluate the athlete's dietary needs and consider how nutritional supplementation and diet manipulation have been proposed to improve performance. In chapter 14, "Body Weight, Body Composition, and Sport," we address the issues of assessing body composition, relating body composition to sport performance, and the use of weight standards. We also determine the most effective means for successfully losing excess body fat while maintaining fat-free body mass, enabling an athlete to achieve optimal competitive weight and probably improve performance. In chapter 15, "Ergogenic Aids and Sport," we discuss various pharmacological, hormonal, and physiological agents that have been proposed to improve performance. We examine the potential benefits, proven effects, and health risks associated with their use.

CHAPTER 12

TRAINING FOR SPORT

In the never-ending quest for performance perfection, many athletes devote as much time as possible to training throughout the year, believing that the more they train, the better they will perform. For others, the end of the competitive season marks the beginning of a period of rest and relaxation, during which training abruptly halts. These athletes often assume that when the season begins again, they still will be well conditioned. Athletes who are injured and undergo immobilization while they heal might worry that the improvements they have trained so hard to achieve will be lost by the time they are allowed to restart training. Yet none of these beliefs are totally correct. Athletes who relentlessly push harder could eventually see their performance suffer, not improve. Likewise, both seasonal athletes who take a break from training and injured athletes who have a limb immobilized will suffer some loss of training gains, but most will be able to rebound quickly.

In this chapter, we present the scientific basis for optimizing the training program. We consider what constitutes optimal training, the effects of both too much training and too little training, and how to regain training losses that might occur from a temporary interruption of training. We will learn that more is not always better as we explore the complexities of optimizing training to maximize performance.

Throughout his college career, Eric had trained at swimming 4 h each day, covering as much as 13.7 km (8.5 mi) per day. Despite this effort, his performance time for the 200-yd (183-m) butterfly event had not improved since his freshman year. With a best performance of 2 min 15 s for the event, he was seldom given a chance to compete because several teammates could perform the event in less than 2 min 5 s. During Eric's senior year, his coach made a major change in the team's training plan. The swimmers trained only 2 h per day and swam an average of 4.5 to 4.8 km (2.8-3.0 mi) per day. Suddenly Eric's performance began to improve. After 3 months, his time dropped to 2 min 10 s, still not good enough to make him a major contender. But as a reward for Eric's improvement, the coach chose him to swim the 200-yd butterfly event at the conference championship meet, which was preceded by 3 weeks of reduced training of only 1.6 km (1 mi) per day. Subsequently, with less training than in previous years and well rested after the taper, Eric was able to make the finals of the event at the championship meet. His preliminary time was 2 min 1 s. In the finals, he improved even further, posting a third-place finish with a time of 1 min 57.7 s, an impressive performance indicating a swimmer who performed better with "less" training.

Repeated days and weeks of training can be considered positive stress because training improves your capacity for energy production, tolerance of physical stress, and exercise performance. The major physical changes associated with training occur in the first 6 to 10 weeks. The magnitude of these adaptations depends on the volume and intensity of exercise performed during training, which has led many coaches and athletes to believe erroneously that the athlete who undertakes the greatest volume (quantity) and intensity (quality) of training will be the best performer. Unfortunately, we often mistakenly consider quantity and quality of training to be synonymous. Too often, training sessions are judged by the total volume (e.g., distance run, cycled, or swum) performed in each training session, leading coaches to design training programs that are not specific for the sport and often impose unrealistic demands on the athlete.

The rate at which an individual adapts to training is limited and cannot be forced beyond the body's capacity for development. Too much training can reduce the potential for improvement and in some cases can cause a breakdown in the adaptation process, eventually reducing performance.

A person's rate of adaptation to training is limited and cannot be forced beyond his or her body's capacity for development. Each individual responds differently to the same training stress, so what might be excessive training for one person might be well below the capacity of another. Therefore, it is important that trainers recognize and account for individual differences when designing training programs.

Although the volume of work performed in training is an important stimulus for physical conditioning, there needs to be a proper balance between volume and intensity. Training can be overdone, leading to chronic fatigue, illness, overtraining syndrome, and performance decrements. In contrast, proper rest and reaching the appropriate balance between training volume and intensity can enhance performance. Much effort has been directed toward determining the appropriate volume and intensity required to achieve optimal adaptation. Exercise physiologists have tested numerous training regimens to determine both the minimal and maximal stimuli needed for cardiovascular and muscular improvements. Let's now examine those factors that can affect your response to a given training program, looking first at a model for optimizing the training stimulus.

Optimizing Training— A Model

All well-designed training programs incorporate the principle of progressive overload. In general, this principle holds that to maximize the benefits of training, the training stimulus must be progressively increased as the body adapts to the current stimulus. Your body responds to training by adapting to the stress of the training stimulus. If the amount of stress remains constant, you will eventually adapt to that level of stimulation, and your body won't need further adaptation. The only way to continue to improve with training is to progressively increase the training stimulus, or stress. When this concept is carried too far, the training may become excessive, pushing the body beyond its ability to adapt. Such excessive training, with too high a volume or intensity, produces no additional improvement in conditioning or performance and can lead to performance decrements and chronic fatigue. Nevertheless, some coaches and athletes believe that maximal improvements with training can be achieved only with excessive training. For example, many swimmers train for 4 to 6 h/day, believing that these overload periods hasten their adaptation or raise their fitness to levels that can't be achieved with lower volume or less intense training. Conversely, if the volume or the intensity of training is too low, optimal performance will not be achieved. Thus, the coach and athlete face the challenge of determining the optimal training stimulus for that particular athlete, recognizing that what works for one athlete might not work for another.

A model demonstrating the continuum of training stages that a competitive athlete might go through during a full year is provided in figure 12.1. This model is based on the principle of periodization, which was described in the introductory chapter and which is illustrated and described in figure 12.2. (It was also discussed in chapter 3 as it applies to resistance training.) In this model, **undertraining** represents the type of training an athlete would undertake between competitive seasons or during active

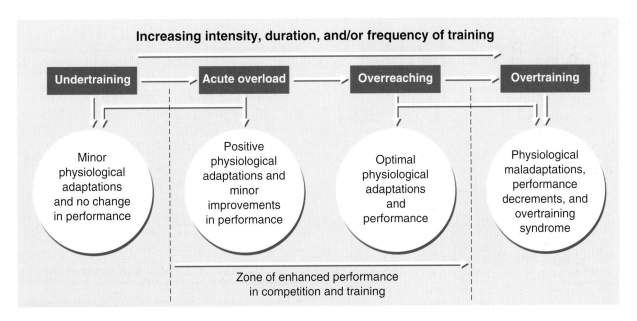

Increasing intensity, duration, and/or frequency of training

| Undertraining | Acute overload | Overreaching | Overtraining |

Minor physiological adaptations and no change in performance

Positive physiological adaptations and minor improvements in performance

Optimal physiological adaptations and performance

Physiological maladaptations, performance decrements, and overtraining syndrome

Zone of enhanced performance in competition and training

▲ **Figure 12.1**   Model of the continuum of training stages in a periodized training mesocycle.

Adapted, by permission, from L.E. Armstrong and J.L. VanHeest, 2002, "The unknown mechanism of the overtraining syndrome," *Sports Medicine* 32(1): 185-209.[1]

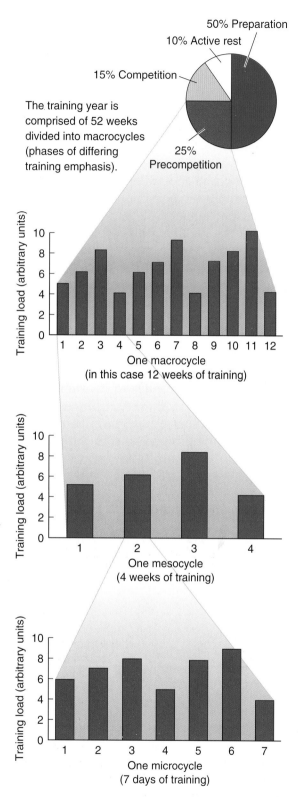

▲ Figure 12.2   The structure of a periodized training program.

Adapted, by permission, from R.W. Fry, A.R. Morton, and D. Keast, 1991, "Overtraining in athletes: An update," *Sports Medicine* 12: 32-65.[13]

rest. Generally, physiological adaptations will be minor, and there will be no improvement in performance. **Acute overload** represents what might be considered an "average" training load, where the athlete is stressing the body to the extent necessary to improve both physiological function and performance. **Overreaching** is a relatively new term that refers to a brief period of heavy overload without adequate recovery, thus exceeding the athlete's adaptive capacity. There will be a performance decrement, but it will be relatively short, from several days to several weeks. Finally, **overtraining** refers to that point where an athlete starts to experience physiological maladaptations and chronic performance decrements. This generally leads to the **overtraining syndrome.**[1]

In the following sections, we discuss overreaching, overtraining, and the overtraining syndrome. However, first, we discuss excessive training, which doesn't fit as nicely in the model described in figure 12.1. **Excessive training** refers to training that is well above what is needed for peak performance but does not strictly meet the criteria for either overreaching or overtraining.

Excessive Training

With excessive training, either or both volume of training and intensity of training are increased to extreme levels. The "more is better" philosophy drives the training schedule. For many years, athletes were undertrained. As coaches and athletes became bolder and started to push the envelope by increasing both training volume and intensity, they found that athletes responded well, and world records began to tumble. However, you can only take this philosophy so far. At a certain point, performance begins to either plateau or decline. Let's take a look at some examples of this.

Most of the research on excessive training has been conducted on swimmers. For that reason, the material in the next two subsections is derived from research conducted on swimmers, but the information applies to most other forms of training.

Volume of Training

Training volume can be increased by increasing either the duration or the frequency of training bouts. The question is, does increased volume translate into increased performance? Research shows that swim training 3 to 4 h/day, 5 or 6 days each week, provides no greater benefits than training only 1 to 1.5 h/day.[7] In fact, such excessive training has been shown to significantly decrease muscular strength and sprint swimming performance.

Few studies have compared the physical conditioning and performance benefits of single versus multiple daily training sessions.[6, 22, 33] Studies conducted thus far reveal no scientific evidence that multiple daily training sessions enhance fitness and performance more than a single daily session. This is illustrated by the data in figure 12.3, which show the responses of two groups of swimmers who trained once per day (group 1) or twice per day (group 2) for a period of 6 weeks during a 25-week training program.[6] All swimmers began the program following the same training regimen: one time per day. But during weeks 5 through 11, group 2 increased its training to twice per day. After 6 weeks on the different regimens, both groups returned to the once-daily program. All the swimmers' heart rates and blood lactate values decreased dramatically with the initiation of training, and no significant differences were seen in the two groups' results in response to the change in training volume. The swimmers who trained twice per day showed no additional improvements over those who trained only once per day. In fact, their blood lactate levels (figure 12.3a) and heart rates (figure 12.3b) appeared to be somewhat higher for the same fixed-pace swim, although the differences were not statistically significant.

To determine the influence of long-term, excessive training, performance improvements of swimmers who trained twice daily for a total distance of more than 10,000 m (10,936 yd) per day (the LS, or long-swim, group) were compared with improvements in those who swam approximately half that distance in a single session each day (the SS, or short-swim, group).[7] The results are indicated in figure 12.4.

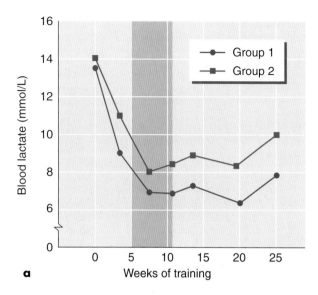

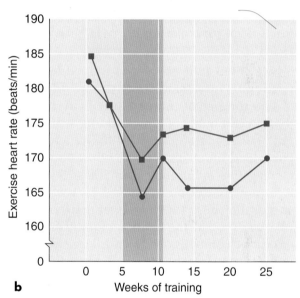

▲ **Figure 12.3**   Changes in swimmers' (*a*) blood lactate levels and (*b*) heart rates during a standardized 366-m (400-yd) swim during 25 weeks of training. Between the 5th and 11th weeks, one group trained twice per day (group 2), whereas the other group trained once each day (group 1).

Changes in performance time for the 100-yd (91-m) front crawl were examined over a 4-year period for both groups. The LS swimmers and SS swimmers both experienced an identical average improvement of 0.8% per year. Similar findings also were observed for competitors in other events, such as the 200-, 500-, and 1,650-yd front crawl. Elite swimmers who trained with greater volume—twice each day—showed no greater improvements in performance than

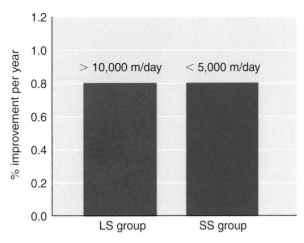

▲ **Figure 12.4**   Percentages of performance improvement for collegiate male swimmers who trained at more than 10,000 m/day (LS group) and those who swam about 5,000 m/day (SS group) over a 4-year period. Note that despite training only about half the distance, the SS group had an average annual improvement equal to that of the LS group.

less talented swimmers who trained less and with only one session per day.

The concept of training specificity implies that several hours of daily training won't provide the adaptations needed for athletes who participate in events of short duration. Most competitive swimming events last less than 2 min. How can training for 3 to 4 h/day at speeds that are markedly slower than competitive pace prepare the swimmer for the maximal efforts of competition? Such a large training volume prepares the athlete to tolerate a high volume of training but likely does little to benefit actual performance.

> The need for long daily workouts (high volume) is being seriously questioned by researchers. For certain sports, it appears that training volume could be reduced significantly, possibly by as much as one half in some sports, without reducing the benefits and with less risk of overtraining the athletes to the point of decreased performance. The principle of training specificity suggests that low-intensity, high-volume training for sprint-type athletes does not improve performance.

Intensity of Training

Training intensity refers to both the relative force of muscle action (i.e., resistance training) and the relative stress placed on the cardiovascular system (i.e., aerobic training). Because we discussed resistance training in chapter 3, discussion in this section is limited to aerobic and anaerobic training.

We typically relate the intensity of effort to the capacity to generate energy or to a percentage of the person's $\dot{V}O_2$max. As training intensity increases, greater demands are placed on the aerobic system, and this stimulates improvements in oxygen transport and oxidative metabolism. Studies have shown that, for most people, training intensities between 50% and 90% $\dot{V}O_2$max significantly improve aerobic capacity. As intensity is increased to energy levels that exceed $\dot{V}O_2$max, anaerobic capacity and strength improve, but there is less improvement in aerobic capacity. The precise relationship between training intensity and aerobic, anaerobic, and strength gains has not been well defined.

We must also remember the strong interaction between training intensity and training volume: As intensity is reduced, training volume must be increased to achieve adaptation. Training at very high intensities requires substantially less training volume, but the adaptations that occur will be significantly different from those achieved with low-intensity, high-volume training.

Attempts to perform large amounts of high-intensity training can have negative effects on adaptation. The energy needs of high-intensity exercise place greater demands on the glycolytic system, rapidly depleting muscle glycogen. If such training is attempted too often, such as daily, the muscles can become chronically depleted of their energy reserves, and the person might demonstrate signs of chronic fatigue or overtraining (discussed in the next section).

▶ Optimal training involves following a model that incorporates the principles of periodization, in that the body needs to systematically go through stages of undertraining, acute overload, and overreaching to maximize performance.

▶ Excessive training refers to training that is done with an unnecessarily high volume, intensity, or both. It provides little or no additional improvements in conditioning or performance and can lead to chronic fatigue and decreased performance.

▶ Training volume can be increased by increasing either or both the duration and frequency of training bouts. But numerous studies have shown no significant differences in improvement between athletes who train with typical training volumes and those who train with twice the volume (training conducted twice the duration or twice a day instead of once a day).

▶ Training intensity determines the specific adaptations that occur in response to the training stimulus. High-intensity, low-volume training can be tolerated only for brief periods, so, although this type of training does increase muscular strength in resistance training and total body speed and anaerobic capacity in high-intensity interval training, this training will provide little or no improvement in aerobic capacity. Conversely, low-intensity, high-volume training stresses the oxygen transport and oxidative metabolism systems, causing greater gains in aerobic capacity but having little or no effect on muscular strength, anaerobic capacity, or total body speed.

▶ Training intensities of between 50% and 90% $\dot{V}O_2$max markedly improve aerobic capacity for most people.

Overreaching

In contrast to excessive training, overreaching is a systematic attempt to intentionally overstress the body, allowing the body to adapt even more to the training stimulus, above and beyond that attained during a period of acute overload. Like overtraining, there is a decrement in performance but it is short term, lasting only several days to several weeks. Furthermore, overreaching is accompanied by both increased physiological function and, subsequently, increased performance. Obviously, this is the critical phase of training—on the edge, leading either to improved physiological function and performance or, if one goes too far, overtraining. Although with overreaching the period of full recovery from training takes several days to several weeks, if one steps over the line and overtrains the body, recovery can take many months or, in some cases, years. So the key to overreaching is to push the athlete hard enough to accomplish the desired positive physiological and performance improvements but to avoid going into the stage of overtraining. As we will see, this is not an easy task!

Overtraining

Despite hard overload training, athletes may experience an unexplained decline in performance and physiological function that extends over weeks, months, or years. This condition is termed overtraining, a condition that has been attributed to both psychological and physiological causes. Although the precise causes for this breakdown in performance and physiological function are not fully understood, overtraining often seems to be associated with periods of overreaching. When the training load is too intense or the volume of training exceeds the body's ability to adequately recover and adapt, the body experiences more catabolism (breakdown) than anabolism (buildup).

Athletes experience various levels of fatigue during repeated days and weeks of training, so not all fatigue-producing situations can be classified as overtraining (as we noted previously with overreaching). Fatigue that often follows one or more exhaustive training sessions usually is corrected by a few days of reduced training or rest and a carbohydrate-rich diet. Overtraining, on the other hand, is characterized by a sudden decline in performance and physiological function that cannot be remedied by a few days of reduced training or rest and dietary manipulation.

Effects of Overtraining: The Overtraining Syndrome

Most of the symptoms that result from overtraining, collectively referred to as the overtraining syndrome, are subjective and identifiable only after the individual's performance and physiological function have suffered. Unfortunately, these symptoms can be highly individualized, which can make it very difficult for athletes, trainers, and coaches to recognize that performance decrements are brought on by overtraining. Usually, the first indication of the overtraining syndrome is a decline in physical performance with continued training (see figure 12.5). The athlete can sense a loss in muscular strength, coordination, and maximal working capacity and generally feels fatigued. Other primary signs and symptoms of the overtraining syndrome include[1]

- change in appetite and body weight loss;
- sleep disturbances;
- irritability, restlessness, excitability, anxiousness;
- loss of motivation;
- lack of mental concentration; and
- feelings of depression.

Physiological changes also indicate the presence of the overtraining syndrome. We address these later in this section.

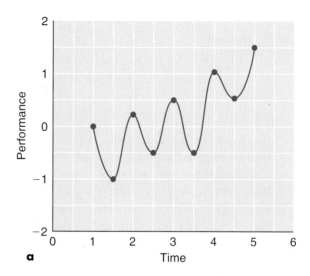

a

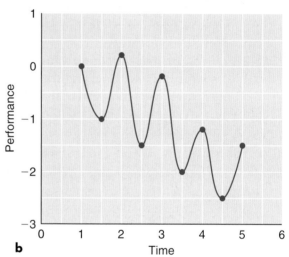

b

▲ **Figure 12.5**   Typical pattern of the expected improvement in performance with acute overload and overreaching (*a*) in contrast to the pattern seen with overtraining (*b*).

Reprinted, by permission, from M.L. O'Toole, 1998, Overreaching and overtraining in endurance athletes.   In *Overtraining in sport*, edited by R.B. Krieder, A.C. Fry and M.L. O'Toole (Champaign, IL: Human Kinetics), 10, 13.[26]

Few athletes are undertrained, but unfortunately, many are overtrained, often erroneously believing that more training always produces more improvement. As their performance declines with overtraining, they train even harder in an effort to compensate. We cannot overstress the importance of designing training programs to include both rest and variation in the training intensity and volume to avoid overtraining and chronic fatigue.

The underlying causes of overtraining syndrome are often a combination of emotional and physiological factors. Hans Selye[28] noted that a person's stress tolerance can break down as often from a sudden increase in anxiety as from an increase in physical distress. The emotional demands of competition, the desire to win, the fear of failure, unrealistically high goals, and others' expectations can be sources of intolerable emotional stress. Because of this, overtraining is typically accompanied by a loss of competitive desire and a loss of enthusiasm for training. Furthermore, Armstrong and VanHeest[1] made the important observation that the overtraining syndrome and clinical depression involve remarkably similar signs and symptoms, brain structures, neurotransmitters, endocrine pathways, and immune responses, suggesting that they have similar etiologies.

> The symptoms of overtraining syndrome are highly individualized and subjective, so they cannot be universally applied. The presence of one or more of these symptoms is sufficient to alert the coach or trainer that an athlete might be overtrained.

The physiological factors responsible for the detrimental effects of overtraining are not fully understood. However, many abnormal responses have been reported that suggest that overtraining is associated with alterations in the neurological, hormonal, and immune systems. Although a cause-and-effect relationship between these changes and the symptoms of overtraining has not been clearly established, these symptoms often can help determine whether an individual is overtrained. In the following discussion, we focus on some of the observed changes associated with overtraining and on potential causes of the overtraining syndrome.

Autonomic Nervous System Overtraining

Some studies suggest that overtraining is associated with abnormal responses in the autonomic nervous system. Physiological symptoms accompanying the decline in performance often reflect changes in the neural or endocrine systems that are controlled by either the sympathetic or the parasympathetic nervous systems. Sympathetic overtraining can lead to

- increased resting heart rate,
- increased blood pressure,
- loss of appetite,
- decreased body mass,
- sleep disturbances,
- emotional instability, and
- elevated basal metabolic rate.

Other studies suggest that the parasympathetic nervous system might be dominant in some cases of overtraining.[19] In these cases, athletes show the same performance failures but have markedly different responses than those with sympathetic overtraining. Signs of parasympathetic overtraining include

- early onset of fatigue,
- decreased resting heart rate,
- rapid heart rate recovery after exercise, and
- decreased resting blood pressure.

Some of the symptoms associated with autonomic nervous system overtraining are also seen in people who are not overtrained. For this reason, we cannot always assume that the presence of these symptoms confirms overtraining. Of the two conditions, symptoms of sympathetic overtraining are the most frequently observed.[19] However, there really is not strong scientific evidence to support the autonomic nervous system overtraining theory, although the autonomic nervous system definitely is affected by overtraining.[32]

Hormonal Responses to Overtraining

Measurements of various blood hormone levels during periods of overreaching suggest that marked disturbances in endocrine function accompany the excessive stress. As shown in

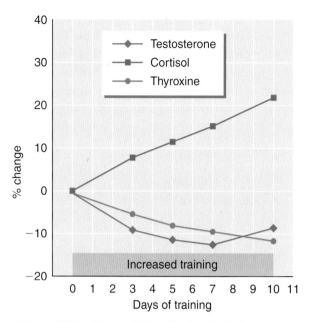

▲ Figure 12.6 Changes in blood levels of testosterone, cortisol, and thyroxine during a period of intensified training. During the 10-day period shown here, the swimmers increased their training from about 4,000 m/day to 8,000 m/day. These data show that cortisol levels increased in response to the added stress, whereas testosterone and thyroxine showed a remarkable decline during this period.

figure 12.6, when athletes increase their training 1.5- to 2-fold, their blood levels of thyroxine and testosterone usually decrease and their blood levels of cortisol increase.[17, 18] The ratio of testosterone to cortisol is thought to regulate anabolic processes in recovery, so a change in this ratio is considered an important indicator, and perhaps a cause, of the overtraining syndrome.[19] Decreased testosterone coupled with increased cortisol might lead to more protein catabolism than anabolism in the cells. Other research, however, suggests that although cortisol levels increase with overreaching and the early stages of overtraining, both resting and exercise cortisol levels generally are decreased in the overtraining syndrome.[1] Overtrained athletes often have higher blood levels of urea, and because urea is produced by the breakdown of protein, this indicates increased protein catabolism. This mechanism is thought to be responsible for the loss in body mass seen in overtrained athletes.

Resting blood levels of epinephrine and norepinephrine are elevated during periods of intensified training.[17] These two hormones elevate heart rate and blood pressure. Some suggest that the blood levels of these hormones should be measured to confirm overtraining. Unfortunately, measurement of these hormones is expensive, complex, and time-consuming, so this is not a test that can be widely used.

Acute overload training and overreaching often produce most of the same endocrine changes reported in overtrained athletes. For this reason, measuring these and other hormones might not provide valid confirmation of overtraining. Athletes whose hormone levels appear abnormal may simply be experiencing the normal effects of hard training. These endocrine changes simply might reflect the stress of training rather than a breakdown in the adaptive process.

Armstrong and VanHeest[1] proposed that the various stressors associated with the overtraining syndrome act primarily through the hypothalamus. They postulated that these stressors activate the following two predominant hormonal axes involved in the body's response to stressors:

- the sympathetic–adrenal medullary axis (SAM), involving the sympathetic branch of the autonomic nervous system; and
- the hypothalamic–pituitary–adrenocortical axis (HPA)

This is illustrated in figure 12.7*a*. Figure 12.7*b* illustrates the brain and immune system interactions with these two axes.

A major role for cytokines in the overtraining syndrome recently has been proposed, providing support for the Armstrong and VanHeest model in figure 12.7.[31] Elevated circulating cytokines result from skeletal muscle, bone, and joint trauma associated with overtraining.

Immunity and Overtraining

Your immune system provides a line of defense against invading bacteria, parasites, viruses,

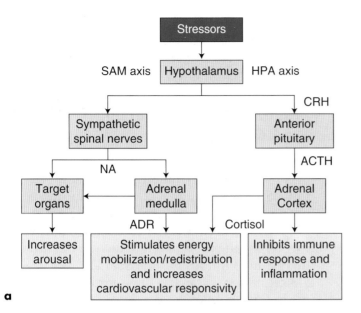

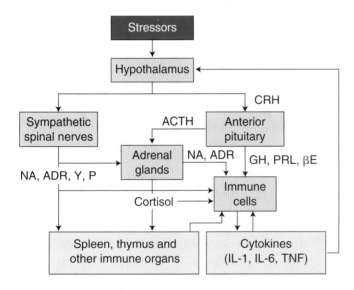

◀ **Figure 12.7** (*a*) The role of the hypothalamus and the sympathetic–adrenal medullary (SAM) and hypothalamic–pituitary–adrenocortical (HPA) axes as possible mediators of the overtraining syndrome. (*b*) The brain–immune system interactions with this model, with the cytokines playing a potentially major role in mediating overtraining.

Symbols: ACTH = adrenocorticotropin; ADR = adrenaline (epinephrine); CRH = corticotropin-releasing hormone; GH = growth hormone; IL-1 = interleukin-1; IL-6 = interleukin 6; NA = noradrenaline (norepinephrine); P = substance P; PRL = prolactin; TNF = tumor necrosis factor; Y = neuropeptide Y; βE = β-endorphin

Adapted, by permission, from L.E. Armstrong and J.L. VanHeest, 2002, "The unknown mechanism of the overtraining syndrome," *Sports Medicine* 32: 185-209.[1]

and tumor cells. This system depends on the actions of specialized cells (such as lymphocytes, granulocytes, and macrophages) and antibodies. These primarily eliminate or neutralize foreign invaders that might cause illness (pathogens). Unfortunately, one of the most serious consequences of overtraining is the negative effect it has on the body's immune system. In fact, from the model proposed in figure 12.7, compromised **immune function** is potentially a major factor in the initiation of the overtraining syndrome.

Recent studies confirm that excessive training suppresses normal immune function, increasing the overtrained athlete's susceptibility

The Overtraining, Chronic Fatigue, and Fibromyalgia Syndromes

Chronic fatigue syndrome is very similar to the overtraining syndrome.[30] Likely, there is considerable overlap between the two. Furthermore, there is considerable overlap between chronic fatigue syndrome and the **fibromyalgia syndrome.** Chronic fatigue and fibromyalgia, however, do occur in nonathletes and in individuals who are not actively exercising. Other than that, there are many similarities in symptoms across the three syndromes. These similarities can include chronic fatigue at rest and during exercise, psychological distress, immune system dysfunction, hormonal dysfunction, HPA axis dysfunction, and neurotransmitter dysfunction. Furthermore, all three syndromes are difficult to diagnose, and generally the specific cause or causes remain unknown.

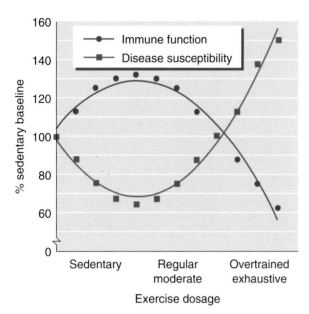

▲ **Figure 12.8**   The inverted J-shaped model of the relationship between amount of exercise and immune function. This model suggests that moderate exercise may lower the risk of infection or disease, whereas overtraining may increase the risk.

Data from D.C. Nieman, 1997.[25]

to infections.[20, 25, 34] This is illustrated in figure 12.8. Numerous studies show that short bouts of intense exercise can temporarily impair the immune response, and successive days of heavy training can amplify this suppression.[25] Several investigators have reported an increased incidence of illness following a single, exhaustive exercise bout. Such immune suppression is characterized by abnormally low levels of both lymphocytes and antibodies. Invading organisms or substances are more likely to cause illness when these levels are low. Also, intense exercise during illness might decrease your ability to fight off the infection and increase your risk of even greater complications.[24]

> Overtraining syndrome appears to be associated with depressed immune function. This places the athlete at an increased risk for infection and disease.

Not all athletes demonstrate a decreased immune function during training, however. Recent studies with young (10- to 12-year-

old) female gymnasts have shown that despite training 22 h/week, their immune responses were similar to an age-matched group of untrained girls.[11]

Predicting the Overtraining Syndrome

We must remember that the underlying cause or causes of the overtraining syndrome are not fully known, although it is likely that physical or emotional overload, or a combination of the two, might trigger this condition. Trying not to exceed an athlete's stress tolerance by regulating the amount of physiological and psychological stress experienced during training is difficult. Most coaches and athletes use intuition to determine training volume and intensity, but few can accurately assess the true impact of a workout on the athlete. No preliminary symptoms warn athletes that they are on the verge of becoming overtrained. By the time coaches realize that they have pushed an athlete too hard, it is often too late. The damage done by repeated days of excessive training or stress can be repaired only by days, and in some cases weeks, of reduced training or complete rest.

Numerous investigators have tried to identify markers of the overtraining syndrome in its early stages by using assorted physiological and psychological measurements. A list of potential markers is provided in table 12.1. Unfortunately, none has proven totally effective. It is often difficult to determine whether the measurements obtained are related to overtraining or whether they simply reflect normal responses to heavy training. In the following sections, we examine several of these physiological markers that have been proposed as potential means for diagnosing the overtraining syndrome.

Blood Enzyme Levels

Measurements of blood enzyme levels have been used with only limited success to diagnose overtraining syndrome. Such enzymes as creatine kinase (CK), lactate dehydrogenase (LDH), and serum glutamic oxalic transaminase (SGOT) are important in muscle energy

Table 12.1

Potential Markers of Overreaching (OR), Overtraining (OT), and the Overtraining Syndrome (OTS)

Marker	Response	Potential marker for OR	OT	OTS
Physiological and psychological				
HRrest and HRmax	Decreased		✓	✓
HRsubmax and $\dot{V}O_2$submax	Increased	✓		✓
$\dot{V}O_2$max	Decreased			✓
Anaerobic metabolism	Impaired		✓	
Basal metabolic rate	Increased			✓
RERsubmax, max	Decreased		✓	✓
Nitrogen balance	Negative			✓
Nerve excitability	Increased			✓
Sympathetic nervous response	Increased			✓
Psychological mood states	Altered	✓		
Risk of infection	Increased	✓		
Blood				
Hematocrit and hemoglobin	Decreased		✓	
Leukocytes and immunophenotypes	Decreased		✓	
Serum iron and ferritin	Decreased		✓	
Serum electrolyte levels	Decreased			✓
Serum glucose and free fatty acids	Decreased		✓	
Plasma lactate concentration, submax, max	Decreased		✓	✓
Ammonia	Increased		✓	✓
Serum testosterone and cortisol	Decreased	✓		
ACTH, growth hormone, prolactin	Decreased			✓
Catecholamines, rest, night	Decreased			✓
Creatine kinase	Increased			✓

Note. HR = heart rate; RER = respiratory exchange ratio; ACTH = adrenocorticotropic hormone.

Adapted from Armstrong and VanHeest, *Sports Medicine,* 2002.[1]

production. These enzymes generally are confined to the inside of cells, so the presence of large amounts of these enzymes in the blood suggests that muscle cell membranes have suffered some damage, allowing the enzymes to escape. Following periods of heavy training, blood levels of such enzymes have been reported to be 2 to 10 times above normal. Recent studies support the idea that these changes might reflect varied degrees of muscle tissue breakdown. As noted in chapter 3, tissue from the leg muscles of marathon runners has shown remarkable damage to the muscle fibers after training and marathon competition, and the onset and timing of these muscle changes paralleled the degree of muscle soreness experienced by the runners.

The performance effects of muscle damage are not fully understood, but experts generally agree that muscle damage might be partly responsible for the localized pain, tenderness, and swelling associated with muscle soreness. Still, no evidence links this condition to overtraining syndrome. Researchers suspect that blood enzyme levels increase and muscle fiber damage occurs frequently during eccentric exercise, regardless of the state of training. For these reasons, and because measuring blood enzyme levels is both difficult and expensive, blood enzyme levels do not appear to be suitable indicators of overtraining syndrome.

Oxygen Consumption

As athletes become overtrained, they often show a loss of skill, which decreases their performance efficiency at submaximal rates of exercise. As their movements become less efficient, their submaximal oxygen consumption typically increases. Because of this, measurement of oxygen consumption during a standardized exercise test can be used to monitor the athlete's loss of skill during overtraining. For example, figure 12.9 compares the submaximal oxygen consumption rate of a college cross country runner during the early season (when he performed well) and in late season (when he experienced symptoms of overtraining). You can see that while the runner experienced the symptoms of overtraining, his submaximal ox-

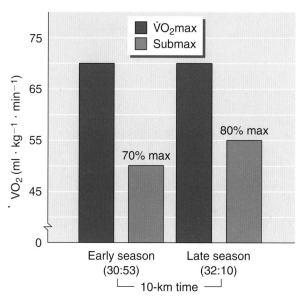

▲ **Figure 12.9**   Oxygen consumption and 10-km times of a cross country runner during the early season (when he was performing well) and in late season (when he was experiencing symptoms of overtraining). Note that his $\dot{V}O_2$max did not change throughout the season, but the oxygen needed to run at a submaximal pace of 3.7 min/km (6 min/mi) was increased. As a result, the percentage of $\dot{V}O_2$max used during this submaximal run increased from 70% in the early season to 80% when he was running poorly.

ygen consumption rate increased by 10%; thus, although $\dot{V}O_2$max did not change, he was exercising at 80% of his $\dot{V}O_2$max in the overtrained state compared with 70% of his $\dot{V}O_2$max in his normal trained state. Unfortunately, such tests are of little value to the coach or athlete because they are too complex, time-consuming, and impractical for field use.

Heart Rate

Current technology allows the coach to monitor the athlete's heart rate response during a standardized exercise bout. For example, the data presented in figure 12.10 illustrate a runner's heart rate response during a 1-mi (1.6-km) run performed at a fixed pace of 6 min/mi (3.7 min/km), or 10 mph (16 km/h). This response was monitored when the runner was untrained (UT), after the runner had trained (TR), and during a period when the runner demonstrated symptoms of overtraining syndrome (OT). This figure shows that heart

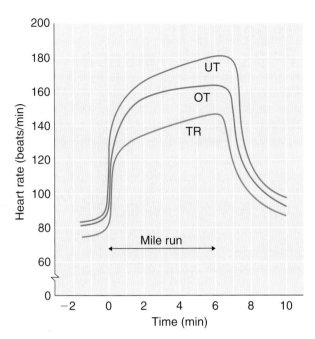

▲ **Figure 12.10**   A runner's heart rate responses during a fixed-pace treadmill run at 10 mph (16 km/h) performed before training (UT), after training (TR), and when the runner showed symptoms of overtraining (OT).

▲ **Figure 12.11**   A runner outfitted with a heart rate monitor. The chest strap picks up and transmits electrical impulses from the heart to the memory device worn on the wrist. After the run, the recordings are played back for interpretation.

rate was higher when the runner was in the overtrained state than when the runner was responding well to training. Similar findings have been reported for swimmers.[6]

The advantage of this test is that it provides an easily obtained, objective measurement of the athlete's cardiovascular response to a given rate of work. Heart rates are relatively simple to record (see figure 12.11) and provide immediate information to the athlete and coach. Such a test provides an objective way to monitor training and can provide a warning signal of overtraining syndrome.[29]

> The best predictor of overtraining syndrome appears to be heart rate response to a standardized bout of work. Performance decrements are also good indicators.

Treating the Overtraining Syndrome

Although the causes for performance deterioration with the overtraining syndrome are not clear, anecdotal reports suggest that training

intensity or speed is a more potent stressor than training volume. Recovery from overtraining syndrome is possible with a marked reduction in training intensity or complete rest. Although most coaches recommend a few days of easy training, overtrained athletes require considerable time for full recovery, and this might necessitate the total cessation of training for a period of weeks or months. In some cases, counseling might be needed to help the athletes cope with other stress in their lives that might contribute to this condition.

The best way to minimize the risk of overtraining is to follow periodization training procedures, alternating easy, moderate, and hard periods of training as discussed earlier in this chapter. Although individual tolerance varies tremendously, even the strongest athletes have periods when they are susceptible to the overtraining syndrome. As a rule, 1 or 2 days of intense training should be followed by an equal number of easy, aerobic training days. Likewise, a week or two of hard training should be followed by a week of reduced effort with little or no emphasis on anaerobic exercise.

Endurance athletes (such as swimmers, cyclists, and runners) must pay particular attention to their carbohydrate intake. Repeated days of hard training gradually reduce muscle glycogen. Unless these athletes consume extra carbohydrate during these periods, their muscle and liver glycogen reserves can be depleted. As a consequence, the most heavily recruited muscle fibers are not able to generate the energy needed for exercise.

▶ Overtraining stresses the body beyond its capacity to adapt, decreasing performance and physiological capacity.

▶ The symptoms of overtraining syndrome are subjective, vary from individual to individual, and many also accompany regular training, which makes prevention or diagnosis of the overtraining syndrome difficult.

▶ Possible explanations for the overtraining syndrome include changes in the functioning of the divisions of the autonomic nervous system, altered endocrine responses, and suppressed immune function.

▶ Many potential signs and symptoms of overtraining might be used to diagnosis overtraining in its earliest stages. At this time, the heart rate response to a fixed-pace exercise bout appears to be the easiest and most specific technique.

▶ Overtraining syndrome is treated by a marked reduction in training intensity or complete rest, for periods of weeks or months. Prevention can best be accomplished by using periodization training procedures that vary training intensity and volume and, for endurance athletes, by ensuring adequate carbohydrate intake to meet energy needs.

Tapering for Peak Performance

Peak performance requires maximal physical and psychological tolerance for the stress of the activity. But periods of intense training reduce muscular strength, decreasing athletes' performance capacity. For this reason, to compete at their peak, many athletes reduce their training intensity and volume before a major competition to give their bodies and minds a break from the rigors of intense training. This practice is referred to as tapering. The taper period, during which intensity and volume are reduced, should provide adequate time for healing of tissue damage caused by intense training and for the body's energy reserves to be fully replenished. Research suggests that, for example, the taper period for swimmers might need to last for at least 2 weeks to maximize performance.

The most notable change during the taper period is a marked increase in muscular strength, which explains at least part of the performance improvement that occurs. It is difficult to determine whether strength improvements result from changes in the muscles' contractile mechanisms or improved muscle fiber recruitment. However, examination of individual muscle fibers taken from swimmers' arms before and after 10 days of intensified training showed that the fast-twitch (FT) fibers exhibit a significant reduction in their maximal shortening velocity.[12] This change has been attributed to changes in the fibers' myosin molecules. In these cases, the myosin in the FT fibers became more like that in the slow-twitch (ST) fibers. We assume from this finding that such changes in the muscle fibers cause the power loss that swimmers and runners experience during prolonged periods of intense training. We can also assume that the recovery of strength and power that occurs with tapering might be linked to modifications of the muscles' contractile mechanisms. Tapering also allows time for the muscle to repair any damage incurred during intense training and for the energy reserves (i.e., muscle and liver glycogen) to be restored.

Tapering for competition is crucial to your best performance. Training tears down the body, so reduced training volume and intensity, coupled with quality rest, are needed to allow your body to repair itself and to restore its energy reserves for competition.

Although tapering is widely practiced in a variety of sports, many coaches fear that reduced training for such a long period before a major competition will decrease conditioning and impair performance. But numerous studies clearly show that this fear is unwarranted. Developing optimal $\dot{V}O_2$max initially requires a considerable amount of training, but once it has been developed, much less training is needed to maintain it at its highest level. In fact, the training level of $\dot{V}O_2$max can be maintained even when training frequency is reduced by two thirds.[14]

Runners and swimmers who reduce their training by about 60% for 15 to 21 days show no losses in $\dot{V}O_2$max or endurance performance.[5, 15] One study found that swimmers' blood lactate levels after a standard swim were lower after a taper period than before. More important, the swimmers experienced a 3.1% improvement in performance as a result of the reduced training and demonstrated a 17.7% to 24.6% increase in arm strength and power.[5]

In a study of distance runners, those runners who went through a 7-day taper decreased their running time in a 5-km time trial by 3% compared with those who did not taper. Submaximal oxygen uptake while running at 80% $\dot{V}O_2$max was decreased by 6% in those who tapered, indicating a greater economy of effort. These results are illustrated in figure 12.12. Blood lactate levels at 80% of $\dot{V}O_2$max were unchanged, as were $\dot{V}O_2$max and leg extension peak force.[16]

Unfortunately, little information is available to demonstrate the influence of tapering on

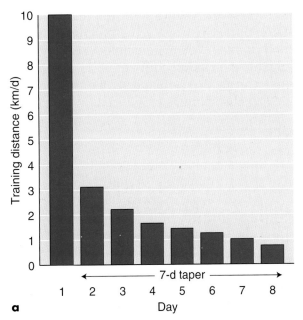

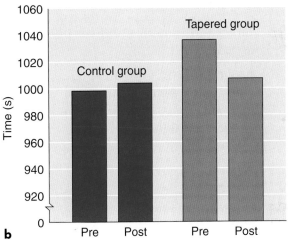

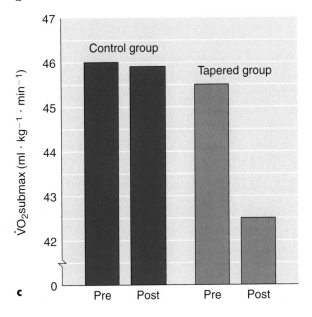

▶ **Figure 12.12**   Effect of a 7-day taper (*a*) on 5-km time trials (*b*) and $\dot{V}O_2$submax (*c*) in two groups of runners who either did or did not (control) taper.

Data from Houmard et al. 1994.[16]

performance in team sports and in long-duration endurance events such as cycling and marathon running. Before guidelines can be offered for athletes in these sports, research is needed to demonstrate that similar benefits can be generated by such periods of reduced training.

▶ Many athletes decrease their training intensity and volume before a competition to increase strength, power, and performance capacity that accompany high-intensity training. This practice is called tapering.

▶ Muscular strength increases significantly during the tapering period.

▶ Tapering allows time for the muscle to repair any damage incurred during intense training and for the energy reserves (i.e., muscle and liver glycogen) to be restored.

▶ Less training is needed to maintain previous gains than was originally needed to attain them, so tapering does not decrease conditioning.

Detraining

What happens to highly conditioned athletes who have fine-tuned their performance skills to a peak level but find that the competitive season and daily training have come to a sudden end? Most athletes in team sports go into physical hibernation at this time. Many have been working 2 to 5 h each day to perfect their skills and improve their physical condition, and they welcome the opportunity to completely relax, purposely avoiding any strenuous physical activity. But how does physical inactivity affect highly trained athletes?

Detraining is defined as the partial or complete loss of training-induced adaptations in response to either the cessation of training or a substantial decrement in the training load.[23] Some of our knowledge about physical detraining comes from clinical research with patients who have been forced into inactivity because of injury or surgery. Athletes generally agree that

suffering the pain of an injury is bad enough, but the situation is even worse when it forces them to stop training. Most fear that all they have gained through hard training will be lost during a period of inactivity. But recent studies reveal that a few days of rest or reduced training will not impair and might even enhance performance. Yet at some point, training reduction or complete inactivity will decrease physiological function and performance.

In the following sections, we examine physiological responses to detraining. We look at specific areas of concern to the athlete: muscular strength and power; muscular endurance; speed, agility, and flexibility; and cardiorespiratory endurance.

Muscular Strength and Power

When a broken limb is immobilized in a rigid cast, changes begin immediately in both the bone and the surrounding muscles. Within only a few days, the cast that was applied tightly around the injured limb is loose. After several weeks, a large space separates the cast and the limb. Skeletal muscles undergo a substantial decrease in size, known as atrophy, when they remain inactive. This is accompanied by considerable loss of muscular strength and power. Total inactivity leads to rapid losses, but even prolonged periods of reduced activity lead to gradual losses that eventually can become quite significant.

Similarly, research confirms that muscular strength and power both are reduced once athletes stop training. But these changes are relatively small during the first few months. In one study, no strength loss was noted 4 weeks after completion of a 3-week resistance training program.[6] In another investigation, only 45% of the original strength gained from a 12-week training program had been lost when the subjects, who did no further training, were reevaluated 1 year later.[21]

A study with collegiate swimmers revealed that even with up to 4 weeks of inactivity, terminating training did not affect arm or shoulder strength.[3] No strength changes were seen in these swimmers, whether they spent 4 weeks at complete rest or whether they reduced their training frequency to one or three sessions per

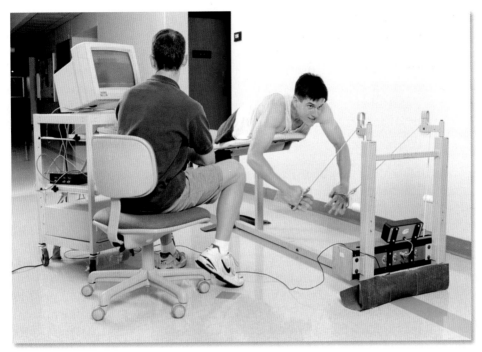

▲ **Figure 12.13**   Measurements of arm strength and power using a biokinetic swim bench.

week. But swimming power was reduced by 8% to 13.5% during the 4 weeks of reduced activity, whether the swimmers underwent complete rest or merely reduced training frequency.

This study's measurement techniques for strength and power differed in a significant respect. The swimmers' strength was measured on land, by using the semi-accommodating swim bench (see figure 12.13), but their power was measured in the water, by using tethered swimming, which allows the swimmers to use more natural actions. (Tethered swimming is discussed in the introduction of this book.)

The results of these measurements, depicted in figure 12.14, suggest that the less specific land measurements of muscular strength might not have accurately reflected the performance loss that the swimmers experienced. Although muscular strength might not have diminished during the 4 weeks of reduced training, the swimmers might have lost their ability to apply force during swimming, probably attributable to a loss of skill. As swimming coaches would say, the swimmers appeared to have "lost their feel" for the water.

The physiological mechanisms responsible for the loss of muscular strength as a conse-

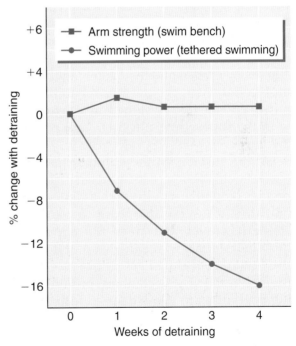

▲ **Figure 12.14**   Changes in arm strength and swimming power during 4 weeks of detraining. Arm strength was assessed from performance on a swim bench (nonspecific), whereas swimming power was assessed from performance of tethered swimming (specific).

quence of either immobilization or inactivity are not clearly understood. Muscle atrophy causes a noticeable decrease in muscle mass and water content, which could partly account for a loss in development of maximal muscle fiber tension. When muscles aren't used, the frequency of their neurological stimulation is reduced, and normal fiber recruitment is disrupted. Thus, part of the strength loss associated with detraining could result from an inability to activate some muscle fibers.

Research indicates that after training is terminated, an athlete can retain gained muscular strength and power for periods of up to 6 weeks. By continuing to train once every 10 to 14 days, athletes usually can maintain strength gains for much longer periods. Evidently, muscle requires minimal stimulation to retain the strength, power, and size gained during training.

This retention of muscular strength, power, and size has extremely important implications for the injured athlete. The athlete can save much time and effort during rehabilitation by performing even a low level of exercise with the injured limb, starting in the first few days of recovery. Simple isometric actions are very effective for rehabilitation because their intensity can be graded and they don't require joint movement. Any program of rehabilitation, however, must be designed in cooperation with the supervising physician and physical therapist.

On the surface, all these findings seem to conflict with the observations noted earlier that periods of total inactivity, such as limb immobilization, cause sizable losses in muscular strength, power, and mass. But apparently most individuals who either stop or decrease their training regimen get sufficient exercise through walking, stair climbing, pushing, pulling, and lifting to allow them to retain much of the strength they previously gained through strength training. With immobilization, though, virtually no activation occurs to stimulate the contractile processes. Strength and mobility are lost rapidly under such conditions.

Muscular Endurance

Muscular endurance performance decreases after only 2 weeks of inactivity. At this time,

not enough evidence is available to determine whether this performance decrement results from changes in the muscle or from changes in cardiovascular capacity. In this section, we examine muscle changes that are known to accompany detraining and that could decrease muscular endurance.

The localized muscle adaptations that occur during periods of inactivity are well documented, but knowledge of the exact role that these changes play in the loss of muscular endurance still eludes us. We know from postsurgery cases that after a week or two of cast immobilization, the activities of oxidative enzymes such as succinate dehydrogenase (SDH) and cytochrome oxidase decrease by 40% to 60%. Data collected from swimmers, shown in figure 12.15, indicate that the muscles' oxidative potential decreases much more rapidly than the subjects' maximal oxygen uptake with detraining. Reduced oxidative enzyme activity would be expected to impair muscular endurance, and this most likely relates to submaximal endurance capacity rather than to maximal aerobic capacity, or $\dot{V}O_2$max.

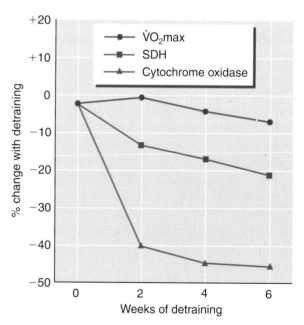

▲ **Figure 12.15** Percentage decreases in $\dot{V}O_2$max, muscle succinate dehydrogenase (SDH) activity, and cytochrome oxidase activity during 6 weeks of detraining. These interesting findings suggest that the muscles experience a decline in metabolic potential, although tests of $\dot{V}O_2$max show little change over this period of detraining.

In contrast, when athletes stop training, the activities of their muscles' glycolytic enzymes, such as phosphorylase and phosphofructokinase, change little, if at all, for at least 4 weeks. In fact, Coyle and colleagues[9] observed no change in glycolytic enzyme activities with up to 84 days of detraining compared with a nearly 60% decrease in the activities of various oxidative enzymes. This means that with detraining, the muscles' capacity for anaerobic performance is maintained longer than their capacity for aerobic performance. This might at least partly explain why performance times in sprint events are unaffected by a month or more of inactivity, but the ability to perform longer endurance events may decrease significantly with as little as 2 weeks of detraining.

One notable change in the muscle during detraining is a change in its glycogen content. Endurance-trained muscle tends to increase its glycogen storage. But 4 weeks of detraining have been shown to decrease muscle glycogen by 40%.[4] Figure 12.16 illustrates the decrease in muscle glycogen accompanying 4 weeks of detraining in competitive collegiate swimmers and in untrained subjects. The untrained people showed no change in muscle glycogen content after 4 weeks of inactivity, but the swimmers' values decreased until they were about equal to those of the untrained people. This indicates that the trained swimmers' improved capacity for muscle glycogen storage was reversed during detraining.

Measurements of blood lactate and pH after a standard work bout have been used to assess the physiological changes that accompany training and detraining. For example, a group of collegiate swimmers were required to perform a standard-paced, 200-yd (183-m) swim at 90% of their seasonal best following 5 months of training and then to repeat this test at the same absolute pace once a week for the following 4 weeks of detraining. The results are shown in table 12.2 (on page 396). Blood lactate levels, taken immediately after this standard swim, increased from week to week during a month of inactivity. At the end of the fourth week of detraining, the swimmers' acid-base balance was significantly disturbed. This was reflected by a significant increase in blood lactate levels and a significant decrease in the levels of bicarbonate (a buffer). These findings support the theory that the muscles' oxidative and anaerobic energy systems change slowly and are probably unaffected by only a few days of rest. Only during periods of complete inactivity (immobilization) do such changes impair performance in the first week or two.

Muscle fiber composition does not appear to change during short periods of inactivity. But some clinical cases have reported dramatic changes in the percentages of ST and FT fibers, with a shift toward more FT fibers, in athletes who have undergone immobilization following surgery. These findings have not been replicated, however, so we cannot draw valid conclusions about the effects of detraining on muscle fiber composition.

Another structural change that has been proposed as a possible reason for the decrease in muscular endurance involves blood flow to the muscles. Although not all investigators agree, there is evidence to suggest that muscle capillary supply may decrease during detraining. This would impair oxygen delivery to the muscles, decreasing their oxidative potential. However, these findings are inconclusive.

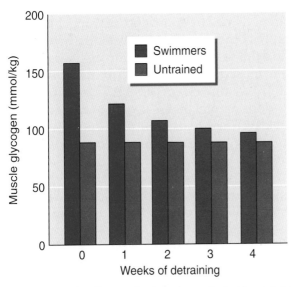

▲ **Figure 12.16** Changes in glycogen content of the deltoid muscle in competitive swimmers during 4 weeks of detraining. Muscle glycogen returned almost to the untrained level at the end of this period.

Table 12.2

Blood Lactate, pH, and Bicarbonate (HCO_3^-) in Eight Collegiate Swimmers Undergoing Detraining

Measurement	Weeks of detraining			
	0[a]	1[b]	2	4
Lactate (mmol/L)	4.2	6.3	6.8	9.7[c]
pH	7.259	7.237	7.236	7.183[c]
HCO_3^- (mmol/L)	21.1	19.5[c]	16.1[c]	16.3[c]
Swim time (s)	130.6	130.1	130.5	130.0

Note. Measurements were taken immediately after a fixed-pace swim.

[a]The values at week 0 represent the measurements taken at the end of 5 months of training.

[b]The values for weeks 1, 2, and 4 are the results obtained after 1, 2, and 4 weeks of detraining, respectively.

[c]Significant difference from the value at the end of training.

Speed, Agility, and Flexibility

Training produces less improvement in speed and agility than it does in strength, power, muscular endurance, flexibility, and cardiorespiratory endurance. Consequently, losses of speed and agility that occur with inactivity are relatively small. Also, peak levels of both can be maintained with a limited amount of training. But this does not imply that the track sprinter can get by with training only a few days a week. Success in actual competition relies on factors other than basic speed and agility, such as correct form, skill, and the ability to generate a strong finishing sprint. Many hours of practice are required to tune performance to its optimal level, but most of this time is spent developing performance qualities other than speed and agility.

Flexibility, on the other hand, is lost rather quickly during inactivity, so it must be worked on throughout the year. Stretching exercises should be incorporated into both in-season and off-season training programs. But during the off-season, many athletes tend to ignore flexibility training because flexibility can be regained so rapidly. Although flexibility can be reestablished in little time, the athlete should maintain the desired flexibility level year-round. Reduced flexibility has been proposed to increase athletes' susceptibility to serious injury.

Cardiorespiratory Endurance

The heart, like other muscles in the body, is strengthened by endurance training. Inactivity, on the other hand, can substantially decondition the heart and the cardiovascular system. The most dramatic example of this is seen in a study conducted on subjects undergoing long periods of total bed rest; they weren't allowed to leave their beds, and physical activity was kept to an absolute minimum.[27] Cardiovascular function was assessed while the subjects performed at a constant rate of work both before and after the 20-day period of bed rest. The cardiovascular effects that accompanied bed rest included

- a considerable increase in submaximal heart rate,
- a 25% decrease in submaximal stroke volume,

- a 25% reduction in maximal cardiac output, and
- a 27% decrease in maximal oxygen consumption.

The reductions in cardiac output and $\dot{V}O_2$max appear to result from reduced stroke volume, which is probably attributable to a combined decrease in heart volume, total blood volume, plasma volume, and ventricular contractility.

It is interesting that the two most highly conditioned subjects in this study (the two who had the highest $\dot{V}O_2$max values) experienced greater decrements in $\dot{V}O_2$max than the three less fit people, as shown in figure 12.17. Furthermore, the untrained subjects regained their initial conditioning levels (before bed rest) in the first 10 days of reconditioning, but the well-trained subjects needed about 40 days for full recovery. This suggests that highly trained individuals cannot afford long periods with little or no endurance training. The athlete who totally abstains from physical training at the completion of the season will experience great difficulty getting back into physical condition when the new season begins.

Studies have shown that impairment of cardiovascular function following a few weeks of detraining is caused largely by a reduction in plasma volume, which in turn diminishes the stroke volume of the heart. In one study,[8] 2 to 4 weeks of reduced activity following months of training for cycling and running resulted in a 9% decrease in blood volume and a 12% decrease in both stroke volume and plasma volume. As a result, $\dot{V}O_2$max decreased by 5.9%. After the subjects were detrained, they were infused with a dextran (sugar) solution to expand their blood volume until it exceeded their trained level. As shown in table 12.3 on page 398, this improved cardiovascular function and $\dot{V}O_2$max, although it proved to be of little benefit to endurance performance.

Studies also have revealed changes in trained subjects' endurance performance during periods of inactivity. This reduction in cardiorespiratory endurance is much greater than the reductions of strength, power, and muscular endurance for the same period of inactivity. Drinkwater and Horvath[10] studied seven female track athletes at the end of the competitive season and again 3 months after formal training ended. During that 3-month period, the athletes participated in typical physical activities of their age group, in-

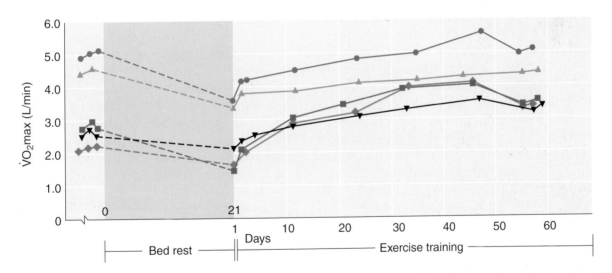

▲ **Figure 12.17**   Changes in $\dot{V}O_2$max with 20 days of bed rest for five individual subjects. Note that the subjects who were least fit (lowest $\dot{V}O_2$max values) at the start of bed rest showed smaller decrements with inactivity and greater gains when they trained after bed rest. Highly fit individuals, on the other hand, were far more affected by the period of inactivity.

Adapted, by permission, from B. Saltin et al., 1968, "Response to submaximal and maximal exercise after bed rest and training," *Circulation* 38(7): 75.

Table 12.3

Effects of Detraining and Blood Volume Expansion

Parameter	Normal blood volume		Expanded blood volume
	Trained	Detrained	Detrained
Blood volume (ml)	5,177	4,692[b]	5,412
Stroke volume (ml/beat)[a]	166	146[b]	164
$\dot{V}O_2$max (L/min)	4.42	4.16[b]	4.28
Exercise time to exhaustion (min)	9.13	8.44	8.06[c]

[a] Stroke volume measured during submaximal exercise.

[b] Denotes a significant difference from the trained (normal blood volume) and detrained (expanded blood volume) values.

[c] Denotes a significant difference from the trained (normal blood volume) value.

Adapted from Coyle et al. 1986.[8]

cluding required physical education. At the end of the 3 months, $\dot{V}O_2$max in the subjects had decreased an average of 15.5%. The new $\dot{V}O_2$max levels were similar to those found in nonathletic girls of the same age.

Inactivity can significantly reduce $\dot{V}O_2$max. How much activity is needed to prevent such considerable losses of physical conditioning? Although a decrease in training frequency and duration reduces aerobic capacity, the losses are significant only when frequency and duration are reduced by two thirds of the regular training load. However, training intensity apparently plays a more crucial role in maintaining aerobic power during periods of reduced training. Studies by Hickson and coworkers[14] suggest that training intensity must be at least 70% $\dot{V}O_2$max to maintain training-induced improvements in $\dot{V}O_2$max. In subjects who previously trained for 10 weeks, as little as a one-third reduction in training intensity for 15 weeks significantly decreased

- $\dot{V}O_2$max,
- long-term endurance (at 80% $\dot{V}O_2$max until exhaustion), and
- cardiac size.

Short-term (4- to 8-min) endurance and body composition, however, were not changed by this one-third reduction in training intensity.

From these and other studies, it is apparent that cardiorespiratory endurance capacity is lost rapidly after the cessation of formal endurance training. Although complete bed rest provides the most dramatic decreases, even periods of light activity or formal endurance training at a frequency of only once or twice a week are not sufficient to prevent the loss of cardiovascular conditioning. Thus, athletes must try to maintain their endurance capacity during the off-season, because once endurance capacity is lost, regaining peak levels takes considerable time. Likewise, an injured athlete should get back into some modified form of endurance exercise as soon as possible to minimize loss of cardiorespiratory endurance capacity.

Your body rapidly loses many of the benefits of training if training is discontinued. Some minimal level of training is necessary to prevent these losses. Research indicates that at least three training sessions per week at an intensity of at least 70% $\dot{V}O_2$max are needed to maintain aerobic conditioning.

▶ Detraining is defined as the partial or complete loss of training-induced adaptations in response to either the cessation of training or a substantial decrement in the training load. The effects of stopping training are quite minor compared with those from immobilization. In general, the greater the gains during training, the greater the losses during detraining simply because the well-trained person has more to lose than the untrained person.

▶ Detraining causes muscle atrophy, which is accompanied by losses in muscular strength and power. However, muscles require only minimal stimulation to retain these qualities during periods of reduced activity.

▶ Muscular endurance decreases after only 2 weeks of inactivity. Possible explanations for this are

  1. decreased oxidative enzyme activity,
  2. decreased muscle glycogen storage,
  3. disturbance of the acid–base balance, or
  4. decreased blood supply to the muscles.

▶ Detraining losses in speed and agility are small, but flexibility seems to be lost quickly.

▶ With detraining, losses of cardiorespiratory endurance are much greater than losses of muscular strength, power, and endurance over the same time period.

▶ To maintain cardiorespiratory endurance, training must be conducted at least three times per week, and training intensity should be at least 70% of $\dot{V}O_2$max.

Retraining

Recovery of conditioning after a period of inactivity, known as **retraining,** is affected by your fitness level and how long you were inactive.

As mentioned earlier, the most highly trained individuals typically experience the greatest loss of conditioning from detraining. Because of this, these people also take considerably longer to regain their initial fitness levels than subjects who are less trained.

Two to three weeks of detraining have been shown to cause the following decrements in highly trained subjects:

- Muscle oxidative enzyme activities decreased by 13% to 24%.
- Performance time decreased by about 2% to 5%.
- $\dot{V}O_2$max decreased by about 4%.

Following 15 days of retraining, only $\dot{V}O_2$max had returned to its original trained level. Oxidative enzyme activities did not improve, and although performance time showed some improvement, it still remained 2% to 5% below the trained level. This suggests that even short periods of detraining significantly change physiological capacity in highly trained subjects and that a longer period of retraining is necessary for these trained individuals to regain their conditioning.[27]

As noted earlier, muscles that have been immobilized in a cast, whether for a few days or for many weeks, lose much of their strength, power, and endurance. After the cast is removed, most people can't begin activity immediately because they lack sufficient joint mobility. Regaining the range of joint motion is a relatively slow process, often taking several months for full recovery.

Several procedures have been proposed to speed the recovery of muscle function following immobilization. For example, when patients who had undergone surgical reconstruction of their anterior cruciate ligaments were given a cast that allowed some movement (20-60° range), full recovery of the knee's range of motion occurred in 4 weeks of retraining. But with an immovable cast, 16 weeks were needed to regain normal motion. The movable cast resulted in minimal reduction of muscle fiber cross-sectional area and no reduction in oxidative enzyme activities.

Other studies have revealed effective ways to limit the reduction in muscle aerobic capacity following cast immobilization. Compared with strength training alone, 20 to 60 min of daily cycling following cast removal leads to greater gains in muscle aerobic capacity and improved knee flexibility. Also, electrical stimulation of the muscles while they are immobilized prevents the usual decrease in their oxidative capacity and also can prevent muscle fiber atrophy.

▶ Retraining is the recovery of conditioning after a period of inactivity. It is affected by a person's fitness level and the duration and extent of the inactivity.

▶ The time needed for retraining can be reduced in cases of cast immobilization if the cast allows some range of movement.

▶ Electrical stimulation of the muscles prevents the usual decrease in muscle oxidative capacity and can prevent muscle fiber atrophy.

▶ The earlier an individual can resume active motion after immobilization or inactivity, the quicker the recovery of muscle function.

In Closing . . .

In this chapter we have examined how the quantity of training can affect your performance. We saw that too much training, either in the form of excessive training or overtraining, can actually impair performance. Then we looked at the effects of too little training—detraining—as a result of either inactivity or immobilization after an injury. We saw that with detraining, many of the gains achieved during regular training are quickly lost, especially cardiovascular endurance. Finally, we briefly considered the process of retraining, during which you try to regain what you have lost through detraining.

Now that we have dispelled the myth that more training always means better performance, in what other ways can athletes try to optimize their performance? In the next chapter, we turn our attention to sport nutrition.

▶ Key Terms

acute overload
anabolism
catabolism
chronic fatigue syndrome
detraining
excessive training
fibromyalgia syndrome
immune function
overreaching
overtraining
overtraining syndrome
retraining
tapering
taper period
undertraining

▶ Study Questions

1. What causes overtraining? How can it be identified? What is the suggested treatment for overtraining?

2. What physiological changes occur during the taper period that can be credited with improvements in performance?

3. What alterations occur in strength, power, and muscular endurance with physical detraining?

4. What changes take place in the muscle during periods of inactivity? During total muscle immobilization (casting)?

5. What alterations occur in speed, agility, and flexibility with physical detraining?

6. What changes occur in the cardiovascular system as one becomes deconditioned?

7. During periods of reduced training, what factors (frequency, intensity, or duration) must be stressed to prevent a decline in long-term endurance and aerobic capacity?

8. How can the negative effects of muscle immobilization be reduced?

▷ References

1. Armstrong, L.E., & VanHeest, J.L. (2002). The unknown mechanism of the overtraining syndrome. *Sports Medicine, 32,* 185-209.

2. Costill, D.L. (1986). *Inside running: Basics of sports physiology.* Indianapolis: Benchmark.

3. Costill, D.L. (1998). *Training adaptations for optimal performance.* Paper presented at the VIII International Symposium on Biomechanics and Medicine of Swimming, June 28, University of Jyväskylä, Finland.

4. Costill, D.L., Fink, W.J., Hargreaves, M., King, D.S., Thomas, R., & Fielding, R. (1985). Metabolic characteristics of skeletal muscle during detraining from competitive swimming. *Medicine and Science in Sports and Exercise, 17,* 339-343.

5. Costill, D.L., King, D.S., Thomas, R., & Hargreaves, M. (1985). Effects of reduced training on muscular power in swimmers. *Physician and Sportsmedicine, 13*(2), 94-101.

6. Costill, D.L., Maglischo, E., & Richardson, A. (1991). *Handbook of sports medicine: Swimming.* London: Blackwell.

7. Costill, D.L., Thomas, R., Robergs, R.A., Pascoe, D.D., Lambert, C.P., Barr, S.I., & Fink, W.J. (1991). Adaptations to swimming training: Influence of training volume. *Medicine and Science in Sports and Exercise, 23,* 371-377.

8. Coyle, E.F., Hemmert, M.K., & Coggan, A.R. (1986). Effects of detraining on cardiovascular responses to exercise: Role of blood volume. *Journal of Applied Physiology, 60,* 95-99.

9. Coyle, E.F., Martin, W.H., III, Sinacore, D.R., Joyner, M.J., Hagberg, J.M., & Holloszy, J.O. (1984). Time course of loss of adaptations after stopping prolonged intense endurance training. *Journal of Applied Physiology, 57,* 1857-1864.

10. Drinkwater, B.L., & Horvath, S.M. (1972). Detraining effects in young women. *Medicine and Science in Sports, 4,* 91-95.

11. Eliakim, A., Kodesh, E., Gavrieli, R., Radnay, J., Ben-Tovim, T., Yarom, T., & Falk, B. (1997). Cellular and humoral immune response to exercise among gymnasts and untrained girls. *International Journal of Sports Medicine, 18,* 208-212.

12. Fitts, R.H., Costill, D.L., & Gardetto, P.R. (1989). Effect of swim-exercise training on human muscle fiber function. *Journal of Applied Physiology, 66,* 465-475.

13. Fry, R.W., Morton, A.R., & Keast, D. (1991). Overtraining in athletes: An update. *Sports Medicine, 12,* 32-65.

14. Hickson, R.C., Foster, C., Pollock, M.L., Galassi, T.M., & Rich, S. (1985). Reduced training intensities and loss of aerobic power, endurance, and cardiac growth. *Journal of Applied Physiology, 58,* 492-499.

15. Houmard, J.A., Costill, D.L., Mitchell, J.B., Park, S.H., Hickner, R.C., & Roemmish, J.N. (1990). Reduced training maintains performance in distance runners. *International Journal of Sports Medicine, 11,* 46-51.

16. Houmard, J.A., Scott, B.K., Justice, C.L., & Chenier, T.C. (1994). The effects of taper on performance in distance runners. *Medicine and Science in Sports and Exercise, 26,* 624-631.

17. Kirwan, J.P., Costill, D.L., Flynn, M.G., Mitchell, J.B., Fink, W.J., Neufer, P.D., & Houmard, J.A. (1988). Physiological responses to successive days of intense training in competitive swimmers. *Medicine and Science in Sports and Exercise, 20,* 255-259.

18. Kirwan, J.P., Costill, D.L., Houmard, J.A., Mitchell, J.B., Flynn, M.G., & Fink, W.J. (1990). Changes in selected blood measures during repeated days of intense training and carbohydrate control. *International Journal of Sports Medicine, 11,* 362-366.

19. Kuipers, H., & Keizer, H.A. (1988). Overtraining in elite athletes: Review and directions for the future. *Sports Medicine, 6,* 79-92.

20. Mackinnon, L.T. (1989). Exercise and natural killer cells: What is the relationship? *Sports Medicine, 7,* 141-149.

21. McMorris, R.O., & Elkins, E.C. (1954). A study of production and evaluation of muscular hypertrophy. *Archives of Physical Medicine and Rehabilitation, 35,* 420-426.

22. Mostardi, R., Gandee, R., & Campbell, T. (1975). Multiple daily training and improvement in aerobic power. *Medicine and Science in Sports, 7,* 82.

23. Mujika, I., & Padilla, S. (2000). Detraining: Loss of training-induced physiological and performance adaptations. Part 1 and Part 2. *Sports Medicine, 30,* 79-87, 145-154.

24. Nieman, D.C. (1994). Exercise, infection, and immunity. *International Journal of Sports Medicine, 15,* S131-S141.

25. Nieman, D.C. (1997). Immune response to heavy exertion. *Journal of Applied Physiology, 82,* 1385-1394.

26. O'Toole, M.L. (1998). Overreaching and overtraining in endurance athletes. In R.B. Kreider, A.C. Fry, & M.L. O'Toole (Eds.), *Overtraining in sport* (pp. 10, 13). Champaign, IL: Human Kinetics.

27. Saltin, B., Blomqvist, G., Mitchell, J.H., Johnson, R.L., Jr., Wildenthal, K., & Chapman, C.B. (1968). Response to submaximal and maximal exercise after bed rest and training. *Circulation, 38*(Suppl. 7).

28. Selye, H. (1956). *The stress of life.* New York: McGraw-Hill.

29. Sharp, R.L., Vitelli, C.A., Costill, D.L., & Thomas, R. (1984). Comparison between blood lactate and heart rate profiles during a season of competitive swim training. *Journal of Swimming Research, 1,* 17-20.

30. Shepherd, R.J. (2001). Chronic fatigue syndrome: An update. *Sports Medicine, 31,* 167-194.

31. Smith, L.L. (2000). Cytokine hypothesis of overtraining: A physiological adaptation to excessive stress? *Medicine and Science in Sports and Exercise, 32,* 317-331.

32. Uusitalo, A.L.T. (2001). Overtraining: Making a difficult diagnosis and implementing targeted treatment. *Physician and Sportsmedicine, 29,* 35-50.

33. Watt, E., Buskirk, E., & Plotnicki, B. (1973). A comparison of single versus multiple daily training regimens: Some physiological considerations. *Research Quarterly, 44,* 119-123.

34. Woods, J.A., Davis, J.M., Smith, J.A., & Nieman, D.C. (1999). Exercise and cellular innate immune function. *Medicine and Science in Sports and Exercise, 31,* 57-66.

▷ Selected Readings

Bompa, T.O. (1983). *Theory and methodology of training.* Dubuque, IA: Kendall/Hunt.

Busso, T., Benoit, H., Bonnefoy, R., Feasson, L., & Lacour, J.-R. (2002). Effects of training frequency on the dynamics of performance response to a single training bout. *Journal of Applied Physiology, 92,* 572-580.

Costill, D.L., Flynn, M.G., Kirwan, J.P., Houmard, J.A., Mitchell, J.B., Thomas, R., & Park, S.H. (1988). Effects of repeated days of intensified training on muscle glycogen and swimming performance. *Medicine and Science in Sports and Exercise, 20,* 249-254.

Costill, D.L., Hinrichs, D., Fink, W.J., & Hoopes, D. (1988). Muscle glycogen depletion during swimming interval training. *Journal of Swimming Research, 4(1),* 15-18.

Haggmark, T., Eriksson, E., & Jansson, E. (1986). Muscle fiber type changes in human skeletal muscle after injuries and immobilization. *Orthopedics, 9,* 181-185.

Henriksson, J., & Reitman, J.S. (1977). Time course of changes in human skeletal muscle succinate dehydrogenase and cytochrome oxidase activities and maximal oxygen uptake with physical activity and inactivity. *Acta Physiologica Scandinavica, 99,* 91-97.

Hickson, R.C., Kanakis, J.C., Moore, A.M., & Rich, S. (1981). Effects of frequency of training, reduced training and retraining on aerobic power and left ventricular responses. *Medicine and Science in Sports and Exercise, 13,* 93.

Hooper, S.L., & Mackinnon, L.T. (1995). Monitoring overtraining in athletes. *Sports Medicine, 20,* 321-327.

Houmard, J.A., Costill, D.L., Mitchell, J.B., Park, S.H., Fink, W.J., & Burns, J.M. (1990). Testosterone, cortisol, and creatine kinase levels in male distance runners during reduced training. *International Journal of Sports Medicine, 11,* 41-45.

Kirwan, J.P., Costill, D.L., Mitchell, J.B., Houmard, J.A., Flynn, M.G., Fink, W.J., & Beltz, J.D. (1988). Carbohydrate balance in competitive runners during successive days of intense training. *Journal of Applied Physiology, 65,* 2601-2606.

Kreider, R.B, Fry, A.C., & O'Toole, M.L. (Eds.). (1998). *Overtraining in sport.* Champaign, IL: Human Kinetics.

Lehman, M., Foster, C., & Keul, J. (1993). Overtraining in endurance athletes: A brief review. *Medicine and Science in Sports and Exercise, 25(7),* 854-862.

Mackinnon, L.T., & Hooper, S.L. (1996). Plasma glutamine and upper respiratory tract infection during intensified training in swimmers. *Medicine and Science in Sports and Exercise, 28,* 285-290.

Mackinnon, L.T., Hooper, S.L., Jones, S., Gordon, R.D., & Bachmann, A.W. (1997). Hormonal, immunological, and hematological responses to intensified training in elite swimmers. *Medicine and Science in Sports and Exercise, 29,* 1637-1645.

McCully, K.K., Sisto, S.A., & Natelson, B.H. (1996). Use of exercise for treatment of chronic fatigue syndrome. *Sports Medicine, 21,* 35-48.

Morgan, W.P., Brown, D.R., Raglin, J.S., O'Connor, P.J., & Ellickson, K.A. (1987). Psychological monitoring of overtraining and staleness. *British Journal of Sportsmedicine, 21,* 107-114.

Morgan, W.P., Costill, D.L., Flynn, M.G., Raglin, J.S., & O'Connor, P.J. (1988). Mood disturbance following increased training in swimmers. *Medicine and Science in Sports and Exercise, 20,* 408-414.

Mujika, I., & Padilla. S. (2001). Cardiorespiratory and metabolic characteristics of detraining in humans. *Medicine and Science in Sports and Exercise, 33,* 413-421.

Mujika, I., & Padilla, S. (2001). Muscular characteristics of detraining in humans. *Medicine and Science in Sports and Exercise, 33,* 1297-1303.

Nieman, D.C., & Pedersen, B.K. (1999). Exercise and immune function. *Sports Medicine, 27,* 73-80.

Pate, R.R., Hughes, R.D., Chandler, J.V., & Ratliffe, J.L. (1978). Effects of arm training on retention of training effects derived from leg training. *Medicine and Science in Sports, 10,* 71-74.

Perhonen, M.A., Franco, F., Lane, L.D., Buckey, J.C., Blomqvist, C.G., Zerwekh, J.E., Peshock, R.M., Weatherall, P.T., & Levine, B.D. (2001). Cardiac atrophy after bed rest and spaceflight. *Journal of Applied Physiology, 91,* 645-653.

Sherman, W.M., Plyley, M.J., Pearson, D.R., Habansky, A.J., Vogelgesang, D.A., & Costill, D.L. (1983). Isokinet-ic rehabilitation after meniscectomy: A comparison of two methods of training. *Physician and Sportsmedicine, 11,* 121-133.

Urhausen, A., & Kindermann, W. (2002). Diagnosis of overtraining: What tools do we have? *Sports Medicine, 32,* 95-102.

CHAPTER 13

NUTRITION AND SPORT

overview

The intense effort and energy expenditure of sports training and competition place unusual demands on the diet of athletes. Athletes in some sports such as swimming and distance running can have trouble balancing their energy intake to the caloric demands of training. For these reasons, the major dietary concern of many athletes is the amount of food they consume rather than what they eat. Yet, in their quest for success, most athletes have at some time searched for a magic food that will produce a winning performance. Unfortunately, diet manipulations typically are based on testimonials from more successful performers, poorly designed research studies, unsupported commercial advertising claims, and misinterpretation of nutritional research. Few areas of exercise science are more fraught with fads and fraud than the field of sport nutrition. Too often this leads to unfounded nutritional practices.

In this chapter, we examine the substances we ingest and their importance beyond their role in bioenergetics. We focus on the optimal composition of the diet and the special dietary needs of athletes. And we carefully examine how nutrition can affect performance, focusing on the potential benefits of various nutrients and dispelling many myths.

outline

In 1970, we (DLC) conducted a study in an effort to determine why athletes who train or compete intensely on repeated days gradually become chronically fatigued. Trained marathon runners were asked to run on a treadmill for 2 h on 3 successive days at a pace equivalent to their best marathon performance. During this period they ate a normal mixed diet containing 50% of calories from carbohydrate, 35% from fat, and 15% from protein. On the average, the runners covered about 32 km (20 mi) in the 2 h, becoming more and more fatigued with each succeeding day. By the third day, none of the runners could maintain the pace of the previous days, and all had to terminate the run before completing the 2-h effort. Why did they become chronically fatigued? Muscle biopsy data revealed that their glycogen levels were extremely low, suggesting that the diet was inadequate to meet the energy needed for exercise. Could the runners' diet have been changed to prevent this fuel deficit? In the following discussion, we focus on the role of nutrition for training and optimal performance.

Optimal physical performance requires a careful dietary balance of the essential nutrients. The U.S. government has established standards for optimal nutrient intake that are termed Recommended (Daily) Dietary Allowances, or RDA. The RDA of a substance is an estimate of the intake adequate to maintain good health. RDA values are guidelines to help people of average activity levels gauge their diets.

The nutritional needs of very active athletes can exceed the RDAs considerably. Individual caloric needs are quite variable, depending on the athlete's size, sex, and sport choice. Some athletes have been reported to need as many as 12,000 kcal per day! Also, some competitive sports require adherence to rigid weight standards. Athletes who participate in these sports must closely monitor their weight and thus their caloric intake. Too often, this leads to nutritional abuses, drug use, dehydration, and serious health risks. In addition, the dietary tactics used by some athletes to achieve excessive weight loss are of increasing concern because of the potential association with eating disorders, such as anorexia nervosa and bulimia nervosa.

A person's diet should contain a relative balance of carbohydrate, fat, and protein. Of the total calories consumed, the recommended balance for most people is

- carbohydrate: 55% to 60%,
- fat: no more than 30% (less than 10% saturated), and
- protein: 10% to 15%.

Interestingly, this recommended percentage distribution of total calories consumed appears to be optimal for both athletic performance and health. A similar distribution of caloric intake is recommended for the prevention of cardiovascular disease, cancer, and obesity.

Although all foods ultimately can be broken down to carbohydrate, fat, or protein, these nutrients are not all that the body needs. Let's look at the six classes of nutrients.

Six Nutrient Classes

The energy from the foods we eat is essential to our ability to sustain physical activity, but we rely on foods for much more than energy. Food can be categorized into six classes of nutrients, each with specific functions in the body:

- Carbohydrate
- Fat (lipid)
- Protein
- Vitamins
- Minerals
- Water

In the following discussion, we examine the physiological importance to the athlete of each class of nutrients.

In Transition: From the RDAs to the DRIs

A major revision of the United States Recommended Daily Allowance (RDA) guidelines is in progress. In the 1950s, the Food and Nutrition Board of the National Academy of Sciences adopted the concept of the RDA. The latest edition was released in 1989. The RDAs provide estimates of safe and adequate daily dietary intakes and estimated minimum requirements for selected vitamins and minerals. The RDAs will be replaced by new recommendations called Dietary Reference Intakes (DRIs). The DRIs reflect a joint effort between the United States and Canada to provide dietary intake recommendations grouped by nutrient function and classification. Seven nutrient groups have been established:

- Calcium, vitamin D, phosphorus, magnesium, and fluoride
- Folate and other B vitamins
- Antioxidants (e.g., vitamins C and E, selenium)
- Macronutrients (e.g., protein, fat, carbohydrates)
- Trace elements (e.g., iron, zinc)
- Electrolytes and water
- Other food components (e.g., fiber, phytoestrogens)

Although recommendations have been made for some of these groups, the final DRIs for all groups are not yet available. When available, they will include at least three different reference values:

- Estimated Average Requirement—the intake value estimated to meet the requirement for 50% of an age- and sex-specific group
- Recommended Dietary Allowance—the intake value sufficient to meet the nutrient requirements of nearly all individuals in a specific group
- Tolerable Upper Intake Level—the maximum level of daily nutrient intake that is unlikely to pose risks of adverse health effects to almost all individuals in a specific group.

Carbohydrate

A **carbohydrate** (CHO) is classified as either monosaccharide, disaccharide, or polysaccharide. Monosaccharides are the simple one-unit sugars (such as glucose, fructose, and galactose) that cannot be reduced to a simpler form. Disaccharides (such as sucrose, maltose, and lactose) are composed of two monosaccharides. For example, sucrose (table sugar) consists of glucose and fructose. Oligosaccharides are short chains of 3 to 10 monosaccharides linked together. Polysaccharides contain starch and fiber and are commonly referred to as complex carbohydrates. Simple carbohydrates refer to carbohydrates derived from processed foods or foods high in sugar.

All carbohydrates must be broken down to monosaccharides before the body can use them.

Carbohydrate serves many functions in the body:

- It is a major energy source, particularly during high-intensity exercise.
- Its presence regulates fat and protein metabolism.
- The nervous system relies exclusively on carbohydrate for energy.
- Muscle and liver glycogen are synthesized from carbohydrate.

Major sources of carbohydrate include grains, fruits, vegetables, milk, and concentrated sweets.

Refined sugar, syrup, and cornstarch are nearly pure carbohydrates. Many concentrated sweets such as candy, honey, jellies, molasses, and soft drinks contain few if any other nutrients.

Carbohydrate Consumption and Glycogen Storage

Your body stores excess carbohydrate, primarily in your muscles and liver, as glycogen. Because of this, your carbohydrate consumption directly influences your muscle glycogen storage and your ability to train and compete in endurance events. As shown in figure 13.1, athletes who trained intensely and ate a low-carbohydrate diet (40% of total calories) experienced a day-to-day decrease in muscle glycogen.[17, 19] When these athletes consumed a high-carbohydrate diet (70% of total calories), their muscle glycogen levels recovered almost completely within the 22 h between training bouts. In addition, athletes perceive training as easier when their muscle glycogen is maintained throughout a workout.

Early studies demonstrated that when men eat a diet containing a normal amount of carbohydrate (about 55% of total calories ingested), their muscles store about 100 mmol of glycogen per kilogram of muscle. One study showed that diets containing less than 15% carbohydrate led to storage of only 53 mmol/kg, but carbohydrate-rich diets (60-70% CHO) led to storage of 205 mmol/kg. When subjects exercised to exhaustion at 75% of their maximal oxygen uptake, their exercise times were proportional to the amount of muscle glycogen stored before the test, as shown in figure 13.2.

> **fyi** Carbohydrate is the primary fuel source for most athletes and should constitute at least 50% of their total caloric intake. For endurance athletes, carbohydrate intake as a percentage of total caloric intake should be even higher: 55% to 65%.

More recent studies have shown that glycogen storage replacement is not determined

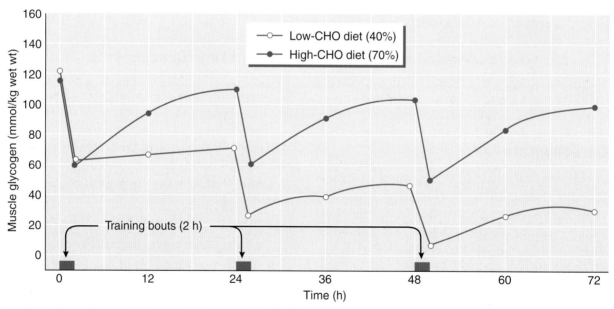

▲ **Figure 13.1** The influence of dietary carbohydrates (CHO) on muscle glycogen stores during repeated days of training. Note that when a low-CHO diet was consumed, muscle glycogen gradually declined over the 3 days of study, whereas the CHO-rich diet was able to return the glycogen to near normal each day.

Adapted, by permission, from D.L. Costill and J.M. Miller, 1980, "Nutrition for endurance sport: Carbohydrate and fluid balance," *International Journal of Sports Medicine* 1(1): 2-14.[19]

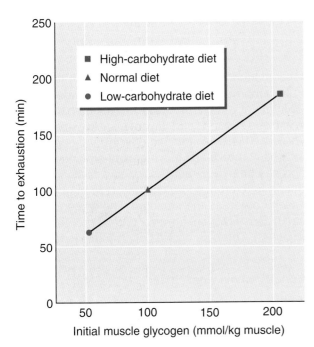

▲ **Figure 13.2** The relationship between preexercise muscle glycogen content and exercise time to exhaustion. The exercise time to exhaustion and muscle glycogen were nearly four times greater when the subjects ate a carbohydrate-rich diet than when the diet was composed mostly of fat and protein.

Adapted from Åstrand 1967.[3]

ing the amount of glucose available for resynthesizing muscle glycogen. In addition, some evidence suggests that eccentrically exercised muscle is less sensitive to insulin, which would limit muscle fiber uptake of glucose. Perhaps future studies will more fully explain why eccentric-type activities delay glycogen storage. But for now, we can only observe that glycogen recovery from various forms of exercise can differ and that this should be considered for optimal diet, training, and competition.

When athletes eat only as much food as hunger dictates, they often fail to consume enough carbohydrate to compensate for the amount used during training or competition. This imbalance between glycogen use and carbohydrate intake might explain in part why some athletes become chronically fatigued and need 48 h or more to restore normal muscle glycogen levels. Athletes who train exhaustively on successive days require a diet rich in carbohydrate to reduce the heavy, tired feeling associated with muscle glycogen depletion.

Carbohydrate Type

Simple carbohydrates (sugars) are absorbed from the digestive system quickly. Because of this, ingestion of simple carbohydrates causes hyperglycemia (elevated blood glucose level). Insulin then helps move the glucose from the blood into the cells. When carbohydrate intake is higher than can be used or stored within the cells, the excess carbohydrate is converted to fat. This in turn can elevate the blood concentration of triglycerides and cholesterol (both are fat derivatives), which are associated with a higher risk of heart disease. Complex carbohydrates, such as starch, require more time for complete breakdown, so they produce a slower and smaller increase in blood glucose. Because of this, complex carbohydrates have less impact on blood lipid levels.

These effects of carbohydrate on blood lipid levels were observed in relatively inactive subjects. But in endurance athletes, most carbohydrate consumed is used for glycogen storage.

simply by carbohydrate intake. Exercise with an eccentric (muscle-lengthening) component, such as running and weightlifting, can induce some muscle damage and impair glycogen resynthesis. In these situations, muscle glycogen levels can appear quite normal during the first 6 to 12 h after exercise, but glycogen resynthesis slows or stops completely as muscle repair begins.

The precise cause for this response is unknown, but conditions in the muscle could inhibit muscle glucose uptake and glycogen storage. For example, within 12 to 24 h after intense eccentric exercise, damaged muscle fibers are infiltrated with inflammatory cells (leukocytes, macrophages) that remove cellular debris resulting from damage to the cells' membranes. This repair process can require a significant amount of the blood glucose, reduc-

The Glycemic Index

It has long been known that the rapid increase in blood sugar levels (hyperglycemia) with the intake of carbohydrate usually is associated with simple carbohydrates, such as glucose, sucrose, fructose, and high-fructose corn syrup. However, this is not always the case. Scientists have discovered that the glycemic response (i.e., increase in blood sugar) to carbohydrate intake varies considerably for both simple and complex carbohydrates. This led to the use of what has been termed the glycemic index of foods. The ingestion of white bread leads to a high and prolonged increase in blood sugar. It is used as a standard and has been arbitrarily assigned a glycemic index of 100. The glycemic response of all other foods is referenced against the response of white bread, using 50 g of both the test food and white bread as the standard. The glycemic index (GI) is calculated as follows: GI = 100 × (blood glucose response over 2 h to 50 g of test food/blood glucose response over 2 h to 50 g white bread). Three categories of glycemic index have been established:

- High glycemic index foods (GI >85) such as soft drinks, whole wheat bread, raisins, honey/syrups, potatoes, carrots, and ice cream
- Moderate glycemic index foods (GI 60-85) such as pastry, pita bread, white rice, orange, popcorn, banana, and low fat ice cream
- Low glycemic index foods (GI <60) such as spaghetti, milk, grapefruit, beans, apples, pears, peanuts, and yogurt[40]

Before exercise, low glycemic index foods would be preferred to reduce the likelihood of hyperinsulinemia. However, high glycemic index foods should be an advantage during exercise by helping maintain blood glucose levels.[54] This should also be the case during recovery from intense and prolonged exercise, as the higher blood sugar level could increase muscle and liver glycogen storage.

Their blood lipid levels change very little with carbohydrate ingestion because training depletes their glycogen reserves, which in turn triggers increased glycogen synthesis.

With all of this information, we might expect that altering the relative amounts of simple and complex carbohydrates in an athlete's diet would affect the rate and quantity of glycogen formation. Tests of this theory, however, are inconclusive. Because of conflicting reports, any potential benefits from the preferential use of either simple or complex carbohydrates for muscle glycogen replacement are unclear.

Carbohydrate Intake and Performance

As noted earlier, muscle glycogen provides a major source of energy during exercise. Because muscle glycogen depletion has been shown to be a major cause of fatigue and ultimate exhaus-tion in events lasting more than an hour, efforts to load the muscle with extra glycogen before starting exercise have been considered ergogenic for performance. Early studies demonstrated that men who ate a carbohydrate-rich diet for 3 days stored nearly twice their normal amounts of muscle glycogen.[3] When they were asked to exercise to exhaustion at 75% of $\dot{V}O_2$max, their exercise times significantly increased (see figure 13.2). This practice, called **glycogen loading**, is widely used by distance runners, cyclists, and other athletes who must perform exercise for several hours. We discuss this practice in greater detail later in this chapter.

Blood glucose levels become low (hypogly-cemia) during exhaustive long-distance run-ning and cycling, and this might contribute to fatigue. Several studies have shown that subjects' performances improve when they are given carbohydrate feedings during exercise lasting 1 to 4 h.[23] Comparisons of subjects who received carbohydrate feedings and those who received placebos revealed no performance differences during the early phase of the exercise, but during the final stage of the experiments, performance was greatly improved with carbohydrate feedings, as shown in figure 13.3.[31]

Although we don't fully understand how carbohydrate feedings improve performance, most scientists believe that maintaining blood glucose near normal levels allows the muscles to obtain more energy from blood glucose. Carbohydrate feedings during exercise do not spare muscle glycogen use. Instead, they may preserve liver glycogen, enabling the exercising muscles to rely more on blood glucose for energy late in the exercise.[21] Endurance performance (more than 1 h) can be enhanced when carbohydrate is consumed within 5 min before the exercise begins, more than 2 h before exercise (such as during the precompetition meal), and at frequent intervals during the activity.

An athlete should not ingest carbohydrate foods during the period 15 to 45 min before

exercise because it could cause hypoglycemia shortly after the exercise begins, which could lead to early exhaustion by depriving the muscle of one of its energy sources. As shown in figure 13.4, carbohydrate ingested during that period stimulates insulin secretion, elevating insulin when the activity begins.[18] In response, glucose uptake by the muscles reaches an abnormally high rate, leading to hypoglycemia. Not everyone experiences this reaction, but sufficient evidence indicates that simple carbohydrates (those that cause a large increase in blood insulin) should be avoided in the period 15 to 45 min before exercise.

Why don't carbohydrate feedings during exercise produce the same hypoglycemic effects observed with preexercise feedings? Sugar feedings during exercise result in smaller increases in both blood glucose and insulin, lessening the threat of an overreaction that leads to a sudden decrease in blood glucose. This finer control of blood glucose during exercise

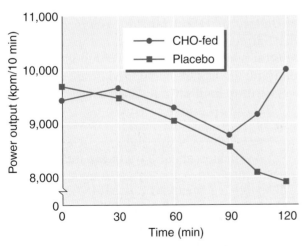

▲ **Figure 13.3** The influence of drinking carbohydrate (CHO) and placebo (flavored water) solutions on 2-h cycling performance. The solutions were taken every 15 min during the exercise. Note the increase in work output from 90 min to 120 min with carbohydrate feedings.

Adapted, by permission, from J.L. Ivy, D.L. Costill, W.J. Fink, and R.W. Lower, 1979, "Influence of caffeine and carbohydrate feedings on endurance performance," *Medicine and Science in Sports and Exercise* 11: 6-11.

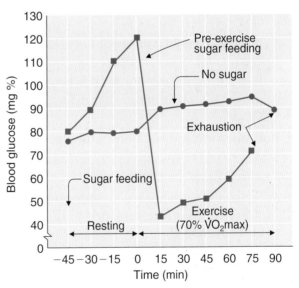

▲ **Figure 13.4** The effects of preexercise carbohydrate (sugar) feeding on blood glucose levels during exercise. Note the decrease in blood glucose to hypoglycemic levels with sugar feeding 45 min before exercise. Also, subjects during the sugar-feeding trial were unable to complete the full 90 min at 70% of $\dot{V}O_2$max, achieving only 75 min.

Adapted, by permission, from D.L. Costill, E. Coyle, G. Dalsky, W. Evans, W.W. Fink, and D. Hoopes, 1977, "Efffects of elevated plasma FFA and insulin on muscle glycogen usage during exercise," *Journal of Applied Physiology* 43: 695-699.

might be caused by increased muscle fiber permeability that decreases the need for insulin, or insulin-binding sites may be altered during muscular activity. Regardless of the cause, carbohydrate intake during exercise appears to supplement the carbohydrate supply needed for muscular activity.

Fat

Fat, also termed lipid, is a class of organic compounds with limited water solubility. It exists in the body in many forms, such as triglycerides, free fatty acids (FFA), phospholipids, and sterols. The body stores most fat as triglycerides, composed of three molecules of fatty acids and one molecule of glycerol. Triglycerides are our most concentrated source of energy.

Dietary fat, especially cholesterol and triglycerides, plays a major role in cardiovascular disease (chapter 20), and excessive fat intake also has been linked to other diseases such as cancer, diabetes, and obesity. But despite the negative publicity, fat serves many vital functions in the body:

- It is an essential component of cell membranes and nerve fibers.
- It is a primary energy source, providing up to 70% of our total energy in the resting state.
- It supports and cushions vital organs.
- All steroid hormones in the body are produced from cholesterol.
- Fat-soluble vitamins gain entry into, are stored in, and are transported through the body via fat.
- Body heat is preserved by the insulating subcutaneous fat layer.

The most basic unit of fat is the fatty acid, which is the part used for energy production. Fatty acids occur in two forms: saturated and unsaturated. Unsaturated fats contain one (monounsaturated) or more (polyunsaturated) double bonds between carbon atoms, and each double bond takes the place of two hydrogen atoms. A saturated fatty acid possesses no double bonds, so it has the maximum amount of hydrogen bound to the carbons. Excessive

saturated fat consumption is a risk factor for numerous diseases.

Fats derived from animal sources generally contain more saturated fatty acids than fats derived from plants. Fats that are more highly saturated tend to be solids at room temperature, whereas less saturated fats tend to be liquid. The tropical oils are notable exceptions: Palm, palm kernel, and coconut oil are plant-derived fats that are liquid at room temperature but are very high in saturated fat. And although many vegetable oils are low in saturated fats, they are often used in foods as hydrogenated shortening. The process of hydrogenation adds hydrogen to the fat, increasing its saturation. Saturated fat contents of some common fats are shown in figure 13.5.

Fat Consumption

Fat can enhance food's palatability by absorbing and retaining flavors and by affecting the food's texture. For this reason, it is quite common in our diets. The dietary fat intake for both men and women was as high as 45% of total calories consumed in 1965 but decreased to 34% in 1995. Most likely this decrease is attributable to the recent media attention on the health risks of dietary fat. Most nutritionists recommend that fat consumption should not exceed 30% of total calories consumed. Guidelines accompanying the USDA's new food guide pyramid advise us to limit saturated fats to less than 10% of total caloric intake.

> Fat constitutes about 34% of the total calories consumed in the typical American diet. Although this figure has decreased from 45% of total calories over the past 30 years, it is still higher than the recommendation of 30% or less, which many believe is essential for good health, disease prevention, and optimal athletic performance.

Fat Intake and Performance

For the athlete, fat is especially important as an energy source. Muscle and liver glycogen stores in the body are limited, so the use of fat

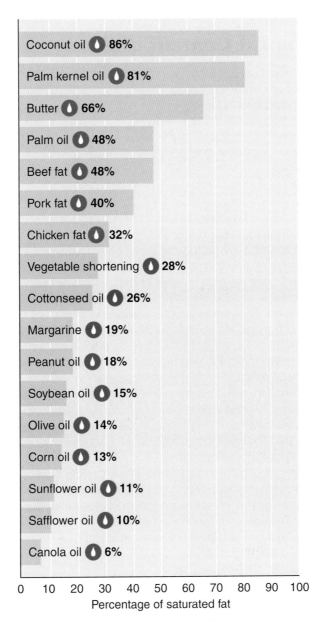

▲ **Figure 13.5** Percentages of saturated fat in some common fats and oils.

Adapted from E. Tribole, 1992, *Eating on the run*, 2nd ed. (Champaign, IL: Leisure Press) 202, 203. Permission by author.

(or FFA) for energy production can delay exhaustion. Clearly, any change that allows the body to use more fat would be an advantage, particularly for endurance performance. In fact, one adaptation that occurs in response to endurance training is an increased ability to use fat as an energy source. Unfortunately, merely eating fat does not stimulate the muscles to burn fat. Instead, eating fatty foods tends only to elevate plasma triglycerides, which then must be broken down before the FFA can be used for energy production. To increase the use of fat, the FFA levels in the blood, not the triglyceride levels, must be increased. Highly trained athletes can adapt to a high-fat diet. In one study, endurance athletes were fed a high-fat diet (70% of total calories) for 5 consecutive days. Fat oxidation was increased and carbohydrate oxidation decreased during exercise at 70% of $\dot{V}O_2$max, but performance during a time trial was not improved.[11]

Calculating the Fat Content of Foods

We're told that reading labels on our foods is important so that we can make wise nutritional choices, but these labels are often confusing. This is especially true with fat content. The fat content of a food can be calculated in terms of

- the weight of the fat (grams of fat),
- its percentage of the food's total weight,
- the kilocalories it provides, or
- its percentage of the total calories.

We are advised to restrict our fat intake to less than 30% of our total calories, but food labels don't always give this information. With some simple math, you can do your own calculations. The table that follows compares 8-oz (236.6-ml) servings of four types of milk. The first column shows the weight of the serving and the second shows what percentage of the milk's weight is water. The third column indicates the total number of kilocalories per serving (in human nutrition, the capitalized Calorie you see on food labels is actually a kilocalorie).

The last four columns are measurements of fat content. First we see the actual weight of the fat; for example, 8 oz (236.6 ml) of whole milk contains 8.15 g of fat. The next column gives the percentage of the total weight that is fat. This shows us that whole milk is only 3.3% fat. That sounds great, because we are trying to keep fat consumption down to less than 30%, but 3.3% is the percentage of the total weight, not of the total calories.

Recall from chapter 4 that 1 g of fat contains 9 kcal of energy. Multiply the 8.15 g by 9 kcal/g and you find that fat accounts for 73.4 kcal in 8 oz of whole milk. Now comes the final calculation: What percentage of the total calories is this? Simply dividing the kilocalories from fat by the total kilocalories in the milk (73.4/150) reveals that 48.9% of the total kilocalories in whole milk is from fat. This far exceeds our target of less than 30%!

Perhaps math is not your strong point, especially when walking the aisles at the grocery store and reading tiny print on confusing food labels. An easier method can be used to calculate the total percentage of calories from fat. Most foods list both the total kilocalories per serving and the grams of fat per serving. To keep your fat intake below 30% of your total calories, select foods that have no more than 3 g of fat per 100 kilocalories, because

3 g of fat per 100 kcal $\times$ 9 kcal/g of fat

$= 27$ kcal of fat per 100 kcal

$= 27\%$ total kilocalories from fat

Milk type (8 oz)	Weight (g)	H_2O (% weight)	Energy (kcal)	Fat (g)	Fat (% weight)	Fat (kcal)	Fat (% kcal)
Whole	244	88	150	8.15	3.3	73.4	48.9
2% low-fat	244	89	121	4.78	2.0	43.0	35.6
1% low-fat	244	90	102	2.54	1.0	22.9	22.4
Skim	245	91	86	0.44	0.2	4.0	4.6

Protein

Protein is a class of nitrogen-containing compounds formed by amino acids. Protein serves numerous functions in our bodies:

- It is the major structural component of the cell.
- It is used for growth, repair, and maintenance of body tissues.
- Hemoglobin, enzymes, and many hormones are produced from protein.
- It is one of the three primary buffers in the control of acid–base balance.
- Normal blood osmotic pressure is maintained by proteins in the plasma.
- Antibodies for disease protection are formed from protein.
- Energy can be produced from protein.

Twenty amino acids have been identified as necessary for human growth and metabolism (table 13.1). Of these, 11 (for children) or 12 (for adults) are termed nonessential amino acids, meaning that our bodies synthesize them, so we don't rely on dietary intake for their supply. The remaining eight or nine are termed essential amino acids because we cannot synthesize them; thus, they are an essential part of our daily diets. Absence of one of these essential amino acids from the diet precludes formation of any proteins that contain that amino acid, and thus any tissue requiring those proteins cannot be maintained.

A dietary protein source that contains all the essential amino acids is called a complete protein. Meat, fish, poultry, eggs, and milk are examples. The proteins in vegetables and grains are called incomplete proteins because they do not supply all the essential amino acids. This concept is important for people on vegetarian diets (discussed later in this chapter).

Protein Consumption

Protein accounts for approximately 5% to 15% of the total calories consumed per day in the United States. Many experts believe this is two to three times the actual amount needed. The RDAs for protein are shown in table 13.2.

Table 13.1

Essential and Nonessential Amino Acids

Essential	Nonessential
Isoleucine	Alanine
Leucine	Arginine
Lysine	Asparagine
Methionine	Aspartic acid
Phenylalanine	Cysteine
Threonine	Glutamic acid
Tryptophan	Glutamine
Valine	Glycine
Histidine (children)[a]	Proline
	Serine
	Tyrosine
	Histidine (adult)*

[a]Histidine is not synthesized in infants and young children, so it is an essential amino acid for children but not for adults.

Table 13.2

Protein Requirements (RDA) for Male and Female Teens and Adults

	Male RDA (g)	Female RDA (g)
Teen	45	46
Adult	58-63	44-50

Note. Recommended Daily Allowance (RDA) is based on 1989 standards by the National Research Council.

The RDA depends on individual body weight and composition. Men typically require more protein than women because men generally weigh more and have greater muscle mass.

In general, though, an allowance of 0.8 g per kilogram of body weight is considered appropriate for adults.

Protein Intake and Performance

Should athletes who are training for strength and endurance increase their protein intake? Amino acids are the body's building blocks, so protein is essential for the growth and development of body tissues. For many years, protein supplementation was believed essential for athletes. In fact, muscle was once thought to consume itself as fuel for its own actions, so protein supplementation was considered necessary to prevent muscle wasting. Over the years, nutritionists and physiologists have argued against the need for supplementing proteins for optimal sport performance. It was generally believed that the RDA of 0.8 g of protein per kilogram of body weight each day would adequately meet the demands of hard training.

Recent studies using metabolic-tracer and nitrogen-balance technologies have shown that the overall protein and specific amino acid requirements are higher for individuals in training than for normally active people.[39, 52] The role of protein differs for endurance and strength training athletes. It appears that strength training individuals need up to 2.1 times the RDA, or about 1.7 g of protein per kilogram of body weight per day, whereas athletes engaging in endurance training need 1.2 to 1.6 g of protein per kilogram of body weight per day.[52] Whereas endurance exercise places greater demand on protein as an auxiliary fuel, strength training requires additional amino acids as the building blocks for muscle development.

Is it necessary to supplement athletes' diets to optimize their intake of protein? Because most athletes consume a large number of calories each day, it is possible to obtain the additional protein by consuming as little as 10% of total calories as protein. Despite the belief that if a little extra protein is good, then diets extremely high in protein or specific amino acids must be better, there is no scientific evidence that protein diets exceeding 1.7

▶ Carbohydrates are sugars and starches. They exist in the body as monosaccharides, disaccharides, oligosaccharides, and polysaccharides. All carbohydrates must be broken down into monosaccharides before the body can use them as a fuel.

▶ Muscle glycogen loading by eating a diet rich in carbohydrate offers major benefits to performance.

▶ Fats, or lipids, exist in the body as triglycerides, free fatty acids, phospholipids, and sterols. They are stored primarily as triglycerides, which are the body's most concentrated energy source. A triglyceride molecule can be broken down into one glycerol and three fatty acid molecules. Only the free fatty acids are used by the body for energy production.

▶ Although fat is a major energy source, dietary attempts to elevate free fatty acids have been only partially successful.

▶ The smallest unit of protein is an amino acid. All proteins must be broken down to amino acids before the body can use them. Only the nonessential amino acids can be synthesized in our bodies. The essential amino acids must be attained through our diets. Protein is not a primary energy source in our bodies, but it can be used for energy production during endurance exercise.

▶ The current RDA for protein (0.8 g/kg per day) may be too low for athletes involved in intense resistance training (1.7 g/kg per day) or for endurance athletes (1.2-1.6 g/kg per day) during early or heavy periods of training. However, extremely high protein diets offer no additional benefits and could offer a health risk to normal kidney function.

g/kg per day provide an additional advantage. In fact, some health risks might be associated with excessive protein intake because it places greater demands on the kidneys to excrete the unused amino acids. A diet containing 10% or at most 15% of calories from protein should be adequate for most athletes, unless their total energy intake is deficient.[30] For example, a 100-kg (220-lb) bodybuilder with a 4,500 kcal/day intake containing 15% protein would consume 675 kcal of protein, or about 165 g per day. Thus, the bodybuilder's total protein intake would be 1.65 g/kg per day, over twice the RDA.

Vitamins

Vitamins are a group of unrelated organic compounds that perform specific functions to promote growth and maintain health. We need them in relatively small quantities, but without them we could not use the other nutrients we ingest. Vitamins act primarily as catalysts in chemical reactions. They are essential for energy release, tissue building, and metabolic regulation. Vitamins can be classified into one of two major categories: fat soluble or water soluble. The fat-soluble vitamins, A, D, E, and K, are absorbed from the digestive tract bound to lipids (fats). These vitamins are stored in the body, so excessive intake can cause toxic accumulations. The B-complex vitamins and vitamin C are water soluble. They are absorbed from the digestive tract along with water. Any excess of these vitamins is excreted, mostly in the urine, but vitamin toxicity has been reported with some of these. Table 13.3 on the following two pages lists the various vitamins and the RDA, good dietary sources, major functions, and symptoms of deficiencies and toxicities for each.

Most vitamins have some function important to the athlete:

- Vitamin A is crucial for normal growth and development because it plays an integral role in bone development.
- Vitamin D is essential for intestinal absorption of calcium and phosphorus and thus for bone development and strength.

By regulating calcium absorption, this vitamin also has a key role in neuromuscular function.

- Vitamin K is an intermediate in the electron transport chain, making it important for oxidative phosphorylation.

Of all the vitamins, though, only the B-complex vitamins and vitamins C and E have been extensively investigated for their potential to facilitate athletic performance. In the following sections, we briefly consider these vitamins.

B-Complex Vitamins

The B-complex vitamins were once thought to be a single vitamin. Now more than a dozen B-complex vitamins have been identified. These vitamins' essential roles in cellular metabolism cannot be overemphasized. Among their diverse functions, they serve as cofactors in various enzyme systems involved in the oxidation of food and the production of energy. Consider just a few examples. Vitamin B_1 (thiamin) is needed for the conversion of pyruvic acid to acetyl coenzyme A. Vitamin B_2 (riboflavin) becomes flavin adenine dinucleotide (FAD), which acts as a hydrogen acceptor during oxidation. Vitamin B_3 (niacin) is a component of nicotinamide adenine dinucleotide phosphate (NADP), a coenzyme in glycolysis. Vitamin B_{12} has a role in amino acid metabolism and is also needed for the production of red blood cells, which transport oxygen for oxidation. The B-complex vitamins have such a close interrelationship that a deficiency in one can impair utilization of the others. Symptoms of deficiencies vary with the vitamins involved.

Several studies have shown that supplementation of one or more of the B-complex vitamins facilitates performance. However, most researchers agree that this is only true if the individual being studied suffers a preexisting B-complex deficiency.[6, 9] Creating a deficiency in one or more of the B-complex vitamins usually impairs performance, but this is reversed when the deficiency is corrected with supplementation. No compelling evidence supports supplementation when there is no deficiency.

Table 13.3

Vitamin Requirements for Adult Men and Women

Vitamin	Fat- or water-soluble	Source	Function	Symptoms of deficiency	RDA Men	RDA Women
A (retinol)	F	Provitamin carotene in yellow and green vegetables; preformed in liver, egg yolk, butter, and milk	Necessary for rhodopsin synthesis, normal health of epithelial cells, bone and tooth growth	Rhodopsin deficiency, night blindness, retarded growth, skin disorders, and increased infection risk	1,000 μg	800 μg
B$_1$ (thiamine)	W	Yeast, grains, and milk	Involved in carbohydrate and amino acid metabolism; necessary for growth	Beriberi—muscle weakness (including cardiac muscle), neuritis, and paralysis	1.5 mg	1.1 mg
B$_2$ (riboflavin)	W	Green vegetables, liver, wheat germ, milk, and eggs	Component of FAD involved in citric acid cycle	Eye disorders and skin cracking, especially at corners of the mouth	1.7 mg	1.3 mg
Pantothenic acid (part of B$_2$ complex)	W	Liver, yeast, green vegetables, grains, and intestinal bacteria	Constituent of coenzyme A, glucose production from lipids and amino acids, and steroid hormone synthesis	Neuromuscular dysfunction and fatigue	4-7 mg	4-7 mg
B$_3$ (niacin)	W	Fish, liver, red meat, yeast, grains, peas, beans, and nuts	Component of NAD; involved in glycolysis and citric acid cycle	Pellagra—diarrhea, dermatitis, and mental disturbance	19 mg	15 mg
B$_6$ (pyridoxine)	W	Fish, liver, yeast, tomatoes, and intestinal bacteria	Involved in amino acid metabolism	Dermatitis, retarded growth, and nausea	2.0 mg	1.6 mg
Folic acid	W	Liver, green leafy vegetables, and intestinal bacteria	Nucleic acid synthesis, hematopoiesis	Macrocytic anemia (enlarged red blood cells)	200 μg	180 μg

Vitamin C

Vitamin C (ascorbic acid) is common in our foods, but deficiencies can occur in people who smoke, use oral contraceptives, have surgery, or run a fever. This vitamin is important for the formation and maintenance of collagen, a

Vitamin	Fat- or water-soluble	Source	Function	Symptoms of deficiency	RDA Men	RDA Women
B₁₂ (cyano-cobalamin)	W	Liver, red meat, milk, and eggs	Necessary for erythrocyte production, some nucleic acid and amino acid metabolism	Pernicious anemia and nervous system disorders	2.0 μg	2.0 μg
C (ascorbic acid)	W	Citrus fruits, tomatoes, and green vegetables	Collagen synthesis; general protein metabolism	Scurvy—defective bone formation and poor bone healing	60 mg	60 mg
D (chole-calciferol, ergosterol)	F	Fish liver oil, enriched milk and eggs; provitamin D converted by sunlight to cholecalciferol in the skin	Promotes calcium and phosphorus use, normal growth, and bone and teeth formation	Rickets—poorly developed, weak bones, osteomalacia; bone reabsorption	10 μg	10 μg
E (α-tocopherol)	F	Wheat germ; cottonseed, palm, and rice oils; grain; liver; and lettuce	Prevents catabolism of certain fatty acids; may prevent miscarriage	Muscular dystrophy and sterility	10 mg	8 mg
H (biotin) often considered part of the B-vitamin group	W	Liver, yeast, eggs, and intestinal bacteria	Fatty acid and purine synthesis; movement of pyruvic acid into citric acid cycle	Mental and muscle dysfunction, fatigue, and nausea	Not known; 0.3-1.0 mg	Not known; 0.3-1.0 mg
K (phylloquinone)	F	Alfalfa, liver, spinach, vegetable oils, cabbage, and intestinal bacteria	Required for synthesis of a number of clotting factors	Excessive bleeding due to retarded blood clotting	65-80 μg	65-80 μg

Note. Recommended Daily Allowance (RDA) is based on 1989 standards set by the National Research Council. FAD = flavin adenine dinucleotide; NAD = nicotinamide adenine dinucleotide.

crucial protein found in connective tissue, so it is essential for healthy bones, ligaments, and blood vessels. Vitamin C also functions in

- the metabolism of amino acids;
- synthesis of some hormones, such as the catecholamines (epinephrine and norepinephrine) and the anti-inflammatory corticoids; and
- promoting iron absorption from the intestines.

Many people also believe that vitamin C assists healing, combats fever and infection, and prevents or cures the common cold. Although evidence to date is inconclusive, the role of vitamin C in the fight against disease is an area of major interest.

Vitamin C supplementation has produced equivocal findings in the research conducted to date. However, those who have reviewed this area generally agree that, even with the increased requirements of training, vitamin C supplementation does not improve performance when no deficiency exists. As noted in the sidebar below, it has been suggested that vitamins, including vitamin C, also may function as antioxidants to combat the cellular damage created by the metabolically generated free radicals.

Vitamin E

Vitamin E is stored in muscle and fat. This vitamin's functions are not well established, although it is known to enhance the activity of vitamins A and C by preventing their oxidation. Indeed, the most important role of vitamin E is its action as an antioxidant. It disarms free radicals (highly reactive molecules) that could otherwise severely damage cells and disrupt metabolic processes. Exercise has been shown to cause DNA damage to the cell, whereas supplementing the intake of vitamin E (3×800 mg) reduced DNA damage. In addition, with the supplementation of vitamin E for 14 days before an exhaustive run, exercise-induced damage was reduced.[28] However, investigators found no benefit of 30 days of vitamin E supplementation on the muscle damage that resulted from 240 maximal isokinetic eccentric knee flexion/extension actions (24 sets of 10 repetitions each) when compared with a placebo control condition.[5]

Vitamin E has received considerable media attention over the years as a potential miracle vitamin that might prevent or alleviate a number of medical conditions, such as rheumatic fever, muscular dystrophy, coronary artery disease, sterility, menstrual disorders, and spontaneous abortions. It also has been suggested that vitamin E supplements may prevent lung damage from many of the pollutants that we inhale. Such claims generally lack supporting scientific evidence; however, recent reports suggest that people who take more than 400 mg of vitamin E per day are at less risk of coronary artery disease.

Many athletes consume supplementary doses of vitamin E. It is postulated to benefit performance through its relationship with oxygen use and energy supply. However, reviews generally have concluded that vitamin E supplementation does not improve athletic performance.

Minerals

A number of inorganic substances known as minerals are essential for normal cellular functions. Minerals account for approximately 4% of your body weight. Some are present in high concentrations in your skeleton and teeth, but

Free Radicals and Antioxidants

Most of the oxygen consumed during aerobic exercise is used in the mitochondria for oxidative phosphorylation and is reduced to water. However, a small number of univalently produced oxygen intermediates, termed **free radicals,** may leak out of the electron transport chain.[12] Laboratory studies have shown that free radical generation increases after acute exercise, which has been theorized to coincide with oxidative tissue damage.[35] Because these free radicals are highly reactive, they are theorized to modulate muscle function and accelerate the fatigue process. Fortunately, under normal conditions, muscle fibers are equipped with antioxidant enzymes that serve as an efficient defense system to prevent the damaging accumulation of free radicals. In addition, the dietary intake of antioxidants, such as vitamin E and β-carotene, also directly traps free radicals, preventing them from interfering with cellular function. Some researchers suggest that these dietary supplements may help block the negative effects of exercise-induced free radical release.[37] Consequently, the importance of antioxidant vitamins has become a popular topic of discussion and study in the fields of nutrition and cellular biology.

minerals are also found throughout your body, in and around every cell, dissolved in your body's fluids. They can be present either as ions or combined with various organic compounds. Mineral compounds that can dissociate into ions in the body are called **electrolytes.**

By definition, **macrominerals** are those of which your body needs more than 100 mg per day. **Microminerals,** or **trace elements,** are those needed in smaller amounts. Table 13.4 on pages 422-423 lists the 17 essential minerals, their major functions, symptoms of deficiencies and excesses, and their RDAs.

Mineral intake is less likely to be supplemented by athletes than vitamin intake. Far less concern has been shown by athletes for their mineral status, possibly because far fewer claims have been made about the performance-enhancing qualities of specific minerals. Of the minerals, calcium and iron have been most frequently investigated.

Calcium

Calcium is the most abundant mineral in your body, constituting approximately 40% of the total mineral content. Calcium is well known for its importance in building and maintaining healthy bones, and that is where most of it is stored. But it is also essential for nerve impulse transmission. Calcium plays major roles in enzyme activation and regulation of cell membrane permeability, both important for metabolism. And this mineral is also essential for normal muscle function: Recall from chapter 1 that calcium is stored in the sarcoplasmic reticulum of muscles and released when the muscle fibers are stimulated. It is required for formation of the actin–myosin cross-bridges that cause the fibers to contract.

Sufficient calcium intake is critical to our health. If we do not consume enough calcium, it will be removed from its storage sites in the body, especially the bones. This condition is called osteopenia. It weakens the bones and can lead to osteoporosis, a common problem in postmenopausal women. Unfortunately, few studies have been conducted on calcium supplementation; these suggest that supplementation is of no value in the presence of an adequate (RDA) dietary intake of calcium.

Phosphorus

Phosphorus is closely linked to calcium. It constitutes approximately 22% of your body's total mineral content. About 80% of this phosphorus is combined with calcium (calcium phosphate), providing strength and rigidity to the bones. Phosphorus is an essential part of metabolism, cell membrane structure, and the buffering system (to maintain constant blood pH). As we saw in chapter 4, phosphorus plays a major role in bioenergetics: It is an essential component of adenosine triphosphate.

Iron

Iron—a micromineral—is present in the body in relatively small amounts (35-50 mg per kilogram of body weight). It plays a crucial role in oxygen transportation: Iron is required for the formation of both hemoglobin and myoglobin. Hemoglobin, located in the red blood cells, binds with oxygen in the lungs and then transports it to the body tissues via the blood. Myoglobin, found in muscle, combines with oxygen and stores it until needed.

Iron deficiency is prevalent throughout the world. By some estimates, as much as 25% of the world's population is iron deficient. The major problem associated with this condition is iron-deficiency anemia, in which hemoglobin levels are reduced, decreasing the blood's oxygen-carrying capacity. This causes fatigue, headaches, and other symptoms. Iron deficiency is a more common problem in women than in men because both menstruation and pregnancy cause iron losses that must be replenished. This problem is compounded by the fact that women generally consume less food, and thus less iron, than men.

Iron has received much attention in the research literature. Women are considered anemic only when their hemoglobin concentration is below 12 g per 100 ml of blood. But in the United States, 22% of all women between the ages of 17 and 44 are thought to be iron deficient despite having normal hemoglobin values. Studies generally suggest that 22% to 25% of female athletes and 10% of male athletes are iron deficient. But these numbers may be conservative: Risser and coworkers[45] found that 31% of a group of female varsity athletes at two major U.S. universities

Table 13.4

Mineral Requirements for Adult Men and Women

Mineral	Function	Symptoms of deficiency	RDA Men	RDA Women
Calcium	Bone and teeth formation, blood clotting, muscle activity, and nerve function	Spontaneous nerve discharge and tetany	1,000-1,200 mg	1,000-1,200 mg
Chlorine	Blood acid–base balance; hydrochloric acid production in stomach	Acid–base imbalance	Not established	Not established
Chromium	Associated with enzymes in glucose metabolism	Unknown	50-200 μg	50-200 μg
Cobalt	Component of vitamin B_{12}; erythrocyte production	Anemia	Not established	Not established
Copper	Hemoglobin and melanin production; electron transport system	Anemia and loss of energy	1.5-3.0 mg	1.5-3.0 mg
Fluoride	Provides extra strength in teeth, prevents dental caries	No real pathology	1.5-4.0 mg	1.5-4.0 mg
Iodine	Thyroid hormone production; maintenance of normal metabolic rate	Decrease of normal metabolism	150 μg	150 μg
Iron	Component of hemoglobin; ATP production in electron transport system	Anemia, decreased oxygen transport, and energy loss	10 mg	15 mg
Magnesium	Coenzyme constituent; bone formation; muscle and nerve function	Increased nervous system irritability, vasodilation, and arrhythmias	350 mg	280 mg
Manganese	Hemoglobin synthesis; growth; activation of several enzymes	Tremors and convulsions	2.5-5.0 mg	2.5-5.0 mg
Molybdenum	Enzyme component	Unknown	75-250 μg	75-250 μg
Phosphorus	Bone and teeth formation; important in energy transfer (ATP); component of nucleic acids	Loss of energy and cellular function	1,200 mg	1,200 mg
Potassium	Muscle and nerve function	Muscle weakness, abnormal electrocardiogram, and alkaline urine	Not established	Not established
Selenium	Component of many enzymes	Unknown	70 μg	55 μg

Mineral	Function	Symptoms of deficiency	RDA Men	RDA Women
Sodium	Osmotic pressure regulation; nerve and muscle function	Nausea, vomiting, exhaustion, and dizziness	Not established, probably about 2,500 mg	Not established, probably about 2,500 mg
Sulfur	Component of hormones, several vitamins, and proteins	Unknown	Not established	Not established
Zinc	Component of several enzymes; carbon dioxide transport; necessary for protein metabolism	Deficient carbon dioxide transport and deficient protein metabolism	15 mg	12 mg

Note. Recommended Daily Allowance (RDA) is based on 1989 standards set by the National Research Council.

were iron deficient. These studies indicate that hemoglobin is not the only marker of anemia or necessarily the best. Plasma ferritin levels provide a good marker of the body's iron stores. When values are below 20 to 30 µg/L, this indicates low body iron stores.

When iron supplements are given to those who are iron deficient (i.e., with low plasma ferritin levels), performance measures, particularly aerobic capacity, typically improve.[26] However, supplementation of iron in those who are not deficient appears to have no benefit. In fact, iron supplements can be a health risk, because excess iron is toxic for the liver, and ferritin levels higher than 200 µg/L are associated with an increased risk for coronary artery disease.[13]

Sodium, Potassium, and Chloride

Sodium, potassium, and chloride are distributed throughout all body fluids and tissues. Sodium and chloride are found primarily in the fluid outside of your cells and in your blood plasma, but potassium is located mainly in your cells. This selective distribution of these three minerals establishes the separation of electrical charge across neuron and muscle cell membranes. Thus, these minerals enable neural impulses to control muscle activity (see chapter 2). In addition, they are responsible for maintaining the body's water balance and distribution, normal osmotic equilibrium, acid-base balance (pH), and normal cardiac rhythm.

▶ Vitamins perform numerous functions in our bodies and are essential for normal growth and development. Many are involved in metabolic processes, such as those leading to energy production.

▶ Vitamins A, D, E, and K are fat soluble. These can accumulate to toxic levels in the body. Vitamins C and B-complex are water soluble. Excesses of these are excreted, so toxicity is rarely a problem. Several of the B-complex vitamins are involved in the processes of energy production.

▶ Macrominerals are minerals of which we require more than 100 mg per day. Microminerals (trace elements) are those of which we require smaller amounts.

▶ Minerals are required for numerous physiological processes, such as muscle contraction, oxygen transport, fluid balance, and bioenergetics. Minerals can dissociate into ions, which can participate in numerous chemical reactions. Minerals that can dissociate into ions are called electrolytes.

▶ Vitamins and minerals do not appear to have any performance-enhancing value. Taking them in amounts greater than the RDA will not improve performance and could be counterproductive.

Western diets are replete with sodium, so dietary deficiency is unlikely. However, minerals are lost with sweating, so any condition that causes excessive sweating, such as extreme exertion or exercise in a hot environment, can deplete these minerals. When discussing mineral imbalances, we often focus on deficiencies. However, many of these minerals also have negative effects when taken in excess. In fact, excess potassium can cause heart failure! Individual needs vary, but megadoses are never advisable.

Water

Seldom is water thought of as a nutrient because it has no caloric value. Yet its importance in maintaining life is second only to oxygen's.

Water constitutes about 60% of a typical young man's and 50% of a typical young woman's total body weight. It has been estimated that we can survive losses of up to 40% of our body weight in fat, carbohydrate, and protein. But a water loss of only 9% to 12% of body weight can be fatal.

> A water loss of 9% to 12% of a person's total body weight can lead to death.

The body's fluid compartments are illustrated in figure 13.6. Approximately two thirds of the water in our bodies is contained in our cells and is referred to as **intracellular fluid**. The remainder is outside the cells, referred to as the

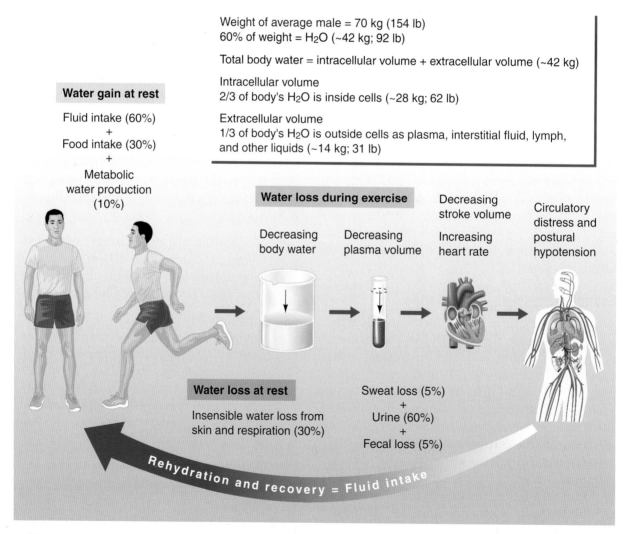

Weight of average male = 70 kg (154 lb)
60% of weight = H_2O (~42 kg; 92 lb)

Total body water = intracellular volume + extracellular volume (~42 kg)

Intracellular volume
2/3 of body's H_2O is inside cells (~28 kg; 62 lb)

Extracellular volume
1/3 of body's H_2O is outside cells as plasma, interstitial fluid, lymph, and other liquids (~14 kg; 31 lb)

Water gain at rest

Fluid intake (60%)
+
Food intake (30%)
+
Metabolic water production (10%)

Water loss during exercise

Decreasing body water → Decreasing plasma volume → Decreasing stroke volume / Increasing heart rate → Circulatory distress and postural hypotension

Water loss at rest

Insensible water loss from skin and respiration (30%)

Sweat loss (5%)
+
Urine (60%)
+
Fecal loss (5%)

Rehydration and recovery = Fluid intake

▲ **Figure 13.6**   The body's fluid compartments and sources of body water gains and losses at rest and during exercise.

extracellular fluid. Extracellular fluid includes the interstitial fluid surrounding the cells, the blood plasma, lymph, and some other fluids.

Water plays several critical roles in exercise:

- Red blood cells carry oxygen to your active muscles via the blood plasma, which is primarily water.
- Nutrients such as glucose, fatty acids, and amino acids are transported to your muscles by blood plasma.
- Carbon dioxide and other metabolic wastes leave the cells and then enter the blood plasma to be cleared from your body.
- Hormones that regulate metabolism and muscular activity during exercise are transported by the blood plasma to their targets.
- Body fluids contain buffering agents to maintain proper pH when lactate is being formed.
- Water facilitates the dissipation of body heat that is generated during exercise.
- Blood plasma volume is a major determinant of blood pressure and thus cardiovascular function.

In the next sections, we more closely examine the role of water in exercise and performance.

▶ Water is our most important nutrient. We would die much more quickly if deprived of water than we would if deprived of any other nutrient.

▶ Water is found in the intracellular compartment (inside the cells) and the extracellular compartment (outside the cells). Extracellular fluids include blood plasma, lymph, interstitial fluid, and other body fluids.

▶ Among its most important functions, water provides transportation between and delivery to the body's different tissues, regulates body temperature, and maintains blood pressure for proper cardiovascular function.

Water and Electrolyte Balance

For optimal performance, the body's water and electrolyte contents should remain relatively constant. Unfortunately, this doesn't always happen during exercise. In the next sections, we examine water content and electrolyte balance, how exercise affects them, and the impact on performance when water or electrolyte balance is disturbed.

Water Balance at Rest

Under normal resting conditions, our body water content is relatively constant: Our water intake equals our water output. About 60% of our daily water intake is obtained from the fluids we drink and about 30% is from the foods we consume. The remaining 10% is produced in our cells during metabolism (recall from chapter 4 that water is a by-product of oxidative phosphorylation). Metabolic water production varies from 150 to 250 ml per day, depending on the rate of energy expenditure: Higher metabolic rates produce more water. The total daily water intake from all sources averages about 33 ml per kilogram of body weight per day. For a 70-kg (154-lb) person, average intake is 2.31 L per day.

Water output, or water loss, occurs from four sources:

- Evaporation from the skin
- Evaporation from the respiratory tract
- Excretion from the kidneys
- Excretion from the large intestine

Human skin is permeable to water. Water diffuses to the skin's surface, where it evaporates into the environment. In addition, the gases we breathe are constantly being humidified by water as they pass through our respiratory tracts. These two types of water loss (from the skin and respiration) occur without our sensing them. Thus, they are termed insensible water losses. Under cool, resting conditions, these losses account for about 30% of daily water loss.

The majority of water loss—60% when at rest—occurs from our kidneys, which excrete water and waste products as urine. Under resting conditions, the kidneys excrete about 50 to 60 ml of water per hour. Another 5% of the water is lost by sweating (although this is often considered along with insensible water loss), and the remaining 5% is excreted from the large intestine in the feces. The sources of water intake and water output at rest are depicted in figure 13.6 on page 424.

Water Balance During Exercise

Water loss accelerates during exercise, as seen in table 13.5. Your body's ability to lose the heat generated during exercise depends primarily on the formation and evaporation of sweat. As your body's temperature increases, sweating increases in an effort to prevent overheating (see chapter 10). But at the same time, more water is produced during exercise because of increased oxidative metabolism. Unfortunately, the amount produced even during the most intense effort has only a small impact on the **dehydration**, or water loss, that results from heavy sweating. During an hour of intense effort, for example, a 70-kg (154-lb) person might metabolize about 245 g of carbohydrate. This would produce about 636 g of water. During that same period, however, sweat losses could exceed 1,500 ml, approximately 10 times more than generated metabolically. Nevertheless, the water produced during oxidative metabolism helps minimize, if only to a small degree, the dehydration that occurs during exercise.

> **fyi** During a marathon race, a runner's muscles can produce nearly 500 ml of water over 2 to 3 h.

In general, the amount of sweat produced during exercise is determined by

- environmental temperature, radiant heat load, humidity, and air velocity;
- body size; and
- metabolic rate.

These three factors influence the body's heat storage and temperature. Heat is transferred from warmer areas to cooler ones, so heat loss from the body is impaired by high environmental temperatures, radiation, high humidity, and still air. Body size is important because large individuals generally need more energy to do a given task, so they usually have higher metabolic rates and produce more heat. But they also have more surface area (skin), which allows more sweat formation and evaporation.

As exercise intensity increases, so does the metabolic rate. This increases body heat production, which in turn increases sweating. To

Table 13.5

Comparison of Water Loss From the Body at Rest in a Cool Environment and During Prolonged Exhaustive Exercise

Source of loss	Resting		Prolonged exercise	
	ml/h	% total	ml/h	% total
Insensible loss				
skin	14.6	15	15	1.1
respiration	14.6	15	100	7.5
Sweating	4.2	5	1,200	90.6
Urine	58.3	60	10	0.8
Feces	4.2	5	—	0.0
Total	95.9	100	1,325	100

conserve water during exercise, blood flow to the kidneys decreases in an attempt to prevent dehydration, but like the increase in metabolic water production, this too may be insufficient. During high-intensity exercise under environmental heat stress, sweating and respiratory evaporation can cause rapid losses of as much as 2 to 3 L of water per hour.[16] (Chapter 10 contains additional information about body water losses during exercise in warm environments.)

> **fyi** During an event such as the marathon, sweating and water loss from respiration may reduce body water content by 6% to 10%, despite fluid intake during the event.

Dehydration and Exercise Performance

Even minimal changes in your body's water content can impair endurance performance. Without adequate fluid replacement, a subject's exercise tolerance shows a pronounced decrease during long-term activity because of water loss through sweating. The impact of dehydration on the cardiovascular and thermoregulatory systems is quite predictable. Fluid loss decreases plasma volume. This decreases blood pressure, which in turn reduces blood flow to the muscles and skin. In an effort to overcome this, heart rate increases. Because less blood reaches the skin, heat dissipation is hindered, and the body retains more heat. Thus, when a person is dehydrated by 2% of body weight or more, both heart rate and body temperature are elevated during exercise.

As you might expect, these physiological changes will decrease exercise performance. Figure 13.7 illustrates the effects of an approximate 2% decrease in body weight attributable to dehydration from the use of a diuretic on a distance runner's performance in 1,500-m, 5,000-m, and 10,000-m time trials on an outdoor track.[2] The dehydration condition resulted in plasma volume decreases of between 10% and 12%. Although the average $\dot{V}O_2$max did not differ between the normally hydrated and dehydrated trials, mean running velocity decreased by 3% in the 1,500-m run and by more than 6%

in the 5,000- and 10,000-m runs. The greater the duration of the performance, the greater the expected decline in performance for the same degree of dehydration. These trials were conducted in relatively cool weather. The higher the temperature, humidity, and radiation, the greater the expected decrement in performance for the same degree of dehydration. The decrement in performance would be progressively greater with greater levels of dehydration.[46]

The effect of dehydration on performance in muscular strength, muscular endurance, and anaerobic types of activities is not as clear. Decrements have been seen in some studies, whereas other studies have reported no change in performance.[46] Wrestlers and other weight-category athletes commonly dehydrate to get a weight advantage during the weigh-in for a competition. Most rehydrate after the weigh-in before the competition and experience only small decrements in performance. A summary of the effects of dehydration on exercise performance is shown in table 13.6.

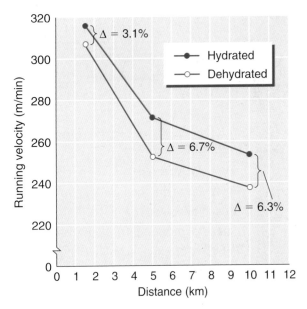

▲ **Figure 13.7** The decline in running velocity (meters per minute) with dehydration of about 2% of body weight for 1,500-m, 5,000-m, and 10,000-m time trials compared with the normally hydrated condition.

Reprinted, by permission, from L.E. Armstrong, D.L. Costill, and W.J. Fink, 1985, "Influence of diuretic-induced dehydration on competitive running performance," *Medicine and Science in Sports and Exercise* 17: 456-461.

Table 13.6

Alterations in Physiological Function and Performance From Dehydration

Variables	Dehydration	Rehydration
Physiological function		
Cardiovascular		
Blood volume/plasma volume	↓	↓[a]
Cardiac output	↓	?
Stroke volume	↓	?
Heart rate	↑	?
Metabolic		
Aerobic capacity—$\dot{V}O_2$max	↔,↓	↔[a]
Anaerobic power—Wingate test	↔,↓	↔,↓
Anaerobic capacity—Wingate test	↔,↓	↔,↓
Blood lactate, peak value	↓	↓[a]
Buffer capacity of the blood	↓	?
Lactate threshold, velocity at LT	↓	?
Muscle and liver glycogen	↓	↓
Blood glucose during exercise	possible ↓	?
Protein degradation with exercise	possible ↑	?
Thermoregulation and fluid balance		
Electrolytes, muscle and blood	↓	↔
Exercise core temperature	↑	?
Sweat rate	↓, delayed onset	?
Skin blood flow	↓	?
Performance		
Muscular strength	↔,↓	↔,↓
Muscular endurance	↔,↓	↔,↓
Muscular power	?	↓[b]
Speed of movement	?	?
Run time to exhaustion	↓	?
Total work performed	↓	↓[a]
Wrestling simulation tests[a]	↓	↔,↓[b]

Note. Data for this table were derived from the following reviews:

Fogelholm (1994)[25]; Horswill (1994)[29]; Keller, Tolly, and Freeson (1994)[38]; and Opplinger et al. (1996)[44].

↓ = decrease; ↑ = increase; ↔= no known change or return to normal values; ? = unknown; LT = lactate threshold.

[a]From Burge, Carey, and Payne (1993).[10]

[b]From Oopik et al. (1996).[43]

> ▶ Water balance depends on electrolyte balance, and vice versa.

> ▶ At rest, water intake equals water output. Water intake includes water ingested from foods and fluids and produced as a metabolic by-product. The majority of water output at rest is generated by the kidneys, but water also is lost from the skin, from the respiratory tract, and in the feces.

> ▶ During exercise, metabolic water production increases as the metabolic rate increases.

> ▶ Water loss during exercise increases because, as heat in the body increases, more water is lost in sweat. Sweat becomes the primary avenue for water loss during exercise. In fact, the kidneys decrease their excretion in an effort to prevent dehydration.

> ▶ When dehydration reaches 2% of body weight, aerobic endurance performance is notably impaired. Heart rate and body temperature increase in response to dehydration.

Thus far, we have examined only the effects of dehydration on performance. In addition to the body water lost during endurance events, many nutrients, especially minerals, escape with sweat. In the following discussion, we examine the effects of heavy sweating not only on water balance but also on the electrolyte balance of body tissues.

Electrolyte Balance During Exercise

As mentioned earlier, normal body function depends on a balance between water and electrolytes. We have discussed the effects of water loss on performance. Now we can turn our attention to the effects of the other component of this delicate balance: electrolytes. When large amounts of water are lost from the body, such as during exercise, the balance between water and electrolytes can be disrupted quickly. In the next sections, we examine the effects of exercise on electrolyte balance. Our focus is on the two major routes for electrolyte loss: sweating and urine production.

Electrolyte Loss in Sweat

Human sweat is a filtrate of blood plasma, so it contains many substances found there, including sodium (Na^+), chloride (Cl^-), potassium (K^+), magnesium (Mg^{2+}), and calcium (Ca^{2+}). Although sweat tastes salty, it contains far fewer minerals than the plasma and other body fluids. In fact, sweat is 99% water.

Sodium and chloride are the predominant ions in sweat and blood. As indicated in table 13.7 on page 430, the concentrations of sodium and chloride in sweat are roughly one third those found in plasma and five times those found in muscle. Each of these three fluids' **osmolarity**, which is the ratio of solutes (such as electrolytes) to fluid, is also shown. Sweat's electrolyte concentration can vary considerably between individuals. It is strongly influenced by the rate of sweating, the state of training, and the state of heat acclimatization.

At the elevated rates of sweating reported during endurance events, sweat contains large amounts of sodium and chloride but little potassium, calcium, and magnesium.

Based on estimates of the athlete's total body electrolyte content, such losses would lower the body's sodium and chloride content by only about 5% to 7%. Total body levels of potassium and magnesium, two ions principally confined to the insides of cells, would decrease by about 1%. These losses probably have no measurable effect on an athlete's performance.

As electrolytes are lost in sweat, the remaining ions are redistributed among the body tissues. Consider potassium. It diffuses from active muscle fibers as they contract, entering the extracellular fluid. The increase this causes in extracellular potassium levels does not equal the amount of potassium that is released from active muscles, because potassium is taken up by inactive muscles and other tissues while the active muscles are losing it. During recovery, intracellular potassium levels normalize quickly.

Table 13.7

Electrolyte Concentrations and Osmolarity in Sweat, Plasma, and Muscle of Men Following 2 h of Exercise in the Heat

	Electrolytes (mEq/L)				Osmolarity (mOsm/L)
Site	Na⁺	Cl⁻	K⁺	Mg²⁺	
Sweat	40-60	30-50	4-6	1.5-5	80-185
Plasma	140	101	4	1.5	295
Muscle	9	6	162	31	295

Note. mEq/L = milliequivalents per liter (thousandths of 1 g of solute per 1 L of solvent).

Some researchers suggest that these muscle potassium disturbances during exercise might contribute to fatigue by altering the membrane potentials of neurons and muscle fibers, making it more difficult to transmit impulses.

Electrolyte Loss in Urine

In addition to clearing wastes from the blood and regulating water levels, the kidneys also regulate the body's electrolyte content. Urine production is the other major source of electrolyte loss. At rest, electrolytes are excreted in the urine as necessary to maintain homeostatic levels, and this is the primary route for electrolyte loss. But as your body's water loss increases during exercise, your urine production rate decreases considerably in an effort to conserve water (the mechanisms involved were discussed in chapter 5). Consequently, with very little urine being produced, electrolyte loss by this avenue is minimized.

The kidneys play another role in electrolyte management. If, for example, a person eats 250 mEq of salt (NaCl), the kidneys will normally excrete 250 mEq of these electrolytes to keep the body NaCl content constant. Heavy sweating and dehydration, however, trigger the release of the hormone aldosterone from the adrenal gland. This hormone stimulates renal reabsorption of sodium. Consequently, the body retains more sodium than usual during the hours and days after a prolonged exercise

▶ The loss of large amounts of water can disrupt electrolyte balance, although electrolytes are rather dilute in sweat, which is 99% water.

▶ Electrolyte loss during exercise occurs primarily with water loss from sweating. Sodium and chloride are the most abundant electrolytes in sweat.

▶ At rest, excessive electrolytes are excreted in the urine by the kidneys. But urine production declines tremendously during exercise, so little electrolyte loss occurs by this route.

▶ Dehydration causes the hormone aldosterone to promote renal retention of Na⁺ and Cl⁻, increasing their concentrations in the blood. This triggers thirst in an effort to make us consume more fluid to replace what has been lost.

bout. This elevates the body's sodium content and increases the osmolarity of the extracellular fluids.

This increased sodium content triggers thirst, compelling the person to consume more water, which is then retained in the extracellular compartment. The increased water consumption reestablishes normal osmolarity in the extracellular fluids but leaves these fluids expanded, which dilutes the other substances present there. This expansion of the extracellular fluids has no negative effects and is temporary. In fact, this is one of the major mechanisms for the increase in plasma volume that occurs with training and with acclimatization to exercise in the heat. Fluid levels return to normal within 48 to 72 h after exercise, providing there are no subsequent exercise bouts.

Replacement of Body Fluid Losses

Your body loses more water than electrolytes when you are sweating heavily. This raises the osmotic pressure in your body fluids because your electrolytes become more concentrated. Your need to replace body water is greater than your need for electrolytes because only by replenishing your water content can the electrolytes return to normal concentrations. But how does your body know when this is necessary?

Thirst

When you feel thirsty, you drink. The thirst sensation is regulated by your hypothalamus. It triggers thirst when the plasma's osmotic pressure is increased. Unfortunately, your body's thirst mechanism doesn't precisely gauge your state of dehydration. You don't sense thirst until well after dehydration begins. Even when you are dehydrated, you might desire fluids only at intermittent intervals.

The control of thirst is not fully understood. When permitted to drink water as their thirst dictates, people can require 24 to 48 h to completely replace water lost through heavy sweating. In contrast, dogs and burros can drink up to 10% of their total body weight within the first few minutes after exercise or heat exposure, replacing all lost water. Because of our sluggish drive to replace body water and to prevent chronic dehydration, we are advised to drink more fluid than our thirst dictates. Because of the increased water loss during exercise, it is imperative that athletes' water intake be sufficient to meet their bodies' needs, and it is

essential that they rehydrate during and after an exercise bout.

Benefits of Fluids During Exercise

Drinking fluids during prolonged exercise, especially during hot weather, has obvious benefits. Water intake will minimize dehydration, increases in body temperature, cardiovascular stress, and declines in performance.

As seen in figure 13.8, when subjects became dehydrated during several hours of treadmill running in the heat (40° C, or 104° F) without fluid replacement, their heart rates increased steadily throughout the exercise.[4] When they were deprived of fluids, the subjects became exhausted and couldn't complete the 6-h exercise. Ingesting either water or a saline solution in amounts equal to weight loss prevented dehydration and kept subjects' heart rates lower. Even warm fluids (near body temperature) provide some protection against overheating, but cold fluids enhance body cooling because

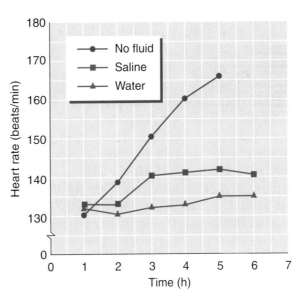

▲ **Figure 13.8**   Effects of 6 h of treadmill running in the heat on heart rate. Subjects received no fluid, a saline solution, or water. Subjects deprived of fluids became exhausted and couldn't complete the 6-h exercise.

Data from S.I.Barr et al., 1991, "Fluid replacement during prolonged exercise: Effects of water, saline, or no fluid," *Medicine and Science in Sports and Exercise* 23: 811-817.

some of the deep body heat is used to warm cold drinks to body temperature.

Hyponatremia

Fluid replacement is beneficial, but can too much of a good thing be bad? In 1981, the first two cases of hyponatremia were reported in endurance athletes.[42] Hyponatremia is clinically defined as a blood sodium concentration below the normal range of 136 to 143 mmol/L. Symptoms of hyponatremia appear in stages: weakness, disorientation, seizures, and coma if the condition is not reversed. How likely is hyponatremia to occur?

The processes that regulate fluid volumes and electrolyte concentrations are highly effective, so consuming enough water to dilute plasma electrolytes is difficult under normal circumstances. Marathoners who lose 3 to 5 L of sweat and drink 2 to 3 L of water maintain normal plasma concentrations of sodium, chloride, and potassium. Distance runners who run 25 to 40 km (15.5-24.9 mi) per day in warm weather and do not salt their food don't develop electrolyte deficiencies. And subjects maintained normal electrolyte levels even when they consumed only 30% as much potassium as they normally would while losing 3 to 4 L of sweat daily for 8 consecutive days.[16]

Some research has suggested that during ultramarathon running (more than 42 km, or 26.1 mi), athletes can experience hyponatremia. A case study of two runners who collapsed after an ultramarathon race (160 km, or 100 mi) in 1983 revealed that their blood sodium concentrations had decreased from a normal value of 140 mEq/L to values of 123 and 118 mEq/L.[27] One of the runners experienced a grand mal seizure; the other became disoriented and confused. Examining the runners' fluid intakes and estimating their sodium intakes during the run suggested that they had diluted their sodium contents by consuming fluids that contained too little sodium. A study by Barr and colleagues,[4] however, showed that when subjects consumed more than 7 L of plain water during 6 h of exercise in the heat, their plasma sodium concentration decreased only negligibly, by about 3.9 mmol/L.

The ideal solution to prevent hyponatremia would be to replace water at the exact rate that it is being lost or to add sodium to the ingested fluid. The problem with the latter approach is that sport drinks that contain no more than 25 mmol/L of sodium are apparently too weak to prevent sodium dilution, but stronger concentrations cannot be tolerated. Exercise hyponatremia appears to be the result of a fluid overload, underreplacement of sodium losses, or both.[41] Only a small number of cases have been reported. Thus, it is probably inappropriate to form conclusions from this information to design a fluid replacement regimen for people who must exercise for long periods in the heat.

▶ Our need to replace lost body fluid is greater than our need to replace lost electrolytes.

▶ Our thirst mechanism does not exactly match our hydration state, so we should consume more fluid than we are aware that we need.

▶ Water intake during prolonged exercise reduces the risk of dehydration and optimizes our bodies' cardiovascular and thermoregulatory functions.

▶ In some cases, drinking too much fluid with too little sodium has led to hyponatremia (low plasma levels of sodium), which can cause confusion, disorientation, and even seizures.

The Athlete's Diet

Athletes place considerable demands on their bodies every day they train and compete. Their bodies must be as finely tuned as possible. This, by necessity, must include optimal nutrition. Too often, athletes spend considerable time and effort perfecting skills and attaining top physical condition while ignoring proper nutrition and sleep. Performance deterioration often can be traced to poor nutrition.

Unfortunately, we know very little about the eating habits of athletes. To gain insight into their practices, the diets of a group of highly trained distance runners were recorded during a period of training and during the 3 days before a marathon. The results are shown in table 13.8 on page 434. The 22 runners (11 men, 11 women) had running experience ranging from recreational to international competition. The findings revealed little dietary difference between elite and average runners. Diet did not appear to determine success or failure among these performers. Although specific diet items varied, when food was analyzed for percentage of fat, protein, and carbohydrate or for vitamin and mineral content, the overall differences were small.

It is interesting that the runners came very close to meeting the RDAs. These runners ate diets containing 50% carbohydrate, 36% fat, and 14% protein. We might at first consider the carbohydrate intake of these runners to be low, knowing the need for a high-carbohydrate diet when training for distance running. But these runners actually ate more than enough carbohydrate to meet the energy needs of training. Their total calorie intake was nearly 50% higher than expected for nontraining people of similar size (about 65.8 kg, or 145 lb), so their total carbohydrate intake was well above average.

Most of the runners studied did not use vitamin supplements, contrary to some recent surveys about the habits of runners. Still, these runners consumed adequate amounts of most vitamins and minerals, at least equal to the RDA. Unless a runner's vitamin intake falls well below the RDA for an extended period of time, no effects on performance are expected. Although diets rich in simple carbohydrates tend to be deficient in some of the B-complex vitamins, only two runners consumed too little vitamin B_{12}. Participants also obtained ample dietary fiber, another item associated with good health.

In the last 3 days before a marathon, the subjects changed both their training and their eating habits. They reduced their daily distance from an average of 13.7 km (8.5 mi) to 3.7 km (2.3 mi). In an attempt to load their muscles with glycogen, the runners increased their daily caloric intake from 3,012 kcal during training to a premarathon average of 3,730 kcal.

During the 3-day prerace period, the marathoners had reduced their running distance, so on average they were burning only about 2,526 kcal per day while eating 3,730 kcal per day. The daily surplus of 1,204 kcal for 3 days could result in storage of an extra 0.45 kg (1 lb) of unnecessary and unproductive fat (0.45 kg of fat contains 3,500 kcal). Several athletes ate more than 5,000 kcal a day, nearly twice their rate of caloric expenditure during that period; such overeating might theoretically hurt their performances. For performance, however, eating a bit too much food, principally carbohydrates, is probably better than risking not being fully loaded with muscle and liver glycogen at the time of competition.

These results are from only one study of highly trained distance runners. They are not meant to reflect the typical diet of all athletes. Athletes in different sports might eat differently.

Vegetarian Diet

In an effort to eat a healthy diet and increase their carbohydrate intake, many athletes have adopted vegetarianism. Vegans are strict vegetarians who eat only food from plant sources. Lactovegetarians also consume dairy products. Ovovegetarians add eggs to their vegetable diets, and lacto-ovovegetarians eat plant foods, dairy products, and eggs.

Can athletes survive on a vegetarian diet? The answer is a qualified yes. Athletes who are strict vegans must be very careful in selecting the plant foods they eat to provide a good balance of the essential amino acids, sufficient calories, and adequate sources of vitamin A, riboflavin, vitamin B_{12}, vitamin D, calcium, zinc, and iron. Adequate iron intake is of particular concern in women vegetarian athletes because of the lower bioavailability of iron in plant-based diets and because of women's greater risk for anemia and low iron stores. Some professional athletes have noted significant deterioration in athletic performance

Table 13.8

A Comparison of 22 Runner's Diets With the Recommended Daily Allowance (RDA)

Diet composition	Runners' average	RDA
Calories (kcal/day)	3,012	(2,000)
Carbohydrate (g)	375	(250)
Protein (g)	112	(70)
Saturated fats (g)	42	(26)
Unsaturated fats (g)	64	(54)
Total fat (g)	122	(66-100)
Cholesterol (mg)	377	(300)
Fiber (g)	7	(3-6)
Vitamin A (IU)	10,814	5,000
Vitamin B_1 (mg)	1.9	1.5
Vitamin B_2 (mg)	2.5	1.7
Vitamin B_6 (mg)	2.2	2.0
Vitamin B_{12} (μg)	3.8	2.0
Folic acid (μg)	230	200
Niacin (mg)	27.3	19.0
Pantothenic acid (mg)	5.3	4-7
Vitamin C (mg)	205	60
Vitamin E (mg)	5.2	10
Iron (mg)	25	15
Potassium (g)	4.3	—[a]
Calcium (mg)	1,300	1,200
Magnesium (mg)	400	350
Phosphorus (mg)	200	800-1,200
Sodium (mg)	2,600	(2,500)[b]

Note. Figures shown in parentheses represent estimates of the average values in the American Diet, which may or may not be healthy.

[a]RDA is not established.

[b]RDA is not established; this is an estimate.

after switching to strict vegetarian diets. The problem usually is traced to unwise selection of foods. Including milk and eggs in the diet decreases the risk of nutritional deficiencies. Anyone contemplating switching to a vegetarian diet should either read authoritative material on the subject written by qualified nutritionists or consult a registered dietitian.

Precompetition Meal

For years, athletes have eaten the traditional steak dinner several hours before competition. This practice might have originated from the early belief that muscle consumes itself to fuel its own activity and that steak would provide the necessary protein to counteract this loss. But we now know that steak is probably the worst food an athlete could eat before competing. Steak contains a high percentage of fat, which requires several hours for full digestion; during competition, the digestive system would compete with the muscles for the available blood supply. Also, nervous tension is typically high before a big competition, so even the choicest steak cannot truly be enjoyed at this time. The steak would be more satisfying and less likely to disturb performance if the athlete were to eat it either the night before or after the competition. But if steak is out, what should the athlete eat before competing?

Although the meal ingested a few hours before competition might contribute little to muscle glycogen stores, it can ensure a normal blood glucose level and prevent hunger. This meal should contain only about 200 to 500 kcal and consist mostly of carbohydrate foods that are easily digested. Foods such as cereal, juice, and toast are digested rather quickly and won't leave the athlete feeling full during competition. In general, this meal should be consumed at least 2 h before competition. The rates at which food is digested and nutrients are absorbed into the body are quite individual, so timing the precompetition meal might depend on prior experience. In one study of endurance cyclists, a prolonged cycling exercise trial to

exhaustion at 70% of the subject's $\dot{V}O_2$max was performed under two different conditions, with 14 days between trials: 100 g of carbohydrate breakfast fed 3 h before exercise (Fed) and no feeding before exercise (Fasted). Subjects tested under the Fed condition exercised 136 min before reaching exhaustion compared with 109 min in the fasted trial, indicating the importance of the precompetition meal.[47]

A liquid precompetition meal might be less likely to result in nervous indigestion, nausea, vomiting, and abdominal cramps. Such feedings are commercially available and generally have been found useful both before and between events. As with any precompetition feeding, however, a liquid meal should be avoided in the final hour before competition. Finding time for athletes to eat is often difficult when they must perform in multiple preliminary and final events. Under these circumstances, a liquid feeding that is low in fat and high in carbohydrate might be the only solution.

Muscle Glycogen Replacement and Loading

Earlier in this chapter, we established that different diets can markedly influence muscle glycogen stores and that endurance performance depends largely on these stores. The theory is that the greater the amount of glycogen stored, the better the potential endurance performance because fatigue will be delayed. Thus, an athlete's goal is to begin an exercise bout or competition with as much stored glycogen as possible.

Based on muscle biopsy studies conducted in the mid-1960s, Åstrand[3] proposed a plan to help runners store the maximum amount of glycogen. This process is known as glycogen loading. According to Åstrand's regimen, athletes should prepare for competition by completing an exhaustive training bout 7 days before the event. For the next 3 days, they should eat fat and protein almost exclusively to deprive the muscles of carbohydrate, which increases the activity of glycogen synthase, an enzyme responsible for glycogen synthesis. Athletes should then eat a carbohydrate-rich diet for the remaining 3 days before the event. Because glycogen synthase activity is increased, increased carbohydrate intake results in greater muscle glycogen storage. Training intensity and volume during this 6-day period should be markedly reduced to prevent additional muscle glycogen depletion, thus maximizing liver and muscle glycogen reserves.

This regimen has been shown to elevate muscle glycogen stores to twice the normal level, but it is somewhat impractical for most highly trained competitors. During the 3 days of low carbohydrate intake, athletes generally find training difficult. They are also often irritable and unable to perform mental tasks, and they typically show signs of low blood sugar, such as muscle weakness and disorientation. In addition, the exhaustive depletion bouts of exercise performed 7 days before the competition have little training value and can impair glycogen storage rather than enhance it. This depletion exercise also exposes athletes to possible injury or overtraining.

Considering these limitations, many propose that the depletion exercise and the low carbohydrate aspects of Åstrand's regimen should be eliminated. Instead, according to Sherman and colleagues[51], the athlete should simply reduce training intensity a week before competition and eat a normal, mixed diet containing 55% of the calories from carbohydrate until 3 days before the competition. For these days, training should be reduced to a daily warm-up of 10 to 15 min of activity, accompanied by a carbohydrate-rich diet. Following this plan, as seen in figure 13.9, glycogen will be elevated to nearly 200 mmol/kg of muscle, the same level attained with Åstrand's regimen, and the athlete will be better rested for competition.

It is possible to increase carbohydrate stores rapidly after even a very short near–maximal intensity bout of exercise. In a study of seven endurance athletes, scientists found that cycling for 150 s at 130% of $\dot{V}O_2$peak followed by 30 s of all-out cycling and 24 h of high-carbohydrate intake was sufficient to nearly double muscle glycogen stores in just one day.[24]

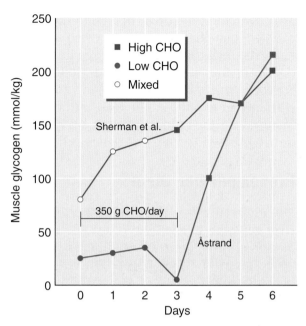

▲ **Figure 13.9**   Two regimens for muscle glycogen loading. In one regimen, the subjects were depleted of muscle glycogen (day 0) then ate a low-carbohydrate (CHO) diet for 3 days. At that time they switched to a CHO-rich diet, which caused muscle glycogen to increase to about 185 mmol/kg. In the other dietary regimen, the subjects ate a normal, mixed diet and reduced their training volume for the first 3 days and then changed to a high-CHO diet and further reduction in training volume for 3 days, which also resulted in muscle glycogen of 185 mmol/kg.

Data from P.-O. Åstrand, 1979, Nutrition and physical performance. In *Nutrition and the world food problem*, edited by M. Rechcigl (Basel, Switzerland: S. Karger); and W.M. Sherman, et al., 1981, "Effects of exercise-diet manipulation on muscle glycogen and its subsequent utilization during performance, *International Journal of Sport Medicine* 2: 1-15.

A diet rich in carbohydrates is critical to the success of endurance athletes. Furthermore, carbohydrate loading is a very effective technique for increasing both muscle and liver glycogen stores. The additional stored carbohydrate provides the critical energy source for improved endurance performance.

Diet is also important in preparing the liver for the demands of endurance exercise. Liver glycogen stores decrease rapidly when a person is deprived of carbohydrates for only 24 h, even when at rest. With only 1 h of strenuous exercise, liver glycogen decreases by 55%. Thus, hard training combined with a low-carbohydrate diet can empty the liver glycogen stores. A single carbohydrate meal, however, quickly restores liver glycogen to normal. Clearly, a carbohydrate-rich diet in the days preceding competition will maximize the liver glycogen reserve and minimize the risk of hypoglycemia during the event.

Water is stored in the body at a rate of about 2.6 g of water for each gram of glycogen. Consequently, an increase or decrease in muscle and liver glycogen generally produces a change in body weight of from 0.45 to 1.36 kg (1-3 lb). Some scientists have proposed that muscle

> ▶ From one study on men and women runners, most ate a nutritionally sound diet, meeting the RDA for most nutrients, but the composition of the diet was different from that expected, with 50% of the total calories coming from carbohydrate, 36% from fat, and 14% from protein.
>
> ▶ Many athletes have adopted vegetarian diets and appear to perform well. However, careful consideration must be given to protein sources and to consuming adequate levels of iron, zinc, calcium, and several vitamins.
>
> ▶ The precompetition meal should be taken no sooner than 2 h before competition, and it should be low in fat and easily digestible. A liquid precompetition meal low in fat and high in carbohydrate has advantages.
>
> ▶ Carbohydrate loading greatly increases muscle glycogen content, which, in turn, increases endurance performance.
>
> ▶ After endurance competition or training, it is important to consume substantial carbohydrate to replace the glycogen used during activity. The body seems primed to replace glycogen during the first few hours after training or competing because glycogen synthase levels are at their peak.

and liver glycogen stores can be monitored by recording the athlete's early morning weight immediately after rising: after emptying the bladder but before eating breakfast. A sudden decrease in weight might reflect a failure to replace glycogen, a deficit in body water, or both.

Athletes who must train or compete in exhaustive events on successive days should replace muscle and liver glycogen stores as rapidly as possible. Although liver glycogen can be depleted totally after 2 h of exercise at 70% $\dot{V}O_2$max, it is replenished within a few hours when a carbohydrate-rich meal is consumed. Muscle glycogen resynthesis, on the other hand, is a slower process, taking several days to return to normal after an exhaustive exercise bout such as the marathon (see figure 13.10).[7, 50] Studies in the late 1980s revealed that muscle glycogen resynthesis was most rapid when individuals were fed at least 50 g (about 0.7 g/kg body weight) of glucose every 2 h after the exercise.[8, 34] Feeding subjects more than

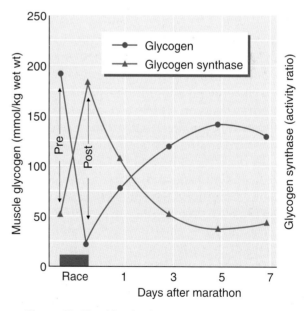

▲ **Figure 13.10** Muscle glycogen resynthesis is a slow process, requiring several days to restore normal muscle glycogen storage following exhaustive exercise. Note that when muscle glycogen decreases with hard exercise (race), muscle glycogen synthase is markedly elevated. This triggers the muscle to store glycogen when carbohydrates are eaten, returning glycogen synthase to the baseline level.

The Zone Diet

In the mid-1990s, many athletes were attracted to a new diet proposed to enhance athletic performance, touted in a popular book written by Dr. Barry Sears, *The Zone*.[48] *The Zone* diet argues against the high-carbohydrate diet typically advocated for the athlete and the general population. The Zone diet is centered on the premise that you should take in 1.8 to 2.2 g of protein per kilogram of fat-free mass (i.e., the weight or mass of all nonfat tissue in the body; see chapter 14). The diet approximates a 40% carbohydrate, 30% fat, and 30% protein proportion of total calories consumed. However, for athletes, a much higher percentage of calories from fat is recommended.[49] Supposedly, this low-carbohydrate diet promotes a more favorable insulin to glucagon ratio, ultimately improving oxygen delivery to exercising muscle.[14]

Although many anecdotal stories supported the performance-enhancing qualities of the Zone diet, its efficacy has yet to be clearly established by well-designed research studies. In fact, a wealth of data in the sport nutrition literature strongly argue against such a diet. The diet promotes an unnecessarily high intake of protein and an extremely low intake of carbohydrate. Furthermore, if the diet is pushed to the extreme, the percentage of total calories from fat increases. So, until controlled research studies support the claims made for this diet, the athlete should follow the dietary recommendations that have been proposed in this chapter, recommendations that have the support of many studies over a number of years.[1]

this amount did not appear to accelerate the replacement of muscle glycogen. During the first 2 h after exercise, the rate of muscle glycogen resynthesis is 7% to 8% per hour (7-8 mmol · kg of muscle^{-1} · h^{-1}), which is somewhat faster than the normal rate of 5% to 6% per hour.[33] Thus, an athlete recovering from an exhaustive endurance event should ingest sufficient carbohydrate as soon after exercise as is practical. It also was postulated that adding protein and amino acids to the carbohydrate ingested during the recovery period would enhance muscle glycogen synthesis above that achieved with carbohydrate alone. Several subsequent studies were unable to substantiate this,[36, 53] but a more recent study has clearly demonstrated that a carbohydrate-protein supplement is more effective for replenishing muscle glycogen after exercise than a carbohydrate supplement alone.[32]

Designing Sport Drinks

We mentioned earlier that ingesting carbohydrate solutions during exercise can benefit performance by ensuring adequate fuel for energy production and adequate fluid for rehydration. Now that we have discussed the nutrient needs of athletes, we can consider what types of feedings are best during exercise. As we have seen, adequate carbohydrate intake is essential for maintaining athletes' energy levels. For this reason, the sport drink industry has focused on carbohydrate solutions. Let's examine some of the factors that must be considered when designing sport drinks to maximize performance.

Carbohydrate Type

Although the body relies on glucose for energy production, is glucose the best sugar to include in sport drinks? Other sugar molecules may empty from the stomach faster than glucose. For example, earlier studies with solutions of maltodextrin (complex chains of glucose) showed that a 5 g per 100 ml of solution emptied from the stomach faster than a glucose solution of similar concentration. However, subsequent research has not supported this finding, leading most investigators to conclude that the rates of **gastric emptying** for these two carbohydrate forms differ little, if at all. But fructose might leave the stomach faster than other carbohydrates. Fructose at concentrations below 200 mmol/L, when given alone, caused little or no slowing of gastric emptying, whereas some other forms of sugar, such as sucrose, maltose, galactose, and lactose, might inhibit gastric emptying. So, along with concentration, the type of carbohydrate in the solution is important. Most commercial sport drinks contain mixtures of glucose, sucrose, fructose, high-fructose corn syrup, and maltodextrins.

Carbohydrate Concentration

In general, carbohydrate solutions empty more slowly from the stomach than either water or a weak sodium chloride (salt) solution.[20, 22] Research suggests that a solution's caloric content, a reflection of its concentration, might be a major determinant of how quickly it empties from the stomach and is absorbed in the intestine. Rich solutions remain in the stomach longer than either water or weak solutions. As illustrated in figure 13.11, increasing the glucose concentration of a drink drastically reduces the gastric emptying rate. For example, 400 ml of a weak glucose solution (139 mmol/L) is almost completely emptied from the stom-

ach in 20 min, but emptying a similar volume of a strong glucose solution (834 mmol/L) can require nearly 2 h.

However, when even a small amount of a strong glucose drink leaves the stomach, it can contain more sugar than a larger amount of a weaker solution simply because of its higher concentration. Table 13.9 shows that despite

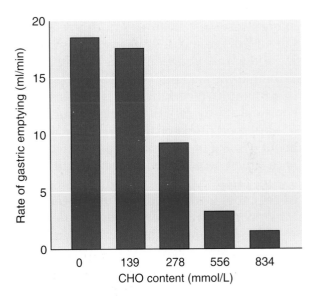

▲ **Figure 13.11** The relationship between a solution's carbohydrate (CHO) concentration and the rate of gastric emptying.

Table 13.9

Composition of Water and Glucose Solutions Before and 20 Min After Ingestion

Variables	Water		5 g of glucose per 100 ml		10 g of glucose per 100 ml	
	Before	Residue	Before	Residue	Before	Residue
Osmolarity (mOsm/L)	23	87	266	245	532	434
Sodium (mEq/L)	0.7	7.9	1.5	18.6	1.9	14.5
Potassium (mEq/L)	0.10	4.11	0.11	5.21	0.10	3.63
Glucose (g/100 ml)	0.0	0.0	5.0	3.3	10.0	6.5
pH	4.76	2.05	3.50	2.29	3.46	2.40
Gastric secretion[a] (ml)	—	32	—	52	—	65

[a]Gastric secretion denotes the volume of secretion calculated to be present in the residue.

their slower rates of gastric emptying, strong sugar solutions deliver more glucose per minute to the intestine than weak ones. If an athlete is trying to prevent dehydration, however, this would deliver less water and thus be counterproductive.

Results of early studies suggested that sport drinks should have less than 2.5 g of sugar per 100 ml of water to speed their passage through the stomach. Unfortunately, such a small amount of carbohydrate contributes little to energy reserves. Even if you drank 200 ml (about 7 oz) of such a drink every 15 min during a long run, you would take in only 20 g of carbohydrate per hour. Recent studies suggest that to improve performance, athletes should consume at least 50 g of sugar per hour.

Most sport drinks on the market contain only about 6 to 8 g of sugar per 100 ml. An endurance athlete would need to drink about 625 to 833 ml (21-28 oz) of these drinks every hour to get enough carbohydrates to be beneficial. Most people can drink only about 270 to 450 ml (9-15 oz) per hour during exercise. Thus, only drinks containing at least 11 g of carbohydrate per 100 ml should be of any performance value. The sport drinks currently on the market fall far short of this. Besides, such a rich mixture might be delayed in the stomach, draw water from the stomach's lining, and cause an uncomfortable feeling of fullness. Furthermore, many studies have documented enhanced endurance performance with solutions containing 10 g or less of carbohydrate per 100 ml of fluid, so this higher concentration is likely not necessary.[15]

Rehydration With Sport Drinks

Just adding fluid to the body during exercise lessens the risk of serious dehydration. But some research indicates that adding glucose to rehydrative beverages, aside from supplying an energy source, also might stimulate both water and sodium absorption. Recall that when sodium is retained, it causes more water to be retained. Furthermore, some exercise physiologists believe that sodium is required for glucose transport.

These beliefs are used to justify adding sodium to sport drinks, but the interactions of glucose and sodium are still unconfirmed. Studies on the topic have examined intestinal content through a tube that bypasses the stomach. Thus, these studies ignore the normal gastric contributions of sodium and other ions to the ingested solutions, so the addition of sodium to sport drinks remains controversial. Still other people have suggested that the addition of amino acids to a glucose and electrolyte solution will enhance its absorption, but again, this remains unconfirmed.

What Works Best

Athletes will not drink solutions that taste bad. Unfortunately, we all have different taste preferences. To further confound the issue, what tastes good before and after a long, hot bout of exercise will not necessarily taste good during the event. A recent study tested the taste preferences of runners and cyclists during 60 min of exercise. Most of the 50 subjects chose a drink with a light flavor and no strong aftertaste. Nearly all the commercial sport drinks failed this criterion. In another study, runners ran on a treadmill for 90 min and then recovered while seated for an additional 90 min in an environmental chamber at a temperature of 32° C (86° F), 50% humidity. Three trials were conducted, two with two different sport drinks (6% and 8% carbohydrate) and one with water. Subjects were encouraged to drink throughout each trial. The volume consumed during exercise was similar for all three drinks, but during recovery, the runners drank about 55% more of each of the two sport drinks than water.[55]

So what should the athlete drink during training and competition? Under the extreme stress of hot weather, water is the primary need. Although a good case can be made for plain water, most agree that definite nutritional benefits can be gained by adding carbohydrate to the solution, and the total volume of fluid consumed does not differ from that of water alone. The inclusion of 4 to 8 g of carbohydrate per 100 ml should not compromise the delivery of water to the body tissues. Consuming 100 to 150 ml of solution every 10 or 15 min should reduce the risk of dehydration and hyperthermia and provide a partial energy supplement.

In events lasting less than an hour, the need for fluids is very small because dehydration is not very significant in these events; the body's carbohydrate stores are sufficient to sustain the activity for this period.

> Sport drinks appear to have benefits in addition to those provided by plain water. The addition of carbohydrate to sport drinks provides an important energy source, and the formulation for optimizing taste likely will result in greater fluid consumption, thus delaying dehydration.

Clearly, how to design the best drink for rehydration is still debatable. In light of the commercial competition surrounding sport drinks, the debate over the ideal exercise drink will probably continue for some time.

> ► In designing sport drinks, there is no clear advantage of using one type of carbohydrate over another.
>
> ► The carbohydrate concentration of a sport drink generally should not exceed 8% to maximize both sugar and fluid intake.
>
> ► Various ideas have been proposed about which solution would be best absorbed from the gastrointestinal tract. Plain water is good; adding carbohydrate is probably even better. But to date no ideal solution has been identified.

In Closing . . .

In this chapter we examined the nutritional needs of the athlete, considering the importance of the six nutrient classes to exercise and athletic performance. We also discussed several ways in which athletes try to use nutritional supplementation for ergogenic purposes. We dispelled the myth of the value of the precompetition steak dinner and explored the effectiveness of commercial sport drinks. Now that we have a thorough knowledge of the importance of a balanced diet, we turn our

attention to another aspect of the athlete's diet. In the next chapter, we consider the effects of body weight on athletic performance.

▷ Key Terms

carbohydrate
dehydration
electrolyte
essential amino acid
extracellular fluid
fat
free radical
gastric emptying
glycogen loading
hyponatremia
intracellular fluid
macromineral
micromineral (trace element)
nonessential amino acid
osmolarity
protein
thirst mechanism
trace element
vitamin

▷ Study Questions

1. What are the six categories of nutrients?
2. What role does dietary fat play in endurance performance?
3. What is an appropriate protein allowance for a normally active adult man? for a woman?
4. Discuss the value of using protein supplements to enhance performance in strength and endurance events.
5. Which vitamins are most likely to be deficient in the athlete's diet?
6. How does dehydration affect exercise performance? What effect does dehydration have on exercise heart rate and body temperature?
7. How does the body regulate electrolyte balance during acute exercise and chronic exercise?

8. Describe the recommended precompetition meal.

9. Describe the methods used to maximize muscle glycogen storage (glycogen loading).

10. Describe the proper dietary regimen to glycogen-load the muscle before an exhaustive event lasting 3 to 4 h.

11. Discuss the value of consuming carbohydrate during and after endurance exercise.

12. List the factors that regulate the rate of gastric emptying. Which of these appear to have the greatest impact on gastric emptying during exercise?

13. What characteristics or components should the ideal sport drink have?

14. What foods can be considered as ergogenic aids to performance? In which events will the athlete benefit from consuming these foods?

▷ References

1. American College of Sports Medicine, American Dietetic Association and Dietitians of Canada. (2000). Nutrition and athletic performance. Joint position statement. *Medicine and Science in Sports and Exercise, 32,* 2130-2145.

2. Armstrong, L.E., Costill, D.L., & Fink, W.J. (1985). Influence of diuretic-induced dehydration on competitive running performance. *Medicine and Science in Sports and Exercise, 17,* 456-461.

3. Åstrand, P.-O. (1967). Diet and athletic performance. *Federation Proceedings, 26,* 1772-1777.

4. Barr, S.I., Costill, D.L., & Fink, W.J. (1991). Fluid replacement during prolonged exercise: Effects of water, saline or no fluid. *Medicine and Science in Sports and Exercise, 23,* 811-817.

5. Beaton, L.J., Allan, D.A., Tarnopolsky, M.A., Tiidus, P.M., & Phillips, S.M. (2002). Contraction-induced muscle damage is unaffected by vitamin E supplementation. *Medicine and Science in Sports and Exercise, 34,* 798-805.

6. Belko, A.Z. (1987). Vitamins and exercise—An update. *Medicine and Science in Sports and Exercise, 19,* S191-S196.

7. Blom, P., Costill, D.L., & Vøllestad, N.K. (1987). Exhaustive running: Inappropriate as a stimulus of muscle glycogen super-compensation. *Medicine and Science in Sports and Exercise, 19,* 398-403.

8. Blom, P., Vøllestad, N.K., & Costill, D.L. (1986). Factors affecting changes in muscle glycogen concentration during and after prolonged exercise. *Acta Physiologica Scandinavica, 128*(Suppl. 556), 67-74.

9. Bruce, R., Ekblom, B., & Nilsson, I. (1985). The effect of vitamin and mineral supplements and health foods on physical endurance and performance. *Proceedings of the Nutrition Society, 44,* 283-295.

10. Burge, C.M., Carey, M.F., & Payne, W.R. (1993). Rowing performance, fluid balance, and metabolic function following dehydration and rehydration. *Medicine and Science in Sports and Exercise, 25,* 1358-1364.

11. Burke, L.M., Hawley, J.A., Angus, D.J., Cox, G.R., Clark, S.A., Cummings, N.K., Desbrow, B., & Hargreaves, M. (2002). Adaptations to short-term high-fat diet persist during exercise despite high carbohydrate availability. *Medicine and Science in Sports and Exercise, 34,* 83-91.

12. Chance, B., Sies, H., & Boveris, A. (1979). Hydroperoxide metabolism in mammalian organs. *Physiological Reviews, 59,* 527-605.

13. Chatard, J.-C., Mujika, I., Guy, C., & Lacour, J.-R. (1999). Anaemia and iron deficiency in athletes. *Sports Medicine, 27,* 229-240.

14. Cheuvront, S.N. (1999). The Zone diet and athletic performance. *Sports Medicine, 27,* 213-228.

15. Coombes, J.S., & Hamilton, K.L. (2000). The effectiveness of commercially available sports drinks. *Sports Medicine, 29,* 181-209.

16. Costill, D.L. (1977). Sweating: Its composition and effect of body fluids. *Annals of the New York Academy of Science, 301,* 160-174.

17. Costill, D.L., Bowers, R., Branam, G., & Sparks. K. (1971). Muscle glycogen utilization during prolonged exercise on successive days. *Journal of Applied Physiology, 31,* 834-838.

18. Costill, D.L., Coyle, E., Dalsky, G., Evans, W., Fink W., & Hoopes D. (1977). Effects of elevated plasma FFA and insulin on muscle glycogen usage during exercise. *Journal of Applied Physiology, 43*(4), 695-699.

19. Costill, D.L., & Miller, J.M. (1980). Nutrition for endurance sport: Carbohydrate and fluid balance. *International Journal of Sports Medicine, 1*(1), 2-14.

20. Costill, D.L., & Saltin, B. (1974). Factors limiting gastric emptying during rest and exercise. *Journal of Applied Physiology, 37,* 679-683.

21. Coyle, E.F. (1995). Substrate utilization during exercise in active people. *American Journal of Clinical Nutrition, 61*(Suppl.), 968S-979S.

22.   Coyle, E.F., Costill, D.L., Fink, W.J., & Hoopes, D.G. (1978). Gastric emptying rates for selected athletic drinks. *Research Quarterly,* **49,** 119-124.

23.   Coyle, E.F., Hagberg, J.M., Hurley, B.F., Martin, W.H., Ehsani, A.A., & Holloszy, J.O. (1983). Carbohydrate feeding during prolonged strenuous exercise can delay fatigue. *Journal of Applied Physiology,* **55,** 230-235.

24.   Fairchild, T.J., Fletcher, S., Steele, P., Goodman, C., Dawson, B., & Fournier, P.A. (2002). Rapid carbohydrate loading after a short bout of near maximal-intensity exercise. *Medicine and Science in Sports and Exercise,* **34,** 980-986.

25.   Fogelholm, M. (1994). Effects of body weight reduction on sports performance. *Sports Medicine,* **18,** 249-267.

26.   Friedmann, B., Weller, E., Mairbäurl, H., & Bärtsch, P. (2001). Effects of iron repletion on blood volume and performance capacity in young athletes. *Medicine and Science in Sports and Exercise,* **33,** 741-746.

27.   Frizzell, R.T., Lang, G.H., Lowance, D.C., & Lathan, S.R. (1986). Hyponatremia and ultramarathon running. *Journal of the American Medical Association,* **255,** 772-774.

28.   Hartmann, A., Nieb, A.M., Grünert-Fuchs, M., Poch, B., & Speit, G. (1995). Vitamin E prevents exercise-induced DNA damage. *Mutation Research,* **346,** 195-202.

29.   Horswill, C.A. (1994). Physiology and nutrition for wrestling. In D.R. Lamb, H.G. Knutten, & R. Murray (Eds.), *Physiology and nutrition for competitive sport* (Vol. 7, pp. 131-174). Carmel, IN: Cooper.

30.   Houston, M.E. (1992). Protein and amino acid needs of athletes. *Nutrition Today* (Sept.-Oct.), 36-38.

31.   Ivy, J.L., Costill, D.L., Fink, W.J., & Lower, R.W. (1979). Influence of caffeine and carbohydrate feeding on endurance performance. *Medicine and Science in Sports and Exercise,* **11,** 6-11.

32.   Ivy, J.L., Goforth, H.W., Damon, B.M., McCauley, T.R., Parsons, E.C., & Price, T.B. (2002). Early postexercise muscle glycogen recovery is enhanced with a carbohydrate-protein supplement. *Journal of Applied Physiology,* **93,** 1337-1344.

33.   Ivy, J.L., Katz, A.L., Cutler, C.L., Sherman, W.M., & Coyle, E.F. (1988). Muscle glycogen synthesis after exercise: Effect of time of carbohydrate ingestion. *Journal of Applied Physiology,* **64,** 1480-1485.

34.   Ivy, J.L., Lee, M.C., Brozinick, J.T., Jr., & Reed, M.J. (1988). Muscle glycogen storage after different amounts of carbohydrate ingestion. *Journal of Applied Physiology,* **65,** 2018-2023.

35.   Jackson, M.L., Edwards, R.H.T., & Symons, M.C.R. (1985). Electron spin resonance studies of intact mammalian skeletal muscle. *Biochimica et Biophysica Acta,* **847,** 185-190.

36.   Jentjens, R.L.P.G., Van Loon, L.J.C., Mann, C.H., Wagenmakers, A.J.M., & Jeukendrup, A.E. (2001). Addition of protein and amino acids to carbohydrates does not enhance postexercise muscle glycogen synthesis. *Journal of Applied Physiology,* **91,** 839-846.

37.   Ji, L.L. (1995). Exercise and oxidative stress: Role of the cellular antioxidant system. *Exercise and Sport Sciences Reviews,* **23,** 135-166.

38.   Keller, H.L., Tolly, S.E., & Freedson, P.S. (1994). Weight loss in adolescent wrestlers. *Pediatric Exercise Science,* **6,** 211-224.

39.   Lemon, P.W.R. (1995). Do athletes need more dietary protein and amino acids? *International Journal of Sport Nutrition,* **5,** S39-S61.

40.   Manore, M., & Thompson, J. (2000). *Sport nutrition for health and performance.* Champaign, IL: Human Kinetics.

41.   Montain, S.J., Sawka, M.N., & Wenger, C.B. (2001). Hyponatremia associated with exercise: Risk factors and pathogenesis. *Exercise and Sport Sciences Reviews,* **29,** 113-117.

42.   Noakes, T.D., Norman, R.J., Buck, R.H., Godlonton, J., Stevenson, K., & Pittaway, D. (1990). The incidence of hyponatremia during prolonged ultraendurance exercise. *Medicine and Science in Sports and Exercise,* **22,** 165-170.

43.   Oopik, V., Paasuke, M., Sikku, T., Timpmann, S., Medijainen, L., Ereline, J., Smirnova, T., Gapejeva, E. (1996). Effect of rapid weight loss on metabolism and isokinetic performance capacity: A case study of two well trained wrestlers. *Journal of Sports Medicine and Physical Fitness,* **36,** 127-131.

44.   Oppliger, R., Case, H., Horswill, C., Landry, G., & Shelter, A. (1996). Weight loss in wrestlers: An American College of Sports Medicine position stand. *Medicine and Science in Sports and Exercise,* **28,** ix-xii.

45.   Risser, W.L., Lee, E.J., Poindexter, H.B.W., West, M.S., Pivarnik, J.M., Risser, J.M.H., & Hickson, J.F. (1988). Iron deficiency in female athletes: Its prevalence and impact on performance. *Medicine and Science in Sports and Exercise,* **20,** 116-121.

46.   Sawka, M.N., & Pandolf, K.B. (1990). Effects of body water loss on physiological function and exercise performance. In C.V. Gisolfi & D.R. Lamb (Eds.), *Fluid homeostasis during exercise, Vol. 3, Perspectives in exercise science and sports medicine* (pp. 1-30). Carmel, IN: Benchmark Press.

47.   Schabort, E.J., Bosch, A.N., Weltan, S.M., & Noakes, T.D. (1999). The effect of a preexercise meal

on time to fatigue during prolonged cycling exercise. *Medicine and Science in Sports and Exercise, 31,* 464-471.

48. Sears, B. (1995). *The Zone.* New York: Harper Collins.

49. Sears, B. (2000) The Zone diet and athletic performance (letter). *Sports Medicine, 29,* 289-291.

50. Sherman, W.M., Costill, D.L., Fink, W.J., Hagerman, F.C., Armstrong, L.E., & Murray, T.F. (1983). Effect of a 42.2-km footrace and subsequent rest or exercise on muscle glycogen and enzymes. *Journal of Applied Physiology, 55,* 1219-1224.

51. Sherman, W.M., Costill, D.L., Fink, W.J., & Miller, J.M. (1981). Effect of exercise diet manipulation on muscle glycogen and its subsequent utilization during performance. *International Journal of Sports Medicine, 2,* 114-118.

52. Tarnopolsky, M. (1999). Protein metabolism in strength and endurance activities. In D.R. Lamb & R. Murray (Eds.), *The metabolic basis of performance in exercise and sport: Vol. 12, Perspectives in exercise science and sports medicine* (pp. 125-157). Carmel, IN: Cooper.

53. Van Hall, G., Shirreffs, S.M., & Calbet, J.A.L. (2000). Muscle glycogen resynthesis during recovery from cycle exercise: No effect of additional protein ingestion. *Journal of Applied Physiology, 88,* 1631-1636.

54. Walton, P., & Rhodes, E.C. (1997). Glycaemic index and optimal performance. *Sports Medicine, 23,* 164-172.

55. Wilmore, J.H., Morton, A.R., Gilbey, H.J., & Wood, R.J. (1998). Role of taste preference on fluid intake during and after 90 min of running at 60% of V̇O₂max in the heat. *Medicine and Science in Sports and Exercise, 30,* 587-595.

▷ Selected Readings

American College of Sports Medicine. (1996). Exercise and fluid replacement: Position stand. *Medicine and Science in Sports and Exercise, 28*(1), i-vii.

Bergstrom, J., Hermansen, L., Hultman, E., & Saltin, B. (1967). Diet, muscle glycogen and physical performance. *Acta Physiologica Scandinavica, 71,* 140-150.

Bergstrom, J., & Hultman, E. (1967). A study of the glycogen metabolism during exercise in man. *Scandinavian Journal of Clinical Laboratory Investigation, 19,* 218-228.

Bishop, N.C., Blannin, A.K., Walsh, N.P., Robson, P.J., & Gleeson, M. (1999). Nutritional aspects of immunosuppression in athletes. *Sports Medicine, 28,* 151-176.

Burke, E.R., & Berning, J.R. (1996). *Training nutrition: The diet and nutrition guide for peak performance.* Carmel, IN: Cooper.

Burke, L.M., & Read, R.S.D. (1993). Dietary supplements in sport. *Sports Medicine, 15*(1), 43-65.

Coggan, A.R., & Coyle, E.F. (1987). Reversal of fatigue during prolonged exercise by carbohydrate infusion or ingestion. *Journal of Applied Physiology, 63,* 2388-2395.

Coggan, A.R., & Swanson, S.C. (1992). Nutritional manipulations before and during endurance exercise: Effects on performance. *Medicine and Science in Sports and Exercise, 24,* S331-S335.

Costill, D.L. (1988). Carbohydrates for exercise: Dietary demands for optimal performance. *International Journal of Sports Medicine, 9,* 1-18.

Costill, D.L., Cote, R., & Fink, W. (1982). Dietary potassium and heavy exercise: Effects on muscle water and electrolytes. *American Journal of Clinical Nutrition, 36,* 266-275.

Coyle, E.F. (1991). Timing and method of increased carbohydrate intake to cope with heavy training, competition, and recovery. *Journal of Sports Sciences, 9,* 29-52.

Coyle, E.F., Coggan, A.R., Hemmert, M.K., & Ivy, J.L. (1986). Muscle glycogen utilization during prolonged strenuous exercise when fed carbohydrates. *Journal of Applied Physiology, 61,* 165-172.

Economos, C.D., Bortz, S.S., & Nelson, M.E. (1993). Nutritional practices of elite athletes: Practical recommendations. *Sports Medicine, 16,* 381-399.

Eichner, E.R. (1992). Sports anemia, iron supplements, and blood doping. *Medicine and Science in Sports and Exercise, 24,* S315-S318.

Foster, C., Costill, D.L., & Fink, W.J. (1979). Effects of pre-exercise feedings on endurance performance. *Medicine and Science in Sports, 11,* 1-5.

Hargreaves, M., Dillo, P., Angus, D., & Febbraio, M. (1996). Effect of fluid ingestion on muscle metabolism during prolonged exercise. *Journal of Applied Physiology, 80,* 363-366.

Hawley, J.A., Brouns, F., & Jeunkendrup, A. (1998). Strategies to enhance fat utilisation during exercise. *Sports Medicine, 25,* 241-257.

Hawley, J.A., Schabort, E.J., Noakes, T.D., & Dennis, S.C. (1997). Carbohydrate-loading and exercise performance: An update. *Sports Medicine, 24,* 73-81.

Hermansen, L., Hultman, E., & Saltin, B. (1967). Muscle glycogen during prolonged severe exercise. *Acta Physiologica Scandinavica, 71,* 129-139.

Kirwan, J.P., O'Gorman, D., & Evans, W. (1998). A moderate glycemic meal before endurance exercise can enhance performance. *Journal of Applied Physiology, 84,* 53-59.

Kozlowski, S., & Saltin, B. (1964). Effect of sweat loss on body fluids. *Journal of Applied Physiology*, **19**, 1119-1124.

Maughan, R.J., & Murray, R. (2001). *Sports drinks: Basic science and practical aspects*. Boca Raton: CRC Press.

McConell, G., Kloot, K., & Hargreaves, M. (1996). Effect of timing of carbohydrate ingestion on endurance exercise performance. *Medicine and Science in Sports and Exercise,* **28**, 1300-1304.

McDonald, R., & Keen, C.L. (1988). Iron, zinc and magnesium nutrition and athletic performance. *Sports Medicine,* **5**, 171-184.

Mitchell, J.B., Costill, D.L., Houmard, J.A., Fink, W.J., Pascoe, D.D., & Pearson, D.R. (1989). Influence of carbohydrate dosage on exercise performance and glycogen metabolism. *Journal of Applied Physiology,* **67**, 1843-1949.

Mitchell, J.B., Costill, D.L., Houmard, J.A., Flynn, M.G., Fink, W.J., & Beltz, J.D. (1988). Effects of carbohydrate ingestion on gastric emptying and exercise performance. *Medicine and Science in Sports and Exercise,* **20**, 110-115.

Nielsen, B., Sjøgaard, G., Ugelvig, J., Knudsen, B., & Dohlmann, B. (1986). Fluid balance in exercise dehydration and rehydration with different glucose-electrolyte drinks. *European Journal of Applied Physiology,* **55**, 318-325.

Piehl, K. (1974). Time course for refilling of glycogen stores in human muscle fibres following exercise-induced glycogen depletion. *Acta Physiologica Scandinavica,* **90**, 297-302.

Robergs, R.A. (1991). Nutrition and exercise determinants of postexercise glycogen synthesis. *International Journal of Sport Nutrition,* **1**, 307-337.

Saltin, B., & Hermansen, L. (1967). Glycogen stores and prolonged severe exercise. In G. Blix (Ed.), *Nutrition and physical activity* (pp. 32-46). Uppsala, Sweden: Almqvist & Wiksells.

Sen, C.K. (2001). Antioxidants in exercise nutrition. *Sports Medicine,* **31**, 891-908.

Sherman, W.M. (1992). Recovery from endurance exercise. *Medicine and Science in Sports and Exercise,* **24**, S336-S339.

Shirreffs, S.M., & Maughan, R.J. (2000). Rehydration and recovery of fluid balance after exercise. *Exercise and Sport Sciences Reviews,* **28**, 27-32.

Short, K.R., Sheffield-Moore, M., & Costill, D.L. (1997). Glycemic and insulinemic responses to multiple pre-exercise carbohydrate feedings. *International Journal of Sport Nutrition,* **7**, 128-137.

Tarnopolsky, M.A., MacDougall, J.D., & Atkinson, S.A. (1988). Influence of protein intake and training status on nitrogen balance and lean body mass. *Journal of Applied Physiology,* **64**, 187-193.

Tsintzas, K., & Williams, C. (1998). Human muscle glycogen metabolism during exercise: Effect of carbohydrate supplementation. *Sports Medicine,* **25**(1), 7-23.

Tsintzas, O.K., Williams, C., Wilson, W., & Burrin, J. (1996). Influence of carbohydrate supplementation early in exercise on endurance running capacity. *Medicine and Science in Sports and Exercise,* **28**, 1373-1379.

Vellar, O.D. (1968). Studies on sweat losses of nutrients: I. Iron content of whole body sweat and its association with other sweat constituents, serum iron levels, hematological indices, body surface area and sweat rate. *Scandinavian Journal of Clinical Laboratory Investigation,* **1**, 157-167.

Venkatraman, J.T., & Pendergast, D.R. (2002). Effect of dietary intake on immune function in athletes. *Sports Medicine,* **32**, 323-337.

Wilmore, J.H. (2000). Weight category sports. In R. Maughan (Ed.), *Nutrition in sport* (pp. 637-645). London: Blackwell Scientific.

CHAPTER 14

BODY WEIGHT,
BODY COMPOSITION,
AND SPORT

overview

An athlete's body size, build, and composition play major roles in determining athletic success. Of primary concern is the athlete's fat mass and fat-free mass. The ideal body type varies with each sport. The endurance runner strives for leanness, minimizing the load carried during a distance run, but the sumo wrestler tries hard to maximize body weight, because the tradition of this sport dictates that "bigger is better." Athletes of various weights can participate successfully in some sports, such as football, depending on the position they play. Yet other sports, such as wrestling, set rigid weight standards that often force athletes to shed large amounts of weight in a short time. Many athletes turn to crash diets, fasting, or the current fad diet, too often with little or no regard for the overall effects on health or performance.

In this chapter, we focus on body composition and how it affects physical performance. We discuss the significance of fat-free mass and relative body fat, and then we put these into perspective by looking at the ideal ranges found in elite performers. We explore the use of weight standards and the medical problems that are common when athletes use unrealistic methods to try to "make weight." Finally, we examine the proper way to plan weight loss, considering the composition of the body, so that a weight goal can be met and maintained without impairing performance.

outline

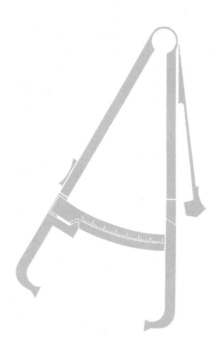

A former major league baseball player made minimum salary during his first few years in the majors. Early preseason polls projected his team to finish the season at the bottom of its division, but the team ended up in the World Series. This player became one of the best at his position in the National League during that season, and once the World Series was over, he asked management for a substantial salary increase ($75,000 in the mid-1970s). Management agreed to his salary demands contingent on his loss of 25 lb (11 kg)! The player refused to lose the weight, so the parties were deadlocked.

The team physician suggested sending the player to a major university for an accurate body composition assessment, and both parties agreed. A hydrostatic weighing was performed, and the results showed that the player had less than 6% body fat, with a total of only 11 lb (5 kg) of fat! Because 3% to 4% body fat is necessary for survival, this player had only 4 to 5 lb (about 2 kg) of fat to lose, and that loss wasn't advised because he was already at the lower range of acceptable body fat levels recommended for athletes. Management was satisfied, this player received his salary increase, and he did not have to lose weight.

This athlete's weight was well above the weight range recommended for his height, and he had a peculiar gait commonly referred to as a waddle. The combination of being overweight by the standard height–weight charts and having a waddle led management to demand the 25-lb (11-kg) weight loss. Had the athlete agreed to management's demand, he would have likely destroyed his career as a professional athlete. How many athletes have been faced with a similar situation? How many gave in?

Coaches and athletes today are acutely aware of the importance of achieving and maintaining optimal body weight for peak performance in sports. Appropriate size, build, and body composition are critical to success in almost all athletic endeavors. Compare the specific performance requirements of the 152-cm, 45-kg (5-ft, 100-lb) Olympic gymnast and those of the 206-cm, 147-kg (6-ft 9-in., 325-lb) defensive lineman in professional football. Body shape, size, and composition largely are predetermined by the genes inherited from one's parents. But this doesn't mean athletes should dismiss these components of their physical profile, believing that nothing can be done to change or improve them. Although size and body build can be altered only slightly, body composition can change substantially with diet and exercise. Resistance training can substantially increase muscle mass, and a sound diet combined with vigorous exercise can significantly decrease body fat. Such changes can be of major importance in achieving optimal athletic performance.

Body Build, Body Size, and Body Composition

What are the differences between body build, body size, and body composition? **Body build** refers to **morphology**, or the form and structure of the body. Most scientific systems for classifying body build have identified three major components (predominant somatotypes):

- Muscularity (mesomorphy)
- Linearity (ectomorphy)
- Fatness (endomorphy)

Each athlete's build is a unique combination of these three components. Athletes in certain sports usually exhibit a predominance of one component over the other two. The bodybuilder exhibits primarily muscularity, the 218-cm (7-ft 2-in.) basketball center who weighs only 82 kg (180 lb) exhibits linearity, and the sumo wrestler exhibits fatness. Most athletes are more balanced between muscularity and linearity, but muscularity tends to dominate in male athletes.[5]

Body size refers to the height and mass (weight) of an individual. Body size often is categorized as short or tall, large or small, heavy or light. Distinctions in these categories can vary depending on the specific performance requirements, so body size must be considered relative to the specific sport, the athlete's position, or the type of event. For example, among men a height of 190.5 cm (6 ft 3 in.) would be short for a professional basketball player but tall for a long-distance runner. Similarly, in professional football, a weight of 104 kg (230 lb) would be heavy for a quarterback, just right for a linebacker, but light for an offensive tackle.

Body composition refers to the body's chemical composition. Figure 14.1 illustrates four models of body composition. The first two divide the body into its various chemical or anatomical components; the last two simplify body composition into two components. The major difference between these last two models is the terminology *lean body mass* and *fat-free mass*. Behnke originally proposed the concept of **lean body mass**, defined to include fat-free mass and essential fat—the amount of fat necessary for survival.[10] Although this model is conceptually sound, it presents measurement problems: It isn't possible to differentiate between essential and nonessential fat. Consequently, most scientists have adopted the two-component model that includes fat mass and fat-free mass, which is the model used in this book. **Fat mass** is often discussed in terms of **relative body fat**, which is the percentage of the total body mass that is composed of fat. **Fat-free mass** simply refers to all body tissue that is not fat.

> The fat-free mass is composed of all of the body's nonfat tissue, including bone, muscle, organs, and connective tissue.

▶ Body build refers to morphology and usually is evaluated in terms of three components: muscularity, linearity, and fatness.

▶ Body size refers to the body's height and mass.

▶ Body composition refers to the body's chemical composition. The model we use considers two components: fat mass and fat-free mass.

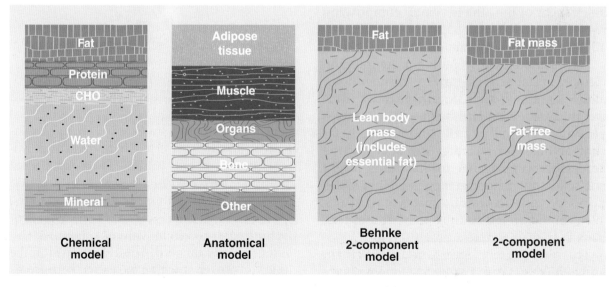

| Chemical model | Anatomical model | Behnke 2-component model | 2-component model |

▲ **Figure 14.1**   Four models of body composition. There is a slight difference in the fat component across the four models. In the anatomical model, adipose tissue includes fat and the cellular matrix and water within adipose tissue. In the Behnke model, the lean body mass includes some fat, which Behnke termed *essential fat*. Thus, fat in his model is total fat less essential fat.

Adapted, by permission, from J.H. Wilmore, 1992, Body weight and body composition. In Brownell, Rodin, and Wilmore (Eds.) *Eating, body weight, and performance in athletes: Disorders of modern society* (Philadelphia, PA: Lippincott, Williams, and Wilkins), 77-93.

Assessing Body Composition

Assessment of body composition provides additional information beyond the basic measures of height and weight to both the coach and the athlete. As an example, if the center fielder of a major league baseball team is 188 cm (6 ft 2 in.) tall and weighs 91 kg (200 lb), is he at his ideal playing weight? Knowing that 4.5 kg (10 lb) of a total weight of 91 kg (200 lb) is fat weight and that the remaining 86.5 kg (190 lb) is fat-free weight provides considerably more insight than weight alone. In this example, only 5% of his body weight is fat, which is about as low as any athlete should go, as discussed in our chapter opening. Armed with this knowledge, both athlete and coach realize that this athlete's body composition is ideal. They should not be concerned with weight loss, even though standard height–weight charts indicate that the athlete is overweight. However, another baseball player of the same height and weight who has 23 kg (50 lb) of fat would be 25% fat. This would constitute a serious weight problem for an elite athlete: He would be overfat. In most sports, the higher the percentage of body fat, the poorer the performance. An accurate assessment of the athlete's body composition provides valuable insight into the weight that allows optimal performance.

Appropriate goal weights for competition cannot be established simply by considering an athlete's present weight. An athlete can be overweight by the standard height–weight tables yet have a normal or below-normal body fat content. Likewise, an athlete can be within the acceptable weight range for his or her height but be overfat. Standard height–weight tables do not provide accurate estimates of what athletes should weigh because they do not take body composition into account.

This was clearly established in Welham and Behnke's classic 1942 study relating body composition to athletics.[33] They investigated the body compositions of a group of professional football players. Of these 25 professional athletes, 17 had been classified as physically unqualified for military duty or for first-class insurance because of their weight. But of these 17 "overweight" players, 11 were found to have very low levels of body fat. Their overweight condition was the result of excess fat-free mass, not fat mass. The average relative body fat for these athletes was only 9.3%. This demonstrates that knowing the composition of an athlete's weight is more important than relying on standard height–weight tables. But how do you determine an athlete's body composition?

> Although total body size and weight are important for most athletes, an athlete's body composition is generally of greater concern. Being overweight is usually not a problem, but being overfat typically has a negative impact on athletic performance. Standard height–weight tables do not provide accurate estimates of what an athlete should weigh because they do not take into account the composition of the weight. An athlete can be overweight according to these tables yet have very little body fat.

Densitometry

Densitometry involves measuring the density of the athlete's body. Density (D) is defined as mass (M) divided by volume (V):

$$D_{body} = M_{body} \div V_{body}$$

The mass of the body is the athlete's scale weight. Body volume can be obtained by several different techniques, but the most common is hydrostatic weighing, in which the athlete is weighed while totally immersed in water. The difference between the athlete's scale weight and underwater weight, when corrected for the density of water, equals the body's volume. This volume must be further corrected to account for air trapped in the body. The amount of air trapped in the gastrointestinal tract is difficult to measure, but fortunately it is a small volume (about 100 ml) and is usually ignored. The gas trapped in the lungs, however, must be measured because its volume is generally large, averaging 1,500 ml in young adult men and 1,200 ml in young adult women, depending on size.

Figure 14.2 illustrates the hydrostatic weighing technique and provides sample data for two

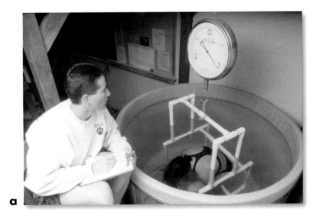

◀ **Figure 14.2**   (a) The underwater weighing technique. (b) Applying the technique to two professional football players of the same weight but different body compositions.

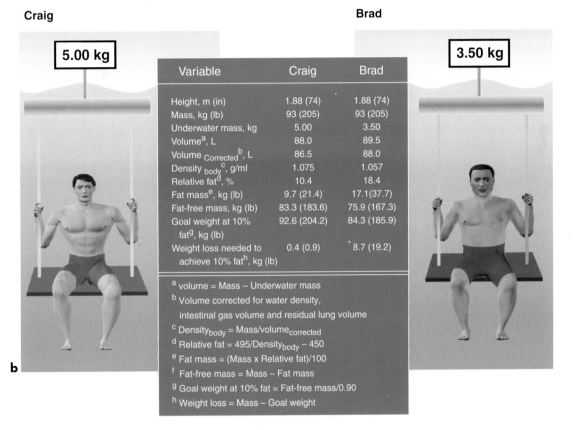

Craig

Brad

5.00 kg

3.50 kg

Variable	Craig	Brad
Height, m (in)	1.88 (74)	1.88 (74)
Mass, kg (lb)	93 (205)	93 (205)
Underwater mass, kg	5.00	3.50
Volume[a], L	88.0	89.5
Volume Corrected[b], L	86.5	88.0
Density body[c], g/ml	1.075	1.057
Relative fat[d], %	10.4	18.4
Fat mass[e], kg (lb)	9.7 (21.4)	17.1(37.7)
Fat-free mass, kg (lb)	83.3 (183.6)	75.9 (167.3)
Goal weight at 10% fat[g], kg (lb)	92.6 (204.2)	84.3 (185.9)
Weight loss needed to achieve 10% fat[h], kg (lb)	0.4 (0.9)	8.7 (19.2)

[a] volume = Mass − Underwater mass
[b] Volume corrected for water density, intestinal gas volume and residual lung volume
[c] Density$_{body}$ = Mass/volume$_{corrected}$
[d] Relative fat = 495/Density$_{body}$ − 450
[e] Fat mass = (Mass × Relative fat)/100
[f] Fat-free mass = Mass − Fat mass
[g] Goal weight at 10% fat = Fat-free mass/0.90
[h] Weight loss = Mass − Goal weight

professional football defensive backs of identical height and weight but substantially different body compositions. The table in the figure shows the calculations of their body composition, demonstrating that Brad is nearly twice as fat as Craig. The team goal for defensive backs is a range of 8% to 12% body fat. If the target was 10% body fat, Craig would have to lose only 0.4 kg (less than 1 lb), but Brad would need to lose 8.7 kg (nearly 20 lb).

Densitometry has long been the technique of choice for assessing body composition. New techniques typically are compared against densitometry to determine their accuracy. However, densitometry has its limitations. If body weight, underwater weight, and lung volume during underwater weighing are measured correctly, the resulting body density value is accurate. Densitometry's major weakness is in the conversion of body density to an estimate of relative body fat.[18]

Accurate estimates of the densities of fat mass and fat-free mass are required when the two-component model of body composition is used. The equation most often used to convert body density to an estimate of relative body fat is the standard equation of Siri:

$$\% \text{ body fat} = (495 \div D_{body}) - 450$$

This equation assumes that the densities of the fat mass and the fat-free mass are relatively constant in all people. Indeed, the density of fat at different sites is very consistent in the same individual and relatively consistent between people. The value generally used is 0.9007 g/cm³. But determining the density of the fat-free mass (D_{FFM}), which the equation of Siri assumes is 1.100, is more problematic. To determine this density, we must make two assumptions:

1. The density of each tissue constituting the fat-free mass is known and remains constant.

2. Each tissue type represents a constant proportion of the fat-free mass (e.g., we assume that bone always represents 17% of the fat-free mass).

Exceptions to either of these assumptions cause error when we convert body density to relative body fat, and this error can be substantial. Unfortunately, the density of the fat-free mass varies considerably between people.

This point is illustrated by considering the data from three athletes, shown in table 14.1.[34] If Edna, Vicki, and Susan each had the same total body density of 1.060, using the standard equation of Siri we would estimate that the relative body fat for each woman is 17.0%. However, this equation would be appropriate only for Edna, because she is the only one with a D_{FFM} of 1.100. If new equations were developed for Vicki and Susan by using their correct fat-free mass densities, Vicki's relative body fat would be 21.8% (her D_{FFM} is 1.115) and Susan's would be 11.5% (her D_{FFM} is 1.085). Thus, for the same body density, which is all that can be measured, the true relative body fat in these athletes varied from 11.5% to 21.8%. Although this example seems to exaggerate the point, the point itself is valid.

As just mentioned, the density of a mature, fully grown person's fat-free mass generally is assumed to be 1.100 g/cm³. Primarily because of known differences in bone mass, bone density, and total body water, evidence supports using lower values for children, women, and the elderly and higher values for certain racial and athletic groups.[14, 22, 27] In our example, Vicki is a mature 17-year-old African-American gymnast and Susan is an 11-year-old white distance runner. The D_{FFM} values obtained from the sum of the proportional densities for Vicki (1.115) and Susan (1.085) are almost identical to the values that have been established for African-American and adolescent populations. Research is

Table 14.1

Differences in the Density of Fat-Free Mass (D_{FFM}) in Three Female Athletes

Body tissue	Edna			Vicki			Susan		
	D_T	%	D_P	D_T	%	D_P	D_T	%	D_P
Muscle	1.065	46	0.490	1.065	41	0.437	1.065	46	0.490
Bone	1.350	17	0.229	1.350	22	0.297	1.260	17	0.214
Remainder	1.030	37	0.381	1.030	37	0.381	1.030	37	0.381
D_{FFM}			1.100			1.115			1.085

Note. D_T = density of the tissue; % = percent contribution of this tissue to the total-fat-free mass; D_P = proportional density of the tissue ($D_T \times$ %); D_{FFM} = density of the fat-free mass, which is the sum of the proportional densities.

progressing in this area. Population-specific equations are now available for select populations, improving the accuracy of converting body density to relative body fat values.[10] Specifically, equations have been derived for women,[14] children,[14] and African-American men[27] and women.[22] However, this research needs to be expanded to additional populations. As an example, it has been shown that the D_{FFM} of men who resistance train is below that used in the standard equations, such as the Siri equation.[21] Thus, the fat-free mass of this group of athletes will be underestimated and the fat mass overestimated when we use these standard equations. Furthermore, more recent research has questioned whether there is a true difference in the D_{FFM} of blacks and whites, both athletes and nonathletes alike.[9, 20, 32]

Inaccuracies in densitometry largely reflect the variation in the density of the fat-free mass from one individual to another. Density of fat-free mass is affected by age, sex, and race.

Other Laboratory Techniques

Many other laboratory techniques are available for assessing body composition. These include radiography, computed tomography (CT), magnetic resonance imaging (MRI), hydrometry (for measuring total body water), total body electrical conductivity, and neutron activation. Most of these techniques are complex and require expensive equipment. None of them are likely to be used for assessing athletic populations, so we will not discuss them further in this chapter; however, we do discuss computed tomography in chapter 21. These techniques have been reviewed extensively by others.[3, 17, 19] Two other techniques hold considerable promise—dual-energy X-ray absorptiometry and air displacement. Let's take a brief look at each of these.

Dual-energy X-ray absorptiometry (DXA) evolved from the earlier single- and dual-photon absorptiometry techniques used between 1963 and 1984.[15] The earlier techniques were used to estimate regional bone mineral content and bone mineral density, primarily in the spine, pelvis, and femur. The new DXA technique (see figure 14.3) allows the quantification of not only bone but also soft tissue com-position. Furthermore, it is not limited to regional estimates but can provide total body es-

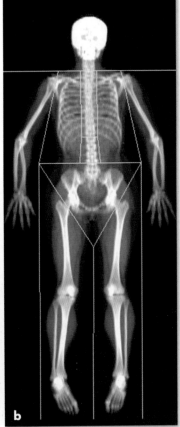

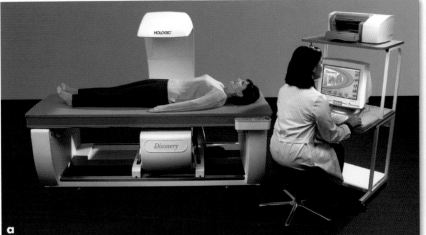

▲ **Figure 14.3**   The dual-energy X-ray absorptiometry (DXA) machine used to estimate bone density and bone mineral content as well as total body composition (fat mass and fat-free mass): (*a*) the machine, (*b*) a regional scan of the body.

timates. Research to date suggests that DXA provides precise and reliable estimates of body composition. The advantage of DXA over the underwater weighing technique includes the ability to estimate bone density and bone mineral content in addition to fat and fat-free mass. Furthermore, it is a passive technique where the subject simply lies on a table during the scan, as opposed to having to submerge him- or herself underwater multiple times.

Air plethysmography is a densitometric technique. Volume is determined by air displacement rather than by water immersion. This technique, developed in the early 1900s, was used largely in research laboratories up until the 1990s, when a commercial model became available that is now widely used (see figure 14.4).[7] The principle of operation is rather simple. You have a closed chamber of room air at atmospheric pressure, which has a known volume. You then open the chamber door, enter the chamber, sit in a fixed position, and then close the chamber door, forming an air-tight seal. The new volume of the air in the chamber is determined, which is then subtracted from the total volume of the chamber to provide an estimate of your volume.

- Total volume = subject's volume + remaining volume
- Subject's volume = total volume − remaining volume

Although this is a relatively simple technique for the subject, it requires considerable accuracy in controlling for changes in temperature, gas composition, and the subject's breathing while in the chamber. Studies have confirmed the accuracy of this technique under most conditions. It appears to provide a relatively precise measurement of body volume. Just like the underwater weighing technique, you can obtain relatively accurate measurements of total body volume and, thus, obtain accurate estimates of total body density. However, you still must use the subject's body density in an equation to estimate relative body fat, recognizing the uncertainties of the D_{FFM} for that subject.

Field Techniques

Several field techniques are also available for assessing body composition. These techniques are more accessible than laboratory techniques

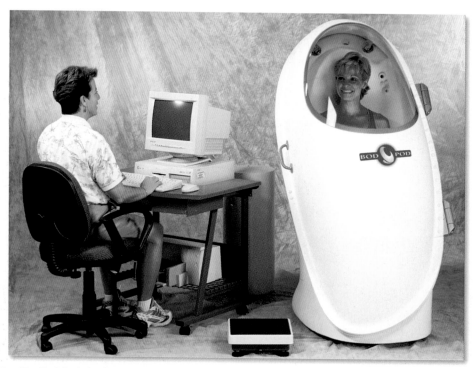

▲ **Figure 14.4**   The Bod Pod air plethysmography device uses the air displacement technique to estimate total body volume.
Reprinted with permission of Life Measurement, Inc.

because the equipment is less costly and cumbersome, so they can be used more easily by the coach, the trainer, or even the athlete, outside the laboratory.

Skinfold Fat Thickness

The most widely applied field technique involves measuring the skinfold fat thickness (see figure 14.5) at one or more sites and using the values obtained to estimate body density, relative body fat, or fat-free mass. It generally is recommended that the sum of the measurements from three or more skinfold sites be used in a quadratic, curvilinear equation to estimate body density.[25] A curvilinear equation more accurately describes the relationship between the sum of skinfold measurements and body density than a linear equation does. Linear equations underestimate the density of lean people, which causes overestimation of body fat. Just the opposite happens for obese people:

Body density is overestimated, and body fat is underestimated. Skinfold fat thickness measurements that use quadratic equations provide reasonably accurate estimates of total body fat or relative fat, with correlations ranging from .90 to .96.

> **fyi** Laboratory techniques such as densitometry and DXA provide reasonable estimates of true body composition: relative body fat, fat mass, and fat-free mass. Multiple skinfold fat thickness measurements used in an equation appropriate for the population being assessed also provide a good estimate of body composition.

Bioelectric Impedance

Bioelectric impedance is a simple procedure introduced in the 1980s that takes just 5 min to perform. Four electrodes are attached to the body at the ankle, the foot, the wrist, and the back of the hand, as shown in figure 14.6. An undetectable current is passed through the distal electrodes (hand and foot). The proximal electrodes (wrist and ankle) receive the current flow. Electrical conduction through the tissues between the electrodes depends on the water and electrolyte distribution in that tissue. Fat-free mass contains almost all the body water

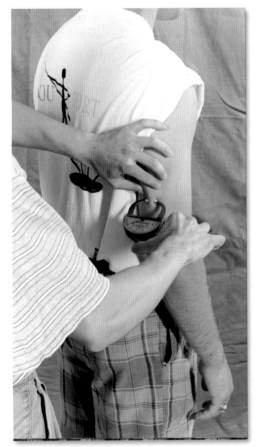

▲ **Figure 14.5** Measuring skinfold fat thickness at the triceps skinfold site.

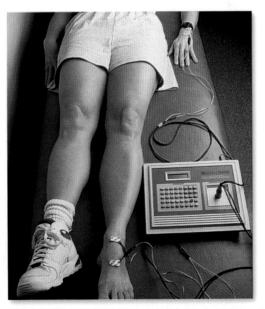

▲ **Figure 14.6** The bioelectric impedance technique for assessing relative body fat.

and the conducting electrolytes, so conductivity is much greater in the fat-free mass than in the fat mass. The fat mass has a much greater impedance, meaning that it is much more difficult for the current to flow through the fat mass. Thus, the amount of current flow through the tissues reflects the relative amount of fat contained in that tissue.

With the bioelectric impedance technique, measurements of the impedance, the conductivity, or both are transformed into estimates of relative body fat. Estimates of relative body fat based on bioelectric impedance highly correlate with body fat measurements obtained through hydrostatic weighing ($r \cong .90$-$.94$). However, the relative body fat in lean athletic populations tends to be overestimated with bioelectric impedance because of the nature of the equations used. More sport-specific equations are being developed, and a newer technique, multifrequency bioelectric impedance spectroscopy, could possibly improve the accuracy of measurement in these lean athletic populations.

Body Composition and Sport Performance

Many athletes believe that they must be big to be good in their sport because size traditionally has been associated with performance quality in certain sports (e.g., football and basketball): The bigger the athlete, the better the performance. But big does not always mean better. In certain other sports, smaller and lighter are considered to be better for performance (e.g., gymnastics, figure skating, and diving). Yet this can be taken to extremes, compromising the health of the athlete. In the following sections, we consider how performance can be affected by body composition.

Fat-Free Mass

Rather than be concerned with total body size or weight, most athletes should be concerned specifically with fat-free mass. Maximizing fat-free mass is desirable for athletes involved in activities that require strength, power, and muscular endurance. But increased fat-free mass is likely to be undesirable for the endurance athlete, such as a distance runner, who must move his or her total body mass horizontally for extended periods. A higher fat-free mass is an additional load that must be carried and might impair the athlete's performance. This might also be true for the high jumper, long jumper, triple jumper, and pole-vaulter, who must maximize their vertical or horizontal distances or both. Additional weight, even though it is active fat-free mass, could decrease rather than facilitate performance in these events.

> ▶ Knowing a person's body composition is more valuable for predicting performance potential than knowing merely height and weight.
>
> ▶ Densitometry is one of the best methods for assessing body composition and has long been considered the most accurate, although it does carry certain risks of error. It involves calculating the density of the athlete's body by dividing body mass by body volume, which is typically determined by hydrostatic weighing or air displacement. Body composition can be calculated, although there is some margin of error.
>
> ▶ The density of the fat-free mass generally is assumed to be 1.100 g/cm³ for a fully mature person, but lower values are suggested for children, women, and the elderly and higher values for racial and athletic subgroups.
>
> ▶ DXA, originally developed for estimating bone density and bone mineral content, is now capable of providing accurate estimates of total body composition—fat mass and fat-free mass.
>
> ▶ Field techniques for assessing body composition include measuring skinfold fat thickness and bioelectric impedance. These techniques are less costly and more accessible for the athlete and the coach than are laboratory techniques.

Techniques eventually will be available to estimate not only athletes' fat mass and fat-free mass but also their potential for increasing their fat-free mass. Such techniques would allow athletes to design training programs that would develop their fat-free mass to this projected maximum while maintaining their fat mass at relatively low levels. Combining resistance training with the ingestion of carbohydrate, or carbohydrate and protein, during recovery from resistance training appears to be effective for increasing the fat-free mass.[12] This routine appears to stimulate the release of the anabolic hormones.

Relative Body Fat

Relative body fat is a major concern of athletes. Adding more fat to the body just to increase the athlete's weight and overall size is generally detrimental to performance. Many studies have shown that the higher the percentage of body fat, the poorer the person's performance. This is true of all activities in which the body weight must be moved through space, such as running and jumping. (It is less important for more stationary activities, such as archery and shooting.) In general, leaner athletes perform better.

One study determined the relationship between body fat and performance in young men.[26] The results, shown in table 14.2, clearly indicated that degree of fatness had a great influence on the performance of four fitness and athletic ability tests. Other studies have clearly shown that body fatness is associated with poorer performance on tests of

- speed,
- endurance,
- balance and agility, and
- jumping ability.

Endurance athletes try to minimize their fat stores because excess weight is proven to impair their performance. Both absolute fat and relative body fat can profoundly influence running performance in highly trained distance runners. Less fat generally leads to better performance. Male runners normally have much less relative body fat than female runners; this is thought to be one of the most important reasons for the differences in running performance between elite male and female distance runners.[35] This premise was confirmed in a study of male and female runners who, when matched by their 15-mi (24-km) road-race times, did not differ in relative body fat.[24]

In another study, male and female runners performed both submaximal and maximal runs on the treadmill and an all-out 12-min

Table 14.2

The Effect of Relative Body Fat on Selected Performance Tests in Young Men

Performance test	Subject's level of fatness (% body fat)		
	Low (<10%)	Moderate (10-15%)	High (>15%)
75-yd dash (s)	9.8	10.1	10.7
220-yd dash (s)	29.3	31.6	35.0
Standing long jump[a] (ft)	23.8	22.7	20.2
Sit-ups in 2 min	43.4	41.6	36.2

Note. The men in this study were classified according to three levels of fatness: low, moderate, and high.

[a]Sum of three trials.

Adapted from Riendeau et al (1958).[26]

run.[6] Each male runner performed his tests under normal conditions and also while carrying external weight added to the trunk to simulate the relative body fat of the female runner to whom he had been matched. With the added weight, the metabolic cost of submaximal exercise was increased, and maximal oxygen uptake was reduced. Furthermore, the large performance differences noted between the men and women in the unweighted tests were reduced considerably when the men ran with the added weight.

> Excessive body fat is associated with decreased athletic performance in activities where the body mass must be moved through space. Speed, endurance, balance, agility, and jumping ability are all negatively affected by a high level of fatness.

Heavyweight weightlifters might be exceptions to the general rule that less fat is better. These athletes add large amounts of fat weight just before competition under the premise that the additional weight will lower their center of gravity and give them a greater mechanical advantage in lifting. Research has not yet confirmed the value of this additional fat weight. The sumo wrestler is another notable exception to the theory that overall size is not the major determinant of athletic success. In this sport, the larger individual has a decided advantage, but even so, the wrestler with the higher fat-free mass should have the best overall success.

Performance in swimming also seems to be an exception to this general rule. In one study, the relationship between body fatness and swimming performance was determined in 284 competitive female swimmers, ages 12 to 17 years.[30] An examination of the subjects' best times in their best events and in the 100-yd (91-m) freestyle revealed that swim performance was unrelated to relative body fat and only slightly related to fat-free mass. Body fat might provide some advantage to the swimmer because it improves buoyancy, which can reduce body drag in the water and reduce the metabolic cost of staying on the surface of the water.

▶ The ideal body composition varies with different sports, but in general, the less fat mass, the greater the performance.

▶ Maximizing fat-free mass is desirable for athletes in sports that require strength, power, and muscular endurance but could be a hindrance to endurance athletes, who must be able to move their total body mass for extended periods, and jumpers, who must move their body mass vertically or horizontally for distance.

▶ The degree of fatness has more influence on performance than does total body weight. In general, the greater the relative body fat, the poorer the performance. Possible exceptions include heavyweight weightlifters, sumo wrestlers, and swimmers.

Weight Standards

Coaches and athletes alike are always looking for that winning edge. Once a coach or an athlete hits on something that works and improves performance, the word quickly spreads. The widespread use of anabolic steroids is a classic example. What started out in the late 1940s and early 1950s as experimentation among a small number of bodybuilders and weightlifters has spread throughout the world of sport so that now the majority of the elite athletes in certain sports are habitual steroid users. A similar phenomenon is the emphasis on leanness in athletes.

The elite athlete has long been esteemed for representing the most desirable physical and physiological characteristics for performance in a sport or activity. Theoretically, the elite athlete's genetic foundation and years of intense training have combined to provide the ultimate athletic profile for that sport. These elite athletes set the standards toward which others aspire.

In the 1970s, we (JHW) had the opportunity to test many of the United States' elite female track and field athletes.[35] In addition to treadmill testing, each athlete underwent a body composition assessment. The results of these assessments are presented in figure 14.7. If we look just at the distance runners, many of the best were below 12% body fat. The two top distance runners had only about 6% fat. One of these had won six consecutive international cross country championships, and the other held what was then the best time in the world for the marathon. From these results, we could be tempted to suggest that any female distance runner should have between 6% and 12% relative body fat if she has world-class aspirations. However, one of the best distance runners in the United States at that time, who was within 2 years of taking over the top spot, had a relative body fat of 17%. Furthermore, one of the women in this study had a relative body fat of 37%, and she set the best time in the world for the 50-mi (80-km) run within 6 months of her evaluation! More than likely, neither of these women would have gained an advantage if she had been forced to decrease her weight to achieve 12% body fat or lower.

Body weight has been a general concern in several sports for many years. Over the past 10 to 15 years, this concern has become more widespread, and most sports have now adopted weight standards intended to ensure that athletes have the optimal body size and composition for maximal performance. Unfortunately, this is not always the result.

Inappropriate Use of Weight Standards

Weight standards have been seriously abused. Coaches have seen that athletes' performances generally improve as body weight decreases. This has led some coaches to adopt the philosophy that if small weight losses improve performance a little, then major weight losses should improve it even more. Not only coaches are guilty of making this assumption: Athletes and their parents also are drawn into this way of thinking. As an example, a university athlete, considered one of the best in the United States in her sport, had dieted and exercised down to such a low body weight that her relative body fat was less than 5%. If anyone who joined the team appeared to be leaner, she would work even harder to reduce her weight and fat content. This woman's athletic performance began to deteriorate, and she started to develop injuries that never seemed to heal. She was eventually diagnosed with anorexia nervosa (chapter 18) and underwent professional treatment. But her career as an elite athlete was over.

Making Weight: Risks With Severe Weight Loss

Male athletes are not immune to problems associated with excessive weight loss and disordered eating. The medical and scientific communities have been concerned over problems in male athletes associated with making weight. The major concern has centered around the sport of wrestling. In 1986, a survey was conducted that involved 63 collegiate wrestlers representing 15 teams at the Eastern Intercollegiate Wrestling Association Championships.[31] These wrestlers began wrestling at an average age of 10.9 years. They started cutting weight by the age of 13.5 years and lost weight an

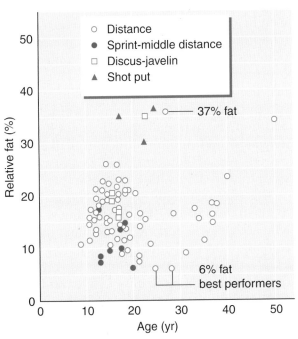

▲ **Figure 14.7** Relative body fat in elite female track and field athletes.
Data from Wilmore et al. 1977.[35]

average of 15 times during a normal season. The average for the most weight lost at any one time was 7.2 kg (15.8 lb). For one specific championship meet, the average weight loss by these wrestlers was 4.4 kg (9.7 lb) in less than 3 days. Wrestlers typically make weight by combining food restriction, fluid deprivation, thermal dehydration, and increased activity. The potential health hazards of such practices led the American College of Sports Medicine to publish a position statement on weight loss in wrestlers in 1976, which was revised in 1996.[1] The influence of dehydration on physiological function and performance is illustrated in the previous chapter in table 13.6.

Many schools, districts, or state-level organizations organize other sports, in addition to wrestling, on the basis of size, with weight as the predominant factor. Athletes in these sports often attempt to achieve the lowest possible weight to gain an advantage over opponents. In so doing, many athletes have jeopardized their health. In the following sections, we examine a few of the consequences of severe weight loss in athletes, both male and female.

Dehydration

Fasting or very low calorie diets lead to large amounts of weight loss, primarily through dehydration. As we discussed in chapter 13, for every gram of carbohydrate stored, there is an obligatory gain of 2.6 g of water. When you use carbohydrates for energy, you lose that water. Thus, with fasting and very low calorie diets, carbohydrate stores are substantially depleted during the first few days. This results in a substantial loss of weight attributable to the loss of body water.

Furthermore, athletes trying to make weight might exercise in rubberized sweat suits, sit in steam and sauna baths, chew on towels to lose saliva, and minimize their fluid intake. Such severe water losses compromise kidney and cardiovascular function and are potentially dangerous. Weight losses of 2% to 4% of the athlete's weight through dehydration can impair performance. The consequences of weight loss through dehydration, discussed in chapter 13, include

- decreased blood volume and blood pressure,
- decreased submaximal and maximal stroke volume and maximal cardiac output,
- decreased blood flow to and through the kidneys,
- increased submaximal heart rate,
- decreased aerobic and anaerobic capacity,
- decreased performance, and
- impaired thermoregulation.

Chronic Fatigue

Pushing body weight too low can have major repercussions. When weight drops below a certain optimal level, the athlete is likely to experience performance decrements and increased incidence of illness and injury. The performance decrements can be attributable to many factors, including chronic fatigue that often accompanies major weight losses. The causes of this fatigue have not been established, but there are several likely possibilities.

The symptoms of an athlete who is chronically underweight (below optimal competitive weight) mimic those seen with overtraining (chapter 12). Both neural and hormonal components are involved in the phenomenon of overtraining. In most cases, the sympathetic nervous system appears to be inhibited, and the parasympathetic system dominates. In addition, the hypothalamus does not function normally, and immune function is likely impaired. These alterations lead to a cascade of symptoms that include chronic fatigue.[2, 13]

This chronic fatigue also could be attributed to substrate depletion. Energy for almost all athletic activities is derived predominantly from carbohydrate. Carbohydrate also represents the smallest source of stored energy. The combined carbohydrate storage in muscle, liver, and extracellular fluid accounts for approximately 2,000 kcal of stored energy. When athletes are training hard and are not eating an adequate diet (when they are deficient either in total calories or in total carbohydrate calories), their carbohydrate energy stores become depleted. Most important to the athlete, liver and muscle glycogen levels decrease, which in turn reduce blood glucose levels. The combined effect of these decreases can be chronic fatigue and considerable declines in performance.[29] In addition, under these conditions the body also uses its protein stores as an energy substrate for exercise. This can, over time, gradually deplete muscle protein.[4]

During the 1990s, we started to see athletes diagnosed with chronic fatigue syndrome. This syndrome might or might not be related to what we have termed chronic fatigue. At this time, very little is known about chronic fatigue syndrome, although it does appear to involve immune system dysfunction. Patients have incapacitating fatigue, and the symptoms may vary in severity over time but generally last for months or years. The symptoms include prolonged, debilitating fatigue; sore throat; muscle tenderness or pain (myalgia); and cognitive dysfunction.

Eating Disorders

The constant attention given to achieving and maintaining a prescribed weight goal, particularly if the weight goal is inappropriate, can lead to disordered eating. A high proportion of athletes, particularly females, have disordered eating. This term can simply refer to restricting food intake to levels that are well below energy expenditure, but disordered eating can also involve pathological behaviors to control body weight, such as self-induced vomiting and laxative abuse. Disordered eating can lead to clinical eating disorders, such as anorexia nervosa or bulimia nervosa. These disorders have become prevalent among female athletes. Each has strict criteria for diagnosis that set it apart from disordered eating in general.

Obtaining accurate estimates of the prevalence of eating disorders is difficult, if not impossible, particularly in athletic populations. Those with eating disorders are not likely to disclose their problem. In fact, most deny they have a problem. However, a significant number of indicators suggest a high prevalence in select athletic populations. More than 90% of people with eating disorders are women. Among athletes, those in appearance sports (such as

gymnastics, figure skating, diving, and dancing) and in endurance sports (such as running and swimming) seem at greatest risk. On certain teams, particularly in these appearance and endurance sports, the prevalence of eating disorders might approach or even exceed 50% at the elite or world-class level. Athletes and coaches must realize the potential link between weight standards and eating disorders. The issue of eating disorders in athletes is a major focus of chapter 18.

A female athlete who is prone to disordered eating is open to a triad of disorders that are likely interrelated: anorexia nervosa or bulimia nervosa, menstrual dysfunction, and bone mineral disorders. This group of disorders is now referred to as the female athlete triad.[23] This is discussed in much more detail in chapter 18.

> Eating disorders appear to be prevalent among elite female athletes. It is important to establish weight standards that maximize performance but minimize the risk of initiating an eating disorder.

Menstrual Dysfunction

Menstrual dysfunction, or abnormal menstruation, is widely recognized in female athletes. High prevalences of oligomenorrhea (infrequent or scant menstrual flow), amenorrhea (cessation of menstrual flow), and delayed menarche (first period) have been associated with sports that emphasize low body weight or low body fat. The combination of caloric restriction and a vegetarian diet is common among women endurance athletes. Substantial weight loss induced by either or both of these is associated with a shortened luteal phase and menstrual dysfunction.[28] Most likely, menstrual dysfunction is the body's natural adaptation to a prolonged energy deficit, where energy intake remains below energy expenditure over long periods of time.[11, 16] This is discussed more fully in chapter 18.

A strong link exists between anorexia nervosa and menstrual dysfunction. In fact, amenorrhea is one of the strict criteria necessary for the diagnosis of anorexia nervosa in females. A similar relationship has not yet been established with bulimia (see chapter 18), but an increasing number of athletes are found to be both bulimic and amenorrheic.

Bone Mineral Disorders

Bone mineral disorders are recognized as a potentially serious consequence of menstrual dysfunction.[8] The link between the two was first reported in 1984. Now a number of scientists are researching the relationship between athletic-induced amenorrhea and low bone mineral content or density. Past studies suggested that bone density increases with the resumption of normal menses (menstruation), but more recent observations suggest that the amount of bone that is regained might be limited and that bone density might remain well below normal even with reestablishment of normal menstrual function.[8] The long-term consequences of chronically low bone densities in athletic populations have not yet been established.

Establishing Appropriate Weight Standards

The potential for abuse of weight standards is clearly established. If standards are not set appropriately, athletes could be pushed well below optimal body weight. Thus, it is critically important to properly set weight standards.

Body weight standards should be based on an athlete's body composition. Once body composition has been determined, the amount of fat-free mass is used to estimate what the athlete should weigh at a specific relative body fat. Consider an example in which the goal is to get a 72.6-kg (160-lb) female swimmer with 25% body fat down to 18% body fat, as shown in table 14.3. We know that her goal weight will consist of 18% fat and 82% fat-free mass, so to estimate her weight goal at 18% body fat, we divide her fat-free mass (54.4 kg, or 120 lb) by 82%, which is the fraction of her weight goal that is to be represented by her fat-free mass. This calculation gives us a goal weight of 66.3 kg (146 lb), so this woman needs to lose 6.3 kg (14 lb).

Thus, establishing weight standards should translate into establishing standards of relative body fat for each sport and, where appropriate, for each event within a sport. With this in mind, what is the recommended relative body fat for an elite athlete in any given sport? For each sport, an optimal range of values for relative body fat should be established, outside of which the athlete's performance is likely impaired. And because fat distribution shows definite sex differences, the weight standards should be sex specific. Representative ranges for men and women in various sports are presented in table 14.4. In most cases, these values represent the elite athletes in those sports.

Table 14.3

Computing a Weight Goal for Performance for a Female Swimmer

Parameter	Measure
Weight	72.6 kg (160 lb)
Relative fat	25%
Fat weight	18.2 kg (40 lb) (weight × 25%)
Fat-free weight	54.4 kg (120 lb) (weight − fat weight)
Relative fat goal	18% (= 82% fat-free)
Weight goal	66.3 kg (146 lb) (fat-free weight ÷ 82%)
Weight loss goal	6.3 kg (14 lb)

Table 14.4

Ranges of Relative Body Fat Values for Male and Female Athletes in Various Sports

Group or sport	% fat	
	Men	**Women**
Baseball/softball	8-14	12-18
Basketball	6-12	10-16
Bodybuilding	5-8	6-12
Canoeing/kayaking	6-12	10-16
Cycling	5-11	8-15
Fencing	8-12	10-16
Football	6-18	—
Golf	10-16	12-20
Gymnastics	5-12	8-16
Horse racing (jockey)	6-12	10-16
Ice/field hockey	8-16	12-18
Orienteering	5-12	8-16
Pentathlon	—	8-15
Racquetball	6-14	10-18
Rowing	6-14	8-16
Rugby	6-16	—
Skating	5-12	8-16
Skiing (alpine and Nordic)	7-15	10-18
Ski jumping	7-15	10-18
Soccer	6-14	10-18
Swimming	6-12	10-18
Synchronized swimming	—	10-18
Tennis	6-14	10-20
Track and field, field events	8-18	12-20
Track and field, running events	5-12	8-15
Triathlon	5-12	8-15
Volleyball	7-15	10-18
Weightlifting	5-12	10-18
Wrestling	5-16	—

The recommended values might not be appropriate for all athletes who engage in a specific activity. The existing techniques for measuring body composition include inherent errors, as we discussed earlier. Even with the better laboratory techniques, measurement of body density can introduce a 1% to 3% error, and an even greater error is associated with converting that density to relative body fat. In addition, we must understand the concept of individual variability. Not every male distance runner will have his best performance at 6% body fat. Some will improve performance with slightly lower values. Others won't be able to get down to such low relative fat values, or they will find that their performance starts to decline before they reach the suggested values. For these reasons, a range of values should be set for males and females in specific activities, recognizing individual variability, methodological error, and sex differences.

▶ Many sports enforce weight standards with the goal of ensuring that the athletes are of optimal body size for participation. Unfortunately, athletes often turn to questionable, ineffective, or even dangerous methods of weight loss to reach their weight goal.

▶ Severe weight loss in athletes can cause potential health problems, such as dehydration, chronic fatigue, disordered eating, menstrual dysfunction, and bone mineral disorders.

▶ The chronic fatigue symptoms that often accompany severe weight loss mimic those of overtraining. This fatigue also can be caused by substrate depletion.

▶ Body weight standards should be based on body composition. Thus, these standards should emphasize relative body fat rather than total body mass.

▶ For each sport, a range of values should be established, recognizing the importance of individual variation, methodological error, and sex differences.

It is important to establish realistic weight standards for athletes. This is generally best accomplished by using a range of relative fat values that are considered acceptable for the sport and the athlete's age and sex.

Achieving Optimal Weight

Many athletes discover that they are considerably above their assigned playing weight with only a few weeks remaining before they report to training camp. Consider a 25-year-old professional football player who realizes his weight is 9 kg (20 lb) above his playing weight in the previous season. He must lose this excess weight by the start of the preseason training camp, a mere 4 weeks away. Failure will cost him a fine of $500 per day for each pound over his assigned weight. Exercise alone is of little value because he would need 9 to 12 months to lose this much weight through that means. Will this athlete accomplish his goal?

Avoiding Fasting and Crash Dieting

Our football player must lose 2.3 kg (5 lb) per week for the next 4 weeks, so he decides to embark on a crash diet, selecting whatever diet is in vogue, knowing that a person can lose 2.7 to 3.6 kg (6-8 lb) per week with such diets. He is not unique. Many athletes find themselves overweight and out of shape because of overeating and reduced activity during the off-season, and they typically wait until the last few weeks before their reporting dates to attack the problem. In our example, the football player might be able to shed 9 kg (20 lb) in 4 weeks with his crash diet. But much of this weight loss would be from body water and very little from stored fat. Several studies have reported that substantial weight losses occur with very low calorie diets (500 kcal per day or less), but of the weight lost, more than 60% comes from the body's fat-free tissue and less than 40% from fat depots.

Although much of our football player's weight is lost from water stores, a substantial amount of protein is lost as well. Also, most crash diets are based on a major reduction in carbohydrate intake. This reduced intake is insufficient to supply the body's needs for carbohydrate, and as a result the body's carbohydrate stores become depleted. Because water storage accompanies carbohydrate storage, the water stores are also reduced as the carbohydrate stores diminish. With every gram of carbohydrate used by the body, approximately 2.6 g of water is lost. The total body glycogen content is about 800 to 1,000 g, so depletion of the stored glycogen would result in a loss of approximately 2,080 to 2,600 g of water—about 2 to 2.6 kg (4.6-5.7 lb).

In addition, the body relies more heavily on free fatty acids for energy, because its carbohydrate stores are depleted. As a result, ketone bodies, a by-product of fatty acid metabolism, accumulate in the blood, causing a condition known as ketosis. This condition further increases water loss. Much of this water loss occurs during the first week of the diet.

Figure 14.8 illustrates these changes in body water, fat, protein, and carbohydrate during 30 days of fasting. Carbohydrate depletion occurs within about 3 days. Protein content decreases significantly in the same time period. The body's water content decreases abruptly during the first 3 days. The rate of fat loss, although it decreases somewhat in the second and third day, remains relatively unchanged for the remainder of the fasting period.

It is impossible for most people to lose more than 1.8 kg (4 lb) of fat per week, even under conditions of total fasting. This can be easily demonstrated. Fat (adipose tissue) has the caloric equivalent of 3,500 kcal per 0.45 kg (1 lb), so a deficit of 3,500 kcal is needed to lose 0.45 kg of fat. The resting metabolic rate of the professional football player in our previous example would be approximately 2,500 kcal per day. So if he fasted, he would have a 2,500-kcal deficit per day. The most he could lose each day would be about 0.32 kg (0.7 lb). However, research indicates that total body metabolism is reduced by 20% to 25% during fasting. A 20% reduction in this athlete's resting metabolic rate would lower his total deficit to only 2,000

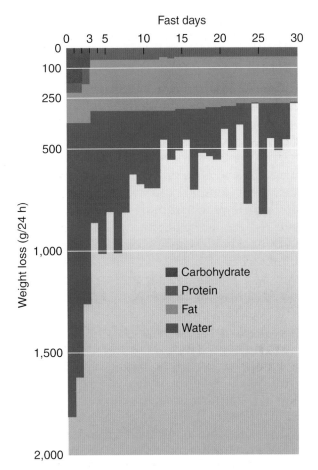

▲ **Figure 14.8**   The composition of weight lost with 30 days of fasting.

kcal per day, which would result in a maximal weight loss of only about 0.26 kg (0.6 lb) of fat per day. In 1 week of total fasting, this would result in a loss of only 1.8 kg (4 lb)! By increasing his activity level, our football player could increase this predicted rate of fat weight loss. But he would find it impossible to train very hard with no food intake and an almost total depletion of his glycogen stores. Also, few people can tolerate the discomfort associated with prolonged periods of fasting. The rapid weight losses experienced with crash diets are quickly regained, probably because when a balanced diet is substituted for the low-calorie, low-carbohydrate diet, the water that was lost is quickly regained as the carbohydrate stores are replenished.

Optimal Weight Loss: Decreasing Fat Mass and Increasing Fat-Free Mass

The sensible approach to reducing body fat stores is to combine moderate dietary restriction with increased exercise. Appetite is delicately balanced with the body's actual caloric needs. If you reduce your dietary intake by a mere 100 kcal each day (one slice of buttered bread), for example, you will lose about 4.5 kg (10 lb) in 1 year, assuming that your activity level remains constant. But if you add to this an additional modest loss of 0.1 to 0.2 kg (0.25-0.5 lb) per week attributable to increased activity (such as a 3-day/week jogging program), your total weight loss would be about 10 to 16 kg (23-36 lb) in a single year, and most of this weight loss would be from stored body fat.

When athletes exceed the upper end of the weight range for their sport, they should work toward achieving the upper-end goal weight slowly, losing no more than 0.5 to 1 kg (less than 2.2 lb) per week. Losing more weight than that per week leads to losses in fat-free mass, which is usually not the desired outcome. When the upper limit of the range is reached, further weight loss should be undertaken only with close supervision of the coach, athletic trainer, or team physician. This weight loss should be achieved at an even slower rate—less than 0.5 kg (1.1 lb) per week—to ensure that performance is not negatively affected. The rate of this loss should be reduced still more if performance is affected or if medical symptoms are noted.

Decreasing caloric intake by 200 to 500 kcal per day will allow weight losses of about 0.5 kg (1.1 lb) per week, particularly if combined with a sound exercise program. This is a realistic goal, and such losses add up to a substantial weight loss over time. When trying to reduce weight, athletes should consume their total daily calories over at least three meals per day. Many athletes make the mistake of eating only one or two meals per day, skipping breakfast, lunch, or both, and then consuming a large dinner. Research in animals has shown that, given the same number of total calories, the animals that eat their daily food ration in one or two meals gain more weight than those that nibble their ration throughout the day. Human research is less clear.

The purpose of weight-loss programs is to lose body fat, not fat-free mass. Because of this, the combination of diet and exercise is the preferred approach. Combining increased activity with caloric reduction prevents any significant loss of fat-free mass. In fact, body composition can be significantly altered with physical training. Chronic exercise can increase fat-free mass and decrease fat mass. The magnitude of these changes varies with the type of exercise used for training. Resistance training promotes gains in fat-free mass, and both resistance training and endurance training promote loss of fat mass. To lose weight, athletes should combine a moderate resistance and endurance training program with modest caloric restriction.

> Athletes who are above their weight standard should lose weight gradually, not more than 0.5 to 1 kg (about 1-2 lb) per week to preserve their fat-free mass. This should be accomplished by integrating a good diet containing 200 to 500 kcal less than their daily energy expenditure with a reasonable increase in resistance and endurance activities.

As a final point, a balanced diet is, of course, essential to ensure that the athlete receives all necessary vitamins and minerals. Vitamin supplementation might or might not be necessary: Results of research thus far are in conflict. But if the nutritional adequacy of the diet is at all questionable, a simple multivitamin that meets the RDA for that person is suggested.

In Closing . . .

In this chapter, we discussed the importance of body composition, the way in which fat-free mass and relative body fat affect performance, and the best method for athletes to lose weight

- When severe (very low calorie) diets are followed, much of the weight loss that occurs is from water, not fat.

- Most severe diets limit carbohydrate intake, depleting carbohydrate stores. Water is lost along with the carbohydrates, exacerbating the problem of dehydration. Also, the increased reliance on free fatty acids can lead to ketosis, which further increases water loss.

- The combination of diet and exercise is the preferred approach to optimal weight loss.

- Athletes should lose no more than about 0.5 to 1.0 kg (1.1-2.2 lb) per week until reaching the upper end of the desired weight range. After that, weight loss should be less than 0.5 kg (1 lb) per week until goal weight is reached. More rapid weight losses result in loss of fat-free mass. Weight loss at the recommended rate can be accomplished by reducing dietary intake 200 to 500 kcal per day, especially when combined with a sound exercise program.

- For fat loss, moderate resistance and endurance training is most effective. Resistance training also promotes gains in fat-free mass.

(body fat) to optimize their body composition for their chosen activity. We now understand how optimizing body composition can enhance performance. In the next chapter, we investigate various substances and phenomena that have been purported to enhance performance.

▷ Key Terms

air plethysmography
bioelectric impedance
body build
body composition
body density
body size
chronic fatigue syndrome
densitometry
dual-energy X-ray absorptiometry (DXA)
fat-free mass
fat mass
female athlete triad
hydrostatic weighing
lean body mass
morphology
relative body fat
skinfold fat thickness

▷ Study Questions

1. Differentiate between body build, body size, and body composition.
2. What tissues of the body constitute the fat-free mass?
3. What is densitometry? How is it used to assess the body composition of the athlete? What is the major weakness of densitometry with respect to its accuracy?
4. What are several field techniques for estimating body composition? What are their strengths and weaknesses?
5. What is the relationship of relative leanness and fatness to performance in sport?
6. Which is more important to sports performance, body fat or body weight? Why?
7. What guidelines should be used to determine the athlete's goal weight?
8. What are potential problems associated with a fixation on body weight that is too low?
9. Why should the athlete avoid crash diets?
10. What is the lowest weight an athlete should be allowed to attain?
11. How much weight should an overweight athlete lose per week to maximize fat loss and minimize fat-free mass loss?

▷ References

1.   American College of Sports Medicine. (1996). ACSM position stand on weight loss in wrestlers. *Medicine and Science in Sports and Exercise,* **28**(6), ix-xii.

2.   Barron, J.L., Noakes, T.D., Levy, W., Smith, C., & Millar, R.P. (1985). Hypothalamic dysfunction in overtrained athletes. *Journal of Clinical Endocrinology and Metabolism,* **60**, 803-806.

3.   Brodie, D.A. (1988). Techniques for measurement of body composition (parts I and II). *Sports Medicine,* **5**, 11-40, 74-98.

4.   Butterfield, G. (1991). Amino acids and high protein diets. In D.R. Lamb & M.H. Williams (Eds.), *Ergogenics—Enhancement of performance in exercise and sport* (pp. 1-27). Dubuque, IA: Brown & Benchmark.

5.   Carter, J.E.L., Aubry, S.P., & Sleet, D.A. (1982). Somatotypes of Montreal Olympic athletes. In J.E.L. Carter (Ed.), *Physical structure of Olympic athletes* (pp. 53-80). New York: Karger.

6.   Cureton, K.J., & Sparling, P.B. (1980). Distance running performance and metabolic responses to running in men and women with excess weight experimentally equated. *Medicine and Science in Sports and Exercise,* **12**, 288-294.

7.   Dempster, P., & Aitkens, S. (1995). A new air displacement plethysmograph for measuring human body composition. *Medicine and Science in Sports and Exercise,* **27**, 1686-1691.

8.   Drinkwater, B.L., Bruemner, B., & Chesnut, C.H. (1990). Menstrual history as a determinant of current bone density in young athletes. *Journal of the American Medical Association,* **263**, 545-548.

9.   Evans, E.M., Prior, B.M., Arngrimsson, S.A., Modlesky, C.M., & Cureton, K.J. (2001). Relation of bone mineral density and content to mineral content and density of the fat-free mass. *Journal of Applied Physiology,* **91**, 2166-2172.

10.   Grande, F., & Keys, A. (1980). Body weight, body composition and calorie status. In R.S. Goodhart & M.E. Shils (Eds.), *Modern nutrition in health and disease* (6th ed., p. 16). Philadelphia: Lea & Febiger.

11.   Harber, V.J. (2000) Menstrual dysfunction in athletes: An energetic challenge. *Exercise and Sport Sciences Reviews,* **28**, 19-23.

12.   Houston, M.E. (1999). Gaining weight: The scientific basis of increasing skeletal muscle mass. *Canadian Journal of Applied Physiology,* **24**, 305-316.

13.   Kuipers, H., & Keizer, H.A. (1988). Overtraining in elite athletes: Review and directions for the future. *Sports Medicine,* **6**, 79-92.

14.   Lohman, T.G. (1986). Applicability of body composition techniques and constants for children and youths. *Exercise and Sport Sciences Reviews,* **14**, 325-357.

15.   Lohman, T.G. (1996). Dual energy X-ray absorptiometry. In A.F. Roche, S.B. Heymsfield, & T.G. Lohman (Eds.), *Human body composition* (pp. 63-78). Champaign, IL: Human Kinetics.

16.   Loucks, A.B., Verdun, M., & Heath, E.M. (1998). Low energy availability, not stress of exercise, alters LH pulsatility in exercising women. *Journal of Applied Physiology,* **84**, 37-46.

17.   Lukaski, H.C. (1987). Methods for the assessment of human body composition: Traditional and new. *American Journal of Clinical Nutrition,* **46**, 537-556.

18.   Martin, A.D., & Drinkwater, D.T. (1991). Variability in the measures of body fat: Assumptions or technique? *Sports Medicine,* **11**, 277-288.

19.   McCrory, M.A., Gomez, T.D., Bernauer, E.M., & Molé, P.A. (1995). Evaluation of a new air displacement plethysmograph for measuring human body composition. *Medicine and Science in Sports and Exercise,* **27**, 1686-1691.

20.   Millard-Stafford, M.L., Collins, M.A., Modlesky, C.M., Snow, T.K., & Rosskopf, L.B. (2001). Effect of race and resistance training status on the density of fat-free mass and percent fat estimates. *Journal of Applied Physiology,* **91**, 1259-1268.

21.   Modlesky, C.M., Cureton, K.J., Lewis, R.D., Prior, B.M., Sloniger, M.A., & Rowe, D.A. (1996). Density of the fat-free mass and estimates of body composition in male weight trainers. *Journal of Applied Physiology,* **80**, 2085-2096.

22.   Ortiz, O., Russell, M., Daley, T.L., Baumgartner, R.N., Waki, M., Lichtman, S., Wang, J., Pierson, Jr., R.N., & Heymsfield, S.B. (1992). Differences in skeletal muscle and bone mineral mass between black and white females and their relevance to estimates of body composition. *American Journal of Clinical Nutrition,* **55**, 8-13.

23.   Otis, C.L., Drinkwather, B., Johnson, M., Loucks, A., & Wilmore, J. (1997). The female athlete triad. *Medicine and Science in Sports and Exercise,* **29**(5), i-ix.

24.   Pate, R.R., Barnes, C., & Miller, W. (1985). A physiological comparison of performance-matched female and male distance runners. *Research Quarterly for Exercise and Sport,* **56**, 245-250.

25.   Pollock, M.L., & Jackson, A.S. (1984). Research progress in validation of clinical methods of assessing body composition. *Medicine and Science in Sports and Exercise,* **16**, 606-613.

26.   Riendeau, R.P., Welch, B.E., Crisp, C.E., Crowley, L.V., Griffin, P.E., & Brockett, J.E. (1958). Relationships

of body fat to motor fitness test scores. *Research Quarterly, 29,* 200-203.

27.   Schutte, J.E., Townsend, E.J., Hugg, J., Shoup, R.F., Malina, R.M., & Blomqvist, C.G. (1984). Density of lean body mass is greater in blacks than in whites. *Journal of Applied Physiology,* **56,** 1647-1649.

28.   Shangold, M., Rebar, R.W., Wentz, A.C., & Schiff, I. (1990). Evaluation and management of menstrual dysfunction in athletes. *Journal of the American Medical Association, 263,* 1665-1669.

29.   Sherman, W.M. (1991). Carbohydrate feedings before and after exercise. In D.R. Lamb & M.H. Williams (Eds.), *Ergogenics—Enhancement of performance in exercise and sport* (pp. 87-117). Dubuque, IA: Brown & Benchmark.

30.   Stager, J.M., & Cordain, L. (1984). Relationship of body composition to swimming performance in female swimmers. *Journal of Swimming Research,* **1,** 21-26.

31.   Steen, S.N., & Brownell, K.D. (1990). Patterns of weight loss and regain in wrestlers: Has the tradition changed? *Medicine and Science in Sports and Exercise,* **22,** 762-768.

32.   Visser, M., Gallagher, D., Deurenberg, P., Wang, J., Pierson, R.N., & Heymsfield, S.B. (2001). Density of fat-free body mass: Relationship with race, age, and level of body fatness. *American Journal of Physiology,* **272,** E781-E787.

33.   Welham, W.C., & Behnke, A.R. (1942). The specific gravity of healthy men. *Journal of the American Medical Association,* **118,** 498-501.

34.   Wilmore, J.H. (1992). Body weight and body composition. In K.D. Brownell, J. Rodin, & J.H. Wilmore (Eds.), *Eating, body weight, and performance in athletes: Disorders of modern society* (pp. 77-93). Philadelphia: Lea & Febiger.

35.   Wilmore, J.H., Brown, C.H., & Davis, J.A. (1977). Body physique and composition of the female distance runner. *Annals of the New York Academy of Sciences,* **301,** 764-776.

▶ Selected Readings

Behnke, A.R., & Wilmore, J.H. (1974). *Evaluation and regulation of body build and composition.* Englewood Cliffs, NJ: Prentice Hall.

Brownell, K.D., Rodin, J., & Wilmore, J.H. (Eds.). (1992). *Eating, body weight, and performance in athletes: Disorders of modern society.* Philadelphia: Lea & Febiger.

Brownell, K.D., Steen, S.N., & Wilmore, J.H. (1987). Weight regulation practices in athletes: Analysis of metabolic and health effects. *Medicine and Science in Sports and Exercise,* **19,** 546-556.

Fleck, S.J. (1983). Body composition of elite American athletes. *American Journal of Sports Medicine,* **11,** 398-403.

Fogelholm, M. (1994). Effects of bodyweight reduction on sports performance. *Sports Medicine,* **18,** 249-267.

Heyward, V.H. (1996). Evaluation of body composition: Current issues. *Sports Medicine,* **22,** 146-156.

Heyward, V.H., & Stolarczyk, L.M. (1996). *Applied body composition assessment.* Champaign, IL: Human Kinetics.

Horswill, C.A. (1992). When wrestlers slim to win: What's a safe minimum weight? *Physician and Sportsmedicine,* **20**(9), 91-101.

Kohrt, W.M. (1995). Body composition by DXA: Tried and true? *Medicine and Science in Sports and Exercise,* **27,** 1349-1353.

Lehman, M., Foster, C., & Keul, J. (1993). Overtraining in endurance athletes: A brief review. *Medicine and Science in Sports and Exercise,* **25**(7), 854-862.

Lohman, T.G. (1992). *Advances in body composition assessment.* Champaign, IL: Human Kinetics.

Roche, A.F., Heymsfield, S.B., & Lohman, T.G. (Eds.). (1996). *Human body composition.* Champaign, IL: Human Kinetics.

Rosen, L.W., McKeag, D.B., Hough, D.O., & Curley, V. (1986). Pathogenic weight control behavior in female athletes. *Physician and Sportsmedicine,* **14**(1), 79-86.

Sinning, W.E. (1985). Body composition and athletic performance. In D.H. Clarke & H.M. Eckert (Eds.), *Limits of human performance* (pp. 45-56). Champaign, IL: Human Kinetics.

Wilmore, J.H. (1983). Body composition in sport and exercise: Directions for future research. *Medicine and Science in Sports and Exercise,* **15,** 21-31.

Wilmore, J.H. (1992). Body weight standards and athletic performance. In K.D. Brownell, J. Rodin, & J.H. Wilmore (Eds.), *Eating, body weight, and performance in athletes: Disorders of modern society* (pp. 315-329). Philadelphia: Lea & Febiger.

Wilmore, J.H. (2000). Weight category sports. In R. Maughan (Ed.), *Nutrition in sport* (pp. 637-645). London: Blackwell Scientific.

CHAPTER 15

ERGOGENIC AIDS AND SPORT

The skill levels of athletes in various sports improve from year to year. Athletic records reach new heights, and the margin between success and failure in the world of sport becomes smaller. Consequently, both coaches and athletes look for that slight edge that might ensure victory. Many turn to potential ergogenic aids—substances or phenomena that enhance performance. Some of these potential aids can benefit performance, but others can have deadly consequences. In this chapter, we examine various pharmacological, hormonal, physiological, and nutritional agents that have been proposed to have ergogenic properties.

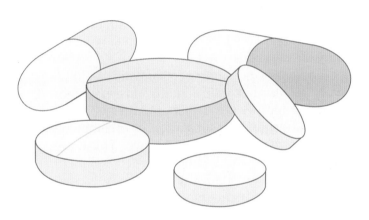

The Winter Olympics held in Salt Lake City, Utah, in 2002 provided hours of entertainment—seeing athletes performing at their very best. Some went home with medals, whereas others went home with the satisfaction of knowing that they had performed their best and had been good enough to qualify as Olympic athletes. A few won medals only to have their performances later questioned as a result of positive drug tests, suggesting that they had used agents banned by the International Olympic Committee (IOC). Johann Muehlegg, German-born cross-country skier competing for Spain, won three gold medals, but a random urine test found traces of darbepoietin. Although not on the IOC's list of banned substances, darbepoietin has properties similar to erythropoietin (EPO), which is on the banned substance list. Both drugs increase the number of red blood cells in the body, increasing the blood's oxygen-carrying capacity. This increase has been shown to increase both $\dot{V}O_2max$ and endurance performance, as we see later in this chapter.

In the never-ending quest for glory, athletes often are willing to try anything to improve their performance. For some, a special diet can be the deciding factor. Others might rely on stress reduction or hypnosis to alter their psychological states. Still others might try certain drugs or hormones.

Substances or phenomena that improve an athlete's performance are referred to as **ergogenic aids**. The variety of potential ergogenic aids is immense. Here are just a few examples:

- Athletes in a number of sports have taken anabolic steroids, hoping to increase muscle mass and strength and thus improve strength-related performance.

- Endurance athletes have loaded up on carbohydrates a few days before competition to pack extra glycogen into their leg muscles.

- Athletes have tried hypnosis to help them through emotional or psychological problems.

- Even the cheers of the home crowd can place the home team at a distinct advantage.

The effects of proposed **ergogenic** substances often are shrouded in myth. Most athletes have received tips about ergogenic aids from a friend or coach and assume that the information is accurate, but this is not always the case. Some athletes experiment with substances, hoping for even a slight performance improvement

regardless of possible harmful consequences. In the quest for performance perfection, a concern only with maximizing performance coupled with a lack of knowledge about ergogenic substances can lead an athlete to make unwise decisions.

The list of possible ergogenic aids is long, but the number that actually possess ergogenic properties is much shorter. In fact, some allegedly ergogenic substances or phenomena actually can impair performance. These are usually drugs, and Eichner[25] has termed them **ergolytic** drugs. Ironically and tragically, several ergolytic agents have been promoted as ergogenic aids.

An ergogenic aid is any substance or phenomenon that enhances performance. An ergolytic agent is one that has a detrimental effect on performance. Some substances generally thought to be ergogenic are actually ergolytic.

Table 15.1 provides a selected listing of substances and agents proposed to have ergogenic properties; the table also lists proposed mechanisms of action by which ergogenic aids might work. These have been studied in sufficient depth to establish their efficacy.[29] Many other substances have been proposed but not adequately researched.

This chapter focuses on pharmacological agents, hormones, physiological agents, and nutritional agents. More general nutritional practices are addressed in chapter 13. Psychological

Table 15.1

Proposed Ergogenic Aids and Mechanisms Through Which They Might Work

Agent	Act on heart, blood, circulation, and aerobic endurance	Increase oxygen delivery	Supply fuel for muscle and general muscle function	Act on muscle mass and strength	Result in weight loss or weight gain	Counteract or delay onset or sensation of fatigue	Counteract central nervous system inhibition	Aid in relaxation and stress reduction
Pharmacological								
Alcohol	✓		✓					✓
Amphetamines	✓					✓	✓	
β-blockers	✓							✓
Caffeine	✓		✓			✓		
Cocaine and marijuana	✓							✓
Diuretics	✓				✓			
Nicotine	✓							✓
Hormones								
Anabolic steroids				✓	✓			
Human growth hormone				✓	✓			
Oral contraceptives								✓
Physiological								
Bicarbonate loading						✓		
Blood doping	✓	✓				✓		
Erythropoietin	✓	✓				✓		
Oxygen	✓	✓				✓		
Phosphate loading	✓	✓				✓		
Nutritional								
Amino acids	✓		✓	✓	✓	✓		
Creatine			✓	✓	✓	✓		
L-Carnitine			✓			✓		

Adapted from E.L. Fox, R.W. Bowers, and M.L. Foss 1988[29], *The physiological bases of physical education and athletics* (Philadelphia: Saunders College Publishing), 632. Copyright 1988 The McGraw Hill Companies. Adapted by permission of The McGraw Hill Companies.

phenomena and mechanical factors are beyond the scope of this book but are reviewed in depth in Williams' book *Ergogenic Aids in Sport*.[60]

Researching Ergogenic Aids

Assume a professional athlete consumes a particular substance several hours before game time and then has a successful performance. The athlete likely will attribute the success to this substance, even though there is no proof that ingesting the substance will ensure other athletes similar success.

Anyone can claim that a certain substance is ergogenic—and many substances have been so labeled strictly because of speculation—but before a substance can be legitimately classified as ergogenic, it must be proven to improve performance. Unfortunately, science, with its carefully controlled investigations, does not have all the answers. Still, scientific studies in this area are essential to differentiate between a true ergogenic response and a pseudoergogenic response, in which performance improves simply because the athlete expects improvement.

Placebo Effect

As we discussed in the introductory chapter, the phenomenon by which your expectations of a substance determine your body's response to it is known as the placebo effect. This effect can seriously complicate the study of ergogenic qualities because researchers must be able to distinguish between the placebo effect and true responses to the substance being tested.

The placebo effect was clearly demonstrated in one of the earliest studies of anabolic steroids.[5] Fifteen male athletes who had been involved in heavy weightlifting for the previous 2 years volunteered for a weight-training experiment using anabolic steroids. They were told that those who made the greatest strength gains over a preliminary 4-month weight-training period would be selected for the second phase of the study, in which they would receive anabolic steroids.

Following the initial period, 8 of these 15 subjects were randomly selected to enter the treatment phase. Only six of these subjects passed all medical screening tests and were allowed to continue to the treatment phase. This phase consisted of a 4-week period in which the subjects were told that they would receive 10 mg per day of Dianabol (an anabolic steroid), when in fact they received a placebo—an inactive substance typically provided in a form identical to the genuine drug.

Strength data were collected over the last 7 weeks of the pretreatment training period and over all 4 weeks of the treatment (placebo) period (see figure 15.1). Even though the subjects were experienced weightlifters, they continued to gain impressive amounts of strength during the pretreatment training period. However, strength gains while subjects were taking the placebo were substantially greater than during the pretreatment period! The group improved an average of 11 kg during the 7-week pretreatment period but improved 45 kg during the 4-week treatment (placebo) period. This represents an average gain in strength of 1.6 kg per week during the pretreatment training period and 11.3 kg per week during the placebo period, a more than seven times greater increase in the rate of strength gain during the placebo (supposed steroid) period over the pretreatment training period. Furthermore, placebos are inexpensive, risk free, and legal for use in sport.

One of the authors of this textbook (JHW) repeatedly witnessed the placebo effect while conducting a large series of studies investigating the effects of β-blocking drugs on the ability to perform single bouts of exercise or to train aerobically. The Human Subjects Committee, a committee mandated by the federal government to oversee all research conducted with human subjects in the United States, requires that all human subjects receive a full disclosure of the risks associated with any experimental intervention so that they can provide informed consent before participating. Before the start of each study, a cardiologist presented a comprehensive background on β-blocking drugs to each subject, including the drugs' significance in treating various cardiovascular diseases and

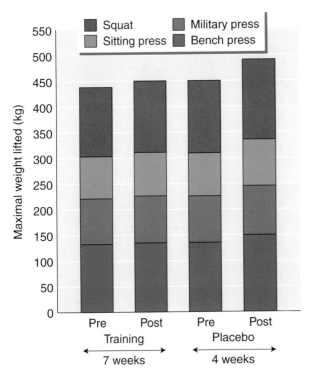

▲ **Figure 15.1**   The placebo effect on muscular strength gains. The increase in total strength and strength in each of four maximum lifts over the last 7 weeks of an intense 4-month pretreatment training period is compared with strength increases during a subsequent 4-week treatment period in which the subjects took placebos that they thought were anabolic steroids and continued intense resistance training. Data from Ariel and Saville.[5]

potential side effects associated with their use. It was amazing to note that over the course of 6 years of study, the most serious side effects almost always appeared in the subjects taking the placebo.

When evaluating a substance for possible ergogenic qualities, researchers must remember that witnessing an ergogenic effect does not necessarily prove that a substance is truly ergogenic. All studies of potential ergogenic substances must include a placebo group so that researchers can compare actual responses resulting from the test substance with those resulting from a placebo.

Although the placebo effect has a psychological origin, the body's physical response to a placebo is not merely imagined: It is quite real. This clearly illustrates how effective our mental state can be in altering our physical state.

Limitations of Research

To satisfy the scientific community, scientists often rely on laboratory techniques to evaluate the efficacy of any potential ergogenic aid. Often, however, scientific studies cannot provide absolutely clear answers to the questions under study. With elite athletes, success is defined in fractions of a second or in millimeters. Laboratory tests are often unable to detect such subtle differences in performance.

Scientists can be greatly limited by the accuracy of their equipment or techniques. All research methods have some margin of error. If the results fall within that margin of error, the researcher cannot be certain that the result is an effect of the substance being tested. The results might reflect limitations of the research methodology. Unfortunately, because of measurement error, individual differences, and the day-to-day variability of subjects' responses, a potential ergogenic aid must exert a major effect before scientific tests can prove that it is ergogenic.

The testing situation can also limit accuracy. Performance in a laboratory is considerably different from performance in the usual athletic environment, so laboratory results won't always accurately reflect natural athletic results. Yet an advantage of laboratory testing is that the environment can be carefully controlled, which isn't always possible in field studies conducted in the athlete's usual environment, where several uncontrollable variables—such as temperature, humidity, wind, and distractions—can affect the results. Thorough testing of a potential ergogenic aid should include both field and laboratory studies.

Realizing that science is limited in its ability to determine the efficacy of a substance, we can now examine some proposed ergogenic aids. We consider substances in four classes:

- Pharmacological agents
- Hormonal agents
- Physiological agents
- Nutritional agents

Pharmacological Agents

Numerous **pharmacological agents**, or drugs, have been suggested as having ergogenic properties. The IOC, the United States Olympic Committee (USOC), the International Amateur Athletic Federation (IAAF), and the National Collegiate Athletic Association (NCAA) all publish extensive lists of banned substances, most of which are pharmacological agents. Each athlete, coach, athletic trainer, and team physician must know which drugs are prescribed for and taken by the athlete, and they must check these drugs periodically against the listing of banned substances because the list changes frequently. (The USOC has a drug education hotline that provides up-to-date information: 1-800-233-0393.) Athletes have been disqualified and have had to relinquish medals, ribbons, awards, and prizes after testing positive for a banned substance. In many cases, the medication had been used legitimately to treat a known medical condition.

We review here only drugs for which a research base has been established. Many other drugs have been touted as being ergogenic, but controlled studies must yet be conducted to determine if they are effective. The drugs we discuss are

- amphetamines,
- β-blockers,
- caffeine,
- diuretics, and
- recreationally used drugs.

Amphetamines

Amphetamine and its derivatives are central nervous system (CNS) stimulants. They are also considered sympathomimetic amines, which means that their activity mimics that of the sympathetic nervous system. For many years they have been used as appetite suppressants in medically supervised weight-loss programs. During World War II, army troops used amphetamines to combat fatigue and to improve endurance. The drugs soon found their way into the athletic arena, where they were considered stimulants with possible ergogenic properties.

Proposed Ergogenic Benefits

Athletes have found amphetamines readily available even though they are prescription

Medical Needs Versus Banned Substance

During the 1972 Summer Olympic Games in Munich, a 16-year-old swimmer, Rick DeMont, won the 400-m freestyle event, but his gold medal was taken away when officials discovered traces of ephedrine, a banned substance, in his blood. DeMont, an asthmatic, had taken drugs for allergies and also was receiving weekly shots for his allergies. Ironically, DeMont had declared these drugs on his medical statement, but the information was not provided to the appropriate authorities at the IOC. The USOC, in December 2001, admitted that it had not acted properly in its handling of DeMont's medical information. His name has now been cleared with the USOC and now the IOC must decide if he should get his medal back. As a final note, it is still not clear if the more popular allergy and asthma medications used today have ergogenic properties when administered to nonasthmatic individuals.

This real-life event illustrates the importance for all those who work with athletes—their physicians, trainers, coaches—and athletes themselves being aware of banned substance rules and regulations. Mistakes can be made by anyone, but the athlete pays the price!

drugs. Amphetamines are used by athletes for many reasons. Psychologically, the drugs are thought to increase concentration and mental alertness. Their stimulating effect decreases mental fatigue. Athletes anticipate more energy and motivation and often feel more competitive when using amphetamines. The drugs also produce a state of euphoria, which is part of their attraction as so-called recreational drugs. Often athletes who use amphetamines report a heightened sense of ability, which they feel spurs them to higher performance levels.

In terms of actual performance, amphetamines are thought to help athletes run faster, throw farther, jump higher, and delay the onset of total fatigue or exhaustion. Athletes who use these drugs expect virtually every aspect of performance to be enhanced.

Proven Effects

Generally, for any physiological, psychological, or performance variable that has been investigated, some studies show that amphetamines have no effect, others demonstrate an ergogenic effect, and still others indicate an ergolytic effect. As potent central nervous system stimulants, amphetamines do increase your state of arousal, which leads to a sense of increased energy, self-confidence, and faster decision making. People who take amphetamines experience

- decreased sense of fatigue,
- increased systolic and diastolic blood pressure,
- increased heart rate,
- redistribution of blood flow to skeletal muscles,
- elevation of blood glucose and free fatty acids, and
- increased muscle tension.[37]

Do these effects aid physical performance? Although studies are not in total agreement, the more recent studies, which have used better experimental designs and controls, show that amphetamines can enhance skills that are important in athletic performance, specifically

- speed;
- strength, power, and muscular endurance;
- concentration; and
- fine motor coordination.[18, 37, 59]

The results of one of the better controlled studies are presented in table 15.2. on page 478.[16] Subjects in this study were tested three times under each of two conditions: a placebo and 15 mg of the amphetamine Dexedrine per 70 kg of body weight, both administered 2 h before testing. Following amphetamine administration, significant increases were found in

- knee extension strength,
- acceleration during a 30-yd (27-m) sprint,
- time to exhaustion during maximal treadmill testing,
- peak lactate response following the maximal treadmill test, and
- maximum heart rate.

Even though the time to exhaustion on the treadmill was increased, there were no differences in aerobic power.[16] The increased time to exhaustion likely reflected the subjects' psychological response to the drug, making them feel more energetic, delaying mental fatigue, and thus allowing them to push harder and work longer. This conclusion is supported by the increase in peak lactate.

As mentioned previously, though, such laboratory tests might not accurately duplicate conditions encountered in field situations. Also, athletes might be consuming far greater doses of amphetamines than allowed in controlled research studies. Future studies must take these factors into account.

Amphetamines can improve performance in certain sports or activities, but, in addition to being illegal, these drugs carry risks that far outweigh their benefits. Amphetamines can be addictive, and they mask important signals that the body sends out to inform us when we are in potentially dangerous situations.

Table 15.2

Physical and Physiological Performance Changes With the Use of Amphetamines

Variable	Placebo	Drug	Mean difference	% Difference
Elbow flexion strength (N)	681	724	43	6.3
Knee extension strength (N)	1,264	1,550	286	22.6[a]
Leg power (W)	623	642	19	3.0
Peak speed (s/9.1 m)	1.11	1.11	0	0.0
Acceleration for 30-yd run (m/s^2)	2.89	3.00	0.11	3.8[a]
Aerobic power $\dot{V}O_2$max (L/min)	3.96	3.97	0.01	0.3
Treadmill time to exhaustion (s)	427	446	19	4.4[a]
Peak lactate (mmol/L)	13.3	14.4	1.1	8.3[a]
Maximum heart rate (beats/min)	191	195	4	2.1[a]

Note. Six subjects were tested on six consecutive Fridays, three times under placebo and three times under amphetamine conditions. Values represent the average of the three trials under each condition.

[a] A statistically significant difference indicating a better performance while on amphetamines.

Adapted, by permission, from J.V. Chandler and S.N. Blair, 1980, "The effect of amphetamines on selected physiological components related to athletic success," *Medicine and Science in Sports and Exercise* 12: 65-69.

Risks of Amphetamine Use

Deaths have been attributed to excessive amphetamine use. Because heart rate and blood pressure are increased, amphetamine users place greater stress on their cardiovascular systems. These drugs can trigger cardiac arrhythmias in some susceptible individuals. Also, rather than delay the onset of fatigue, amphetamines likely delay the sensation of fatigue, enabling the athletes to push dangerously beyond normal limits to the point of circulatory failure. Deaths have occurred when athletes have pushed themselves far beyond the normal point of exhaustion.

Amphetamines can be psychologically addictive because of the euphoria and energized feelings they cause. But the drugs also can be physically addictive if taken regularly, and a person's tolerance to them builds with continued use, requiring increasingly larger doses over time to obtain the same effects. Amphetamines also can be toxic. Extreme nervousness, acute anxiety, aggressive behavior, and insomnia are frequently mentioned side effects of regular use.

β-Blockers

The sympathetic nervous system influences bodily functions through adrenergic nerves: those that use norepinephrine as their neurotransmitter. Neural impulses traveling through these nerves trigger the release of norepinephrine, which crosses the synapses and binds to adrenergic receptors at the target cells. These adrenergic receptors are classified into two groups: α-adrenergic receptors and β-adrenergic receptors.

β-adrenergic blockers, or **β-blockers**, are a class of drugs that block the β-adrenergic receptors, preventing binding of the neurotransmitter norepinephrine. This greatly reduces the effects of stimulation by the sympathetic nervous system. β-blockers generally are prescribed for the treatment of hypertension, angina pectoris, and certain cardiac arrhythmias. They also are prescribed as a preventive treatment for migraine headaches, to reduce the symptoms of anxiety and stage fright, and for initial recovery from heart attacks.

Proposed Ergogenic Benefits

Because the sympathetic response gears the body up for physical activity (through the fight-or-flight mechanism), it is difficult to understand why athletes might turn to β-blockers as ergogenic aids. β-blocker use in sport has been limited mostly to sports where anxiety and tremor could impair performance. When a person stands on a force platform (a highly sophisticated device that measures mechanical forces), measurable body movement is detected each time the heart beats. This movement is sufficient to affect a shooter's aim. Accuracy in shooting sports improves if the rifle or pistol can be shot or the arrow released between heartbeats. β-blockers can slow a shooter's heart rate, allowing more time to stabilize the aim before shooting or releasing before the next heartbeat.

β-blockers also have been postulated to enhance physiological adaptations to endurance training.[61] Research has shown that chronic use of β-blocking drugs increases the body's number of β-receptors. It is theorized that endurance training while taking β-blocking drugs would increase an athlete's number of β-receptors, allowing a greater sympathetic response after the drugs are discontinued.

Proven Effects

β-blockers decrease the effects of sympathetic nervous system activity. This is well illustrated by the marked reduction in maximum heart rate with β-blocker administration. It is not unusual for a 20-year-old male athlete with a normal maximum heart rate of 190 beats/min to have a maximum heart rate of only 130 beats/min when taking β-blocking drugs. Resting and submaximal heart rates also are reduced by these drugs. Several studies have confirmed improved scores in shooting sports as a result of this heart rate reduction when subjects used β-blockers. Because of this, the

IOC, the USOC, and the NCAA have banned the use of β-blockers for these sports.

The body contains two types of β-adrenergic receptors: β-1 and β-2. Nonselective β-blockers affect both types of receptors, but β-1 selective blockers primarily affect only the β-1 receptors. β-1 receptors are located mainly in the heart, so a β-1 selective blocker decreases the heart's rate and contractility. β-2 receptors are located in blood vessels, the lungs, the liver, skeletal muscle, and the intestines. Because nonselective β-blockers block both receptor types, they have greater overall effects than selective ones: They can affect blood flow, airflow, and metabolism. If an athlete must take a β-blocking drug for a medical condition, β-1 selective blockers usually are preferred because they have fewer negative effects on performance.

Laboratory studies have shown that β-blocking drugs reduce

- maximal oxygen uptake, particularly in highly trained individuals;
- maximal ventilatory capacity, because airflow through the airways is reduced;
- submaximal and maximal heart rate;
- maximal cardiac output, because the stroke volume cannot increase enough to compensate for the reduced heart rate; and
- blood pressure, because cardiac output is reduced.[64]

The negative results of these laboratory studies have been confirmed by controlled studies during actual competition. In a study of long-distance runners, 10-km race times were greatly affected by β-blocking drugs.[3] Under both control and placebo conditions, the runners averaged 35.8 min to complete the 10-km race. When using a nonselective β-blocker, the runners averaged 41.0 min (14.5% longer), but when using a selective β-1 blocker, the runners averaged 39.2 min (9.5% longer).

Finally, β-blocking drugs appear to have little influence on strength, power, and local muscular endurance (in activities that elicit fatigue in less than 2 min).[61] Thus, depending on the type of performance desired, β-blockers can be ergogenic (accuracy in shooting sports), ergolytic (decreased aerobic capacity), or without effect (strength, power, and local muscular endurance).

β-blockers can enhance performance in shooting types of sports and therefore have been banned. However, they are ergolytic with respect to endurance performance.

Risks of β-Blocker Use

Most risks from β-blockers are associated with prolonged use, not isolated incidents of use as in athletics. β-blockers can induce bronchospasm in people with asthma. They can cause cardiac failure in people who have underlying problems with cardiac function. In people with bradycardia, these drugs can lead to heart block. The decreased blood pressure they cause can result in lightheadedness. Some people with type 2 diabetes can become hypoglycemic because β-blockers increase insulin secretion. These drugs, through their various effects, can cause pronounced fatigue, which can inhibit athletic performance and decrease motivation.

Caffeine

Caffeine, one of the most widely consumed drugs in the world, is found in coffee, tea, cocoa, soft drinks, and various other foods.[18] This drug is also common in several over-the-counter medications, often even in simple aspirin compounds. Caffeine is a central nervous system stimulant, and its effects are similar to those noted previously for amphetamines, although weaker. The caffeine contents of some common products are listed in table 15.3.

Proposed Ergogenic Benefits

As with amphetamines, caffeine generally is touted as improving alertness, concentration, reaction time, and energy levels. People taking the drug often feel stronger and more competitive. They believe that they can perform longer before the onset of fatigue and that if they are fatigued beforehand, the fatigue is reduced.

Proven Effects

Because of its effects on the central nervous system, caffeine

- increases mental alertness,
- increases concentration,
- elevates mood,
- decreases fatigue and delays its onset,
- decreases reaction time (i.e., faster response),
- enhances catecholamine release,
- increases free fatty acid mobilization, and
- increases the use of muscle triglycerides.

In terms of ergogenic properties, caffeine initially was studied for potential effects that could benefit endurance activities. The first studies, conducted by Costill, Ivy, and their colleagues,[19, 38] demonstrated marked improvements in endurance performance when competitive cyclists ingested a caffeinated beverage compared with a placebo beverage. Caffeine increased endurance times in fixed-pace work bouts and decreased times in fixed-distance races.

Although several studies were unable to replicate these results, the most recent studies have demonstrated substantial ergogenic effects of caffeine ingestion in recreational cyclists and highly trained distance runners.[34, 54] It is now generally concluded that caffeine does improve endurance performance. It was initially postulated that this improvement was the result of an increased mobilization of free fatty acids, sparing muscle glycogen for later use. But the actual mechanisms by which caffeine improves endurance performance appear to be more complex.[33] It is now well documented that caffeine lowers your perception of effort at a given rate of work, potentially allowing you to perform at a higher rate of work for the same perceived effort. Cellular mechanisms within skeletal muscle also are being explored.[22]

Caffeine can enhance performance in endurance sports at doses well under the limit of sport governing bodies. However, an athlete may experience a negative response, in which case caffeine can be considered ergolytic.

Caffeine also might improve performance in sprint and strength types of activities.[4, 17] Unfortunately, fewer studies have investigated

Table 15.3

Caffeine Contents of Some Common Products[a]

Substance	Caffeine (mg)
28 g (1 oz) baking chocolate	45
57 g (2 oz) chocolate candy	45
237 ml (8 oz) chocolate milk	48
355 ml (12 oz) Mello Yellow	51
355 ml (12 oz) Mountain Dew	54
355 ml (12 oz) cola beverages	32-65
177 ml (6 oz) instant coffee	54-75
177 ml (6oz) iced tea	70-75
177 ml (6 oz) hot tea (strong)	65-107
Standard dose of some aspirin products (see labels)	30-128
177 ml (6 oz) automatic perk coffee	125
177 ml (6 oz) automatic drip coffee	181
Standard dose No Doz, Vivarin	100-200
Standard dose Dexatrim, Dietac	200
Standard dose Prolamine	280

[a]Recommended caffeine intake is less than 250 mg per day

Adapted from E. Tribole, 1992, *Eating on the run,* 2nd ed. (Champaign, IL: Leisure Press), 202. By permission of author.

this area, but caffeine might facilitate calcium exchange at the sarcoplasmic reticulum and increase the activity of the sodium–potassium pump, better maintaining the muscle membrane potential.

Risks of Caffeine Use

In people who are not accustomed to using caffeine, who are sensitive to it, or who consume high doses, caffeine can produce nervousness, restlessness, insomnia, and tremors. Caffeine also acts as a strong diuretic, increasing an athlete's risk for dehydration and heat-related illness when performing in hot environments. It can disrupt normal sleep patterns, contributing to fatigue. Caffeine is also physically addictive; abrupt discontinuation of caffeine intake can

result in severe headache, fatigue, irritability, and gastrointestinal distress.

Diuretics

Diuretics affect the kidneys, increasing urine production. Used appropriately, they rid the body of excess fluid and frequently are prescribed to control hypertension and reduce edema (water retention) associated with congestive heart failure or other conditions.

Proposed Ergogenic Benefits

Diuretics generally are used as ergogenic aids for weight control. For decades, diuretics have been used by some jockeys, wrestlers, and gymnasts to keep their weight down. More recently, they have been used by anorexics and bulimics for weight loss.[65]

Some athletes who are taking banned drugs also have turned to diuretics but not to enhance their performance. Because diuretics increase fluid loss, these athletes hope that the extra fluid in their urine will dilute the concentration of banned drugs, thus decreasing the likelihood that the banned substances will be detected during drug testing. This practice and other means of altering the urine in an effort to escape drug detection are called masking.[59]

Proven Effects

Diuretics lead to significant weight loss, but no evidence suggests any other potentially ergogenic effects. In fact, several side effects make diuretics ergolytic. The fluid loss results primarily from losses in extracellular fluid, including plasma. For athletes, particularly those who depend on moderate to high levels of aerobic endurance, this reduction in plasma volume reduces maximal cardiac output, which in turn reduces aerobic capacity and impairs performance.

Risks of Diuretic Use

In addition to reducing plasma volume, diuretics also hinder thermoregulation. As internal body heat increases, more blood must be diverted to the skin so that the heat can be lost to the environment. However, when blood plasma volume is diminished, as with diuretic use, more blood must be kept in the central regions to maintain central venous blood pressure and adequate blood supply and blood pressure to the vital organs. Thus, less blood is available to be shunted to the skin, and heat loss is impaired.

Electrolyte imbalance also can occur. Many diuretics cause fluid loss by ensuring electrolyte loss. A diuretic called furosemide inhibits sodium reabsorption in the kidneys, thus allowing more of it to be excreted in the urine. Because fluid follows the sodium, more fluid also will be excreted. Electrolyte imbalances can occur with losses of either sodium or potassium. These imbalances can cause fatigue and muscle cramping. More serious imbalances

can lead to exhaustion, cardiac arrhythmias, and even cardiac arrest. Some athletes' deaths have been attributed to electrolyte imbalances caused by diuretic use.

▶ Amphetamines are CNS stimulants that increase mental alertness, elevate mood, decrease the sense of fatigue, and produce euphoria.

▶ Studies indicate that amphetamines can increase strength, acceleration, maximum heart rate, lactate responses during exhaustive exercise, and time to exhaustion.

▶ Amphetamines elevate both heart rate and blood pressure and can trigger cardiac arrhythmias. Excessive use of these drugs has been blamed for some deaths, and the drugs can be both psychologically and physically addictive.

▶ β-blockers block β-adrenergic receptors, preventing neurotransmitter binding.

▶ β-blockers slow the resting heart rate, which is a distinct advantage for shooters who try to release the arrow or squeeze the trigger between heartbeats to minimize the slight tremor associated with each beat. But these drugs impair endurance performance, reducing $\dot{V}O_2$max in highly trained people because cardiac output is reduced (stroke volume cannot fully compensate for the reduced heart rate).

▶ β-blockers cause bradycardia and can even cause heart block, hypotension, bronchospasm, pronounced fatigue, and decreased motivation. Selective β-blockers have fewer side effects than nonselective blockers and usually would be prescribed for an athlete with a medical need for β-blocking drugs.

▶ Caffeine, one of the most widely consumed drugs in the world, is also a CNS stimulant, and its effects are similar to those of amphetamine but weaker.

▶ Caffeine increases mental alertness and concentration, elevates mood, decreases fatigue and delays its onset, increases catecholamine release and mobilization of free fatty acids, and is proposed to increase muscle use of free fatty acids to spare glycogen.

▶ Caffeine can cause nervousness, restlessness, insomnia, tremors, and diuresis. Diuresis increases susceptibility to heat injury.

▶ Diuretics affect the kidneys, increasing urine production and excretion. They often are used by athletes for weight reduction or maintenance and also by those trying to mask use of other drugs during drug testing.

▶ Weight loss is the only proven ergogenic effect of diuretics, but this weight loss is primarily from the extracellular fluid compartment, including blood plasma. This leads to dehydration, which can impair thermoregulation and cause electrolyte imbalances.

Recreationally Used Drugs

A class of drugs referred to as "recreational drugs" have been widely used by athletes for both recreation as well as for their potential ergogenic properties. None of these have been shown to have ergogenic properties and most are ergolytic in nature. We briefly discuss alcohol, cocaine, marijuana, and nicotine.

Alcohol

Alcohol consumption is the number one drug problem in the United States today. Furthermore, its use among adolescents is high, and young athletes may be more likely to abuse alcohol than nonathletes. Alcohol can be classified as a food or nutrient because it provides energy (7 kcal/g), but it also can be considered an antinutrient because it can interfere with the metabolism of other nutrients. Alcohol is also correctly classified as a drug because of its depressant effects on the CNS. However,

psychologically, alcohol ingestion appears to result in a two-part response: an initial sensation of excitement, followed by depressive effects.[61]

Proposed Ergogenic Benefits Some athletes use alcohol primarily for its psychological effects. It is thought to improve self-confidence and calm the nerves. Some athletes believe alcohol reduces inhibitions and makes them more alert. Physiologically, some people view alcohol as a good carbohydrate source. It also has been touted as a means for reducing pain and muscle tremor. Its reputation for reducing tremor and anxiety made alcohol a potential ergogenic aid for shooting sports, but these sports have banned it.

Proven Effects Unfortunately, little is known about the influence of varying amounts of alcohol on athletic performance. Although alcohol intoxication causes erratic and unpredictable performance, the influence of small amounts of alcohol just before or during a contest is not well understood.

In the absence of field studies of alcohol use during competition, laboratory studies have been conducted to observe the effects of small and moderate doses of alcohol on psychomotor skills. Studies suggest that most psychomotor functions associated with sport performance are impaired by alcohol, not improved.[61] Although athletes can feel more alert and self-confident, their reaction time, coordination, movement, and thinking are all impaired. Small amounts of alcohol impair psychomotor skills, yet athletes are unaware of these changes and often believe that their performance has improved. Well-controlled research studies also consistently support the conclusion that alcohol ingestion has no ergogenic effects on physiological function and performance.

Risks of Alcohol Use More important than its lack of ergogenic qualities, alcohol has many ergolytic features. Alcohol is a poor source of carbohydrates and, as mentioned earlier, is an antinutrient. Its depressant effects on the CNS dull pain sensation, but pain indicates injury, and physical activity while injured always carries a great risk of increasing the extent of injury. In fact, injury incidence is higher in athletes who drink compared with those who don't drink.[47] Although muscle tremor and anxiety might be reduced, the accompanying impairment of psychomotor skills offsets any advantage that the athlete might gain.

Alcohol suppresses the release of antidiuretic hormone (ADH), causing your body to excrete more water in the urine. This can, in turn, transiently decrease your blood pressure and cause dehydration. This can have serious consequences during athletic performance, especially in hot environments. Alcohol also causes peripheral vasodilation (dilation of blood vessels in the skin). Loss of body heat through the blood vessels in your skin can trigger hypothermia in cold environments if more heat is lost from your body than is desirable.

Cocaine

Little is known about the influence of **cocaine** on athletic performance. Cocaine acts as a CNS stimulant. Cocaine also can be characterized as a sympathomimetic drug, and its actions are very similar to those of amphetamines.

Cocaine blocks the reuptake of norepinephrine and dopamine (two major neurotransmitters) by the neurons after release. Recall that norepinephrine is released by sympathetic nerves, including those supplying the heart. Both norepinephrine and dopamine are used in the brain. By blocking their reuptake, cocaine potentiates these neurotransmitters' effects throughout the body.

Proposed Ergogenic Benefits   Although cocaine use has become far too common in athletics, most use is recreational. But some athletes believe that cocaine is an ergogenic aid. The drug creates an intense euphoria that is thought to increase self-confidence and motivation. Like amphetamines, cocaine masks both fatigue and pain, increases alertness, and makes the athlete feel energetic.

Proven Effects   Because of the dangers involved with cocaine use, few studies have been conducted on its ergogenic properties. If we consider only the well-controlled studies, no evidence indicates that cocaine has any ergogenic properties, regardless of its similarities to amphetamine.

Risks of Cocaine Use   Athletes must recognize that even if athletic performance could benefit from cocaine, the risks associated with its use far outweigh any benefits. Several comprehensive reviews of the research literature conclude that tremendous health risks and no known performance benefits are associated with cocaine use.[15, 18, 43, 59] Deaths of some prominent sports figures have been directly attributed to cocaine use. By drastically increasing the stimulation of the heart, cocaine puts a tremendous strain on even a healthy heart. Because these effects occur so rapidly, the heart is suddenly faced with tremendous stress and can go into cardiac arrest. With the added stress of physical performance, the risk of death is greatly increased.

Marijuana

Marijuana is another so-called recreational drug. Like alcohol, it can elicit both stimulant and depressant effects.[61] It acts primarily on the central nervous system, but its mode of action is poorly understood.

Proposed Ergogenic Benefits   Marijuana has not been proposed as ergogenic. In fact, it generally is considered ergolytic.[25] However, use of this drug is quite common, especially among younger athletes, so it must be considered in terms of its effects on performance. Many who use marijuana seek the sense of euphoria and relaxation it produces. Like alcohol, it is often a means of escape or a way to reduce tension. Many youths deem its use appropriate merely because their peers are smoking it.

Proven Effects   Marijuana impairs the performance of motor skills. Of major concern in athletes is the "amotivational syndrome" seen frequently in marijuana users. This syndrome is characterized by apathy, impaired judgment, loss of ambition, and an inability to carry out long-term plans.

Risks of Marijuana Use   Health risks of marijuana use are still under investigation. Personality changes are noted with just a few marijuana cigarettes. Short-term memory is impaired, which has led to concerns that marijuana might cause permanent brain damage. High intake of the drug can cause hallucinations and psychotic-like behavior. Also, studies have reported decreased circulating testosterone levels in users.[61] For more detail, see the excellent reviews by Wadler and Hainline[59] and Williams.[61]

Another concern with marijuana is how the drug is ingested. It is usually smoked, although it can be eaten. The serious health consequences associated with cigarette smoking are common knowledge. Are these same health problems associated with marijuana smoking? Do other health problems result from the combustion of different chemicals in marijuana? These questions are still unanswered.

Nicotine

Athletes have used **nicotine** as a stimulant. The most familiar nicotine form is cigarettes, and fortunately fewer people now smoke. But among athletes, the smokeless forms—chewing tobacco (chew), snuff (dip), and compressed tobacco (plug)—are still popular, and their use appears to be increasing.

As with other recreational drugs, some athletes turn to nicotine for possible ergogenic effects. Others are addicted and use nicotine daily, so its effects carry over into their athletic performance.

Proposed Ergogenic Benefits   Nicotine is a stimulant. Some athletes believe that it makes them more alert and better able to concentrate. Yet, paradoxically, the drug also is reported to have a calming effect, opposite that of a stimulant. For this reason, many athletes also use it to soothe jittery nerves.

Proven Effects   Nicotine generally has been found to be detrimental or of little value to athletic performance. In general, smokers have demonstrated lower $\dot{V}O_2$max values than nonsmokers; such lower values are as-

sociated with increased carbon monoxide binding to hemoglobin, which reduces oxygen transport capacity. Nicotine from cigarettes or smokeless tobacco increases heart rate, blood pressure, and autonomic reactivity. Other changes noted after nicotine use include vasoconstriction, decreased peripheral circulation, increased secretion of antidiuretic hormone and catecholamines, and increased blood lipid levels, plasma glucose, glucagon, insulin, and cortisol. The effects on performance parameters have not been studied adequately to draw any conclusions.

Risks of Nicotine Use   Nicotine has serious long-term health effects. The drug is highly addictive, which is why so many people who first try tobacco with their peers find themselves years later with a serious habit that is difficult to break.

Many of nicotine's risks relate to the method of ingestion. Smokeless forms are known to cause cancers of the mouth, pharynx, and larynx, and smoking is linked to several cancers, most notably lung cancer. Smokers are often more susceptible to respiratory infections, because cigarette smoke paralyzes the cilia in the respiratory tract, which sweep particulate matter away from the lungs. When the cilia are paralyzed, this cleansing action is diminished or halted, and particles can settle into the alveoli, blocking or damaging them. Smoking also can lead to emphysema. Furthermore, smoking leads to serious cardiovascular changes. It raises blood cholesterol levels and promotes atherosclerosis, which can directly lead to myocardial infarction or stroke. A smoker's risk of a heart attack is twice that of a nonsmoker, and smoking is the main risk factor for sudden cardiac death.

Another effect is impaired circulation to the extremities. Smoking is the major factor contributing to peripheral vascular disease, in which the blood vessels to the extremities are constricted. The American Heart Association reports that this disease is found almost exclusively in smokers. This places athletes who smoke at a much higher risk of frostbite when performing in cold environments.

▶ Alcohol consumption is the number one drug problem in the United States today. Alcohol is correctly classified as a drug because of its depressant effects on the CNS.

▶ Alcohol is used by athletes primarily for its psychological effects. It is thought to improve self-confidence, calm nerves, reduce anxiety and inhibitions, increase mental alertness, and reduce pain and muscle tremor.

▶ Most measures of psychomotor function are impaired by alcohol consumption, not improved, and no improvements in physiological function have been observed.

▶ Alcohol use can negatively affect the athlete's health as well as performance.

▶ Cocaine is a CNS stimulant. Although cocaine is not generally considered ergogenic, some athletes associate the euphoria it creates with increased self-confidence and motivation. There is no evidence that cocaine is in any way ergogenic.

▶ Cocaine is extremely addictive. It has tremendous potential for triggering major psychological disorders and has numerous undesirable physiological effects, most involving heart function, that can lead to death.

▶ Marijuana acts on the CNS and can elicit both stimulant and depressant effects.

▶ Marijuana has not been proposed to have ergogenic qualities. It is in fact ergolytic. It impairs performance that requires hand–eye coordination, fast reaction time, motor coordination, tracking ability, and perceptual accuracy.

▶ Marijuana use can lead to personality changes, short-term memory impairment, hallucinations, and psychotic-like behavior. When smoked, it might pose the same risks as cigarette smoking.

▶ Nicotine is a stimulant, ingested either by smoking or in smokeless forms: chewing tobacco, snuff, and compressed tobacco. Some athletes believe that nicotine makes them more alert and better able to concentrate yet also more calm.

▶ Nicotine is generally detrimental to performance. It causes several changes in cardiovascular, metabolic, respiratory, and hormonal function that can impair both submaximal and maximal performance.

▶ Proven risks of nicotine use include various forms of cancer and cardiovascular disease.

fyi Many pharmacological agents don't have ergogenic properties, yet some athletes believe that they do. Several substances are banned not because they are ergogenic but because their use carries high risks. Such bans are intended to keep athletes from trying harmful substances with the erroneous notion that they will enhance performance, when in fact some of these substances can be lethal.

Hormonal Agents

The use of **hormonal agents** as ergogenic aids in competitive athletics began in the late 1940s or early 1950s. Anabolic steroids were the hormones most frequently used by athletes between the 1950s and the 1980s. During the last half of the 1980s, a new potential ergogenic aid emerged with the introduction of synthetic human growth hormone, and women began experimenting with oral contraceptives (birth control pills) to see if manipulating their menstrual cycles could facilitate athletic performance.

Although numerous scientific studies have been conducted on anabolic steroids and sport, much less is known about the effects of human growth hormone and birth control pills on sport performance. Both anabolic steroids and human growth hormone are banned for all sports, and the medical risks associated with their use are high. Medical risks also are associated with taking oral contraceptives, but their use in sport is unregulated at this time.

We now examine the three major hormone groups being used (or abused) by athletes today:

- Anabolic steroids
- Human growth hormone
- Oral contraceptives

Anabolic Steroids

Androgenic-anabolic steroids, commonly referred to simply as **anabolic steroids**, are nearly identical to the male sex hormones (see chapter 5). The anabolic (building) properties of these steroid hormones accelerate growth by increasing the rate of bone maturation and the development of muscle mass. For years, anabolic steroids have been given to youngsters with delayed growth patterns to normalize their growth curves. The development of synthetic steroids has allowed alteration of the natural chemical composition of these hormones to reduce their androgenic (masculinizing) properties and increase their anabolic effects on muscle.

Proposed Ergogenic Benefits

Theoretically, steroid administration will increase fat-free mass and strength. Consequently, an athlete who depends on muscle size, body size, or strength might be tempted to take steroids. Early claims that aerobic capacity improves with anabolic steroid use also caught the attention of endurance athletes. Anabolic steroids also have been postulated to facilitate recovery from exhaustive training bouts, allowing athletes to train hard on subsequent days. This potential benefit has stirred the interest of athletes from almost all sports. The potential for anabolic steroid use among athletes is very high, and this continues to be a major problem in sport.

Proven Effects

The limitations of scientific research have been apparent in the study of the effects of anabolic steroids. Results of early investigations were almost evenly divided. Many studies found no significant change in body size or physical performance attributable to taking steroids, yet many other studies found steroids to have considerable positive influence on increasing muscle mass and strength.

One basic problem with almost all research conducted in this area to date is the inability to observe in the research laboratory the effects of the drug dosages being used in the athletic world. Some athletes are estimated to be taking 10 times the recommended maximum daily dosage or more. Obviously, it would be unethical to design a study using dosages that exceed the recommended maximum. However, some researchers have been able to observe athletes both when the athletes are taking high doses of steroids and when they are off the drug. Let's examine what the research shows about the effects of steroids on performance.

Muscle Mass and Strength   In one of the first studies to observe athletes who were taking steroids on their own, the effects of relatively high doses were observed in seven male weightlifters.[36] Two treatment periods, each lasting 6 weeks, were separated by a 6-week interval without treatment. Half the subjects received a placebo during the first treatment period and the steroid during the second treatment period. The other half received the medications in reverse order: steroid first, then placebo. When the data from all subjects were analyzed, results showed that while on the steroid, the weightlifters had significant increases in

- body mass and fat-free mass,
- total body potassium and total body nitrogen (markers of fat-free mass),
- muscle size, and
- leg strength.

These increases did not occur during the placebo period. Results of this study are summarized in figure 15.2 on page 488.

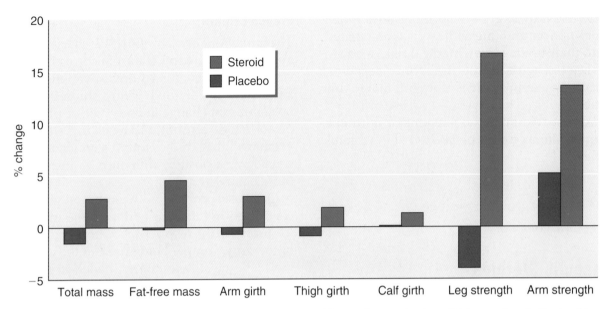

▲ Figure 15.2   Percentage changes in body size, body composition, and strength when athletes used anabolic steroids and a placebo.

Adapted, by permission, from G.R. Hervey et al., 1981, "Effects of methandienone on the performance and body composition of men undergoing athletic training," *Clinical Science*, 60: 457-461.[36]

In a second study, Forbes[28] observed body composition changes in a professional bodybuilder and a competitive weightlifter. Both were on self-prescribed high doses of steroids. The bodybuilder had been on the high dose for 140 days and the weightlifter for 125 days. Fat-free body mass increased an average of 19.2 kg (42.3 lb), and fat mass decreased almost 10 kg (22 lb).

Forbes plotted the results of a number of different studies that used different dosages (figure 15.3). He concluded that only minimal increases of 1 to 2 kg (2.2-4.4 lb) in fat-free body mass occur with low doses of anabolic steroids. But with high doses, fat-free body mass increases markedly. His results show a threshold level for steroid doses, with only very high doses resulting in substantial increases in fat-free body mass.

A third study looked at supraphysiological doses of testosterone on muscle size and strength in men who were nonathletes but were experienced with weightlifting.[8] Forty men completed the study and were assigned to one of the following groups: placebo with no exercise, placebo with exercise, testosterone with no exercise, and testosterone with exercise. The men received either 600 mg of testosterone

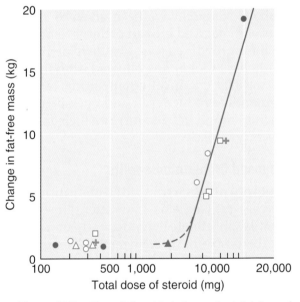

▲ Figure 15.3   The relationship between the total dose of steroid (mg/day) and the change in the fat-free mass (FFM) in kilograms. The symbols represent different anabolic steroid drugs. Steroid dose is plotted logarithmically.

Reprinted from *Metabolism*, vol. 34, G.B. Forbes, "The effect of anabolic steroids on lean body mass: The dose response curve," pp. 571-573, Copyright 1985, with permission from Elsevier.

enanthate or placebo intramuscularly each week for 10 weeks. The exercise groups strength-trained 3 days per week for 10 weeks. Body composition was measured by underwater weighing, muscle size by magnetic resonance imaging, and strength by the 1-repetition maximum technique. The testosterone and exercise group showed the largest increases in fat-free mass, muscle area, and strength, whereas the placebo and no-exercise group remained unchanged (figure 15.4). The placebo and exercise group increased strength, quadriceps area, and fat-free mass, and the testosterone and no-exercise group increased squat strength and quadriceps and triceps muscle areas. This is one of the best designed studies conducted on steroids and resistance exercise because it used placebo and no-exercise groups.

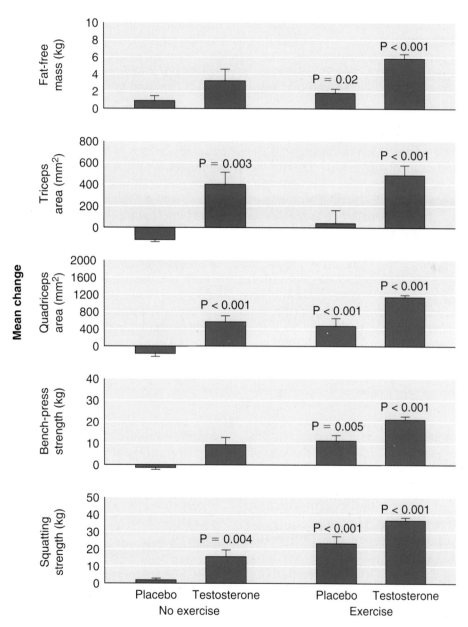

▲ **Figure 15.4**   Changes in fat-free mass and quadriceps and triceps muscle areas from magnetic resonance imaging, and changes in squat and bench press strength over the 10 weeks of placebo or testosterone, with or without exercise training.

"Andro"—Fuel for Power Hitters?

During the 1998 Major League Baseball season, Mark McGuire blasted 70 home runs to eclipse the previous home run record of 61 set by Roger Maris in 1961. During the season, McGuire admitted to using androstenedione ("Andro"), a precursor to testosterone, which is marketed to increase testosterone levels and, subsequently, skeletal muscle mass. Sales of Andro increased markedly in response to McGuire's revelation. Does Andro work? King and his colleagues at Iowa State University were the first to investigate the combined effects of Andro and resistance training in twenty 19- to 29-year-old men.[39] Ten subjects were assigned to the Andro and resistance-training group and 10 to the placebo and resistance-training group. Andro had no effect on serum testosterone levels, and the gains in strength and muscle mass were the same for the Andro and placebo groups. However, there was an unexpected finding: Andro increased serum estradiol and estrone concentrations! Other studies have substantiated these findings in young, middle-aged, and older men.[10, 24] Research is now looking at alternate ways of getting Andro into the blood other than through oral ingestion, which can lead to hepatic breakdown of the ingested androgens.

Dehydroepiandrosterone (DHEA) is another steroid hormone, as is its conjugate sulfate (DHEA-S). Bloodborne DHEA can be converted to androstenedione or androstenediol, which can then be converted to testosterone.[12] DHEA has been proposed to increase muscle mass. Brown and colleagues[12] investigated the potential benefit of DHEA in increasing both serum testosterone and strength in 19 young men participating in an 8-week resistance training program, 9 receiving DHEA and 10 receiving the placebo. As with the Andro study, there were no changes in serum testosterone concentrations as a result of taking DHEA, and the changes in strength and muscle mass were the same for the DHEA and placebo groups.

Major League Baseball players must have anticipated the results of these studies on Andro and DHEA. The cover of the June 3, 2002, issue of *Sports Illustrated* announced, "Special Report—Steroids in Baseball: Confessions of an MVP." From this issue, it appears that a substantial number of professional baseball players have skipped over Andro and DHEA and have gone straight to anabolic steroids. Although the actual percentage of players using steroids is unknown, some estimate that it is as great as 50%. Curt Schilling, Arizona Diamondback pitcher and Co-Most Valuable Player of the 2001 World Series, was quoted as saying, "Guys look like Mr. Potato Head, with six or seven body parts that just don't look right."

Cardiorespiratory Endurance Several early studies reported increases in $\dot{V}O_2$max with the use of anabolic steroids. These results were consistent with the known effects of steroid administration on increasing red blood cell production and total blood volume. However, in these studies, $\dot{V}O_2$max was estimated indirectly. In later, better controlled studies, $\dot{V}O_2$max was measured directly, and anabolic steroids produced no benefit. However, none of the studies investigating anabolic steroid use and improvement in aerobic capacity involved trained endurance athletes.

Recovery From Training The theory that anabolic steroids facilitate recovery from high-intensity training is attractive. A major concern in training elite athletes today is how to reduce the negative physiological and psychological effects associated with high-intensity training, enabling the athlete to continue to train at peak levels day after day. At this time, though, no available data about steroid use support this idea.

Anabolic steroid use does increase muscle mass and strength, which can improve performance in strength-type sports or activities. Aerobic endurance appears to be unaffected by steroid use. Steroid use in sport is illegal and is banned in most sports. Furthermore, the health risks can be considerable.

Risks of Anabolic Steroid Use

Although use of anabolic steroids can be beneficial for certain types of athletic performance, several major issues must be addressed. It is neither moral nor ethical for athletes to use drugs to improve their chances in competition. Most athletes feel that it is wrong for their competitors to artificially improve their performance. Yet many of these same athletes feel compelled to use steroids in an effort to compete with the other athletes in their sport or event who are chronic steroid users. Fair competition is not possible if you are the only athlete in a particular competition who has remained steroid free.

How widespread is this problem? Although in the past steroid use was a problem primarily in male-dominated sports, many female athletes are now taking steroids to increase their fat-free body mass, strength, and performance. Steroid use among athletes also has worked its way down to the high school and junior high school levels. Furthermore, steroid use among adolescent nonathletes who simply want to look good has also sharply increased.

> **fyi** At one time, it was estimated that 80% of all weightlifters, shot-putters, discus throwers, and javelin throwers of national caliber were using anabolic steroids, and this was considered by many to be a conservative number.

Although the pressures on the athlete are great, are the potential gains worth the possible risks associated with steroid use? These drugs are illegal, and athletes who use them risk being banned from their sports. More important, however, are the medical risks associated with steroid use, especially with the massive doses often used by athletes. Steroid use by people who are not physically mature can lead to early closure of the epiphyses of the long bones, so final stature can be reduced. Use of anabolic steroids suppresses the secretion of gonadotropic hormones, which control the development and function of the gonads (testes and ovaries). In males, decreased gonadotropin secretion can cause testicular atrophy, decreased secretion of testosterone, and a reduced sperm count. Decreased testosterone can lead to enlargement of the male breasts. In females, gonadotropins are required for ovulation and secretion of estrogens, so a decrease in these hormones disrupts those processes and menstruation. In addition, these hormonal disturbances in females can lead to masculinization: breast regression, enlargement of the clitoris, deepening of the voice, and growth of facial hair.

Prostate gland enlargement in males is another possible side effect of steroid use. Liver damage from a form of chemical hepatitis brought on by steroid use has also been identified, and it can lead to liver tumors.

Cardiomyopathy (diseased heart muscle) has been reported in chronic steroid users. It is suspected that steroid use was at least partially responsible for the medical condition of one former offensive lineman in the National Football League who was on the waiting list for a heart transplant. Scientists have found markedly depressed high-density–lipoprotein cholesterol (HDL-C) levels—reductions of 75% or more—in athletes on even moderate steroid doses. HDL-C has antiatherogenic properties, meaning it prevents the development of atherosclerosis. Low levels of it are associated with a high risk for both coronary artery disease and heart attack (see chapter 20). Furthermore, low-density–lipoprotein cholesterol (LDL-C), which has atherogenic properties, appears to be increased with steroid use.

Substantial personality changes have occurred with steroid use. The most notable change is a marked increase in aggressive behavior, or "roid rage." Some teens have become extremely violent and have attributed these drastic mood changes to steroid use. Evidence also suggests that drug dependency can result from steroid use. As a final point of great concern, neither scientists nor physicians know the potential long-term effects of chronic steroid use. One study of male mice that received four different anabolic steroids of the kinds and doses taken by athletes found that their life span was markedly decreased.[11] It is critical to develop a large database of former steroid users who can be followed over their life span. We have to remember that most diseases begin many years before symptoms develop. It is possible that the more serious health risks of steroid use won't be apparent for 20 or 30 years.

Recent reviews provide more detail on the potential ergogenic effects and health risks associated with anabolic steroid use.[6, 48, 59, 69] Most governing bodies of sports likely to be affected by anabolic steroid use have developed educational materials for their athletes in hopes of preventing steroid use. Also, national governing bodies for most sports have instituted aggressive year-round testing programs in which athletes are tested randomly for steroid use. Interestingly, Major League Baseball, which is now confronting the issue of steroid use, has no formal policy or testing program in place.

> ▶ Anabolic steroids are more appropriately termed androgenic-anabolic steroids because in their natural state they include both androgenic (masculinizing) and anabolic (building) properties. Synthetic steroids have been designed to maximize the anabolic effects while minimizing androgenic effects.
>
> ▶ Anabolic steroids have been proposed to increase muscle mass, strength, and endurance capacity and to facilitate recovery from exhaustive training bouts.
>
> ▶ Anabolic steroids can increase muscle mass and strength, but this effect is dose dependent. They do not increase endurance capacity, and their ability to facilitate recovery from exhaustive exercise has not been proven.
>
> ▶ Andro and DHEA, precursors of testosterone, have been proposed to have ergogenic properties—increased muscle mass and strength—but research has not supported the claims.
>
> ▶ Potential risks are associated with use of anabolic steroids, including personality changes, "roid rage," testicular atrophy in men, reduced sperm count in men, breast enlargement in men, breast regression in women, prostate gland enlargement in men, masculinization in women, menstrual cycle disruption in women, liver damage, and cardiovascular disease.

Human Growth Hormone

For years, the medical treatment for hypopituitary dwarfism has been administration of **human growth hormone (hGH)**, a hormone secreted by the anterior pituitary gland. Before 1985, this hormone was obtained from cadaver pituitary extracts, and the supply was limited. Since the introduction of genetically engineered human growth hormone in the mid-1980s, availability is no longer an issue, although the cost is still high.

During the 1980s, realizing this hormone's numerous functions, athletes started investigating human growth hormone as a possible substitute for or complement to their use of anabolic steroids. As drug testing for anabolic steroids became more sophisticated, athletes were looking for an alternative for which there was no drug test at the time. Growth hormone appeared to be the designer drug for athletes who wanted to increase their strength and muscle mass.

Proposed Ergogenic Benefits

Growth hormone (GH) has five functions of interest to athletes:

- Stimulation of protein and nucleic acid synthesis in skeletal muscle
- Stimulation of bone growth (elongation) if bones are not yet fused (important to young athletes)
- Increase in lipolysis, leading to an increase in free fatty acids and an overall decrease in body fat
- Increase in blood glucose levels
- Enhancement of healing after musculoskeletal injuries

Athletes turned to this hormone thinking that it would increase muscle development and thus increase fat-free mass. Often GH is used with anabolic steroids to maximize the anabolic effects.

Proven Effects

Research on whether GH aids healing is inconclusive at this time, but several studies have reported other effects. One study involved healthy men, 61 to 81 years of age.[50] Following a 6-month treatment period in which 12 men received GH three times a week, researchers found that

- fat-free body mass increased by 9%,
- fat mass decreased by 14%, and
- lumbar vertebral bone density increased by 2%.

A control group of nine men showed no change in any of these measurements over the same period. Resistance training, or any other form of exercise, was not a part of this study's design.

In another study, young men were randomly assigned to one of two groups.[67] Both underwent resistance training, but one group was given a placebo while the other received GH. Following a 12-week training period, the GH group showed the greatest increases in

- fat-free body mass,
- total body water,
- whole-body protein synthesis rate, and
- whole-body protein balance (rate of synthesis–breakdown).

When researchers looked at specific muscles, however, the two groups did not differ significantly. Muscle size, muscle strength, and the rate of muscle protein synthesis for the quadriceps were not any greater in the GH group than in the placebo group, indicating that resistance exercise alone resulted in similar increases in these areas, irrespective of GH use.

In a study of experienced weightlifters who were placed on GH treatment for 14 days, the investigators found that GH did not alter either the rate of muscle protein synthesis or the rate of whole-body protein breakdown, the two factors that would promote an increase in muscle mass.[68]

Some athletes also take other drugs and certain amino acid supplements to stimulate GH release from the pituitary. To date, little evidence suggests that this practice is effective.

> It is unclear if human growth hormone has ergogenic properties in young, healthy athletes. However, major health risks are associated with the use of growth hormone.

Risks of Growth Hormone Use

As with steroids, potential medical risks are associated with growth hormone use. Acromegaly can result from taking GH after the bones have fused. This disorder results in bone thickening, which causes broadening of the hands, feet, and face; skin thickening; and soft tissue growth. Internal organs typically enlarge. Ultimately, the victim suffers muscle and joint weakness and often heart disease. Cardiomyopathy is the most common cause of death with GH use. Glucose intolerance, diabetes, and hypertension also can result from GH use.

Oral Contraceptives

Oral contraceptives (birth control pills) contain synthetic versions of natural estrogens and progesterones. They function as contraceptives by preventing ovulation.

Proposed Ergogenic Benefits

Oral contraceptives have been proposed as potential ergogenic aids because they can control the athlete's menstrual cycle. Many female athletes find that their performance is unaffected by their menstrual cycle, but others notice a difference (see chapter 18). Those who have cyclic fluctuations in their performance frequently suffer from premenstrual syndrome (PMS), experiencing emotional or physical symptoms, or often both, 3 to 5 days before menstruation. Many experience dysmenorrhea (difficult or painful menstruation).

Proven Effects

Shangold[51] stated that it is rarely advisable or necessary to manipulate an athlete's menstrual cycle to improve her performance. However, she also reported that it might be appropriate for those few elite female athletes who perform better in the follicular (early cycle) phase than at other times to consider using oral contraceptives to regulate the cycle for special events of great importance. This can be accomplished by administering low-dose oral contraceptives continuously (not cyclically) for several months before the competitive event, continuing until 10 days before the competition. Withdrawal bleeding can be expected within 3 days of discontinuation. This assures the athlete of a predictable bleeding pattern and leaves her with low levels of both estrogen and progesterone at the time of the event. This approach is relatively safe.

> Oral contraceptives can help regulate the timing of the menstrual cycle, but their affect on athletic performance has yet to be determined.

Risks of Oral Contraceptive Use

Though the use of oral contraceptives is quite widespread, these drugs are not without medical risks. Some of the risks include

- nausea,
- weight gain,
- fatigue,
- hypertension,
- liver tumors,
- blood clots,
- stroke, and
- heart attack.

The risk of the last three increases tremendously for women who also smoke. It is important to understand that these risks are unrelated to whether you are an athlete. These risks are associated with oral contraceptives for any user.

fyi Human growth hormone stimulates synthesis of protein and nucleic acid in skeletal muscle, stimulates bone growth, increases lipolysis (thus decreasing body fat), increases blood glucose levels, and enhances healing of musculoskeletal injuries.

▶ Growth hormone has not been studied extensively for its potential ergogenic effects. The limited research available supports its ability to increase fat-free mass and decrease fat mass, but GH might have little or no effect on increasing muscle mass and strength.

▶ Risks associated with GH use include acromegaly, hypertrophy of internal organs, muscle and joint weakness, diabetes, hypertension, and heart disease.

▶ Oral contraceptives have been proposed as ergogenic aids for female athletes because of their ability to regulate the menstrual cycle.

▶ Little research at this time supports the use of oral contraceptives for ergogenic purposes, although the drugs may be beneficial for elite female athletes who suffer from PMS or dysmenorrhea.

▶ Risks associated with oral contraceptives include nausea, weight gain, fatigue, hypertension, liver tumors, blood clots, stroke, and heart attack.

Physiological Agents

Many **physiological agents** have been proposed as ergogenic aids. The goal of using these agents is to improve the body's physiological response during exercise. An athlete typically adds to a substance that occurs naturally in the body to try to improve performance. The reasoning is that if natural levels of a substance are beneficial to performance, higher levels should be even better. Several physiological agents have been proven effective but generally only under very specific conditions or for certain events or sports.

As with hormonal agents, many athletes consider use of these substances to be more ethical than use of pharmacological agents because these substances are found naturally in the body. Athletes also often assume that because these substances are normally found in the body, they must be safe at any level. Unfortunately, our bodies can be very unforgiving, and this assumption can be fatal.

We will look at only a few examples of physiological agents being used as ergogenic aids:

- Blood doping
- Erythropoietin
- Oxygen supplementation
- Bicarbonate loading
- Phosphate loading

Blood Doping

Although altering blood composition in any way can be considered blood doping, the term has taken on a more specific meaning. **Blood doping** refers to any means by which a person's total volume of red blood cells is increased. This often is accomplished by transfusion of red blood cells, previously donated by either the recipient (autologous transfusions) or someone else with the same blood type (homologous transfusions). Blood doping also includes the use of erythropoietin, but this is discussed separately in the next section.

Proposed Ergogenic Benefits

Because oxygen is carried through the body bound to hemoglobin, it seems logical that increasing the number of red blood cells available to ferry the oxygen to the tissues could benefit performance. Increasing the number of oxygen carriers should increase your blood's oxygen-carrying capacity, allowing more oxygen to be delivered to your active tissues. If this happens, your aerobic endurance and thus your performance could be substantially improved. That is the premise underlying blood doping.

Proven Effects

Ekblom and coworkers[27] created quite a stir in the sports world in the early 1970s. In a landmark study, they withdrew between 800 and 1,200 ml of blood from their subjects, refrigerated the blood, then reinfused the red blood cells into those subjects about 4 weeks later. Results showed a considerable improvement in $\dot{V}O_2max$ (9%) and treadmill performance time (23%) after reinfusion. Over the next few years, several studies con-

firmed these original findings, but several others failed to demonstrate an ergogenic effect.

Thus, the research literature was divided on the effectiveness of blood doping until a major breakthrough occurred in 1980 as a result of a study by Buick and colleagues.[13] Eleven highly trained distance runners were tested at different times during the study: (1) before blood withdrawal, (2) following blood withdrawal after allowing the body adequate time to reestablish normal red blood cell levels but before reinfusion of the removed blood, (3) following a sham reinfusion of 50 ml of saline (a placebo), (4) following reinfusion of 900 ml of the subject's own blood that originally had been withdrawn and preserved by freezing, and (5) after the elevated red blood cell levels had returned to normal. The sham trial was identical to the blood reinfusion trial, except that a small volume of saline was infused rather than whole blood. This trial served as a placebo to rule out any psychological effects.

As shown in figure 15.5 on page 496, the researchers found a substantial increase in $\dot{V}O_2max$ and treadmill running time to exhaustion after the reinfusion of the red blood cells and no change after the sham reinfusion. This increase in $\dot{V}O_2max$ persisted for up to 16 weeks, but the increase in treadmill time decreased within the first 7 days.

Maximizing the Benefits Why was the Buick study a major breakthrough? Gledhill[30, 31] helped explain the controversy arising from the early studies. Many early studies that found no improvement with blood doping had reinfused only small volumes of red blood cells, and the reinfusion was conducted within 3 to 4 weeks of the blood withdrawal. First, it appears that you must withdraw and reinfuse 900 ml or more of whole blood to have an effect. Increases in $\dot{V}O_2max$ and performance are not as great when smaller volumes are used. In fact, some studies using smaller volumes failed to find any differences.

Second, it appears that you must wait for at least 5 to 6 weeks and possibly as long as 10 weeks before reinfusion. This is based on the time it takes your body to reestablish your blood's prewithdrawal hematocrit.

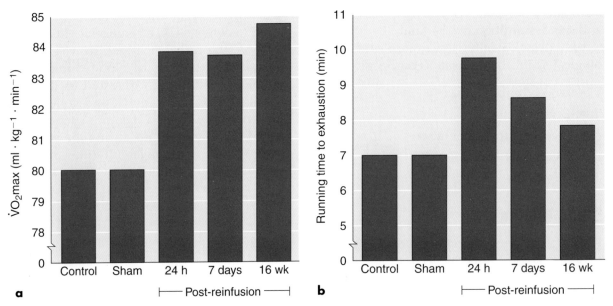

▲ Figure 15.5 Changes in $\dot{V}O_2$max and running time to exhaustion following reinfusion of red blood cells.
Adapted, by permission, from F.J. Buick, N. Gledhill, A.B. Froese, L. Spriet, and E.C. Meyers, 1980, "Effect of induced erythocythemia on aerobic work capacity," *Journal of Applied Physiology* 48: 636-642.

Finally, researchers who conducted early studies refrigerated the withdrawn blood. Maximal storage time for blood under refrigeration is approximately 5 weeks. Furthermore, when blood is refrigerated, approximately 40% of the red blood cells are destroyed or lost. Later studies used frozen storage. Freezing allows an almost unlimited storage time, and only about 15% of the red blood cells are lost.

Gledhill[30, 31] concluded that blood doping significantly improves $\dot{V}O_2$max and endurance performance when done under optimal conditions:

- A minimum of 900 ml of blood reinfusion
- A 5- to 6-week minimum interval between withdrawal and reinfusion
- Blood storage by freezing

He also showed that these improvements are the direct result of the blood's increased hemoglobin content, not increased cardiac output caused by an expanded plasma volume.

Blood Doping and Endurance Performance Does an increase in $\dot{V}O_2$max and treadmill time as a result of blood doping translate into improved endurance performance? Several studies have addressed this issue. One study observed 5-mi (8-km) treadmill run times in a group of 12 experienced distance runners.[63] Their times were checked before and after saline (placebo) infusion and before and after blood infusion. The 5-mi (8-km) run times on the treadmill were significantly faster following blood infusion, but this difference became significant only over the last 2.5 mi (4 km). The blood infusion trials were 33 s (3.7%) faster over the last 2.5 mi (4 km) and 51 s (2.7%) faster over the full 5 mi (8 km) than the placebo trials.

A second study looked at 3-mi (4.8-km) run times in a group of six trained distance runners and reported a decrease of 23.7 s following blood doping.[32] This decrease was significantly different from the runners' blind control trials. Subsequent studies confirmed improvements in distance running and cross-country skiing performance with blood doping.[53] Figure 15.6 illustrates the improvement in run time with blood doping for distances of up to 11 km (6.8 mi).

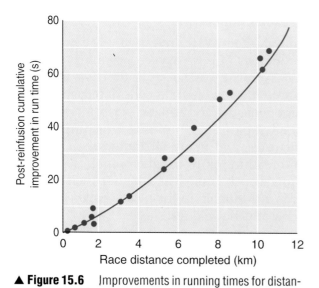

▲ **Figure 15.6**   Improvements in running times for distances of up to 11 km (6.8 mi) following reinfusion of red blood cells from two units of blood preserved by freezing. Values on the *y* axis reflect the reduction in time to run a specific distance on the *x* axis. As an example, for a 10-km (6.2-mi) race, you could expect to run 60 s faster after reinfusion.

Adapted, by permission, from L.L Spriet, 1991, Blood doping and oxygen transport. In *Ergogenics—Enhancement of performance in exercise and sport,* edited by D.R. Lamb and M.H. Williams (Dubuque, IA: Brown & Benchmark), 213-242. Copyright 1991 Cooper Publishing Group, Carmel, IN.

Risks of Blood Doping

Although this procedure is relatively safe in the hands of competent physicians, it has inherent dangers. Adding more red blood cells into the cardiovascular system can overload it, causing the blood to become too viscous, which could lead to clotting and possibly heart failure. With autologous blood transfusions, in which the recipient receives his or her own blood, mislabeling of the blood could occur. With homologous blood transfusions, in which blood is received from a matched donor, several other complications can occur. The reinfused blood could be mismatched. An allergic reaction could be triggered. The athlete may experience chills, fever, and nausea. The athlete also risks contracting hepatitis or acquired immune deficiency syndrome (AIDS). The potential risks of blood doping, even without considering the legal, moral, and ethical issues involved, outweigh any potential benefits.

Erythropoietin

As mentioned in the last section, erythropoietin falls into the class of blood doping, but because the mechanism is somewhat different, we examine it separately. **Erythropoietin** is a naturally occurring hormone produced by the kidneys. It stimulates red blood cell production. In fact, this hormone is responsible for the elevated red blood cell production seen when training at altitude; training in the presence of a lower partial pressure of oxygen stimulates erythropoietin release.

Human erythropoietin can now be cloned through genetic engineering, so it is widely available. This hormone increases the hematocrit substantially when administered to patients with renal failure.

Proposed Ergogenic Benefits

Theoretically, if administered to athletes, human erythropoietin would have the same effects as the reinfusion of red blood cells. The goal of its use is to increase the red blood cell volume, thus increasing the blood's oxygen-carrying capacity.

Proven Effects

Erythropoietin's ability to increase oxygen capacity was demonstrated in 1991 when the first study was conducted of the effects of subcutaneous injections of low doses of human erythropoietin on maximal treadmill time and $\dot{V}O_2$max.[26] The study involved moderately trained to well-trained subjects. Six weeks after erythropoietin administration, the following effects were observed:

- Both hemoglobin concentration and hematocrit increased 10%.
- $\dot{V}O_2$max increased 6% to 8%.
- Time to exhaustion on the treadmill increased 13% to 17%.

Seven of the 15 subjects had been through a previous study of red blood cell reinfusion conducted 4 months earlier. Increases in $\dot{V}O_2$max and treadmill time were almost identical in both studies, and these improvements were attributed directly to the increase in hemoglobin.

Risks of Erythropoietin Use

Serious consequences can arise from erythropoietin use. Up to 18 deaths among competitive cyclists, reported between 1987 and 1990, were alleged to be linked to erythropoietin use, but this has not been confirmed.[1]

The outcome of erythropoietin use is less predictable than that of red blood cell reinfusion. Once the hormone has been put into the body, no one can predict how much red blood cell production will occur. This places the athlete at great risk of substantial increases in blood viscosity, which could lead to clotting or heart failure.

> Blood doping and erythropoietin can enhance aerobic capacity and the performance of aerobic sports or activities. This occurs through an increase in the oxygen-carrying capacity of the blood, attributable to increased red blood cells and to the increase in total blood volume that results from blood doping and the use of erythropoietin.

Oxygen Supplementation

You're watching a professional football game on television. The star running back breaks loose for a 35-yd (32-m) touchdown run, struggles back to the bench, grabs a face mask, and starts breathing 100% oxygen to facilitate his recovery. How much does he gain by using this oxygen supplementation instead of just breathing ordinary air?

Proposed Ergogenic Benefits

Obviously, the purpose of taking in oxygen is to increase the oxygen content of the blood, as with blood doping. Blood doping attempts to do this by increasing the oxygen-carrying capacity of the blood; **oxygen supplementation** tries to achieve this directly by providing more oxygen. By increasing the available oxygen, athletes hope to fend off fatigue for longer periods. This technique also has been suggested as a means to speed recovery between exercise bouts.

Proven Effects

Initial attempts to investigate the ergogenic properties of pure oxygen began in the early 1900s, but it was not until the 1932 Olympic Games that oxygen was considered a potential ergogenic aid for athletic performance. That year, Japanese swimmers won impressive victories, and many attributed their success to breathing pure oxygen before competing. However, it is unclear whether their success was attributable to their use of oxygen or to their athletic abilities.

Of historical note, one of the first studies to observe the effects of breathing oxygen on performance was conducted by Sir Roger Bannister, a physician–scientist who is world renowned for his research in neurological disorders.[7] As an athlete, Dr. Bannister was the first person in the world to break the 4-min mile barrier.

Oxygen can be administered immediately before competition, during competition, during recovery from competition, or at any combination of these times.

Oxygen breathing before exercise has a limited effect on performance of that exercise bout. The total amount of work or the rate of work (exercise intensity) can be increased by breathing oxygen if the bout is of short duration and occurs within a few seconds after the athlete breathes oxygen. During these short bouts, submaximal work can be performed at a lower pulse rate. However, no improvement occurs unless the exercise follows within seconds of breathing oxygen.

For exercise bouts exceeding 2 min or when more than 2 min lapse between oxygen breathing and actual performance, oxygen supplementation's influence is greatly diminished. This simply reflects the limits of the body's oxygen-storage potential: Extra oxygen dissipates rapidly; it is not stored.

When oxygen is administered during exercise, definite performance improvements occur. The total amount of work performed and the rate of work performed increase substantially. Likewise, submaximal work is performed more efficiently with lower physiological cost to the individual. Peak blood lactate levels are depressed following exhaustive exercise that the subject performs while breathing oxygen, even though considerably more work can be performed.

Studies have been unable to demonstrate any clear advantage to oxygen breathing during the recovery period. Recovery does not seem to be facilitated, nor does subsequent performance improve. In an unpublished study in our laboratory (JHW), subjects performed an exhaustive, all-out ride on a cycle ergometer for 60 s. They then immediately began breathing a gas mixture of either pure oxygen or air for a 2-min recovery period. This was immediately followed by a second all-out 60-s ride. Neither the 2-min recovery between the exhaustive bouts nor the total work accomplished in the second bout (see figure 15.7) was facilitated by oxygen breathing. A similar study with professional soccer players running on a treadmill also found no improvements in recovery or subsequent performance on the second exhaustive bout as a result of oxygen breathing.[66]

From a practical standpoint, oxygen administration before exercise would have little value because of the relatively short time that oxygen stores remain elevated. The nature of most sports doesn't allow an athlete to go immediately from oxygen breathing into competition. Regardless of the ergogenic effects of oxygen intake during performance, administration during exercise has limited value for obvious reasons: During which sports or events could you carry a cylinder of oxygen without significant restriction of movement?

The recovery period seems the only practical time to administer oxygen, but it would be worthwhile only if oxygen administration were known to speed the recovery process, allowing the athlete to reenter the contest more fully recovered. However, such an effect has not been substantiated by research.

> Oxygen supplementation can increase aerobic performance but only if administered during exercise, which is not practical in sport.

Risks of Oxygen Supplementation

At this time, no known risks are associated with oxygen supplementation. More research must be conducted to determine its safety. However, oxygen is flammable, so oxygen equipment should never be near any heat source or flame, nor should it be allowed near anyone who is smoking.

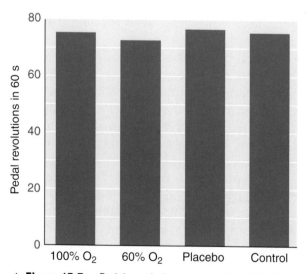

▲ **Figure 15.7**   Pedal revolutions accomplished during a second 60-s all-out work bout on a cycle ergometer. This second bout was performed 2 min after an initial, identical work bout. Either oxygen (60% and 100%) or room air was administered between bouts on a double-blind basis.

▶ Blood doping refers to an artificial increase in a person's total volume of red blood cells. It has been proposed to improve endurance performance by increasing the blood's oxygen-carrying capacity.

▶ Early studies produced conflicting results, but more recent work found major increases in maximal oxygen uptake, time to exhaustion, and actual performance in cross-country skiing and distance running as a result of blood doping.

▶ Risks associated with blood doping include major complications such as blood clotting, heart failure, and administration of mislabeled blood and its potential consequences, such as transfusion reactions, transmission of hepatitis, and transmission of the virus that causes AIDS.

> ▶ Erythropoietin is the naturally occurring hormone that stimulates red blood cell production. It has been proposed as an ergogenic aid under the premise that increasing the number of red blood cells increases the blood's oxygen-carrying capacity.
>
> ▶ Few studies have been conducted on the role of erythropoietin in enhancing aerobic endurance performance, but one study demonstrated increased maximal oxygen consumption and increased time to exhaustion.
>
> ▶ Because we cannot predict the magnitude of the body's response to erythropoietin administration, it is very dangerous. The hormone can lead to death if red blood cells are overproduced, because increased blood viscosity can cause clotting and heart failure.
>
> ▶ Oxygen administration during exercise improves performance but is too cumbersome to be practical. Administration before or immediately after exercise has not been proven ergogenically effective.
>
> ▶ No serious risks are associated with oxygen breathing.

Bicarbonate Loading

Recall from chapter 8 that bicarbonates are an important part of the buffering system necessary to maintain the acid–base balance of body fluids. Scientists naturally began to investigate whether performance in highly anaerobic events, in which large amounts of lactic acid are formed, could be improved by enhancing the body's buffering capacity through elevation of the blood's bicarbonate concentrations, a process called **bicarbonate loading**.

Proposed Ergogenic Benefits

By ingesting agents that increase the bicarbonate concentrations in the blood plasma, such as sodium bicarbonate (baking soda), subjects can increase blood pH, making the blood more alkaline. It was proposed that increasing plasma bicarbonate levels would provide additional buffering capacity, allowing higher concentrations of lactate in the blood. Theoretically, this could delay the onset of fatigue in short-term, all-out anaerobic work, such as all-out sprinting.

Proven Effects

Oral intake of sodium bicarbonate elevates plasma bicarbonate concentrations. However, this has little effect on intracellular concentrations of bicarbonate in muscle. Therefore, the potential benefits of bicarbonate ingestion were thought to be limited to anaerobic bouts of exercise lasting longer than 2 min, because bouts less than 2 min would be too brief to allow many hydrogen ions (H^+, from the lactic acid) to diffuse out of the muscle fibers into the extracellular fluid where they could be buffered.

In 1990, however, Roth and Brooks[49] described a cell membrane lactate transporter that operates in response to the pH gradient. Increasing the extracellular buffering capacity by ingesting bicarbonate increases the extracellular pH, which in turn increases transport of lactate from the muscle fiber via this membrane transporter to the blood plasma and other extracellular fluids. This should improve anaerobic performances even for events briefer than 2 min.

Although the theory proposing bicarbonate ingestion as an ergogenic aid for anaerobic performance is sound, the research literature is, again, conflicting. However, Linderman and Fahey,[42] in their review of the research literature, found several important patterns in the research that had been conducted that might explain these conflicts. They concluded that bicarbonate ingestion had little or no effect on performances of less than 1 min or of more than 7 min, but for performances between 1 and 7 min, the ergogenic effects were evident. Furthermore, they found that the dose was important. Most studies that used a dose of 300 mg/kg of body mass showed a benefit, whereas most studies of lower dosage showed little or no benefit. Thus, it appears that bicarbonate ingestion of 300 mg/kg of body mass can enhance the performance of all-out, maximal anaerobic activities of 1- to 7-min duration.

An example of a study supporting these conclusions is illustrated in figure 15.8. In this study, blood bicarbonate concentrations were artificially elevated by bicarbonate ingestion before and during five sprint-cycling bouts, each lasting 1 min (see figure 15.8a).[20] Performance on the final trial improved by 42%! This elevation in blood bicarbonate levels reduced the concentration of free H[+] both during and after exercise (see figure 15.8b), thereby elevating blood pH. The authors concluded that in addition to improving buffering capacity, the extra bicarbonate appeared to speed the removal of H[+] ions from the muscle fibers, thereby reducing the decrease in intracellular pH. Six years later, in 1990, Roth and Brooks[49] reported the presence of a lactate transporter in the muscle cell membrane, as described earlier, which works precisely as Costill et al.[20] had originally postulated.

Risks of Bicarbonate Loading

Although sodium bicarbonate has long been used as a remedy for indigestion, many authors studying bicarbonate loading have reported severe gastrointestinal discomfort, including diarrhea, cramps, and bloating, from high doses of bicarbonate. These symptoms can be prevented by ingesting as much water as desired and by dividing the total bicarbonate dosage of at least 300 mg/kg of body mass into five equal parts over a 1- to 2-h period.[42] Also, several studies have found sodium citrate to have similar effects on buffering capacity and performance without gastrointestinal discomfort.[40, 45]

Phosphate Loading

Since the early 1900s, scientists have been interested in the possibility of increasing the dietary consumption of phosphorus to improve cardiovascular and metabolic function during exercise. Several of the early studies suggested that phosphate loading, which involves ingestion

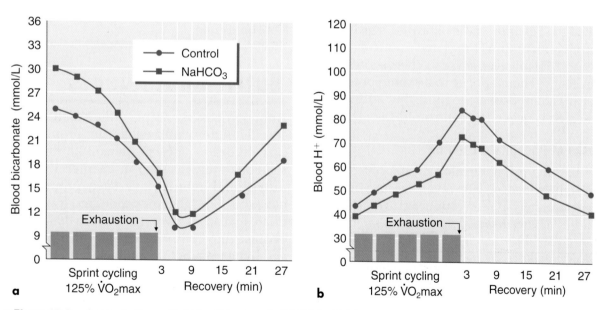

▲ **Figure 15.8**   Concentrations of (a) blood bicarbonate (HCO₃⁻) and (b) blood hydrogen ion (H⁺) before, during, and after five sprint-cycling bouts with and without ingestion of sodium bicarbonate (NaHCO₃). The fifth sprint was performed to exhaustion. The elevated blood HCO₃⁻ concentrations caused an attenuation in the elevation of blood H⁺, a smaller decline in blood pH, a 42% increase in performance to exhaustion during the fifth sprint, and faster recovery after the sprints.

Adapted, by permission, from D.L. Costill, F. Verstappen, H. Kuipers, E. Janssen, and W. Fink, 1984, "Acid–base balance during repeated bouts of exercise: Influence of HCO₃⁻," *International Journal of Sports Medicine* 5: 228-231. [20]

of sodium phosphate as a dietary supplement, is an effective ergogenic aid.

Proposed Ergogenic Benefits

Phosphate loading has been proposed to have numerous potential benefits during exercise. These include elevation of extracellular and intracellular phosphate levels, which would increase the availability of phosphate for oxidative phosphorylation and phosphocreatine synthesis, thus improving your body's energy production capacity. It is also thought to enhance 2,3-diphospho-glycerate (2,3-DPG) synthesis in the red blood cells. This substance facilitates the release of oxygen from the red blood cells. This increase in 2,3-DPG would shift the hemoglobin–oxygen dissociation curve to the right, permitting greater oxygen unloading in the active muscles.[59] Phosphate loading also has been proposed to improve the cardiovascular response to exercise, improve the body's buffering capacity, and consequently improve endurance capacity and performance.

Proven Effects

Only a few studies have been conducted to determine the ergogenic benefits of phosphate loading. Unfortunately, the results are divided. Several studies showed significant improvements in $\dot{V}O_2$max and time to exhaustion.[14, 41, 55] However, several others showed no effects.[9, 23] There appear to be some potential benefits to phosphate loading, but additional research is needed to confirm this.

Risks of Phosphate Loading

At this time, no known risks are associated with phosphate loading. However, because insufficient research has been conducted to date, more studies are needed to determine its safety.

▶ Bicarbonate is an important component of the body's buffering system, needed to maintain normal pH by neutralizing excess acid.

▶ Bicarbonate loading is proposed to increase the blood's alkalinity, thus increasing the buffering capacity so that more lactate can be cleared. This delays the onset of fatigue.

▶ Bicarbonate ingestion of at least 300 mg/kg of body weight can delay fatigue and increase performance in all-out bouts of exercise lasting more than 1 min but less than 7 min.

▶ Bicarbonate loading can cause gastrointestinal distress, including cramping, bloating, and diarrhea.

▶ Ingestion of sodium phosphate has been postulated to improve general cardiovascular and metabolic functioning. During exercise, phosphate loading has been proposed to elevate phosphate levels throughout the body, which would increase the potential for oxidative phosphorylation and phosphocreatine synthesis, enhance oxygen release to the cells, improve cardiovascular response to exercise, improve the body's buffering capacity, and improve endurance capacity.

▶ Little research at this time supports the use of phosphate loading as an ergogenic aid. Existing research is conflicting, and the risks of phosphate loading are largely unknown.

Nutritional Agents

Although basic concepts of nutrition and the specific performance-enhancing properties of carbohydrates, fats, proteins, vitamins, and minerals are discussed in chapter 13, many nutritional agents have been proposed to have specific ergogenic properties. It is appropriate to discuss several of these in this chapter because they have received so much publicity and hype from both manufacturers and users. Most of these nutritional agents have not been adequately researched, however, so our discussion of each is brief.

Amino Acids

Specific amino acids, or groups of amino acids, have been proposed to have special ergogenic properties. **L-tryptophan,** an essential amino acid, has been proposed to increase aerobic endurance performance through its effects on the CNS, acting as an analgesic and delaying fatigue. L-tryptophan is the first precursor of serotonin, a potent CNS neurotransmitter. Although an initial study of L-tryptophan indicated dramatic increases in endurance performance, subsequent studies have been unable to confirm these results, showing no improvement in endurance performance.

Branched-chain amino acids (BCAA)—leucine, isoleucine, and valine—have been postulated to work in combination with L-tryptophan to delay fatigue, primarily through CNS mechanisms. There is convincing evidence that exercise-induced increases in the plasma free tryptophan/BCAA ratio are associated with increased brain serotonin and the onset of fatigue during prolonged exercise.[21] Theoretically, increasing the BCAA would reduce the ratio and delay the onset of fatigue. One study observed the time to exhaustion on a cycle ergometer at 70% to 75% of $\dot{V}O_2$max under conditions that either increased tryptophan levels, increased

BCAA levels, or reduced BCAA levels, all substantially altering the tryptophan/BCAA ratio.[58] Exercise time to exhaustion was not different among treatments (see figure 15.9). This study and others call into question the efficacy of supplementing either tryptophan or BCAA to improve endurance performance.[21]

Others have postulated that supplementation of specific amino acids increases serum growth hormone release from the anterior pituitary gland. This also has not been clearly substantiated by research. There is, however, evidence that supplementation with a metabolite of leucine (β-hydroxy-β-methylbutyrate, or HMB) does increase fat-free mass and strength. It acts by decreasing the breakdown of protein that occurs with resistance training.[46] The benefits, however, might be limited to untrained individuals.[52]

L-Carnitine

Long-chain fatty acids are the major source of energy in the body, and fatty acid oxidation provides energy both at rest and during exercise. **L-carnitine** is important in fatty acid metabolism because it assists in the transfer of fatty acids from the cytosol (the fluid portion of the cytoplasm, exclusive of organelles) across the inner mitochondrial membrane for β-oxidation. This membrane is normally impermeable to long-chain fatty acids, so the availability of L-carnitine may be a limiting factor for the rate of fatty acid oxidation.

It has been theorized that by increasing the availability of L-carnitine, you might facilitate the oxidation of lipids. By relying more on fat as an energy source, you could possibly spare glycogen and increase aerobic endurance capacity. The studies of L-carnitine are mixed: Several show indirect evidence of increased fat oxidation with L-carnitine supplementation, but most show no effect on fat oxidation when using either indirect or direct estimates. Studies have shown that L-carnitine supplementation doesn't increase muscle storage of carnitine, enhance fatty acid oxidation, spare glycogen, or postpone fatigue during exercise, nor has it been shown unequivocally to improve athletes' performance.[35]

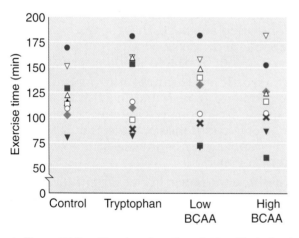

▲ **Figure 15.9**   Time to exhaustion of 10 subjects (each represented by a different symbol) on a cycle ergometer at 70% to 75% of $\dot{V}O_2$max. Each subject was tested under control conditions and after each of three treatments that increased tryptophan, reduced branched chain amino acid (BCAA), or increased BCAA. Exercise time to exhaustion was not significantly different between control and treatment conditions.

Adapted, by permission, from G. van Hall et al., 1995, "Ingestion of branched-chain amino acids and tryptophan during sustained exercise in man: Failure to affect performance," *Journal of Physiology* 486: 789-794.

Creatine

The use of **creatine** as an ergogenic aid has been widespread among athletes, ranging from the recreational to the professional. The primary use of creatine is based on its role in skeletal muscle, where approximately two thirds of its total is in the form of phosphocreatine (PCr). Increasing the creatine content of skeletal muscle through creatine supplementation is theorized to increase muscle PCr levels, thus enhancing the adenosine triphosphate (ATP)–PCr energy system by better maintaining muscle ATP levels. This, in turn, theoretically would enhance peak power production during intense exercise and possibly facilitate recovery from high-intensity exercise. Creatine also serves as a buffer, helping to regulate acid–base balance, and is involved in the oxidative metabolic pathways.

Because of the popularity of creatine supplementation in the 1990s and the wide-ranging claims of its ergogenic properties, the American College of Sports Medicine published a Consensus Statement on "The Physiological and Health Effects of Oral Creatine Supplementation" in 2000.[2] A group of eminent exercise and sport scientists reviewed the existing research literature on creatine and performance at that time to develop a consensus on the actual ergogenic benefits of creatine. From their review, they concluded the following:

- Creatine supplementation can increase muscle phosphocreatine content, but not in all individuals.
- The combination of creatine with a large amount of carbohydrate might increase muscle uptake of creatine.
- Exercise performance in short periods of intense, high power output activity can be enhanced, particularly with repeated bouts, consistent with the role of PCr in this type of activity.
- Maximal isometric strength, the rate of maximal force production, and maximal aerobic capacity are not enhanced by creatine supplementation.
- Creatine supplementation leads to weight gain within the first few days, likely at-tributable to water accumulation with creatine uptake in the muscle.
- In combination with resistance training, creatine supplementation is associated with greater gains in strength, possibly associated with the increased ability to train at higher levels.
- The high expectations for performance enhancement exceed the true ergogenic benefits.

Drs. Melvin Williams, Richard Kreider, and David Branch published an excellent book in 2000, which summarized all that was known about creatine supplementation from their extensive review of the basic science and clinical research literature.[62] They carefully reviewed all aspects of creatine supplementation and concluded the following:

Muscle creatine levels: Although not everyone responds to supplementation, most will increase their muscle creatine stores by about 20%, particularly when creatine is ingested with carbohydrate.

Anaerobic power: In activities that rely predominantly on the ATP-PCr energy system, 50 studies found enhanced performance and 42 reported no effect. The response was quite variable, dependent on the type of activity, and unclear with respect to sport performance.

Anaerobic endurance: Improvements in performance were found in 12 of 22 studies. With the exception of swimming performance, anaerobic endurance of more than 30 s up to 150 s appears to be increased with creatine supplementation.

Aerobic endurance: In 7 of 16 studies, aerobic endurance performance was improved with creatine supplementation, but the improvement was generally for aerobic endurance of relatively short duration.

Body mass and composition: Most studies have observed increases in total body mass in males, but there is not good agreement on the source of the gain in mass. Contractile protein might explain part of the gain in total mass, but it is likely that water retention explains most of the gains. When coupled with resistance training, creatine supplementation appears to increase fat-free mass.

So, it appears that there is potential for ergogenic benefits from creatine supplementation. Furthermore, there appears to be little risk in supplementing creatine, particularly at the smaller doses, with the possible exception of weight gain, which would not be desirable for some athletes.

> Creatine supplementation appears to have some ergogenic benefits, particularly for increasing the creatine content of skeletal muscle and for improving performance in intense, short-duration maximal exercise bouts of between 30 and 150 s. The evidence is less clear concerning its ability to increase muscle mass and strength.

Other Nutritional Agents

Other nutritional agents are proposed to have ergogenic properties, but there is insufficient research to clearly document the actual effect of these agents. Glycerol, released in the breakdown of triglycerides, has been postulated to be a gluconeogenic substrate, thus reducing the utilization of muscle and liver glycogen stores and enhancing aerobic endurance performance. Although limited data from rats support the reduced muscle and liver glycogen storage utilization effects of glycerol, human data are lacking. Glycerol also has been postulated to have an ergogenic effect by combining glycerol and water and thus allowing the body to become hyperhydrated. This water-loading effect would allow the body to better survive exercise in the heat, improving thermoregulation and delaying dehydration. The limited research to date supports the water-loading effect and suggests improved aerobic endurance performance. Along these same lines, limited data suggest that saline ingestion before exercise may expand the plasma volume and improve thermoregulation during exercise in the heat.[44]

Pyruvate has been promoted as an ergogenic agent for improving aerobic capacity, promoting weight and fat weight loss, serving as an antioxidant, and lowering blood lipids. Although the research on pyruvate supplementation is limited, there appears to be no basis to support any of these claims.[56]

- ▶ Although amino acid supplementation, particularly L-tryptophan and BCAA, has been proposed to have ergogenic properties, little evidence supports this proposal.

- ▶ Although L-carnitine is important in fatty acid metabolism, most studies show that supplementation doesn't increase muscle storage of carnitine, enhance fatty acid oxidation, spare glycogen, or delay fatigue during exercise.

- ▶ Creatine supplementation has been shown to increase muscle creatine levels and can increase performance in anaerobic power, anaerobic endurance, and aerobic endurance types of activities, although the results of these performance studies are mixed.

- ▶ Glycerol might have ergogenic properties, facilitating water storage and improving exercise performance in the heat through improved thermoregulation and reduced dehydration. Pyruvate supplementation has no proven benefits.

In Closing . . .

In this chapter, we reviewed some common substances and procedures thought to have ergogenic properties. All athletes must recognize the legal, ethical, moral, and medical consequences of using any ergogenic agent. The list of banned substances increases almost daily. Athletes who use banned substances risk disqualification from a particular competition, and they can be banned from competition in their sport for a year or more. In their quest for the perfect performance, athletes can easily get caught up in the hype surrounding various substances and the purported benefits they might bestow. Unfortunately, too many athletes are blinded by ambition and do not consider the consequences of their actions until their careers have been jeopardized or their health seriously impaired.

We have discussed pharmacological, hormonal, physiological, and some specific nutritional ergogenic aids. In the next part, we shift our focus away from athletes in general to the unique characteristics of younger, older, and female athletes within the broader categories of growth and development, aging, and sex differences in exercise performance. We begin in chapter 16 by examining special considerations in the child and adolescent.

▷ Key Terms

alcohol
amino acid
amphetamine
anabolic steroid
β-blocker
bicarbonate loading
blood doping
branched-chain amino acid (BCAA)
caffeine
cocaine
creatine
diuretic
ergogenic
ergogenic aid
ergolytic
erythropoietin
hormonal agent
human growth hormone (hGH)
L-carnitine
L-tryptophan
marijuana
nicotine
nutritional agent
oral contraceptive
oxygen supplementation
pharmacological agent
phosphate loading
physiological agent
placebo
placebo effect

▷ Study Questions

1. What is the meaning of the term *ergogenic aid*? What is an ergolytic effect?
2. Why is it important to include control groups and placebos when studying the ergogenic properties of any substance or phenomenon?
3. How does the use of alcohol in moderate or large doses affect athletic performance?
4. What is presently known about the use of amphetamines in athletic competition? What are the potential risks of using amphetamines?
5. Under what circumstances might β-blockers be ergogenic aids? What are some of their ergolytic properties?
6. How might caffeine improve athletic performance?
7. What is known about cocaine and marijuana as ergogenic aids?
8. Are diuretics ergogenic? What are some risks associated with their use?
9. What are the effects of anabolic steroid use on athletic performance? What are some of the medical risks of steroid use?
10. What is known about human growth hormone as a potential ergogenic aid? What are the risks associated with its use?
11. How are oral contraceptives used as potential ergogenic aids?
12. What is blood doping? Does blood doping improve athletic performance?
13. How is erythropoietin theorized to benefit performance?
14. How beneficial is breathing oxygen before the start of competition, during competition, and during the recovery from competition?
15. What are the potential ergogenic properties of bicarbonate, phosphate, and creatine?

▷ References

1. American College of Sports Medicine Position Stand. (1996). The use of blood doping as an ergogenic aid. *Medicine and Science in Sports and Exercise,* **28**(6), i-xii.

2. American College of Sports Medicine Consensus Statement. (2000). The physiological and health effects of oral creatine supplementation. *Medicine and Science in Sports and Exercise,* **32**, 706-717.

3. Anderson, R.L., Wilmore, J.H., Joyner, M.J., Freund, B.J., Hartzell, A.A., Todd, C.A., & Ewy, G.A. (1985). Effects of cardioselective and nonselective beta-adrenergic blockade on the performance of highly trained runners. *American Journal of Cardiology,* **55**, 149D-154D.

4. Anselme, F., Collomp, K., Mercier, B., Ahma didi, S., & Préfaut, C. (1992). Caffeine increases maximal anaerobic power and blood lactate concentration. *European Journal of Applied Physiology,* **65**, 188-191.

5. Ariel, G., & Saville, W. (1972). Anabolic steroids: The physiological effects of placebos. *Medicine and Science in Sports and Exercise,* **4**, 124-126.

6. Bahrke, M.S., Yesalis, C.E., & Wright, J.E. (1990). Psychological and behavioural effects of endogenous testosterone levels and anabolic-androgenic steroids among males: A review. *Sports Medicine,* **10**, 303-337.

7. Bannister, R.G., & Cunningham, D.J.C. (1954). The effects on the respiration and performance during exercise of adding oxygen to the inspired air. *Journal of Physiology,* **125**, 118-137.

8. Bhasin, S., Storer, T.W., Berman, N., Callegari, C., Clevenger, B., Phillips, J., Bunnell, T.J., Tricker, R., Shirazi, A., & Casaburi, R. (1996). The effects of supraphysiologic doses of testosterone on muscle size and strength in normal men. *New England Journal of Medicine,* **335**, 1-7.

9. Bredle, D.L., Stager, J.M., Brechue, W.F., & Farber, M.O. (1988). Phosphate supplementation, cardiovascular function, and exercise performance in humans. *Journal of Applied Physiology,* **65**, 1821-1826.

10. Broeder, C.E., Quindry, J., Brittingham, K., Panton, L., Thomson, J., Appakondu, S., Bruel, K., Byrd, R., Douglas, J., Earnest, C., Mitchell, C., Olson, M., Roy, T., & Yarlagadda, C. (2000). The Andro Project: Physiological and hormonal influences of androstenedione supplementation in men 35 to 65 years old participating in a high-intensity resistance training program. *Archives of Internal Medicine,* **160**, 3093-3104.

11. Bronson, F.H., & Matherne, C.M. (1997). Exposure to anabolic-androgenic steroids shortens life span of male mice. *Medicine and Science in Sports and Exercise,* **29**, 615-619.

12. Brown, G.A., Vukovich, M.D., Sharp, R.L., Reifenrath, T.A., Parsons, K.A., & King, D.S. (1999). Effect of oral DHEA on serum testosterone and adaptations to resistance training in young men. *Journal of Applied Physiology,* **87**, 2274-2283.

13. Buick, F.J., Gledhill, N., Froese, A.B., Spriet, L., & Meyers, E.C. (1980). Effect of induced erythrocythemia on aerobic work capacity. *Journal of Applied Physiology,* **48**, 636-642.

14. Cade, R., Conte, M., Zauner, C., Mars, D., Peterson, J., Lunne, D., Hommen, N., & Packer, D. (1984). Effects of phosphate loading on 2,3-diphosphoglycerate and maximal oxygen uptake. *Medicine and Science in Sports and Exercise,* **16**, 263-268.

15. Cantwell, J.D., & Rose, F.D. (1986). Cocaine and cardiovascular events. *Physician and Sportsmedicine,* **14**(11), 77-88.

16. Chandler, J.V., & Blair, S.N. (1980). The effect of amphetamines on selected physiological components related to athletic success. *Medicine and Science in Sports and Exercise,* **12**, 65-69.

17. Collomp, K., Ahma didi, S., Chatard, J.C., Audran, M., & Préfaut, C. (1992). Benefits of caffeine ingestion on sprint performance in trained and untrained swimmers. *European Journal of Applied Physiology,* **64**, 377-380.

18. Conlee, R.K. (1991). Amphetamine, caffeine, and cocaine. In D.R. Lamb & M.H. Williams (Eds.), *Ergogenics—Enhancement of performance in exercise and sport* (pp. 285-325). Dubuque, IA: Brown & Benchmark.

19. Costill, D.L., Dalsky, G.P., & Fink, W.J. (1978). Effects of caffeine ingestion on metabolism and exercise performance. *Medicine and Science in Sports,* **10**, 155-158.

20. Costill, D.L., Verstappen, F., Kuipers, H., Janssen, E., & Fink, W. (1984). Acid–base balance during repeated bouts of exercise: Influence of HCO_3^-. *International Journal of Sports Medicine,* **5**, 228-231.

21. Davis, J.M. (1995). Carbohydrates, branched-chain amino acids, and endurance: The central fatigue hypothesis. *International Journal of Sport Nutrition,* **5**, S29-S38.

22. Dodd, S.L., Herb, R.A., & Powers, S.K. (1993). Caffeine and exercise performance: An update. *Sports Medicine,* **15**, 14-23.

23. Duffy, D.J., & Conlee, R.K. (1986). Effects of phosphate loading on leg power and high intensity treadmill exercise. *Medicine and Science in Sports and Exercise, 18,* 674-677.

24. Earnest, C.P. (2001). Dietary androgen "supplements." *Physician and Sportsmedicine, 29*(5), 63-70.

25. Eichner, E.R. (1989). Ergolytic drugs. *Sports Science Exchange, 2*(15), 1-4.

26. Ekblom, B., & Berglund, B. (1991). Effect of erythropoietin administration on maximal aerobic power. *Scandinavian Journal of Medicine and Science in Sports, 1,* 88-93.

27. Ekblom, B., Goldbarg, A.N., & Gullbring, B. (1972). Response to exercise after blood loss and reinfusion. *Journal of Applied Physiology, 33,* 175-180.

28. Forbes, G.B. (1985). The effect of anabolic steroids on lean body mass: The dose response curve. *Metabolism, 34,* 571-573.

29. Fox, E.L., Bowers, R.W., & Foss, M.L. (1988). *The physiological basis for exercise and sport* (4th ed.). Madison, WI: Brown & Benchmark.

30. Gledhill, N. (1982). Blood doping and related issues: A brief review. *Medicine and Science in Sports and Exercise, 14,* 183-189.

31. Gledhill, N. (1985). The influence of altered blood volume and oxygen transport capacity on aerobic performance. *Exercise and Sport Sciences Reviews, 13,* 75-93.

32. Goforth, H.W., Jr., Campbell, N.L., Hodgdon, J.A., & Sucec, A.A. (1982). Hematologic parameters of trained distance runners following induced erythrocythemia [Abstract]. *Medicine and Science in Sports and Exercise, 14,* 174.

33. Graham, T.E. (2001) Caffeine and exercise: Metabolism, endurance and performance. *Sports Medicine, 31,* 785-807.

34. Graham, T.E., & Spriet, L.L. (1991). Performance and metabolic responses to a high caffeine dose during prolonged exercise. *Journal of Applied Physiology, 71,* 2292-2298.

35. Heinonen, O.J. (1996). Carnitine and physical exercise. *Sports Medicine, 22,* 109-132.

36. Hervey, G.R., Knibbs, A.V., Burkinshaw, L., Morgan, D.B., Jones, P.R.M., Chettle, D.R., & Vartsky, D. (1981). Effects of methandienone on the performance and body composition of men undergoing athletic training. *Clinical Science, 60,* 457-461.

37. Ivy, J.L. (1983). Amphetamines. In M.H. Williams (Ed.), *Ergogenic aids in sport* (pp. 101-127). Champaign, IL: Human Kinetics.

38. Ivy, J.L., Costill, D.L., Fink, W.J., & Lower, R.W. (1979). Influence of caffeine and carbohydrate feedings on endurance performance. *Medicine and Science in Sports and Exercise, 11,* 6-11.

39. King, D.S., Sharp, R.L., Vukovich, M.D., Brown, G.A., Reifenrath, T.A., Uhl, N.L., & Parsons, K.A. (1999). Effect of oral androstenedione on serum testosterone and adaptations of resistance training in young men: A randomized controlled trial. *Journal of the American Medical Association, 281,* 2020-2028.

40. Kowalchuk, J.M., Maltais, S.A., Yamaji, K., & Hughson, R.L. (1989). The effect of citrate loading on exercise performance, acid-base balance and metabolism. *European Journal of Applied Physiology, 58,* 858-864.

41. Kreider, R.B., Miller, G.W., Williams, M.H., Somma, C.T., & Nasser, T.A. (1990). Effects of phosphate loading on oxygen uptake, ventilatory anaerobic threshold, and run performance. *Medicine and Science in Sports and Exercise, 22,* 250-256.

42. Linderman, J., & Fahey, T.D. (1991). Sodium bicarbonate ingestion and exercise performance: An update. *Sports Medicine, 11,* 71-77.

43. Lombardo, J.A. (1986). Stimulants and athletic performance: Cocaine and nicotine. *Physician and Sportsmedicine, 14*(12), 85-91.

44. Luetkemeier, M.J., Coles, M.G., & Askew, E.W. (1997). Dietary sodium and plasma volume levels with exercise. *Sports Medicine, 23,* 279-286.

45. McNaughton, L.R. (1990). Sodium citrate and anaerobic performance: Implications of dosage. *European Journal of Applied Physiology, 61,* 392-397.

46. Nissen, S., Sharp, R., Ray, M., Rathmacher, J.A., Rice, D., Fuller, J.C., Connelly, A.S., & Abumrad, N. (1996). Effect of leucine metabolite β-hydroxy-β-methylbutyrate on muscle metabolism during resistance-exercise training. *Journal of Applied Physiology, 81,* 2095-2104.

47. O'Brien, C.P., & Lyons, F. (2000). Alcohol and the athlete. *Sports Medicine, 29,* 295-300.

48. Pärssinen, M., & Seppälä, T. (2002). Steroid use and long-term health risks in former athletes. *Sports Medicine, 32,* 83-94.

49. Roth, D.A., & Brooks, G.A. (1990). Lactate transport is mediated by a membrane-bound carrier in rat skeletal muscle sarcolemmal vesicles. *Archives of Biochemistry and Biophysics, 279,* 377-385.

50. Rudman, D., Feller, A.G., Nagraj, H.S., Gergans, G.A., Lalitha, P.Y., Goldberg, A.F., Schlenker, R.A., Cohn, L., Rudman, I.W., & Mattson, D.E. (1990). Effects of human growth hormone in men over 60 years old. *New England Journal of Medicine, 323,* 1-6.

51. Shangold, M.M. (1988). Gynecologic concerns in exercise and training. In M. Shangold & G. Mirkin (Eds.), *Women and exercise: Physiology and sports medicine* (pp. 186-194). Philadelphia: Davis.

52. Slater, G.J., & Jenkins, D. (2002). β-hydroxy-β-methylbutyrate (HMB) supplementation and the promotion of muscle growth and strength. *Sports Medicine, 30*, 105-116.

53. Spriet, L.L. (1991). Blood doping and oxygen transport. In D.R. Lamb & M.H. Williams (Eds.), *Ergogenics—Enhancement of performance in exercise and sport* (pp. 213-242). Dubuque, IA: Brown & Benchmark.

54. Spriet, L.L., MacLean, D.A., Dyck, D.J., Hultman, E., Cederblad, G., & Graham, T.E. (1992). Caffeine ingestion and muscle metabolism during prolonged exercise in humans. *American Journal of Physiology, 262*, E891-E898.

55. Stewart, I., McNaughton, L., Davies, P., & Tristram, S. (1990). Phosphate loading and the effects on $\dot{V}O_2$max in trained cyclists. *Research Quarterly for Exercise and Sport, 61*, 80-84.

56. Sukala, W.R. (1998). Pyruvate: Beyond the marketing hype. *International Journal of Sport Nutrition, 8*, 241-249.

57. Tribole, E. (1992). *Eating on the run* (2nd ed.). Champaign, IL: Leisure Press.

58. van Hall, G., Raaymakers, J.S.H., Saris, W.H.M., & Wagenmakers, A.J.M. (1995). Ingestion of branched-chain amino acids and tryptophan during sustained exercise in man: Failure to affect performance. *Journal of Physiology, 486*, 789-794.

59. Wadler, G.I., & Hainline, B. (1989). *Drugs and the athlete.* Philadelphia: Davis.

60. Williams, M.H. (Ed.). (1983). *Ergogenic aids in sport.* Champaign, IL: Human Kinetics.

61. Williams, M.H. (1991). Alcohol, marijuana, and beta blockers. In D.R. Lamb & M.H. Williams (Eds.), *Ergogenics—Enhancement of performance in exercise and sport* (pp. 331-369). Dubuque, IA: Brown & Benchmark.

62. Williams, M.H., Kreider, R.B., & Branch, J.D. (2000). *Creatine: The power supplement.* Champaign, IL: Human Kinetics.

63. Williams, M.H., Wesseldine, S., Somma, T., & Schuster, R. (1981). The effect of induced erythrocythemia upon 5-mile treadmill run time. *Medicine and Science in Sports and Exercise, 13*, 169-175.

64. Wilmore, J.H. (1988). Exercise testing, training, and beta-adrenergic blockade. *Physician and Sportsmedicine, 16*(12), 45-52.

65. Wilmore, J.H. (1991). Eating and weight disorders in the female athlete. *International Journal of Sport Nutrition, 1*, 104-117.

66. Winter, F.D., Snell, P.G., & Stray-Gundersen, J. (1989). Effects of 100% oxygen on performance of professional soccer players. *Journal of the American Medical Association, 262*, 227-229.

67. Yarasheski, K.E., Campbell, J.A., Smith, K., Rennie, M.J., Holloszy, J.O., & Bier, D.M. (1992). Effect of growth hormone and resistance exercise on muscle growth in young men. *American Journal of Physiology, 262*, E261-E267.

68. Yarasheski, K.E., Zachwieja, J.J., Angelopoulos, T.J, & Bier, D.M. (1993). Short-term growth hormone treatment does not increase muscle protein synthesis in experienced weight lifters. *Journal of Applied Physiology, 74*, 3073-3076.

69. Yesalis, C.E. (Ed.). (2000). *Anabolic steroids in sport and exercise* (2nd ed.). Champaign, IL: Human Kinetics.

▷ Selected Readings

Armsey, T.D., Jr., & Green, G.A. (1997). Nutrition supplements: Science vs. hype. *Physician and Sportsmedicine, 25*(6), 77-92.

Bahrke, M.S., & Yesalis, C.E. (2002). *Performance-Enhancing Substances in Sport and Exercise.* Champaign, IL: Human Kinetics.

Catlin, D.H., & Murray, T.H. (1996). Performance-enhancing drugs, fair competition, and Olympic sport. *Journal of the American Medical Association, 276*, 231-237.

Clarkson, P.M., & Thompson, H.S. (1997). Drugs and sport: Research findings and limitations. *Sports Medicine, 24*, 366-384.

Demant, T.W., & Rhodes, E.C. (1999). Effects of creatine supplementation on exercise performance. *Sports Medicine, 28*, 49-60.

Eichner, E.R. (1997). Ergogenic aids: What athletes are using—and why. *Physician and Sportsmedicine, 25*(4), 70-83.

Heigenhauser, G.J.F., & Jones, N.L. (1991). Bicarbonate loading. In D.R. Lamb & M.H. Williams (Eds.), *Ergogenics—Enhancement of performance in exercise and sport* (pp. 183-207). Dubuque, IA: Brown & Benchmark.

Huie, M.J. (1996). The effects of smoking on exercise performance. *Sports Medicine, 22*, 355-359.

Mesa, J.L.M., Ruiz, J.R., Gonzalez-Gross, M.M., Sainz, A.G., & Garzon, M.J. (2002). Oral creatine supplementation and skeletal muscle metabolism in physical exercise. *Sports Medicine, 32,* 903-944.

Morgan, W.P. (Ed.). (1972). *Ergogenic aids and muscular performance.* New York: Academic Press.

Reents, S. (2000). *Sport and exercise pharmacology.* Champaign, IL: Human Kinetics.

Rogol, A.D. (1989). Growth hormone: Physiology, therapeutic use, and potential for abuse. *Exercise and Sport Sciences Reviews, 17,* 353-377.

Tarnopolsky, M.A. (1994). Caffeine and endurance performance. *Sports Medicine, 18,* 109-125.

Tremblay, M.S., Galloway, S.D., & Sexsmith, J.R. (1994). Ergogenic effects of phosphate loading: Physiological fact or methodological fiction? *Canadian Journal of Applied Physiology, 19,* 1-11.

Yarasheski, K.E. (1994). Growth hormone effects on metabolism, body composition, muscle mass, and strength. *Exercise and Sport Sciences Reviews, 22,* 285-312.

P A R T V I

Age and Sex Considerations in Sport and Exercise

From the previous parts of this book, we have gained a good understanding of the general principles of exercise and sport physiology. Now we turn our attention to how these principles are specifically applied to children and adolescents, older individuals, and females. In chapter 16, "Children and Adolescents in Sport and Exercise," we examine the processes of human growth and development and how different stages affect a young person's physiological capacity and performance. We also consider how these stages of growth and development might alter our strategies for training young athletes for competition. In chapter 17, "Aging in Sport and Exercise," we discuss how exercise capacity and sport performance change as we age, asking to what extent this change is attributable to physiological aging and how much change might be attributable to an increasingly sedentary lifestyle. We discover the important role that training can play in minimizing the loss of performance capacity and physical conditioning that accompanies aging. In chapter 18, "Sex Differences in Sport and Exercise," we examine potential differences between women's and men's responses to exercise and training and the extent to which these differences are biological. We also focus on physiological and clinical issues specific to female athletes, including menstrual function, pregnancy, osteoporosis, and the high prevalence of eating disorders in female athletes.

CHAPTER 16

CHILDREN AND ADOLESCENTS IN SPORT AND EXERCISE

overview

Throughout the previous chapters of this book, we have examined the body's physiological responses to acute bouts of exercise and its adaptations to training and the environment. But the focus has been on the adult. For many years, it was assumed that children and adolescents responded and adapted identically to adults, but few scientists actually had studied children and adolescents. There is now a cadre of researchers—pediatric exercise physiologists—who focus on the exercise responses and adaptations of the child and adolescent. As a result, we now have a much better understanding and appreciation of both the differences and similarities between adults and children and adolescents, which we discuss in this chapter.

With respect to sport, competition in youth sports has grown considerably over the last three decades and is now a major part of the world of sport. Girls and boys compete in sports such as Little League softball, Bobby Sox softball, Little League baseball, Pop Warner football, gymnastics, minibike racing, soccer, swimming, track and field, and long-distance running. With this growing interest in age-group competition, many questions have been raised. Is competition physically or psychologically harmful for the preadolescent? Should preadolescents be allowed to compete in such activities as long-distance running or strength training?

We address these and similar questions in this chapter. We begin by considering the processes of growth and development, and then we examine how these processes affect responses and adaptations to exercise. Finally, we discuss motor ability and sports performance and special issues in the training of young athletes.

outline

At age 16, Bulgarian weightlifter Naim Suleimanov lifted 170 kg (375 lb) from the floor to full arms' length overhead. He was able to perform this incredible feat at a body weight of only 56 kg (123 lb)! In other words, he was able to lift more than three times his body weight, something that no one else had ever been able to do. He was considered at the time to be the best competitive weightlifter in the world. We are discovering that children and adolescents have tremendous physiological capacity in many areas of sport, but we must be cautious in our expectations and training techniques because this is a time of rapid change and possibly increased risk for injury—anatomical, physiological, and emotional.

Growth, development, and maturation are terms used to describe changes that occur in the body starting at conception and continuing through adulthood. **Growth** refers to an increase in the size of the body or any of its parts. **Development** refers to differentiation of cells along specialized lines of function (e.g., organ systems), so it reflects the functional changes that occur with growth. Finally, **maturation** refers to the process of taking on adult form and becoming fully functional, and it is defined by the system or function being considered. For example, skeletal maturity refers to having a fully developed skeletal system in which all bones have completed normal growth and ossification, whereas sexual maturity refers to having a fully functional reproductive system. The state of a child's or adolescent's maturity can be defined by

- chronological age,
- skeletal age, and
- stage of sexual maturation.

Throughout this chapter we refer to the child and the adolescent. The period of life from birth to the start of adulthood is generally divided into three phases: infancy, childhood, and adolescence. **Infancy** is defined as the first year of life. **Childhood** spans the period of time between the end of infancy (the first birthday) and the beginning of **adolescence** and is usually broken into early childhood (preschool) and middle childhood (elementary school). The period of adolescence is more difficult to define in chronological years, because it varies in both its onset and termination. Its onset generally is defined as the onset of **puberty**, when secondary sex characteristics develop and sexual reproduction becomes possible, and its termination as the completion of growth and development processes, such as attaining adult height. For most girls, adolescence ranges from 8 to 19 years and for most boys from 10 to 22 years.[17]

With the increasing popularity of youth sport and an emphasis on increasing children's physical fitness, we must understand the physiological aspects of growth and development. Children and adolescents must not be regarded as mere miniature versions of adults. They are unique at each stage in their development. The growth and development of their bones, muscles, nerves, and organs largely dictate their physiological and performance capacities. As children's size increases, so do almost all of their functional capacities. This is true of motor ability, strength, cardiovascular and respiratory function, and aerobic and anaerobic capacity. In the following sections, we examine age-related changes in a child's physical abilities.

Body Composition: Growth and Development of Tissues

To understand the physical capabilities of children and the potential impact that sport activity can have on young athletes, we must first consider the physical state of their bodies. In this section, we examine growth and development of selected body tissues.

Height and Weight

Specialists in the field of growth and development have spent considerable time analyzing the changes in height and weight that accompany growth. These two variables are most useful when we examine their rates of change. Change in height is assessed in terms of centimeters per year and change in weight in terms of kilograms per year. Figure 16.1*a* shows that height increases rapidly during the first 2 years of life. In fact, the child reaches about 50% of adult height by age 2. After this, height increases at a progressively slower rate throughout childhood; thus, there is a decline in the rate of its change. Just before puberty, the rate of change

in height increases markedly, followed by an exponential decrease in rate until full height is attained at a mean age of about 16.0 years in girls and 18.0 years in boys. The peak rate of growth in height occurs at approximately 11.4 years in girls and 13.4 years in boys. Figure 16.1*b* reveals the same overall trend for the rate of increase in weight. As with height, the peak rate of growth in body weight occurs at approximately 12.5 years in girls and 14.5 years in boys—slightly later than height.

> Girls mature physiologically about 2 years earlier than boys do.

Bone

Bones, joints, cartilage, and ligaments form the body's structural support. Bones provide points of attachment for the muscles, protect delicate tissues, and act as reservoirs for calcium and phosphorus, and some are involved in blood cell formation. Early in fetal development, the bones begin to develop from cartilage. Some flat bones, such as those of the skull, develop from fibrous membranes, but the vast majority of bones instead develop from hyaline cartilage. During fetal development, as well as during the initial 14 to 22 years of life, membranes and cartilage are transformed into bone through the process of **ossification,** or bone formation. The average ages at which the different bones in our bodies complete ossification differ widely, but bones typically begin to fuse in the preteens, and all are fused by the early 20s. On average, girls achieve full bone maturity several years before boys.

The structure of mature long bones is complex. Bone is a living tissue that requires essential nutrients, so it receives a rich blood supply. Bone consists of cells distributed throughout a matrix or lattice-type arrangement, and it is dense and hard because of deposits of lime salts, mainly calcium phosphate and calcium carbonate. For this reason, calcium is an essential nutrient, particularly during periods of bone growth and in the later years of life when bone tends to become brittle because of bone mineral loss associated with aging. Bones

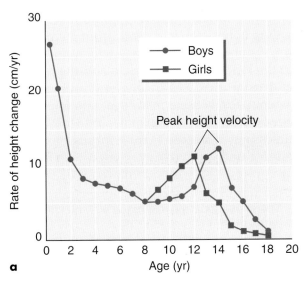

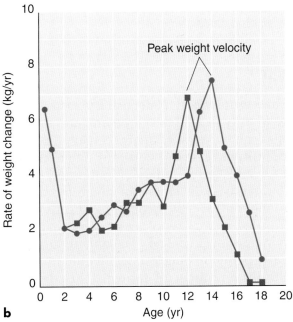

▲ **Figure 16.1**   Changes with age in the rate of increase in (*a*) height and (*b*) weight.

also store calcium. When our blood calcium level is high, excess calcium can be deposited in our bones for storage, and when calcium levels are too low, bone is resorbed, or broken down, to release calcium into the blood. When injury occurs or when extra stress is placed on a bone, more calcium is deposited. Thus, throughout life, our bones are constantly changing.

Exercise is essential for proper bone growth. Although exercise has little or no influence on bone lengthening, it does increase bone width and bone density by depositing more mineral in the bone matrix, which increases the bone's strength. There is preliminary evidence indicating that the prepubertal years may be the most opportune time to lay down bone in response to an exercise stimulus.[3]

> Exercise, along with an adequate diet, is essential for proper bone growth. Exercise affects primarily bone width, density, and strength but has little or no effect on length.

> ▶ Growth in height is very rapid during the first 2 years of life, with a child reaching 50% of adult stature by age 2. After that, the rate is slower throughout childhood until a marked increase occurs near puberty.
>
> ▶ The peak rate of height growth occurs at age 11.4 in girls and age 13.4 in boys. Full height is typically attained at age 16.0 in girls and age 18.0 in boys.
>
> ▶ Growth in weight follows the same trend as height. The peak rate of weight increase occurs at age 12.5 in girls and age 14.5 in boys.

Muscle

From birth through adolescence, the body's muscle mass steadily increases, along with the youngster's weight. In males, the total muscle mass increases from 25% of total body weight at birth to about 50% or more in the adult. Much of this gain occurs when the muscle development rate peaks at puberty.

This peak corresponds to a sudden, almost 10-fold increase in testosterone production. Girls don't experience such rapid acceleration of muscle growth at puberty, but their muscle mass does continue to increase, although much more slowly than boys', to about 40% of their total body weight as adults. This rate difference is largely attributed to hormonal differences at puberty (see chapter 18).

Increases in muscle mass with age appear to result primarily from hypertrophy (increase in size) of existing fibers, with little or no hyperplasia (increase in fiber number). This hypertrophy results from increases in the myofilaments and myofibrils. Increases in muscle length as young bones elongate result from increases in the number of sarcomeres (which are added at the junction of the muscle and the tendon) and from increases in the length of existing sarcomeres. Muscle mass peaks in females at age 16 to 20 years and in males at 18 to 25 years, unless it is increased further through exercise, diet, or both.

> The increase in muscle mass with growth and development is accomplished primarily by hypertrophy of individual muscle fibers through increases in their myofilaments and myofibrils. Muscle length increases through the addition of sarcomeres and by increases in the length of existing sarcomeres.

Fat

Fat cells form and fat deposition starts in these cells early in fetal development, and this process continues indefinitely thereafter. Each fat cell can increase in size at any age from birth to death. Early studies investigating fat cell and fat mass development suggested that the number of fat cells becomes fixed early in life. This led many scientists to believe that maintaining a low total body fat content during the early period of development would minimize the total number of fat cells that develop, greatly reducing the likelihood of obesity as an adult. But subsequent studies provided evidence suggesting that the number of fat cells can continue to increase throughout life.[4]

The most recent evidence suggests that as fat is added to the body, existing fat cells continue to fill with fat to a certain critical volume. Once these cells are filled to this point, new fat cells are formed. In light of this evidence, it is important to maintain good dietary and exercise habits throughout life!

The amount of fat that accumulates with growth and aging depends on

- your diet,
- your exercise habits, and
- heredity.

Heredity is unchangeable, but you can alter both diet and exercise to either increase or decrease your fat stores.

> Fat storage occurs by increasing the size of existing fat cells and by increasing the number of fat cells. It appears that as existing fat cells become full, they signal the need for the development of new fat cells.

At birth, 10% to 12% of total body weight is fat. At **physical maturity**, the fat content reaches approximately 15% of total body weight for males and approximately 25% for females. This sex difference, like that seen in muscle growth, is primarily attributable to hormonal differences. When girls reach puberty, their estrogen levels increase, which promotes the deposition of body fat. The trend of body fat increasing with age is shown in figure 16.2, which illustrates the relationship between subcutaneous fat (measured at the triceps and subscapular sites) and age for boys and girls from ages 2 to 18 years. The amount of subcutaneous fat is representative of total body fat. Figure 16.3 illustrates the changes in percent body fat, fat mass, and fat-free mass for both males and females from ages 8 to 20 years.[17] It is important to realize that both fat mass and

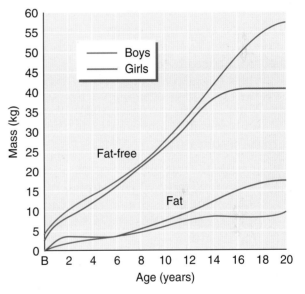

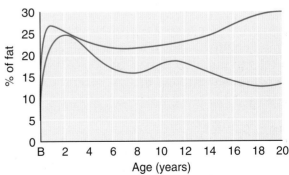

▲ **Figure 16.3** Changes in percent fat, fat mass, and fat-free mass for females and males from birth to 20 years of age.

Reprinted, by permission, from R.M. Malina and C. Bouchard, 1991, *Growth, maturation, and physical activity* (Champaign, IL: Human Kinetics), 97.

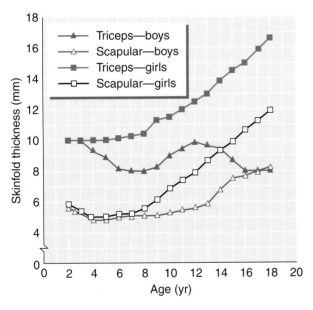

▲ **Figure 16.2** Changes in skinfold fat thickness at the triceps brachii and subscapular sites in boys and girls from ages 2 to 18 years.

Data from the NHANES-I, National Center for Health Statistics.

fat-free mass increase during this time period, so an increase in absolute fat does not necessarily mean an increase in relative fat.

Nervous System

As children grow, they develop better balance, agility, and coordination as their nervous systems develop. Myelination of the nerve fibers must be completed before fast reactions and skilled movement can occur because conduction of an impulse along a nerve fiber is considerably slower if myelination is absent or incomplete (chapter 2). **Myelination** of the cerebral cortex occurs most rapidly during childhood but continues well beyond puberty. Although practicing an activity or skill can improve performance to a certain extent, the full development of that activity or skill depends on full maturation (and myelination) of the nervous system. The development of strength is also likely influenced by myelination.

> ▶ Muscle mass increases steadily along with weight gain from birth through adolescence.
>
> ▶ In boys, the rate of muscle-mass increase peaks at puberty, when testosterone production increases dramatically. Girls do not experience this sharp increase in muscle mass.
>
> ▶ Muscle mass increases in boys and girls result primarily from fiber hypertrophy with little or no hyperplasia.
>
> ▶ Muscle mass peaks in girls between ages 16 and 20 and in boys between ages 18 and 25, although it can be further increased through diet and exercise.
>
> ▶ Fat cells can increase in size and number throughout life.
>
> ▶ The amount of fat accumulation depends on diet, exercise habits, and heredity.
>
> ▶ At physical maturity, the body's fat content averages 15% in males and 25% in females. The differences are caused

> primarily by higher testosterone levels in males and higher estrogen levels in females.
>
> ▶ Balance, agility, and coordination improve as children's nervous systems develop.
>
> ▶ Myelination of nerve fibers must be complete before fast reactions and skilled movements are fully developed because myelination speeds the transmission of electrical impulses.

Physiological Responses to Acute Exercise

The function of almost all physiological systems improves until full maturity is reached or shortly before. After that, physiological function plateaus for a period of time before starting to decline with advancing age. In this section, we focus on some of the changes in children and adolescents that accompany growth and development, including the following:

- Strength
- Cardiovascular and respiratory function
- Metabolic function, including aerobic capacity, running economy, and anaerobic capacity

Strength

Strength improves as muscle mass increases with age. Peak strength usually is attained by age 20 in women and between ages 20 and 30 in men. The hormonal changes that accompany puberty lead to marked increases in strength in pubescent males because of the increased muscle mass noted before. Brooks and Fahey[6] also observed that the extent of development and the performance capacity of muscle depend on the relative maturation of the nervous system. High levels of strength, power, and skill are impossible if the child has not reached neural maturity.

Myelination of many motor nerves is incomplete until sexual maturity, so the neural control of muscle function is limited before that time.

Figure 16.4 illustrates changes in leg strength in a group of boys from the Medford Boys' Growth Study.[7] The boys were followed longitudinally from age 7 to 18. The rate of strength gain increased noticeably around age 12, the typical age for onset of puberty. Similar longitudinal data for girls are not available. Cross-sectional data, however, indicate that girls experience a more gradual increase in strength and do not exhibit a marked change in their rate of strength gain with puberty,[11] as shown in figure 16.5.

Cardiovascular and Respiratory Function

Cardiovascular function undergoes considerable change as children grow and age. Let's consider some of these changes during submaximal and maximal exercise.

Rest and Submaximal Exercise

Blood pressure while at rest and during submaximal levels of exercise is lower in children than in adults but progressively increases to adult values during the late teen years. Blood pressure is also directly related to body size: Larger people generally have higher blood pressures. In children, blood flow to active muscles during exercise can be greater for a given volume of muscle than in adults because children have less peripheral resistance.

Recall that cardiac output is the product of heart rate and stroke volume. A child's smaller heart size and total blood volume result in a lower stroke volume, both at rest and during exercise, than in an adult. In an attempt to compensate for this, the child's heart rate response to a given rate of submaximal work (such as on a cycle ergometer), where the absolute oxygen requirement is the same, is higher than an adult's. As the child ages, heart size and blood volume increase along with body size.

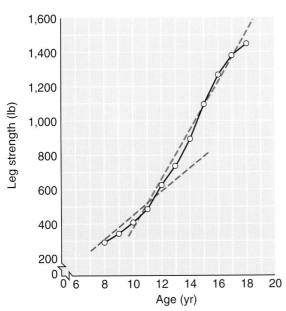

▲ **Figure 16.4** Gains with age in leg strength of young boys followed longitudinally over 12 years. Note the increase in the slope of the curve from 12 to 16 years of age.

Data from H.H. Clarke, 1971, *Physical and motor tests in the Medford boys' growth study* (Englewood Cliffs, NJ: Prentice-Hall).

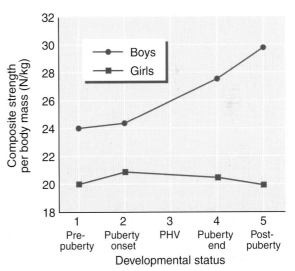

▲ **Figure 16.5** Changes in strength with developmental status in boys and girls. Strength is expressed as a composite static strength score from several strength testing sites, and the data are expressed per kilogram of body mass to account for differences in size between boys and girls. PHV = peak height velocity.

Reprinted, by permission, from K. Froberg and O. Lammert, 1996, "Development of muscle strength during childhood." In *The child and adolescent athlete* (London: Blackwell Publishing Company) 28.

Consequently, stroke volume also increases, as body size increases, for the same absolute rate of work.

However, a child's higher submaximal heart rate cannot completely compensate for the lower stroke volume. Because of this, the child's cardiac output is also somewhat lower than the adult's for the same absolute rate of work or same oxygen consumption. To maintain adequate oxygen uptake during these submaximal levels of work, the child's arterial-mixed venous oxygen difference, or a-$\bar{v}O_2$ difference, increases to further compensate for the lower stroke volume. The increase in a-$\bar{v}O_2$ difference is most likely attributable to increased blood flow to the active muscles—a greater percentage of the cardiac output goes to the active muscles.[29] These submaximal relationships are illustrated in figure 16.6, in which the responses of a 12-year-old boy are compared with those of a fully mature man.

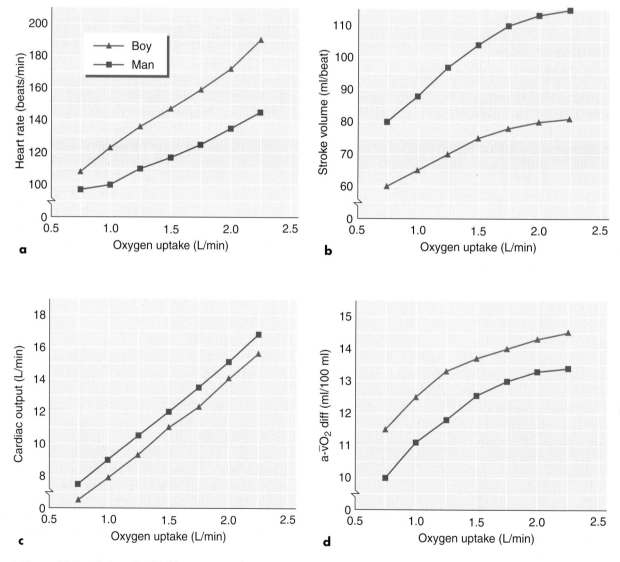

▲ **Figure 16.6**   Submaximal (*a*) heart rate, (*b*) stroke volume, (*c*) cardiac output, and (*d*) arterial-venous oxygen difference, or a-$\bar{v}O_2$ difference, in a 12-year-old boy and a fully mature man at the same rates of oxygen uptake.

Data from O. Bar-Or, 1983, *Pediatric sports medicine for the practitioner: From physiologic principles to clinical applications* (New York: Springer-Verlag).

Maximal Exercise

Maximum heart rate (HRmax) is higher in children than in adults but decreases linearly as children age. Children under age 10 frequently have maximum heart rates exceeding 210 beats/min, whereas the average 20-year-old has a maximum heart rate of approximately 195 beats/min. Results of cross-sectional studies suggest that maximum heart rate decreases by slightly less than 1 beat/min per year. Longitudinal studies, however, suggest that maximum heart rate decreases only 0.5 beats/min per year. Longitudinal studies, in which the same people are followed over time, generally provide more accurate estimates of the true changes.

During maximal levels of exercise, as also seen with submaximal exercise, the child's smaller heart and blood volume limit the maximal stroke volume that he or she can achieve. Again, the elevated maximum heart rate cannot fully compensate for this, leaving the child with a lower maximal cardiac output than the adult. This limits the child's performance at high absolute rates of work (e.g., pedaling at 100 W on a cycle ergometer or trying to achieve the same absolute $\dot{V}O_2$) because the child's capacity for oxygen delivery is less than an adult's. However, for high relative rates of work in which the child is responsible for moving only his or her body mass (e.g., running on a treadmill at the same speed with no grade), this lower maximal cardiac output is not as serious a limitation. In running, for example, a 25-kg (55-lb) child requires (in direct proportion to body size) considerably less oxygen than a 90-kg (198-lb) man would require, yet the rate of oxygen consumption per kilogram of body weight is about the same for both.

Heart size is directly related to body size, so children have smaller hearts than adults. As a result of this and a smaller blood volume, children have a smaller stroke volume capacity. A child's higher maximum heart rate can only partially compensate for this lower stroke volume capacity, and thus maximal cardiac output is lower than that of an equally trained adult.

Lung function changes markedly with growth. All lung volumes increase until growth is complete. Peak flow rates follow the same pattern. The changes in these volumes and flow rates are matched by the changes in the highest ventilation that can be achieved during exhaustive exercise, which is referred to

▶ Strength improves as muscle mass increases with age.

▶ Gains in strength with growth also depend on neural maturation because neuromuscular control is limited until myelination is complete, usually around sexual maturity.

▶ Blood pressure is directly related to body size: It is lower in children than in adults but increases to adult levels in the late teen years, both at rest and during exercise.

▶ During both submaximal and maximal exercise, a child's smaller heart and blood volume result in a lower stroke volume than in adults. In partial compensation, a child's heart rate is higher than an adult's for the same rate of work or $\dot{V}O_2$.

▶ Even with increased heart rate, a child's cardiac output remains less than an adult's. In submaximal exercise, an increase in the a-$\bar{v}O_2$ difference ensures adequate oxygen delivery to the active muscles. But at maximal work rates, oxygen delivery limits performance in activities other than those in which the child merely needs to move his or her body mass, such as in running.

▶ All lung volumes increase until physical maturity, primarily because of increasing body size.

▶ Until physical maturity, maximal ventilatory capacity and maximal expiratory ventilation increase in direct proportion to the increase in body size during maximal exercise.

as maximal expiratory ventilation ($\dot{V}_E$max), or maximal minute ventilation. $\dot{V}_E$max increases with age until physical maturity and then decreases with aging. For example, cross-sectional data show that $\dot{V}_E$max averages about 40 L/min for 4- to 6-year-old boys and increases to 110 to 140 L/min at full maturity. Girls follow the same general pattern, but their absolute values are considerably lower post-puberty, primarily because of their smaller body size. These changes are associated with the growth of the pulmonary system, which parallels the general growth patterns of children. As body size increases with growth and development, so does lung size and function.

Metabolic Function

Metabolic function also changes as the child and adolescent grow larger, as you would expect from the changes that we have just reviewed in muscle mass and strength and cardiorespiratory function. We look first at aerobic capacity.

Aerobic Capacity

The purpose of the basic cardiovascular and respiratory adaptations that occur in response to varying levels of exercise (rates of work) is to accommodate the exercising muscles' need for oxygen. Thus, increases in cardiovascular and respiratory function that accompany growth suggest that aerobic capacity ($\dot{V}O_2$max) similarly increases. In 1938, Robinson[20] demonstrated this phenomenon in a cross-sectional sample of boys and men ranging in age from 6 to 91 years. He found that $\dot{V}O_2$max peaks between ages 17 and 21 and then decreases linearly with age. Other studies subsequently confirmed these observations. Studies of girls and women have shown essentially the same trend, although in females the decrease begins at a much younger age, generally age 12 to 15 (refer to chapter 18), probably attributable to an earlier assumption of a sedentary lifestyle. The changes in $\dot{V}O_2$max with age, expressed in liters per minute, are illustrated in figure 16.7a.

Expressing $\dot{V}O_2$max relative to body weight (ml · kg^{-1} · min^{-1}) provides a considerably different picture, as shown in figure 16.7b. Values change little in boys from age 6 to young adulthood. For girls, however, little change occurs from age 6 to 13, but after age 13, aerobic capacities show a gradual decrease. Although these observations are of general interest, they might not accurately reflect the development

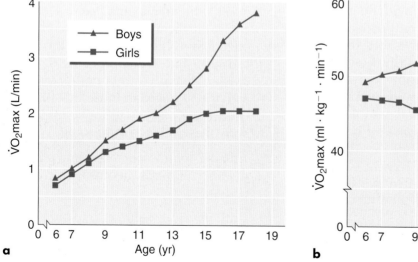
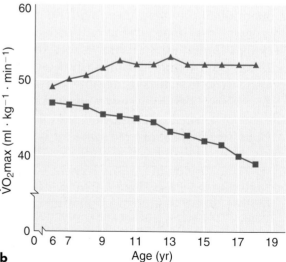

▲ **Figure 16.7** Changes in maximal oxygen uptake with age. Values are expressed (*a*) as absolute values in L/min and (*b*) relative to body weight in ml · kg^{-1} · min^{-1}.

of the cardiorespiratory system as children grow and their physical activity levels change. Several concerns have been raised about the validity of using body weight to account for changes in the size of the cardiorespiratory and metabolic systems, as when dividing absolute values by body weight, for example, $\dot{V}O_2$ per kilogram.

First, although $\dot{V}O_2$max values expressed relative to body weight remain relatively stable or decline with age, endurance performance steadily improves. The average 14-year-old boy can run the mile (1.6 km) almost twice as fast as the average 5-year-old boy, yet their $\dot{V}O_2$max values expressed relative to body weight are similar.[23] Second, although the increases in $\dot{V}O_2$max that accompany endurance training in children are relatively small compared with adults, the performance increases in these children are relatively large. Body weight might not be the most appropriate variable to use in scaling $\dot{V}O_2$max values for differences in body size in a young, growing child. The relationships between $\dot{V}O_2$max, body dimensions, and system functions during growth are extraordinarily complex.[24] This is discussed in greater detail later in this chapter.

> Aerobic capacity ($\dot{V}O_2$max), when expressed in liters per minute, is lower in children than in adults at similar levels of training. This is attributable primarily to the child's lower maximal cardiac output capacity. When $\dot{V}O_2$max values are expressed to reflect the differences in body size between children and adults, there is little or no difference in aerobic capacity.

Running Economy

How do growth-related changes in aerobic capacity affect a child's performance? For any activity that requires a fixed rate of work, such as cycling on an ergometer, the child's lower $\dot{V}O_2$max limits endurance performance. But as noted earlier, for activities where body weight is the major resistance to movement, such as distance running, children should not be at a disadvantage because their $\dot{V}O_2$max values

expressed relative to body weight are already at or near adult values.

Yet children cannot maintain a running pace as fast as adults because of basic differences in economy of effort. At a given speed on a treadmill, a child will have a substantially higher submaximal oxygen consumption when expressed relative to body weight than an adult. Even if the child's lactate threshold occurred at the same relative oxygen consumption as the adult (at the same percentage of their respective $\dot{V}O_2$max values), the child would be running at a much slower pace. Also, as children age, their legs lengthen, their muscles become stronger, and their running skills improve. Running economy increases, and this improves their distance-running pace, even if the children are not training and if their $\dot{V}O_2$max values don't increase.[8, 14]

Rowland[23] hypothesized that the following factors, which change with growth and development, explain at least in part the lower running economy in children and its improvement with maturation:

- Stride frequency
- Gait mechanics
- Musculotendinous elastic energy storage
- Surface area to body mass ratio
- Changes in body composition
- Thermal responses to exercise
- Substrate utilization
- Anaerobic capacity
- Ventilatory efficiency

Of these factors, only stride frequency thus far has been proven to be important (based on studies with children ages 8-20 years). It is also possible that scaling oxygen consumption to body weight is inappropriate during growth and development, as discussed in the previous section.[21]

Anaerobic Capacity

Children have a limited ability to perform anaerobic-type activities. This is demonstrated in several ways. Children cannot achieve adult

► As pulmonary and cardiovascular function improve with continued development, so does aerobic capacity.

► $\dot{V}O_2$max, expressed in liters per minute, peaks between ages 17 and 21 years in males and between 12 and 15 years in females, after which it steadily decreases.

► When $\dot{V}O_2$max is expressed relative to body weight, it plateaus in males from ages 6 to 25 years before it begins to decline. In females, the decline in $\dot{V}O_2$max is small from ages 6 to 12 years but begins a more substantial decline at about age 13. However, expressing $\dot{V}O_2$max relative to body weight might not provide an accurate estimate of aerobic capacity. Such $\dot{V}O_2$max values do not reflect the significant gains in endurance performance capacity that are noted with both maturation and training.

► The child's lower $\dot{V}O_2$max value (L/min) limits endurance performance unless body weight is the major resistance to movement, such as in distance running.

► When expressed relative to body weight, a child's $\dot{V}O_2$max is similar to an adult's, yet in activities such as distance running, a child's performance is far inferior to adult performance.

► Running economy is lower in children compared with adults, when $\dot{V}O_2$ is expressed relative to body weight. The only factor that has been found to explain this difference is the difference between children and adults in stride frequency for the same fixed-pace run.

concentrations of lactate in either muscle or blood for maximal and supramaximal rates of exercise, which indicates a lower glycolytic capacity. The lower lactate levels might reflect a lower concentration of phosphofructokinase, the key rate-limiting enzyme of anaerobic gly-

colysis. Lactate dehydrogenase activity also seems to be lower in children. However, lactate threshold, when expressed as a percentage of $\dot{V}O_2$max, does not appear to be a limiting factor in children because children's lactate thresholds are similar to, if not somewhat higher than, those of similarly trained adults. Also, children's resting levels of adenosine triphosphate (ATP) and phosphocreatine (PCr) are similar to those of adults, so activities of less than 10 to 15 s should not be compromised. Thus, only activities that tax the anaerobic glycolytic system—those from 15 s up to 2 min in duration—will be lower.[5]

> Anaerobic capacity is lower in children than adults, which might simply reflect children's lower concentration of the key rate-limiting enzyme phosphofructokinase or lactate dehydrogenase.

Children cannot achieve high respiratory exchange ratios during maximal or exhaustive exercise. Maximal respiratory exchange ratios in children are seldom above 1.10 and are sometimes below 1.00, but adult ratios are usually more than 1.10 and often greater than 1.15. This indicates that less carbon dioxide is produced in children for the same oxygen consumption, which in turn indicates less buffering of lactate.

Anaerobic mean and peak power output, as determined by the Wingate anaerobic power test (a 30-s, all-out maximal effort on a cycle ergometer), is also lower in children than in adults.[12] Figure 16.8 illustrates the results of a similar cycle ergometer anaerobic power test that is potentially a better discriminator of peak power output capacity.[26] In this figure, peak power is statistically adjusted for body weight to account for differences in body size when we compare values for preteenagers, teenagers, and adults. This figure demonstrates the very low peak power outputs for preteenagers (9-10 years of age) compared with both teenagers (14-15 years of age) and adults (mean age of 21 years). Teenagers were much closer to the values for adults than the values for the preteenagers. Again, these values were adjusted for body size, so they should accurately reflect anaerobic power.

Bar-Or[1] summarized the development of both the aerobic and anaerobic characteristics of boys and girls from ages 9 through 16, using 18 years of age as the criterion for 100% of the adult value. The changes with age are shown in figure 16.9. Aerobic power is represented by the child's $\dot{V}O_2max$, whereas anaerobic power is represented by the child's performance on the Margaria step-running test (a field test). Maximal energy expenditure per kilogram represents the maximal energy-generating capacities of the aerobic and anaerobic systems, scaled to body weight to account for body size differences with growth. Notice that aerobic fitness remains constant for the boys but declines for the girls from 12 to 16 years. Nine- to 12-year-old girls have a higher aerobic capacity than the 18-year-old reference adult value; thus, their

values are 110% of the adult value. For both boys and girls, anaerobic capacity increases from 9 through 15 years of age.

▶ Children's ability to perform anaerobic activities is limited. A child has a lower glycolytic capacity, possibly because of a limited amount of phosphofructokinase or lactate dehydrogenase.

▶ Children have lower lactate concentrations in both blood and muscle at maximal and supramaximal rates of work.

▶ Children cannot attain high respiratory exchange ratios during maximal or exhaustive exercise, suggesting less lactate production.

▶ Anaerobic mean and peak power outputs are lower in children than in adults, even when scaled for body mass.

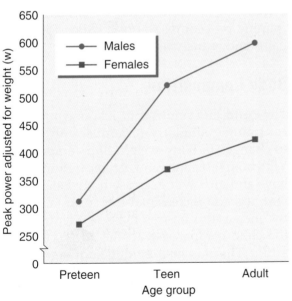

▲ **Figure 16.8** Optimal peak power output (anaerobic power) adjusted for body mass in preteenagers (9-10 years old), teenagers (14-15 years old), and adults (mean age of 21 years). These values represent anaerobic power independent of body size.

Data from Santos et al., 2002.[26]

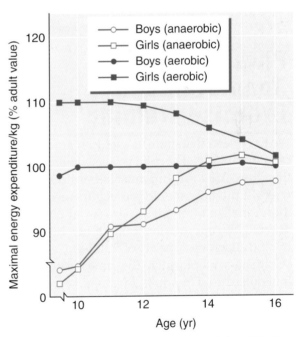

▲ **Figure 16.9** Development of aerobic and anaerobic characteristics in boys and girls ages 9 to 16 years. Values are expressed as a percentage of adult values (values at age 18).

Adapted, by permission, from O. Bar-Or, 1983, *Pediatric sports medicine for the practitioner: From physiologic principles to clinical applications* (New York: Springer-Verlag).

Scaling Physiological Data to Account for Size Differences

Throughout this chapter and previous chapters, we have discussed the need to express physiological data relative to the size of the individual. In chapter 4, when we introduced the concept of $\dot{V}O_2$max, we mentioned that values normally are expressed relative to body mass, by dividing the absolute $\dot{V}O_2$max (expressed in L/min) by body weight (ml · kg^{-1} · min^{-1}). Many scientists believe that dividing by body mass alone does not adequately account for size differences. This becomes a major issue when we compare children's values with those of adults, or men's with women's, as we see in chapter 18.

Strong cases have been made for scaling $\dot{V}O_2$, cardiac output, stroke volume, and other size-related physiological variables relative to body surface area, measured in square meters, or relative to weight, expressed to the 0.67 power or 0.75 power (wt$^{0.67}$ or wt$^{0.75}$). For years, cardiologists have expressed heart volumes relative to body surface area. Recent research suggests that using body surface area (ml · m^{-2} · min^{-1}) or wt$^{0.75}$ (ml · kg$^{-0.75}$ · min^{-1}) provides the best means by which to express the data to reduce the effect of body size.[21] One study followed young boys longitudinally from 12 to 20 years of age; one group remained untrained but active, and the other group trained.[28] There was little or no increase with run training in $\dot{V}O_2$max expressed in ml · kg^{-1} · min^{-1}, whereas submaximal $\dot{V}O_2$ expressed in the same manner decreased with age, suggesting no change in aerobic capacity but an improvement in running economy. When these same data were expressed in ml · kg$^{-0.75}$ · min^{-1}, the boys who were training showed increased aerobic capacity with increased training and age but no change in running economy. This latter finding intuitively makes more sense, supporting the use of wt$^{0.75}$ as the best way of expressing the data.

Physiological Adaptations to Exercise Training

We have seen that, indeed, children are not just miniature adults. The child is physiologically distinct from the adult and must be considered differently. But how should these differences affect individualized training programs for children? Training can improve the strength, aerobic capacity, and anaerobic capacity of children. Generally, youngsters adapt well to the same type of training routine used by adults. But training programs for children and adolescents should be designed specifically for each age group, keeping in mind the developmental factors associated with that age. In this section, we look at training-induced changes in each of the following:

- Body composition
- Strength
- Aerobic capacity
- Anaerobic capacity

Then, where appropriate, we discuss proper training procedures to optimize performance gains and reduce the risk of injury.

Body Composition

The child and adolescent respond to physical training similarly to adults with respect to changes in body weight and composition. With both resistance and aerobic training, both boys and girls will decrease body weight and fat mass and increase fat-free mass, although the increase in fat-free mass is attenuated in the child compared with the adolescent and adult. There is also evidence of significant bone growth as a result of exercise training, above that seen with normal growth. In fact, Bass[3] suggested that the prepubertal years may be the most opportune time to increase bone mass because of increases in bone density and periosteal expansion of cortical bone.

We shall see in chapter 21 that there is presently an epidemic of obesity in the United States, Canada, and much of Europe. This is true not only in adults but in children and adolescents as well. Physical training and an active lifestyle are critical throughout the growing years to maintain a healthy body composition and establish a life-long habit of exercise and activity.

Strength

For many years, the use of resistance training to increase muscular strength and endurance in prepubescent and adolescent boys and girls was highly controversial. Boys and girls were discouraged from using free weights for fear that they might injure themselves and prematurely stop the growth process. Furthermore, many scientists speculated that resistance training would have little or no effect on the muscles of prepubescent boys because their levels of circulating androgens were still low.

Studies on animals suggest that heavy resistance exercise can lead to stronger, broader, and more compact bones. But these studies have not contributed much to our understanding of the benefits or risks associated with this form of activity for humans because it is nearly impossible to load these animals to the same extent as youngsters can be loaded. Fortunately, several studies have been conducted in which both prepubescent and adolescent children have participated in resistance training. From these studies, Kraemer and Fleck[13] concluded that the risk of injury is very low. In fact, resistance training might offer some protection against injury, for example, by strengthening the muscles that cross a joint. Still, a conservative approach is recommended in prescribing resistance exercise for children, particularly preadolescents.

Several studies conducted in the mid-1980s demonstrated that prepubescent boys and girls can participate safely in resistance training and gain substantial strength. In one study, prepubescent boys and girls took part in a 9-week progressive resistance training program.[27] They exercised 25 to 30 min per day, 3 days each week. Their mean strength increase was 42.9%, compared with a 9.5% increase in a nontraining control group. The increase in the nontraining control group was expected, attributable to

normal growth over this period of time. In a second study, 16 prepubescent males between ages 6 and 11 participated in a 14-week strength training program using isokinetic techniques with hydraulic resistance, whereas another 10 boys served as nontraining controls.[30] Isokinetic strength increased between 18% and 37% in the training group, but little or no change was observed in the control group. The authors believed that only one reported injury was related to the strength-training routine. The injured boy missed only three training sessions. Interestingly, an additional six subjects reported injuries from activities of normal daily living, independent of the strength training program. None of the subjects demonstrated any damage to the epiphyses, bones, or muscles as a result of strength training.

In a final study,[33] prepubescent, pubescent, and postpubescent males underwent a 9-week resistance training program. All three groups had significant strength gains.[18] Researchers hypothesized that the pubescent group would experience the greatest strength gains because testosterone levels increase dramatically during this period, but this was not the case. In fact, the prepubescent group made greater gains than the pubescent group in several of the strength tests.

How are these increases in strength accomplished? The mechanisms allowing strength changes in children are similar to those for adults, with one minor exception: Prepubescent strength gains are accomplished largely without any changes in muscle size.[25] A comprehensive study of the mechanisms responsible for strength increases in prepubescent boys concluded that the likely determinants of the strength gains achieved are

- improved motor skill coordination,
- increased motor unit activation, and
- other undetermined neurological adaptations.[19]

Strength gains in the adolescent result primarily from neural adaptations and increases in both muscle size and specific tension. Kraemer and Fleck[13] provided a model that integrates various developmental factors that affect a person's potential for muscle strength adaptations with resistance training. This model is illustrated in figure 16.10. In this model, strength is influenced by the amount of fat-free mass, testosterone concentrations (in males), the extent of nervous system development, and the differentiation of fast-twitch (FT) and slow-twitch (ST) muscle fibers. As previously mentioned, the early gains in strength up through puberty are largely the result of changes in neuromuscular patterns.

> Prepubescent children can improve their strength with resistance training. These strength gains are attributable largely to neurological factors, with little or no change in the size of the muscle.

For actual training programs, resistance training for children should be prescribed in much the same way as for adults. Specific guidelines (see sidebar) were established at a workshop in 1985 by a group representing eight different professional organizations: the American Orthopaedic Society for Sports Medicine, the American Academy of Pediatrics, the American College of Sports Medicine, the National Athletic Trainers Association, the National Strength and Conditioning Association, the President's Council on Physical Fitness and Sports, the U.S. Olympic Committee, and the Society of Pediatric Orthopaedics. Also, Kraemer and Fleck[13] have established basic guidelines for the progression of resistance exercise in children, which are presented in table 16.1 on page 530. Further information on resistance training program designs for children is available.[10, 13, 25] Any youth resistance training program must be carefully supervised by competent instructors who have been trained specifically to work with children. Furthermore, resistance training should be only one part of a more comprehensive fitness program for this age group. In the next sections, we consider the value of aerobic and anaerobic training.

Aerobic Capacity

Do prepubescent boys and girls benefit from aerobic training to improve their cardiorespiratory systems? This also has been a highly controversial area because several early studies indicated that training prepubescent children did not change their $\dot{V}O_2$max values.[22] Interestingly, even without significant increases in $\dot{V}O_2$max, the running performance of the children studied did improve substantially. They could run a fixed distance faster following the training program. More recent studies have found small increases in aerobic capacity with training in prepubescent children, but these increases are less than would be expected for adolescents or adults—about 5% to 15% in children compared with about 15% to 25% in adolescents and adults.

More substantial changes in $\dot{V}O_2$max appear to occur once children have reached puberty, although the reason for this is unknown. Because stroke volume appears to be the major limitation to aerobic performance in this age group, it is quite possible that further increases in aerobic capacity depend on heart growth. Also, as discussed earlier in this chapter, scaling of these variables is an issue. The study of Sjödin and Svedenhag[28] presented in the sidebar on page 526 clearly establishes this as a key factor.

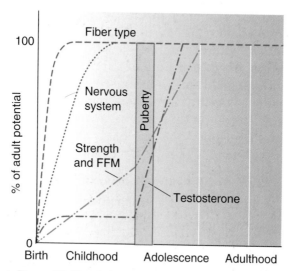

▲ **Figure 16.10**　A theoretical interactive model that integrates various developmental factors that are related to the potential for muscle strength adaptations with training.

Reprinted, by permission, from W.J. Kraemer et al., 1989, "Resistance training and youth," *Pediatric Exercise Science* 1(4): 342.

Resistance Training Recommendations for Prepubescent Children

Resistance training equipment

1. Equipment should be of appropriate design to accommodate the prepubescent's size and degree of maturity.
2. Equipment should be cost effective.
3. Equipment should be safe, free of defects, and inspected frequently.
4. Equipment should be located in an uncrowded area free of obstructions with adequate lighting and ventilation.

Program considerations

1. A preparticipation physical exam is mandatory.
2. The child must have the emotional maturity to accept coaching and instruction.
3. There must be adequate supervision by coaches who are knowledgeable about strength training and the special problems of prepubescents.
4. Strength training should be a part of a comprehensive program designed to increase motor skills and level of fitness.
5. Strength training should be preceded by a warm-up period and followed by a cool-down.
6. Emphasis should be on dynamic concentric actions.
7. All exercises should be carried through a full range of motion.
8. Competition is prohibited.
9. No maximal lift should ever be attempted.

Prescribed program

1. Training is recommended two or three times a week for 20- to 30-min periods.
2. No resistance should be applied until proper form is demonstrated. Six to 15 repetitions equal one set; one to three sets per exercise should be done.
3. Weight or resistance is increased in 0.5- to 1.4-kg (1- to 3-lb) increments after the prepubescent does 15 repetitions in good form.

Anaerobic Capacity

Anaerobic training appears to improve children's anaerobic capacity. Following training, children have

- increased resting levels of phosphocreatine, ATP, and glycogen;
- increased phosphofructokinase activity; and
- increased maximal blood lactate levels.[1, 9]

Ventilatory threshold, a noninvasive marker of lactate threshold, also has been reported to increase with endurance training in 10- to 14-year-old boys.[15]

When you design aerobic and anaerobic training programs for children and adolescents, it appears that standard training principles for adults can be applied. Children and adolescents have not been well studied, but what we do know suggests that they can be trained in a similar manner as adults. Again, because children and adolescents are not adults, it is prudent to be conservative to reduce the risk of injury, overtraining, and loss of interest in sport. The approach outlined earlier for resistance training

Table 16.1

Basic Guidelines for Resistance Exercise Progression in Children

Age (years)	Considerations
7 or younger	Introduce basic exercises with little or no weight; teach exercise techniques.
	Develop the concept of a training session.
	Progress from body weight calisthenics, partner exercises, and lightly resisted exercises.
	Keep volume low.
8-10	Gradually increase the number of exercises and training volume.
	Practice exercise technique in all lifts; keep exercises simple.
	Start gradual, progressive loading of exercises, carefully monitoring toleration to the exercise stress.
11-13	Teach all basic exercise techniques, emphasizing technique.
	Continue progressive loading of each exercise.
	Introduce more advanced exercises with little or no resistance.
14-15	Progress to more advanced youth programs in resistance exercise.
	Add sport-specific components.
	Emphasize exercise techniques.
	Increase volume.
16 or older	Move child to entry-level adult programs after all background knowledge has been mastered and a basic level of training experience has been gained.

Note. If a child of any age begins a program with no previous experience, start the child at previous levels and move him or her to more advanced levels as exercise toleration, skill, amount of training time, and understanding permit.

Reprinted, by permission, from W.J. Kraemer and S.J. Fleck, 1993, *Strength training for young athletes* (Champaign, IL: Human Kinetics), 5.

is a good model to use for aerobic and anaerobic training. This is also an appropriate time in life to focus on learning a variety of motor skills by exploring a number of activities and sports.

Motor Ability and Sport Performance

As shown in figure 16.11 on page 532, the motor ability of boys and girls generally increases with age for the first 17 years, although girls tend to plateau at about the age of puberty for most items tested. These improvements result primarily from development of the neuromuscular and endocrine systems and secondarily from the children's increased activity.

The plateau observed in the girls at puberty is likely explained by three factors. First, as mentioned earlier, the increase in estrogen levels at puberty, or in the estrogen/testosterone ratio,

▶ Body composition changes with training in children and adolescents are similar to those seen in adults—loss of total body weight and fat mass and increase in fat-free mass.

▶ Animal studies suggest that resistance training can lead to stronger, broader, denser bones.

▶ The risk of injury from resistance training in young athletes is relatively low, and the programs they should follow are much like those for adults.

▶ Strength gains achieved from resistance training in preadolescents result primarily from improved motor skill coordination, increased motor unit activation, and other neurological adaptations. Unlike adults, preadolescents who resistance train experience little change in muscle size.

▶ Aerobic training in preadolescents does not alter $\dot{V}O_2max$ as much as would be expected for the training stimulus, possibly because $\dot{V}O_2max$ depends on heart size. But endurance performance improves with aerobic training.

▶ A child's anaerobic capacity increases with anaerobic training.

▶ In general, growth and maturation rates and processes are probably not altered significantly by training.

leads to increased fat deposition. Performance tends to decrease as fat increases. Second, girls have less muscle mass. Finally, and probably of greater importance, around puberty many girls assume a much more sedentary lifestyle than boys. This is largely a matter of social conditioning, as boys are encouraged to be more active and athletic than girls. As girls become less active, their motor abilities tend to plateau. This trend appears to be changing because of changing social attitudes and more opportunities for sport and activity now available for girls (see chapter 18).

Sport performance in children and adolescents improves with growth and maturation, as can be seen for age group records in sports such as swimming and track and field. Figure 16.12 on page 533 illustrates the improvement in American records for the following age groups:

• 10 years and under
• 11 to 12 years
• 13 to 14 years
• 15 to 16 years
• 17 to 18 years

The figure gives values for the 100-m and 400-m swim and the 100-m and 1,500-m run. These events were selected because they represent both a predominantly anaerobic event in swimming and running (100-m swim and run) and a predominantly aerobic activity (400-m swim and 1,500-m run). Both anaerobic and aerobic performance improved progressively with increasing age groups, with the exception of the 1,500-m run for 17- and 18-year-old girls. Similar age-group records for weightlifting do not appear to be available, because weightlifting competition is organized by weight in broad classifications such as 16 and under, 17 to 20 years of age, and then adult classifications. On the basis of normal strength gains with growth and development, it is assumed that weightlifting records would increase markedly from late childhood through adolescence, particularly in boys.

▶ Motor ability generally increases for the first 18 years of life, although in girls it tends to plateau around puberty. This plateau is probably attributable to increased estrogen levels (which promote greater fat deposition), less muscle mass, and a more sedentary lifestyle.

▶ Sport performance improves dramatically through childhood and adolescence.

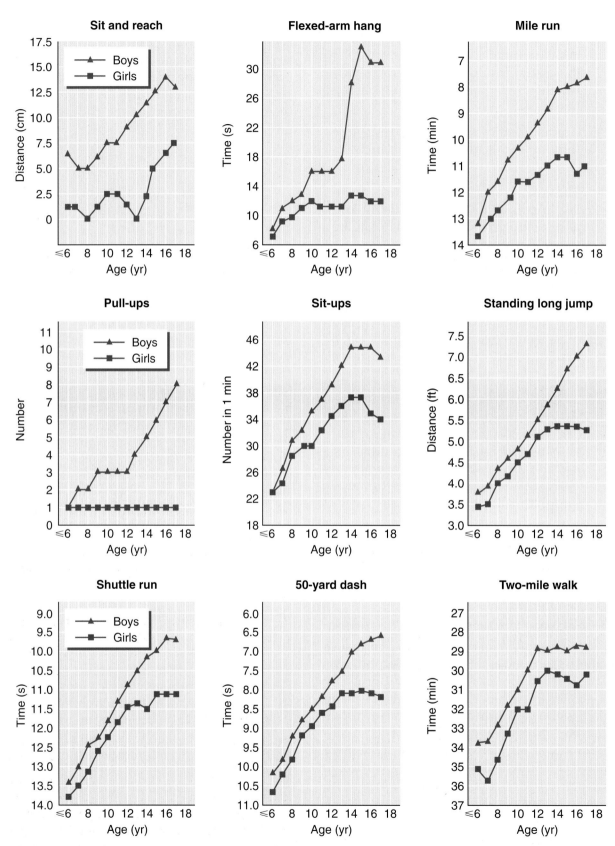

▲ Figure 16.11 Changes in motor ability from the ages of 6 years to 17 years.
Data from the President's Council on Physical Fitness and Sports, 1985.

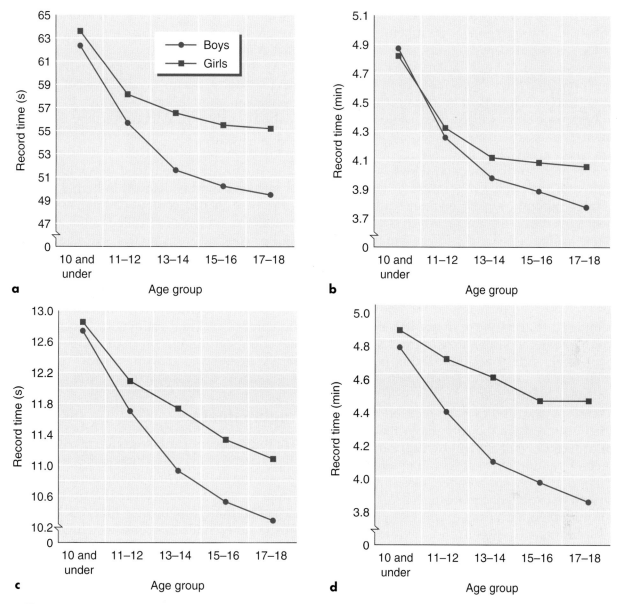

▲ **Figure 16.12**   United States national record performances for boys and girls from 10 years of age and under up through 17 to18 years of age in the following swimming and running events: (*a*) 100-m swim; (*b*) 400-m swim; (*c*) 100-m run; and (*d*) 1,500-m run. Records were obtained from USA Track & Field (as of December 2001; http://www.usatf.org) and United States Swimming (as of November 2001; http://www.usaswimming.org).

Special Issues

During the period of growth and development from childhood through adolescence, several special issues need to be addressed:

- Thermal stress
- Growth and maturation with training
- Youth fitness

Thermal Stress

Laboratory experiments suggest that children are more susceptible to heat- and cold-induced illness or injury than adults. But the number of reported cases of thermal illness or injury has not supported this theory. A major concern is the child's apparently lower capacity when exercising in the heat to dissipate heat through evaporation. Children appear to rely much more on convection and radiation, which are enhanced through greater peripheral vasodilation.[2] Compared with adults, children have a greater ratio of body surface area to mass, meaning that they have more skin surface area from which to gain or lose heat for each kilogram of body weight. In a cool environment, this is an advantage, because children are better able to lose heat through radiation, convection, and conduction. However, once the environmental temperature exceeds the skin's temperature, children more readily gain heat from the environment, which is a distinct disadvantage. A child's lower capacity for evaporative heat loss is largely the result of a lower sweating rate. Individual sweat glands in children form sweat more slowly and are less sensitive to increases in the body's core temperature than those in adults. Although young boys can acclimatize to exercise in the heat, their rate of acclimatization is slower than that of adults. Acclimatization data are not available for girls.

Only a few studies have focused on children exercising in the cold. From the limited information available, children appear to have greater conductive heat loss than adults, because of a larger surface area/mass ratio. This should be expected to place them at higher risk for hypothermia.

Few studies have been conducted on children for either heat or cold stress, and conclusions from the studies sometimes have been contradictory. More research is needed in this area to determine the risks faced by children who exercise in the heat and cold. In the meantime, a conservative approach is advisable. Children may be at an increased risk of heat- and cold-related injuries compared with adults.[2]

Growth and Maturation With Training

Many people have wondered what effect physical training has on growth and maturation. Does hard physical training slow down or accelerate normal growth and development? In a comprehensive review of this area, Malina made some interesting and relevant observations.[16] Regular training has no apparent effect on growth in height. It does, however, affect weight and body composition, as discussed earlier in this chapter.

As for maturation, the age at which peak height velocity occurs generally is not affected by regular training, nor is the rate of skeletal maturation. But the data concerning the influence of regular training on indexes of sexual maturation are not as clear. Although some data suggest that menarche (the initial onset of menstruation) is delayed in highly trained girls, these data are confounded by a number of factors that weren't controlled in the analysis. Malina concluded his review with the following statement: "Responses of the developing individual to the physical activity of regular training are probably not sufficient to alter genotypically programmed growth and maturation processes. Thus, training has no apparent effect on stature and on maturation as ordinarily assessed in growth studies" (p. 261).[16] Menarche is discussed in chapter 18.

Youth Fitness

There has been a major concern with what many perceive to be a reduction in activity levels among children and adolescents in North America and Europe. Over the past several decades, there has been a marked increase in the time children spend watching television and playing computer and video games. Although many children are active in sport and recreational activities, most are not. To further complicate this situation, many states within the United States have either greatly reduced or eliminated the physical education requirement in their public schools. From this, it would seem that our youth must be less fit and more fat. Unfortunately, we have very few data to support this contention, largely because neither good longitudinal nor cross-sectional data are available. We need to track fitness much more closely, because when activity levels decrease below a certain level and fat levels increase above a certain level, children and adolescents are at increased risk for chronic disability and disease, as we will see in part VII. An active lifestyle must be promoted very early in life!

▶ Laboratory studies suggest that children are more susceptible to injury or illness from thermal stress, but the number of reported cases does not support this.

▶ Children are capable of less evaporative heat loss than adults because children sweat less (less sweat is produced by each active sweat gland).

▶ Young boys acclimatize to heat more slowly than adults do. Data on this topic are not available for girls.

▶ Children appear to have greater conductive heat loss than adults, which should place children at greater risk for hypothermia in cold environments.

▶ Until more is known about children's susceptibility to thermal stress, a conservative approach should be used for children who exercise in temperature extremes.

▶ Physical training appears to have little or no negative effect on normal growth and development.

▶ Physical training can reduce body fat stores and increase fat-free body mass

▶ It appears that today's children and adolescents are increasingly more inactive. This is associated with increasing fatness and increased risk for chronic diseases in adulthood.

In Closing . . .

In this chapter, we have discussed children and young athletes. We have seen how children gain more control of movements as their body systems grow and develop. We have seen how their developing systems can sometimes limit performance capacities and how training can improve children's performances.

We have seen that, in general, the ability to perform increases as children approach physical maturity. But as people move beyond the point of physical maturity, their physiological functioning begins to decline. Having considered the developmental process, we are now ready to consider the aging process. How is performance affected as we move beyond our physiological prime? This is our focus in the next chapter as we turn our attention to aging and the older athlete.

▶ Key Terms

adolescence

childhood

development

growth

infancy

maturation

myelination

ossification

physical maturity

puberty

▶ Study Questions

1. What is the major concern when a bone that has not reached full growth breaks?

2. At what ages does fat-free body mass reach its peak rate of growth in males and in females?

3. What typical changes occur in fat cells with growth and development?

4. How does pulmonary function change with growth?

5. What changes occur in stroke volume for a fixed rate of work as a child grows? What factors explain these changes?

6. What changes occur in cardiac output for a fixed rate of work as a child grows? What factors explain these changes?

7. What changes occur in submaximal and maximal heart rate as a child grows?

8. Why does absolute aerobic or cardiorespiratory endurance capacity increase from age 6 to 20?

9. How do children differ from adults with respect to thermoregulation?

10. How dangerous is resistance training for children? What advice would you give to children if they wanted to improve their strength? Can they improve strength, and if so, how does this occur?

11. What happens to aerobic capacity as a prepubescent child trains aerobically?

12. What happens to anaerobic capacity as a prepubescent child trains anaerobically?

13. How do physical activity and regular training affect the growth and maturation processes?

▶ References

1. Bar-Or, O. (1983). *Pediatric sports medicine for the practitioner: From physiologic principles to clinical applications.* New York: Springer-Verlag.

2. Bar-Or, O. (1989). Temperature regulation during exercise in children and adolescents. In C.V. Gisolfi & D.R. Lamb (Eds.), *Perspectives in exercise science and sports medicine: Youth, exercise and sport* (pp. 335-362). Carmel, IN: Benchmark Press.

3. Bass, S.L. (2000). The prepubertal years: A uniquely opportune stage of growth when the skeleton is most responsive to exercise? *Sports Medicine, 30,* 73-78.

4. Bjorntorp, P. (1986). Fat cells and obesity. In K.D. Brownell & J.P. Foreyt (Eds.), *Handbook of eating disorders: Physiology, psychology, and treatment of obesity, anorexia, and bulimia* (pp. 88-98). New York: Basic Books.

5. Boisseau, N., & Delamarche, P. (2000). Metabolic and hormonal responses to exercise in children and adolescents. *Sports Medicine, 30,* 405-422.

6. Brooks, G.A., Fahey, T.D., White, T.P. and Baldwin, K.M. (2000). *Exercise physiology: Human bioenergetics and its applications,* 3rd ed. Mountain View, CA: Mayfield.

7. Clarke, H.H. (1971). *Physical and motor tests in the Medford boys' growth study.* Englewood Cliffs, NJ: Prentice Hall.

8. Daniels, J., Oldridge, N., Nagle, F., & White, B. (1978). Differences and changes in $\dot{V}O_2$ among young runners 10 to 18 years of age. *Medicine and Science in Sports and Exercise, 10,* 200-203.

9. Eriksson, B.O. (1972). Physical training, oxygen supply and muscle metabolism in 11-13-year-old boys. *Acta Physiologica Scandinavica,* (Suppl. 384), 1-48.

10. Fleck, S.J., & Kraemer, W.J. (1997). *Designing resistance training programs,* 2nd ed. Champaign, IL: Human Kinetics.

11. Froberg, K., & Lammert, O. (1996). Development of muscle strength during childhood. In O. Bar-Or (Ed.), *The child and adolescent athlete* (p. 28). London: Blackwell.

12. Inbar, O., & Bar-Or, O. (1986). Anaerobic characteristics in male children and adolescents. *Medicine and Science in Sports and Exercise, 18,* 264-269.

13. Kraemer, W.J., & Fleck, S.J. (1993). *Strength training for young athletes.* Champaign, IL: Human Kinetics.

14. Krahenbuhl, G.S., Morgan, D.W., & Pangrazi, R.P. (1989). Longitudinal changes in distance-running performance of young males. *International Journal of Sports Medicine, 10,* 92-96.

15. Mahon, A.D., & Vaccaro, P. (1989). Ventilatory threshold and $\dot{V}O_2$ max changes in children following endurance training. *Medicine and Science in Sports and Exercise, 21,* 425-431.

16. Malina, R.M. (1989). Growth and maturation: Normal variation and effect of training. In C.V. Gisolfi & D.R. Lamb (Eds.), *Perspectives in exercise science and sports medicine: Youth, exercise and sport* (pp. 223-265). Carmel, IN: Benchmark Press.

17. Malina, R.M., & Bouchard, C. (1991). *Growth, maturation, and physical activity.* Champaign, IL: Human Kinetics.

18. Pfeiffer, R.D., & Francis, R.S. (1986). Effects of strength training on muscle development in prepubescent, pubescent, and postpubescent males. *Physician and Sportsmedicine, 14*(9), 134-143.

19. Ramsay, J.A., Blimkie, C.J.R., Smith, K., Garner, S., MacDougall, J.D., & Sale, D.G. (1990). Strength training effects in prepubescent boys. *Medicine and Science in Sports and Exercise, 22,* 605-614.

20. Robinson, S. (1938). Experimental studies of physical fitness in relation to age. *Arbeitsphysiologie, 10,* 251-323.

21. Rogers, D.M., Olson, B.L., & Wilmore, J.H. (1995). Scaling for the $\dot{V}O_2$-to-body size relationship among children and adults. *Journal of Applied Physiology, 79,* 958-967.

22. Rowland, T.W. (1985). Aerobic response to endurance training in prepubescent children: A critical analysis. *Medicine and Science in Sports and Exercise, 17,* 493-497.

23. Rowland, T.W. (1989). Oxygen uptake and endurance fitness in children: A developmental perspective. *Pediatric Exercise Science, 1,* 313-328.

24. Rowland, T.W. (1991). "Normalizing" maximal oxygen uptake, or the search for the holy grail (per kg). *Pediatric Exercise Science, 3,* 95-102.

25. Sale, D.G. (1989). Strength training in children. In C.V. Gisolfi & D.R. Lamb (Eds.), *Perspectives in exercise science and sports medicine: Youth, exercise and sport* (pp. 165-216). Carmel, IN: Benchmark Press.

26. Santos, A.M.C., Welsman, J.R., De Ste Croix, M.B.A., & Armstrong, N. (2002). Age- and sex-related differences in optimal peak power. *Pediatric Exercise Science, 14,* 202-212.

27. Sewall, L., & Micheli, L.J. (1986). Strength training for children. *Journal of Pediatric Orthopaedia Strabismus, 6,* 143-146.

28. Sjödin, B., & Svedenhag, J. (1992). Oxygen uptake during running as related to body mass in circumpubertal boys: A longitudinal study. *European Journal of Applied Physiology, 65,* 150-157.

29. Turley, K.R., & Wilmore, J.H. (1997). Cardiovascular responses to treadmill and cycle ergometer exercise in children and adults. *Journal of Applied Physiology, 83,* 948-957.

30. Weltman, A., Janney, C., Rians, C.B., Strand, K., Berg, B., Tippitt, S., Wise, J., Cahill, B.R., & Katch, F.I. (1986). The effects of hydraulic resistance strength training in pre-pubertal males. *Medicine and Science in Sports and Exercise, 18,* 629-638.

▷ Selected Readings

Armstrong, L.E., & Maresh, C.M. (1995). Exercise-heat tolerance of children and adolescents. *Pediatric Exercise Science, 7,* 239-252.

Armstrong, N., & Welsman, J.R. (1994). Assessment and interpretation of aerobic fitness in children and adolescents. *Exercise and Sport Sciences Reviews, 22,* 435-476.

Armstrong, N., & Welsman, J.R. (2000). Development of aerobic fitness during childhood and adolescence. *Pediatric Exercise Science, 12,* 128-149.

Åstrand, I. (1967). *Aerobic work capacity: Its relation to age, sex, and other factors* (Monograph No. 15). New York: American Heart Association.

Bailey, D.A., Faulkner, R.A., & McKay, H.A. (1996). Growth, physical activity, and bone mineral acquisition. *Exercise and Sport Sciences Reviews, 24,* 233-266.

Bar-Or, O. (Ed.). (1996). *The child and adolescent athlete.* London: Blackwell.

Beunen, G., & Thomis, M. (2000). Muscular strength development in children and adolescents. *Pediatric Exercise Science, 12,* 174-197.

Blimkie, C.J.R. (1993). Resistance training during pre-adolescence: Issues and controversies. *Sports Medicine, 15,* 389-407.

Borer, K.T. (1995). The effects of exercise on growth. *Sports Medicine, 20,* 375-397.

Cheung, L.W.Y., & Richmond, J.B. (Eds.). (1995). *Child health, nutrition, and physical activity.* Champaign, IL: Human Kinetics.

Cronk, C.E., & Roche, A.F. (1982). Race and sex-specific reference data for triceps and subscapular skin folds and weight/stature. *American Journal of Clinical Nutrition, 35,* 347-354.

Falk, B. (1998). Effects of thermal stress during rest and exercise in the paediatric population. *Sports Medicine, 25,* 221-240.

Ganley, T. (2000). Exercise and children's health. *Physician and Sportsmedicine, 28*(2), 85-92.

Gisolfi, C.V., & Lamb, D.R. (Eds.). (1989). *Perspectives in exercise science and sports medicine: Youth, exercise and sport.* Carmel, IN: Benchmark Press.

Guo, S.S., Chumlea, W.C., Roche, A.R., & Siervogel, R.M. (1997). Age- and maturity-related changes in body composition during adolescence into adulthood: The Fels Longitudinal Study. *International Journal of Obesity, 21,* 1167-1175.

Krahenbuhl, G.S., Skinner, J.S., & Kohrt, W.M. (1985). Developmental aspects of maximal aerobic power in children. *Exercise and Sport Sciences Reviews, 13,* 503-538.

Malina, R.M. (Ed.). (1988). *Young athletes: Biological, psychological, and educational perspectives.* Champaign, IL: Human Kinetics.

Malina, R.M. (1994). Physical growth and biological maturation of young athletes. *Exercise and Sport Sciences Reviews, 22,* 389-433.

McCann, D.J., & Adams, W.C. (2002). A dimensional paradigm for identifying the size-independent cost of walking. *Medicine and Science in Sports and Exercise, 34,* 1009-1017.

Rowland, T.W. (1990). *Exercise and children's health.* Champaign, IL: Human Kinetics.

Rowland, T.W. (1996). *Developmental exercise physiology.* Champaign, IL: Human Kinetics.

Rutenfranz, J., Mocellin, R., & Klimt, F. (Eds.). (1986). *Children and exercise XII.* Champaign, IL: Human Kinetics.

Shephard, R.J. (1992). Effectiveness of training programmes for prepubescent children. *Sports Medicine, 13,* 194-213.

Stull, G.A., & Eckert, H.M. (Eds.). (1986). *Effects of physical activity on children.* Champaign, IL: Human Kinetics.

Svedenhag, J. (1995). Maximal and submaximal oxygen uptake during running: How should body mass be accounted for? *Scandinavian Journal of Medicine and Science in Sports, 5,* 175-180.

Turley, K.R. (1997). Cardiovascular responses to exercise in children. *Sports Medicine, 24*(4), 241-257.

Vaccaro, P., & Mahon, A. (1987). Cardiorespiratory responses to endurance training in children. *Sports Medicine, 4,* 352-363.

Van Praagh, E. (2000). Development of anaerobic function during childhood and adolescence. *Pediatric Exercise Science, 12,* 150-173.

Weltman, A. (1989). Weight training in prepubertal children: Physiologic benefit and potential damage. In O. Bar-Or (Ed.), *Advances in pediatric exercise sciences* (pp. 101-129). Champaign, IL: Human Kinetics.

Williams, C.A. (1997). Children's and adolescents' anaerobic performance during cycle ergometry. *Sports Medicine, 24*(4), 227-240.

CHAPTER 17

AGING IN SPORT AND EXERCISE

The number of adult men and women over 40 years of age who exercise regularly or participate in competitive sports has increased dramatically over the past 30 years. Many of these older competitors, often termed Masters or Senior athletes, engage in competition for recreation and fitness, whereas others train with the same enthusiasm and intensity as Olympians. Opportunities are now available for older athletes to compete in activities ranging from marathon running to weightlifting. The success and the standards of performance set by many older athletes are exceptional and often difficult to explain. However, although these older athletes exhibit strength and endurance capacities that are far greater than those of untrained people of similar age, even the most highly trained older person experiences a decline in performance after the fourth or fifth decade of life.

What physiological changes occur during aging that affect exercise performance? Does intense physical activity pose any health risks for aging athletes? How trainable are middle-aged and older adults? We will attempt to answer these questions in this chapter. We begin by examining body size and composition changes in older adults, and then we consider age-related changes in the physiological responses to acute exercise and chronic adaptations to long-term training. Finally, we look at how aging affects the older athlete's performance and address several special issues unique to the aging population.

Few athletes continue to compete at a national level into middle and old age. One exception was Clarence DeMar, who won his seventh Boston Marathon at age 42, placed 7th at age 50, and was 78th in a field of 153 runners at age 65. In all, he ran more than 1,000 distance races, including more than 100 marathons between 1909 and 1957, a period when it was not popular to exercise or engage in competition as an older adult. His performances at the Boston Marathon alone spanned 48 years, from age 20 to 68. DeMar's last race in 1957, at age 68, was 15 km (9.3 mi), which he ran despite advanced intestinal cancer and a colostomy. His best time for the Boston Marathon was 2:29:42 at age 36. Thereafter, his time gradually slowed to 3:58:37 at age 66.

In modern societies, the level of voluntary physical activity begins to decline soon after people reach adult maturity. In many ways, we try to eliminate all forms of stress from our lives, including muscular effort. Technology has made virtually every aspect of life less physically demanding. Voluntary participation in strenuous physical activity on a regular basis is an unusual pattern of behavior that is not observed in most aging animals. Studies have shown that humans and other animals tend to decrease their physical activity as they grow older. As shown in figure 17.1, rats that were allowed to eat at will ran an average of more than 4,000 m per week in the early months of life but covered less than 1,000 m per week during their final months.

Thus, older men and women who choose to participate in competitive sports or to train

exhaustively do not follow natural human or animal behavior patterns. Why do some older individuals choose to remain physically active when the natural tendency is to become sedentary? The psychological factors that motivate these older athletes to compete are not clearly defined, but these athletes' goals probably do not differ substantially from those of their younger counterparts.

Considering the importance of exercise for maintaining muscle and cardiorespiratory health, it is not surprising that adult inactivity can lead to deterioration of one's capacity and tolerance for strenuous effort. Because of this, distinguishing between the effects of aging and those of reduced activity is difficult when studying lifelong changes in physiological function and physical performance. How does aging affect our body size and composition, physiological responses and adaptations to exercise, and sport performance? These are the focus of this chapter.

Body Size and Composition

As we age, we tend to lose height and gain weight, as illustrated in figure 17.2.[51] The reduction in height generally starts at about 35 to 40 years of age and is primarily attributable to compression of the intervertebral disks and poor posture early in aging. At about the age of 40 to 50 years in women, and 50 to 60 years in men, osteoporosis becomes a factor. **Osteoporosis** refers to a severe loss of bone mass with deterioration of the microarchitecture of bone, leading to increased risk of bone fracture. Poor diet and exercise habits throughout the lifespan

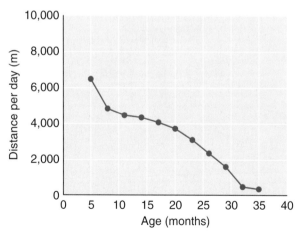

▲ **Figure 17.1**   Voluntary running activity in rats throughout life.

Adapted, by permission, from J.O. Holloszy, 1997, "Mortality rate and longevity of food-restricted exercising male rats: A reevaluation," *Journal of Applied Physiology* 82: 399-403.

also contribute. The gain in weight occurs between the ages of 25 and 45 and is largely attributable to both a decrease in physical activity levels and poor nutrition. Beyond the age of 45, weight stabilizes for about 10 to 15 years and then decreases as the body loses bone calcium and muscle mass. Many people over 65 to 70 years of age tend to lose their appetite and thus don't consume sufficient calories to maintain body weight. An active lifestyle, however, tends to help better regulate appetite so that caloric intake more closely approximates caloric expenditure, thereby maintaining weight.

With aging, beyond 20 years of age, we tend to gain fat. This is largely attributable to three factors that occur as we get older: diet, physical inactivity, and reduced ability to mobilize fat. As one might anticipate, the body fat content of physically active older people, including older athletes, is significantly lower than that of age-matched sedentary people. However, older athletes have substantially more body fat than younger competitors.

Fat-free mass decreases progressively in both men and women starting at about the age of 30 to 40. This results primarily from decreased muscle and bone mass, with muscle having the greatest effect because it constitutes about 50% of the fat-free mass. **Sarcopenia** is the term used to describe the loss of muscle mass associated with the aging process. **Osteopenia** is a companion term used to describe the loss in bone mass with aging. Figure 17.3 illustrates the changes in muscle mass with aging in a cross-sectional study of 468 men and women, aged 18 to 88 years.[29] You can see that there is almost no decline in muscle mass until about age 45, at which time the rate of decline increases. The rate of decline is greater in men than women. Obviously, a decline in activity levels is a major cause of this decline in muscle mass with aging, but there are other factors. It is now known that the rate of muscle protein synthesis is reduced as we age. When we compare 60- to 80-year-olds with 20-year-olds, muscle protein synthesis rate is lower by 30% or more in the older people.[21] The reduction in muscle protein synthesis rate in older people is likely associated with declines in growth hormone and insulin-like growth factor-1.[21]

There is also a significant decrease in bone mineral, starting at about age 30 to 35 in women

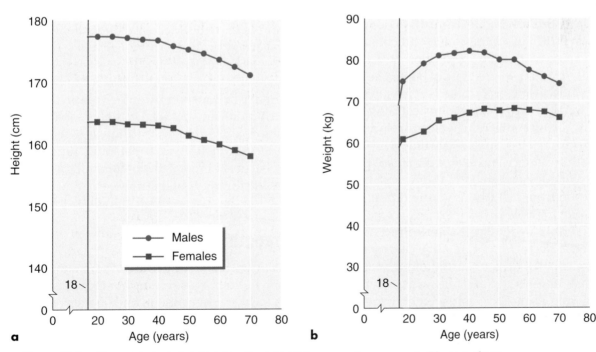

▲ **Figure 17.2**   Changes in body height (*a*) and weight (*b*) in men and women up to 70 years of age.
Reprinted, by permission, from W.W. Spirduso, 1985, *Physical Dimensions of Aging* (Champaign, IL: Human Kinetics), 59.

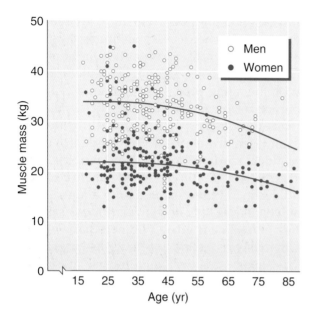

▲ Figure 17.3 Changes in muscle mass with aging in 468 men and women 18 to 88 years of age. The rate of decline is greater in men than in women and is steeper at about 45 years of age.

Adapted, by permission, from I. Janssen et al., 2000, "Skeletal muscle mass and distribution in 468 men and women aged 18-88 yr.," *Journal of Applied Physiology* 89: 81-88.[28]

and in age 45 to 50 in men. Throughout the life cycle, bone is constantly being formed or synthesized by osteoblasts and resorbed by the osteoclasts. Early in life, resorption occurs at a slower rate than synthesis, allowing bone mass to increase. With aging, resorption exceeds synthesis, resulting in a net loss of bone. The loss of both muscle and bone mass is at least partially attributable to decreased physical activity.

These differences in weight, relative (%) body fat, fat mass, and fat-free mass with aging, in both sedentary and exercise-trained populations, are demonstrated in figure 17.4.[35] These data are from a study of young (18-31 years) and older (58-72 years) men and women who were either sedentary or endurance-trained athletes. Body weight, relative body fat, and fat mass were higher in the older sedentary groups, whereas fat-free mass was lower. Similar trends, except for body weight, were noted for the endurance-trained athletes. However, the endurance-trained athletes had much lower total body weight, relative body fat, and fat mass values and similar fat-free mass values.

With age, body fat content increases, while at the same time fat-free mass decreases. Much of these changes can be attributed to the reduction in general activity levels that occurs with aging and poor nutrition.

▶ Body weight tends to increase with aging, whereas body height decreases.

▶ Body fat increases with age, primarily because of increased caloric intake, decreased physical activity, and a reduced ability to mobilize fat.

▶ Beyond age 45, fat-free mass decreases, primarily because of decreased muscle and bone mass, both resulting at least partly from decreased activity.

▶ Training can help attenuate these changes in body composition.

Physiological Responses to Acute Exercise

As we age, muscular and cardiovascular endurance and muscular strength tend to decrease, with the extent of decrease dependent on our level of physical activity and genetics. As activity level declines, which appears to be a natural phenomenon in both animals and humans, these reductions in physiological function are much more substantial. Let's now look at how aging affects strength, cardiovascular and respiratory function, and metabolic function. It will become obvious that maintaining an active lifestyle throughout life attenuates the losses in function that we see with aging.

Strength

The level of strength needed to meet the daily demands of living remains unchanged throughout life. However, a person's maximal strength, generally well above the daily demands early in life, decreases steadily with aging. Eventually, strength declines to the point where simple activities become chal-

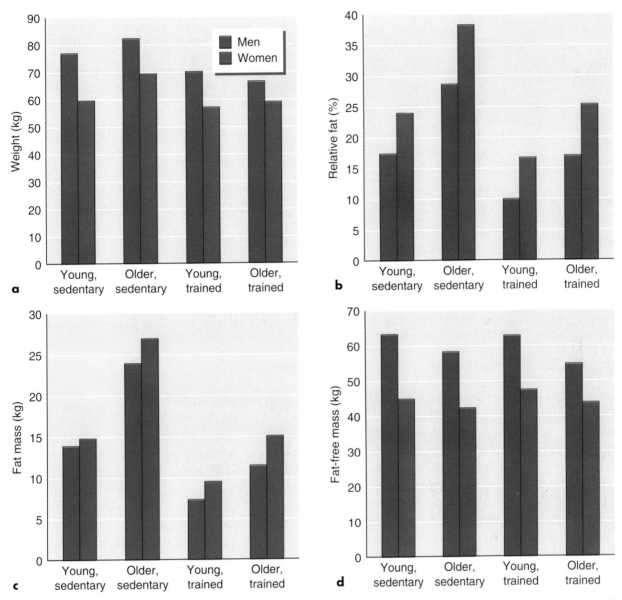

▲ **Figure 17.4**   Differences in total body weight (*a*), relative body fat (*b*), fat mass (*c*), and fat-free mass (*d*) in young and older men and women, sedentary and endurance trained.

Adapted, by permission, from W.M. Kohrt et al., 1992, "Body composition of healthy sedentary and trained, young and older men and women," *Medicine and Science in Sports and Exercise* 24: 832-837.[35]

lenging. For example, the ability to stand up from a sitting position in a chair starts to be compromised at age 50, and before age 80 this task becomes impossible for some people (see figure 17.5*a* on page 544). As a further example, opening the cap on a jar that has a set resistance is a task that can easily be accomplished by 92% of men and women in the age range of 40 to 60. After age 60, the failure rate for this task increases dramatically. By the age of 71 to 80, only 32% can open the jar.[48]

Figure 17.5*b* describes leg strength changes with aging in men. Knee extension strength in normally active men and women starts to decrease rapidly by age 40. But strength training the knee extensor muscles enables older men to perform better at age 60 than most normally active men at half that age. Similar declines in knee extensor and flexor strength were noted in a cross-sectional study of Japanese men and women, aged 20 to 84 years. Peak torque, a measure of strength, was determined

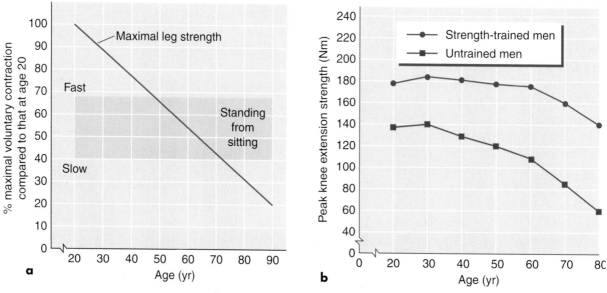

▲ **Figure 17.5** (a) The ability to stand from a sitting position is compromised at age 50, and by age 80 this task becomes impossible for some people. (b) Changes in peak knee extension strength in untrained and trained men at various ages. Note that older men (e.g., 60-80 years) who strength train can have knee extension strength equal to or greater than individuals who are only a third their age.
Data for 17.5b from Human Performance Laboratory, Ball State University.

at various speeds of muscle action by using an isokinetic testing device (see chapter 3, figure 3.2). The reduction in strength with aging was highly correlated with the reduction in the cross-sectional area of the involved muscles.[1]

Age-related losses of muscle strength result primarily from the substantial loss of muscle mass that accompanies aging or decreased physical activity, as we discussed earlier in this chapter. Figure 17.6 shows a computed tomographic (CT) scan of the upper arms of three 57-year-old men of similar body weight (about 78-80 kg, or 172-176 lb). Note that the

untrained subject had substantially less muscle and more fat than the others. The swim-trained subject had less fat and a markedly larger triceps muscle than the untrained subject, but his biceps muscle, which is seldom used during swimming, was not much different. However, both of these muscles were enlarged in the strength-trained subject. The differences between these three men are likely attributable to a combination of genetics and their volume and type of training.

We now know that aging has a marked effect on total muscle mass and strength, but what

Untrained   Swim-trained   Strength-trained

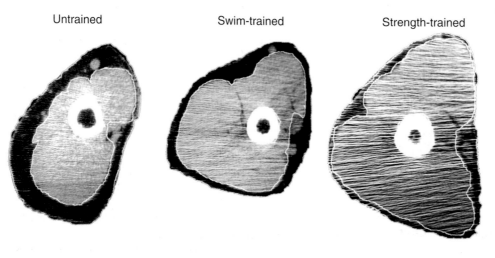

◀ **Figure 17.6** Computed tomography scans of the upper arms of three 57-year-old men of similar body weights. The scans show bone (dark center surrounded by white ring), muscle (striated grey area), and subcutaneous fat (dark perimeter). Note the difference in the muscle areas of the untrained man, swim-trained man, and the strength-trained man.

about fiber type? There are conflicting results about the effects of aging on slow-twitch (ST) and fast-twitch (FT) fibers. Cross-sectional studies that have examined the entire vastus lateralis (quadriceps) muscle in 15- to 83-year-old subjects postmortem have suggested that fiber type remains unchanged throughout life.[30] However, results from longitudinal studies conducted over a 20-year period indicate that the amount or intensity of activity or perhaps both might play an important role in fiber-type distribution with aging.[54, 55] Muscle biopsy samples from the gastrocnemius (calf) muscles of a group of previously elite distance runners obtained in 1970 through 1974 and again in 1992 demonstrated that the runners who had decreased their activity (fitness trained) or become sedentary (untrained) had a significantly greater proportion of ST fibers than when they were 18 to 22 years younger (figure 17.7). Those who remained highly trained had no change. Although some of the elite runners who still competed in distance running (highly trained) showed a small increase in the percentage of

ST fibers, on average these highly trained runners showed no change in their calf muscle fiber composition in the 18 to 22 years of this study.

It has been suggested that the apparent increase in ST fibers is probably attributable to an actual decrease in the number of FT fibers, resulting in a greater proportion of ST fibers. Although the precise cause of this loss of FT fibers is unclear, it has been suggested that the number of FT motor neurons decreases during aging, which eliminates innervation of these muscle fibers. This is possibly caused by death of the motor neurons in the spinal cord.[12] Most fibers from these dead motor neurons gradually atrophy and eventually are absorbed by the body. Surviving motor neurons, however, can develop axonal sprouts and reinnervate some of the muscle fibers from the dead motor neurons.[12] This increases the size of the remaining motor units, in that there are now more muscle fibers per motor neuron.

Documentation from numerous investigations has shown that a decrease in both the number and size of muscle fibers occurs with aging. Research indicates that approximately 10% of the total number of muscle fibers are lost per decade after age 50.[38] This explains in part the muscle atrophy that occurs as we get older. Additionally, it appears that the size of both ST and FT fibers decreases with aging. It was thought that training might reduce the loss in muscle mass observed during aging, but recent studies suggest that endurance training (distance running) may have little impact on the aging decline in muscle mass.[55] Strength training, on the other hand, reduces muscle atrophy in aging adults and can, in fact, cause older individuals to increase their muscle cross-sectional area.[37]

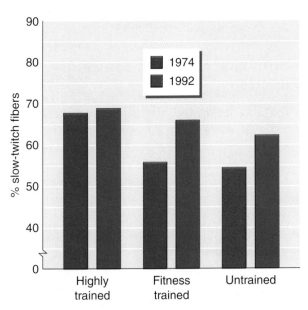

▲ **Figure 17.7**   Changes in muscle fiber-type composition of the gastrocnemius in elite distance runners who remained highly trained, stayed fitness-trained, or became untrained during the 18 to 22 years between tests. Note that the runners who continued to compete showed almost no change in percentage of slow-twitch fibers, whereas the less fit and untrained individuals experienced an increase in the percentage of slow-twitch fibers.

Strength is reduced with aging. This is the result of decreases in both physical activity and muscle mass, the latter largely the result of a reduction in protein synthesis with aging and the loss of FT motor units. Whereas endurance training does little to prevent the aging loss in muscle mass, strength training can maintain or increase the muscle fiber cross-sectional area in older men and women.

Research has shown that aging is accompanied by substantial changes in the nervous system's capacity to process information and to activate muscles. Specifically, aging affects the ability to detect a stimulus and process the information to produce a response. Simple and complex movements are slowed with aging, although people who remain physically active are only slightly slower than younger, active individuals. Studies of motor unit activation have indicated that the maximal discharge rate in adults older than 67 years was less than in 21- to 33-year old adults.[32] It was concluded that the strength reductions experienced by older adults are caused, at least in part, by an impaired ability to fully drive the surviving motor units. This could be largely attributable to the remodeling of motor units, which reduces the number of FT motor units and increases the number of muscle fibers in ST motor units. This theory is supported by another study, where motor unit firing rates and contractile properties of the tibialis anterior muscle of young (~20 years) and old (~80 years) men were compared. The older men were characterized as having lower firing rates and longer twitch contraction durations, whereas the younger men had relatively higher firing rates and shorter contraction times.[7]

These neuromuscular changes during aging are at least partially responsible for decreased strength and endurance, but active participation in exercise and sport tends to lessen the impact of aging on performance. This doesn't mean that regular physical activity can arrest biological aging, but an active lifestyle can markedly reduce many of the decrements in physical work capacity.

Saltin[47] noted that despite the loss of muscle mass in active aging men, the quality of the remaining muscle mass is well maintained. The number of capillaries per unit area is similar in young and old endurance runners. Oxidative enzyme activities in the muscles of endurance-trained older athletes are only 10% to 15% lower than in endurance-trained young athletes. Thus, oxidative capacity of skeletal muscle of endurance-trained older runners is only slightly less than that of young elite runners, which suggests that aging has little effect on skeletal muscle's adaptability to endurance

training. With this in mind, we continue to explore the effects of aging on physical performance, shifting our focus to cardiovascular and respiratory function.

▶ Maximal strength decreases steadily with aging.

▶ Age-related losses of strength result primarily from a substantial loss of muscle mass.

▶ In general, normally active people experience a shift toward a higher percentage of ST muscle fibers as they age, possibly attributable to a reduction in FT fibers.

▶ The total number of muscle fibers and the fiber cross-sectional area decrease with age, but training appears to lessen at least the change in fiber area.

▶ Aging also appears to slow the nervous system's ability to detect a stimulus and to process the information to produce a response.

▶ Training cannot arrest the process of biological aging, but it can lessen the impact of aging on performance.

Cardiovascular and Respiratory Function

What are the underlying physiological causes for the decrease in cardiorespiratory endurance with aging? To a large extent, changes in endurance performance that accompany aging can be attributed to decrements in both central and peripheral circulation. Changes in respiratory function likely play a lesser role. In this section, we see how aging effects both the cardiovascular and the respiratory systems.

Cardiovascular Function

Cardiovascular function, similar to muscle function, declines as we age. One of the most notable changes that accompany aging is a decrease in maximum heart rate (HRmax). Whereas children's values frequently exceed 200 beats/min, the average 60-year-old has an HRmax of approximately 160 beats/min. HRmax is estimated to decrease slightly less than 1 beat/min per year as we age. For years, the average HRmax for any age was estimated from the equation, HRmax = 220 − age. However, Tanaka and his associates discovered a more accurate equation:[52]

$$HRmax = [208 − (0.7 × age)]$$

This new equation seems to be appropriate for all people and is not influenced by the person's sex or activity level. The old equation tended to overestimate the HRmax of children and young adults and to underestimate the HRmax of older adults. When the old equation (HRmax = 220 − age) is used, individual values can deviate by ±20 beats/min or more from the predicted value. For example, the old equation predicts that a 60-year-old would have an HRmax of 160 beats/min, but this individual's actual HRmax might be as low as 140 beats/min or as high as 180 beats/min. Similar data are not yet available for the new equation. We will see in chapter 19 that overestimates and underestimates make a big difference when used in prescribing exercise.

> The new equation used to estimate HRmax is HRmax = [208 − (0.7 × age)]. However, prediction equations estimate only the average value for a given age.

The reduction in HRmax with age appears to be similar in both sedentary and highly trained adults. At age 50, for example, normally active men have the same HRmax values as former and still-active distance runners of the same age. This reduction in HRmax might be attributable to morphological and electrophysiological alterations in the cardiac conduction system, specifically in the sinoatrial (SA) node and in the bundle of His, which could slow cardiac conduction.[36] It is well known that the heart's intrinsic rate (its natural rate without nervous or hormonal stimulation) decreases with age. Apparent down-regulation of the β-1 receptors in the heart also decreases the heart's sensitivity to catecholamine stimulation.

Maximal stroke volume (SVmax) and cardiac output (Q̇max) also appear to decrease with age. Studies of endurance runners have

shown that the lower V̇O$_2$max values observed in older athletes result from a reduction in maximal cardiac output, despite the fact that heart volumes of older athletes are similar to those of young athletes. Saltin[47] reported that 51-year-old orienteers (distance runners) have a maximal cardiac output that is about 5 L/min (21%) lower than young orienteers. This difference is attributable to the older athletes' lower HRmax and SVmax values (recall that cardiac output = heart rate × stroke volume). Lower stroke volumes in older athletes are caused primarily by increased peripheral resistance. But when compared with sedentary men of the same age, these active older orienteers had markedly higher V̇O$_2$max values, primarily because they had greater stroke volumes and thus also greater Q̇max values than their sedentary peers.

Peripheral blood flow, such as to the legs, decreases with aging, even though capillary density in the muscles is unchanged. Studies reveal a 10% to 15% reduction in leg blood flow to the exercising muscles in middle-aged athletes at any given work rate when compared with well-trained young athletes (figure 17.8).[31, 57] But the reduced blood flow to the legs of these

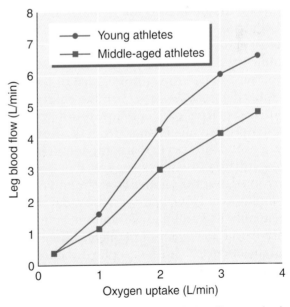

▲ **Figure 17.8**   Leg blood flow during cycling exercise in young and middle-aged orienteers.

Adapted, by permission, from B. Saltin, 1986, The aging endurance athlete. In *Sports medicine for the mature athlete,* edited by J.R. Sutton and R.M. Brock (Indianapolis: Benchmark Press). Copyright 1986 Cooper Publishing Group, Carmel, IN.

middle-aged and older endurance runners during submaximal exercise was apparently compensated for by a greater arterial-venous oxygen difference, or a-$\bar{v}O_2$ difference (more oxygen is extracted by the muscles). As a result, although the blood flow is different, oxygen uptake by the exercising muscles is similar at a given submaximal work intensity in the different age groups. This was confirmed in a study of endurance-trained men, which compared subjects aged 22 to 30 years with subjects 55 to 68 years. Leg blood flow, vascular conductance, and femoral venous oxygen saturation were each 20% to 30% lower in the older men at each submaximal work rate.[45]

Why, then, do maximal cardiac output and $\dot{V}O_2$max decrease with age? One explanation is that aging increases peripheral resistance. With age, arteries and arterioles begin to lose their elasticity and become less capable of vasodilation. This increases peripheral resistance, and as a result, blood pressure increases at rest and during exercise. Although older athletes of both sexes have slightly lower mean arterial pressures than most sedentary men, they still have more peripheral resistance than younger athletes, which limits their stroke volume capacity.

Thus, the gradual decline in maximal cardiac output among older athletes appears to be the result of restrictions placed on both the heart's pumping capacity and peripheral blood flow. It is hard to determine how much of the age-related decreases in stroke volume, cardiac output, and peripheral blood flow result from the aging process alone, apart from **cardiovascular deconditioning** that accompanies reduced activity. Recent studies suggest that both are involved, but the relative contribution of each is unknown. However, even the older athlete generally trains at a lower volume and intensity than the 20-year-old athlete. Aging alone might decrease cardiovascular function and endurance less than the deconditioning that accompanies inactivity, decreased activity, or decreased intensity of training. These declines in cardiovascular function with aging are largely responsible for the declines observed in $\dot{V}O_2$max, which are of similar magnitude. This is discussed later in this chapter.

> ▶ Much of the decline in endurance performance associated with aging can be attributed to decrements in central and peripheral circulation.
>
> ▶ Maximum heart rate decreases slightly less than 1 beat/min per year as we age. The average HRmax for a certain age can be estimated by the following equation: HRmax = [208 − (0.7 × age)].
>
> ▶ Maximal stroke volume and cardiac output also appear to decrease with age. Stroke volume can be well maintained in older athletes who have continued to train, but it will still be less than in younger athletes.
>
> ▶ Peripheral blood flow also decreases with age; in trained older athletes, however, this is offset by an increased submaximal a-$\bar{v}O_2$ difference.
>
> ▶ It is unclear how much of the decrease in cardiovascular function with aging is attributable to physical aging alone and how much is attributable to deconditioning because of decreased activity. However, many studies indicate that these changes are minimized in older athletes who continue to train, which seems to indicate that inactivity might play a larger role than physical aging.

Respiratory Function

Lung function changes considerably in sedentary people with aging. Both **vital capacity (VC,** the total volume of air expelled after maximal inhalation) and **forced expiratory volume in 1 s (FEV$_{1.0}$,** the greatest volume of air exhaled in 1 s) decrease linearly with age, starting at age 20 to 30. Whereas these decrease, **residual volume (RV,** the amount that cannot be exhaled) increases, and the **total lung capacity (TLC)** remains essentially unchanged. As a result, the ratio of the residual volume to total lung capacity (RV/TLC) increases, meaning that less air can be exchanged. In our early 20s, residual volume accounts for 18% to 22% of the total lung capacity, but this increases to 30% or more as we reach age 50. Smoking appears to accelerate this increase.

These changes are matched by changes in maximal ventilatory capacity during exhaustive exercise. **Maximal expiratory ventilation** ($\dot{V}_E$**max**) increases during growth until you achieve physical maturity, and then it decreases with age. $\dot{V}_E$max values average about 40 L/min for 4- to 6-year-old boys, increase to 110 to 140 L/min for fully mature men, and then decrease to 70 to 90 L/min for 60- to 70-year-old men. Females follow the same general pattern, although their absolute values are considerably lower at each age, primarily because of smaller body size. Figure 17.9 illustrates the change in $\dot{V}_E$max with age.

The changes in pulmonary function as adults get older are probably the result of several factors. The most important of these is loss of elasticity of the lung tissue and chest wall as we age, which increases the work involved in breathing. The resulting stiffening of the chest wall appears to be responsible for most of the reduction in lung function. But despite all these changes, the lungs still hold a remarkable reserve and maintain an adequate diffusion capacity to permit maximal exertion.

Endurance training in older adults reduces the extent of loss of elasticity from the lungs and chest wall. As a result, endurance-trained older athletes have only slightly decreased pulmonary ventilation capacities. Decreased aerobic capacity among these older athletes cannot be attributed to changes in pulmonary ventilation. Also, during strenuous exercise, both normally active older people and athletes can maintain near-maximal arterial oxygen saturation (97% saturation).[48] Thus, neither changes in the lungs nor in the blood's oxygen-carrying capacity appear to be responsible for the observed decrease in $\dot{V}O_2$max reported in aging athletes.

Rather, the primary limitation is apparently linked with oxygen transport to the muscles. As discussed earlier in this chapter, aging decreases maximum heart rate and stroke volume, which lower maximal cardiac output and blood flow to the exercising muscles. In addition, maximal a-$\bar{v}O_2$ difference is lower in older people than in younger people, suggesting that less oxygen is extracted by our muscles as we age.

> ▶ Both vital capacity and forced expiratory volume decrease linearly with age. Residual volume increases, and total lung capacity remains unchanged. This increases the RV/TLC ratio, meaning that less air can be exchanged in the lung with each breath.
>
> ▶ Maximal expiratory ventilation also decreases with age.
>
> ▶ Pulmonary changes that accompany aging are primarily caused by a loss of elasticity in the lung tissue and the chest wall. However, aging athletes have only slightly decreased pulmonary ventilation capacity. For them, the primary limiter of $\dot{V}O_2$max appears to be decreased oxygen transport to the muscles. Furthermore, maximal a-$\bar{v}O_2$ difference decreases, indicating that less oxygen is extracted by their muscles.

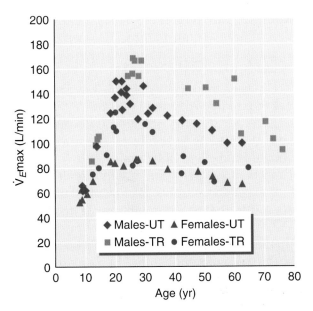

▲ **Figure 17.9**   Changes in maximal expiratory ventilation with age in untrained (UT) and trained (TR) males and females.

Metabolic Function

In investigating the effect of aging on metabolic function, we focus on two key variables—$\dot{V}O_2$max and lactate threshold.

$\dot{V}O_2$max

When we study how $\dot{V}O_2$max changes with aging, there are several important issues to consider. First, we must decide how to express the $\dot{V}O_2$max values. Do you use values expressed in liters per minute (L/min) or liters per minute per kilogram of body weight to adjust for size ($ml \cdot kg^{-1} \cdot min^{-1}$)? In some cases, $\dot{V}O_2$max expressed in liters per minute does not decrease much over a 10- to 20-year period, but when the same subjects' values are expressed relative to body weight, there is a relatively large decrease. This apparent discrepancy in results is simply attributable to the subjects gaining substantial amounts of body weight during the 10 to 20 years between the initial test and the final test. Which is the correct value to use? For exercise where transporting the body mass is not an issue, such as in stationary cycling, using liters per minute is appropriate. Most activities require movement of the body mass through space, where it is more appropriate to express the values per unit of body weight ($ml \cdot kg^{-1} \cdot min^{-1}$).

The second issue relates to whether change values should be expressed in absolute values (L/min or $ml \cdot kg^{-1} \cdot min^{-1}$) or as a percentage increase or decrease, where

% change = [(final value − initial value)/initial value] × 100.

This might seem like a small point, but it isn't. As an example, a 30-year-old man has an initial $\dot{V}O_2$max of 50 $ml \cdot kg^{-1} \cdot min^{-1}$, and at the age of 50 years his $\dot{V}O_2$max has decreased to 40 $ml \cdot kg^{-1} \cdot min^{-1}$. A 60-year-old man has an initial $\dot{V}O_2$max of 35 $ml \cdot kg^{-1} \cdot min^{-1}$, and at the age of 80 years his $\dot{V}O_2$max has decreased to 25 $ml \cdot kg^{-1} \cdot min^{-1}$. In this example, both men have decreased their $\dot{V}O_2$max by 10 $ml \cdot kg^{-1} \cdot min^{-1}$ in a 20-year period, or a decline of 0.5 $ml \cdot kg^{-1} \cdot min^{-1}$ per year. However, the younger man has had a decrease of 20% (10/50 = 0.20, or 20%) over 20 years, or 1% per year, where the older man has had a decrease of 28.6% (10/35 = 0.286, or 28.6%), or 1.4% per year. Although the two men have identical decreases in $\dot{V}O_2$max when expressed in $ml \cdot kg^{-1} \cdot min^{-1}$, the older man has a substantially greater decrease when ex-

pressed as a percentage decrease. Many studies will report both the absolute ($ml \cdot kg^{-1} \cdot min^{-1}$) and relative (%) decrease. With this in mind, let's consider changes in $\dot{V}O_2$max with aging, looking first at changes in normally active people and then at changes in highly trained endurance athletes.

> The decrease in $\dot{V}O_2$max with aging and inactivity is largely explained by decreases in HRmax, maximal stroke volume (SVmax), and maximal a-$\overline{v}O_2$ difference. The decrease in HRmax is attributable largely to decreases in the heart's intrinsic rate but also could be caused by decreases in sympathetic nervous system activity and alterations in the cardiac conduction system. The decrease in SVmax is attributable primarily to increased total peripheral resistance from reduced compliance in the arteries with aging and to possible reductions in left ventricular contractility. The decrease in VO_2max with aging is therefore apparently a function of reduced blood flow to the active muscles, which is associated with the reduction in maximal cardiac output attributable to a reduction in SVmax and HRmax.

Normally Active People   The first studies of aging and physical fitness were performed by Sid Robinson[46] in the late 1930s. He demonstrated that $\dot{V}O_2$max in normally active men declined steadily from age 25 to age 75 (table 17.1). His cross-sectional data show that aerobic capacity declines an average of 0.44 $ml \cdot kg^{-1} \cdot min^{-1}$ per year up to age 75, which is about 1% per year or 10% per decade. For women between the ages of 25 to 56 years, Irma Åstrand in 1960 showed a decline of 0.38 $ml \cdot kg^{-1} \cdot min^{-1}$ per year, or 0.9% per year.[3] A 1987 review of 11 cross-sectional studies on men, most under age 70, found that the average rate of decrease in $\dot{V}O_2$max was 0.41 $ml \cdot kg^{-1} \cdot min^{-1}$ per year.[6] In this same review, analysis of six cross-sectional studies of women resulted in an average rate of decline of 0.30 $ml \cdot kg^{-1} \cdot min^{-1}$ per year.[6] This review did not publish the average rate of decline expressed as a percentage of the subjects' initial $\dot{V}O_2$max values. In the mid-1990s, a large cross-sectional study of changes

Table 17.1

Changes in $\dot{V}O_2$max Among Normally Active Men

Age (years)	$\dot{V}O_2$max (ml · kg^{-1} · min^{-1})	% change from 25 years
25	47.7	—
35	43.1	−9.6
45	39.5	−17.2
52	38.4	−19.5
63	34.5	−27.7
75	25.5	−46.5

Data from Robinson, 1938.[46]

in $\dot{V}O_2$max with aging was conducted at the NASA/Johnson Space Center in Houston, Texas. This study included 1,499 men and 409 women, all of whom were healthy and had performed a maximal treadmill test to exhaustion during which $\dot{V}O_2$ was directly measured.[27, 28] The authors reported a decline in $\dot{V}O_2$max of 0.46 ml · kg^{-1} · min^{-1} per year in men (1.2% per year) and 0.54 ml · kg^{-1} · min^{-1} per year in women (1.7% per year).

Unfortunately, few longitudinal studies have been conducted in this area. Studies that have reexamined normally active men at various stages of their lives reveal a wide range of values for the decline in aerobic capacity.[4, 6, 10, 40] At least part of these variations can be attributed to the subjects' different activity levels and ages at the beginning of the studies. Nevertheless, the rate of decline in $\dot{V}O_2$max generally is agreed to be approximately 10% per decade or 1% per year (−0.4 ml · kg^{-1} · min^{-1} per year) in relatively sedentary men. The results are similar for women, although only a few subjects have been studied.

> $\dot{V}O_2$max decreases by about 10% per decade with aging, starting in the midteens for women and in the mid-20s for men. This decrease is largely associated with a decrease in cardio-respiratory function.

Older Athletes   One of the most notable long-term studies of distance runners and aging was conducted by D.B. Dill and his colleagues from the Harvard Fatigue Laboratory.[11] Don Lash, world record holder for the 2-mi run (8 min 58 s) in 1936, was among those studied by the Harvard group. Although few of the former runners continued to train after leaving college, Lash was still running about 45 min per day at age 49. Despite this activity, his $\dot{V}O_2$max had declined from 81.4 ml · kg^{-1} · min^{-1} at age 24 to 54.4 ml · kg^{-1} · min^{-1} at age 49, a 33% decline. Runners who did not continue to train during middle age showed much larger declines. On the average, their aerobic capacities declined by about 43% from age 23 to age 50 (from 70 to 40 ml · kg^{-1} · min^{-1}). These data suggest that prior training offers little advantage to endurance capacity in later life unless a person continues to engage in some form of vigorous activity. In addition, there are large individual differences in the rate of decline in physical prowess with aging, and genetics is a major contributor.

More recent longitudinal studies of older male runners and rowers have reported a decline in aerobic capacity and cardiovascular function and changes in muscle fiber composition with aging.[17, 33, 44, 55, 56, 58, 59] These athletes were studied for 20 to 28 years, during which time some continued to train for competition, whereas others became quite sedentary. Those athletes who continued training hard experienced a 5% to 6% decline in $\dot{V}O_2$max per decade (0.5-0.6% per year).[17, 44, 55, 56] On the other hand, elite runners who stopped training experienced nearly a 15% decline in aerobic capacity per decade (1.5% per year), the combined effect of deconditioning and aging.[55, 56]

Fewer studies have been published on women, but the results show the same trends. In one study of 86 male and 49 female Masters endurance runners, the authors observed both cross-sectional and longitudinal (approximately 8.5 years) changes in $\dot{V}O_2$max with age.[22] Their results are illustrated in figure 17.10 on page 552. The average rate of decline, as indicated by the cross-sectional data regression line, was 0.47 ml · kg^{-1} · min^{-1} per year in men (0.8% per year) and 0.44 ml · kg^{-1} · min^{-1} in women (0.9% per year). However, from this figure, you can see

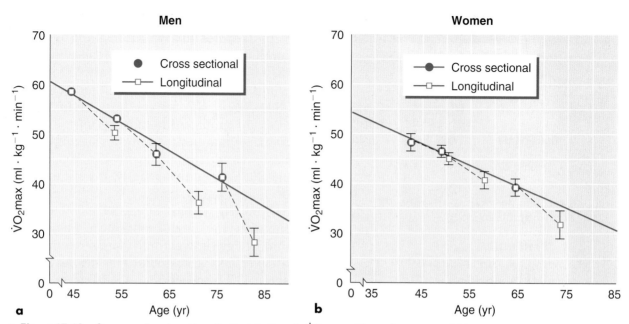

▲ **Figure 17.10**   Cross-sectional and longitudinal declines in V̇O₂max with age in a group of 86 male and 49 female Masters endurance athletes.

From Hawkins, S.A., Marcell, T.J., Jaque, S.V., and Wiswell, R.A., 2001, "A longitudinal assessment of change in V̇O₂max and maximal heart rate in master athletes," *Medicine and Science in Sports and Exercise*, 33: 1744-1750.[22]

that the longitudinal changes are greater than the cross-sectional changes, particularly for the older ages. In a cross-sectional study of sedentary women (*n* = 2,256), active women (*n* = 1,717), and endurance-trained women (*n* = 911) aged 18 to 89 years, V̇O₂max declined by 0.35 ml · kg⁻¹ · min⁻¹ per year in the sedentary women (1.2% per year), 0.44 ml · kg⁻¹ · min⁻¹ per year in the active women (1.1% per year), and 0.62 ml · kg⁻¹ · min⁻¹ per year in the endurance-trained women (1.2%).[14] A 7-year longitudinal study of endurance-trained and sedentary women, who were in their 50s at the start of the study, showed much larger declines—0.84 ml · kg⁻¹ · min⁻¹ per year in the endurance-trained women (1.8% per year) and 0.40 ml · kg⁻¹ · min⁻¹ per year for the sedentary women (1.5% per year).[13] However, this study included only 8 sedentary women and 16 endurance-trained women.

A recent 25-year follow-up study reexamined highly competitive, older male distance runners.[55, 56] These men were initially tested at 18 to 25 years of age. During the interval between testing sessions, the runners trained at about the same relative intensity as they did when they were younger. As a consequence, their V̇O₂max

values (L/min) declined only 3.6% over the 25-year period,[56] as shown in table 17.2. Although their maximal oxygen uptake decreased from 69.0 to 64.3 ml · kg⁻¹ · min⁻¹, this is a decrease of only 0.19 ml · kg⁻¹ · min⁻¹ per year or 0.3% per year, and most of that change was attributable to a 2.1-kg increase in body weight.

This rate of decrease in these older runners' V̇O₂max values is significantly less than that of either sedentary people or those who fitness train at levels and intensities below those of these older runners. One of these runners performed a 4 min 11 s mile and a 2:29 marathon in 1992 at the age of 46! Both of these performances were significantly faster than his best in 1966. Similar findings have been reported for other athletes who continue to train with the same relative intensity and volume as they did in college.

Are these performances exceptions to the natural rules of aging? Can other athletes reduce the effects of aging on their endurance by continuing to train intensely? Much depends on the training adaptability of the individual athlete, a factor that might be determined as much by heredity as by training regimen.

Table 17.2

Changes in Aerobic Capacity and Maximal Heart Rates With Aging in a Group of 10 Highly Trained Masters Distance Runners

| Age (years) | Weight (kg) | $\dot{V}O_2max$ | | HRmax (beats/min) |
		(L/min)	(ml · kg⁻¹ · min⁻¹)	
21.3 (±1.6)	63.9 (±2.2)	4.41 (±0.09)	69.0 (±1.4)	189 (±6)
46.3 (±1.3)	66.0 (±0.6)	4.25 (±0.05)	64.3 (±0.8)	180 (±6)

Note. Values are mean ± *SE.*

The effects of aging and training on $\dot{V}O_2max$ in men are summarized in figure 17.11. Although the number of studies on women is much smaller, a similar trend would be expected. Note that although hard training reduces the normal aging-related decline in $\dot{V}O_2max$, aerobic capacity still declines. Thus, it appears that highly intense training has a slowing effect on the rate of loss in aerobic capacity during the early and middle years of adult life (e.g., 30-50 years) but less effect after 50 years of age.

From the results of all of these studies, we see that $\dot{V}O_2max$ declines with age and that the rate of decline is approximately 1% per year. Many factors influence this rate of decline, including the following:

- General activity level
- Intensity of training
- Volume of training
- Increased body weight and body fat mass, and decreased fat-free mass
- Age range, with older individuals experiencing greater declines

There is not universal agreement on which of these factors are most important.

> It is often difficult to differentiate between the results of biological aging and physical inactivity. A natural deterioration in physiological function occurs with aging, but this is compounded by the fact that most people also become more sedentary as they age.

Lactate Threshold

Few studies have investigated the changes in lactate threshold, or anaerobic threshold derived from ventilatory variables, with aging. Lactate threshold was determined in a cross-sectional study of a group of Masters endurance runners, 40 to 70+ years of age, 111 men and 57 women.[61] Lactate threshold expressed as a percentage of $\dot{V}O_2max$ (LT-%$\dot{V}O_2max$) provides the best marker relative to endurance running performance (see

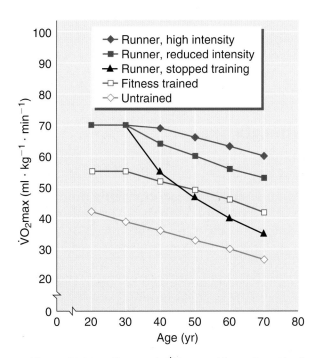

▲ **Figure 17.11**   Changes in $\dot{V}O_2max$ with age for trained and untrained men.

chapter 4). Interestingly, LT-%$\dot{V}O_2$max did not differ between men and women, but it did increase with age. Another study reported similar results in 152 untrained men and 146 untrained women.[43] This is somewhat of a paradox, because both $\dot{V}O_2$max and LT-%$\dot{V}O_2$max are considered to be important determinants of aerobic endurance performance, with better performance being associated with higher values of both of these variables. With aging, performance decreases, $\dot{V}O_2$max decreases, and yet LT-%$\dot{V}O_2$max increases. Furthermore, LT-%$\dot{V}O_2$max was found to be unrelated to distance running performance, which is just the opposite of what has been found in younger runners. At this time, there is no clear physiological explanation for this paradox.

> ▶ Aerobic capacity generally decreases by about 10% per decade or 1% per year in relatively sedentary men and women.
>
> ▶ Similar results have been found for highly trained endurance athletes, although there is a much wider variation in results of different studies.
>
> ▶ Studies of older athletes and less active people of the same age group indicate that the decrease in $\dot{V}O_2$max is not strictly a function of age. Athletes who continue to train have significantly smaller decreases in $\dot{V}O_2$max as they age, particularly if they train at a high intensity.
>
> ▶ Lactate threshold, expressed as a percentage of $\dot{V}O_2$max, unexpectedly increases with aging. The significance of this is not understood.

Physiological Adaptations to Exercise Training

Despite the decrements in body composition and exercise performance associated with aging, middle-aged and older athletes are capable of exceptional performances. Furthermore, those who train for general fitness appear to experience changes in body composition and gains in muscular strength and endurance similar to those of young adults. Let's look at these changes in more detail.

Body Composition

With both resistance training and aerobic training, older men and women will reduce body weight, relative body fat, and fat mass. Furthermore, they can increase their fat-free mass, but this is more likely with resistance training than with aerobic training. Men appear to experience greater changes in body composition than women, but the reason for this has not been clearly established.

The most significant changes in body composition result from a combination of diet and exercise, with a modest reduction in caloric intake (250-500 kcal/day) being the preferred approach. A more substantial reduction in caloric intake (>500 kcal/day) is likely to result in a loss of fat-free mass as well as fat mass. This is not desirable, as a loss in fat-free mass is associated with a reduction in resting metabolic rate, thus decreasing the rate of weight and fat loss. Exercise that increases fat-free mass will likely increase resting metabolic rate, which would increase the rate of weight loss. It appears that older adults will get changes in body composition with exercise training similar to younger adults.[60]

Strength

As noted earlier in this chapter, the loss of strength with aging is likely the result of a combination of the natural aging process and reduced physical activity that produces a decline in muscle mass and function. Although it is difficult to compare the adaptations to strength training of younger and older people, aging appears neither to impair the ability to improve muscle strength nor to prevent muscle hypertrophy. For example, when older men (ages 60-72) strength trained for 12 weeks at 80% of their 1-repetition maximum (1RM) for extension and flexion of both knees, their extension strength increased by 107% and flexion strength increased by 227%.[15] These improvements were attributed to muscle hypertrophy, as determined from midthigh CT scans. Biopsies of the vastus lateralis muscle (in the quadriceps) revealed that the cross-sectional area of ST fibers increased by 33.5% and that of FT fibers by 27.6%. In another study of older untrained men (aged 64 years), a 16-week resistance training program resulted in major increases in strength (50% for leg extension strength, 72% for leg press strength, and 83% for half squat strength) and an increase in the mean cross-sectional area of all major muscle fiber types (46% for ST, 34% for FT_a, and 52% for FT_b).[18, 24]

A study of older women (average age 72 years) who performed an aerobic-resistance program for 50 weeks found a 6% increase in leg strength at the end of the period. This was accompanied by a significant increase (29%) in cross-sectional area of only the FT fibers.[9] In another study of older women (average age 64 years), 21 weeks of resistance training resulted in a 37% increase in leg extensor maximal force development, a 29% increase in leg extension 1RM, an increase in the cross-sectional area of the extensor muscles, and a 22% to 36% increase in ST, FT_a, and FT_b muscle fiber areas.[20]

Another study investigated the changes in leg strength, chair rise time, and muscle fiber type and fiber composition in 60- to 75-year-old men and women who performed heavy resistance and power training using squats twice a week for 24 weeks.[19] The 1RM for strength increased by 26% in women and 35% in men, whereas the time to perform three stand-ups in quick succession from a 40-cm chair (chair rise time) decreased by 24% in women and 25% in men. There was an increase in muscle fiber cross-sectional area for ST and FT_a in women and for ST, FT_a, and FT_b for men. Although the percentage of ST fibers remained unchanged, the percentage of FT_a fibers increased and FT_b fibers decreased in both men and women.

> Many people believe that the degree of strength gains and muscle hypertrophy might be less in older women than in older men, but there are too few data to support such a claim.

Aerobic and Anaerobic Capacity

Recent studies have shown that improvements in $\dot{V}O_2$max with training are similar for younger (ages 21-25) and older (ages 60-71) men and women.[34, 41] Although the pretraining $\dot{V}O_2$max values were, on the average, lower for the older subjects, the absolute increases of 5.5 to 6.0 ml · kg^{-1} · min^{-1} were similar in both groups. Additionally, older men and women experienced similar increases in $\dot{V}O_2$max, averaging 21% for men and 19% for women, when they trained for 9 to 12 months by walking, running, or both about 4 mi (6 km) per day. This research indicates that endurance training produces similar gains in aerobic capacity in healthy people throughout the age range of 20 to 70 years, and this adaptation is independent of age, sex, and initial fitness level. However, this does not mean that endurance training can enable older athletes to achieve the performance standards established by younger athletes.

The precise mechanisms that trigger the body's adaptations to training at any age are not fully understood, so we don't know if improvements from training are achieved in the same way throughout life. For example, much of the improvement in $\dot{V}O_2$max seen in younger subjects is associated with an increase in maximal cardiac output. But older subjects show significantly greater gains in muscle oxidative enzyme activities, which suggests that peripheral factors in older subjects' muscles might play a greater role in aerobic adaptations to training than in younger subjects.

Very little is known about the trainability of anaerobic capacity in older people. We saw earlier in this chapter that lactate threshold, expressed as a percentage of a person's $\dot{V}O_2$max, increases with aging and is not associated with endurance running performance. In young and middle-aged adults, LT-%$\dot{V}O_2$max is the best predictor of endurance performance—running,

> **fyi** The ability to adapt to training was once thought to greatly decrease with aging. Recent studies, however, in which older subjects trained at relatively high intensities, indicate that older people have considerable ability to increase their endurance capacity and strength with training.

cycling, swimming, and cross-country skiing. Presently, there is no logical explanation for this difference between older adults and young and middle-aged adults.

> ▶ Older adults appear to get the same bene-fits from exercise training as younger and middle-aged adults relative to changes in body weight, relative body fat, fat mass, and possibly fat-free mass.
>
> ▶ It appears that aging does not impair a person's ability to increase muscle strength or muscle hypertrophy. Indi-vidual muscle fibers also have the ability to increase in size.
>
> ▶ Endurance exercise training produces similar gains in healthy people, regardless of their age, sex, or initial level of fitness.
>
> ▶ With endurance training, older indi-viduals show greater improvement in their muscles' oxidative enzyme activi-ties, whereas improvement in younger people is largely attributable to increased maximal cardiac output.

Sport Performance

World and national records in running, swim-ming, cycling, and weightlifting suggest that we are in our physical prime during our 20s or early 30s. If we use a cross-sectional ap-proach, comparing these records with national and world records for older athletes in these events allows us to examine the effects aging has on the best performers. Unfortunately, we have little longitudinal information about the effects of aging on performance because few studies have enabled us to follow physical performance in selected individuals over the span of their athletic careers. In the following sections, we consider how aging affects certain types of sport performance.

Running Performance

In 1954, Roger Bannister, a 21-year-old medical student, stunned the sporting world when he became the first person to run the mile (1.61 km) in less than 4 min (3 min 59.4 s). Today, the record for the mile is more than 16 s faster than Bannister's record, a gap that would have placed him more than 100 m behind today's record holder. In 1954, it would have seemed inconceivable that a sub-4-min mile could have been accomplished by someone over the age of 30 years. Yet today there are several runners over the age of 40 years who have equaled or surpassed Bannister's best performance.

Although older runners have achieved some exceptional records, running performance in general declines with age, and the rate of this decline appears to be independent of distance. Longitudinal studies of elite distance runners indicate that despite a high level of training, performance in events from the mile (1.61 km) to the marathon (42 km) declines at a rate of about 1.0% per year from the age of 27 to 47 years.[55,56] It is interesting to note that world records for both 100-m and 10-km runs also decrease by about 1% per year from age 25 to age 60,[8] as shown in figure 17.12. Beyond age 60, however, the records for men slow by nearly 2% per year. A sprint-running test of 560 wom-en between ages 30 and 70 revealed a steady decrease in maximal running velocity of 8.5% per decade (0.85% per year).[42] The patterns of change are about the same in both sprint- and endurance-running performances.

Swimming Performance

A retrospective study of freestyle performances at the U.S. Masters Swimming championships between 1991 and 1995 revealed that both men's and women's performances in the 1,500 m declined steadily from age 35 to about 70 years, after which swimming times slowed at a faster rate.[53] However, the rate and magnitude of the declines in both 50-m and 1,500-m per-formances with age were found to be greater for women than for men.

As shown in figure 17.13, U.S. Masters na-tional records in the 100-m freestyle decrease by about 1% per year for both men and women from age 25 to age 65 to 75. Because success in

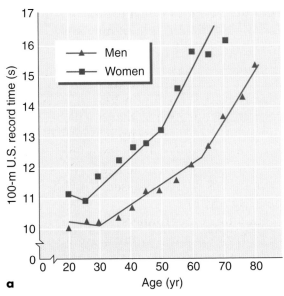

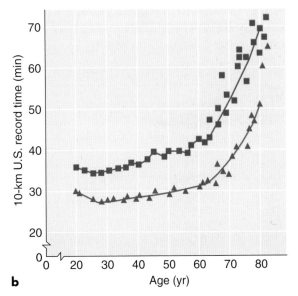

▲ **Figure 17.12** Change with age in men's and women's world records for (*a*) 100-m and (*b*) 10-km runs. Note that these running records slow at a much faster rate after the age of 50 to 60 years.

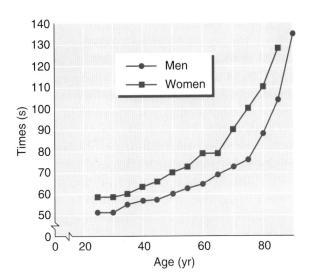

▲ **Figure 17.13** Change with age in 100-m freestyle swimming world records among Masters-level competitors.

this sport depends on skill as well as on strength and endurance, some U.S. Masters swimmers have achieved their personal best performances at 45 to 50 years of age. As an example, the data in table 17.3 on page 558 illustrate a male swimmer's best performances at age 20 and at age 50. Despite a 30-year lapse in swimming training, this swimmer was able to achieve his best performances when he resumed training at age 50. Although the precise reasons for these

improvements are unknown, we can logically assume that they are the combined result of improvements in swimming technique, training methods, and swimming facilities, with little decrease in physiological capacity.

Cycling Performance

As with other strength and endurance sports, record-setting cycling performances are generally achieved in the age range of 25 to 35. Male and female cyclists' records for 40-km (24.9-mi) races decrease at about the same rate with age, an average of 20 s (approximately 0.6%) per year. The U.S. national cycling records for 20 km (12.4 mi) show a similar pattern for both men and women. For this distance, speed decreases by about 12 s (approximately 0.7%) per year from age 20 to nearly age 65.

Weightlifting

In general, maximal muscle strength is achieved between the ages of 25 and 35. Beyond that age range, as shown in figure 17.14, men's records for the sum of three power lifts decline at a steady rate of about 12.1 kg (26.7 lb), approximately 1.8%, per year. Of course, as with other measurements of human perfor-

Table 17.3

Freestyle Swimming Performances at Age 20 and 50 Years for a Male Masters Swimmer

Distance (m)	Best performances (s)		Improvement (%)
	20 years	50 years	
50	27.2	26.5	2.6
100	62.7	60.3	3.8
200	147.8	137.7	6.8
400	318.8	288.9	9.4
1,500	1,403.0	1,227.0	12.5

Note. The best freestyle swimming (front crawl stroke) times were achieved at age 50, despite the fact that the swimmer was an accomplished collegiate swimmer at age 18 to 21 years. It is also interesting to mention that this swimmer trained by swimming about 1,500 m per day at age 20 and 2,500 m per day at age 50.

Data from Ball State University, Human Performance Laboratory.

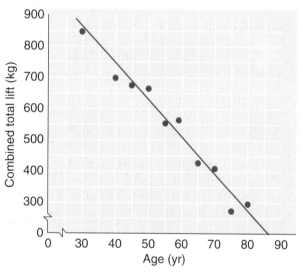

▲ **Figure 17.14** Changes in U.S. National Masters power-lifting records with age among male weightlifters. The values reported are combined totals for the squat, bench press, and dead lift.

> As we age, peak performances in both endurance and strength events decrease by about 1% to 2% per year, starting between ages 25 and 35.

mance, individual strength performances vary considerably. Some individuals, for example, exhibit greater strength at age 60 than people half their age.

Most athletic performances decline at a steady rate during middle and advanced ages. These decreases result from decrements in both muscular and cardiovascular endurance and strength as was discussed earlier in this chapter.

> ▶ Records in running, swimming, cycling, and weightlifting indicate that we are in our physical prime during our 20s and early 30s.
>
> ▶ In all of these sports, performance generally declines with aging beyond our physical prime.
>
> ▶ Most athletic performances decline steadily during middle and older age, primarily because of decrements in endurance and strength.

Special Issues

As we age, we need to consider several special issues that can directly affect us when exercising or performing various sport activities. We look briefly at two environmental stresses, altitude and heat, and then investigate the issue of longevity, injury, and risk of death resulting from exercise and sport.

Environmental Stress

Because a variety of physiological control processes become less effective with aging, we can logically assume that older people are less tolerant of environmental stress than their younger counterparts. In the following discussion, we compare the responses of younger and older adults to exposure to altitude and heat stress. Unfortunately, no specific studies have examined the tolerance of older athletes to these environmental conditions, so responses in only healthy, untrained adults can be reported here.

Exposure to Altitude

With the strenuous demands of exercise at moderate or high altitude, we might assume that normally active older adults would be at a disadvantage during exposure to the hypobaric conditions of altitude. Surprisingly, the opposite might be true. Many anecdotal stories have detailed the mountain-climbing exploits of people who range in age from 70 to 90 years.[5] Although most of these efforts have been at less than 4,500 m (14,764 ft), a 52-year-old American was able to climb to the top of Mount Everest (8,848 m, or 29,028 ft).

As we saw in chapter 11, acute altitude sickness is the major problem confronting most trained and untrained mountain climbers. Onset of acute altitude sickness occurs within 6 to 96 h after ascent and is characterized by such symptoms as headache, insomnia, loss of appetite, nausea, dizziness, and lassitude. A small percentage of cases rapidly progress into life-threatening high-altitude pulmonary edema (HAPE) or cerebral edema (HACE). Surprisingly, people under age 20 are more prone to HAPE than are older people.[26, 49] The incidence of HAPE is generally about 50 cases per 100,000 people, but for those under age 14 it is 140 per 100,000.[49] Some evidence indicates that we might tend to outgrow this predisposition to develop HAPE, so increased age alone should not deter healthy adults from activities at altitude. In fact, aging might provide some protection against the symptoms of acute altitude sickness and HAPE.

Exposure to Heat

Exposure to heat stress presents a problem for older people. Considerable evidence indicates that older adults are more susceptible to fatal heat injuries than are younger people.[2, 23] Measurements of heat stress in older and younger subjects show that aging reduces thermal tolerance. Even when people are matched for body size, body composition, $\dot{V}O_2max$, and degree of acclimatization, these age-related differences still persist. Both at rest and during submaximal exercise, older subjects develop a higher internal body temperature when exposed to heat than their younger counterparts. Part of the explanation for this is that older adults produce less sweat, decreasing their capacity for heat loss via evaporation.

> Aging does not appear to reduce our capacity to perform normal activity at high altitude. In fact, it might enhance our capacity! However, aging reduces our ability to adapt to exercise in the heat. This is largely because sweating capacity decreases as we age.

Most of these observations were made with normally active subjects. Thus, we can't determine what influence various lifestyles and activity levels might have had on these results. Unfortunately, no comparable data exist for highly trained young and old athletes. Knowing the positive influence that regular exercise, heat exposure, and endurance training have on heat tolerance, we can only assume that the negative effects of environmental heat stress would be reduced in older athletes.

> ▶ Some cases of acute altitude sickness progress to life-threatening conditions of high-altitude pulmonary edema or high-altitude cerebral edema. But these conditions are more common in young people. Aging might provide some protection against acute altitude sickness and high-altitude pulmonary or cerebral edema.
>
> ▶ Aging reduces thermal tolerance partly because aging reduces sweat production; consequently, less heat can be lost via evaporation.

Longevity and Risk of Injury and Death

Regular physical activity is an important contributor to good health, so we can logically ask, Does training throughout adulthood affect longevity? Because the aging rate in rats is more rapid than in humans, they have been used as subjects in studies conducted to determine the influence of chronic exercise (training) on longevity (the length of one's life). A study by Goodrick[16] demonstrated that rats who exercised freely lived about 15% longer than sedentary rats. But an investigation at Washington University in St. Louis showed no significant increase in the life span of rats who voluntarily ran on an exercise wheel.[25] More of the active rats lived to old age, but on the average, they still died at the same age as their sedentary counterparts. It is interesting that the rats that had restricted food intake and maintained a lower body weight lived 10% longer than the freely eating, sedentary rats.

Of course, we cannot directly apply these findings to humans, but these results raise some interesting questions that might be relevant to our health and longevity. Although it is true that an endurance exercise program can reduce a number of the risk factors associated with cardiovascular disease, only limited information supports the contention that you will live longer if you exercise regularly. Data collected from the alumni at Harvard University and the University of Pennsylvania and from participants at the Aerobic Center in Dallas suggest that there is a decrease in mortality rate and a small increase in longevity (about 2 years) among people who remain physically active throughout life. Perhaps future longitudinal studies will shed more light on the relationship between lifelong exercise and longevity.

What about the risk of injury and death from exercise as we get older? Studies show that as you get older you are at a greater risk for injuries involving tendons, cartilage, and bone. The most common orthopedic injuries include rotator cuff tears, quadriceps tendon ruptures, Achilles tendon ruptures, degenerative meniscus tears, focal articular cartilage defects and injuries, and stress fractures.[39] Furthermore, when injuries occur, the healing process is usually prolonged and complete recovery can take up to a full year.[39] On the other hand, increasing the strength and endurance of elderly people reduces their risk of falls and related injury.

The risk of death during exercise appears to be no higher in the older athlete compared with the younger and middle-aged athlete. However, the risk in older people who are not regularly active appears to be increased.[50] Importantly, an active lifestyle does, in fact, reduce the risk of death from many chronic diseases, something that we discuss in chapters 19 through 21.

> ▶ An active lifestyle appears to be associated with a small increase in longevity. Just as important, an active lifestyle leads to a higher quality of life!
>
> ▶ There is an increased risk of injury from exercise as you age, and injuries tend to be slower to heal.
>
> ▶ The risk of death during exercise is not increased in those who are regularly active but is increased in those who seldom exercise.

In Closing . . .

In this chapter we examined the effects of aging on physical performance. We evaluated changes in cardiorespiratory endurance and strength with age. We considered the effect of aging on body composition, which we know can affect performance. And yet, in the course of our discussion, it became clear that much of the change that occurs with aging is to a great extent attributable to the inactivity that often accompanies aging. When older people participate in training, most of the changes associated with aging are lessened and the resulting degree of change is similar to that seen in young and middle-aged adults. Thus, we have dispelled many of the myths about the capacity for physical activity of older people.

In the next chapter, we turn our attention to females, a group that is often considered less capable of physical activity than males. We consider the physiology of girls and women, the impact of this physiology on athletic ability, how performances of female athletes compare with those of male athletes, and special issues associated with being female.

▷ Key Terms

cardiovascular deconditioning

forced expiratory volume in 1 s (FEV$_{1.0}$)

longevity

maximal expiratory ventilation ($\dot{V}_E$max)

osteopenia

osteoporosis

peripheral blood flow

residual volume (RV)

sarcopenia

total lung capacity (TLC)

vital capacity (VC)

▷ Study Questions

1. Describe the changes in strength and endurance performance records with aging.

2. What cardiovascular changes occur during aging? How do these changes affect maximal oxygen uptake?

3. Describe the changes in $\dot{V}O_2$max with age. How do trained individuals differ from untrained individuals?

4. How does the respiratory system change with aging? What happens to vital capacity, FEV$_{1.0}$, residual volume, RV/TLC, and $\dot{V}_E$max?

5. Describe the changes in HRmax with age. How does training alter this relationship?

6. How does aging affect maximal stroke volume and maximal cardiac output? What mechanisms potentially can explain these changes?

7. What muscular changes occur with aging? How do they affect athletic performance?

8. How does training alter the biology of aging?

9. Differentiate between biological aging and physical inactivity.

10. What influence do aging and training have on body composition?

11. Describe the trainability of the older individual for both strength and endurance.

▷ References

1. Akima, H., Kano, Y., Enomoto, Y., Ishizu, M., Okada, M., Oishi, Y., Katsuta, S., & Kuno, S.Y. (2001). Muscle function in 164 men and women aged 20-84 yr. *Medicine and Science in Sports and Exercise, 33,* 220-226.

2. Applegate, W.B., Runyan, J.W., Brasfield, L., Williams, M.L., Konigsberg, C., & Fauche, C. (1981). Analysis of the 1980 heat wave in Memphis. *Journal of the American Geriatrics Society, 29,* 337-342.

3. Åstrand, I., (1960). Aerobic work capacity in men and women with special reference to age. *Acta Physiologica Scandinavica, 49*(Suppl. 169), 1-92.

4. Åstrand, I., Åstrand, P.-O., Hallback, I., & Kilbom, A. (1973). Reduction in maximal oxygen intake with age. *Journal of Applied Physiology, 35,* 649-654.

5. Balcomb, A.C., & Sutton, J.R. (1986). Advanced age and altitude illness. In J.R. Sutton & R.M. Brock (Eds.), *Sports medicine for the mature athlete* (213-224). Indianapolis: Benchmark Press.

6. Buskirk, E.R., & Hodgson, J.L. (1987). Age and aerobic power: The rate of change in men and women. *Federation Proceedings, 46,* 1824-1829.

7. Connelly, D.M., Rice, C.L., Roos, M.R., & Vandervoort, A.A. (1999). Motor unit firing rates and contractile properties in tibialis anterior of young and old men. *Journal of Applied Physiology, 87,* 843-852.

8. Costill, D.L. (1986). *Inside running: Basics of sports physiology.* Indianapolis: Benchmark Press.

9. Cress, M.E., Thomas, D.P., Johnson, J., Kasch, F.W., Cassens, R.G., Smith, E.L., & Agre, J.C. (1991). Effect of training on $\dot{V}O_2$max, thigh strength, and muscle morphology in septuagenarian women. *Medicine and Science in Sports and Exercise, 23,* 752-758.

10. Dill, D.B., Alexander, W.C., Myhre, L.G., Whinnery, J.E., & Tucker, D.M. (1985). Aerobic capacity of D.B. Dill, 1928-1984 [Abstract]. *Federation Proceedings, 44,* 1013.

11. Dill, D.B., Robinson, S., & Ross, J.C. (1967). A longitudinal study of 16 champion runners. *Journal of Sports Medicine and Physical Fitness, 7,* 4-27.

12. Enoka, R.M., Burnett, R.A., Graves, A.E., Kornatz, K.W., & Laidlaw, D.H. (1999). Task- and age-dependent variations in steadiness. *Progress in Brain Research, 123,* 389-395.

13. Eskurza, I., Donato, A.J., Moreau, K.L., Seals, D.R., & Tanaka, H. (2002). Changes in maximal aerobic capacity with age in endurance-trained women: 7-yr follow-up. *Journal of Applied Physiology, 92,* 2303-2308.

14. Fitzgerald, M.D., Tanaka, H., Tran, Z.V., & Seals, D.R. (1997). Age-related declines in maximal aerobic capacity in regularly exercising vs. sedentary women: a meta-analysis. *Journal of Applied Physiology, 83,* 160-165.

15. Frontera, W.R., Meredith, C.N., O'Reilly, K.P., Knuttgen, W.G., & Evans, W.J. (1988). Strength conditioning in older men: Skeletal muscle hypertrophy and improved function. *Journal of Applied Physiology, 64,* 1038-1044.

16. Goodrick, C.L. (1980). Effects of long-term voluntary wheel exercise on male and female Wistar rats: 1. Longevity, body weight and metabolic rate. *Gerontology, 26,* 22-33.

17. Hagerman, F.C., Fielding, R.A., Fiatarone, M.A., Gault, J.A., Kirkendall, D.T., Ragg, K.E., & Evans, W.J. (1996). A 20-yr longitudinal study of Olympic oarsmen. *Medicine and Science in Sports and Exercise, 28,* 1150-1156.

18. Hagerman, F.C., Walsh, S.J., Staron, R.S., Hikida, R.S., Gilders, R.M., Murray, T.F., Toma, K., & Ragg, K.E. (2000). Effects of high-intensity resistance training on untrained older men. I. Strength, cardiovascular, and metabolic responses. *Journals of Gerontology Series A–Biological Sciences & Medical Sciences, 55,* B336-B346.

19. Häkkinen, K., Kraemer, W.J., Pakarinen, A., Triplett-McBride, T., McBride, J.M., Häkkinen, A., Alen, M., McGuigan, M.R., Bronks, R., & Newton, R.U. (2002). Effects of heavy resistance/power training on maximal strength, muscle morphology, and hormonal response patterns in 60-75-year-old men and women. *Canadian Journal of Applied Physiology, 27,* 213-231.

20. Häkkinen, K., Pakarinen, A., Kraemer, W.J., Häkkinen, A., Valkeinen, H., & Alen, M. (2001). Selective muscle hypertrophy, changes in EMG and force, and serum hormones during strength training in older women. *Journal of Applied Physiology, 91,* 569-580.

21. Hameed, M., Harridge, S.D.R., & Goldspink, G. (2002). Sarcopenia and hypertrophy: A role for insulin-like growth factor-1 and aged muscle? *Exercise and Sport Sciences Reviews, 30,* 15-19.

22. Hawkins, S.A., Marcell, T.J., Jaque, S.V., & Wiswell, R.A. (2001). A longitudinal assessment of change in $\dot{V}O_2$max and maximal heart rate in master athletes. *Medicine and Science in Sports and Exercise, 33,* 1744-1750.

23. Henschel, A., Burton, L., & Morgalies, L. (1969). An analysis of the deaths in St. Louis during July 1966. *American Journal of Public Health, 59,* 2232-2240.

24. Hikida, R.S., Staron, R.S., Hagerman, F.C., Walsh, S., Kaiser, E., Shell, S., & Hervey, S. (2000). Effects of high-intensity resistance training on untrained older men. II. Muscle fiber characteristics and nucleo-cytoplasmic relationships. *Journals of Gerontology Series A–Biological Sciences & Medical Sciences, 55,* B347-B354.

25. Holloszy, J.O. (1997). Mortality rate and longevity of food-restricted exercising male rats: A reevaluation. *Journal of Applied Physiology, 82,* 399-403.

26. Hultgren, H.N., & Marticorena, E.M. (1978). High altitude pulmonary edema: Epidemiologic observations in Peru. *Chest, 74,* 372-376.

27. Jackson, A.S., Beard, E.F., Wier, L.T., Ross, R.M., Stuteville, J.E., & Blair, S.N. (1995). Changes in aerobic power of men, ages 25-70 yr. *Medicine and Science in Sports and Exercise, 27,* 113-120.

28. Jackson, A.S., Wier, L.T., Ayers, G.W., Beard, E.F., Stuteville, J.E., & Blair, S.N. (1996). Changes in aerobic power of women, ages 20-64 yr. *Medicine and Science in Sports and Exercise, 28,* 884-891.

29. Janssen, I., Heymsfield, S.B., Wang, Z., & Ross, R. (2000). Skeletal muscle mass and distribution in

468 men and women aged 18-88 yr. *Journal of Applied Physiology, 89,* 81-88.

30.   Johnson, M.A., Polgar, J., Weihtmann, D., & Appleton, D. (1973). Data on the distribution of fiber types in thirty-six human muscles: An autopsy study. *Journal of Neurological Science, 1,* 111-129.

31.   Jorfeldt, L., & Wahren, J. (1971). Leg blood flow during exercise in man. *Clinical Science, 41,* 459-473.

32.   Kamen, G., Sison, S.V., Duke Du, C.C., & Patten, C. (1995). Motor unit discharge behavior in older adults during maximal-effort contractions. *Journal of Applied Physiology, 79,* 1908-1913.

33.   Kasch, F.W., Boyer, J.L., Van Camp, S., Verity, L.S., & Wallace, J.P. (1995). Cardiovascular changes with age and exercise. *Scandinavian Journal of Medicine and Science in Sports, 5,* 147-151.

34.   Kohrt, W.M., Malley, M.T., Coggan, A.R., Spina, R.J., Ogawa, T., Ehsani, A.A., Bourey, R.E., Martin, W.H., III, & Holloszy, J.O. (1991). Effects of gender, age, and fitness level on response of VO_2 max to training in 60-71 yr olds. *Journal of Applied Physiology, 71,* 2004-2011.

35.   Kohrt, W.M., Malley, M.T., Dalsky, G.P., & Hollo-szy, J.O. (1992). Body composition of healthy sedentary and trained, young and older men and women. *Medicine and Science in Sports and Exercise, 24,* 832-837.

36.   Lakatta, E.G. (1979). Alterations in the cardiovascular system that occur in advanced age. *Federation Proceedings, 38,* 163-167.

37.   Lexell, J., Downham, D.Y., Larson, Y., Bruhn, E., & Morsing, B. (1995). Heavy-resistance training in older Scandinavian men and women: Short- and long-term effects on arm and leg muscles. *Scandinavian Journal of Medicine and Science in Sports, 5,* 329-341.

38.   Lexell, J., Taylor, C.C., & Sjostrom, M. (1988). What is the cause of the aging atrophy? Total number, size, and proportion of different fiber types studied in whole vastus lateralis muscle from 15- to 83-year-old men. *Journal of Neurological Science, 84,* 275-294.

39.   Maharam, L.G., Bauman, P.A., Kalman, D., Skol-nik, H., & Perle, S.M. (1999). Masters athletes: Factors affecting performance. *Sports Medicine, 28,* 273-285.

40.   MacKeen, P.C., Rosenberger, J.L., Slater, J.S., Nicholas, W.C., & Buskirk, E.R. (1985). A 13-year follow-up of a coronary heart disease risk factor screening and exercise program for 40- to 59-year-old men: Exercise habit maintenance and physiologic status. *Journal of Cardiopulmonary Rehabilitation, 5,* 510-523.

41.   Meredith, C.N., Frontera, W.R., Fisher, E.C., Hughes, V.A., Herland, J.C., Edwards, J., & Evans, W.J. (1989). Peripheral effects of endurance training in young and old subjects. *Journal of Applied Physiology, 66,* 2844-2849.

42.   Meusel, H. (1984). Health and well-being for older adults through physical exercises and sport—Outline of the Giessen model. In B. McPherson (Ed.), *Sport and aging* (pp. 107-115). Champaign, IL: Human Kinetics.

43.   Paterson, D.H., Cunningham, D.A., Koval, J.J., & St. Croix, C.M. (1999). Aerobic fitness in a population of independently living men and women aged 55-86 years. *Medicine and Science in Sports and Exercise, 31,* 1813-1820.

44.   Pollock, M.L., Mengelkoch, L.J., Graves, J.E., Lowenthal, D.T., Limacher, M.C., Foster, C., & Wilmore, J.H. (1997). Twenty-year follow-up of aerobic power and body composition of older track athletes. *Journal of Applied Physiology, 82,* 1508-1516.

45.   Proctor, D.N., Shen, P.H., Dietz, N.M., Eickhoff, T.J., Lawler, L.A., Ebersold, E.J., Loeffler, D.L., & Joyner, M.J. (1998). Reduced leg blood flow during dynamic exercise in older endurance-trained men. *Journal of Applied Physiology, 85,* 68-75.

46.   Robinson, S. (1938). Experimental studies of physical fitness in relation to age. *Arbeitsphysiologie, 10,* 251-323.

47.   Saltin, B. (1986). The aging endurance athlete. In J.R. Sutton & R.M. Brock (Eds.), *Sports medicine for the mature athlete* (59-80). Indianapolis: Benchmark Press.

48.   Saltin, B. (1990). *Aging, health and exercise performance* (Provost Lecture Series). Muncie, IN: Ball State University.

49.   Scoggin, E.H., Meyers, T.M., Reeves, J.T., & Grover, R.F. (1977). High-altitude edema in young adults of Leadville, Colorado. *New England Journal of Medicine, 297,* 1269-1272.

50.   Shephard, R.J. (1997). *Aging, physical activity, and health.* Champaign, IL: Human Kinetics.

51.   Spirduso, W.W. (1995). *Physical dimensions of aging.* Champaign, IL: Human Kinetics.

52.   Tanaka, H., Monahan, K.D., & Seals, D.R. (2001). Age-predicted maximal heart rate revisited. *Journal of the American College of Cardiology, 37,* 153-156.

53.   Tanaka, H., & Seals, D. (1997). Age and gender interactions in physiological functional capacity: Insight from swimming performance. *Journal of Applied Physiology, 82,* 846-851.

54.   Trappe, S.W., Costill, D.L., Fink, W.J., & Pearson, D.R. (1995). Skeletal muscle characteristics among distance runners: A 20-yr follow-up study. *Journal of Applied Physiology, 78,* 823-829.

55.   Trappe, S.W., Costill, D.L., Goodpaster, B.H., & Pearson, D.R. (1996). Calf muscle strength in former elite distance runners. *Scandinavian Journal of Medicine and Science in Sports, 6,* 205-210.

55.   Trappe, S.W., Costill, D.L., Vukovich, M.D., Jones, J., & Melham, T. (1996). Aging among elite distance runners: A 22-yr longitudinal study. *Journal of Applied Physiology, 80,* 285-290.

57.   Wahren, J., Saltin, B., Jorfeldt, L., & Pernow, B. (1974). Influence of age on the local circulatory adaptation to leg exercise. *Scandinavian Journal of Clinical Laboratory Investigation, 33,* 79-86.

58.   Widrick, J.J., Trappe, S.W., Blaser, C.A., Costill, D.L., & Fitts, R.H. (1996). Isometric force and maximal shortening velocity of single muscle fibers from elite master runners. *American Journal of Physiology, 271*(40), C666-C675.

59.   Widrick, J.J., Trappe, S.W., Costill, D.L., & Fitts, R.H. (1996). Force-velocity and force power properties of single muscle fibers from elite master runners and sedentary men. *American Journal of Physiology, 271*(40), C676-C683.

60.   Wilmore, J.H., Després, J.-P., Stanforth, P.R., Mandel, S., Rice, T., Gagnon, J., Leon, A.S., Rao, D.C., Skinner, J.S., & Bouchard, C. (1999). Alterations in body weight and composition consequent to 20 wk of endurance training: the HERITAGE Family Study. *American Journal of Clinical Nutrition, 70,* 346-352.

61.   Wiswell, R.A., Jaque, S.V., Marcell, T.J., Hawkins, S.A., Tarpenning, K.M., Constantino, N., & Hyslop, D.M. (2000). Maximal aerobic power, lactate threshold, and running performance in master athletes. *Medicine and Science in Sports and Exercise 32,* 1165-1170.

▷ Selected Readings

Bemben, M.G. (1998). Age-related alterations in muscular endurance. *Sports Medicine, 25,* 259-269.

Blair, S.N., Kohl, H.W., III, Paffenbarger, R.S., Clark, D.G., Cooper, K.H., & Gibbons, L.W. (1989). Physical fitness and all-cause mortality: A prospective study of healthy men and women. *Journal of the American Medical Association, 262,* 2395-2401.

Coggan, A.R., Spina, R.J., Rogers, M.A., King, D.S., Brown, M., Nemeth, P.M., & Holloszy, J.O. (1990). Histochemical and enzymatic characteristics of skeletal muscle in master athletes. *Journal of Applied Physiology, 68,* 1896-1901.

Daley, M.J., & Spinks, W.L. (2000). Exercise, mobility and aging. *Sports Medicine, 29,* 1-12.

Drinkwater, B.L., Bedi, J.F., Loucks, A.B., Roche, S., & Horvath, S.M. (1982). Sweating sensitivity and capacity of women in relation to age. *Journal of Applied Physiology, 53,* 671-676.

Hurley, B.F., & Roth, S.M. (2000). Strength training in the elderly. *Sports Medicine, 30,* 249-268.

Kenny, W.L., & Hodgson, J.L. (1987). Heat tolerance, thermoregulation and aging. *Sports Medicine, 4,* 446-456.

Larsson, L. (1978). Morphological and functional characteristics of the aging skeletal muscle in man. *Acta Physiologica Scandinavica,* (Suppl. 457), 36.

Lindle, R.S., Metter, E.J., Lynch, N.A., Fleg, J.L., Fozard, J.L., Tobin, J., Roy, T.A., & Hurley, B.F. (1997). Age and gender comparisons of muscle strength in 654 women and men aged 20-93 yr. *Journal of Applied Physiology, 83,* 1581-1587.

Ogawa, T., Spina, R.J., Martin, W.H., III, Kohrt, W.M., Schechtman, K.B., Holloszy, J.O., & Ehsani, A.A. (1992). Effects of aging, sex, and physical training on cardiovascular responses to exercise. *Circulation, 86,* 494-503.

Orlander, J., & Aniansson, A. (1979). Effects of physical training on skeletal muscle metabolism and ultrastructure in 70 to 75-year-old men. *Acta Physiologica Scandinavica, 109,* 149-154.

Paffenbarger, R.S., Hyde, R.T., Wing, A.L., & Hsieh, C.-C. (1986). Physical activity, all-cause mortality, and longevity of college alumni. *New England Journal of Medicine, 314,* 605-613.

Pollock, M.L., Miller, H.S., & Wilmore, J.H. (1974). Physiological characteristics of champion American track athletes 40 to 75 years of age. *Journal of Gerontology, 29,* 645-649.

Porter, M.M., Vandervoort, A.A., & Lexell, J. (1995). Aging of human muscle: Structure, function and adaptability. *Scandinavian Journal of Medicine and Science in Sports, 5,* 129-142.

Proctor, D.N., & Joyner, M.J. (1997). Skeletal muscle mass and the reduction of $\dot{V}O_2$max in trained older subjects. *Journal of Applied Physiology, 82,* 1411-1415.

Rogers, M.A., Hagberg, J.M., Martin, W.H., Ehsani, A.A., & Holloszy, J.O. (1990). Decline in $\dot{V}O_2$max with aging in master athletes and sedentary men. *Journal of Applied Physiology, 68,* 2195-2199.

Roubenoff, R. (2001). Origins and clinical relevance of sarcopenia. *Canadian Journal of Applied Physiology, 26,* 78-89.

Saltin, B., & Grimby, G. (1968). Physiological analysis of middle-aged and old former athletes. *Circulation, 38,* 1104-1115.

Seals, D.R., Hagberg, J.M., Hurley, B.F., Ehsani, A.A., & Holloszy, J.O. (1984). Endurance training in older men and women: I. Cardiovascular responses to exercise. *Journal of Applied Physiology: Respiratory, Environmental, and Exercise Physiology, 57,* 1024-1029.

Shephard, R.J. (1984). Physiological aspects of sport and physical activity in the middle and later years of life. In B. McPherson (Ed.), *Sport and aging* (pp. 37-43). Champaign, IL: Human Kinetics.

Spirduso, W.W. (1975). Reaction and movement time as a function of age and physical activity level. *Journal of Gerontology,* **30,** 435.

Sutton, J.R., & Brock, R.M. (1986). *Sports medicine for the mature athlete.* Indianapolis: Benchmark Press.

Tanaka, H., DeSouza, C.A., Jones, P.P., Stevenson, E.T., Davy, K.P., & Seals, D.R. (1997). Greater rate of decline in maximal aerobic capacity with age in physically active vs. sedentary healthy women. *Journal of Applied Physiology,* **83,** 1947-1953.

von Dobeln, W. (1957). Human standard and maximal metabolic rate in relation to fat-free body mass. *Acta Physiologica Scandinavica,* (Suppl. 126), 37-79.

Young, K., & Young, J.H. (1985). *Running records by age: 1985.* Tucson, AZ: National Running Data Center.

CHAPTER 18

SEX DIFFERENCES IN SPORT AND EXERCISE

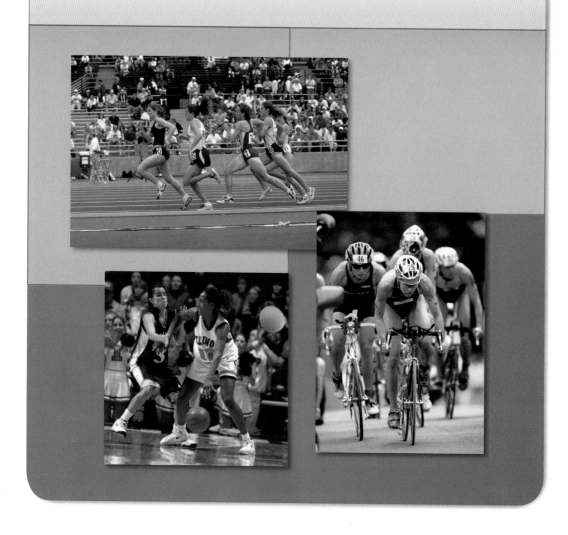

overview

In the not-so-distant past, young girls typically were encouraged to play with dolls, play house, and play dress-up, while young boys climbed trees, raced against each other, and became active in various sports. The underlying notion was that boys were meant to be active and athletic but girls were weaker and less well suited to physical activity. Physical education classes furthered this idea by having the girls exercise differently than the boys—by running shorter distances and performing modified push-ups. Less physical activity was expected of girls, and as they progressed through school, most girls could not compete on an equal basis with boys of their own age, even if given the opportunity. In athletics, girls and women were not allowed to run in long-distance races, and basketball was limited to half court, with each team having only offensive or defensive players.

Now more athletic activities are accessible to girls and women than in the past, and the results have been amazing! Their performances in sport parallel those of boys and men, with performance differences of 15% or less for most sports and events. This has led researchers to ask how much of the difference in the performance capabilities of females and males is attributable to biological differences. In this chapter, we try to answer this question. We probe the similarities and differences between females and males in body size and composition and in physiological responses to acute exercise and to training. We also consider sex differences in athletic performance. Finally, we review several issues important to female athletes, including menstruation and menstrual dysfunction, pregnancy, osteoporosis, and eating disorders. We also look at how men's and women's responses to heat, cold, and altitude differ. We consider how all these factors affect the performance of female athletes.

outline

Girls and women were prohibited from running any race longer than 800 m until the 1960s. They were also barred from official participation in the marathon until 1970. Both of these restrictions resulted from a misconception that women were physiologically unsuited for endurance activity. Yet at the 1984 Los Angeles Olympic Games, American runner Joan Benoit won the gold medal in the first-ever Olympic marathon for women with a time of 2:24:52. Her time would have won 11 of the previous 20 men's Olympic marathons![31]

Another myth that is falling by the wayside is that pregnant women should not exercise. The following story appeared in the *Austin American Statesman* on June 17, 1995. "When Sue Olsen lines up today for Grandma's Marathon in St. Paul, Minn., along the North Shore of Lake Superior, she will be precisely 16 days short of the due date for her first child. 'I get a lot of advice both ways,' Olsen, 38, said with a laugh. 'There are people who think I'm crazy and shouldn't be out there, and there are a lot of supportive people.' . . . Olsen's husband will drive parallel to the course with a cellular phone in his car, prepared to whisk her to the hospital if the situation arises." Although many would question the wisdom of this, the point is that she was able to complete the marathon in 4 h. The following weekend she completed a 24-h race, delivering a baby boy the next day.

On the basis of world records in 2002, women's performances lag behind men's. This is illustrated in table 18.1. Do these performance differences result from biological differences? Or do they reflect social and cultural restrictions placed on females during preadolescent and adolescent development? Or do they reflect that, throughout history, the total number of women who have competed in these events is smaller than the total number of men who have competed in them? Our focus in this chapter is the extent to which biological differences between females and males affect performance capacity. Let's begin by considering basic physical differences and their impact on performance.

Table 18.1

Selected Men's and Women's World Records Through 2002

Event	Men	Women	Difference
Track and field			
100 m	9.79 s	10.49 s	7.2%
1,500 m	3:26.00 min:s	3:50.46 min:s	11.9%
10,000 m	26:22.75 min:s	29:31.78 min:s	11.9%
High jump	2.45 m	2.09 m	14.7%
Long jump	8.95 m	7.52 m	16.0%
Swimming			
100-m freestyle	47.84 s	53.77 s	12.4%
400-m freestyle	3:40.17 min:s	4:03.85 min:s	10.8%
1,500-m freestyle	14:34.56 min:s	15:52.10 min:s	8.9%

Body Size and Composition

Until age 12 to 14—around puberty—boys and girls do not differ substantially in height, weight, girth, bone width, or skinfold thickness.

A study of 609 normal boys and girls ages 7.5 to 20.5 found no **sex-specific differences** in fat-free mass (FFM) prior to adolescence when FFM was expressed per unit of height.[22] At ages 12 to 13, the ratio of FFM to height in females begins to plateau, but in males it continues to increase to age 20. Female FFM peaks at age 15 to 16, but male FFM doesn't peak until age 18 to 20. The peak FFM attained by females is 72% of the FFM attained by males. These changes in FFM with age are illustrated in figure 18.1.

> Major differences in body size and composition between girls and boys do not start to appear until puberty.

Body density data are somewhat inconsistent with these findings. Females typically have lower total body density values at all

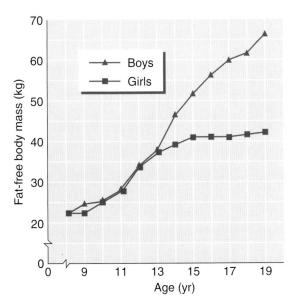

▲ **Figure 18.1**   Sex differences in fat-free body mass changes with age.
Data from G.B. Forbes, 1972, "Growth of the lean body mass in man," *Growth* 36: 325-338.

ages, including preadolescence, which would normally indicate a higher relative body fat. But from age 7 to age 25 years, the density of the fat-free mass in females is consistently lower than in males.[35] The calculations used to determine relative body fat typically assume that these densities are the same in both sexes. As a result, most existing data on females in this age range overestimate their true relative body fat.

At puberty, the body composition of the sexes begins to differ markedly, primarily because of endocrine changes. Before puberty, the anterior pituitary gland does not secrete gonadotropic hormones: follicle-stimulating hormone (FSH) and luteinizing hormone (LH). These hormones stimulate the gonads (ovaries and testes). During puberty, however, the anterior pituitary begins to secrete both of these hormones. In females, when sufficient quantities of FSH and LH are secreted, the ovaries develop and estrogen secretion begins. In males, these same hormones trigger development of the testes and, in turn, testosterone secretion. **Testosterone** increases bone formation, which leads to larger bones, and it increases protein synthesis, which leads to increased muscle mass. As a result, adolescent males are larger and more muscular than females, and these characteristics continue into adulthood. At full maturity, men not only have a greater muscle mass, but the distribution of the muscle mass differs from that of women. Men carry a higher percentage of their muscle mass in the upper body compared with women (42.9% vs. 39.7%).[33]

Estrogen also has a significant influence on body growth by broadening the pelvis, stimulating breast development, and increasing fat deposition, particularly in the thighs and hips. This increase in fat deposition in the thighs and hips is the result of increased lipoprotein lipase activity in these areas. This enzyme is considered the gatekeeper for storing fat in adipose tissue. **Lipoprotein lipase** is produced in the fat cells (adipocytes) but is bound to the walls of the capillaries, where it exerts its influence on the chylomicrons, which are the major transporters of triglycerides in the blood. When lipoprotein lipase activity in any area of the body is high, chylomicrons are trapped and their triglycerides are hydrolyzed and transported into the adipocytes in that area for storage.

Estrogen also increases the growth rate of bone, allowing the final bone length to be reached within 2 to 4 years following the onset of puberty. As a result, females grow very rapidly for the first few years following puberty and then cease to grow. Males have a much longer growth phase, allowing them to attain a greater height. Because of these differences, compared with fully mature males, fully mature females are on average nearly

- 13 cm (5 in.) shorter,
- 14 to 18 kg (30-40 lb) lighter in total weight,
- 18 to 22 kg (40-50 lb) lighter in FFM,
- 3 to 6 kg (7-13 lb) heavier in fat mass, and
- 6% to 10% higher in relative body fat.

Anthropometric measurements at maturity differ substantially between the sexes. Women have narrower shoulders, broader hips, and smaller chest diameters and tend to have more fat in the hips and lower body, whereas men carry more fat in the abdomen and upper body. Table 18.2 provides anthropometric (body-measurement) data for young and middle-aged men and women.[48, 49, 67, 68] From this table you can see how both age and sex affect these measurements.

Table 18.2

Anthropometric Measurements for Young and Middle-Aged Men and Women

Site	Women			Men		
	Young[a] $n = 128$	Young[b] $n = 83$	Middle-aged[b] $n = 60$	Young[a] $n = 133$	Young[b] $n = 95$	Middle-aged[b] $n = 84$
Skinfolds (mm)						
Scapula	13.2	15.3	17.3	14.1	13.9	20.2
Triceps	12.8	18.8	22.2	7.9	13.6	18.5
Midaxillary	10.7	13.3	16.9	11.7	15.5	24.8
Chest	—	14.0	14.0	—	11.4	20.6
Superiliac	17.2	15.3	17.3	19.3	15.2	22.0
Abdominal	15.1	22.8	29.6	16.0	20.6	30.0
Thigh	31.8	28.8	33.1	14.9	17.4	22.2
Knee	7.0	17.4	17.3	5.3		
Circumferences (cm)						
Head	55.0	—	—	57.5	—	—
Neck	31.8	—	—	38.5	—	—
Shoulders	101.9	99.7	100.9	117.0	112.5	114.8
Chest	85.2	84.6	87.1	97.4	91.4	96.3

Site	Women			Men		
	Young[a] n = 128	Young[b] n = 83	Middle-aged[b] n = 60	Young[a] n = 133	Young[b] n = 95	Middle-aged[b] n = 84
Circumferences (cm) *(continued)*						
Bust	87.8	87.7	90.8	—	—	—
Abdomen	75.3	75.0	82.7	84.0	81.0	91.1
Hips	95.9	93.1	97.5	96.9	94.4	98.4
Thigh	57.0	56.5	57.6	58.0	57.1	59.0
Knee	36.1	—	—	37.7	—	—
Calf	35.1	33.9	34.4	37.6	36.5	36.9
Ankle	21.1	20.8	20.8	22.7	22.1	22.1
Deltoid	30.7			36.3		
Biceps, flexed	27.2	27.0	28.6	33.2	32.6	34.0
Biceps, extended	25.0			29.1		
Forearm	23.5	23.8	24.4	27.6	28.3	29.2
Wrist	14.9	14.8	15.1	17.0	16.7	17.4
Diameters (cm)						
Head length	19.0	—	—	19.9	—	—
Head width	14.9	—	—	15.5	—	—
Biacromial	36.5	36.8	36.7	40.4	41.1	41.5
Bideltoid	42.1	41.4	41.8	47.6	46.9	47.4
Chest	25.8	27.8	28.6	29.3	31.8	33.0
Bi-iliac	28.4	29.9	31.2	28.4	29.6	31.4
Bitrochanteric	32.1	34.0	35.3	32.9	33.6	35.1
Knee	8.9	9.3	9.6	9.5	9.8	10.1
Ankle	6.3	—	—	7.1	—	—
Elbow	6.0	—	—	7.0	—	—
Wrist	4.9	5.1	5.2	5.6	5.9	6.0
Arm span	165.8	—	—	181.7	—	—
Foot length	24.1	—	—	26.7	—	—
Hand length	17.3	—	—	19.1	—	—

[a]Data from Wilmore and Behnke (1969 and 1970).[67, 68]

[b]Data from Pollock et al. (1975 and 1976).[48, 49]

With aging, both women and men tend to accumulate fat and lose fat-free body mass starting in their mid-20s. In one of the few longitudinal studies conducted, FFM was found to decrease by approximately 3 kg (6.6 lb) per decade, or 0.3 kg (0.66 lb) per year.[23] These data show slightly greater losses than cross-sectional data, which indicate a loss in FFM of 0.1 to 0.2 kg (0.3-0.5 lb) per year. This loss is associated with lower levels of physical activity and of testosterone. Apparently, the increase in total body fat is associated with a general decline in physical activity without an equal decrease in caloric intake. Table 18.3 illustrates the changes with aging in relative body fat for both sexes.

The average difference in relative body fat between young women and men ages 18 to 24 is about 6% to 10% (20-25% for women vs. 13-16% for men). At first, this difference was thought to reflect sex-specific differences in fat deposit (namely, the breasts, hips, and thighs). But female athletes, particularly distance runners, can be exceptionally lean—well below the relative fat value for the average young woman and even below that for the average young man. Many of the better female runners are below 10% body fat (chapter 14). Such low values could result from either or both a genetic predisposition toward leanness and the high weekly training distance these women run, sometimes exceeding 160 km (100 mi) per week. Thus, we know that women can maintain fat stores well below what is considered normal for their age. In fact, there is increasing concern that some women are becoming too lean. This will be discussed in greater detail later in this chapter.

Table 18.3

Relative Body Fat Values for Average, Untrained Women and Men of Various Ages

Age group (years)	Relative body fat %	
	Women	Men
15-19	20-24	13-16
20-29	22-25	15-20
30-39	24-30	18-26
40-49	27-33	23-29
50-59	30-36	26-33
60-69	30-36	29-33

▶ Until puberty, girls and boys do not differ significantly in most measurements of body size and composition.

▶ At puberty, because of the influences of estrogen and testosterone, body composition begins to change markedly. Estrogen causes increased fat deposition in females, particularly in the hips and thighs, and an increased rate of bone growth, such that bones in females reach their final length earlier than in males.

▶ Although women tend to accumulate more body fat than men, research shows that some female athletes are exceptionally lean.

Fat Deposition: Why the Hips and Thighs?

Many women are constantly fighting fat deposition on the thighs and hips, but they are usually fighting a losing battle. Lipoprotein lipase activity is very high and lipolytic activity (fat breakdown) is low in the hips and thighs of women compared with women's other fat storage areas and with the hips and thighs of men. This results in a rapid storage of fat in women's thighs and hips, and the decreased lipolytic activity makes it difficult for women to lose fat from these areas. During the last trimester of pregnancy and throughout lactation, the activity of lipoprotein lipase decreases and lipolytic activity increases dramatically, which suggests that fat is stored in the hips and thighs for reproductive purposes.

Physiological Responses to Acute Exercise

When females and males are exposed to an acute bout of exercise, whether an all-out run to exhaustion on the treadmill or a single attempt to lift the maximal weight possible, responses differ between the sexes. Differences between children and adolescent boys and girls were discussed in chapter 16. Here we briefly discuss these differences in adults, focusing on the following types of responses:

- Strength
- Cardiovascular and respiratory
- Metabolic

Strength

In terms of strength, women typically have been regarded as the weaker sex. In previous studies, women have been found to be 40% to 60% weaker than men in upper body strength but only 25% to 30% weaker in lower body strength. Because of the considerable size difference between the sexes, several studies have expressed strength either relative to body weight (absolute strength ÷ body weight) or relative to FFM, as a reflection of muscle mass (absolute strength ÷ FFM). When lower body strength is expressed relative to body weight, women are still 5% to 15% weaker than men, but when expressed relative to FFM, this difference disappears. This suggests that the innate qualities of muscle and its mechanisms of motor control are similar for women and men. In fact, this was confirmed by computed tomography scans of the upper arms and thighs of physical education majors of both sexes and of male bodybuilders.[57] Although both groups of men had much greater absolute levels of strength than the women, no differences between the groups were found when strength was expressed per unit of muscle cross-sectional area for both knee extensor muscles (figure 18.2a) and elbow flexor muscles (figure 18.2b).

Although differences in upper body strength are reduced somewhat when expressed relative to total body weight and FFM, substantial differences remain. There are at least two possible explanations for this. Women have a higher percentage of their muscle mass in the

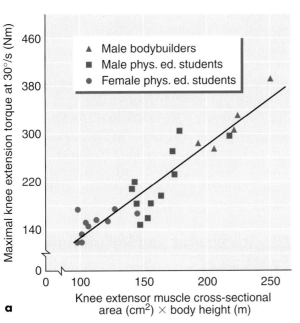

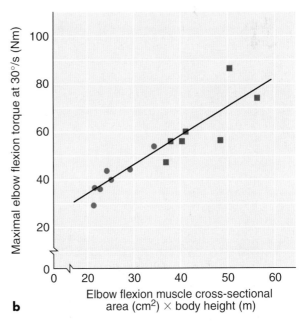

▲ **Figure 18.2**   No sex differences in strength (a), maximal knee extension torque; (b), maximal elbow flexion torque) are seen when strength is expressed per unit of muscle cross-sectional area.

Reprinted, by permission, from P. Schantz et al., 1983, "Muscle fibre type distribution, muscle cross-sectional area and maximal voluntary strength in humans," *Acta Physiologica Scandinavica* 117: 219-226.

lower body when compared with men. In addition and probably related to this muscle mass distribution, women use the muscle mass of their lower bodies much more than they use their upper body muscle mass, particularly when compared with use patterns in men. Some average-sized women have remarkable strength, exceeding that of an average man. This indicates the importance of neuromuscular recruitment and synchronization of motor unit firing in the ultimate determination of strength (chapter 3).

Muscle biopsies have become more common among female athletes. From these biopsy data, we have learned that men and women in the same sport or event have similar fiber-type distributions, as shown in figure 18.3, although men in one study reached greater extremes (greater than 90% slow-twitch [ST] or greater than 90% fast-twitch [FT]). As you can see from figure 18.3, when the vastus lateralis muscle was biopsied, male distance and sprint runners varied between approximately 15% and 85% ST fiber-type distribution, compared with female

distance and sprint runners, whose distributions were between approximately 25% and 75%.[55] However, different results were found in two studies of elite female and male distance runners.[12, 21] In these elite runners, the extremes for percentages of ST fibers were similar (90-96% for women and 92-98% for men), even though the mean values were different: Women had a mean value of 69% ST fibers compared with 79% for the men. However, the women had much smaller fiber areas for both ST and FT fibers (mean values of less than 4,500 μm^2 in women and greater than 8,000 μm^2 in men). Despite smaller fiber size in women, capillarization is similar between men and women.[50]

> For the same amount of muscle, there are no differences in strength between the sexes, although women have smaller muscle fiber cross-sectional areas than men and typically less muscle mass than men.

Research indicates that women have a greater resistance to fatigue compared with men. Fatigue usually is tested by having subjects maintain a constant force output at a given percentage of their maximal voluntary static action. As an example, women would be able to maintain a constant force output at 50% of their maximal static action for a longer period of time than men at the same 50% of their maximal static action. Men, being stronger, will have to apply a greater absolute amount of force to achieve the same 50% relative force. The reason for this greater resistance to fatigue is not yet known but could be related to the amount of muscle mass recruited, substrate utilization, muscle fiber type, and neuromuscular activation.[29]

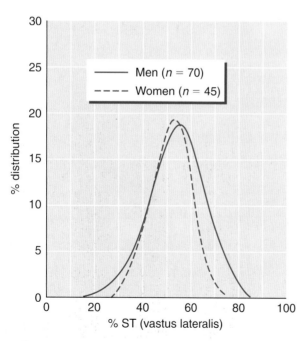

▲ **Figure 18.3** Distribution of slow-twitch (ST) fibers in the vastus lateralis muscle in male and female runners.

Adapted, by permission, from B. Saltin et al., 1977, "Fiber types and metabolic potentials of skeletal muscles in sedentary man and endurance runners," *Annals of the New York Academy of Sciences* 301: 3-29.

Cardiovascular and Respiratory Function

When placed on a cycle ergometer, where the power output can be precisely controlled independent of body weight—50 W, as an example—women generally have a higher heart rate (HR) response for any absolute level of submaximal exercise. However, maximum heart rate (HRmax) is generally the same in both sexes. Cardiac output ($\dot{Q}$) for the same absolute submaximal power output is nearly identical in women and men. Because of this (recall that cardiac output is the product of heart rate and stroke volume), the higher HR response in women is a compensation for a lower stroke volume (SV), which results primarily from at least three factors:

- Women have smaller hearts and therefore smaller left ventricles because of their smaller body size and possibly lower testosterone concentrations.
- Women have a smaller blood volume, which also is related to their size.
- The average woman may be less aerobically active and therefore less aerobically conditioned.

When power output is controlled to provide the same relative level of exercise, usually expressed as a fixed percentage of maximal oxygen uptake ($\dot{V}O_2max$), women's heart rates are still slightly elevated compared with men's, and their stroke volumes are markedly lower. At 60% $\dot{V}O_2max$, for example, a woman's cardiac output, stroke volume, and oxygen consump-

tion are generally less than a man's, and her heart rate is slightly higher. These differences also are seen at maximal levels of exercise with the exception of HRmax.

These relationships between HR, SV, and $\dot{Q}$ for the same absolute power output (50 W) and the same relative power output (60% $\dot{V}O_2max$) are illustrated in figure 18.4. These data were derived from the HERITAGE Family Study.[70] Interestingly, when we compare these same relationships in 7- to 9-year old boys and girls, there are no sex differences.[64]

> For the same rate of work, trained women generally have cardiac outputs similar to comparably trained men, but this is achieved through higher heart rates and lower stroke volumes. Women's lower stroke volume is due to their smaller left ventricle and lower blood volume, both the result of women's smaller body size.

Women also have less potential for increasing their arterial-venous oxygen difference, or a-$\bar{v}O_2$ difference. This is likely attributable to their lower hemoglobin content, which results in lower arterial oxygen content and reduced

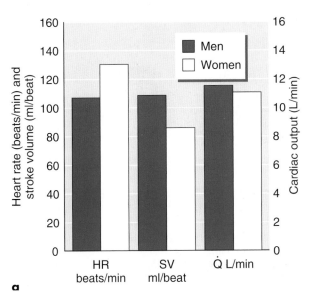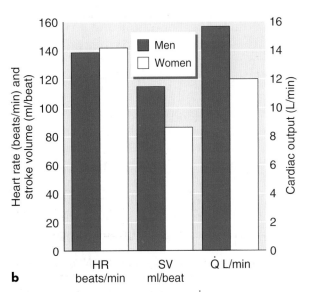

▲ Figure 18.4   Comparison of submaximal heart rate (HR), stroke volume (SV) and cardiac output ($\dot{Q}$) between men and women at (*a*) the same absolute power output (50 W) and (*b*) the same relative power output (60% $\dot{V}O_2max$).

Data from HERITAGE Family Study, 2001.[70]

muscle oxidative potential. Lower hemoglobin content is an important contributor to sex-specific differences in $\dot{V}O_2$max because less oxygen is delivered to the active muscle for a given volume of blood.

The differences between men's and women's respiratory responses to exercise are largely attributable to body size differences. Breathing frequency when working at the same relative power output differs little between the sexes. However, when we instead consider the same absolute power output, women tend to breathe more rapidly than men, probably because when both are working at the same absolute power output, the woman is working at a higher percentage of her $\dot{V}O_2$max.

Tidal volume and ventilatory volume are generally smaller in women at the same relative and absolute power outputs, up to and including maximal levels. Most highly trained female athletes have maximal ventilatory volumes below 125 L/min, but highly trained men have maximal values of 150 L/min and higher, some exceeding 250 L/min (figure 17.9, page 549). Again, these differences are closely associated with body size.

Metabolic Function

$\dot{V}O_2$max is regarded by most exercise scientists as the single best index of a person's cardiorespiratory endurance capacity. Recall that $\dot{V}O_2$ is the product of cardiac output and a-$\bar{v}O_2$ difference. This means that $\dot{V}O_2$max represents that point during exhaustive exercise where the subject has maximized oxygen delivery and utilization capabilities. The average female tends to reach her peak $\dot{V}O_2$max between ages 12 and 15, but the average male does not reach his peak until ages 17 to 21 (see chapter 16). Beyond puberty, the average woman's $\dot{V}O_2$max is only 70% to 75% of the average man's.

$\dot{V}O_2$max differences between women and men must be interpreted carefully. A classic study published in 1965 found considerable variability in $\dot{V}O_2$max within each sex and considerable overlap of values between sexes.[28] The study involved a group of women and men 20 to 30 years of age. The investigators divided the group into subgroups:

- Elite female athletes
- Female nonathletes
- Elite male athletes
- Male nonathletes

They compared the subjects' physiological responses to submaximal and maximal exercise. Barbara Drinkwater's[17] further analysis of this 1965 study revealed that 76% of the female nonathletes overlapped 47% of the male nonathletes and that 22% of the female athletes overlapped 7% of the male athletes. The relationships are illustrated in figure 18.5. These data demonstrate the importance of looking beyond mean values to consider both the subjects' levels of physical conditioning and the extent of overlap between the groups being compared.

Although the $\dot{V}O_2$max values of females and males are similar until puberty, many comparisons of $\dot{V}O_2$max values of normal nonathletic females and males beyond puberty might not be valid. Such data likely reflect an unfair comparison of relatively sedentary females with relatively active males. Thus, reported differences would reflect the level of conditioning as well as possible sex-specific differences. To

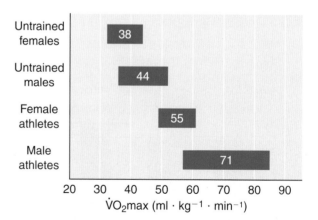

▲ **Figure 18.5**   Range of $\dot{V}O_2$max values (mean ±2 *SD*) for female nonathletes, male nonathletes, elite female athletes, and elite male athletes. The mean $\dot{V}O_2$max value is presented within each box. This figure illustrates that although there can be substantial differences in the average $\dot{V}O_2$max between groups, there can be considerable overlap of one group with another.

Data from L. Hermansen and K.L. Andersen, 1965, "Aerobic work capacity in young Norwegian men and women," *Journal of Applied Physiology* 20: 425-431.

overcome this potential problem, investigators began to examine highly trained female and male athletes, with the assumption that the level of training would be similar for both sexes and would allow a more accurate evaluation of true sex-specific differences.

Saltin and Åstrand[54] compared $\dot{V}O_2$max values of female and male athletes from Swedish national teams. In comparable events, the women had 15% to 30% lower $\dot{V}O_2$max values. However, more recent data from the United States suggest a smaller difference. The $\dot{V}O_2$max values for a group of elite and good female distance runners are compared in figure 18.6 with values for elite male distance runners and for average, nonathletic women and men.[4, 13, 45, 53, 69] The elite female runners had substantially higher values than untrained men and women. Some women's values were even higher than a few of the elite male runners' values, but when you consider the average for each elite group, the women's values were still 8% to 12% lower than those of the elite male runners.

> **fyi** The highest $\dot{V}O_2$max value reported in the literature for a female athlete is 77 ml · kg⁻¹ · min⁻¹, that of a Russian cross-country skier. The highest value for a male athlete was reported in a Norwegian cross-country skier, who achieved a value of 94 ml · kg⁻¹ · min⁻¹![5]

Several studies have attempted to scale $\dot{V}O_2$max values relative to height, weight, FFM, or limb volume in an attempt to more objectively compare women's and men's values. Several of these studies have shown that differences between the sexes disappear when $\dot{V}O_2$max is expressed relative to FFM or active muscle mass, yet some studies continue to demonstrate differences even when adjusted for differences in body fat.

In one study, researchers used a novel approach to this problem.[16] They studied the responses to submaximal and maximal treadmill runs under various conditions in 10 women and 10 men who regularly engaged in distance running. The women exercised only under normal weight conditions, but the men exercised both at normal weight and with external weight added to their trunks so that the total percentage of excess weight, defined as the men's fat weight plus the added weight, equaled the percentage fat of the matched women. Equalizing the sexes for excess weight reduced the amount of mean sex-specific differences in

- treadmill run time (32%),
- the amount of oxygen required per unit of FFM for running at various submaximal speeds (38%), and
- $\dot{V}O_2$max (65%).

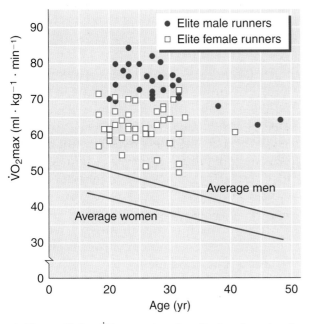

▲ **Figure 18.6** $\dot{V}O_2$max values for elite female and male distance runners compared with average values in untrained women and men.

Data from S. Robinson, 1938, "Experimental studies of physical fitness in relation to age," *Arbeitsphysiologie* 10: 251-323; I. Åstrand, 1960, "Aerobic work capacity in men and women with special reference to age," *Acta Physiologica Scandinavica* 49 (Suppl. 169); D.L. Costill and E. Winrow, 1970, "Maximal oxygen intake among marathon runners," *Archives of Physical Medicine and Rehabilitation* 51: 317-320; M.L. Pollock, 1977, "Submaximal and maximal working capacity of elite distance runners: Part I. Cardiorespiratory aspects," *Annals of the New York Academy of Sciences* 301: 310-322; R.R. Pate, et al., 1987, "Cardiorespiratory and metabolic responses to submaximal and maximal exercise in elite women distance runners," *International Journal of Sports Medicine* 8 (Suppl 2): 91-95; and J.H. Wilmore and C.H. Brown, 1974, "Physiological profiles of women distance runners," *Medicine and Science in Sports* 6: 178-181.

The researchers concluded that women's greater sex-specific essential body fat stores are major determinants of sex-specific differences in metabolic responses to running.

Women have lower hemoglobin levels than men, and this also has been proposed as a factor contributing to their lower $\dot{V}O_2$max values. One study attempted to equate the hemoglobin concentrations of a group of 10 men and 11 women who were active but not highly trained.[14] An amount of blood was withdrawn from the men to equalize their hemoglobin concentrations to those of the women. This significantly reduced the men's $\dot{V}O_2$max values, but these reductions accounted for only a relatively small portion of the sex differences in $\dot{V}O_2$max.

It is also important to understand that a woman's lower cardiac output at maximal rates of work is a limitation to achieving a high $\dot{V}O_2$max value. A woman's smaller heart size and lower plasma volume greatly limit her maximal stroke volume capacity. In fact, the results of several studies have suggested that women have limited ability to increase their maximal stroke volume capacity with high-intensity endurance training. However, more recent studies have shown that young, premenopausal women were able to increase their stroke volume with training identically to men. Furthermore, in the untrained state after artificially increasing their plasma volume with a plasma volume expander and after β-blockade (which reduced the heart rate for a given rate of work, allowing more time to fill the left ventricle), women were able to increase their stroke volume to the same extent as untrained men during an acute bout of exercise.[41, 42]

> Women generally have lower $\dot{V}O_2$max values expressed in ml · kg^{-1} · min^{-1} than men. A part of this difference in $\dot{V}O_2$max values between women and men is related to the extra body fat carried by women and, to a lesser extent, to their lower hemoglobin levels, which result in a lower oxygen content in the arterial blood.

If, instead of looking at $\dot{V}O_2$max, we consider submaximal oxygen consumption ($\dot{V}O_2$), little if any difference is found between women and men for the same absolute power output. But remember that at the same absolute submaximal work rate, women usually are working at a higher percentage of their $\dot{V}O_2$max. As a result, their blood lactate levels are higher, and lactate threshold occurs at a lower absolute power output. Peak blood lactate values are generally lower in active but untrained women than in active but untrained men. Also, limited data suggest that elite female middle-distance and long-distance runners have peak lactate values that are approximately 45% lower than similarly trained elite male runners (8.8 mmol/L vs. 12.9 mmol/L).[45, 47] Such sex differences in peak blood lactate values are unexpected and unexplained.

Lactate threshold values appear to be similar between equally trained men and women if the values are expressed in relative (%$\dot{V}O_2$max), not absolute, terms. Lactate threshold appears to be closely related to the mode of testing and to the individual's state of training. Thus, sex-specific differences are not expected.

> ▶ The innate qualities of muscle and the mechanisms of motor control are similar for women and men.
>
> ▶ Women and men do not differ in lower body strength expressed relative to body weight or to fat-free mass. But women show less upper body strength expressed relative to body weight or fat-free mass than men, largely because more of women's muscle mass is below the waist and women use their lower body muscles more than their upper body muscles.
>
> ▶ At submaximal exercise levels, women have higher heart rates than men, but women's submaximal cardiac outputs are similar for the same rate of work. This indicates that women have lower stroke volumes, primarily because they have smaller hearts, have less blood volume, and are generally less well conditioned than men.

> ▶ Women also have a lower capacity to increase a-$\bar{v}$-O_2 difference, probably because of their lower hemoglobin content, so less oxygen is delivered to their active muscles per unit of blood.
>
> ▶ Differences in women's and men's respiratory responses are primarily attributable to body size differences.
>
> ▶ Beyond puberty, the average woman's $\dot{V}O_2$max is only 70% to 75% that of the average man. However, some of this difference could be attributable to women's less active lifestyles. Research with highly trained athletes reveals that part of this difference is attributable to women's greater fat mass, lower hemoglobin levels, and lower maximal cardiac output.
>
> ▶ Little or no difference in lactate threshold is found between the sexes.

Physiological Adaptations to Exercise Training

As we have seen in earlier chapters, basic physiological function both at rest and during exercise changes substantially with physical training. In this section, we investigate how women adapt to chronic exercise, emphasizing areas in which their responses might differ from men's.

Body Composition

With either cardiorespiratory endurance training or resistance training, both women and men experience

- losses in total body mass,
- losses of fat mass,
- losses of relative fat, and
- gains in FFM.

Women generally gain much less in FFM than men do. With the exception of FFM, the magnitude of the change in body composition appears to be related more to the total energy expenditure associated with the training activities than to the participant's sex. Significantly more FFM is gained in response to strength training than with endurance training, and the magnitude of these responses is much less in women, primarily because of hormonal differences.

Bone and connective tissue undergo alterations with training, but these changes are not well understood. In general, animal studies and limited human studies have found an increase in the density of the weight-bearing long bones. This adaptation appears to be independent of sex, at least in young and middle-aged populations. We discuss some exceptions later in this chapter.

Connective tissue appears to be strengthened with endurance training, and sex-specific differences in this response have not been identified. Higher injury rates in women point to the possibility that women are more susceptible to injury than men while participating in physical activity and sport. This has led to concerns about sex-specific differences in joint integrity and laxity and the strength of ligaments, tendons, and bones. Unfortunately, the research literature contributes little to confirming or denying the validity of such concerns. Where differences in the rate of injury have been observed, the injury could be related more to the level of conditioning than to the participant's sex. Those who are less fit are more prone to injury. This is an extremely difficult area in which to obtain objective data but nevertheless is an important area that needs to be better researched.

Strength

Until the 1970s, prescribing strength training programs for girls and women was not considered appropriate. Women were not believed capable of gaining strength because of their extremely low levels of male anabolic hormones. Paradoxically, many people also generally feared that strength training would masculinize women. During the 1960s and 1970s, however, it became evident that many of the United States' better female athletes were not doing well in international competition, largely because they were weaker than their competitors. Slowly, research demonstrated that women can gain considerably from strength training programs even though strength gains are usually not accompanied by large increases in muscle bulk.

One study compared the training responses of 47 women and 26 men who volunteered to participate in identical progressive resistance weight-training programs. The program was conducted twice each week, 40 min per day, for a total of 10 weeks. The strength gains were as follows:

- Bench-press strength: 29% in women, 17% in men
- Leg-press strength: 30% in women, 26% in men

Muscle girth increased only slightly in the women, whereas the men exhibited classic muscle hypertrophy.[65] Thus, hypertrophy is neither a necessary consequence of nor a prerequisite to gains in muscle strength. Several studies have confirmed these results.

Because of their lower levels of testosterone, women have less total muscle mass. If muscle mass is the major determinant of strength, then women have a distinct disadvantage. But if neural factors are as important or more important than size, women's potential for absolute strength gains is considerable. Also, some women can attain significant muscle hypertrophy. This has been demonstrated in women bodybuilders who have remained free of anabolic steroids. Also, in one study, men and women had similar percentage increases in both strength and muscle hypertrophy following 16 weeks of resistance training.[15] The basic mechanisms allowing greater levels of strength have not yet been clearly defined, so we cannot draw any conclusions at this time. Overall, though, women generally gain substantially less muscle mass in response to a given training stimulus than men do, but some women gain more muscle mass than some men. Like aerobic capacity, discussed earlier, some overlap occurs between females and males.

With all of this in mind, it is of interest to look at the men's and women's world records for weightlifting by weight classification. Fig-

ure 18.7 illustrates these world records for the total weight lifted (the sum of the snatch and the clean and jerk) as of 2002. You can see from this figure that men are considerably stronger at each weight classification. At the 69-kg weight classification, the men lifted 100 kg more than the women did, but men lifted 120 kg more at the higher weight classifications. Part of the difference can be explained by the fact that, for a given body weight, men most likely will have a higher fat-free mass. Furthermore, few women participate in competitive weightlifting. The greater the number of participants, the greater the likelihood of higher world records. Still, the differences are so great that there must be other factors operating.

Women can experience major increases in strength (20-40%) as a result of resistance training, and the magnitude of these changes is similar to that seen in men. These gains are likely attributable more to neural factors, because women's increase in muscle mass is generally small.

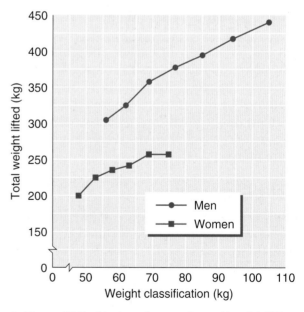

▲ **Figure 18.7** Men's and women's world weightlifting records as of 2002 for the total amount of weight lifted (combination of the snatch and the clean and jerk) by weight category. The highest weight category for both men and women is not included because it is not defined by weight. The men's total weight lifted is considerably more than the women's at the same weight classification.

Cardiovascular and Respiratory Function

Major cardiovascular and respiratory adaptations result from cardiorespiratory endurance training, and these adaptations do not appear to be sex specific. Major increases in maximal cardiac output ($\dot{Q}max$) accompany training. Maximum heart rate usually does not change with training, so this increase in $\dot{Q}max$ is the result of a large increase in stroke volume, which results from two factors. End-diastolic volume (the amount of blood in the ventricles before contraction) increases with training because blood volume increases and venous return is more efficient. In addition, end-systolic volume (the amount of blood remaining in the ventricles after contraction) is reduced with training because the stronger myocardium produces a stronger contraction, ejecting more blood.

At submaximal work rates, cardiac output shows little or no change, although stroke volume is considerably higher for the same absolute rate of work. Consequently, heart rate for any given rate of work is reduced after training. Resting heart rate can be reduced to 50 beats/min or less. Several female distance runners have had resting heart rates below 36 beats/min. This is considered a classic training response and corresponds to an exceptionally high stroke volume.

The increases in $\dot{V}O_2max$ that accompany cardiorespiratory endurance training, which are discussed subsequently, result primarily from the large increases in maximal cardiac output, with only small increases in a-$\bar{v}$-O_2 difference. However, Saltin and Rowell[56] stated that the major limitation to $\dot{V}O_2max$ is oxygen transport to the working muscles. Although cardiac output is important to oxygen transport, these researchers believe that the increases in $\dot{V}O_2max$ that accompany training are primarily attributable to increased maximal muscle blood flow and muscle capillary density. These changes are firmly established in men, and we have no reason to suspect that women differ in this response to training. In fact, one study has demonstrated that endurance-trained women have considerably higher capillary to fiber ratios (1.69 capillaries per fiber) than untrained women (1.11 capillaries per fiber).[32] Elite female distance runners in one study had a mean capillary to fiber ratio of 2.50 capillaries per fiber![12] These values are similar to those reported in men of similar training status. Although women also experience considerable increases in maximal ventilation, reflecting increases in both tidal volume and breathing frequency, these changes are thought to be unrelated to the increase in $\dot{V}O_2max$.

Metabolic Function

With cardiorespiratory endurance training, women experience the same relative increase in $\dot{V}O_2max$ that has been observed in men. They can improve their $\dot{V}O_2max$ by 10% to 50% with cardiorespiratory endurance training. These percentages are similar to improvements seen in men. The magnitude of change noted generally depends on the intensity and duration of the training sessions, the frequency of training, and the length of the study.

> Women can experience major increases in endurance capacity ($\dot{V}O_2max$ increases of 10-50%) with aerobic training.

After cardiorespiratory endurance training, women's oxygen uptake at the same absolute submaximal work rate does not appear to change, although several studies have reported decreases. Women's blood lactate levels are reduced for the same absolute submaximal rates of work, peak lactate levels generally are increased, and the lactate threshold increases with training. Finally, endurance training also improves women's ability to use free fatty acids for fuel, an adaptation that, as we have seen, is very important for glycogen sparing. In fact, it appears that women, during submaximal exercise, obtain a greater percentage of their energy from fat compared with men, in both the untrained and trained state.[30, 63]

From this discussion, we can see that women respond to physical training in the same manner as men. Although the magnitude of their adaptations to training might differ somewhat from those of their male counterparts, the over-all trends appear identical. This is important to remember when prescribing exercise, something that is discussed in chapter 19.

> ▶ With training, women generally gain less fat-free mass than men do, but other changes in body composition seem related more to total energy expenditure than to sex.
>
> ▶ Women can gain considerable strength through strength training, although strength gain usually is not accompanied by large increases in muscle mass.
>
> ▶ Cardiovascular and respiratory changes that accompany cardiorespiratory endurance training do not appear to be sex specific.
>
> ▶ Women experience the same relative increases in $\dot{V}O_2$max that men experience with cardiorespiratory endurance training.
>
> ▶ Women respond to physical training in the same manner as men do.

Sport Performance

Women are outperformed by men in all athletic activities in which performance can be precisely and objectively measured by distance or time. The difference is most pronounced in activities such as the shot put, where high levels of upper body strength are crucial to successful performance. In the 400-m freestyle swim, however, the winning time for women in the 1924 Olympic Games was 16% slower than that for men, but this difference decreased to 11.6% in the 1948 Olympics and to only 6.9% in the 1984 Olympics. The fastest women's 800-m freestyle swimmer in 1979 swam faster than the world record–holding man for the same distance in 1972. From these results, it would appear that the gap between the sexes is narrowing. However, as we see from table 18.1 (page 568), the difference between the men's and women's world record for the 400-m freestyle was 10.8% in 2002. For the 800-m

freestyle it was 8.1%. Unfortunately, making valid comparisons through the years has been difficult because the degree to which an activity has been emphasized and its popularity are not constant, and other factors—such as opportunities to participate, coaching, facilities, and training techniques—have differed considerably between the sexes over the years.

As noted at the beginning of this chapter, large numbers of girls and women didn't start entering competitive sport until the 1970s. Even then, there was an initial reluctance to train women as hard as men. Once girls and women started training as hard as boys and men, their performance improved dramatically. This is illustrated in figure 18.8, which shows world records from 1975 to 2003 for women and men in six running events in track and field. For distances of 400 m through the marathon, women's present world records are consistently 10% to 13% slower than men's. Furthermore, the improvement in women's records, which was initially quite dramatic, is beginning to level off and parallel the curves for men's records.[60]

Special Issues

Although the sexes respond to acute exercise and adapt to chronic exercise in much the same manner, several additional areas that are unique to females must be considered. Specifically, we look at

- menstruation and menstrual dysfunction,
- pregnancy,
- osteoporosis,
- eating disorders, and
- environmental factors.

Menstruation and Menstrual Dysfunction

Two questions of interest to exercising women—particularly female athletes—are, How does the menstrual cycle or pregnancy influence exercise capacity and performance? and How do physical activity and competition

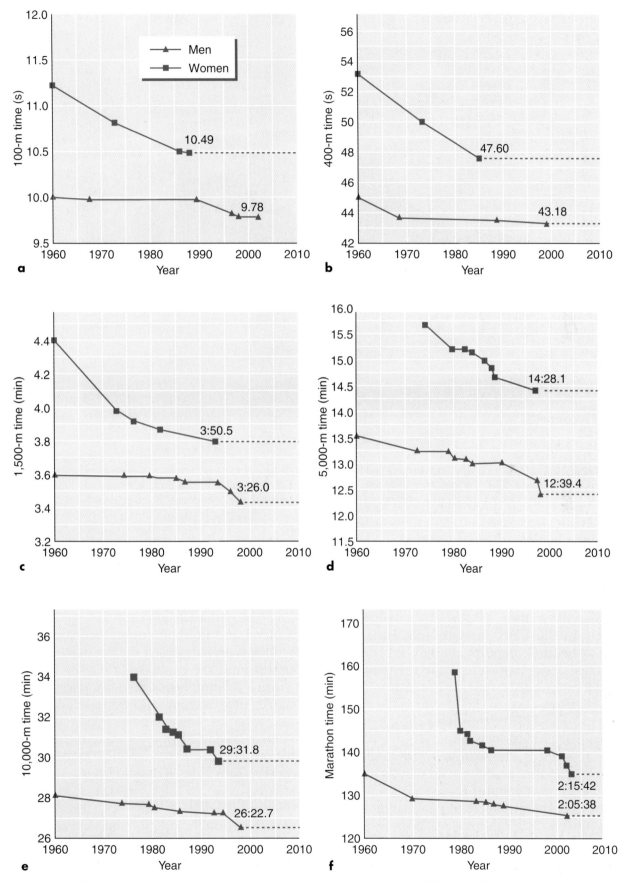

▲ **Figure 18.8** Women's and men's world records in six running events between 1960 and 2002.

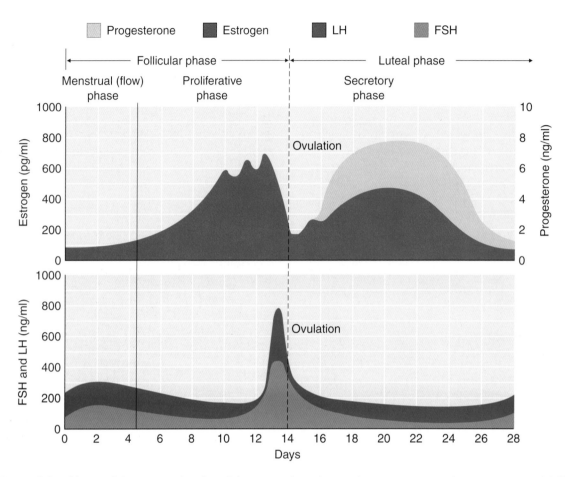

▲ **Figure 18.9**   Phases of the menstrual cycle and the concomitant changes in progesterone and estrogen, top, and follicle-stimulating hormone (FSH) and luteinizing hormone (LH), bottom. For most purposes, the cycle is divided into the follicular phase, which starts with the initiating of bleeding during menstrual flow, and the luteal phase, which starts with ovulation.

influence the menstrual cycle or pregnancy? Let's try to answer these questions, beginning with the relationship between menstruation and physical performance.

The three major phases of the **menstrual cycle** are illustrated in figure 18.9. The first is the menstrual (flow) phase, which lasts 4 to 5 days, during which time the uterine lining (endometrium) is shed and menstrual flow, or bleeding, occurs. The second is the proliferative phase, which prepares the uterus for fertilization and lasts about 10 days. During this phase, the endometrium begins to thicken, and some of the ovarian follicles that house the ova mature. These follicles secrete estrogen. The proliferative phase ends when a mature follicle ruptures, releasing its ovum (ovulation). The menstrual and proliferative phases correspond to the follicular phase of the

ovarian cycle. The third and final phase of the menstrual cycle is the secretory phase, which corresponds to the luteal phase of the ovarian cycle. This phase lasts 10 to 14 days, during which the endometrium continues to thicken, its blood and nutrient supply increases, and the uterus prepares itself for pregnancy. During this time, the empty follicle (now termed a corpus luteum, hence the term *luteal phase*) secretes progesterone, and estrogen secretion also continues. The complete menstrual cycle averages 28 days. However, there is considerable variation in the cycle length among healthy women, from 23 to 38 days in length.

Menstruation and Performance

Alterations in athletic performance experienced during different phases of the menstrual cycle are subject to considerable individual variability. Some women have absolutely no noticeable change in their performance ability at any time during their menstrual cycle, yet others have considerable difficulty in the preflow or early-flow phase, or both. The number of women who report impaired performance during the flow phase is approximately the same as those who experience no difficulty. In fact, some female athletes have reportedly set world records during the flow phase. Adding to the confusion, much of the information available on this topic is based on anecdotes or subjective statements made by athletes during informal surveys.

Very little information is available from well-designed and well-controlled research studies. Several studies have suggested that athletic performance is best during the immediate postflow period up to the 15th day of the cycle (the first day of the cycle corresponds to the initiation of the flow, or **menses**, and ovulation occurs on about the 14th day). However, other studies have reported that performance is best during the flow phase. The confused state of research in this area is illustrated by three studies conducted on swimmers.

- One study showed that swimmers had faster times during the flow phase.[8]
- One study showed that swimmers swam fastest during the postflow phase.[6]
- One study found slightly faster times in the preflow phase, although the differences in times were not statistically significant.[51]

Such divergent results are perplexing. The disagreement between these studies could be attributable to the small number of subjects tested, large individual variations among the swimmers, or inadequately controlled experimental designs. Several studies have been conducted in research laboratories by using physiological measures to gauge performance changes. These generally have found no performance differences in the various phases of the menstrual cycle. Most likely, it depends totally on the individual and her specific conditions at that time.

From currently available information, we can conclude that some women's performance can be affected by the phase of their menstrual cycle, but that many, if not most, women are not affected. Any woman who experiences premenstrual syndrome (PMS) or dysmenorrhea (pain or abdominal cramping with menstruation) will likely not perform as well while she is experiencing symptoms. For these women, some degree of control over their menstrual cycle is possible through the use of low-dose oral contraceptives. This was discussed in chapter 15.

> There appears to be no general pattern concerning the ability of women to achieve their best performances during any specific phase of their menstrual cycle. However, those who experience PMS or dysmenorrhea will likely not perform as well while experiencing symptoms.

Menarche

Menarche, which refers to the first menstrual period, has been reported to be delayed in some young athletes involved in certain sports and activities, such as gymnastics and ballet. The median age of menarche for American girls is 12.8 years. For gymnasts, the median age appears to be closer to 14.5 years. Frisch[24] hypothesized that menarche is delayed 5 months for each year of training before menarche, implying that training causes delayed menarche. Malina,[40] however, postulated that late maturers, such as those with a later menarche, are more likely to be successful in sports such as gymnastics because of their small, lean bodies. This implies that those who naturally experience a later menarche have an advantage in, and thus are likely to be involved in, certain sports rather than that their sport involvement delays menarche.

These opposing viewpoints can be summarized by the following questions: Does intense training to achieve the level of an elite performer delay menarche, or does a later menarche provide an advantage that contributes to the success of the elite performer? Stager and colleagues[61] used computer modeling to analyze this issue and concluded that the age of menarche in athletes is later rather than delayed.

At this time, evidence is insufficient to support the theory that training delays menarche, so Malina's statement that menarche occurs later in the trained athlete is appropriate.

> Menarche appears to come later in highly trained elite athletes in certain sports such as gymnastics. However, there is no strong evidence to support the contention that the intense training for the sport delays menarche.

Menstrual Dysfunction

Female athletes can experience disruptions of their normal menstrual cycle. These disruptions are collectively referred to as **menstrual dysfunction**, of which there are several types. **Eumenorrhea** is the term for normal menstrual function. **Oligomenorrhea** refers to abnormally infrequent or scant menstruation. **Amenorrhea** refers to the absence of menstruation; **primary amenorrhea** refers to the absence of menarche in women 18 years of age and older—women who never began menstruating. Some athletes with previously normal menstrual function have reported the absence of menstruation for months or even years when they have trained and competed intensely in sports such as figure skating, ballet, gymnastics, diving, bodybuilding, cycling, and distance running. This phenomenon is referred to as **secondary amenorrhea.**

The prevalence of secondary amenorrhea and oligomenorrhea among athletes is not well documented but is estimated to be approximately 5% to 40% or higher, depending on the sport or activity and the level of competition. This is considerably higher than the estimated 2% to 3% prevalence for amenorrhea and 10% to 12% prevalence for oligomenorrhea in the general population. The prevalence appears to be greater in those who train many hours each day and in those who train at very high intensities.

Many women who become amenorrheic are relieved to be free of menstruation each month. Most also assume that they have developed a simple but effective form of birth control. However, athletes have become pregnant while

amenorrheic, which indicates that ovulation, and thus fertility, are not always influenced by the absence of menstruation. This point needs to be stressed among female athletes prone to amenorrhea to reduce the possibility of unexpected pregnancy.

For more than 20 years, scientists have been conducting experiments to determine the basic cause of secondary amenorrhea. We are tempted to surmise that high volume and/or high intensity training leads to menstrual dysfunction, but this might not be the case. Some of the factors that have been proposed as potential causes of secondary amenorrhea include the following:

- History of menstrual dysfunction
- Acute effects of stress
- High quantity or intensity of training
- Low body weight or body fat
- Hormonal alterations
- Energy deficit through inadequate nutrition and disordered eating

Let's briefly discuss each of these.

History of Menstrual Dysfunction   The association between secondary amenorrhea and prior menstrual dysfunction was demonstrated in one study that found that 54.5% of amenorrheic runners had a history of irregular menstruation, compared with only 15.5% of eumenorrheic runners and 13.3% of nonrunners. Other studies, however, have been unable to confirm these results. Most likely, a history of menstrual dysfunction can be one of a number of factors, but is probably not the primary factor, in the development of secondary amenorrhea.

Stress   Stress is also a likely factor. One study showed that amenorrheic runners associated more stress with their training than did eumenorrheic runners.[58] However, formal psychological tests failed to detect any differences in the amenorrheic runners' levels of anxiety, depression, and other states that reflect stress.

High Quantity or Intensity of Training   Several studies have shown that quantity of training is associated with secondary amenorrhea. In runners, a direct relationship has been reported between the distance run per week and the prevalence of secondary amenorrhea. High-mileage runners report more amenorrhea. But other studies have been unable to confirm this relationship. Researchers have tried to induce amenorrhea by increasing the quantity of women's training.[7, 9] In one study, none of the runners became amenorrheic, but 18 developed significant menstrual cycle changes, mainly oligomenorrhea.[7] In another study, only 4 of the 28 runners maintained normal menstrual cycles during the increased training period. The most significant menstrual dysfunctions noted were delayed menstruation, abnormal luteal function, and a loss of luteinizing hormone (LH) surge.[9] Within 6 months of completing this study, all runners returned to normal menstrual cycles. Athletically induced amenorrhea is usually reversible during periods of reduced training, injury, or vacation.

The effect of training intensity on menstrual function has not been well documented. Training at a high intensity, because of the high physical stress placed on the body, might be more closely linked with secondary amenorrhea than volume of training. Future studies need to focus on this aspect of training.

Although volume and intensity of training might be important, neither is likely the primary factor leading to secondary amenorrhea. This is discussed in greater detail in the last part of this section.

Low Body Weight or Body Fat   Excessive leanness, undernutrition, or both have long been associated with amenorrhea. Some researchers have suggested that the loss of one third of a woman's body fat or a 10% to 15% decrease in her total body weight (mass) will induce amenorrhea. The reason for this is that androgens are converted into estrogens in adipose tissue, particularly breast and abdominal fat, and this conversion accounts for nearly one third of the estrogen in premenopausal women.[25] Any decrease in adipose tissue influences the storage and metabolism of estrogen. In other words, fat is an important estrogen source, necessary for normal menstrual function. At one time a certain minimal weight for height was believed necessary for

achieving menarche and remaining eumenorrheic.[26] This later evolved into a hypothesis that girls need to attain a minimal relative body fat of 17% to achieve menarche and that females after that need a minimum of 22% relative body fat to maintain normal menstrual function.[59] More recently, however, investigators have challenged this theory. Numerous studies have now been published in which the relative body fat levels of eumenorrheic and amenorrheic athletes were identical. In one study of elite female runners, both amenorrheic and eumenorrheic runners had 10% body fat, very low body fat for female athletes.[71] This strongly suggests that low body weight and body fat are not primary triggers of menstrual dysfunction.

Hormonal Alterations   Numerous hormonal changes occur with acute bouts of exercise (transient changes) as well as with chronic periods of training (long-term changes). Recall from basic physiology the normal endocrine control of the menstrual cycle. We have discussed the importance of the gonadotropic hormones: luteinizing hormone (LH) and follicle-stimulating hormone (FSH). These hormones are released from the anterior pituitary gland in response to gonadotropin-releasing hormone (GnRH), which is produced by the hypothalamus. In athletes with menstrual dysfunction, the normal episodic secretion pattern of LH appears to undergo subtle changes as discussed in the previous section. These changes might be caused by increased secretion of corticotropin-releasing hormone (CRH) from the hypothalamus. This hormone inhibits GnRH release, which in turn inhibits the release of LH and FSH.[34] Cross-sectional studies of amenorrheic athletes have shown abnormal reproductive hormone patterns that suggest that the normal secretion of GnRH by the hypothalamus is disrupted and thus fails to initiate normal hypothalamic–pituitary–ovarian function.[36] It is likely, however, that the hormonal alterations are not the primary cause of menstrual dysfunction but part of a cascade of physiological changes initiated by a primary factor. What is that primary factor? After several decades of research, it appears that an energy deficit is the primary factor, as discussed in the next section.

Energy Deficit: Inadequate Nutrition and Disordered Eating

Current evidence indicates that inadequate nutrition resulting in an energy deficit is the primary cause of secondary amenorrhea. Studies have shown that inadequate intake of calories, where the body is not matching caloric intake to caloric expenditure over an extended period of time, is the primary cause of secondary amenorrhea.

Recent research by Dr. Anne Loucks and her colleagues[37, 38, 39] at Ohio University has clearly demonstrated that simply inducing an energy deficit in eumenorrheic women results in significant hormonal alterations that are associated with amenorrhea. Specifically, reducing caloric intake, with or without the added stress of increased energy expenditure from exercise training, reduces luteinizing hormone pulse frequency and the thyroid hormone triiodothyronine (T_3), both associated with disturbed menstrual function. Exercise training might not be directly associated with menstrual dysfunction at all, other than through its contribution to an energy deficit. An energy deficit, in either the absence or presence of exercise training, is associated with these hormonal alterations. Intense or high-volume training most likely is not associated with menstrual dysfunction as long as energy intake matches or exceeds energy expenditure over days, weeks, and months.

A high percentage of female athletes in endurance and appearance sports experience secondary amenorrhea, in which normal menstrual function is lost for months or even years. This appears to be reversible with reductions in the intensity and volume of training and an increase in caloric intake.

The relationship between disordered eating and menstrual dysfunction is a more recent concern; several studies have shown a strong relationship between the two. In one study, 8 of 13 amenorrheic distance runners reported eating disorders, compared with 0 of 19 eumenorrheic distance runners.[27] In another study, 7 of 9 amenorrheic elite middle- and long-distance runners were diagnosed with either anorexia nervosa, bulimia nervosa, or both, compared with 0 of 5 eumenorrheic runners.[71] Eating disorders are discussed in detail later in this chapter, but we can say that eating disorders generally involve an energy deficit.

> ▶ The effects of different phases of the menstrual cycle on performance are subject to considerable individual variation. In general, the number of women reporting impaired performance during the flow phase is about the same as the number reporting no difficulty. Any woman who experiences PMS or dysmenorrhea is likely to not perform as well while experiencing those symptoms.
>
> ▶ Menarche can occur late in some young athletes in certain sports. However, the most likely explanation for this is that late maturers, because of lean body build, are more likely to participate successfully in these activities, not that these activities cause delayed menarche.
>
> ▶ Female athletes can experience menstrual dysfunction, most often secondary amenorrhea or oligomenorrhea. Current evidence implicates inadequate nutrition, or a prolonged energy deficit, as the primary cause of secondary amenorrhea. In addition, hormonal changes from exercise and training might disrupt GnRH secretion, which is needed to direct the normal cycle. This too, is associated with a prolonged energy deficit.

Pregnancy

What are the effects of exercise during **pregnancy**? Four major physiological concerns are associated with exercise during pregnancy:[73, 74]

1. The acute risk associated with reduced blood flow to the uterus (blood is diverted to the mother's active muscles), leading to fetal hypoxia (insufficient oxygen)

2. Fetal hyperthermia (elevated temperature) associated with the increase in the mother's internal body temperature during prolonged aerobic-type exercise or exercise under conditions of heat stress

3. Reduced carbohydrate availability to the fetus as the mother's body uses more carbohydrate to fuel her exercise

4. The possibility of miscarriage and the final outcome of pregnancy

Let's discuss each of these.

Reduced Uterine Blood Flow and Hypoxia

There have been reports that uterine blood flow in both animals and humans is reduced by 25% or more during moderate to strenuous exercise. The magnitude of the reduction is directly correlated with exercise intensity and duration.[72] Whether this reduction in uterine blood flow leads to fetal hypoxia is less clear. Apparently, an increase in the uterine a-$\bar{v}O_2$ difference at least partially compensates for any reduced blood flow. Increases in fetal heart rate, although not always observed during maternal exercise, have been interpreted as an index of hypoxia in the fetus. Although increased fetal heart rate might reflect hypoxia to a certain degree, it more likely represents the fetal heart's response to increased catecholamine levels in the blood originating from both the fetus and the mother.

Hyperthermia

Fetal hyperthermia is a distinct possibility if the mother's core temperature is elevated substantially during and immediately after exercise. Teratogenic effects (abnormal fetal development) have been documented with chronic exposure to thermal stress in animals, and these effects have been documented with maternal fever in humans. Central nervous system defects are the most common result.[74] Although fetal temperature has been shown to increase with exercise in animal studies, it is unclear whether this increase is sufficient to warrant concern.

Carbohydrate Availability

The potential for reduced carbohydrate availability for the fetus during exercise is also not well understood. We know that endurance athletes who train or compete for long durations reduce both liver and muscle glycogen stores and that blood glucose concentrations also can decrease. But whether this is a potential problem in pregnant women is less clear. One study demonstrated small reductions in blood glucose levels from 5.3 to 4.6 mmol/L following steady-state treadmill running in recreational runners studied at 32 weeks of pregnancy.[11] The significance of this is unclear.

Miscarriage and Pregnancy Outcome

Concerns also have been expressed regarding the potential of exercise to induce miscarriage during the first trimester, to induce premature labor, and to alter the normal course of fetal development. Unfortunately, little information is available concerning the risk for miscarriage and premature labor. Regarding pregnancy outcome, data are scarce and conflicting. Although there are some indications of lighter birth weights and shorter gestation periods, most studies have shown either favorable effects of exercise (such as reduced maternal weight gain, shorter postdelivery hospital stays, and fewer cesarean sections) or no differences between the control and exercise groups.

Although there are several concerns for the health of the fetus during maternal exercise, the risk to the fetus from aerobic exercise during pregnancy appears to be low, particularly if guidelines for exercising during pregnancy are followed.

▶ During exercise, major concerns for the pregnant athlete include the possible risk of fetal hypoxia, fetal hyperthermia, reduced carbohydrate supply to the fetus, miscarriage, premature labor, low birth weight, and abnormal fetal development.

▶ The benefits of a properly prescribed exercise program during pregnancy outweigh the potential risks. Such an exercise program must be coordinated with the woman's obstetrician.

Recommendations for Exercise During Pregnancy

To summarize, exercise during pregnancy can have associated risks (see table 18.4), but the benefits far outweigh the potential risks if caution is taken in designing the exercise program. Wolfe and coworkers[73] provided a set of guidelines that should be followed when prescribing aerobic exercise during pregnancy.

1. Obtain medical clearance before exercise.
2. Non-weight-bearing exercise (e.g., cycling, swimming) is preferable to weight-bearing exercise (e.g., jogging).
3. Exertion levels should be determined on an individual basis.
4. Avoid strenuous exercise during the first trimester.
5. Increases in exercise quantity and quality should be very gradual for previously inactive women.
6. Avoid exercise or positioning of the individual in the supine posture, particularly in late gestation.
7. Avoid exercise in warm and/or humid environments.
8. Drink liquids before and after exercise to ensure adequate hydration.
9. Do not exercise when fatigued, particularly in late gestation.
10. Periodic rest intervals may be helpful to minimize hypoxia or thermal stress to the fetus.
11. Know reasons to discontinue exercise, and consult a physician immediately if they occur.

Table 18.4

Hypothetical Risks and Postulated Benefits of Exercise During Pregnancy

		Hypothetical risks	Postulated benefits
Maternal		Acute hypoglycemia	Increased energy level (aerobic fitness)
		Chronic fatigue	Reduced cardiovascular stress
		Musculoskeletal injury	Prevention of excessive weight gain
			Facilitation of labor
			Faster recovery from labor
			Promotion of good posture
			Prevention of lower back pain
			Prevention of gestational diabetes
			Improved mood state and body image
Fetal		Acute hypoxia	Fewer complications of a difficult labor
		Acute hyperthermia	
		Acute reduction in glucose availability	
		Miscarriage in the first trimester	
		Induction of premature labor	
		Altered fetal development	
		Shortened gestation	
		Reduced birth weight	

Adapted from Wolfe, Hall et al. (1989).[73]

It is important that the pregnant woman coordinate her exercise program with her obstetrician so that sound medical judgment can be used to determine the most appropriate mode, frequency, duration, and intensity of activity.

The American College of Obstetricians and Gynecologists (ACOG) also developed a set of guidelines in 1985, which were subsequently revised in 1994.[1] These guidelines were summarized by Pivarnik[46] as follows:

- Pregnant women can derive health benefits from mild to moderate exercise performed at least 3 days per week.

- Women should avoid supine exercise and motionless standing after the first trimester because this compromises venous return, which, in turn, compromises cardiac output.

- Women should stop exercising when fatigued, should not exercise to exhaustion, and should modify their routines based on maternal symptoms. Weight-bearing exercises under some circumstances may be continued, but non-weight-bearing activities such as cycling or swimming are encouraged to reduce the risk of injury.

- Care should be taken not to participate in sports or exercises where falling, a loss of balance, or blunt abdominal trauma may occur.

- Because pregnancy requires an extra 300 kcal (1,255 kJ) of energy per day, an exercising woman should pay particular attention to diet to ensure that she is receiving adequate calories.

- Heat dissipation is of particular concern in the first trimester, so an exercising woman should wear correct clothing, be sure that her fluid intake is sufficient, and select optimal environmental conditions.

- A woman's regular prepregnancy exercise routine should be resumed gradually postpartum, as pregnancy-associated changes may persist 4 to 6 weeks.

In 2002, ACOG published a short "Committee Opinion" on exercise during pregnancy and the postpartum period, which essentially supported their previous guidelines.[2] In addition, they supported the current recommendation of the Centers for Disease Control and Prevention and the American College of Sports Medicine for nonpregnant individuals, which states that individuals should accumulate 30 min or more per day of moderate exercise on most, if not all, days of the week (see chapter 19). Furthermore, they state that scuba diving should be avoided throughout pregnancy because the fetus is at increased risk for decompression sickness. Also, there is an increased risk when pregnant women exercise at altitudes in excess of 6,000 ft.

Osteoporosis

Maintaining a healthy lifestyle might retard one detrimental aging process that is a major health concern for women: osteoporosis. **Osteoporosis** is characterized by decreased bone mineral content, which causes increased bone porosity (see figure 18.10). **Osteopenia**, as we discovered in chapter 17, refers to a loss of bone mass that occurs with aging. Osteoporosis is a more severe loss of bone mass with deterioration of the microarchitecture of bone leading to skeletal fragility and increased risk of bone fracture.[44] These changes typically begin in the early 30s. The occurrence rate for fractures associated with osteoporosis increases by two to five times after the onset of menopause. Men also experience osteoporosis but to a lesser degree early in life because of a slower rate of bone mineral loss. Much remains to be learned about the etiology of osteoporosis; however, three major contributing factors common to postmenopausal women are

- estrogen deficiency,
- inadequate calcium intake, and
- inadequate physical activity.

Although the first of these is a direct result of menopause, the last two reflect dietary and exercise patterns throughout life.

In addition to postmenopausal women, women with amenorrhea and those with anorexia nervosa also suffer from osteoporosis attributable to insufficient calcium intake, low

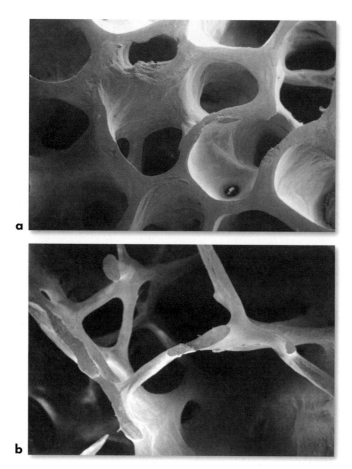

▲ **Figure 18.10**   (*a*) Healthy bone and (*b*) bone showing increased porosity (decreased density, appearing darker) resulting from osteoporosis.

serum estrogen levels, or possibly both. In a study of women with anorexia, the investigators found that the bone densities of anorectics were reduced significantly compared with controls.[52] Cann and associates[10] were the first to report a substantially lower than normal bone mineral content in physically active women classified as having hypothalamic amenorrhea.

In another study, the radial and vertebral bone densities of 14 athletic women (mostly runners) with amenorrhea were compared with those of 14 athletic women with normal menstruation (eumenorrhea).[19] Investigators discovered that physical activity did not protect the group with amenorrhea from significant bone density losses. The amenorrheic group's bone density values at a mean age of 24.9 were equivalent to those of normally active women at a mean age of 51.2. In a follow-up study, increases in vertebral bone mineral density

were found in the women who previously had been amenorrheic but had resumed menstruation.[20] However, their bone mineral densities remained well below the average for their age group, even 4 years after they resumed normal menses.[18]

It generally is assumed that exercise is a positive factor for bone health in that it is associated with an increase in bone mass, or at least with the maintenance of bone mass in older women. Therefore, it is confusing to learn that amenorrheic runners have reduced bone mass. Figure 18.11 attempts to clarify this apparent contradiction. From this figure we see that the bone mineral content of normally menstruating runners tends to be higher than that of normally menstruating nonrunning controls. Furthermore, female runners who are amenorrheic have higher bone mineral contents than untrained women who are amenorrheic. Thus, when we compare women of like menstrual

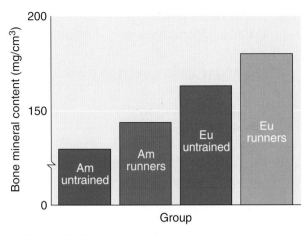

▲ Figure 18.11   Bone mineral content of women runners and untrained women who are amenorrheic (Am) and eumenorrheic (Eu). Note that when women with the same menstrual status are compared, the runners have higher bone mineral content than the untrained women.

Unpublished data from Dr. Barbara Drinkwater.

status, those who are exercising will have the higher bone mineral content. Caution should be used when interpreting data such as those presented in this section, because the results can be confounded by such factors as body composition, age, height, weight, and diet.

> Athletes with secondary amenorrhea are at increased risk for bone mineral loss. This loss does not appear to be totally reversible with the resumption of normal menstrual function.

Although the precise mechanism is unknown, estrogen deficiency appears to play a major role in the development of osteoporosis. In the past, estrogen has been prescribed in an effort to reverse the degenerative effects of osteoporosis, but this therapy can have serious side effects, such as an increased risk of endometrial cancer. Increasing calcium intake to 1.5 to 2.0 g per day also has been proposed for decreasing the risk of osteoporosis. This approach, however, might not be as effective as once hoped, and an increased Recommended Dietary Allowance (RDA) for calcium remains controversial. Certainly, meeting the RDA is very important!

Evidence certainly suggests that increased physical activity and adequate calcium intake combined with adequate caloric intake are a sensible approach to preserving the integrity of bone at any age. However, maintaining normal menstrual function is critical for those who have not reached menopause.

Eating Disorders

Eating disorders are a group of disorders that must meet specific criteria established by the American Psychiatric Association.[3] The two most commonly diagnosed eating disorders are anorexia nervosa and bulimia nervosa. **Disordered eating,** on the other hand, refers to patterns of eating that are not considered normal but don't meet the specific diagnostic criteria for a given eating disorder.

Eating disorders in girls and women became the focus of considerable attention beginning in the 1980s. Men constitute less than 10% of the reported cases. Anorexia nervosa has been considered a clinical syndrome since the late 19th century, but bulimia nervosa was first described in 1976.

Anorexia nervosa is a disorder characterized by

- refusal to maintain more than the minimal normal weight based on age and height,
- distorted body image,
- intense fear of fatness or gaining weight, and
- amenorrhea.

Females from ages 12 to 21 are at greatest risk for this disorder. Its prevalence in this group is likely less than 1%.

Bulimia nervosa, originally termed bulimarexia, is characterized by

- recurrent episodes of binge eating,
- a feeling of lack of control during these binges, and
- purging behavior, which can include self-induced vomiting, laxative use, and diuretic use.

The prevalence of bulimia in the population at greatest risk, again adolescent and young adult females, is generally considered to be less than 5% and possibly closer to 1%.

It is important to realize that a person might exhibit disordered eating and yet not meet the strict diagnostic criteria for either anorexia or bulimia. As an example, the diagnosis of bulimia requires that the individual average a minimum of two binge-eating and purging episodes a week for at least 3 months. What about the person who meets all the criteria, except that bingeing and purging occur only once per week? Although this person cannot technically be diagnosed as having bulimia, her or his eating is certainly disordered and is a potential cause for concern. Thus, the term "disordered eating" has been used to describe those who do not meet the strict criteria for an eating disorder but who do have abnormal eating patterns.

Dr. Jorunn Sundgot-Borgen,[62] an exercise scientist from Norway, developed a third category of disordered eating that she termed *anorexia athletica,* which is characterized by the following:

- An intense fear of gaining weight or becoming fat even though one is underweight (at least 5% less than the expected normal weight for age and height for the general female population)
- A weight loss of at least 5%, which is usually accomplished by a reduction in total energy intake, often with extensive or compulsive exercise
- Reported bingeing, self-induced vomiting, or use of laxatives or diuretics

The prevalence of eating disorders in athletes is not well understood. Numerous studies have used either self-report or at least one of two inventories developed to diagnose disordered eating: the Eating Disorders Inventory (EDI) and the Eating Attitudes Test (EAT). Results have varied because not all studies used the strict standard diagnostic criteria for either anorexia or bulimia. As in the general population, female athletes are typically at a much higher risk than male athletes, and certain sports carry higher risks than others. The high-risk sports can generally be grouped into one of three categories:

1. Appearance sports, such as diving, figure skating, gymnastics, bodybuilding, and ballet
2. Endurance sports, such as distance running and swimming
3. Weight-classification sports, such as horse racing, boxing, and wrestling

Self-reports or inventories do not always provide accurate results. In a study of 110 elite female athletes representing seven sports, EAT results showed that no athlete fell within the disordered eating range of the inventory. But in the subsequent 2-year period, 18 of these athletes received either inpatient or outpatient treatment for eating disorders. In a second study of 14 nationally ranked middle- and long-distance runners who completed the EDI, only 3 were shown to have possible problems with disordered eating, and none were shown to have eating disorders.[71] In follow-up, seven subjects were subsequently diagnosed as having an eating disorder: four with anorexia nervosa, two with bulimia nervosa, and one with both. Eating disorders, by their very nature, are secretive. We cannot realistically expect those with eating disorders to identify themselves, even when anonymity is ensured. For the athlete, this need for secrecy might be heightened by fear that a coach or a parent will learn of the eating disorder and not allow the athlete to compete.

Even though research is limited, it seems appropriate to conclude that athletes (figure 18.12) are at higher risk for eating disorders than the general population. Existing evidence likely does not reflect the seriousness of this problem in athletic populations. Although research data are not yet available, the prevalence might be as high as 50% or more in the specific high-risk athletic populations listed earlier.

> Disordered eating has become a major concern in female athletes. Some researchers have estimated the prevalence to be as high as 50% for elite athletes in certain sports.

▲ **Figure 18.12**   Some female athletes decrease their weight to such dangerously low levels that both health and athletic performance are compromised.

Eating disorders generally are considered to be addictive disorders and are extremely difficult to treat. The physiological consequences are considerable and can include death. Considering this, along with the emotional distress suffered by the athlete, the extraordinary costs of treatment ($5,000-25,000 per month for hospital inpatient treatment), and the effect on those closest to the athlete, eating disorders must be considered among the most serious problems facing female athletes today, paralleling the seriousness of anabolic steroid use in male athletes. The NCAA has become concerned about the problem of eating disorders and in 1990 released a three-part videotape series on eating disorders with accompanying written materials and posters for distribution to athletic administrators, coaches, trainers, and athletes.[43] Among these materials is a list of warning signs designed specifically for athletes. These are presented in table 18.5 on page 596.

When an eating disorder is suspected, it is important to recognize the seriousness of the disorder and refer the athlete to a person specifically trained in dealing with eating disorders.

Most athletic trainers, coaches, and even physicians are not trained to provide professional help to those with serious eating disorders. Most of the athletes who experience eating disorders are very intelligent, come from an upper middle-class or higher socioeconomic level, and are very good at denying that they have a problem. These athletes are unfortunate victims of the unhealthy emphasis on extreme leanness promoted by the media and the challenges of attaining the optimal weight for their sport. Treating eating disorders is extremely difficult, and even the best-trained professionals are not always successful. Some extreme cases end in suicide or premature death from failure of the cardiovascular system. Immediate professional help should be sought for an athlete suspected of having an eating disorder.

Female athletes are at higher risk for disordered eating and eating disorders for several reasons. Perhaps most important, there is tremendous pressure on athletes, particularly female athletes, to get their weight down to very low levels, often below what is appropriate. This weight limit can be imposed by the coach, trainer, or parent or self-imposed by the athlete.

In addition, the personality of the typical female athlete closely matches the profile of the female at high risk for an eating disorder (competitive, perfectionistic, and under the tight control of a parent or other significant figure, such as a coach). Furthermore, the nature of the sport or activity largely dictates those at high risk. As previously mentioned, athletes in three categories are at high risk: appearance sports, endurance sports, and weight-classification

Table 18.5

Warning Signs for Anorexia Nervosa and Bulimia Nervosa

Anorexia nervosa	Bulimia nervosa
Dramatic loss in weight	A noticeable weight loss or gain
A preoccupation with food, calories, and weight	Excessive concern about weight
Wearing baggy or layered clothing	Bathroom visits after meals
Relentless, excessive exercise	Depressed mood
Mood swings	Strict dieting followed by eating binges
Avoiding food-related social activities	Increased criticism of one's body

Note. The presence of one or two of these signs does not necessarily indicate an eating disorder. Diagnosis should be made by appropriate health professionals.

Adapted from a poster distributed by the National Collegiate Athletic Association (1990).[43]

Female Athlete Triad

In the early 1990s, it became apparent that there is a reasonably strong association among

- disordered eating,
- secondary amenorrhea, and
- bone mineral disorders.

This group of disorders has been termed the female athlete triad.[66] From the limited research available at this time, it appears that the triad might start with disordered eating. Over a period of time, the length of which has not been well established and might vary considerably from one athlete to another, an athlete who has disordered eating creating a prolonged energy deficit starts to experience disordered menstrual function, which eventually leads to secondary amenorrhea. Following an additional period of time, again the length of which has not been defined, secondary amenorrhea leads to bone mineral disorders. A number of researchers have become interested in these intriguing relationships, and considerable research is now underway. We should know much more about this important sequence of events when the results of these research efforts are known. In 1997, the American College of Sports Medicine issued a position stand on the female athlete triad, which presents a strong case for the connection of each of the three components of the triad.[44] This report concluded that female athlete triad disorders can decrease physical performance and cause morbidity (illness or disease) and mortality (death). Anyone working with young athletes, particularly female athletes, should be educated about the triad, know how to recognize it, and have a plan of action involving professionals trained in dealing with the female athlete triad to help prevent it, treat it, and reduce its risks.

sports. Added to these are the normal pressures imposed by the media and culture on young women, whether they are athletes or not.

Environmental Factors

Exercise in the heat, in the cold, or at altitude provides additional stress or challenge to the body's adaptive abilities (see chapters 10 and 11). Many early studies indicated that women are less tolerant to heat than men are, particularly when exercising. Much of this difference, however, is the result of lower fitness levels of the women included in these studies, because the men and women were tested at the same absolute rate of work. When the rate of work is adjusted relative to individual $\dot{V}O_2$max values, women's responses are almost identical to men's. Women generally have lower sweat rates for the same exercise and heat stress: Although they possess a larger number of active sweat glands than men do, women produce less sweat per gland.

Research shows that, when exposed to repeated bouts of heat stress, the body undergoes considerable adaptation (acclimatization) that enables it to withstand future heat stress more efficiently. Recent evidence suggests that, after acclimatization, the internal temperature at which sweating and vasodilation begin is similarly lowered in women and men. Also, the sensitivity of the sweating response per unit increase in internal temperature increases by a similar amount in both sexes following both physical training and heat acclimatization. Therefore, any differences noted between the women and men in the early studies can be attributed to initial differences in their physical conditioning and not to their sex.

> Women generally have lower sweat rates than men for the same heat stress, apparently the result of lower sweat production per sweat gland. However, this reduced sweating capacity does not appear to affect women's ability to tolerate heat.

Women have a slight advantage over men during cold exposure because they have more subcutaneous body fat. But their smaller muscle mass is a disadvantage in extreme cold because shivering is the major adaptation for generating body heat. The greater the active muscle mass, the greater the subsequent heat generation. Muscle also provides an additional insulating layer.

Several studies have reported sex differences in response to altitude hypoxia, both at rest and during submaximal exercise. Maximal oxygen consumption decreases during hypoxic work in both sexes, but these decreases do not seem to adversely affect women's ability to work at high altitude. Studies of maximal exercise at altitude demonstrate no difference in response between the sexes.

> ▶ Three major contributing factors to osteoporosis are estrogen deficiency, inadequate calcium intake, and inadequate physical activity.
>
> ▶ Postmenopausal women, amenorrheic women, and those who have anorexia nervosa are at greater risk of osteoporosis.
>
> ▶ Eating disorders, such as anorexia nervosa and bulimia nervosa, are much more common in women than in men and are especially common among athletes in appearance sports, endurance sports, and weight-classification sports. Athletes seem to be at a higher risk for eating disorders than the general population.
>
> ▶ When exercise intensity is adjusted relative to an individual's $\dot{V}O_2$max, women and men respond almost identically to heat stress. Any differences noted are likely attributable to different initial levels of conditioning.
>
> ▶ Because they have more insulating subcutaneous fat, women have a slight advantage over men during cold exposure, but their smaller muscle mass limits their ability to generate body heat.
>
> ▶ Studies indicate that responses during exercise at altitude do not differ in women and men.

In Closing . . .

In this chapter, we discussed sex-specific differences in performance. Most true differences between the sexes result from women's smaller body size, lower fat-free mass, and greater relative and absolute body fat. We also considered how women's relatively more sedentary lifestyle, an artifact from a society that traditionally frowned on women's participation in physical activity, has affected research through the years. Making valid comparisons has been difficult because an event's popularity and other factors—such as opportunities to participate, coaching, facilities, and training techniques—have differed considerably between the sexes over the years. Finally, we have found that female and male athletes are not as different as many people believe.

With this chapter, we conclude our examination of age and sex considerations in sport and exercise. In the next part, we turn our attention from athletics to a different application of exercise physiology: the use of physical activity for health and fitness. We begin with an examination of exercise prescription.

▷ Key Terms

amenorrhea

anorexia nervosa

bulimia nervosa

disordered eating

eating disorder

estrogen

eumenorrhea

lipoprotein lipase

menarche

menses

menstrual cycle

menstrual dysfunction

oligomenorrhea

osteopenia

osteoporosis

pregnancy

primary amenorrhea

secondary amenorrhea

sex-specific difference

teratogenic effect

testosterone

▷ Study Questions

1. How does the body composition of females compare with that of males? How do male and female athletes differ from male and female nonathletes?

2. What is the role of testosterone in the development of strength and fat-free mass?

3. How does women's upper body strength compare with men's? Lower body strength? Fat-free weight? Can women gain strength with resistance training?

4. What differences in $\dot{V}O_2$max exist between average females and males? Between highly trained females and males? What can explain these differences?

5. What cardiovascular differences exist between females and males with respect to submaximal exercise? Maximal exercise?

6. How does the menstrual cycle influence athletic performance?

7. What are some of the possible reasons that some female athletes in intense training stop menstruating for intervals of several months to several years or more?

8. What risks are associated with training during pregnancy? How can these be avoided?

9. What are the effects of amenorrhea on bone mineral? How does exercise training affect bone mineral?

10. What are the two major eating disorders, and what is the level of risk for elite female athletes having these eating disorders? How does this vary by sport?

11. How do women differ from men in their exercise response when exposed to intense heat and humidity? To altitude?

▷ References

1. American College of Obstetricians and Gynecologists. (1994, February). Exercise during pregnancy and the postpartum period. *Technical Bulletin, 189,* 1-5.

2. American College of Obstetricians and Gynecologists Committee Opinion. (2002). Exercise during pregnancy and the postpartum period. *Obstetrics and Gynecology, 99,* 171-173.

3. American Psychiatric Association. (1994). *Diagnostic and statistical manual of mental disorders* (4th ed.). Washington, DC: American Psychiatric Association.

4. Åstrand, I. (1960). Aerobic work capacity in men and women with special reference to age. *Acta Physiologica Scandinavica, 49*(Suppl. 169).

5. Åstrand, P.-O., & Rodahl, K. (1986). *Textbook of work physiology: Physiological bases of exercise* (3rd ed.). New York: McGraw-Hill.

6. Bale, P., & Nelson, G. (1985). The effects of menstruation on performance of swimmers. *Australian Journal of Science and Medicine in Sport, 19,* 19-22.

7. Boyden, T.W., Pamenter, R.W., Stanforth, P., Rotkis, T., & Wilmore, J.H. (1983). Sex steroids and endurance running in women. *Fertility and Sterility, 39,* 629-632.

8. Brooks-Gunn, J., Gargiulo, J.M., & Warren, M.P. (1986). The effect of cycle phase on the adolescent swimmers. *Physician and Sportsmedicine, 14*(3), 182-192.

9. Bullen, B.A., Skrinar, G.S., Beitins, I.Z., von Mering, G., Turnbull, B.A., & McArthur, J.W. (1985). Induction of menstrual disorders by strenuous exercise in untrained women. *New England Journal of Medicine, 312,* 1349-1353.

10. Cann, C.E., Martin, M.C., Genant, H.K., & Jaffe, R.B. (1984). Decreased spinal mineral content in amenorrheic women. *Journal of the American Medical Association, 251,* 626-629.

11. Clapp, J.F., III, Wesley, M., & Sleamaker, R.H. (1987). Thermoregulatory and metabolic responses to jogging prior to and during pregnancy. *Medicine and Science in Sports and Exercise, 19,* 124-130.

12. Costill, D.L., Fink, W.J., Flynn, M., & Kirwan, J. (1987). Muscle fiber composition and enzyme activities in elite female distance runners. *International Journal of Sports Medicine, 8*(Suppl. 2), 103-106.

13. Costill, D.L., & Winrow, E. (1970). Maximal oxygen intake among marathon runners. *Archives of Physical Medicine and Rehabilitation, 51,* 317-320.

14. Cureton, K., Bishop, P., Hutchinson, P., Newland, H., Vickery, S., & Zwiren, L. (1986). Sex differences in maximal oxygen uptake: Effect of equating haemoglobin concentration. *European Journal of Applied Physiology, 54,* 656-660.

15. Cureton, K.J., Collins, M.A., Hill, D.W., & McElhannon, F.M. (1988). Muscle hypertrophy in men and women. *Medicine and Science in Sports and Exercise, 20,* 338-344.

16. Cureton, K.J., & Sparling, P.B. (1980). Distance running performance and metabolic responses to running in men and women with excess weight experimentally equated. *Medicine and Science in Sports and Exercise, 12,* 288-294.

17. Drinkwater, B.L. (1973). Physiological responses of women to exercise. *Exercise and Sport Sciences Reviews, 1,* 125-153.

18. Drinkwater, B.L., Bruemner, B., & Chesnut, C.H., III. (1990). Menstrual history as a determinant of current bone density in young athletes. *Journal of the American Medical Association, 263,* 545-548.

19. Drinkwater, B.L., Nilson, K., Chesnut, C.H., III, Bremner, W.J., Shainholtz, S., & Southworth, M.B. (1984). Bone mineral content of amenorrheic and eumenorrheic athletes. *New England Journal of Medicine, 311,* 277-281.

20. Drinkwater, B.L., Nilson, K., Ott, S., & Chesnut, C.H., III. (1986). Bone mineral density after resumption of menses in amenorrheic athletes. *Journal of the American Medical Association, 256,* 380-382.

21. Fink, W.J., Costill, D.L., & Pollock, M.L. (1977). Submaximal and maximal working capacity of elite distance runners: Part II. Muscle fiber composition and enzyme activities. *Annals of the New York Academy of Sciences, 301,* 323-327.

22. Forbes, G.B. (1972). Growth of the lean body mass in man. *Growth, 36,* 325-338.

23. Forbes, G.B. (1976). The adult decline in lean body mass. *Human Biology, 48,* 161-173.

24. Frisch, R.E. (1983). Fatness and reproduction: Delayed menarche and amenorrhea of ballet dancers and college athletes. In P.E. Garfinkel, P.L. Darby, & D.M. Garner (Eds.), *Anorexia nervosa: Recent developments in research* (pp. 343-363). New York: Liss.

25. Frisch, R.E. (1988). Fatness and fertility. *Scientific American, 255,* 88-95.

26. Frisch, R.E., & McArthur, J.W. (1974). Menstrual cycles: Fatness as a determinant of minimum weight for height necessary for their maintenance or onset. *Science, 185,* 949-951.

27. Gadpaille, W.J., Sanborn, C.F., & Wagner, W.W. (1987). Athletic amenorrhea, major affective disorders, and eating disorders. *American Journal of Psychiatry, 144,* 939-942.

28. Hermansen, L., & Andersen, K.L. (1965). Aerobic work capacity in young Norwegian men and women. *Journal of Applied Physiology, 20,* 425-431.

29. Hicks, A.L., Kent-Braun, J., & Ditor, D.S. (2001). Sex differences in human skeletal muscle fatigue. *Exercise and Sport Sciences Reviews, 29,* 109-112.

30. Horton, T.J., Pagliassotti, M.J., Hobbs, K., & Hill, J.O. (1998). Fuel metabolism in men and women during and after long-duration exercise. *Journal of Applied Physiology, 85,* 1823-1832.

31. Hult, J.S. (1986). The female American runner: A modern quest for visibility. In B.L. Drinkwater (Ed.), *Female endurance athletes* (pp. 1-39). Champaign, IL: Human Kinetics.

32. Ingjer, F., & Brodal, P. (1978). Capillary supply of skeletal muscle fibers in untrained and endurance-trained women. *European Journal of Applied Physiology, 38,* 291-299.

33. Janssen, I., Heymsfield, S.B., Wang, Z., & Ross, R. (2000). Skeletal muscle mass and distribution in 468 men and women aged 18-88 yr. *Journal of Applied Physiology, 89,* 81-88.

34. Keizer, H.A., & Rogol, A.D. (1990). Physical exercise and menstrual cycle alterations: What are the mechanisms? *Sports Medicine, 10,* 218-235.

35. Lohman, T.G. (1986). Application of body composition techniques and constants for children and youths. *Exercise and Sport Sciences Reviews, 14,* 325-357.

36. Loucks, A.B. (1990). Effects of exercise training on the menstrual cycle: Existence and mechanisms. *Medicine and Science in Sports and Exercise, 22,* 275-280.

37. Loucks, A.B., & Heath, E.M. (1994). Dietary restriction reduces luteinizing hormone (LH) pulse frequency during waking hours and increases LH pulse amplitude during sleep in young menstruating women. *Journal of Clinical Endocrinology and Metabolism, 78,* 910-915.

38. Loucks, A.B., & Heath, E.M. (1994). Induction of low-T_3 syndrome in exercising women occurs at a threshold of energy availability. *American Journal of Physiology, 266,* R817-R823.

39. Loucks, A.B., Verdun, M., & Heath, E.M. (1998). Low energy availability, not stress of exercise, alters LH pulsatility in exercising women. *Journal of Applied Physiology, 84,* 37-46.

40. Malina, R.M. (1983). Menarche in athletes: A synthesis and hypothesis. *Annals of Human Biology, 10,* 1-24.

41. Mier, C.M., Domenick, M.A., Turner, N.S., & Wilmore, J.H. (1996). Changes in stroke volume and maximal aerobic capacity with increased blood volume in men and women. *Journal of Applied Physiology, 80,* 1180-1186.

42. Mier, C.M., Domenick, M.A., & Wilmore, J.H. (1997). Changes in stroke volume with β-blockade before and after 10 days of exercise training in men and women. *Journal of Applied Physiology, 83,* 1660-1665.

43. National Collegiate Athletic Association. (1990).

44. Otis, C.L., Drinkwater, B., Johnson, M., Loucks, A., & Wilmore, J. (1997). The female athlete triad. *Medicine and Science in Sports and Exercise, 29*(5), i-ix.

45. Pate, R.R., Sparling, P.B., Wilson, G.E., Cureton, K.J., & Miller, B.J. (1987). Cardiorespiratory and metabolic responses to submaximal and maximal exercise in elite women distance runners. *International Journal of Sports Medicine, 8*(Suppl. 2), 91-95.

46. Pivarnik, J.M. (1994). Maternal exercise during pregnancy. *Sports Medicine, 18,* 215-217.

47. Pollock, M.L. (1977). Submaximal and maximal working capacity of elite distance runners: Part I. Cardiorespiratory aspects. *Annals of the New York Academy of Sciences, 301,* 310-322.

48. Pollock, M.L., Hickman, T., Kendrick, Z., Jackson, A., Linnerud, A.C., & Dawson, G. (1976). Prediction of body density in young and middle-aged men. *Journal of Applied Physiology, 40,* 300-304.

49. Pollock, M.L., Laughridge, E.E., Coleman, B., Linnerud, A.C., & Jackson, A. (1975). Prediction of body density in young and middle-aged women. *Journal of Applied Physiology, 38,* 745-749.

50. Porter, M.M., Stuart, S., Boij, M., & Lexell, J. (2002). Capillary supply of the tibialis anterior muscle in young, healthy, and moderately active men and women. *Journal of Applied Physiology, 92,* 1451-1457.

51. Quadagno, D., Faquin, L., Lim, G.-N., Kuminka, W., & Moffatt, R. (1991). The menstrual cycle: Does it affect athletic performance? *Physician and Sportsmedicine, 19,* 121-124.

52. Rigotti, N.A., Nussbaum, S.R., Herzog, D.B., & Neer, R.M. (1984). Osteoporosis in women with anorexia nervosa. *New England Journal of Medicine, 311,* 1601-1606.

53. Robinson, S. (1938). Experimental studies of physical fitness in relation to age. *Arbeitsphysiologie, 10,* 251-323.

54. Saltin, B., & Åstrand, P.-O. (1967). Maximal oxygen uptake in athletes. *Journal of Applied Physiology, 23,* 353-358.

55. Saltin, B., Henriksson, J., Nygaard, E., & Andersen, P. (1977). Fiber types and metabolic potentials of skeletal muscles in sedentary man and endurance runners. *Annals of the New York Academy of Sciences, 301,* 3-29.

56. Saltin, B., & Rowell, L.B. (1980). Functional adaptations to physical activity and inactivity. *Federation Proceedings, 39,* 1506-1513.

57. Schantz, P., Randall-Fox, E., Hutchison, W., Tyden, A., & Åstrand, P.-O. (1983). Muscle fibre type distribution, muscle cross-sectional area and maximal voluntary strength in humans. *Acta Physiologica Scandinavica, 117,* 219-226.

58. Schwartz, B., Cumming, D.C., Riordan, E., Selye, M., Yen, S.S.C., & Rebar, R.W. (1981). Exercise-associated amenorrhea: A distinct entity? *American Journal of Obstetrics and Gynecology, 141,* 662-670.

59. Sinning, W.E., & Little, K.D. (1987). Body composition and menstrual function in athletes. *Sports Medicine, 4,* 34-45.

60. Sparling, P.B., O'Donnell, E.M., & Snow, T.K. (1998). The gender difference in distance running performance has plateaued: An analysis of world rankings from 1980 to 1996. *Medicine and Science in Sports and Exercise, 30,* 1725-1729.

61. Stager, J.M., Wigglesworth, J.K., & Hatler, L.K. (1990). Interpreting the relationship between age of menarche and prepubertal training. *Medicine and Science in Sports and Exercise, 22,* 54-58.

62. Sundgot-Borgen, J. (1993). Prevalence of eating disorders in elite female athletes. *International Journal of Sport Nutrition, 3,* 29-40.

63. Tarnopolsky, M.A. (2000). Gender differences in substrate metabolism during endurance exercise. *Canadian Journal of Applied Physiology, 25,* 312-327.

64. Turley, K.R., & Wilmore, J.H. (1997). Cardiovascular responses to submaximal exercise in 7- to 9-yr-old boys and girls. *Medicine and Science in Sports and Exercise, 29,* 824-832.

65. Wilmore, J.H. (1974). Alterations in strength, body composition and anthropometric measurements consequent to a 10-week weight training program. *Medicine and Science in Sports, 6,* 133-138.

66. Wilmore, J.H. (1991). Eating and weight disorders in the female athlete. *International Journal of Sport Nutrition, 1,* 104-117.

67. Wilmore, J.H., & Behnke, A.R. (1969). An anthropometric estimation of body density and lean body weight in young men. *Journal of Applied Physiology, 27,* 25-31.

68. Wilmore, J.H., & Behnke, A.R. (1970). An anthropometric estimation of body density and lean body weight in young women. *American Journal of Clinical Nutrition, 23,* 267-274.

69. Wilmore, J.H., & Brown, C.H. (1974). Physiological profiles of women distance runners. *Medicine and Science in Sports, 6,* 178-181.

70. Wilmore, J.H., Stanforth, P.R., Gagnon, J., Rice, T., Mandel, S., Leon, A.S., Rao, D.C., Skinner, J.S., & Bouchard, C. (2001). Cardiac output and stroke volume changes with endurance training: The HERITAGE Family Study. *Medicine and Science in Sports and Exercise, 33,* 99-106.

71. Wilmore, J.H., Wambsgans, K.C., Brenner, M., Broeder, C.E., Paijmans, I., Volpe, J.A., & Wilmore, K.M. (1992). Is there energy conservation in amenorrheic compared to eumenorrheic distance runners? *Journal of Applied Physiology, 72,* 15-22.

72. Wolfe, L.A., Brenner, I.K.M., & Mottola, M.F. (1994). Maternal exercise, fetal well-being and pregnancy outcome. *Exercise and Sport Sciences Reviews, 22,* 145-194.

73. Wolfe, L.A., Hall, P., Webb, K.A., Goodman, L., Monga, M., & McGrath, M.J. (1989). Prescription of aerobic exercise during pregnancy. *Sports Medicine, 8,* 273-301.

74. Wolfe, L.A., Ohtake, P.J., Mottola, M.F., & McGrath, M.J. (1989). Physiological interactions between pregnancy and aerobic exercise. *Exercise and Sport Sciences Reviews, 17,* 295-351.

▷ Selected Readings

American College of Sports Medicine. (1995). ACSM position stand on osteoporosis and exercise. *Medicine and Science in Sports and Exercise, 27*(4), i-vii.

Anderson, J.J.B. (2000). The important role of physical activity in skeletal development: How exercise may counter low calcium intake. *American Journal of Clinical Nutrition, 71,* 1384-1386.

Artal, R., & Sherman, C. (1999). Exercise during pregnancy. *Physician and Sportsmedicine, 27*(8), 51-75.

Artal, R., & Wiswell, R.A. (Eds.). (1986). *Exercise in pregnancy.* Baltimore: Williams & Wilkins.

Åstrand, P.-O. (1952). *Experimental studies of physical working capacity in relation to age and sex.* Copenhagen: Munksgaard.

Beals, K.A., & Manore, M.M. (1994). The prevalence and consequences of subclinical eating disorders in female athletes. *International Journal of Sport Nutrition, 4,* 175-195.

Beck, B.R., & Shoemaker, M.R. (2000). Osteoporosis: Understanding key risk factors and therapeutic options. *Physician and Sportsmedicine, 28*(2), 69-84.

Burrows, M., & Bird, S. (2000). The physiology of the highly trained female endurance runner. *Sports Medicine, 30*, 281-300.

Drinkwater, B.L. (1984). Women and exercise: Physiological aspects. *Exercise and Sport Sciences Reviews, 12*, 21-51.

Drinkwater, B.L. (Ed.). (1986). *Female endurance athletes.* Champaign, IL: Human Kinetics.

Drinkwater, B.L. (Ed.). (2000). *Women in sport.* Oxford, UK: Blackwell Science.

Harber, V.J. (2000). Menstrual dysfunction in athletes: An energetic challenge. *Exercise and Sport Sciences Reviews, 28*, 19-23.

Lebrun, C.M. (1993). Effect of the different phases of the menstrual cycle and oral contraceptives on athletic performance. *Sports Medicine, 16*, 400-430.

Lotgering, F.K., Gilbert, R.D., & Longo, L.D. (1985). Maternal and fetal responses to exercise during pregnancy. *Physiological Reviews, 65*, 1-36.

Loucks, A.B. (2001). Physical health of the female athlete: Observations, effects, and causes of reproductive disorders. *Canadian Journal of Applied Physiology, 26*, S176-S185.

Manore, M.M. (2002). Dietary recommendations and athletic menstrual dysfunction. *Sports Medicine, 32*, 887-901.

McMurray, R.G., & Katz, V.L. (1990). Thermoregulation in pregnancy: Implications for exercise. *Sports Medicine, 10*, 146-158.

McMurray, R.G., Mottola, M.F., Wolfe, L.A., Artal, R., Millar, L., & Pivarnik, J.M. (1993). Recent advances in understanding maternal and fetal responses to exercise. *Medicine and Science in Sports and Exercise, 25*, 1305-1321.

Nattiv, A., Agostini, R., Drinkwater, B., & Yeager, K.K. (1994). The female athlete triad: The inter-relatedness of disordered eating, amenorrhea, and osteoporosis. *Clinics in Sports Medicine, 13*, 405-418.

Otis, C.L., & Goldingay, R. (2000). *The athletic woman's survival guide.* Champaign, IL: Human Kinetics.

Shangold, M.M. (1984). Exercise and the adult female: Hormonal and endocrine effects. *Exercise and Sport Sciences Reviews, 12*, 53-79.

Shangold, M., & Mirkin, G. (Eds.). (1988). *Women and exercise: Physiology and sports medicine.* Philadelphia: Davis.

Shephard, R.J. (2000). Exercise and training in women, Part I: Influence of gender on exercise and training responses. *Canadian Journal of Applied Physiology, 25*, 19-34.

Shephard, R.J. (2000). Exercise and training in women, Part II: Influence of menstrual cycle and pregnancy on exercise responses. *Canadian Journal of Applied Physiology, 25*, 35-54.

Sparling, P.B. (Ed.). (1987). A comprehensive profile of elite women distance runners. *International Journal of Sports Medicine, 8*(Suppl. 2), 71-136.

Wells, C.L. (1991). *Women, sport and performance: A physiological perspective* (2nd ed.). Champaign, IL: Human Kinetics.

PART VII

Physical Activity for Health and Fitness

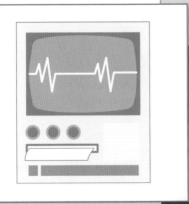

In previous parts of this book, we focused on the physiological bases of physical activity, observing the responses to an acute bout of exercise and the adaptations to chronic training, and on how we can improve our performance in sport-related activities. In part VII, we shift our focus away from athletic performance and turn to a special area of exercise and sport physiology: the use of physical activity for health and fitness. We begin in chapter 19, "Prescription of Exercise for Health and Fitness," where we discuss how to design an exercise program that can improve health and fitness. We consider the essential components, how to tailor the program to an individual's specific needs, and the unique role of physical activity for rehabilitation of people who are ill. In chapter 20, "Cardiovascular Disease and Physical Activity," we examine the major types of cardiovascular disease, their physiological bases, and how physical activity can help prevent these diseases. Finally, in chapter 21, "Obesity, Diabetes, and Physical Activity," we examine the causes of obesity and diabetes, the health risks associated with each, and how physical activity can be used to control both disorders.

CHAPTER 19

PRESCRIPTION OF EXERCISE FOR HEALTH AND FITNESS

overview

Patterns of today's living have channeled the average American into an increasingly sedentary existence. Humans, however, were designed and built for movement. Physiologically, we have not adapted well to this inactive lifestyle. In fact, during what appeared to be a fitness boom in the 1970s and 1980s, fewer than 10% to 20% of adult Americans were exercising at levels that would increase or maintain their aerobic fitness and strength. Yet research has determined that, for most people, an active lifestyle is important for optimal health.

In this chapter, we focus on the principles of exercise prescription. We discuss the importance of obtaining medical clearance before beginning an exercise program and what constitutes an appropriate medical screening. Then we review the components that constitute the exercise prescription. Finally, we examine the components of exercise programs for both healthy people and those with diseases.

outline

Jason Walker, a 35-year-old executive, went in for his annual physical examination with a vow to start a long overdue exercise program. Because of his high blood pressure and one-pack-a-day smoking habit, his physician decided to give Jason a graded exercise test to determine the normality of his electrocardiogram (ECG) during the stress of exercise. As Jason was nearing exhaustion on the treadmill, his doctor noticed changes in the ST segment of his ECG, which are considered indicative of coronary artery disease. The following week, Jason, with fear and trepidation, underwent a coronary arteriogram procedure to check for coronary artery disease. His arteriogram was normal, indicating that his treadmill electrocardiogram was not accurate—it was a false-positive test. Fortunately, Jason had been scared to the point that he stopped smoking and began an exercise program. He is now competing in 10-km (6.2-mi) races, is very fit, and has his blood pressure under control.

The 1990s will be remembered as the decade in which the medical profession formally recognized the fact that physical activity is vital to your body's health. It seems rather ironic that it took this long for clinicians and scientists to reach this conclusion, as Hippocrates (460-377 B.C.), a prominent physician and athlete, had strongly endorsed physical activity and proper nutrition as essential to health more than 2,000 years earlier!

The first acknowledgment from the modern medical profession came in July 1992, when the American Heart Association proclaimed physical inactivity a major risk factor for coronary artery disease, placing it alongside smoking, abnormal blood lipids, and hypertension.[13] In 1994, the Centers for Disease Control and Prevention (CDCP) in collaboration with the American College of Sports Medicine (ACSM) held a press conference to announce to the American public the importance of physical activity as a public health initiative and subsequently published the full text of a consensus statement by a panel of experts in this field in February 1995.[29] The National Institutes of Health (National Heart, Lung, and Blood Institute) released a consensus statement in December 1995, the full text of which was published in 1996, advocating physical activity as important for cardiovascular health.[28] Finally, in July 1996, coinciding with the start of the Olympic Games in Atlanta, the Surgeon General of the United States released a written report on the health benefits of physical activity.[38] This was a landmark report recognizing the importance of physical activity in reducing

the risk for chronic degenerative diseases. (See the sidebar for the major conclusions of this report.)

Much of the research supporting the benefits of physical activity in reducing the risk of developing chronic degenerative disease has come from the field of epidemiology, where large populations are studied and associations between activity levels and disease risk determined. In the year 2000, molecular biologists started waging war on what they termed the "sedentary death syndrome," by forming an action group advocating governmental support for research into the diseases and disorders associated with a sedentary lifestyle. The group, Researchers Against Inactivity-Related Disorders, or RID, has been very effective in gaining the support of top government leaders for basic research into the role of an active lifestyle in preventing or delaying chronic degenerative diseases. Several key scientific articles have been published,[6, 7] and a Web site has been established (http://www.endseds.org).

With the health benefits of an active lifestyle so clearly established, what has been the response of the U.S. population in general? We need to go back a few years to get a proper historical perspective. The seeds for a fitness revolution were planted in the late 1960s with the publication of the book *Aerobics*, written by Dr. Kenneth Cooper (figure 19.1).[11] This book provided a sound medical basis for the importance of exercise, particularly aerobic exercise, to health and fitness. The fitness movement grew throughout the 1970s, possibly peaking in the early 1980s, at which time the media

◄ **Figure 19.1** Dr. Kenneth H. Cooper, founder of the Cooper Institute and author of numerous books on the health-related benefits of maintaining an active lifestyle.

declared that America was in the midst of a fitness boom. This was an exciting time to be in the exercise sciences!

Then, in 1983, came a penetrating article by Kirshenbaum and Sullivan,[21] published in *Sports Illustrated*, that brought everything into proper perspective. The authors questioned the existence of the fitness boom, contending that involvement was basically limited to a small but highly visible segment of the total population. They maintained that the fitness boom included mostly high-income, executive-level, white, college-educated, young to middle-aged adults. The results of a survey by the Pacific Mutual Life Insurance Company confirmed

The Surgeon General's Report on Physical Activity and Health

In July 1996, the U.S. Surgeon General's office released its official report *Physical Activity and Health.*[38] This detailed report resulted in the following major conclusions:

1. People of all ages, both male and female, benefit from regular physical activity.

2. Significant health benefits can be obtained by including a moderate amount of physical activity (e.g., 30 min of brisk walking or raking leaves, 15 min of running, or 45 min of playing volleyball) on most, if not all, days of the week. Through a modest increase in daily activity, most Americans can improve their health and quality of life.

3. Additional health benefits can be gained through greater amounts of physical activity. People who can maintain a regular regimen of activity that is of longer duration or of more vigorous intensity are likely to derive greater benefit.

4. Physical activity reduces the risk of premature mortality in general and of coronary artery disease, hypertension, colon cancer, and diabetes mellitus in particular. Physical activity also improves mental health and is important for the health of muscles, bones, and joints.

5. More than 60% of American adults are not regularly physically active. In fact, 25% of all adults are not active at all.

6. Nearly half of American youths 12 to 21 years of age are not vigorously active on a regular basis. Moreover, physical activity declines dramatically during adolescence.

7. Daily enrollment in physical education classes has declined among high school students from 42% in 1991 to 25% in 1995.

8. Research on understanding and promoting physical activity is at an early stage, but some interventions to promote physical activity through schools, work sites, and healthcare settings have been evaluated and found to be successful.

this analysis.[16] Furthermore, additional national surveys reported that only 15% to 36% of the adult American population participated in regular vigorous activity.[17, 18, 27] Most important, estimates in the mid-1980s suggested that only 7.5% of all adult Americans met the ACSM guidelines for cardiorespiratory fitness development and maintenance.[1, 10]

Despite the disappointing statistics, most Americans are aware that exercise is an integral part of preventive medicine. And yet people often equate exercise with jogging 8 km (5 mi) a day or lifting weights until their muscles can do no more. Many believe that high volume and intensity of exercise training are necessary to attain health-related benefits, yet this is not true. This myth was the primary focus of the CDCP/ACSM report published in 1995,[29] which concluded that significant health benefits can be obtained by including a moderate amount of physical activity, such as 30 min of brisk walking, 15 min of running, or 45 min of playing volleyball, on most, if not all, days of the week. The major emphasis of this report was that through a modest increase in daily activity, most people can improve their health and quality of life. However, the Surgeon General's report[38] emphasized that additional health benefits can be gained through greater amounts of physical activity. Research suggests that people who can maintain a regular regimen of activity that is of longer duration or of more vigorous intensity are likely to derive greater benefit. It is now apparent that the appropriate exercise type and intensity vary, depending on individual characteristics, current fitness level, and specific health concerns. Knowing this, how should people begin exercise programs to improve their general health and fitness? The first step is deciding to take action. The next step is getting medical clearance.

Medical Clearance

Is a medical evaluation really necessary before starting an exercise program? Dr. Per-Olof Åstrand (figure 19.2), eminent Swedish physi-

cian and physiologist, has argued that those individuals who elect to remain sedentary should be required to have a medical evaluation to determine if their bodies can withstand the rigors of a sedentary lifestyle. The medical evaluation is perceived as a significant barrier to starting an exercise program for many people, yet it is useful and important for the following reasons:

- Some people either should not exercise at all or are considered at high risk and should be restricted to exercising only under close medical supervision. A comprehensive medical evaluation will help identify these high-risk individuals.

- The information obtained in a medical evaluation can be used to develop the exercise prescription.

- The values obtained for certain clinical measures, such as blood pressure, body fat content, and blood lipid levels, can be used to motivate the person to adhere to the exercise program.

- A comprehensive medical evaluation, particularly of healthy people, can provide a baseline against which any subsequent changes in health status can be compared.

- Children and adults should establish the habit of periodic medical evaluations because many illnesses and diseases, such as cancer and cardiovascular diseases, can be identified in their earliest stages when the chances of successful treatment are much higher.

Medical Evaluation

Unfortunately, although a comprehensive medical evaluation is useful and desirable before prescribing exercise, not all people will have one. Many people can't afford the cost of such an evaluation, and the medical system is not prepared to provide this service for the total population, even if money were available. Also, medical evaluation before prescribing exercise for a population presumed healthy has not been proven to reduce the medical risks associated with exercise. For

▲ **Figure 19.2**   Dr. Per-Olof Åstrand, eminent Swedish physician and physiologist, bicycling through the woods.

these reasons, guidelines or recommendations have been established that attempt to target the higher risk individuals.[1, 3, 14] People at high risk are those who have two or more risk factors for coronary artery disease (table 19.1 on page 610) or one or more symptoms or signs of cardiopulmonary disease (table 19.2 on page 611).

In general, before beginning any exercise program that includes vigorous activity, the following people should have a complete medical examination by a physician:

- Men over 40 years of age
- Women over 50 years of age
- People of any age who are considered to be at high risk

For moderate-intensity activity, only those with two or more risk factors (table 19.1) or one or more symptoms or signs of cardiopulmonary disease (table 19.2) should have a medical exam.

> Although a general medical evaluation on a regular basis is important and desirable for everyone, it simply is not practical to require this for all people who want to start an exercise program.

The American College of Sports Medicine has published specific recommendations for each phase of the medical evaluation.[1] This document should be consulted whenever there is any question as to what should be included.

The physical examination should include discussion between the physician and patient of the proposed exercise program in case any medical contraindications are associated with the proposed activity. For example, people with hypertension should be cautioned to avoid activities that use isometric actions. Isometric actions tend to increase blood pressure considerably and usually result in the Valsalva maneuver, in which intra-abdominal and intrathoracic pressures increase to the point of

Table 19.1

Coronary Artery Disease Risk Factors for Targeting High-Risk People

Positive risk factors	Defining criteria
Age	Men >45 years; women >55 years or premature menopause without estrogen replacement therapy
Family history	Myocardial infarction or sudden death before 55 years of age in father or other male first-degree relative or before 65 years of age in mother or other female first-degree relative
Current cigarette smoking	
Hypertension	Blood pressure ≥140/90 mmHg, confirmed by measurements on at least two separate occasions, or on antihypertensive medication
Hypercholesterolemia	Total serum cholesterol >200 mg/dl (5.2 mmol/L, if lipoprotein profile is unavailable) or HDL <35 mg/dl (0.9 mmol/L)
Diabetes mellitus	People with insulin-dependent diabetes mellitus (IDDM) who are >30 years of age or who have had IDDM for >15 years, and people with non-insulin-dependent diabetes mellitus (NIDDM) who are >35 years of age
Sedentary lifestyle/ physical inactivity	People comprising the least active 25% of the population, as defined by the combination of sedentary jobs involving sitting for a large part of the day and no regular exercise or active recreational pursuits
Negative risk factors	**Defining criteria**
High serum HDL-C	>60 mg/dl (1.6 mmol/L)

Note. It is common to sum risk factors in making clinical judgments. If high-density–lipoprotein cholesterol (HDL-C) is high, subtract one risk factor from the sum of positive risk factors, because high HDL-C decreases coronary artery disease risk. Obesity is not listed as an independent positive risk factor because its effects are exerted through other risk factors (e.g., hypertension, hyperlipidemia, diabetes). Obesity should be considered as an independent target for intervention.

Adapted, by permission, American College of Sports Medicine, 2000, *ACSM's guidelines for exercise testing and prescription*, 6th ed. (Philadelphia, PA: Lippincott, Williams, and Wilkins), 24.

restricting blood flow through the vena cava, limiting venous return to the heart. Both responses can lead to serious medical complications, such as loss of consciousness or stroke. Also, as we have discussed previously (see chapter 7), even dynamic resistance training can cause very high blood pressure during the activity.

Exercise Electrocardiogram

The **exercise electrocardiogram (ECG)** is obtained while you exercise, usually on either a treadmill or a cycle ergometer, as shown in figure 19.3. The ECG is monitored as you progress from low-intensity exercise, such as slow walking, up to maximal-intensity exercise. Maximal intensity might be brisk walking for an older, deconditioned subject or running up a grade for a younger, fit individual. The rate of work is generally increased every 1 to 3 min until the maximal rate of work is achieved. This progression is referred to as a **graded exercise test** (GXT). The ECG is monitored for arrhythmias and changes in the ST segment; the latter can suggest existing coronary artery disease.

Table 19.2

Major Symptoms or Signs Suggestive of Cardiopulmonary Disease

Pain or discomfort (or other anginal equivalent) in the chest, neck, jaw, arms, or other areas that may be ischemic in nature

Shortness of breath at rest or with mild exertion

Dizziness or syncope

Orthopnea or paroxysmal nocturnal dyspnea

Ankle edema

Palpitations or tachycardia

Intermittent claudication

Known heart murmur

Unusual fatigue or shortness of breath with usual activities

Note. These symptoms must be interpreted in the clinical context in which they appear because they are not all specific for cardiopulmonary or metabolic disease.

Adapted, by permission, American College of Sports Medicine, 2000, *ACSM's guidelines for exercise testing and prescription*, 6th ed. (Philadelphia, PA: Lippincott, Williams, and Williams), 25.

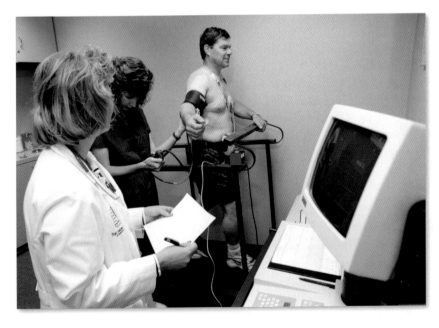

▶ **Figure 19.3** Obtaining an exercise electrocardiogram. Once he steadies himself, the subject is usually asked to refrain from using the handrail during testing.

The exercise ECG is an extremely important part of the medical evaluation because a small but significant percentage of the adult population have abnormalities in ECGs taken during or following exercise, even though they have normal resting ECGs. These abnormalities often indicate the presence of coronary artery disease. Because of this, an exercise test is recommended primarily for disease detection. But the sensitivity, specificity, and predictive value of an exercise test must be considered. **Sensitivity** refers to the exercise test's ability to correctly identify people who have the disease in question, such as coronary artery disease. **Specificity** refers to the test's ability to correctly identify people who don't have the disease. And the **predictive value of an abnormal exercise test** refers to the accuracy with which abnormal test results reflect presence of the disease.

These concepts are illustrated in table 19.3. The total population tested is divided into those who have documented coronary artery disease and those who don't. This is normally determined by a coronary arteriogram, where radiopaque dye (dye that is opaque to X-rays) is injected through a catheter into coronary arteries, allowing visualization of the insides of the arteries. Results from an exercise test are considered either positive, when there have been changes in the ECG that suggest the presence of coronary artery disease, or negative, which implies that no disease was detected. The exercise ECG is not 100% accurate, so some people have a normal, or negative, exercise test, yet have coronary artery disease. These are false-negative tests. There are also people who do not have the disease, as determined by coronary arteriograms, yet who have a positive exercise ECG suggesting disease, as we saw in the example of Jason Walker, presented at the opening of this chapter. These are false-positive tests.

Table 19.3 illustrates how sensitivity, specificity, and the predictive value of an abnormal exercise test are calculated. As an example, a total of 100 subjects have been tested and 6 are true positive, 10 are false positive, 4 are false negative, and 80 are true negative. The sensitivity of the test is $SN = 6/(6 + 4) \times 100\% = 60\%$.

The specificity of the test is $SP = 80/(10 + 80) \times 100\% = 88.9\%$. The predictive value of an abnormal test is $PV = 6/(6 + 10) \times 100\% = 37.5\%$. The prevalence of the disease in this population is 6% (6 true positives ÷ 100 subjects).

Unfortunately, both the sensitivity and the predictive value of an abnormal exercise test for detecting coronary artery disease are relatively low in healthy populations of people who have no symptoms of this disease. Past studies reveal that sensitivity averages from 50% to 80%, indicating that between 50% and 80% of those with coronary artery disease are correctly identified by exercise electrocardiograms as having the disease. Conversely, from 20% to 50% of those with the disease are incorrectly diagnosed as disease-free based on exercise ECGs. Average specificity values range from 80% to 90%, indicating that 80% to 90% of those without disease are correctly identified by the exercise test as being disease free. Unfortunately, this means that 10% to 20% of the population are incorrectly identified as having the disease.

> The sensitivity and predictive value of an abnormal exercise test are generally low in a young, healthy population where there is a low prevalence of coronary artery disease. As a result, the value of using exercise electrocardiography to screen for coronary artery disease in this population is questionable.

Table 19.3

Illustration of Exercise Test Sensitivity (SN), Specificity (SP), and the Predictive Value (PV) of an Abnormal Test with Respect to Coronary Artery Disease (CAD)

Exercise test result	Those with CAD	Those without CAD
Positive (abnormal)	True positive (TP)	False positive (FP)
Negative	False negative (FN)	True negative (TN)
	$SN = [TP/(TP/FN)] \times 100\%$	$SP = [TN/(FP + TN)] \times 100\%$
	$PV = [TP/(TP + FP)] \times 100\%$	

Note. An abnormal exercise test is defined as one in which the ST segment of the electrocardiogram is depressed, suggestive of a myocardial ischemia.

The predictive value of an abnormal exercise test varies considerably with the prevalence of coronary artery disease in the population. This is illustrated in table 19.4, which compares the predictive value of an abnormal test when the prevalence of coronary artery disease per 1,000 people is 5% versus 50%. In both cases, we assume an average sensitivity of 60% and an average specificity of 90% based on past studies. For the population with the 5% prevalence, the predictive value of an abnormal test is only 24.0%. This indicates that only 24% of those who have an abnormal exercise ECG actually have coronary artery disease. In other words, in this population, more than three of every four people identified with abnormal exercise tests will be labeled as diseased but won't actually have identifiable coronary artery disease! But if we consider the population with a prevalence of 50%, the predictive value of an abnormal exercise test is much higher—85.7%.

In a 10-year study of YMCAs in the United States, investigators reported one sudden death during exercise per 3 million person-hours and one cardiac arrest per 2 million person-hours.[26] Person-hours refers to the total amount of time the entire population exercises. For example, if each of 10 people exercised for 500 h, that gives a total of 5,000 person-hours of exercise. Based on the uncertainties associated with positive and negative results from exercise tests in populations that have a low prevalence of disease and a relative scarcity of sudden deaths or cardiac arrests during exercise, the investigators concluded that it is impractical to use exercise testing to prevent significant cardiovascular complications during exercise in asymptomatic individuals.

From this information, we can conclude that exercise testing is of limited value in screening young, apparently healthy individuals before prescribing exercise for them. The accuracy of interpreting the results of the exercise ECG is questionable, particularly in a population with such a low prevalence of disease. Also, the actual risk of death or cardiac arrest during

Table 19.4

Comparison of the Predictive Value of Exercise Testing Between Populations With 5% and 50% Prevalence of Coronary Artery Disease (CAD), Using a Sensitivity of 60% and a Specificity of 90%

Population subgroups	*n* (%) of subjects	*n* (%) with abnormal exercise ECG	*n* (%) with normal exercise ECG
5% prevalence			
Normal	950 (95)	95 (10) FP	855 (90) TN
CAD	50 (5)	30 (60) TP	20 (40) FN
Total	1,000 (100)	125 (12.5)	875 (87.5)
Predictive value = [30/(30 + 95)] × 100 = 24.0%			
50% prevalence			
Normal	500 (50)	50 (10) FP	450 (90) TN
CAD	500 (50)	300 (60) TP	200 (40) FN
Total	1,000 (100)	350 (35)	650 (65)
Predictive value = [300/(300 + 50)] × 100 = 85.7%			

Note. An abnormal exercise test is defined as one in which the ST segment of the electrocardiogram (ECG) is depressed, suggestive of myocardial ischemia. FP = false positive; TP = true positive; FN = false negative; TN = true negative.

exercise is relatively low. Another important consideration is the expense of conducting clinical exercise tests, generally around $150 to $500 per test. Finally, far too few clinical facilities are equipped to conduct these tests to accommodate testing everyone who should be involved in an exercise program. Fortunately, the ACSM and the American Heart Association have recommended this exercise test only for the at-risk groups mentioned earlier in this chapter.

From a medical and legal perspective, however, the question must be raised as to whether a recommendation within a set of na-

▶ Before beginning any exercise program, men over age 40, women over age 50, and anyone who is considered to be at a high risk for coronary artery disease should have a complete medical evaluation.

▶ ACSM guidelines should be followed for each phase of the evaluation, and the physician should be consulted about the proposed exercise activity in case there are any medical contraindications.

▶ Exercise ECGs should be conducted for anyone in one of the previously mentioned high-risk categories. This test can detect undiagnosed coronary artery disease and other cardiac abnormalities.

▶ Test sensitivity refers to the test's ability to correctly identify people with a given disease. Test specificity refers to its ability to correctly identify people who do not have the disease. The predictive value of an abnormal exercise test refers to the accuracy with which the test reflects presence of the disease.

▶ The most recent ACSM guidelines state that a medical examination and exercise test might not be necessary if moderate exercise is undertaken gradually in people without symptoms of cardiopulmonary disease.

tional guidelines is tantamount to a standard of practice in the medical community. The most recent guidelines of the ACSM state that a medical examination and exercise test might not be necessary in people without symptoms if moderate exercise is undertaken gradually, with appropriate guidance, and with no competitive participation.[1] Moderate exercise is defined as exercise that is well within the individual's current capacity and can be sustained comfortably for a prolonged period of time, such as 60 min. In contrast, vigorous exercise is defined as exercise at an intensity greater than 60% of the individual's $\dot{V}O_2max$. This seems like a reasonable compromise, because moderate exercise is associated with considerable health benefits and few risks.[4, 22-24] But, as mentioned earlier, exercise testing offers benefits other than diagnosing coronary artery disease: It can provide valuable physiological data, such as a person's blood pressure response to exercise, and much of the data obtained can be used in formulating the exercise prescription.

Exercise Prescription

The **exercise prescription** involves four basic factors:

1. Mode or type of exercise
2. Frequency of participation
3. Duration of each exercise bout
4. Intensity of the exercise bout

In our discussion, we assume that the goal of the exercise program is to improve aerobic capacity in people who have not been exercising. The information contained in this section is not appropriate for designing training programs for elite endurance athletes or for those who simply wish to gain the health-related benefits of moderate activity but do not wish to improve aerobic capacity. The prescription of resistance training programs is discussed in detail in chapter 3.

Before examining the components of the exercise prescription, we must consider how much exercise is effective. A minimum **thresh-**

Parallel Careers, Lifelong Impact!

Since the early 1970s, considerable progress has been made in providing a research base for better understanding the relationship between an active lifestyle and reduced risk for chronic debilitating disease and for identifying how much and what types of activity promote health. During this time, two exercise scientists have had a particularly significant impact on helping us better understand the relationship between physical activity and disease prevention through their research, advocacy, and professional leadership. Interestingly, both had their roots in the Los Angeles area of Southern California and both were graduate students at the same time studying for their PhD degrees in exercise physiology at the University of Illinois. Dr. William L. Haskell (figure 19.4a), received his undergraduate training at the University of California, Santa Barbara, while Dr. Michael L. Pollock (figure 19.4b) was playing baseball and completing his undergraduate degree at the University of Arizona. Both served in the U.S. Army before completing their PhD degrees.

During their professional careers, both were deeply committed to their research and to their primary professional organization, the American College of Sports Medicine, each serving as president. The two of them were instrumental in developing the original and subsequent ACSM Position Stands on the recommended quantity and quality of exercise needed to promote health and prevent chronic disease. These two scientists had parallel careers and a mutual impact on our understanding of the importance of an active lifestyle in promoting health and preventing chronic disease in today's sedentary society.

◀ **Figure 19.4**   (a) Bill Haskell and (b) Michael Pollock.

old for frequency, duration, and intensity of exercise must be reached before any aerobic benefits are obtained. But, as we have discussed elsewhere, individual responses to any given training program are highly variable, so the threshold required differs from one person to the next. If we use exercise intensity as an example, a position statement by the American College of Sports Medicine for developing and maintaining aerobic capacity recommends a training intensity of 55% or 60% to 90% of one's maximum heart rate (HRmax) or 40% or 50% to 85% of $\dot{V}O_2$max.[2] Although this recommendation is appropriate for most healthy adults, some might improve their aerobic capacities

at, for example, intensities below 40% of their $\dot{V}O_2$max, whereas others would have to exercise at intensities greater than 85% $\dot{V}O_2$max to show improvement. Each individual's threshold for frequency, duration, and intensity must be exceeded to achieve gains in aerobic capacity, and this threshold is likely to increase as aerobic capacity improves.

> A minimal threshold for frequency, duration, and intensity of exercise must be reached to gain aerobic benefits from that exercise. Furthermore, minimal thresholds vary widely, making individualized exercise prescription necessary.

Exercise Mode

The prescribed exercise program should focus on one or more **modes,** or types, of cardiovascular endurance activities. Traditionally, the activities prescribed most frequently are

- walking,
- jogging,
- running,
- hiking,
- cycling,
- rowing, and
- swimming.

Because these activities do not appeal to everyone, alternative activities (see figure 19.5) have been identified that should promote similar cardiovascular endurance gains. Spinning, aerobic dance, box or bench stepping, and most racket sports also have been shown to improve aerobic capacity.

▲ **Figure 19.5**　Many activities promote gains in cardiovascular endurance, including spinning.

For most competitive sport activities, preconditioning with one of the standard endurance activities, such as jogging or cycling, is advisable before undertaking serious competition. Some researchers, clinicians, and practitioners believe that for you to successfully compete in certain sports or activities, a basic preconditioning program is essential to bring you up to the level of conditioning needed for the sport or activity and to reduce your risk of injury. Rather than using the sport or activity to get in shape, you get in shape—or precondition—before participating in that sport or activity. For example, if the desired activity requires a moderate to high level of cardiovascular endurance, such as basketball, you might engage in jogging, aerobic dance, or a cycling program for several months until your endurance capacity increases to the necessary level. At that time, you switch over to the sport. The sport then acts as a maintenance activity through which you maintain your desired fitness level. In some cases, with intense sports you can continue to develop your aerobic fitness level.

> Sport and recreational activities are appropriate for maintaining desirable fitness levels, but they generally are not appropriate for developing fitness in unfit individuals. Relatively unfit individuals should use conditioning activities to reach the desired level of fitness and then switch to the sport or recreational activity.

Individuals should select activities that they enjoy and are willing to continue throughout life. Exercise must be regarded as a lifetime pursuit because, as we saw in chapter 12, the benefits are soon lost if participation stops. Motivation is probably the most important factor in a successful exercise program. Selecting an activity that is fun, provides a challenge, and can produce needed benefits is one of the most crucial tasks in exercise prescription. Other considerations include geographic location, climate, and availability of equipment and facilities. Home exercise has become more common as many people are homebound either because of responsibilities, such as child rearing, or because of weather considerations, such as heat, humidity, cold, rain, ice, and snow. Exercise videos and home exercise equipment have become popular, but they need to be selected carefully to avoid inappropriate exercise or faulty equipment. Potential purchasers should seek professional advice and, when possible, use the video or equipment during a trial period before purchase.

Exercise Frequency

The frequency of exercise participation, although certainly an important factor to consider, is probably less critical than either exercise duration or intensity. Research studies conducted on exercise frequency show that 3 to 5 days per week is an optimal frequency. This does not mean that 6 or 7 days per week won't give additional benefits, but simply for the health-related benefits, the optimal gain is achieved with a time investment of 3 to 5 days per week. Exercise initially should be limited to 3 or 4 days per week and increased up to 5 or more days per week only if the activity is enjoyed and physically tolerated. All too often, a person starts out with great intentions, is highly motivated, and exercises every day for the first few weeks, only to stop from utter fatigue or injury. Obviously, additional days above the 3- to 4-day frequency are beneficial for weight loss, but this level should not be encouraged until the exercise habit is firmly established and the injury risk is reduced.

Exercise Duration

Several studies have demonstrated improvement in cardiovascular conditioning with endurance exercise periods as brief as 5 to 10 min per day. More recent research has indicated that 20 to 30 min per day is an optimal amount. Again, optimal is used here to reflect the greatest return for time invested, and the specified time refers to the time during which you are at your appropriate exercise intensity. Exercise duration cannot be discussed appropriately without also discussing exercise intensity. Similar improvements in aerobic capacity are gained with either a short-duration, high-intensity program or a long-duration, low-intensity

program if the minimal threshold is exceeded for both duration and intensity. Similar benefits are also gained whether the daily endurance training session is conducted in multiple shorter bouts (e.g., three 10-min bouts) or a single long one (e.g., a single 30-min bout).

Exercise Intensity

The intensity of the exercise bout appears to be the most important factor. How hard must you push yourself to gain benefits? Former athletes immediately recall the exhaustive workouts they endured to condition themselves for their sport. Unfortunately, this concept also gets carried over into the exercise programs they pursue for health benefits. Evidence now suggests that a substantial training effect can be accomplished in some people by training at intensities of 45% or less of their aerobic capacities. For most, however, the appropriate intensity appears to be at least 50% to 60% $\dot{V}O_2$max.

As mentioned earlier, evidence now suggests that low to moderate levels of activity, below those needed to increase $\dot{V}O_2$max, produce substantial health benefits. Dr. Ronald LaPorte, an epidemiologist from the University of Pittsburgh, was one of the first to observe that individuals in most studies reporting health-related benefits from regular physical activity were involved in relatively low-intensity exercise.[22] In fact, these benefits occurred at intensities substantially below the levels currently recommended for improving aerobic capacity. Lower intensities appear to have considerable health benefits without changing aerobic capacity. This important point must be researched more fully in the future. Less than 10% of adult Americans are meeting the guidelines mentioned earlier that are proposed for improving aerobic capacity.[10] Promoting lower intensity exercise might greatly increase our sedentary society's participation in activities that will provide health benefits, which should then reduce our nation's healthcare costs.

A hypothetical example in figure 19.6 illustrates the degree of risk for a disease by activity level, from sedentary to highly active. You can see that just moving from the sedentary

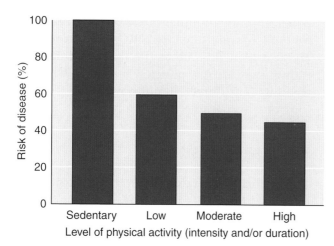

▲ **Figure 19.6** Hypothetical model showing how the risk for a given chronic degenerative disease, such as coronary artery disease, hypertension, or diabetes, is reduced by increasing levels of physical activity. Activity levels are determined on the basis of intensity, duration, or both.

category to the low-level activity category greatly reduces your risk for a given disease. Increasing activity even more will further lower your risk, but the greatest benefits come from moving out of the sedentary category. This was the basis for the CDCP/ACSM statement (see page 608): "Significant health benefits can be obtained by including a moderate amount of physical activity (e.g., 30 minutes of brisk walking or raking leaves, 15 minutes of running, or 45 minutes of playing volleyball) on most, if not all days of the week."

Monitoring Exercise Intensity

Exercise intensity can be quantified on the basis of the training heart rate (THR), the metabolic equivalent (MET), or the rating of perceived exertion (RPE). Let's examine each of these and their strengths and weaknesses in quantifying exercise intensity.

Training Heart Rate

The concept of **training heart rate (THR)** is based on the linear relationship between heart rate and $\dot{V}O_2$ with increasing rates of

▶ The four basic factors in an exercise program are exercise mode, frequency, duration, and intensity. A minimum threshold for the last three must be met to attain any aerobic benefits, and this threshold is quite variable from one individual to another.

▶ The program should include one or more cardiovascular endurance activities. If the activity involves competition, preconditioning with a standard endurance activity is recommended before sport participation begins to bring you up to an appropriate level of fitness.

▶ Activities must be matched with individual needs and likes so that motivation can be maintained.

▶ Optimal exercise frequency is 3 to 5 days of training each week, although greater frequency might provide additional benefits. Exercise should begin with three to four sessions per week and then progress to more if desired.

▶ Exercise duration of 20 to 30 min working at the appropriate intensity is optimal, but the key is reaching the threshold for both duration and intensity.

▶ Exercise intensity appears to be the most important factor. For most people, intensity should be at least 50% to 60% $\dot{V}O_2$max. However, health benefits occur at intensities lower than those needed for aerobic conditioning.

work, as shown in figure 19.7. When you are exercise-tested, your heart rate and $\dot{V}O_2$ values are obtained each minute and plotted against each other. The THR is established by using the heart rate that is equivalent to a set percentage of your $\dot{V}O_2$max. For example, if a training level of 75% $\dot{V}O_2$max is desired, 75% of $\dot{V}O_2$max is calculated ($\dot{V}O_2$max × 0.75), and the heart rate corresponding to this $\dot{V}O_2$ then is selected as the THR. An important point is that the exercise intensity necessary to achieve a given percentage of $\dot{V}O_2$max results in a much higher heart rate than that

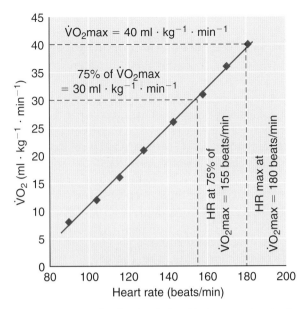

▲ **Figure 19.7** The linear relationship between heart rate and oxygen consumption ($\dot{V}O_2$) with increasing rates of work and the heart rate equivalent to a set percentage (75%) of $\dot{V}O_2$max.

same percentage of HRmax. As an example, a THR that is set at 75% of the $\dot{V}O_2$max represents an intensity of 86% of the HRmax (see figure 19.7).

The Karvonen Method

The training heart rate also can be established by using what is known as the Karvonen concept of maximal heart rate reserve, or the **Karvonen method**.[19] **Maximal heart rate reserve** is defined as the difference between HRmax and the resting heart rate (HRrest):

Maximal heart rate reserve = HRmax − HRrest

With this method, the training heart rate is calculated by taking a given percentage of the maximal heart rate reserve and adding it to the resting heart rate. Let's consider an example. For 75% of maximal heart rate reserve, the equation would be as follows:

$$THR_{75\%} = HRrest + 0.75(HRmax - HRrest)$$

The Karvonen method adjusts the THR so that THR as a specific percentage of the maximal heart rate reserve is nearly identical to the HR equivalent of that same percentage of $\dot{V}O_2$max at moderate to high intensities.[12]

Thus, a THR computed as 75% of maximal heart rate reserve is approximately the same as the heart rate corresponding to 75% of the $\dot{V}O_2$max. However, there is a substantial difference between the two at low intensities.[37]

Training Heart Rate Range

More recently, appropriate exercise intensity has been established by setting a THR range, rather than a single THR value. This is a more sensible approach because exercising at a set percentage of $\dot{V}O_2$max can place you above your lactate threshold, making it difficult for you to train for any extended period. With the THR range concept, low and high values are established that will ensure a training response. You start at the low end of the THR range and progress through the range as you feel comfortable. To illustrate this, using the Karvonen method for establishing the THR, consider the following example. A 40-year-old man has a resting heart rate of 75 beats/min and a maximum heart rate of 180 beats/min, and he is advised to exercise within a THR range of 50% to 75% of his maximal heart rate reserve. His training heart rate range would be as follows:

$$THR_{50\%} = 75 + 0.50 (180 - 75)$$
$$= 75 + 53 = 128 \text{ beats/min}$$

$$THR_{75\%} = 75 + 0.75 (180 - 75)$$
$$= 75 + 79 = 154 \text{ beats/min}$$

This same THR range method can be used by estimating HRmax [208 − (0.7 × age)] without losing much accuracy.

The concept of training heart rate is extremely valuable. Heart rate is highly correlated with the work done by the heart. Heart rate alone is a good index of myocardial oxygen consumption as well as coronary blood flow. By using the training heart rate method for monitoring your exercise intensity, your heart works at the same rate, even though the metabolic cost of the work might vary considerably. As an example, when you exercise at high altitudes or in the heat, your heart rate will be elevated significantly if you attempt to maintain a set rate of work, such as running at a pace of 6 min/km (9 min/mi). With the THR method, you simply train at a lower rate of work under these extreme environmental conditions

Prescribing Exercise Intensity Using the $\dot{V}O_2$ Reserve Method

In the most recent ACSM Position Stand on exercise prescription,[2] a slightly different approach to prescribing exercise intensity was proposed. Exercise intensity is prescribed based on what has been termed the $\dot{V}O_2$ reserve method ($\dot{V}O_2R$). Instead of prescribing exercise at a given percentage of $\dot{V}O_2$max, the prescription is based on a given percentage of the $\dot{V}O_2R$, where $\dot{V}O_2R$ is defined as $\dot{V}O_2$max − $\dot{V}O_2$rest.[36] This also can be thought of as the $\dot{V}O_2$max reserve. As an example, with a $\dot{V}O_2$max of 40 ml · kg⁻¹ · min⁻¹ and a $\dot{V}O_2$rest of 3.5 ml · kg⁻¹ · min⁻¹, $\dot{V}O_2R$ = 40 − 3.5 ml · kg⁻¹ · min⁻¹ = 36.5 ml · kg⁻¹ · min⁻¹. To prescribe an exercise intensity range of between 60% and 75% of $\dot{V}O_2R$, you simply multiply $\dot{V}O_2R$ by 60% and 75%: $\dot{V}O_2R_{60\%}$ = 36.5 ml · kg⁻¹ · min⁻¹ × 0.60 = 21.9 ml · kg⁻¹ · min⁻¹; and $\dot{V}O_2R_{75\%}$ = 36.5 ml · kg⁻¹ · min⁻¹ × 0.75 = 27.4 ml · kg⁻¹ · min⁻¹. The major advantage of using the $\dot{V}O_2R$ technique is that you now have an equivalency between the percentage of the maximal heart rate reserve and the percentage of $\dot{V}O_2$max reserve.

to maintain the same heart rate (THR). This is a much safer approach to monitoring exercise intensity, particularly for high-risk patients in whom the work of the heart must be closely regulated. The THR method also allows for improvement in aerobic capacity with training. As you become better conditioned, your heart rate decreases for the same rate of work, which means that you must perform at a higher rate of work to reach your training heart rate.

It is important to come back to a point that was raised in the first paragraph of this section—the point about lactate threshold. In chapter 4 we learned that lactate threshold was that point as you increase exercise intensity where the rate of production of lactate exceeds the rate of its clearance, resulting in increased blood lactate levels. When you are at an exercise intensity above your lactate threshold, you limit the length of time that you can train comfortably at that intensity. For those just starting a training program, it is important not to exceed the lactate threshold. Having a training heart rate range allows you to set the lower end of the range at an intensity that would be below the expected lactate threshold for an untrained individual. Obviously it would be better to actually measure the lactate threshold so that this range could be more accurately determined. However, this isn't practical because of the difficulty and expense associated with determining lactate threshold.

Heart rate is the preferred method for monitoring exercise intensity because it is highly correlated to the work of the heart (or stress on the heart) and it allows for a progressive increase in the rate of training with improvements in fitness to maintain the same training heart rate. When one is prescribing exercise intensity, it is appropriate to establish a training heart rate range, with exercise starting at the low end of the range and progressing to the upper end of the range over time.

Metabolic Equivalent

Exercise intensity also has been prescribed on the basis of the **metabolic equivalent (MET)** system. The amount of oxygen your body consumes is directly proportional to the energy you expend during physical activity. At rest, your body uses approximately 3.5 ml of oxygen per kilogram (2.2 lb) of body weight per minute (3.5 ml · kg⁻¹ · min⁻¹). This resting metabolic rate is referred to as 1.0 MET.

All activities can be classified by intensity according to their oxygen requirements. An activity that is rated as a 2.0-MET activity would require two times the resting metabolic rate, or 7 ml · kg⁻¹ · min⁻¹, and an activity that is rated at 4.0 METs would require approximately 14 ml · kg⁻¹ · min⁻¹. Some activities and their MET values are presented in table 19.5. These values are only approximations because metabolic

Table 19.5

Selected Activities and Their Respective MET Values

Activity	MET	Activity	MET
Self-care		Beating carpets	4.0
Rest, supine	1.0	*Occupational*	
Sitting	1.0	Sitting at desk	1.5
Standing, relaxed	1.0	Writing	1.5
Eating	1.0	Riding in automobile	1.5
Conversation	1.0	Watch repair	1.5
Dressing and undressing	2.0	Typing	2.0
Washing hands and face	2.0	Welding	2.5
Propulsion, wheelchair	2.0	Radio assembly	2.5
Walking, 4 km/h (2.5 mph)	3.0	Playing musical instrument	2.5
Showering	3.5	Parts assembly	3.0
Walking downstairs	4.5	Bricklaying and plastering	3.5
Walking, 5.6 km/h (3.5 mph)	5.5	Heavy assemble work	4.0
Ambulation, braces and crutches	6.5	Wheeling wheelbarrow, 52 kg, 4 km/h (115 lb, 2/5 mph)	4.0
Housework		Carpentry	5.5
Hand sewing	1.0	Mowing lawn by hand mower	6.5
Machine sewing	1.5	Chopping wood	6.5
Sweeping floor	1.5	Shoveling	7.0
Polishing furniture	2.0	Digging	7.5
Peeling potatoes	2.5	*Physical conditioning*	
Scrubbing, standing	2.5	Level walking, 3.2 km/h, 1 km in 19 min (2 mph, 1 mi in 30 min)	2.5
Washing small clothes	2.5	Level cycling, 8.9 km/h, 1 km in 6 min 44 s (5.5 mph, 1 mi in 10 min 54 s)	3.0
Kneading dough	2.5		
Scrubbing floors	3.0		
Cleaning windows	3.0	Level cycling, 9.7 km/h, 1 km in 6 min 12 s (6 mph, 1 mi in 10 min)	3.5
Making beds	3.0		
Ironing, standing	3.5	Level walking, 4 km/h, 1 km in 15 min (2.5 mph, 1 mi in 24 min)	3.5
Mopping	3.5		
Wringing wash by hand	3.5	Level walking, 4.8 km/h, 1 km in 12 min 30 s (3 mph, 1 mi in 20 min)	4.5
Hanging wash	3.5	Calisthenics	4.5

Table 19.5 (continued)

Selected Activities and Their Respective MET Values

Activity	MET	Activity	MET
Level cycling, 15.6 km/h, 1 km in 3 min 50 s (9.7 mph, 1 mi in 6 min 18 s)	5.0	Playing piano	2.0
		Driving car	2.0
Swimming, crawl, 0.3 m/s (1 ft/s)	5.0	Canoeing, 4 km/h (2.5 mph)	2.5
Level walking, 5.6 km/h, 1 km in 10 min 43 s (3.5 mph, 1 mi in 17 min)	5.5	Horseback riding, walk	2.5
		Volleyball, 6-player recreational	3.0
Level walking, 6.4 km/h, 1 km in 9 min 23 s (4.0 mph, 1 mi in 15 min)	6.5	Billiards	3.0
		Bowling	3.5
Level jogging, 8 km/h, 1 km in 7 min 30 s (5.0 mph, 1 mi in 12 min)	7.5	Horseshoes	3.5
		Golf	4.0
Level cycling, 20.9 km/h, 1 km in 2 min 52 s (13 mph, 1 mi in 4 min 37 s)	9.0	Cricket	4.0
		Archery	4.5
Level running, 12 km/h, 1 km in 5 min (7.5 mph, 1 mi in 8 min)	9.0	Ballroom dancing	4.5
		Table tennis	4.5
Swimming, crawl 0.6 m/s (2 ft/s)	10.0	Baseball	4.5
Level running, 13.7 km/h, 1 km in 4 min 23 s (8.5 mph, 1 mi in 7 min)	12.0	Tennis	6.0
Level running, 16 km/h, 1 km in 3 min 45 s (10.0 mph, 1 mi in 6 min)	15.0	Horseback riding, trot	6.5
		Folk dancing	6.5
Swimming, crawl, 0.8 m/s (2.5 ft/s)	15.0	Skiing	8.0
Swimming, crawl, 0.9 m/s (3.0 ft/s)	20.0	Horseback riding, gallop	8.0
Level running, 19.3 km/h, 1 km in 3 min 6 s (12 mph, 1 mi in 5 min)	20.0	Squash racquets	8.5
Level running, 24.1 km/h, 1 km in 2 min 30 s (15 mph, 1 mi in 4 min)	30.0	Fencing	9.0
		Basketball	9.0
Swimming, crawl, 1.1 m/s (3.5 ft/s)	30.0	Football	9.0
Recreational		Gymnastics	10.0
Painting, sitting	1.5	Handball and paddleball	10.0

efficiency varies considerably from one person to the next, and even in the same individual. Although the MET system is useful as a guideline for training, it fails to account for changes in environmental conditions, and it does not allow for changes in physical conditioning as discussed in the previous section.

Ratings of Perceived Exertion

Ratings of perceived exertion (RPE) also have been proposed for use in prescribing exercise intensity. With this method, individuals subjectively rate how hard they feel that they are working. A given numerical rating corresponds to the perceived relative intensity of exercise. When the RPE scale is used correctly, this system for monitoring exercise intensity has proven very accurate.[8] Using the **Borg RPE scale**[9], which is a rating scale ranging from 6

▶ Exercise intensity can be monitored on the basis of training heart rate, metabolic equivalent, or rating of perceived exertion.

▶ Training heart rate (THR) can be established by using the heart rate equivalent to a certain percentage of $\dot{V}O_2$max. It can also be determined using the Karvonen method, which takes a given percentage of maximal heart rate reserve and adds it to resting heart rate. With this method, the percentage of maximal heart rate reserve used corresponds to approximately the same percentage of $\dot{V}O_2$max when a person is exercising at moderate to high intensities.

▶ A sensible approach is to establish a THR range to work within, instead of a single THR, attempting to estimate the low end at an intensity below lactate threshold.

▶ The amount of oxygen consumed reflects the amount of energy expended during an activity. $\dot{V}O_2$ at rest is about 3.5 ml · kg^{-1} · min^{-1}, which equals 1.0 MET. Activity intensities can be classified by their oxygen requirements as multiples of the resting metabolic rate.

▶ The rating of perceived exertion method requires that a person subjectively rate how difficult the work is, using a numerical scale that is related to exercise intensity. The subject looks at the standard scale to determine the appropriate number.

Table 19.6

The Borg Ratings of Perceived Exertion Scale

Rating	Intensity
6	No exertion at all
7	Extremely light
8	
9	Very light
10	
11	Light
12	
13	Somewhat hard
14	
15	Hard (heavy)
16	
17	Very hard
18	
19	Extremely hard
20	Maximal exertion

Borg RPE scale
© Gunar Borg 1970, 1985, 1984, 1998

From G. Borg, 1998, *Borg's perceived exertion and pain scales* (Champaign, IL: Human Kinetics), 47. By permission of G. Borg.

to 20 (table 19.6), your exercise intensity should be between an RPE of 12 to 13 (somewhat hard) and an RPE of 15 to 16 (hard). Initially, this sounds too simple. However, most people can use the RPE technique very accurately. Studies have shown that when you ask people to select a pace on a treadmill, or a resistance on a cycle ergometer, at a moderate or heavy intensity of exercise (see table 19.7), they are able to select a pace or resistance that gets their heart rates into the appropriate range. This is a more natural way to prescribe exercise and very efficient if the person is able to relate perceptions of intensity accurately.

Table 19.7

Classification of Exercise Intensity Based on 20 to 60 min of Endurance Activity Comparing Three Methods

Relative intensity			
HRmax	$\dot{V}O_2$ max or HRmax reserve	Rating of perceived exertion	Classification of intensity
<35%	<30%	<9	Very light
35-59%	30-49%	10-11	Light
60-79%	50-74%	12-13	Moderate
80-89%	75-84%	14-16	Heavy
≥90%	≥85%	>16	Very heavy

Adapted from Pollock and Wilmore (1990).[32]

Table 19.7 compares the various methods for rating exercise intensity. Let's use them to determine a moderate exercise intensity. As the first column shows, you would want to work within a range of 60% to 79% HRmax. If, instead, you are monitoring intensity by $\dot{V}O_2$max or using the Karvonen method, this HR range is equivalent to 50% to 74% of either $\dot{V}O_2$max or HRmax reserve, as shown in the second column. If you use the rating of perceived exertion, shown in the third column, this is equivalent to an RPE value of 12 to 13. All these values reflect moderate-intensity exercise.

> Physical activity must be considered a lifetime pursuit! The benefits of a sound exercise program are rapidly lost once that program is discontinued.

Exercise Program

Once the exercise prescription has been determined, it is integrated into a total exercise program, which is generally only part of an overall health improvement plan. Individual exercise capacity varies widely even among people of similar ages and physical builds. For this reason, each program must be indi-vidualized, based on results of physiological and medical tests and, if possible, individual needs and interests.

The total exercise program consists of the following activities:

- Warm-up and stretching activities
- Endurance training
- Cool-down and stretching activities
- Flexibility training
- Resistance training
- Recreational activities

Generally, the first three activities are performed three to four times each week. Flexibility training can be included as part of the warm-up, cool-down, and stretching exercises, or it can be done at a separate time during the week. Resistance training is usually done on alternate days when endurance training is not done; however, the two can be combined into the same workout. Now, let's examine each of these activities.

Warm-Up and Stretching Activities

The exercise session should begin with low-intensity, calisthenic-type and stretching exercises (figure 19.8). Such a warm-up period increases both heart rate and breathing, preparing you for the efficient and safe functioning of

▲ **Figure 19.8**   The warm-up period should include stretching and low-intensity activity, such as walking or light jogging.

your heart, blood vessels, lungs, and muscles during the more vigorous exercise that follows. A good warm-up also reduces the amount of muscle and joint soreness that you experience during the early stages of the exercise program and can decrease your risk of injury. An acceptable warm-up would begin with 5 to 10 min of stretching, followed by 5 to 10 min of low-intensity activity using the mode of exercise selected for endurance training. For example, if you train by running, you might start with stretching and follow it with 5 to 10 min of light jogging.

Endurance Training

Physical activities that develop cardiovascular endurance are the heart of the exercise program. They are designed to improve both the capacity and efficiency of your cardiovascular, respiratory, and metabolic systems. These activities also help you control or reduce your body weight. Activities such as walking, jogging, running, cycling, swimming, rowing, aerobic

dancing, box stepping, and hiking are good endurance activities. Sports such as handball, racquetball, tennis, badminton, and basketball also have aerobic potential if they are pursued vigorously. Activities such as golf, bowling, and softball are generally of little value for developing aerobic capacity, but they are fun, have definite recreational value, and may offer health-related benefits. For these reasons, such activities certainly have a place in the overall exercise program.

Cool-Down and Stretching Activities

Every endurance exercise session should conclude with a cool-down period. Cool-down is best accomplished by slowly reducing the intensity of the endurance activity during the last several minutes of your workout. After running, for example, a slow, restful walk for several minutes helps prevent blood from pooling in your extremities. Stopping abruptly after an endurance exercise bout causes blood to pool in your legs and can result in dizziness or fainting. Also, catecholamine levels might be elevated during the immediate recovery period, and this can lead to a fatal heart arrhythmia. After the cool-down period, stretching exercises can be performed to facilitate increased flexibility.

Flexibility Training

Flexibility exercises (see figure 19.9) usually supplement exercises performed during the warm-up or cool-down period and are useful for those who have poor flexibility or muscle and joint problems, such as low back pain. These exercises should be performed slowly. Quick stretching movements are potentially dangerous and can lead to muscle pulls or spasms. At one time it was recommended that these exercises be performed before the endurance conditioning period. However, recently it has been hypothesized that the muscles, tendons, ligaments, and joints are more adaptable and responsive to flexibility exercises when they are done after the endurance conditioning phase. Research has yet to confirm this hypothesis.

Resistance Training

Increasing interest surrounds the use of resistance training as part of a general health and fitness exercise program. Indeed, many health-related benefits can be obtained from resistance training. The American College of Sports Medicine has included resistance training in its recommendations for a general health and fitness program.[2]

Starting a Resistance Training Program

Recall from chapter 3 that the maximum amount of weight you can lift successfully only one time is your 1-repetition maximum, or 1RM. When you begin a resistance training program, you should start with a weight that is exactly one half of your maximal strength, or 1RM, for each lift. You should attempt to lift that weight 10 consecutive times. If you can lift the weight just 10 times before you reach fatigue, this is the proper starting point. If you can do more repetitions, you should go to the next higher weight for your second set. If, instead, you were able to lift the weight fewer than eight times in your first set, the original

▲ **Figure 19.9**   Flexibility exercises should be performed slowly to prevent injury.

weight was too heavy and should be reduced to the next lower weight for your second set.

When a given weight brings you to fatigue by the eighth to tenth repetition in your first set, this is your appropriate starting weight. You should try to achieve as many repetitions as possible during the second and third sets, but the number of repetitions you can complete in these last sets will probably decrease as your muscles become fatigued. You can perform two or three sets of each lift per day, 2 to 3 days per week, for weight-control purposes. Strength gains, however, appear to be fully achieved with just one set per day in previously untrained people.

As your strength increases, the number of repetitions you can complete per set will increase. When you reach 15 repetitions on the first set, you are ready to progress to the next higher weight. This training technique is referred to as progressive resistance exercise. See chapter 3 for further details.

Health Benefits Associated With Resistance Training

Interest has surged in the use of resistance training for promoting improvements in general health. Unfortunately, research findings are not always consistent, and this is likely attributable to differences in the training programs used in different studies. The volume and intensity of the training, the rest interval between sets and exercises, and the selection of exercises all can greatly influence the results of a given study. With this in mind, let's briefly summarize some of the key findings from an extensive review by Stone and coworkers of the health-related benefits of resistance training.[35]

Resistance training can affect cardiorespiratory fitness, specifically the risk factors associated with cardiovascular disease. With resistance training, the heart rate at submaximal rates of exercise generally is reduced, which typically reflects improved cardiorespiratory fitness. However, the resting heart rate has not been shown to be reduced consistently across studies. The heart can be enlarged by resistance training, likely because of increases in the thickness (hypertrophy) of the interven-

tricular septum and the left ventricular wall. As we saw in earlier chapters, this can increase the contractility of the left ventricle and enhance stroke volume. These changes in the myocardium of the left ventricle are theorized to be adaptations to the heart contracting against an increased systemic arterial pressure during resistance training.

Some evidence indicates that resistance training can reduce the resting blood pressure of people with hypertension or borderline hypertension.[20] Studies also have supported the use of resistance training to promote favorable alterations in blood lipid profiles, although the results have not always been consistent from one study to another. When blood lipid changes have been found, they generally reflect a decrease in the ratio either of total cholesterol to high-density–lipoprotein cholesterol (HDL-C) or of low-density–lipoprotein cholesterol (LDL-C) to HDL-C. Resistance training also increases insulin sensitivity and improves glucose tolerance, both of which are important factors in preventing diabetes.[34] And diabetes is not only a disease itself but also a risk factor for cardiovascular disease.

Resistance training reduces the risk of obesity. Good evidence indicates that a program of resistance training increases the participant's fat-free mass and decreases the fat mass. Some scientists speculate that this increased fat-free mass will increase the person's resting metabolic rate because muscle is more metabolically active than fat. This would increase daily caloric expenditure.

Finally, the role of resistance training in preventing osteoporosis is currently under investigation, and preliminary results appear promising. Studies of resistance training by older women suggest that the bone loss associated with menopause can be attenuated, or even reversed, with a resistance training program.

The potential health benefits of resistance training compared with those of aerobic training are summarized in table 19.8 on page 629.[31] This table illustrates that both aerobic and resistance training exercise can provide tremendous health benefits, independently and in combination.

▲ **Figure 19.10**   The recreational activities in which you can participate regularly may depend on your location, but many communities offer a wide variety of programs.

Recreational Activities

Recreational activities (see figure 19.10) are important to any comprehensive exercise program. Although people engage in these activities primarily for enjoyment and relaxation, many recreational activities can also improve health and fitness. Activities such as hiking, tennis, handball, squash, and certain team sports fall into this category. Guidelines for selecting these activities include the following:

- Can you learn or perform the activities with at least a moderate degree of success?

- Do the activities include opportunities for social development, if that is desired?

- Are the costs associated with participation reasonable and within your budget?

- Are the activities varied enough to maintain your continued long-term interest?

Many excellent opportunities exist for people who have no recreational hobbies or activities but who would like to become involved. Local public recreation centers, park districts, YMCAs, YWCAs, and some public schools, community colleges, and universities offer instructional classes in a wide variety of activities at little or no cost. Often the entire family can participate in these classes—an added bonus to a total health-improvement program! Also, the number of commercial fitness centers is rapidly growing, and many now employ trained staff who can properly prescribe exercise programs and help individuals get started.

It is important to emphasize the whole exercise program and not focus on just one or two parts. This will ensure that total-body fitness needs are met.

▶ An exercise session should begin with a warm-up of low-intensity, calisthenic-type and stretching exercises to prepare the cardiovascular, respiratory, and muscle systems to work more efficiently.

▶ Endurance activities should be performed three to four times each week.

▶ Each endurance session should be followed by cool-down and stretching to prevent blood pooling in the extremities and muscle soreness.

▶ Flexibility exercise should be performed slowly, and this phase of the program might be best included immediately after the endurance component.

▶ Resistance training should begin with a weight of one half your 1RM. This is the proper weight if you can lift it about 10 times. If you can lift it more than that, you need more weight, and if you can lift it fewer than eight times, you need less.

▶ Recreational activities should be included in your exercise program for enjoyment and relaxation.

Table 19.8

Comparison of the Effects of Aerobic Endurance Training to Strength Training on Health and Fitness Variables

Variable	Aerobic exercise	Resistance exercise
Bone mineral density	↑↑	↑↑
Body composition		
% fat	↓↓	↓
Lean body mass	↔	↑↑
Strength	↔	↑↑↑
Glucose metabolism		
Insulin response to glucose challenge	↓↓	↓↓
Basal insulin levels	↓	↓
Insulin sensitivity	↑↑	↑↑
Serum lipids		
HDL-C	↑↑	↑↔
LDL-C	↓↓	↓↔
Resting heart rate	↓↓	↔
Stroke volume	↑↑	↔
Blood pressure at rest		
Systolic	↓↓	↔
Diastolic	↓↓	↓↔
$\dot{V}O_2max$	↑↑↑	↑
Endurance performance	↑↑↑	↑↑
Physical function	↑↑	↑↑↑
Basal metabolism	↑	↑↑

Note. HDL-C = high-density–lipoprotein cholesterol; LDL-C = low-density–lipoprotein cholesterol.
↑ = increase; ↓ = decrease; ↔ = little or no change; the more arrows, the greater the change

Reprinted from Pollock, M.L., and Vincent, K.R. (1996).[31]

Exercise and Rehabilitation of People With Diseases

Exercise has become a major component in **rehabilitation programs** for a number of diseases. Cardiopulmonary rehabilitation programs, which began in the 1950s, have become the most visible. Tremendous advances in cardiopulmonary rehabilitation have led to the formation of a professional association, the American Association of Cardiovascular and Pulmonary Rehabilitation, and a professional research journal, the *Journal of Cardiopulmonary Rehabilitation.*

Exercise is also an important part of the rehabilitation of people with

- cancer,
- obesity,
- diabetes,
- renal disease,
- arthritis, and
- cystic fibrosis.

Most recently, emphasis on the use of exercise in the rehabilitation of transplant patients, including those with heart transplants, liver transplants, and kidney transplants, has increased because exercise helps alleviate some drug side effects and improves general health.

> Exercise training has become an extremely important part of rehabilitation programs for a number of diseases. Although the specific physiological mechanisms explaining the benefits of exercise training for each of these diseases have not been clearly defined, exercise training carries many general health benefits that appear to improve the patient's prognosis.

The manner in which exercise is used in the rehabilitation of people with disease is highly specific to the nature and extent of the disease. It is therefore beyond the scope of this chapter to go into specific details for any disease, but many of the references and selected readings listed at the end of this chapter provide more details about establishing exercise programs for those with specific diseases and the clinical values of these programs.[5, 15, 25, 30, 32, 33]

> ▶ Exercise is a vital part of cardiopulmonary rehabilitation and is also essential in rehabilitating patients with such diseases as cancer, obesity, diabetes, renal disease, arthritis, and cystic fibrosis.
>
> ▶ The type and details of the rehabilitation program depend on the patient, the specific disease involved, and its extent.

In Closing . . .

In this chapter, we have seen that the medical community now regards a physically active lifestyle as vital to maintaining good health and reducing the risk of disease. We looked at the importance and practicality of both a medical examination and an exercise electrocardiogram in screening previously sedentary adults before prescribing exercise. We discussed the components of an exercise prescription and methods of monitoring exercise intensity. Finally, we reviewed the components of an exercise program and the role of exercise in rehabilitating patients with disease.

Now that we have seen the importance of exercise in disease prevention, we will look more closely at physical activity as it relates to specific disease states. In the next chapter, we turn our attention to cardiovascular diseases.

▶ Key Terms

Borg RPE scale
exercise electrocardiogram (ECG)
exercise prescription
graded exercise test (GXT)
Karvonen method
maximal heart rate reserve
metabolic equivalent (MET)
mode
predictive value of an abnormal exercise

test
rating of perceived exertion (RPE)
rehabilitation programs
sensitivity
specificity
threshold
training heart rate (THR)

▷ Study Questions

1. How active are adult Americans today? Are we in a fitness revolution?

2. What role does the graded exercise test to exhaustion play in the medical clearance? Is this test essential for all adults?

3. Discuss the concepts of sensitivity and specificity of exercise testing and the predictive value of an abnormal test. Of what value is this information in the establishment of policy mandating who should be exercise tested?

4. How can we get our population to be more active? What levels of exercise do we need to promote to help people gain the health-related benefits associated with exercise?

5. What four factors must be considered in the exercise prescription? Which of these is the most important?

6. Discuss the concept of a minimal threshold for initiating physiological changes with exercise training as it relates to the exercise prescription.

7. Discuss the various ways of monitoring exercise intensity, and give the advantages and disadvantages of each.

8. Describe the components of a good exercise program and their importance in the total program.

9. How do you effectively motivate individuals to maintain regular exercise habits?

▷ References

1. American College of Sports Medicine. (2000). *Guidelines for exercise testing and prescription* (6th ed.). Philadelphia, PA: Lippincott, Williams & Wilkins.

2. American College of Sports Medicine. (1998). The recommended quantity and quality of exercise for developing and maintaining cardiorespiratory and muscular fitness, and flexibility in healthy adults. *Medicine and Science in Sports and Exercise, 30,* 975-991.

3. American Heart Association. (1972). *Exercise testing and training of apparently healthy individuals: A handbook for physicians.* New York: American Heart Association.

4. Blair, S.N., Kohl, H.W., Paffenbarger, R.S., Clark, D.G., Cooper, K.H., & Gibbons, L.W. (1989). Physical fitness and all-cause mortality: A prospective study of healthy men and women. *Journal of the American Medical Association, 262,* 2395-2401.

5. Bloomfield, J., Fricker, P.A., & Fitch, K.D. (1992). *Textbook of science and medicine in sport.* Boston: Blackwell Scientific.

6. Booth, F.W., Chakravarthy, M.V., Gordon, S.E., & Spangenburg, E.E. (2002).Waging war on physical inactivity: Using modern molecular ammunition against an ancient enemy. *Journal of Applied Physiology, 93,* 3-30.

7. Booth, F.W., Gordon, S.E., Carlson, C.J., & Hamilton, M.T. (2000). Waging war on modern chronic disease: Primary prevention through exercise biology. *Journal of Applied Physiology, 88,* 774-787.

8. Borg, G.A.V. (1982). Psychophysical bases of perceived exertion. *Medicine and Science in Sports and Exercise, 14,* 377-381.

9. Borg, G.A.V. (1998). *Borg's perceived exertion and pain scales.* Champaign, IL: Human Kinetics.

10. Caspersen, C.J. (1987). Physical activity and coronary heart disease. *Physician and Sportsmedicine, 15*(11), 43-44.

11. Cooper, K.H. (1968). *Aerobics.* New York: Evans.

12. Davis, J.A., & Convertino, V.A. (1975). A comparison of heart rate methods for predicting endurance training intensity. *Medicine and Science in Sports, 7,* 295-298.

13. Fletcher, G.F., Blair, S.N., Blumenthal, J., Caspersen, C., Chaitman, B., Epstein, S., Falls, H., Froelicher, E.S.S., Froelicher, V.F., & Pina, I.L. (1992). Statement on exercise: Benefits and recommendations for physical activity programs for all Americans. *Circulation, 86,* 340-344.

14. Fletcher, G.F., Froelicher, V.F., Hartley, L.H., Haskell, W.L., & Pollock, M.L. (1990). Exercise standards: A statement for health professionals from the American Heart Association. *Circulation, 82,* 2286-2322.

15. Franklin, B.A., Gordon, S., & Timmis, G.C. (Eds.). (1989). *Exercise in modern medicine.* Baltimore: Williams & Wilkins.

16. Harris, Louis, & Associates, Inc. (1978). *Health maintenance*. Newport Beach, CA: Pacific Mutual Life Insurance.

17. Harris, Louis, & Associates, Inc. (1979). *The Perrier study: Fitness in America*. New York: Perrier-Great Waters of France.

18. Harris, Louis, & Associates, Inc. (1983, October-November). Prevention in America: Steps people take—or fail to take—for better health. *Prevention Magazine*.

19. Karvonen, M.J., Kentala, E., & Mustala, O. (1957). The effects of training heart rate: A longitudinal study. *Annales Medicinae Experimentalis et Biologiae Fenniae*, **35**, 307-315.

20. Kelley, G. (1997). Dynamic resistance exercise and resting blood pressure in adults: A meta-analysis. *Journal of Applied Physiology*, **82**, 1559-1565.

21. Kirshenbaum, J., & Sullivan, R. (1983). Hold on there, America. *Sports Illustrated*, **58**(5), 60-74.

22. LaPorte, R.E., Adams, L.L., Savage, D.D., Brenes, G., Dearwater, S., & Cook, T. (1984). The spectrum of physical activity, cardiovascular disease and health: An epidemiologic perspective. *American Journal of Epidemiology*, **120**, 507-517.

23. Leon, A.S., & Connett, J. (1991). Physical activity and 10.5 year mortality in the Multiple Risk Factor Intervention Trial (MRFIT). *International Journal of Epidemiology*, **20**, 690-697.

24. Leon, A.S., Connett, J., Jacobs, D.R., & Rauramaa, R. (1987). Leisure-time physical activity levels and risk of coronary heart disease and death. *Journal of the American Medical Association*, **258**, 2388-2395.

25. Lowenthal, D.T., Bharadwaja, K., & Oaks, W.W. (1979). *Therapeutics through exercise*. New York: Grune & Stratton.

26. Malinow, M.R., McGarry, D.L., & Kuehl, K.S. (1984). Is exercise testing indicated for asymptomatic active people? *Journal of Cardiac Rehabilitation*, **4**, 376-380.

27. Miller Brewing Co. (1983). *The Miller Lite report on American attitudes towards sports*. Milwaukee: Miller Brewing.

28. National Institutes of Health. Consensus Development Panel on Physical Activity and Cardiovascular Health. (1996). Physical activity and cardiovascular health. *Journal of the American Medical Association*, **276**, 241-246.

29. Pate, R.R., Pratt, M., Blair, S.N., Haskell, W.L., Macera, C.A., Bouchard, C., Buchner, D., Ettinger, W., Heath, G.W., King, A.C., Kriska, A., Leon, A.S., Marcus, B.H., Morris, J., Paffenbarger, R.S., Patrick, K., Pollock, M.L., Rippe, J.M., Sallis, J., & Wilmore, J.H. (1995). Physical activity and public health: A recommendation from the Centers for Disease Control and Prevention and the American College of Sports Medicine. *Journal of the American Medical Association*, **273**, 402-407.

30. Pollock, M.L., & Schmidt, D.H. (Eds.). (1986). *Heart disease and rehabilitation* (2nd ed.). Boston: Houghton Mifflin.

31. Pollock, M.L., & Vincent, K.R. (1996). Resistance training for health. *Research Digest: President's Council on Physical Fitness and Sports*, **2**(8), 1-6.

32. Pollock, M.L., & Wilmore, J.H. (1990). *Exercise in health and disease: Evaluation and prescription for prevention and rehabilitation* (2nd ed.). Philadelphia: Saunders.

33. Skinner, J.S. (Ed.). (1993). *Exercise testing and exercise prescription for special cases: Theoretical basis and clinical application* (2nd ed.). Philadelphia: Lea & Febiger.

34. Smutok, M.A., Reece, C., Kokkinos, P.F., Farmer, C., Dawson, P., Shulman, R., DeVane-Bell, J., Patterson, J., Charabogos, C., Goldberg, A.P., & Hurley, B.F. (1993). Aerobic versus strength training for risk factor intervention in middle-aged men at high risk for coronary heart disease. *Metabolism*, **42**, 177-184.

35. Stone, M.H., Fleck, S.J., Triplett, N.T., & Kraemer, W.J. (1991). Health and performance-related potential of resistance training. *Sports Medicine*, **11**, 210-231.

36. Swain, D.P. (2000). Energy cost calculations for exercise prescription: An update. *Sports Medicine*, **30**, 17-22.

37. Swain, D.P., & Leutholtz, B.C. (1997). Heart rate reserve is equivalent to %$\dot{V}O_2$ reserve, not to %$\dot{V}O_2$max. *Medicine and Science in Sports and Exercise*, **29**, 410-414.

38. U.S. Department of Health and Human Services. (1996). *Physical activity and health: A report of the Surgeon General*. Atlanta: U.S. Department of Health and Human Services, Centers for Disease Control and Prevention, National Center for Chronic Disease Prevention and Health Promotion.

▷ Selected Readings

Birk, T.J., & Birk, C.A. (1987). Use of ratings of perceived exertion for exercise prescription. *Sports Medicine*, **4**, 1-8.

Gordon, N.F., Kohl, H.W., Scott, C.B., Gibbons, L.W., & Blair, S.N. (1992). Reassessment of the guidelines for exercise testing: What alterations to current recommendations are required? *Sports Medicine*, **13**, 293-302.

Graves, J.E., & Franklin, B.A. (Eds.). (2001). *Resistance training for health and rehabilitation*. Champaign, IL: Human Kinetics.

Haskell, W.L. (1994). Health consequences of physical activity: Understanding and challenges regarding dose-response. *Medicine and Science in Sports and Exercise, 26,* 649-660.

Hass, C.J., Feigenbaum, M.S., & Franklin, B.A. (2001). Prescription of resistance training for healthy populations. *Sports Medicine, 31,* 353-364.

Howley, E.T., & Franks, B.D. (1997). *Health fitness instructor's handbook* (3rd ed.). Champaign, IL: Human Kinetics.

Karvonen, M.J., & Vuorimaa, T. (1988). Heart rate and exercise intensity during sport activities: Practical application. *Sports Medicine, 5,* 303-312.

Kell, R.T., Bell, G., & Quinney, A. (2001). Musculoskeletal fitness, health outcomes and quality of life. *Sports Medicine, 31,* 863-873.

King, C.N., & Senn, M.D. (1996). Exercise testing and prescription: Practical recommendations for the sedentary. *Sports Medicine, 21,* 326-336.

Mazzeo, R.S., & Tanaka, H. (2001). Exercise prescription for the elderly: Current recommendations. *Sports Medicine, 31,* 809-818.

Swain, D.P., Abernathy, K.S., Smith, C.S., Lee, S.J., & Bunn, S.A. (1994). Target heart rates for the development of cardiorespiratory fitness. *Medicine and Science in Sports and Exercise, 26*(1), 112-116.

Swain, D.P., Leutholtz, B.C., King, M.E., Haas, L.A., & Branch, J.D. (1998). Relationship between % heart rate reserve and % VO_2 reserve in treadmill exercise. *Medicine and Science in Sports and Exercise, 30*(2), 318-321.

CHAPTER 20

CARDIOVASCULAR DISEASE AND PHYSICAL ACTIVITY

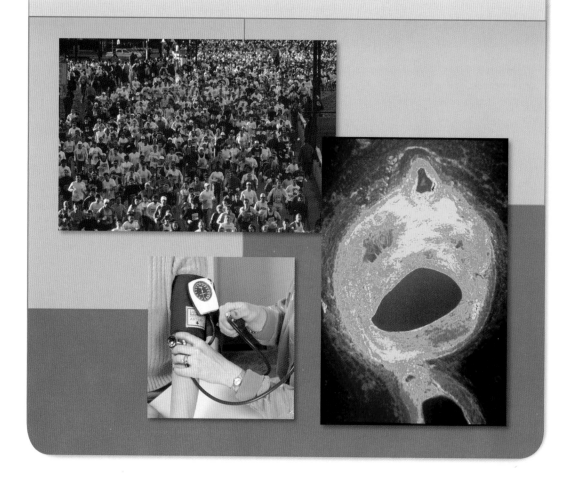

overview

Most of us consider ourselves to be healthy until we experience some overt sign of illness. With chronic degenerative diseases, such as heart disease, most people are unaware that the disease process is smoldering and progressing to the point that it could cause major complications, including death. Fortunately, early detection and proper treatment of various chronic diseases can substantially reduce their severity and often avert disability and death. Even more important, decreasing the risk factors for a disease often can either prevent the disease or delay its onset. To do this, we should

- develop and maintain nutritionally sound dietary habits;
- develop and maintain regular physical activity habits;
- abstain from the use of tobacco and other drugs;
- consume alcohol only in moderation, if at all;
- get adequate rest and sleep; and
- improve our ability to cope with stress.

Although this list is familiar to most of us, physical activity is too often ignored because it takes time and effort. But its importance to our health cannot be overlooked. In this chapter, we consider different types of cardiovascular disease and their pathophysiology, and we explore the impact of physical activity on both the prevention and treatment of these diseases.

outline

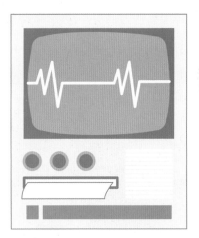

On Saturday afternoon, June 22, 2002, St. Louis Cardinals pitcher Darryl Kile was found dead in his hotel room in Chicago. The Cardinals were in Chicago to play a three-game series against the Chicago Cubs. He was scheduled to start the last game of the series on Sunday night. He was considered one of the Cardinals' best pitchers and a clubhouse leader. Kile, who was only 33 years old, apparently died of coronary atherosclerosis—on autopsy, two of his three major coronary arteries were found to be narrowed by 80% to 90%. Although he had no medical history or symptoms of disease, his father had died of a stroke at the age of 44 years, and Kile had complained of shoulder pain and fatigue during dinner the previous night. This tragedy illustrates the important fact that being an outstanding athlete during youth and young adulthood does not confer lifelong immunity for coronary artery disease (CAD). Although a genetic predisposition to CAD is serious, it doesn't have to result in premature death. Paying close attention to all of the CAD risk factors and knowing how to minimize the risk become extremely important.

Chronic and degenerative diseases of the cardiovascular system are the major cause of serious illness and death in the United States (figure 20.1). Cardiovascular diseases affect nearly 62 million Americans each year, result in nearly 1 million deaths each year, and cost individuals, government, and private industry over $350 billion annually.[4]

From the early 1900s to the mid-1960s, the relative number of heart disease deaths, expressed per 100,000 people, increased threefold. The population of the United States more than doubled during that time, so the absolute number of heart disease deaths increased even more dramatically than the relative rate indicates. It was estimated that in 2003

- more than 1.1 million heart attacks would occur in America,
- about 515,000 Americans would die from heart attacks,
- about one out of every five Americans would have some form of cardiovascular disease, and
- one of every 2.5 deaths in the United States is attributable to cardiovascular disease.

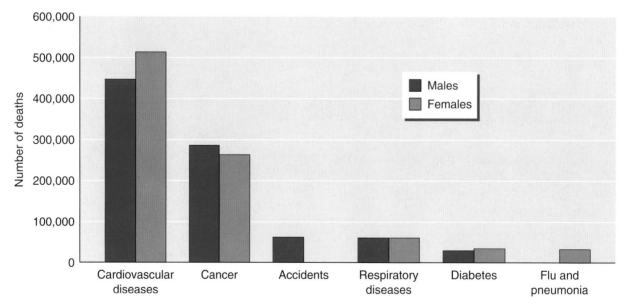

▲ **Figure 20.1** The leading causes of death in the United States in 2002. Accidents were not a leading cause of death for women, and flu and pneumonia were not a leading cause of death for men.
Data from American Heart Association, 2003.

Furthermore, it was estimated that there were

- about 519,000 coronary artery bypass surgeries in 2000,
- about 561,000 percutaneous transluminal coronary angioplasties (PTCA) in 2000, and
- about 2,200 heart transplants in 2001.

Fortunately, the number of deaths from cardiovascular disease and heart attacks has steadily decreased since its peak in the mid-1960s. The reasons for this decline have been heavily debated but likely include the following:

- Better and earlier diagnosis
- Better medical care
- Improved drugs for specific treatment
- Better emergency care and treatment for heart attack victims
- Improved public awareness of symptoms and risk factors
- Increased use of preventive measures, including lifestyle changes to reduce individual risk

Cardiovascular diseases remain the number one cause of death in the United States, accounting for more than 40% of all deaths. However, from 1989 to 1999, there was a 15.6% decrease in the death rate from cardiovascular diseases, and the death rate from heart disease declined by almost 25% during the same period.

Now that we have briefly discussed the extent of this health problem, let's turn our attention to some specific types of cardiovascular disease.

Forms of Cardiovascular Disease

There are several different cardiovascular diseases. In this section, we focus primarily on those that are preventable and that affect the largest number of Americans each year; these are illustrated in figure 20.2.

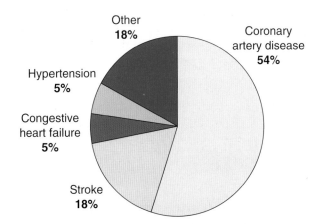

▲ **Figure 20.2** The leading causes of death from cardiovascular diseases.
Data from American Heart Association, 2003.

Coronary Artery Disease

As most humans age, their coronary arteries, which supply the myocardium (heart muscle) itself, become progressively narrower as a result of the formation of fatty **plaque** along the inner wall of the artery, as seen in figure 20.3 on page 638. This progressive narrowing of the arteries in general is referred to as **atherosclerosis,** and when the coronary arteries are involved, it is termed **coronary artery disease (CAD).** As the disease progresses and the coronary arteries become narrower, the capacity to supply blood to the myocardium is progressively reduced. This is what happened to the baseball pitcher described in the beginning of this chapter.

As the narrowing worsens, the myocardium eventually can't receive enough blood to meet all of its needs. When this occurs, the portion of the myocardium (heart muscle) that is supplied by the narrowed arteries becomes ischemic, meaning it suffers a deficiency of blood. Ischemia of the heart usually causes severe chest pain, referred to as *angina pectoris*. This typically is first experienced during periods of physical exertion or stress, when the demands on the heart are greatest.

When blood supply to a part of the myocardium is severely or totally restricted, ischemia

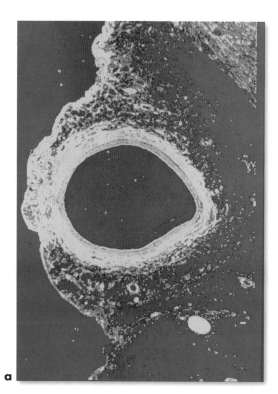 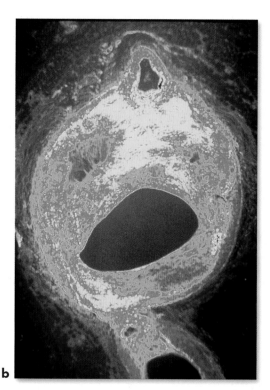

▲ **Figure 20.3**   A comparison of two coronary arteries illustrates the relative internal dimensions of (*a*) a healthy artery versus (*b*) an artery partially blocked by fatty plaque deposits.

can lead to a heart attack, or **myocardial infarction,** because cardiac muscle cells that are deprived of blood for several minutes are also deprived of oxygen, which leads to irreversible damage and necrosis (cellular death). This can lead to mild, moderate, or severe disability or even death, depending on the location of the infarction and the extent of the damage. Sometimes a heart attack is so mild that the victim is unaware that it has occurred. In such cases, the heart attack is discovered weeks, months, or even years later when an electrocardiogram is obtained during a routine medical examination.

Atherosclerosis is not a disease of the aged. Rather, it is more appropriately classified as a pediatric disease because the pathological changes that lead to atherosclerosis begin in infancy and progress during childhood.[28] **Fatty streaks,** or lipid deposits, which are thought to be the probable precursors of atherosclerosis, commonly are found in the aortas of children by age 3 to 5. These fatty streaks start to appear

in the coronary arteries during the early teens, can develop into fibrous plaques during one's 20s, and can progress to complicated lesions during one's 40s and 50s.

The rate at which atherosclerosis progresses is determined largely by genetics and lifestyle factors, including smoking history, diet, physical activity, and stress. For some people, the disease progresses rapidly, with a heart attack occurring at a relatively young age—in their 20s or 30s. For others, the disease progresses very slowly, with few or no symptoms throughout their lives. Most people fall somewhere between these two extremes.

> Atherosclerosis begins in childhood and progresses at different rates, depending primarily on heredity and lifestyle choices.

To illustrate this, a study of combat fatalities from the Korean War revealed that 77% of autopsied American soldiers, average age 22.1, already had some gross evidence of coronary atherosclerosis.[16] The extent of disease ranged from fibrous thickening to complete occlusion of one or more of the main branches of the coronary arteries. The autopsied Korean soldiers, however, were free of the disease. Evidence of coronary atherosclerosis also was found in 45% of the American fatalities from the Vietnam War, and 5% exhibited severe manifestations of the disease.[38]

Hypertension

Hypertension is the medical term for high blood pressure, a condition in which blood pressure is chronically elevated above levels considered desirable or healthy for a person's age and size. Blood pressure depends primarily on body size, so children and young adolescents have much lower blood pressures than adults. For this reason, determining what constitutes hypertension in the growing child and adolescent is difficult. Clinically, hypertension in these groups is defined as blood pressure values above the 90th or the 95th percentile for the youth's age. Hypertension is uncommon during childhood but can appear during midadolescence. For adults, the Joint National Committee on Detection, Evaluation, and Treatment of High Blood Pressure has established guidelines, presented in table 20.1, for **systolic**

blood pressure, which is the highest pressure in the arteries at any time, and for **diastolic blood pressure,** which is the lowest pressure in the arteries at any time.[27]

Hypertension causes the heart to work harder than normal, because it has to expel blood from the left ventricle against a greater resistance. Furthermore, hypertension places great strain on the systemic arteries and arterioles. Over time, this stress can cause the heart to enlarge and the arteries and arterioles to become scarred, hardened, and less elastic. Eventually, this can lead to atherosclerosis, heart attacks, heart failure, stroke, and kidney failure.

In 2003, about 50 million Americans age 6 and older were estimated to have high blood pressure.[4] Approximately 25% of the United States' adult population has high blood pressure. The age-adjusted death rate from hypertension increased 21% from 1990 to 2000, and over 44,000 Americans died as a result of hypertension in 2000. Compared with white Americans, black Americans develop high blood pressure at an earlier age, and it is more severe at any decade of life. Consequently, black Americans have a 1.3 times greater rate of nonfatal stroke, a 1.8 times greater rate of fatal stroke, a 1.5 times greater rate of heart disease deaths, and a 4.2 times greater rate of end-stage renal disease when compared with white Americans.[4] The estimated age-adjusted prevalence of high blood pressure in American adults age 20

Table 20.1

Classification of Blood Pressure for Adults, Age 18 Years and Older

Category	Systolic (mmHg)	Diastolic (mmHg)
Normal	<120	<80
Prehypertension	120-139	80-89
Hypertension		
Stage 1	140-159	90-99
Stage 2	≥160	≥100

From the seventh report of the Joint National Committee on Prevention, Detection, Evaluation, and Treatment of High Blood Pressure, 2003, *Journal of the American Medical Association* 289:2560-2572.[27]

and older was 25.2% for non-Hispanic white males, 20.5% for non-Hispanic white females, 36.7% for non-Hispanic black males, 36.6% for non-Hispanic black females, 24.2% for Mexican American males, and 22.4% for Mexican American females.[4]

> About one of every four adult Americans has hypertension.

Stroke

Stroke, also called *cerebrovascular accident* (CVA), is a form of cardiovascular disease that affects the cerebral arteries, those that supply the brain. Approximately 700,000 strokes occur in the United States each year, resulting in more than 167,000 deaths in 2000.[4] As with coronary artery disease, the death rate from strokes also has decreased significantly in recent years—a 12.3% reduction between 1990 and 2000.

The most common cause of stroke is **cerebral infarction** (see figure 20.4*a*), which typically results from

- cerebral thrombosis, in which a thrombus (blood clot) forms in a cerebral vessel, often at the site of atherosclerotic damage to the vessel;

- cerebral embolism, in which an embolus (an undissolved mass of material, such as fat globules, bits of tissue, or a blood clot) breaks loose from another site in the body and lodges in a cerebral artery; or

- atherosclerosis that leads to narrowing of and damage to a cerebral artery.

In cases of cerebral infarction, blood flow beyond the blockage is restricted, and the part of the brain that relies on that supply becomes ischemic, is oxygen deficient, and can die.

Hemorrhage is the other major cause of stroke. The two major types are cerebral hemorrhage, in which one of the cerebral arteries ruptures in the brain (figure 20.4*b*), and subarachnoid hemorrhage, in which one of the brain's surface vessels ruptures, dumping blood into the space between the brain and the skull. In both cases, blood flow beyond the rupture is diminished because the blood leaves

the vessel at the site of injury. Also, as the blood accumulates outside of the vessel, it puts pressure on the fragile brain tissue, which can alter brain function. Brain hemorrhages often result from aneurysms, which arise from weak spots in the vessel wall that balloon outward, and aneurysms often arise because of hypertension or atherosclerotic damage to the vessel wall.

As with a heart attack, a stroke results in death of the affected tissue. The consequences depend largely on the location and extent of the stroke. Brain damage from a stroke can affect

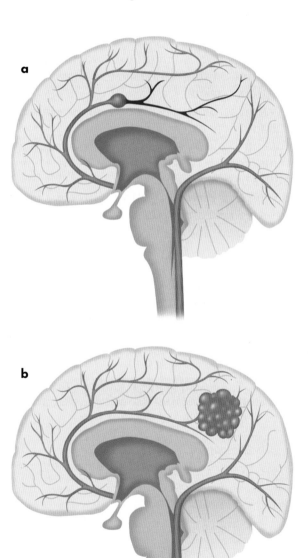

▲ **Figure 20.4** Two major causes of stroke are (*a*) cerebral infarction and (*b*) cerebral hemorrhage.

the senses, speech, body movement, thought patterns, and memory. Paralysis on one side of the body is common, as is the inability to verbalize thoughts. Most effects of a stroke are indicative of the side of the brain that was damaged. These are listed in table 20.2.

Congestive Heart Failure

Congestive heart failure is a clinical condition in which the heart muscle becomes too weak to maintain an adequate cardiac output to meet the body's oxygen demands. This usually results from either damage to or overworking of the heart. Hypertension, atherosclerosis, and heart attack are among the possible causes of this disorder.

When cardiac output is inadequate, blood begins to back up in the veins. This causes excess fluids to accumulate in the body, particularly in the legs and ankles. This fluid accumulation (edema) also can affect the lungs (pulmonary edema), disrupting breathing and causing shortness of breath. Congestive heart failure can progress to the point of irreversible damage to the heart, and the patient becomes a candidate for a heart transplant.

Other Cardiovascular Diseases

Other cardiovascular diseases include peripheral vascular diseases, valvular heart diseases, rheumatic heart disease, and congenital heart disease.

Peripheral vascular diseases involve the systemic arteries and veins, as opposed to the coronary vessels. **Arteriosclerosis** refers to numerous conditions in which the walls of the arteries become thickened, hard, and less elastic. Atherosclerosis is a form of arteriosclerosis. Arteriosclerosis obliterans, in which an artery becomes completely occluded, is another form. Peripheral venous diseases include varicose veins and phlebitis. Varicose veins result from incompetency of the valves in the veins, allowing blood to back up in the veins and causing them to become enlarged, tortuous, and painful. Phlebitis is inflammation of a vein and is also very painful.

Valvular heart diseases involve one or more of the four valves that control the direction of

Table 20.2

The Effects of Brain Damage Resulting from Stroke

	Left brain damage	Right brain damage
Paralysis	Right side	Left side
Deficits	Speech, language	Spatial, perceptual
Behavioral style	Slow, cautious	Quick, impulsive
Memory deficits	Performance	Language

blood flow into and out of the four heart chambers. **Rheumatic heart disease** is one form of valvular heart disease involving a streptococcal infection that has caused acute rheumatic fever, typically in children between ages 5 and 15. Rheumatic fever is an inflammatory disease of the connective tissue and commonly affects the heart, specifically the heart valves. The damage to the valves usually causes difficulty in their opening, hindering blood flow out of that chamber, or difficulty in their closing, allowing blood to flow back into the previous chamber.

Congenital heart disease includes any heart defects that are present at birth, which are also appropriately termed *congenital heart defects*. These defects occur when the heart or the blood vessels near the heart do not develop normally before birth. These include coarctation of the aorta, in which the aorta is abnormally constricted; valvular stenosis, in which one or more heart valves are narrowed; and septal defects, in which the septum separating the right and left sides of the heart is defective, allowing blood from the systemic side to mix with that in the pulmonary side, and vice versa.

Now that we have considered various types of cardiovascular diseases, in the remainder of this chapter we focus on the two major diseases in this category: coronary artery disease and hypertension.

▶ Atherosclerosis is a process in which arteries become progressively narrower. Coronary artery disease is atherosclerosis of the coronary arteries.

▶ When blood flow to the heart is sufficiently blocked, the part of the heart supplied by the diseased artery suffers from lack of blood (ischemia), and the resulting oxygen deprivation can cause myocardial infarction, which results in tissue necrosis.

▶ Atherosclerotic changes in the arteries actually begin in young children, but the extent and progression of this disease process are quite variable.

▶ Hypertension is the clinical term for high blood pressure.

▶ Stroke, or cerebral vascular accident, affects the cerebral arteries so that the part of the brain they supply receives too little blood. The most common cause of stroke is cerebral infarction, usually resulting from a cerebral thrombosis or embolism or from atherosclerosis. Another common cause of stroke is cerebral hemorrhage.

▶ Congestive heart failure is a condition in which the cardiac muscle becomes too weak to maintain an adequate cardiac output, causing blood to back up in the veins.

▶ Peripheral vascular diseases involve systemic, rather than coronary, vessels and include arteriosclerosis, varicose veins, and phlebitis.

▶ Congenital heart disease includes all heart defects present at birth.

Understanding the Disease Process

Pathophysiology refers to the pathology and physiology of a specific disease process or disordered function. Understanding the pathophysiology of a disease gives us insight into how physical activity might affect or alter the disease process. In the following sections, we examine the pathophysiology of coronary artery disease and hypertension.

Pathophysiology of Coronary Artery Disease

How does atherosclerosis develop in the coronary arteries? The walls of the coronary arteries are composed of three distinct tunics or layers, as shown in figure 20.5: the tunica intima (inner layer), the tunica media (middle layer), and the tunica adventitia (outer layer). These are referred to more simply as the intima, the media, and the adventitia. The innermost layer

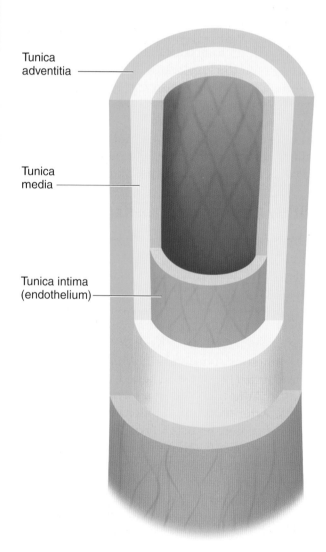

Tunica adventitia

Tunica media

Tunica intima (endothelium)

▲ **Figure 20.5**   The wall of an artery has three layers: tunica intima, tunica media, and tunica adventitia.

of the intima, the endothelium, is formed by a thin lining of endothelial cells that provides a smooth protective coating between the blood flowing through the artery and the intimal layer of the vessel wall. The endothelium provides a protective barrier between toxic substances in the blood and the vascular smooth muscle cells. For vessels larger than 1 mm in diameter, the intima also includes a subendothelial layer, formed from connective tissue. The media consists mainly of the smooth muscle cells, which control the constriction and dilation of the vessel, and elastin. The adventitia is composed of collagen fibers that protect the vessel and anchor it to its surrounding structure.[37] Local injury to endothelial cells appears to initiate the process of atherosclerosis.

This process is depicted in figure 20.6. Early studies of primates have shown that scratching the inner lining of a coronary artery causes the endothelial cells to slough off, exposing the underlying connective tissue. Blood platelets then are attracted to the site of injury and adhere to the exposed connective tissue (see figure 20.6b). These platelets release a substance referred to as **platelet-derived growth factor (PDGF)** that promotes migration of smooth muscle cells from the media into the intima. The intima normally contains few if any smooth muscle cells. A plaque, which is basically composed of smooth muscle cells, connective tissue, and debris, forms at the site of injury (see figure 20.6c). Eventually, lipids in the blood, specifically low-density–lipoprotein cholesterol (LDL-C), are attracted to and deposited in the plaque (see figure 20.6d). This early theory of atherosclerosis evolved from the work of Dr. Russell Ross and his colleagues at the University of Washington.[50]

a
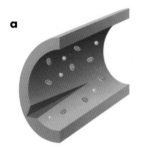

b
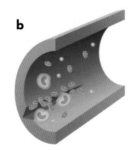

c
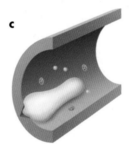

d
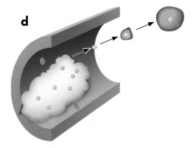

A blood-borne irritant injures the arterial wall, disrupting the endothelial layer and exposing the underlying connective tissue.

Blood platelets and circulating immune cells known as monocytes are then attracted to the site of the injury and adhere to the exposed connective tissue. The platelets release a substance referred to as platelet-derived growth factor (PDGF) that promotes migration of smooth muscle cells from the media to the intima.

A plaque, which is basically composed of smooth muscle cells, connective tissue, and debris, forms at the site of injury.

As the plaque grows, it narrows the arterial opening and impedes blood flow. Lipids in the blood, specifically low-density-lipoprotein cholesterol (LDL-C), are deposited in the plaque. When pieces of the plaque break loose they can start clots that lodge in other parts of the vessel.

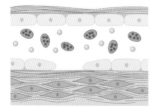
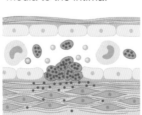
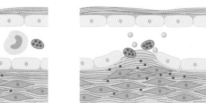
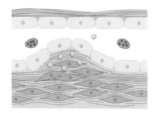

▲ **Figure 20.6** Changes in the arterial wall with injury, illustrating the disruption of the endothelium and the subsequent alterations that lead to atherosclerosis.

More recently, researchers have theorized that monocytes, which are effector cells of the immune system, attach between endothelial cells. These monocytes eventually become foam cells, or macrophages, and form fatty streaks. Smooth muscle cells then accumulate under these foam cells. The endothelial cells then separate or are sloughed off, exposing the underlying connective tissue and allowing platelets to attach to it.[50] In this modification of the original theory, endothelial injury is not always the precipitating event. Injury or disruption of the endothelium can result from hypertension and high blood concentrations of the atherogenic form of cholesterol (LDL-C), among other factors.

An additional hypothesis, proposed in 1989, is referred to as the lipid-infiltration hypothesis and is based primarily on the presence of elevated plasma LDL-C levels and the oxidative modification of LDL-C within the arterial wall.[53] This lipid-infiltration hypothesis interacts well with the response-to-injury hypothesis in explaining various aspects and stages of the atherosclerotic process.

> The process of atherosclerosis appears to begin with injury to or disruption of the endothelial cells lining the intima. This leads to a chain of events that eventually develops into a full-blown atherosclerotic plaque.

Pathophysiology of Hypertension

The pathophysiology of hypertension is not well understood. In fact, it is estimated that 90% or more of those identified with hypertension are classified as having idiopathic hypertension, or hypertension of unknown origin. Idiopathic hypertension, also referred to as essential hypertension, can result from

- genetic factors,
- high sodium intake,
- obesity,
- insulin resistance,
- physical inactivity,
- psychological stress,
- a combination of these factors, or
- other factors yet to be substantiated or determined.

> ▶ Pathophysiology refers to the pathology and physiology of a specific disease process or disordered function.
>
> ▶ Early theories held that coronary artery disease can be initiated by damage to the smooth endothelial lining of the intimal layer of the arterial wall. This damage attracts platelets to the area, which in turn release platelet-derived growth factor (PDGF). PDGF attracts smooth muscle cells, and a plaque, composed of smooth muscle cells, connective tissue, and debris, begins to form. Eventually, lipids are deposited in the plaque.
>
> ▶ More recent research indicates that monocytes, involved with the immune system, can attach between endothelial cells in the intima and begin forming fatty streaks; this then leads to plaque formation. According to this theory, endothelial damage is not necessary for plaque formation.
>
> ▶ The pathophysiology of hypertension is poorly understood.
>
> ▶ More than 90% of people with hypertension have idiopathic, or essential, hypertension, meaning its cause is unknown. Possible causes might be genetic factors, excessive sodium intake, obesity, insulin resistance, physical inactivity, and psychological stress.

Determining Individual Risk

Over the years, scientists have attempted to determine the basic etiology, or cause, of both coronary artery disease and hypertension. Much of our understanding of these two diseases comes from the field of epidemiology, a science that studies the relationships of various factors to a specific disease or disease process. In several studies, selected members of various communities have been observed for extended periods of time. These observations include periodic medical examinations and clinical tests.

Eventually, some of the participants in such studies become diseased and many die. All those who develop heart disease or hypertension or who die from heart attacks or hypertension are grouped accordingly. Then their previous medical and clinical tests are analyzed to determine shared attributes or factors. Although this approach does not define the causal mechanism of the disease, it does provide researchers with valuable insights into the disease process. As identified in long-term longitudinal population studies, the factors that place individuals at risk for disease are referred to as risk factors. Let's examine the risk factors for heart disease and hypertension.

Risk Factors for Coronary Artery Disease

The factors associated with an increased risk for premature development of coronary artery disease can be classified into two groups: those over which a person has no control and those that can be altered through basic changes in lifestyle (see table 20.3). Those that a person cannot control include heredity (family history of coronary artery disease), male gender, and advanced age. Factors that can be controlled or altered include

- cigarette smoking,
- hypertension,
- elevated blood lipids (cholesterol and triglycerides),
- physical inactivity,
- obesity, and
- diabetes and insulin resistance.

Primary risk factors are those that have been shown conclusively to have a strong association with coronary artery disease. These include smoking, hypertension, high blood lipid levels, physical inactivity, and obesity.

Table 20.3

Coronary Artery Disease Risk Factors

Primary risk factors	Secondary risk factors
Smoking	*Treatable*
Hypertension	Diabetes/ hyperinsulinemia
Blood lipids	
High LDL-C	*Untreatable*
Low HDL-C	Heredity
High triglycerides	Male sex
	Advanced age
Physical inactivity	
Obesity	

Note. LDL-C = low-density–lipoprotein cholesterol; HDL-C = high-density–lipoprotein cholesterol.

Physical inactivity was added to this list in July 1992,[18] and obesity was added to the list in the late 1990s. Table 20.4 lists the risk levels associated with actual values for these and other factors.[48] Refer to this table as we discuss each of these risk factors throughout this chapter and the next chapter, which focuses on obesity and diabetes.

> When one or more risk factors for a certain disease are present, an individual is at increased risk for developing the disease or dying from it. The primary risk factors for coronary artery disease are smoking, hypertension, elevated blood lipids, physical inactivity, and obesity.

Lipids and Lipoproteins

The inclusion of elevated blood lipids as a primary risk factor needs to be further defined. For many years, cholesterol and triglycerides were the only lipids observed in these epidemiological studies. The public was confused by conflicting data and opinions about the role

Table 20.4

Risk of Developing Coronary Artery Disease on the Basis of Specific Values for the Various Risk Factors

	Relative level of risk				
Risk factor	**Very low**	**Low**	**Moderate**	**High**	**Very high**
Blood pressure (mmHg)					
Systolic	<110	120	130-140	156-160	>170
Diastolic	<70	76	82-88	94-100	>106
Cigarettes (per day)	Never or 0 in 1 year	5	10-20	30-40	>50
Cholesterol (mg/dl)	<180	<200	220-240	260-280	>300
Cholesterol ÷ HDL	<3.0	<4.0	<4.5	>5.2	>7.0
Triglycerides (mg/dl)	<50	<100	>130	>200	>300
Glucose (mg/dl)	<80	90	100-110	120-130	>140
Body fat (%)					
Men	12	16	25	30	>35
Women	16	20	30	35	>40
Body mass index[a]	<20	20-24	25-29	30-40	>40
Stress tension	Never	Almost never	Occasional	Frequent	Nearly constant
Physical activity (min/week)					
Above 6 kcal/min (5 METs)[b]	240	180-120	100	80-60	<30
Above 60% HRmax reserve	120	90	30	0	0
ECG abnormality (ST depression; mV)[c]	0	0	0.05	0.10	0.20
Family history of premature heart attack (blood relative)[d]	0	0	1	2	3+
Age	<30	40	50	60	>70

Note. HDL = high-density lipoprotein; MET = metabolic equivalent; ECG = electrocardiogram.

[a]Body mass index = weight (kg)/height^2 (m). Risk information adapted from Bray, G.A., 1987, Obesity and the heart, *Modern Concepts of Cardiovascular Disease,* 56: 67-71.

[b]A MET is equal to the oxygen cost at rest. One MET is generally equal to 3.5 ml · kg^{-1} · min^{-1} or 1.2 kcal/min.

[c]Other ECG abnormalities are also potentially dangerous and are not listed here.

[d]Premature heart attack refers to people younger than 60 years.

Adapted from Pollock and Wilmore (1990).[48]

of lipids in the development of atherosclerosis. More recently, scientists have studied the manner in which lipids are transported in the blood. Lipids by themselves are insoluble in blood, so they are packaged with a protein to allow transport through the body. **Lipoproteins** are the proteins that carry the blood lipids. Two classes of lipoproteins of major concern for coronary artery disease are **low-density lipoprotein (LDL)** and **high-density lipoprotein (HDL)**. High levels of **low-density–lipoprotein cholesterol (LDL-C)** and low levels of **high-density–lipoprotein cholesterol (HDL-C)** place a person at extremely high risk of having a heart attack at a relatively young age—under age 60. Conversely, a high level of HDL-C and a low level of LDL-C place a person at an extremely low risk. Yet a third class of lipoproteins is called **very low density lipoproteins (VLDL)**. **Very-low density lipoprotein cholesterol (VLDL-C)** is becoming increasingly implicated as a risk factor for coronary artery disease.

> High levels of HDL-C and low levels of LDL-C place the individual at the lowest risk for coronary artery disease. LDL-C has been implicated in plaque formation, whereas HDL-C is probably involved in plaque regression.

Merely looking at total cholesterol is not sufficient. A person might have a moderately high level of total cholesterol (Total-C) and yet be at a relatively low risk because of a high concentration of HDL-C and low concentration of LDL-C. Conversely, a person might have a moderately low level of total cholesterol and yet be at a relatively high risk because of a high concentration of LDL-C and a low concentration of HDL-C.

Why are these two cholesterol carriers associated with different risk levels? LDL-C is theorized to be responsible for depositing cholesterol in the arterial wall. HDL-C, however, is regarded as a scavenger that removes cholesterol from the arterial wall and transports it to the liver to be metabolized. Because of these very different roles, it is essential to know the specific levels of both of these lipoproteins when determining individual risk. The ratio of Total-C to HDL-C may be the best index of risk for coronary artery disease. Values of 3.0 or less place a person at low risk, but values of 5.0 or greater place a person at high risk. As an example, with a Total-C of 225 mg/dl, an HDL-C of 45 mg/dl would provide a ratio of 5.0 (225 ÷ 45 = 5.0), but an HDL-C of 75 mg/dl would provide a ratio of 3.0 (225 ÷ 75 = 3.0). Others have used the ratio of Total-C to LDL-C or LDL-C to HDL-C to establish the degree of risk. At this time, there is no consensus as to which ratio provides the best estimate of risk.

> The ratio of Total-C to HDL-C is possibly the most accurate lipid index of risk for coronary artery disease, with values of 5.0 or greater indicating increased risk, and values of 3.0 or lower representing low risk.

Scientists are looking at other factors involved in the transportation of lipids in the blood and their possible relationship to atherosclerosis. These include Lipoprotein(a) and apolipoprotein E. As the technology for assessing these factors in the blood improves, additional potential markers are likely to emerge.

Early Detection of Risk Factors

Evidence now suggests that coronary artery disease risk factors can be identified at an early age, and the earlier they are identified, the earlier preventive treatment can begin. In a study of 96 boys ages 8 to 12,[61]

- 19.8% had total cholesterol values above the suggested high-normal value of 200 mg/dl,
- 5.2% exhibited abnormal resting electrocardiograms,
- 37.5% had more than 20% relative body fat, and
- none had elevated blood pressure.

Similar data were reported in a later study of 13- to 15-year-old boys.[60] Both studies are summarized in table 20.5. Those with an elevated risk during childhood generally remain at elevated risk as young adults.

In addition, the results of the Bogalusa Heart Study must be considered. This is a longitudinal study of cardiovascular disease risk factor development from birth through age 39. In 204 of the subjects who died prematurely (primarily from accidents, homicides, or suicides), the scientists found a strong relationship between the risk factors and development of fatty streaks; the greater the number of risk factors,

Table 20.5

Coronary Artery Disease Risk Factor Prevalence in Boys 8 Through 15 Years of Age

Risk factor	Percentage of boys with risk factor	
	8- to 12-year-olds ($n = 96$)	13- to 15-year-olds ($n = 308$)
Blood lipids		
Total cholesterol ($\geq$200 mg/dl)	20.0	11.0
HDL-C ($\leq$36 mg/dl)	No data	14.6
Triglycerides ($\geq$100 mg/dl)	8.4	25.0
Blood pressure		
Systolic (>90th percentile)	0.0	13.0
Diastolic (>90th percentile)	0.0	4.9
Smoking ($\geq$10 cigarettes a day)	0.0	0.0
Diabetes	0.0	1.3
Obesity ($\geq$25% relative body fat)	12.6	14.9
Physical activity ($\dot{V}O_2$max$\leq$42 ml $\cdot$ kg^{-1} $\cdot$ min^{-1})	3.2	18.8
Family history (heart attack at$\leq$60 years of age)	33.7	30.9
Presence of risk factors		
None	36.0	29.9
One	46.0	35.4
Two	14.0	22.1
Three	3.0	10.7
Four	1.0	1.9

Note. Values represent the percentage of the boys with the risk factor. HDL-C = high-density–lipoprotein cholesterol.

the greater the development of aortic and coronary artery fatty streaks.[5]

Risk Factors for Hypertension

The risk factors for hypertension, like those for coronary artery disease, can be classified as ones we can control and ones we cannot. Those we cannot control are heredity (family history of hypertension), advanced age, and race (increased risk for people of African or Hispanic ancestry). Risk factors we can control are

- insulin resistance,
- obesity,
- diet (excess sodium intake),
- use of oral contraceptives,
- stress, and
- physical inactivity.

Although heredity is a risk factor for hypertension, it probably plays a much smaller role than many of the other proposed factors. We must remember that lifestyle factors are often quite similar within a family.

Recently, scientists have shown great interest in a possible link between hypertension, obesity, type 2 diabetes, and coronary heart disease through the common pathway of insulin resistance or impaired insulin action (see sidebar below). But obesity also has been established as an independent risk factor for hypertension. Numerous studies have shown substantial reductions in blood pressure with weight loss in

▶ Risk factors for coronary artery disease that we cannot control are heredity (and family history), male sex, and advanced age. Those that we can control are elevated blood lipids, hypertension, cigarette smoking, physical inactivity, obesity, and diabetes. Primary risk factors are those that have been proven to be strongly associated with the diseases. For coronary artery disease, these are smoking, hypertension, abnormal blood lipids, physical inactivity, and obesity.

▶ LDL-C is thought to be responsible for depositing cholesterol in the arterial walls. VLDL-C is also implicated in the development of atherosclerosis. However, HDL-C acts as a scavenger, removing cholesterol from the vessel walls. Thus, high HDL-C levels provide some degree of protection from coronary artery disease.

▶ The ratio of total cholesterol to HDL-C might be the best indicator of personal risk for coronary artery disease. Values below 3.0 reflect a low risk, but values above 5.0 reflect a high risk.

▶ Risk factors for hypertension that can't be controlled include heredity, advanced age, and race. Those we can control are insulin resistance, obesity, diet (excess sodium), use of oral contraceptives, stress, and physical inactivity.

Metabolic Syndrome

Metabolic syndrome is a term that has been used to link coronary artery disease, hypertension, abnormal blood lipids, type 2 diabetes, and upper body obesity to insulin resistance and hyperinsulinemia. This syndrome also has been referred to as syndrome X and the civilization syndrome. It is not totally clear where the syndrome starts, but it has been observed that upper body obesity is associated with insulin resistance and that insulin resistance is highly correlated with increased risk for coronary artery disease, hypertension, and type 2 diabetes. It appears, however, that obesity is the trigger that leads to a cascade of events leading to the metabolic syndrome. This became a major topic of research in the 1990s and continues to be today. The results of this research should help us better understand the pathophysiology of these diseases and their interrelationships.

hypertensive patients. Also, although sodium intake traditionally has been linked to hypertension, this relationship is likely limited to those who are salt sensitive.

> Although the pathways are complex, it is becoming increasingly clear that hypertension, coronary artery disease, abnormal blood lipids, obesity, and diabetes might be linked through the common pathway of insulin resistance. It is also likely that obesity is the trigger that starts a cascade of events leading to the metabolic syndrome.

Physical inactivity is a risk factor for hypertension. Its role has been conclusively established in epidemiological studies.[2] Furthermore, substantial evidence indicates that increasing physical activity tends to reduce elevated blood pressure.[2, 17, 22, 24, 51, 56, 57]

Reducing Risk Through Physical Activity

The role that physical activity might play in preventing or delaying the onset of coronary artery disease and hypertension has been of major interest to the medical community for many years. In the following sections, we try to unravel this mystery by examining the following areas:

- Epidemiological evidence
- Physiological adaptations with training that might reduce risk
- Risk factor reduction with exercise training

Reducing the Risk of Coronary Artery Disease

Physical activity has been proven effective in reducing the risk of coronary artery disease. In the following sections, we discover what is known about this topic and what physiological mechanisms are involved.

Epidemiological Evidence

Well over 100 research papers have dealt with the epidemiological relationship between physical inactivity and coronary artery disease. Generally, studies have found the risk of heart attack in sedentary male populations to be about two to three times that of men who are physically active in either their jobs or their recreational pursuits.[49] The early studies of Dr.

Exercise Type and Intensity Is Related to CAD Risk

In 2002, a group of scientists from Harvard University reported in the *Journal of the American Medical Association* the results of their epidemiologic study of the relationship of exercise type and intensity to coronary artery disease (CAD) in more than 44,000 men enrolled in the Health Professional's Follow-Up Study.[54] These men were followed every 2 years from 1986 through 1998 to assess potential CAD risk factors, to identify newly diagnosed cases of CAD, and to assess levels of leisure-time physical activity. Men who ran 6 mph or faster for 1 h or more per week had a 42% risk reduction compared with men who didn't run. Men who trained with weights for 30 min or more per week had a 23% risk reduction when compared with men who didn't weight train. Brisk walking for 30 min or more per day was associated with an 18% risk reduction, as was rowing for one or more hours per week. Surprisingly, swimming and cycling were unrelated to risk. This study was the first to show the direct benefits of weight training on CAD risk and that exercise intensity is also a critical consideration, with higher intensities providing greater risk reduction.

J.N. Morris (see figure 20.7) and his colleagues in England in the 1950s were among the first to demonstrate this relationship.[42] In these studies, sedentary bus drivers were compared with active bus conductors who worked on double-decker buses, and sedentary postal workers were compared with active postal carriers who walked their routes. The death rate from coronary artery disease was about twice as high in the sedentary groups as in the active groups. Many studies published over the subsequent 20 years showed essentially the same results: Those who were occupationally sedentary were at twice the risk for death from coronary artery disease than those who were active.

Most of these early epidemiologic studies focused exclusively on occupational activity. Not until the 1970s did researchers start looking at leisure-time activity as well. Again, the studies by Dr. Morris and his colleagues[41, 43] were among the first to observe the relationship between leisure-time activity and the risk of coronary artery disease: The least active people were at two to three times greater risk. Subsequent studies by epidemiologists such as Paffenbarger, Leon, and Blair (figure 20.7) have provided similar results.[6, 8, 15, 35, 36, 46] Physical inactivity approximately doubles the risk of having a fatal heart attack.[49]

Powell and his colleagues at the Centers for Disease Control in Atlanta conducted an extensive review of all epidemiologic studies published on physical inactivity and coronary artery disease up to the mid-1980s.[49] They used stringent criteria for including studies in their analysis, and the quality of each study also was assessed. They found that the average relative risk of coronary artery disease associated with inactivity ranged from 1.5 to 2.4, with a median value of 1.9, meaning that inactive people have about twice the risk of more active people. The researchers found that relative risk from

▲ **Figure 20.7**   Key exercise epidemiologists whose research activities were instrumental in leading the American Heart Association to include physical inactivity as a major risk factor for coronary artery disease: Drs. Steven Blair, Ralph Paffenbarger, Jerry Morris, and Art Leon.

physical inactivity is similar to the risk associated with the three other major risk factors for coronary artery disease. Furthermore, the percentage of the total population of the United States who are physically inactive far exceeds the percentage with the other three major risks: those who smoke, are hypertensive, or have elevated cholesterol levels.[9] This is illustrated in figure 20.8. The results of these epidemiological studies played a major role in leading the American Heart Association in 1992 to declare physical inactivity a primary risk factor for coronary artery disease.

Another important concern was raised in the mid-1980s: What level of physical activity or fitness is necessary to reduce one's risk of coronary artery disease?[32] It was not totally clear from the epidemiological studies what level of fitness or activity was effective. In fact, during the mid-1980s, scientists had just started to differentiate between activity level and fitness, defining personal fitness by a person's $\dot{V}O_2$max. In retrospect, distinguishing these two terms was crucial because a person can be active yet unfit (low $\dot{V}O_2$max) or fit (high $\dot{V}O_2$max) yet inactive. Dr. Ronald LaPorte and his colleagues at the University of Pittsburgh were instrumental in redirecting the thinking and subsequent research in this area.[32] Dr. LaPorte pointed out that, based on various epidemiological studies, the levels of activity associated with a lower risk for coronary artery disease were generally low and certainly not at the level that would increase aerobic capacity. Subsequent studies have supported this.[8, 35, 36] Low levels of activity, such as walking and gardening, can provide considerable benefit by reducing the risk for coronary artery disease. More vigorous exercise likely provides even greater benefit.[34]

> From epidemiological studies, it has been established that physical inactivity doubles the risk of coronary artery disease. However, it is now equally clear that low-intensity activity is sufficient to reduce the risk of this disease. Health benefits do not require high-intensity exercise!

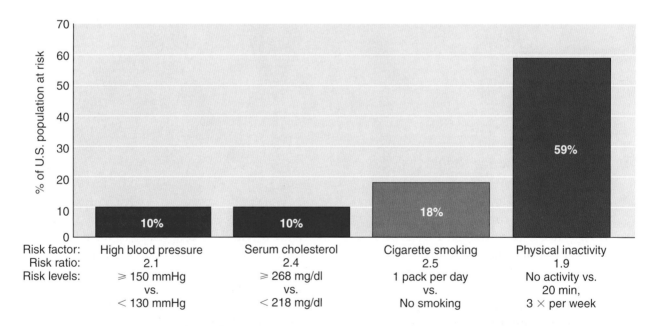

Risk factor:	High blood pressure	Serum cholesterol	Cigarette smoking	Physical inactivity
Risk ratio:	2.1	2.4	2.5	1.9
Risk levels:	≥ 150 mmHg vs. < 130 mmHg	≥ 268 mg/dl vs. < 218 mg/dl	1 pack per day vs. No smoking	No activity vs. 20 min, 3 × per week

▲ **Figure 20.8**   Percentages of the U.S. population at increased risk for coronary artery disease based on the primary risk factors.

Reproduced, with permission, from Caspersen CJ: Physical activity and coronary heart disease. *Physicians Sportsmedicine* 1987; 15(11): 43-44 © The McGraw-Hill Companies.

Physical Activity Versus Physical Fitness: Are Both Important?

In 2001, Dr. Paul Williams of the Lawrence Berkeley National Laboratory published an important article suggesting that both physical fitness, as measured by $\dot{V}O_2$max, and physical activity levels are independent risk factors for coronary artery disease.[59] He conducted a meta-analysis of a number of studies that had been conducted on large populations. He found that both being active and having a high fitness level were independently related to the degree of risk for coronary artery disease (CAD) and cardiovascular diseases (CVD) in general. This is illustrated in figure 20.9. As the percentages for both physical activity and physical fitness increase, going from the lowest percentile to the highest, above about 15% there is a reduction in risk for both CAD and CVD. The benefit of being physically fit had a stronger relationship to reduced risk than being physically active. These findings are controversial among epidemiologists, so there will likely be more studies and dialogue on this issue.[7]

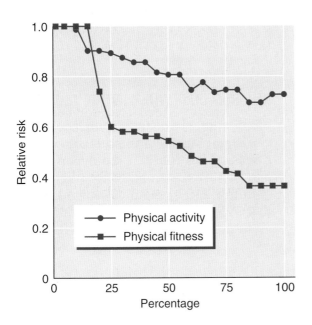

◀ **Figure 20.9** The reduction in relative risk for both coronary artery disease and cardiovascular disease risk with increasing levels of physical activity and physical fitness. This is a dose-response curve, indicating that the higher the dose, the better the response.

Reprinted, by permission, from P.T. Williams, 2001, "Physical fitness and activity as separate heart disease risk factors: A meta-analysis," *Medicine and Science in Sports and Exercise* 33: 754-761.

Training Adaptations That Might Reduce Risk

The importance of regular physical activity in reducing the risk of coronary artery disease becomes apparent when we consider anatomical and physiological adaptations in response to exercise training. For example, as we learned in chapter 9, exercise training causes the heart to hypertrophy, primarily through an increase in left ventricular chamber size but also through increases in left ventricular wall thickness. This adaptation may be important for improved contractility and increased cardiac work capacity.

The capacity of coronary circulation appears to increase with training. Studies have shown that the size of major coronary vessels increases, which implies an increased capacity for blood flow to all regions of the heart. In fact, several studies have demonstrated that the peak flow rate in the major coronary arteries increases following an exercise training program. An important study was conducted at Boston University by Dr. Dieter Kramsch and his associates,[31] who studied the effects of moderate exercise training on the development of coronary artery disease in monkeys. The monkeys were divided into three groups:

1. A control group that ate normal, low-fat monkey chow
2. A nonexercising group that ate an atherogenic (high-fat) diet known to induce heart disease
3. An exercising group that also ate the atherogenic diet

The sedentary group that consumed the atherogenic diet developed atherosclerosis. However, the coronary arteries of the exercising monkeys on this same diet had a greater internal diameter and substantially less atherosclerosis than the sedentary monkeys, as shown in figure 20.10. For this group, the cross-sectional area of the lumen (diameter) of all of the major coronary vessels was two to three times larger than in the sedentary monkeys.

Some evidence also suggests that the heart's collateral circulation improves with exercise training. The collateral circulation is a system of small vessels that branch off the major coronary vessels and are important in providing blood to all regions of the heart, particularly when there are blockages in the major coronary arteries. It is possible, however, that collateral circulation development is more the result of the blockages and compromised circulation than of exercise training.

> Aerobic training produces favorable anatomical and physiological changes that decrease the risk of heart attack, including larger coronary arteries, increased heart size, and increased pumping capacity. Aerobic training also has a favorable effect on most of the other risk factors for coronary artery disease.

Risk Reduction With Exercise Training

Many studies have investigated the role of exercise in altering risk factors associated with heart disease. Let's consider the major risk factors and how exercise might affect them.

Little direct evidence is available to indicate that exercise leads to smoking cessation or reduces the number of cigarettes smoked. However, a

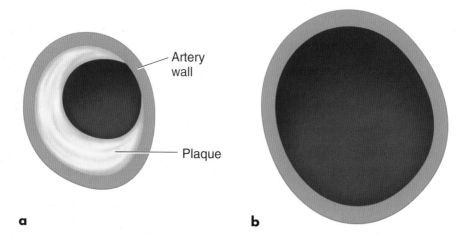

▲ **Figure 20.10**   Comparison of the left main coronary artery in (*a*) sedentary and (*b*) exercising monkeys on atherogenic diets.

great deal of anecdotal information suggests that this is true. Relatively strong data support the effectiveness of exercise in reducing blood pressure in those with mild to moderate hypertension. Endurance training can reduce both systolic and diastolic blood pressures by approximately 10 and 8 mmHg in individuals who have blood pressures of 160 and greater systolic and/or 95 mmHg or greater diastolic and by 6 and 7 mmHg in those who have blood pressures between 140 and 159 mmHg systolic and/or 90 and 95 mmHg diastolic.[2, 17, 22, 23, 24, 51, 56, 57] Exercise even can lead to small reductions in both systolic and diastolic blood pressure (by 3 mmHg each) in those with normal blood pressure.[17] But exercise appears to have little or no effect in those with severe hypertension. The specific mechanisms responsible for the decreases in blood pressure with endurance training have yet to be determined.

Exercise possibly exerts its most beneficial effect on blood lipid levels.[12, 13, 21, 25, 58] Although the decreases in Total-C and LDL-C with endurance training are relatively small (generally less than 10%), there appear to be relatively major increases in HDL-C and major decreases in triglycerides. Cross-sectional studies of athletes and nonathletes alike show unequivocally that people with greater levels of aerobic activity or higher aerobic capacities have higher HDL-C and lower triglyceride levels. Results of longitudinal training studies, however, are much less clear. Many studies have reported increases in HDL-C and decreases in triglycerides from training, yet others have reported little or no change. Several even have reported decreased HDL-C levels. Almost all studies, however, have shown that the ratios of LDL-C to HDL-C and of Total-C to HDL-C are decreased following endurance training. This implies a reduced risk.

Two confounding factors must be considered when we evaluate lipid changes with exercise training because they can have a marked independent effect on such changes. Because plasma lipids are expressed as a concentration (milligrams of lipid per deciliter of blood), any change in plasma volume will affect plasma concentrations independently of the change in total lipid. Recall that training typically increases plasma volume (chapter 9). With this plasma expansion, the absolute amount of HDL-C could increase, yet the HDL-C concentration might not change or even could be lowered. In addition, plasma lipid levels are tightly coupled with changes in body weight. When we evaluate the effects of exercise training, the independent effects that a change in body weight could have on plasma lipids must be considered.

With respect to the remaining risk factors, exercise plays an important role in weight reduction and control and in the control of diabetes. These are discussed in detail in chapter 21. Exercise also has been reported to be effective for stress reduction and control and for reducing anxiety.[30, 47] Some research supports the use of exercise training in the treatment of depression, although the results are not yet conclusive.[11, 40]

Table 20.6 on page 656 provides a list of potential physiological mechanisms by which physical activity might help prevent coronary artery disease. These each are listed in one of four categories. The first two categories address factors that could affect the oxygen supply to the heart and the oxygen demand and work of the heart. The third category lists factors that could affect the actual function of the myocardium, and the last category lists factors that could increase the electrical stability of the myocardium. Most of these factors have been discussed in this or previous chapters.

Reducing the Risk of Hypertension

Physical activity's role in reducing the risk of hypertension has not been as well established as its role in coronary artery disease. As we saw in the last section, exercise training lowers blood pressure in those with moderate hypertension, but the precise mechanisms allowing this reduction are not yet fully known. Let's consider what is known.

Epidemiological Evidence

Very few epidemiological studies have investigated the relationship between physical inactivity and hypertension. In the Tecumseh

Table 20.6

Biological Mechanisms by Which Exercise May Contribute to the Primary or Secondary Prevention of Coronary Artery Disease

Maintain or increase myocardial oxygen supply

Delay progression of coronary atherosclerosis (possible)

    Improve lipoprotein profile (increase HDL-C/LDL-C ratio (probable)

    Improve carbohydrate metabolism (increase insulin sensitivity) (probable)

    Decrease platelet aggregation and increase fibrinolysis (probable)

    Decrease adiposity (usual)

Increase coronary collateral vascularization (unlikely)

Increase epicardial artery diameter (possible)

Increase coronary blood flow (myocardial perfusion) or distribution (possible)

Decrease myocardial work and oxygen demand

Decrease HR at rest and submaximal exercise (usual)

Decrease systolic and mean systemic arterial pressure during submaximal exercise (usual) and at rest (possible)

Decrease cardiac output during submaximal exercise (probable)

Decrease circulating plasma catecholamine levels (decrease sympathetic tone) at rest (probable) and at submaximal exercise (usual)

Increase myocardial function

Increase stroke volume at rest and at submaximal and maximal exercise (likely)

Increase ejection fraction at rest and during exercise (likely)

Increase intrinsic myocardial contractility (possible)

Increase myocardial function resulting from decreased "afterload" (probable)

Increase myocardial hypertrophy[a] (probable)

Increase electrical stability of myocardium

Decrease regional ischemia or ischemia at submaximal exercise (possible)

Decrease catecholamines in myocardium at rest (possible) and at submaximal exercise (probable)

Increase ventricular fibrillation threshold attributable to reduction of cyclic AMP (possible)

Note. Expression of likelihood that effect will occur in an individual participating in endurance training program for 16 weeks or longer at 65% to 80% of functional capacity for 25 min or longer per session (300 kcal) for three or more sessions per week ranges from unlikely, possible, likely, probable, to usual.

HDL-C = high-density–lipoprotein cholesterol; LDL-C = low-density–lipoprotein cholesterol, HR = heart rate; AMP = adenosine monophosphate.

[a]This may not reduce coronary artery disease risk.

Community Health Study, 1,700 males (age 16 and older) completed questionnaires and interviews to provide estimates of their average daily energy expenditures, their peak daily energy expenditures, and the hours they spent in particular activities. The more active men had significantly lower systolic and diastolic blood pressures, irrespective of age.[39] Similar results were found when resting blood pressure was analyzed by fitness level in nearly 3,000 adult men and more than 3,900 adult women tested at the Cooper Clinic in Dallas.[10, 20] The more fit individuals exhibited lower systolic and diastolic blood pressures. In a follow-up of the participants from the Cooper Clinic study, the investigators reported a relative risk of 1.5 for the development of hypertension in people with low levels of fitness compared with highly fit people.[6] From these limited studies, active people and fit people are at reduced risks for developing hypertension. Epidemiological studies also have shown that higher physical activity levels and aerobic fitness are related to a decreased risk for stroke in both men[33] and women.[26]

Training Adaptations That Might Reduce Risk

A number of physiological adaptations that accompany endurance training could affect blood pressure both at rest and during exercise. One of the most important changes associated with endurance training is the previously mentioned plasma volume increase. We might logically assume that any increase in plasma volume would increase blood pressure, particularly because one of the first lines of drug treatment for hypertension is the prescription of a diuretic to reduce total body water and thus plasma volume. However, recall from chapter 9 that trained muscle has a notable increase in capillaries. Also, the venous system in a trained person has a greater capacity, allowing it to contain more blood. For these reasons, the increased plasma volume following exercise training does not increase blood pressure.

Specific mechanisms responsible for reductions in resting blood pressure with endurance training have not been established. Some stud-

ies show that resting cardiac output is reduced and that the body's oxygen demands are met by an increased arterial-mixed venous oxygen difference, or a-$\bar{v}O_2$ difference. But other studies have found cardiac output to remain unchanged. Without a decrease in cardiac output, the observed reductions in resting blood pressure that follow training must result from reductions in peripheral vascular resistance, which may be attributable to an overall reduction of sympathetic nervous system activity. Weight loss also is associated with reductions in blood pressure. Therefore, any weight loss associated with training should contribute to better blood pressure control.

> Aerobic training reduces blood pressure in those who have moderate hypertension but seems to have little effect on those with severe hypertension. The mechanisms by which exercise reduces blood pressure have not been completely determined.

Risk Reduction With Exercise Training

In the previous section on coronary artery disease, we determined that exercise training lowers resting blood pressure in those with moderate hypertension. Even small reductions in resting blood pressure have been noted in normotensive people (i.e., those with normal blood pressure)[29] but not in those with severe hypertension. Hagberg and his colleagues[23] conducted an extensive review of all studies previously conducted in this area and concluded that blood pressure is reduced approximately 11 mmHg systolic and 8 mmHg diastolic, and the reductions appear to be slightly greater in women and in middle-aged people. The reductions are unrelated to the duration of the training program but might be greater in response to low- to moderate-intensity activity compared with higher intensity activity.

Not only does exercise reduce high blood pressure itself in those who are moderately hypertensive, but it also affects other risk factors. Exercise is important in reducing body fat, and it can increase muscle mass, which may be important in reducing blood glucose levels and thus assisting in better glycemic (blood sugar) control. This latter effect could reduce insulin resistance, another risk for hypertension. Exercise training also has been associated with stress reduction.

▶ Epidemiological studies generally have found that the risk of coronary artery disease in sedentary male populations is about two to three times that of men who are physically active, and physical inactivity approximately doubles a person's risk of a fatal heart attack.

▶ The levels of activity associated with a reduced risk for coronary artery disease are generally lower than those needed to increase aerobic capacity.

▶ Physical training improves the heart's contractility, work capacity, and coronary circulation.

▶ Exercise may have its major impact on blood lipid levels. Studies show that endurance training decreases the ratios of LDL-C to HDL-C and of Total-C to HDL-C.

▶ Exercise also can help control blood pressure, weight, and blood glucose levels; can help alleviate stress; and might decrease cigarette smoking.

▶ People who are active and those who are fit have reduced risk for developing hypertension.

▶ Increased plasma volume that accompanies physical training does not increase blood pressure because trained people have more capillaries and greater venous capacity.

▶ Resting blood pressure is decreased by training in people with moderate hypertension, probably attributable to decreased peripheral resistance, but the actual mechanisms are unknown.

▶ Exercise also reduces body fat and blood glucose levels, which could reduce insulin resistance.

Risk of Heart Attack and Death During Exercise

Whenever a person dies while exercising, the incident usually makes newspaper headlines. Deaths during exercise don't happen often, but they are highly publicized. How safe, or how dangerous, is exercise? It was estimated in one study that there will be approximately one death for every 7,620 middle-aged joggers per year.[55] In a second study, the estimate was one death for every 18,000 physically active men.[52] In a third study, the risk was one death per 1.51 million episodes of exercise.[1] Although the risk of death increases during the period of vigorous exercise, habitual vigorous exercise is associated with an overall decreased risk of heart attack.[52] This is illustrated in figure 20.11. In one study, the relative risk of sudden death during exercise in men who rarely exercise was 74.1 compared with a relative risk of 10.9 in men who exercised at least five times per week.[1]

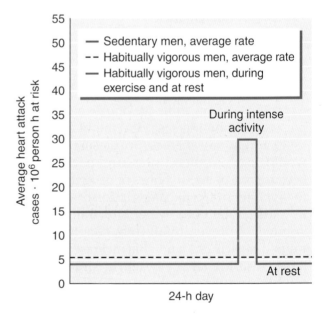

▲ **Figure 20.11**   The risk of primary cardiac arrest during vigorous exercise and at other times throughout a 24-h period, comparing sedentary men with habitually active men.

Data from D.S. Siscovick et al., 1984, "The incidence of primary cardiac arrest during vigorous exercise," *New England Journal of Medicine* 311: 874-877.

Exercise Training and Rehabilitating Patients With Heart Disease

Can active participation in a cardiac rehabilitation program that has a strong aerobic exercise component help a heart attack survivor to either survive a subsequent attack or avoid one altogether? Endurance training leads to many physiological changes that reduce the work or oxygen demand of the heart. As we have seen, many of these are peripheral, not involving the heart directly. To recap, training increases the capillary/muscle fiber ratio and plasma volume. Because of these changes, blood flow increases to the muscles. In some cases, as mentioned earlier, this allows a reduction in cardiac output, with the body's oxygen demands being met by an increased a-$\overline{v}O_2$ difference. Training also possibly can increase or maintain oxygen supply to the heart.

However, significant changes also may occur in the heart itself. Studies of heart disease patients at Washington University in St. Louis have provided dramatic evidence that intense aerobic conditioning not only can substantially change peripheral factors but also can alter the heart itself, possibly increasing blood flow to the heart and increasing left ventricular function.[14]

From our previous discussions in this chapter, it is clear that endurance exercise training can significantly reduce the risk of cardiovascular disease through its independent effect on the individual risk factors for coronary artery disease and hypertension. Favorable changes in blood pressure, lipid levels, body composition, glucose control, and stress have been reported in patients undergoing exercise training for cardiac rehabilitation. We have every reason to believe that these changes are just as important to the health of a patient who has had a heart attack as they are for a presumably healthy person.

Some researchers have tried to determine whether participation in a cardiac rehabilitation program reduces the risk of a subsequent heart attack or of death from a subsequent heart attack. However, it is nearly impossible to design a study to resolve this issue, primarily because it would be necessary to enroll several thousand people into one study to have a large enough sample to prove a statistically significant effect. Consequently, two published reports have combined the results of the most highly controlled of these studies and have used special statistical analyses on the data.[44, 45] Both reports concluded that exercise rehabilitation substantially reduces the risk of death from a subsequent heart attack but has relatively little effect on reducing the risk for recurrence of a nonfatal heart attack.

The evidence that physical activity is important in the rehabilitation of the cardiac patient is sufficiently clear. The American College of Sports Medicine issued a position stand in 1994 that concluded, "most patients with coronary artery disease should engage in individually designed exercise programs to achieve optimal physical and emotional health" (p. iv).[3] They recommended that such programs include a comprehensive preexercise medical evaluation, including a graded exercise test, and an individualized exercise prescription. Such programs should focus on multifactorial risk factor modification by using diet, drugs, and exercise to control blood lipid disorders, diabetes, and hypertension. With an aggressive approach to rehabilitation, it is even possible to see slight regression in the disease and a reduced incidence of rupture of the atherosclerotic plaque (a major cause of heart attack death).[19]

There is an increased risk of heart attack during the actual period of exercise. However, over the course of a 24-h period, those who exercise regularly have a much lower risk of a heart attack than those who do not exercise.

When death during exercise occurs in people age 35 or older, it usually results from a cardiac arrhythmia caused by atherosclerosis of the coronary arteries. On the other hand, those under age 35 are most likely to die from hypertrophic cardiomyopathy (enlarged diseased heart, usually genetically transmitted), congenital coronary artery anomalies, an aortic aneurysm, or myocarditis (inflammation of the myocardium).

▶ Deaths during exercise are rare, although typically highly publicized.

▶ Deaths during exercise in people over age 35 usually are caused by a cardiac arrhythmia resulting from atherosclerosis.

▶ Deaths during exercise in people under age 35 usually are caused by hypertrophic cardiomyopathy, congenital coronary artery anomalies, aortic aneurysm, or myocarditis.

In Closing . . .

In this chapter, we have seen how important physical activity is in reducing the risk for cardiovascular diseases, especially coronary artery disease and hypertension. We discussed the prevalence of these disorders, the risk factors associated with each, and how physical activity can help reduce our personal risks. In the next chapter, we continue examining the effects of exercise on our health as we turn our attention to obesity and diabetes.

▶ Key Terms

arteriosclerosis
atherosclerosis
blood lipid
cerebral infarction
congenital heart disease
congestive heart failure
coronary artery disease (CAD)
diastolic blood pressure
fatty streak
high-density lipoprotein (HDL)
high-density–lipoprotein cholesterol (HDL-C)
hypertension
ischemia

lipoprotein
low-density lipoprotein (LDL)
low-density–lipoprotein cholesterol (LDL-C)
metabolic syndrome
myocardial infarction
pathophysiology
peripheral vascular disease
plaque
platelet-derived growth factor (PDGF)
primary risk factors
rheumatic heart disease
stroke
systolic blood pressure
triglycerides
valvular heart disease
very low density lipoprotein (VLDL)
very low density lipoprotein cholesterol (VLDL-C)

▶ Study Questions

1. What are currently the major causes of death in the United States?

2. What is atherosclerosis, how does it develop, and at what age does it begin?

3. What is hypertension, how does it develop, and at what age does it begin?

4. What is stroke? How does stroke occur? What are the results of stroke?

5. What are the basic risk factors for coronary artery disease? For hypertension?

6. What is the risk of death from coronary artery disease associated with a sedentary lifestyle as compared with an active lifestyle? How has this been established?

7. What are three basic physiological alterations resulting from exercise training that would reduce the risk of death from coronary artery disease?

8. In what ways does endurance exercise training alter risk factors for heart disease?

9. What is a sedentary individual's risk of developing hypertension compared with an active individual's?

10. What are three basic physiological alterations resulting from exercise training that would reduce the risk for developing hypertension?

11. What changes in blood pressure that result from endurance exercise training occur in moderately hypertensive individuals?

12. Of what value is cardiac rehabilitation in treating a patient who has had a heart attack?

13. What is the risk of death with endurance exercise training?

▷ References

1. Albert, C.M., Mittleman, M.A., Chae, C.U., Lee, I.-M., Hennekens, C.H., & Manson, J.E. (2000). Triggering of sudden death from cardiac causes by vigorous exertion. *New England Journal of Medicine, 343,* 1355-1361.

2. American College of Sports Medicine Position Stand. (1993). Physical activity, physical fitness, and hypertension. *Medicine and Science in Sports and Exercise, 25*(10), i-x.

3. American College of Sports Medicine Position Stand. (1994). Exercise for patients with coronary artery disease. *Medicine and Science in Sports and Exercise, 26*(3), i-v.

4. American Heart Association. (2002). *2003 heart and stroke statistics.* Dallas: American Heart Association.

5. Berenson, G.S., Srinivasan, S.R., Bao, W., Newman, W.P., Tracy, R.E., & Wattigney, W.A. (1998). Association between multiple cardiovascular risk factors and atherosclerosis in children and young adults. The Bogalusa Heart Study. *New England Journal of Medicine, 338,* 1650-1656.

6. Blair, S.N., Goodyear, N.N., Gibbons, L.W., & Cooper, K.H. (1984). Physical fitness and incidence of hypertension in healthy normotensive men and women. *Journal of the American Medical Association, 252,* 487-490.

7. Blair, S.N., & Jackson, A.S. (2001). Guest editorial: Physical fitness and activity as separate heart disease risk factors: A meta-analysis. *Medicine and Science in Sports and Exercise, 33,* 762-764.

8. Blair, S.N., Kohl, H.W., Paffenbarger, R.S., Clark, D.G., Cooper, K.H., & Gibbons, L.W. (1989). Physical fitness and all-cause mortality: A prospective study of healthy men and women. *Journal of the American Medical Association, 262,* 2395-2401.

9. Caspersen, C.J. (1987). Physical inactivity and coronary heart disease. *Physician and Sportsmedicine, 15*(11), 43-44.

10. Cooper, K.H., Pollock, M.L., Martin, R.P, White, S.R., Linnerud, A.C., & Jackson, A. (1976). Physical fitness levels vs. selected coronary risk factors: A cross-sectional study. *Journal of the American Medical Association, 236,* 166-169.

11. Dunn, A.L., & Dishman, R.K. (1991). Exercise and the neurobiology of depression. *Exercise and Sport Sciences Reviews, 19,* 41-98.

12. Durstine, J.L., Grandjean, P.W., Davis, P.G., Ferguson, M.A., Alderson, N.L, DuBose, K.D. (2001). Blood lipid and lipoprotein adaptations to exercise. *Sports Medicine, 31,* 1033-1062.

13. Durstine, J.L., & Haskell, W.L. (1994). Effects of exercise training on plasma lipids and lipoproteins. *Exercise and Sport Sciences Reviews, 22,* 477-521.

14. Ehsani, A.A. (1987). Cardiovascular adaptations to endurance exercise training in ischemic heart disease. *Exercise and Sport Sciences Reviews, 15,* 53-66.

15. Ekelund, L.-G., Haskell, W.L., Johnson, J.L., Whaley, F.S., Criqui, M.H., & Sheps, D.S. (1988). Physical fitness as a predictor of cardiovascular mortality in asymptomatic North American men: The Lipid Research Clinics mortality follow-up study. *New England Journal of Medicine, 319,* 1379-1384.

16. Enos, W.F., Holmes, R.H., & Beyer, J. (1953). Coronary disease among United States soldiers killed in action in Korea. *Journal of the American Medical Association, 152,* 1090-1093.

17. Fagard, R.H., & Tipton, C.M. (1994). Physical activity, fitness, and hypertension. In C. Bouchard, R.J. Shephard, & T. Stephens (Eds.), *Physical activity, fitness, and health* (pp. 633-655). Champaign, IL: Human Kinetics.

18. Fletcher, G.F., Blair, S.N., Blumenthal, J., Caspersen, C., Chaitman, B., Epstein, S., Falls, H., Froelicher, E.S.S., Froelicher, V.F., & Pina, I.L. (1992). Statement on exercise: Benefits and recommendations for physical activity programs for all Americans. *Circulation, 86,* 340-344.

19. Franklin, B.A., & Kahn, J.K. (1996). Delayed progression or regression of coronary atherosclerosis with intensive risk factor modification: Effects of diet, drugs, and exercise. *Sports Medicine, 22,* 306-320.

20. Gibbons, L.W., Blair, S.N., Cooper, K.H., & Smith, M. (1983). Association between coronary heart disease risk factors and physical fitness in healthy adult women. *Circulation, 67,* 977-983.

21. Goldberg, L., & Elliot, D.L. (1985). The effect of physical activity on lipid and lipoprotein levels. *Medical Clinics of North America, 69,* 41-55.

22. Hagberg, J.M. (1990). Exercise, fitness, and hypertension. In C. Bouchard, R.J. Shephard, T. Stephens, J.R. Sutton, & B.D. McPherson (Eds.), *Exercise, fitness, and health* (pp. 455-466). Champaign, IL: Human Kinetics.

23. Hagberg, J.M., Park, J.-J., & Brown, M.D. (2000). The role of exercise training in the treatment of hypertension. *Sports Medicine, 30,* 193-206.

24. Hagberg, J.M., & Seals, D.R. (1986). Exercise training and hypertension. *Acta Medica Scandinavica,* (Suppl. 711), 131-136.

25. Haskell, W.L., Leon, A.S., Caspersen, C.J., Froelicher, V.F., Hagberg, J.M., Harlan, W., Holloszy, J.O., Regensteiner, J.G., Thompson, P.D., Washburn, R.A., & Wilson, P.W.F. (1992). Cardiovascular benefits and assessment of physical activity and physical fitness in adults. *Medicine and Science in Sports and Exercise, 24,* S201-S220.

26. Hu, F.B., Stampfer, M.J., Colditz, G.A., Ascherio, A., Rexrode, K.M., Willett, W.C., & Manson, J.E. (2000). Physical activity and risk of stroke in women. *Journal of the American Medical Association, 283,* 2961-2967.

27. Joint National Committee on Prevention, Detection, Evaluation, and Treatment of High Blood Pressure. (2003). The seventh report of the Joint National Committee on Prevention, Detection, Evaluation, and Treatment of High Blood Pressure. *Journal of the American Medical Association, 289,* 2560-2572.

28. Kannel, W.B., & Dawber, T.R. (1972). Atherosclerosis as a pediatric problem. *Journal of Pediatrics, 80,* 544-554.

29. Kelley, G., & Tran, Z.V. (1995). Aerobic exercise and normotensive adults: A meta-analysis. *Medicine and Science in Sports and Exercise, 27,* 1371-1377.

30. Kirkcaldy, B. (1989). Exercise as a therapeutic modality. *Medicine and Sports Science, 29,* 166-187.

31. Kramsch, D.M., Aspen, A.J., Abramowitz, B.M., Kreimendahl, T., & Hood, W.B. (1981). Reduction of coronary atherosclerosis by moderate conditioning exercise in monkeys on an atherogenic diet. *New England Journal of Medicine, 305,* 1483-1489.

32. LaPorte, R.E., Adams, L.L., Savage, D.D., Brenes, G., Dearwater, S., & Cook, T. (1984). The spectrum of physical activity, cardiovascular disease and health: An epidemiologic perspective. *American Journal of Epidemiology, 120,* 507-517.

33. Lee, C.D., & Blair, S.N. (2002). Cardiorespiratory fitness and stroke mortality in men. *Medicine and Science in Sports and Exercise, 34,* 592-595.

34. Lee, I.-M., & Paffenbarger, R.S., Jr. (1996). Do physical activity and physical fitness avert premature mortality? *Exercise and Sport Sciences Reviews, 24,* 135-171.

35. Leon, A.S., & Connett, J. (1991). Physical activity and 10.5 year mortality in the Multiple Risk Factor Intervention Trial (MRFIT). *International Journal of Epidemiology, 20,* 690-697.

36. Leon, A.S., Connett, J., Jacobs, D.R., & Rauramaa, R. (1987). Leisure-time physical activity levels and risk of coronary heart disease and death. *Journal of the American Medical Association, 258,* 2388-2395.

37. Marieb, E.N. (1995). *Human anatomy and physiology* (3rd ed.). Redwood City, CA: Benjamin/Cummings.

38. McNamara, J.J., Molot, M.A., Stremple, J.F., & Cutting, R.T. (1971). Coronary artery disease in combat casualties in Vietnam. *Journal of the American Medical Association, 216,* 1185-1187.

39. Montoye, H.J., Metzner, H.L., Keller, J.B., Johnson, B.C., & Epstein, F.H. (1972). Habitual physical activity and blood pressure. *Medicine and Science in Sports and Exercise, 4,* 175-181.

40. Morgan, W.P. (1994). Physical activity, fitness and depression. In C. Bouchard, R.J. Shephard, & T. Stephens (Eds.), *Physical activity, fitness, and health* (pp. 851-867). Champaign, IL: Human Kinetics.

41. Morris, J.N., Adam, C., Chave, S.P.W., Sirey, C., Epstein, L., & Sheehan, D.J. (1973). Vigorous exercise in leisure-time and the incidence of coronary heart-disease. *Lancet, 1,* 333-339.

42. Morris, J.N., Heady, J.A., Raffle, P.A.B., Roberts, C.G., & Parks, J.W. (1953). Coronary heart-disease and physical activity of work. *Lancet, 265,* 1053-1057, 1111-1120.

43. Morris, J.N., Pollard, R., Everitt, M.G., Chave, S.P.W., & Semmence, A.M. (1980). Vigorous exercise in leisure-time: Protection against coronary heart disease. *Lancet, 2,* 1207-1210.

44. O'Connor, G.T., Buring, J.E., Yusuf, S., Goldhaber, S.Z., Olmstead, E.M., Paffenbarger, R.S., & Hennekens, C.H. (1989). An overview of randomized trials of rehabilitation with exercise after myocardial infarction. *Circulation, 80,* 234-244.

45. Oldridge, N.B., Guyatt, G.H., Fischer, M.E., & Rimm, A.A. (1988). Cardiac rehabilitation after myocardial infarction: Combined experience of randomized clinical trials. *Journal of the American Medical Association, 260,* 945-950.

46. Paffenbarger, R.S., Hyde, R.T., Wing, A.L., & Hsieh, C.-C. (1986). Physical activity, all-cause mortality, and longevity of college alumni. *New England Journal of Medicine, 314,* 605-613.

47. Petruzzello, S.J., Landers, D.M., Hatfield, B.D., Kubitz, K.A., & Salazar, W. (1991). A meta-analysis on the anxiety-reducing effects of acute and chronic exercise: Outcomes and mechanisms. *Sports Medicine, 11,* 143-182.

48. Pollock, M.L., & Wilmore, J.H. (1990). *Exercise in health and disease: Evaluation and prescription for prevention and rehabilitation* (2nd ed.). Philadelphia: Saunders.

49. Powell, K.E., Thompson, P.D., Caspersen, C.J., & Kendrick, J.S. (1987). Physical activity and the incidence of coronary heart disease. *Annual Reviews in Public Health, 8,* 253-287.

50. Ross, R. (1986). The pathogenesis of atherosclerosis—An update. *New England Journal of Medicine, 314,* 488-500.

51. Seals, D.R., & Hagberg, J.M. (1984). The effect of exercise training on human hypertension: A review. *Medicine and Science in Sports and Exercise, 16,* 207-215.

52. Siscovick, D.S., Weiss, N.S., Fletcher, R.H., & Lasky, T. (1984). The incidence of primary cardiac arrest during vigorous exercise. *New England Journal of Medicine, 311,* 874-877.

53. Steinberg, D., Parthasarathy, S., Carew, T.E., Khoo, J.C., & Witztum, J.L. (1989). Beyond cholesterol: Modifications of low-density lipoprotein that increase its atherogenicity. *New England Journal of Medicine, 320,* 915-924.

54. Tanasescu, M., Leitzmann, M.F., Rimm, E.B., Willett, W.C., Stampfer, M.J., & Hu, F.B. (2002). Exercise type and intensity in relation to coronary heart disease in men. *Journal of the American Medical Association, 288,* 1994-2000.

55. Thompson, P.D. (1982). Cardiovascular hazards of physical activity. *Exercise and Sport Sciences Reviews, 10,* 208-235.

56. Tipton, C.M. (1984). Exercise, training, and hypertension. *Exercise and Sport Sciences Reviews, 12,* 245-306.

57. Tipton, C.M. (1991). Exercise training and hypertension: An update. *Exercise and Sport Sciences Reviews, 19,* 447-505.

58. Tran, Z.V., & Weltman, A. (1985). Differential effects of exercise on serum lipid and lipoprotein levels seen with changes in body weight. *Journal of the American Medical Association, 254,* 919-924.

59. Williams, P.T. (2001). Physical fitness and activity as separate heart disease risk factors: A meta-analysis. *Medicine and Science in Sports and Exercise, 33,* 754-761.

60. Wilmore, J.H., Constable, S.H., Stanforth, P.R., Tsao, W.Y., Rotkis, T.C., Paicius, R.M., Mattern, C.M., & Ewy, G.A. (1982). Prevalence of coronary heart disease risk factors in 13- to 15-year-old boys. *Journal of Cardiac Rehabilitation, 2,* 223-233.

61. Wilmore, J.H., & McNamara, J.J. (1974). Prevalence of coronary heart disease risk factors in boys 8 to 12 years of age. *Journal of Pediatrics, 84,* 527-533.

▷ Selected Readings

American Heart Association. (1996). Cardiovascular preparticipation screening of competitive athletes: Exercise, sudden death. *Medicine and Science in Sports and Exercise, 28,* 1445-1452.

Andersen, L.B., & Hippe, M. (1996). Coronary heart disease risk factors in the physically active. *Sports Medicine, 22,* 213-218.

Caspersen, C.J. (1989). Physical activity epidemiology: Concepts, methods, and applications to exercise science. *Exercise and Sport Sciences Reviews, 17,* 423-473.

Franklin, B.A., & Shephard, R.J. (2000). Avoiding repeat cardiac events. *Physician and Sportsmedicine, 28*(9), 31-58.

Harris, S.S., Caspersen, C.J., DeFriese, G.H., & Estes, E.H., Jr. (1989). Physical activity counseling for healthy adults as a primary preventive intervention in the clinical setting. *Journal of the American Medical Association, 261,* 3590-3598.

Laughlin, M.H., & McAllister, R.M. (1992). Exercise training–induced coronary vascular adaptation. *Journal of Applied Physiology, 73,* 2209-2225.

Leon, A.S. (Ed.). (1997). *Physical activity and cardiovascular health: A national consensus.* Champaign, IL: Human Kinetics.

Morris, C.K., & Froelicher, V.F. (1993). Cardiovascular benefits of improved exercise capacity. *Sports Medicine, 16,* 225-236.

Morris, J.N., & Hardman, A.E. (1997). Walking to health. *Sports Medicine, 23,* 306-332.

Nieman, D.C. (1998). *The exercise-health connection.* Champaign, IL: Human Kinetics.

Pollock, M.L., & Schmidt, D.H. (Eds.). (1995). *Heart disease and rehabilitation* (3rd ed.). Champaign, IL: Human Kinetics.

Schaible, T.F., & Scheuer, J. (1985). Cardiac adaptations to chronic exercise. *Progress in Cardiovascular Disease, 27,* 297-324.

Shephard, R.J. (1986). Exercise in coronary heart disease. *Sports Medicine, 3,* 26-49.

Thompson, P.D. (2001). Exercise rehabilitation for cardiac patients. *Physician and Sportsmedicine, 29*(1), 69-75.

Thompson, P.D. (2001). Cardiovascular risks of exercise. *Physician and Sportsmedicine, 29*(4), 33-47.

Wannamethee, S.G., & Shaper, A.G. (2001). Physical activity in the prevention of cardiovascular disease: An epidemiological perspective. *Sports Medicine, 31,* 101-114.

CHAPTER 21

OBESITY, DIABETES, AND PHYSICAL ACTIVITY

overview

While millions of people are dying of starvation each year in most parts of the world, many Americans are dying as an indirect result of overconsumption of food. Billions of dollars are spent each year overfeeding the American public, which in turn leads to the expenditure of billions of dollars more each year on various weight-loss methods. Another common disorder in America is diabetes mellitus, which affects about 15 million Americans. This disorder of carbohydrate metabolism centers on insulin. Interestingly, scientists have established a link between insulin resistance and obesity, coronary artery disease, and hypertension.

As we saw in the previous chapter, exercise is essential for reducing our risk of coronary artery disease and hypertension. In this chapter, we examine obesity and diabetes, keeping in mind the important link among the four disorders, and again explore the role of physical activity in prevention and treatment.

outline

William "the Refrigerator" Perry, defensive lineman for the Chicago Bears professional football team during the 1980s and early 1990s, reported to the 1988 summer training camp at a weight of 170 kg (375 lb), some 25 kg (55 lb) over his mandated playing weight. Although there was an obvious concern regarding his ability to perform on the football field at this excessive weight, of greater concern are the health risks associated with obesity. Chris Taylor, an Iowa State and U.S. Olympic team wrestler, competed at a weight of between 181 and 204 kg (400-450 lb). He died in his sleep at age 29, most likely from obesity-related causes.

A sedentary lifestyle has been associated with an increased risk for two major metabolic and endocrine disorders: obesity and diabetes. Although neither disease by itself represents a major cause of death, both are strongly associated with other diseases that have high mortality rates, such as hypertension, coronary artery disease, and cancer. Furthermore, millions of Americans have obesity, diabetes, or both. The consequences of these diseases are debilitating, and the costs associated with their treatment are high.

In this chapter we focus on obesity and diabetes, discussing their prevalence, their etiology, health problems associated with each disease, and general treatment options. Finally, we consider the role that physical activity can play in prevention and treatment.

Obesity

The terms *overweight* and *obesity* often are used interchangeably, but technically they have different meanings. In this next section, we discuss terminology in some detail.

Terminology and Standards

Overweight is defined as a body weight that exceeds the normal or standard weight for a particular person based on height and frame size. These standard weights, presented in table 21.1, were established in 1959 but are still the most widely used. New weight and height tables were introduced in 1983, but their introduction was controversial because many experts believed the weight allowances were too liberal. Many professional health organizations have refused to accept the newer tables.

Weight values in the standard tables are based solely on population averages. For this reason, a person can be overweight according to these standards and yet have a lower than normal body fat content. For example, football players frequently are found to be overweight according to standard tables, yet they are typically much leaner than people of the same age, height, and frame size who are of normal weight or who are even underweight (see chapter 14). Still other people are within the normal range of body weights for their height and frame size by the standard tables and yet are obese.

Obesity refers to the condition of having an excessive amount of body fat. This implies that the actual amount of body fat or its percentage of total weight must be assessed or estimated (see chapter 14 for assessment techniques). Exact standards for allowable fat percentages have not been established. However, men with more than 25% body fat and women with more than 35% should be considered obese. Men with relative fat values of 20% to 25% and women with values of 30% to 35% should be considered borderline obese.

Most studies of large populations have used measurements of weight alone, weight combined with height, or skinfold thickness to estimate the prevalence of overweight and obesity. Relative weight is a term used to express the percentage by which an individual is either overweight or underweight and is used as an index of obesity as well as overweight. This value generally is determined by dividing a person's weight by the mean weight (from the standard weight tables) for the medium frame category, based on height. The result is then expressed as a percentage. For example, according to table 21.1, a man who weighs 104

Table 21.1

Standard Tables (1959) for Men's and Women's Weight for Height and Frame Size

Men ages 25 years and over					Women ages 25 years and over				
Height w/ shoes 1-in. heels		Desirable weight			Height w/ shoes 2-in. heels		Desirable weight		
Ft	In.	Small frame	Medium frame	Large frame	Ft	In.	Small frame	Medium frame	Large frame
5	2	112-120	118-129	126-141	4	10	92-98	96-107	104-119
5	3	115-123	121-133	129-144	4	11	94-101	98-110	106-122
5	4	118-126	124-136	132-148	5	0	96-104	101-113	109-125
5	5	121-129	127-139	135-152	5	1	99-107	104-116	112-128
5	6	124-133	130-143	138-156	5	2	102-110	107-119	115-131
5	7	128-137	134-147	142-161	5	3	105-113	110-122	118-134
5	8	132-141	138-152	147-166	5	4	108-116	113-126	121-138
5	9	136-145	142-156	151-170	5	5	111-119	116-130	125-142
5	10	140-150	146-160	155-174	5	6	114-123	120-135	129-146
5	11	144-154	150-165	159-179	5	7	118-127	124-139	133-150
6	0	148-158	154-170	164-184	5	8	122-131	128-143	137-154
6	1	152-162	158-175	168-189	5	9	126-135	132-147	141-158
6	2	156-167	162-180	173-194	5	10	130-140	136-151	145-163
6	3	160-171	167-185	178-199	5	11	134-144	140-155	149-168
6	4	164-175	172-190	182-204	6	0	138-148	144-159	153-173

Note: Weights include indoor clothing. For nude weight, deduct 5 to 7 lb for men and 2 to 4 lb for women.

Reprinted courtesy of Metropolitan Life Insurance Company, Statistical Bulletin.

kg (230 lb) at a height of 183 cm (6 ft) would have a relative weight of 142% (104 kg ÷ 73 kg = 142%, or 230 lb ÷ 162 lb = 142%). The figure 73 kg (162 lb) is the mean value for the medium frame category for a 6-ft (183-cm) man (154-170 lb, or 70-77 kg).

Body mass index (BMI) is a frequently used standard to estimate obesity. A person's BMI is determined by dividing body weight in kilograms by the square of body height in meters. As an example, the man who weighs 104 kg (230 lb) and is 183 cm (6 ft) tall would have a BMI of 31 kg/m²; $104 \text{ kg}/(1.83 \text{ m})^2 = 104$ kg/3.35 m² = 31 kg/m². Generally, the BMI is related to body composition. It is highly correlated with relative body fat and probably provides a better estimate of obesity than does relative weight.

In 1997, the World Health Organization proposed a classification system for underweight, overweight, and obesity based solely on BMI values.[40] This classification system was adopted by the National Institutes of Health in 1998 with several modifications (see table 21.2) and has been used widely since 2000.[22] In table 21.2, BMI values have been divided into five categories: underweight, normal weight, overweight, obesity, and extreme obesity. Within the obesity classification, there are two subclassifications, Class I and Class II. Extreme obesity is Class III. The degree of disease risk also is included and is determined by both BMI and waist circumference. A larger waist circumference increases risk for a given BMI category. Waist circumference reflects the role of abdominal visceral fat in increasing your risk for disease. Table 21.3 on pages 670-671 provides a simple way of determining your BMI from height and weight.

This classification system has made a major contribution to our understanding of the true prevalence of overweight and obesity. Prior to the adoption of this system, there was a wide range of estimates of the percentage of adults who were overweight or obese or both. This wide range of estimates, which was the result of studies using different cut-points or standards for overweight and obesity, led to considerable confusion among scientists and the general public about the true prevalence of weight disorders. We can now better understand the true prevalence of overweight and obesity and how it has changed over time.

Prevalence of Overweight and Obesity in the United States

The prevalence of overweight and obesity in the United States has increased dramatically over the past 40 years, as illustrated in figure 21.1. This figure presents data from national surveys of large numbers of men and women representative of the total U.S. population obtained in 1960-1962, 1971-1974, 1976-1980, 1988-1994, and 1999-2000.[10, 11] The percentage overweight represents the percentage of the total population with a BMI of 25.0 to 29.9 (figure 21.1a), and the percentage obese is for those with BMI values of 30.0 or greater (figure 21.1b), consistent with the World Health Organization/National Institutes of Health classification system. Figure 21.1c presents

Table 21.2

Classification of Overweight and Obesity by BMI, Waist Circumference, and Associated Disease Risk[a]

Classification	BMI (kg/m²)	Obesity class	Disease risk (relative to normal weight and waist circumference)	
			Men ≤40 in. (102 cm) Women ≤35 in. (88 cm)	Men >40 in. (102 cm) Women >35 in. (88 cm)
Underweight	<18.5		—	—
Normal[b]	18.5-24.9		—	—
Overweight	25.0-29.9		Increased	High
Obesity	30.0-34.9	I	High	Very high
	35.0-39.9	II	Very high	Very high
Extreme obesity	≥40	III	Extremely high	Extremely high

[a]Disease risk for type 2 diabetes, hypertension, and cardiovascular disease.

[b]Increased waist circumference also can be a marker for increased risk even in persons of normal weight.

Adapted, by permission, from World Health Organization, 1998, "Obesity: Preventing and managing the global epidemic." In *Report of a WHO Consultation on Obesity* (Geneva: WHO).

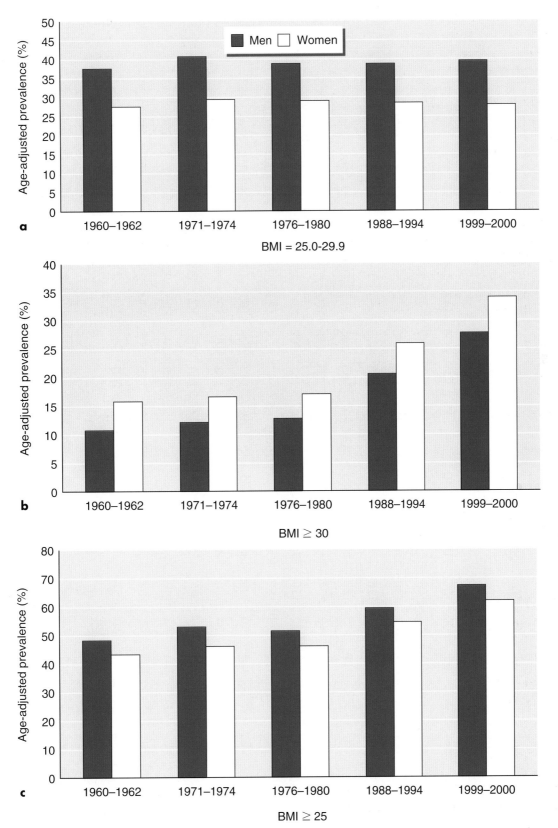

▲ **Figure 21.1** The increasing prevalence of overweight (*a*) and obesity (*b*) and the combination of overweight and obesity (*c*) in the United States from 1960 through 2000.

Adapted from Flegal et al. 1998[10] and Flegal et al. 2002.[11]

Table 21.3

Body Mass Index

BMI Height (inches)	19	20	21	22	23	24	25	26	27	28	29	30	31	32	33	34	35	36
								Body Weight (pounds)										
58	91	96	100	105	110	115	119	124	129	134	138	143	148	153	158	162	167	172
59	94	99	104	109	114	119	124	128	133	138	143	148	153	158	163	168	173	178
60	97	102	107	112	118	123	128	133	138	143	148	153	158	163	168	174	179	184
61	100	106	111	116	122	127	132	137	143	148	153	158	164	169	174	180	185	190
62	104	109	115	120	126	131	136	142	147	153	158	164	169	175	180	186	191	196
63	107	113	118	124	130	135	141	146	152	158	163	169	175	180	186	191	197	203
64	110	116	122	128	134	140	145	151	157	163	169	174	180	186	192	197	204	209
65	114	120	126	132	138	144	150	156	162	168	174	180	186	192	198	204	210	216
66	118	124	130	136	142	148	155	161	167	173	179	186	192	198	204	210	216	223
67	121	127	134	140	146	153	159	166	172	178	185	191	198	204	211	217	223	230
68	125	131	138	144	151	158	164	171	177	184	190	197	203	210	216	223	230	236
69	128	135	142	149	155	162	169	176	182	189	196	203	209	216	223	230	236	243
70	132	139	146	153	160	167	174	181	188	195	202	209	216	222	229	236	243	250
71	136	143	150	157	165	172	179	186	193	200	208	215	222	229	236	243	250	257
72	140	147	154	162	169	177	184	191	199	206	213	221	228	235	242	250	258	265
73	144	151	159	166	174	182	189	197	204	212	219	227	235	242	250	257	265	272
74	148	155	163	171	179	186	194	202	210	218	225	233	241	249	256	264	272	280
75	152	160	168	176	184	192	200	208	216	224	232	240	248	256	264	272	279	287
76	156	164	172	180	189	197	205	213	221	230	238	246	254	263	271	279	287	295

Table 21.3 *(continued)*

BMI	37	38	39	40	41	42	43	44	45	46	47	48	49	50	51	52	53	54
Height (inches)									*Body Weight (pounds)*									
58	177	181	186	191	196	201	205	210	215	220	224	229	234	239	244	248	253	258
59	183	188	193	198	203	208	212	217	222	227	232	237	242	247	252	257	262	267
60	189	194	199	204	209	215	220	225	230	235	240	245	250	255	261	266	271	276
61	195	201	206	211	217	222	227	232	238	243	248	254	259	264	269	275	280	285
62	202	207	213	218	224	229	235	240	246	251	256	262	267	273	278	284	289	295
63	208	214	220	225	231	237	242	248	254	259	265	270	278	282	287	293	299	304
64	215	221	227	232	238	244	250	256	262	267	273	279	285	291	296	302	308	314
65	222	228	234	240	246	252	258	264	270	276	282	288	294	300	306	312	318	324
66	229	235	241	247	253	260	266	272	278	284	291	297	303	309	315	322	328	334
67	236	242	249	255	261	268	274	280	287	293	299	306	312	319	325	331	338	344
68	243	249	256	262	269	276	282	289	295	302	308	315	322	328	335	341	348	354
69	250	257	263	270	277	284	291	297	304	311	318	324	331	338	345	351	358	365
70	257	264	271	278	285	292	299	306	313	320	327	334	341	348	355	362	369	376
71	265	272	279	286	293	301	308	315	322	329	338	343	351	358	365	372	379	386
72	272	279	287	294	302	309	316	324	331	338	346	353	361	368	375	383	390	397
73	280	288	295	302	310	318	325	333	340	348	355	363	371	378	386	393	401	408
74	287	295	303	311	319	326	334	342	350	358	365	373	381	389	396	404	412	420
75	295	303	311	319	327	335	343	351	359	367	375	383	391	399	407	415	423	431
76	304	312	320	328	336	344	353	361	369	377	385	394	402	410	418	426	435	443

Adapted, by permission, from World Health Organization, 1998, "Obesity: Preventing and managing the global epidemic." In *Report of a WHO Consultation on Obesity* (Geneva: WHO).

the data for the combination of those who are overweight and obese (BMI ≥25.0). It is most striking to note that nearly 67% of men and 62% of women in the United States are overweight or obese, and that the prevalence of obesity increased by 62% in men and 52% in women between the 1976-1980 and the 1988-1994 data collection periods[10] and by an additional 34% in men and 31% in women between the 1988-1994 and 1999-2000 periods.[11] Interestingly, the prevalence of overweight remained relatively constant over these same time periods. When we look at these data by race, it is apparent that the problem is much more significant in Mexican-American men and women and in Black women (figure 21.2). These trends are not unique to the United States. Canada, Australia, and most of Europe have seen similar increases, but, with few exceptions, not to the extent seen in the United States.[40]

Unfortunately, this same trend of increasing prevalence of overweight has been reported in U.S. children and adolescents.[23, 32] There was a substantial increase in the prevalence of overweight from the 1976-1980 survey to the most recent 1999-2000 survey in both 6- to 11-year-old children and 12- to 19-year-old adolescents. Figure 21.3 illustrates the trends in the preva-

lence of overweight from 1963 through 2000 in preadolescent and adolescent boys and girls. Because BMI is much less precise for estimating body fat in children and adolescents, scientists typically use the cut-point for BMI of greater than the 95th percentile, a value that likely indicates the child is overfat. Similar to the adult trends shown in figure 21.1, the prevalence of overweight remained relatively constant from 1963 through 1980 but has increased dramatically since 1980.[23] Although there are no significant differences between boys and girls, there are substantial differences by race, with black and Mexican-American children and adolescents having nearly twice the prevalence compared with white children.

> More than 67% of men and nearly 62% of women in the U.S. adult population are overweight or obese, and the prevalence of overweight in children has increased at an alarming rate since 1980.

The average person in the United States will gain approximately 0.45 kg (1 lb) of additional weight each year after age 25. Such a seemingly

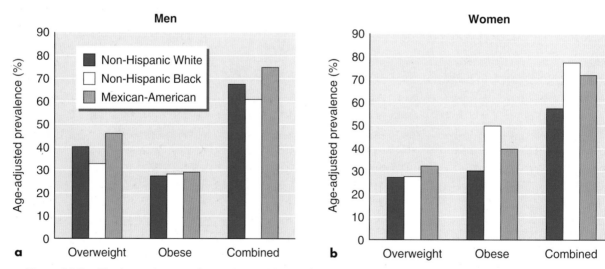

▲ **Figure 21.2** The increasing prevalence of overweight and obesity and the combination of overweight and obesity in (*a*) men and (*b*) women by race.

From Flegal et al., 2002, "Prevalence and trends in obesity among US adults, 1999-2000," *Journal of the American Medical Association* 288: 1723-1727.[11]

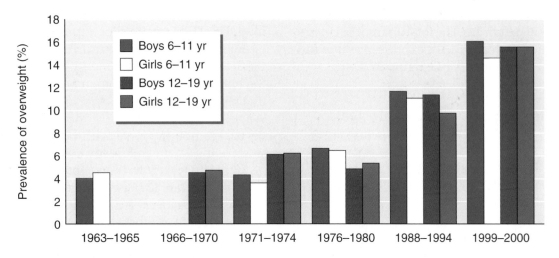

▲ Figure 21.3 The increasing prevalence of overweight (95th percentile) in children and adolescents in the United States from 1962 through 2000.

From Ogden et al., 2002, "Prevalence and trends in overweight among US children and adolescents," *Journal of the American Medical Association* 288: 1728-1732.[23]

small gain, however, results in 14 kg (30 lb) of excess weight by age 55. At the same time, bone and muscle mass decrease by approximately 0.2 kg (0.5 lb) per year because of reduced physical activity. Taking this into account, an average person's body fat actually increases by 0.7 kg (1.5 lb) each year, or a 20-kg (45-lb) fat gain over a 30-year period! It is no wonder that weight loss is an American obsession.

Control of Body Weight

We must understand how body weight is controlled or regulated to better understand how a person becomes obese. Body weight regulation has puzzled scientists for years. The human body takes in an average of about 2,500 kcal per day, or nearly one million kcal per year. The average gain of 0.7 kg (1.5 lb) of fat each year represents an imbalance of only 5,250 kcal per year between energy intake and expenditure (3,500 kcal is the energy equivalent of 0.45 kg, or 1 lb, of adipose tissue). This translates into a surplus of less than 15 kcal per day. Even with a weight gain of 1.5 lb (0.7 kg) of fat per year, the body can balance caloric intake to within one potato chip per day of what is expended! That is truly remarkable.

> The body has the ability to balance energy intake and expenditure to within 10 to 15 kcal per day, about the equivalent of one potato chip!

The body's ability to balance its caloric intake and expenditure to within such a narrow range has led scientists to propose that body weight is regulated around a given set point, similar to the way in which body temperature is regulated. Excellent evidence for this is found in the animal research literature.[17] When animals are force-fed or starved for various periods of time, their weights respectively increase or decrease markedly. But when they go back to their normal eating patterns, they always return to their original weight or to the weight of the control animals (for animals that naturally continue to gain weight throughout their life span).

Similar results have been found in humans, although the number of studies is limited. Subjects placed on semistarvation diets have lost up to 25% of their body weight but regained that weight within months of returning to a normal diet.[18] In a study involving Vermont prisoners, overfeeding resulted in weight gains of 15% to 25%, yet their weights returned to original levels shortly after the experiment ended.[27]

How can the body do this? Let's consider energy expenditure. The total amount of energy expended each day can be expressed as the sum of three components (see figure 21.4):

1. Resting metabolic rate (RMR)
2. The thermic effect of a meal (TEM)
3. The thermic effect of activity (TEA)

Resting metabolic rate (RMR) is your body's metabolic rate early in the morning following an overnight fast and 8 h of sleep. The term basal metabolic rate (BMR) also is used but generally implies that the individual sleeps over in the clinical facility where the metabolic rate measurement will be made. Most research today uses resting metabolic rate. This value, as we learned in chapter 4, represents the minimal amount of energy expenditure needed to support basic physiological processes. It accounts for 60% to 75% of the total energy we expend each day.

The **thermic effect of a meal (TEM)** represents the increase in the metabolic rate that is associated with the digestion, absorption, transport, metabolism, and storage of ingested food. The TEM accounts for approximately 10% of our total energy expenditure each day. This value also includes some energy waste, because the body can increase its metabolic rate above that necessary for food processing and storage. We seldom notice the TEM; however, after a very large holiday meal with family, you will start feeling warm and sleepy, with small beads of sweat forming on your forehead. These changes indicate that your metabolic rate has increased considerably. The TEM component of metabolism might be defective in people with obesity, possibly attributable to a defect in the energy-wastage component, leading to a surplus of calories.

The **thermic effect of activity (TEA)** is simply the energy expended above the resting metabolic rate to accomplish a given task or activity, whether it is combing your hair or running a 10-km race. The TEA accounts for the remaining 15% to 30% of our energy expenditure.

The body adapts to major increases or decreases in energy intake by altering the energy expended by each of these three components—RMR, TEM, and TEA. With fasting or very low calorie diets, all three decrease. The body appears to be attempting to conserve its energy stores. This is dramatically illustrated by decreases in resting metabolic rate of 20% to 30% or more reported within several weeks after patients begin fasting or a very low calorie diet. Conversely, all three components of energy expenditure increase with overeating. In this case, the body appears to be trying to prevent unnecessary storage of the surplus calories. All of these adaptations may be under the control of the sympathetic nervous system and may play a major, if not the primary, role in maintaining weight around a given set point. This remains a most important area for future research.

If the body has a set-point weight, how can we explain the increasing prevalence of overweight and obesity? It appears that the set point can change, at least in the animals that have been most extensively studied. As an example,

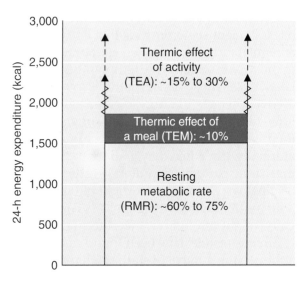

▲ **Figure 21.4** The three components of energy expenditure. See the text for a detailed explanation.

Adapted, by permission, from E.T. Poehlman, 1989, "A review: Exercise and its influence on resting energy metabolism in man," *Medicine and Science in Sports and Exercise* 21: 515-525.

the set-point weight of rats maintained on a high-fat diet over a 6-month period tends to increase: When the rats are placed back on a low-fat diet, they do not go back to their expected weight but stabilize at a much higher weight. For dieting periods of less than 6 months, they do tend to go back to their original set point. The composition of the diet is therefore a prime suspect for increasing set-point weight, as is physical activity level. It is quite possible that an increase in the fat content of the diet and a decrease in physical activity levels allow set-point weight to increase. This could at least partially explain the increasing prevalence of overweight and obesity in the United States today. It is also important to note that people generally consume more calories per day when they are placed on high-fat diets.

Another related factor has evolved during the 1990s—supersizing food portions. Fast food restaurants and many restaurant food chains have started serving much larger portions of food, a trend referred to as *supersizing*. How much difference does it make when you supersize a meal? At several of your fast-food chains, a single meal including a double cheeseburger (~1,000 kcal), supersize French fries (~600 kcal), and a supersize 42-oz soft drink (~400 kcal) provides about 2,000 kcal of energy. For many people, this would be sufficient to supply their total daily caloric needs!

The trends of food availability and food purchasing and preparation in the United States, between 1970 and 1998, reveal that the per capita energy availability increased by 15%.[14] Furthermore, Americans are eating more meals outside the home, relying more heavily on convenience foods, and consuming larger food portions. These trends point to a general increase in daily energy intake over the past 30 years.

> The body attempts to defend its weight when overfed or underfed by increasing or decreasing the three components of energy expenditure: RMR, TEM, and TEA.

▶ Overweight is a body weight that exceeds the standard weight for a certain height and frame size. Obesity refers to having excessive body fat, meaning more than 25% body fat for men and more than 35% body fat for women.

▶ Relative weight refers to the percentage by which a person is either over- or underweight on the basis of standard height and weight tables.

▶ A person's body mass index (BMI) is calculated by dividing body weight in kilograms by the square of height in meters. This value is highly correlated with relative body fat and provides a better estimate of obesity than does relative weight.

▶ Prevalence of obesity and overweight in the United States has increased dramatically over the past 40 years.

▶ The average person gains 0.45 kg (1 lb) per year after age 25 but also loses 0.2 kg (0.5 lb) of fat-free mass per year, meaning a net gain of 0.7 kg (1.5 lb) of fat each year.

▶ Body weight appears to be regulated around a set point.

▶ Daily energy expenditure is reflected by the sum of the resting metabolic rate (RMR), the thermic effect of a meal (TEM), and the thermic effect of activity (TEA). The body adapts to changes in energy intake by adjusting any or all of these components.

Etiology of Obesity

At various times throughout human history, obesity has been thought to be caused by basic hormonal imbalances resulting from failure of one or more of the endocrine glands to properly regulate body weight. At other times, it has been believed that gluttony, rather than glandular malfunction, was the primary cause of obesity. In the first case, a person is perceived as having no control over the situation, and yet in the second, he or she is held directly responsible! Results of recent medical and physiological research show that obesity can be the result of any one or a combination of many factors. Its etiology, or cause, is not as simple as was once believed.

Experimental studies on animals have linked obesity to hereditary (genetic) factors. Studies of humans by Dr. Albert Stunkard and his colleagues at the University of Pennsylvania have shown a direct genetic influence on height, weight, and BMI.[28, 29, 30] A study from Laval University in Quebec provided possibly the strongest evidence yet of a significant genetic component for obesity.[5] The investigators took 12 pairs of young adult male monozygotic (identical) twins and housed them in a closed section of a dormitory under 24-h observation for 120 consecutive days. The subjects' diets were monitored during the initial 14 days to determine their baseline caloric intake. Over the next 100 days, the subjects were fed 1,000 kcal above their baseline consumption for 6 of every 7 days. On the seventh day, the subjects were fed only their baseline diet. Thus, they were overfed by 1,000 kcal per day for 84 of the 100 days. Activity levels were also tightly controlled. At the end of the study period, as shown in figure 21.5, the actual weight gained varied widely, from 4.3 to 13.3 kg (9.5-29.3 lb)—a threefold variation in weight gain for overconsumption of the same calories. However, the response of both twins in any given twin pair was quite similar; the major variations occurred between different twin pairs. Similar results were found for gains in fat mass, percentage body fat, and subcutaneous fat.

> Recent research confirms that there is a significant genetic component in the etiology of obesity. However, it is possible to be obese, attributable basically to lifestyle choices, in the absence of a family history (genetics) of obesity. It is also possible to be relatively lean, even though you have a genetic predisposition to obesity, through proper diet and activity levels.

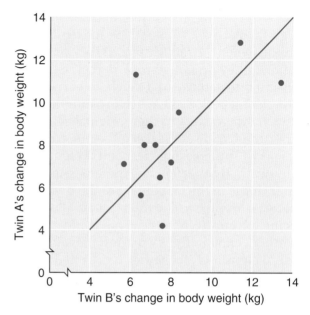

▲ **Figure 21.5** Similarity in weight gains between twins in response to a 1,000-kcal increase in dietary intake for 84 days of a 100-day study. A data point represents the weight gain for each twin in a pair; twin A's value is shown on the *y*-axis, twin B's on the *x*-axis. See the text for further explanation of these data.

Obesity also has been experimentally and clinically linked with both physiological and psychological trauma. Hormonal imbalances, emotional trauma, and alterations in basic homeostatic mechanisms all have been shown to be either directly or indirectly related to the onset of obesity. Environmental factors, such as cultural habits, inadequate physical activity, and improper diets, are major causes of obesity.

Thus, obesity is of complex origin, and the specific causes undoubtedly differ from one person to the next. Recognizing this is important for treating existing obesity and for preventing its onset. To attribute obesity solely to gluttony is unfair and psychologically damaging to people who are concerned about their problem and are attempting to correct it. In fact, several studies have shown that some obese people actually eat less, although they get far less physical activity, than people of the same sex and similar age with average body fat contents.

A Genetic Predisposition to Obesity: The Pima Indians

It has been clearly established that genetics is a major factor in the development of obesity. Dr. Claude Bouchard of Laval University in Quebec has conducted many studies on the heritability of obesity and has concluded that the heritability of fat mass or relative body fat (percentage fat) is about 25% of the age- and sex-adjusted variance.[4] Does having a genetic predisposition to obesity mean that you are destined to be obese? The answer is no! We have learned a great deal from the study of the Pima Indians, who have lived for at least 2,000 years near the Gila River in the Sonora Desert in what is now southern Arizona. Until the turn of this century, the Pima Indians apparently were lean and healthy people who were physically active and ate a healthy diet. As they have moved onto reservations, stopped farming, and started eating a westernized high-fat diet and consuming alcohol, they have become very obese, with a prevalence of obesity of approximately 70%.[19] Associated with their obesity is a high prevalence of diabetes. They are such an unusual population with respect to their extremely high prevalence of obesity and diabetes that the National Institutes of Health established a special research center in Phoenix, Arizona, just to study these Pima Indians and their health issues.

Interestingly, there is another group of Pima Indians who live in northern Mexico. They have stayed active working on farms but use no motorized equipment. Furthermore, they consume a diet high in carbohydrates and low in fat. They have managed to stay relatively lean. The BMIs for these two groups of Indians are seen in figure 21.6. The bottom line is that you can have a predisposition for obesity, but with proper diet and exercise, you can maintain a relatively normal body weight. The Pima Indians have taught us an important lesson.

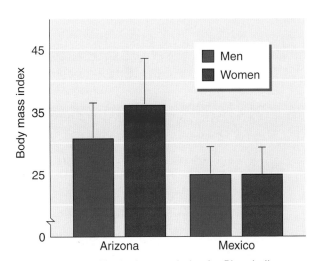

▲ **Figure 21.6**   The body mass index for Pima Indian men and women living in Arizona and in northern Mexico.
From personal communication with Dr. Eric Ravussin, 1997.

Health Problems Associated With Excessive Weight and Obesity

Overweight and obesity are associated with an increased overall rate of death (general excess mortality).[6] This relationship is curvilinear, as shown in figure 21.7. A major increase in risk occurs when the BMI exceeds 30 kg/m², although BMI values between 25.0 and 30.0 are associated with an increased risk for many diseases. Excess mortality associated with obesity and overweight is associated with the following major diseases:

- heart disease,
- hypertension,
- diabetes, and
- certain types of cancer.

With the large increase in the prevalence of obesity in the United States over the past 40 years, it is not surprising to also see a very high prevalence of the metabolic syndrome (chapter 20) in U.S. adults. Data from the third National Health and Nutrition Examination Survey (NHANES-3), conducted between 1988 and

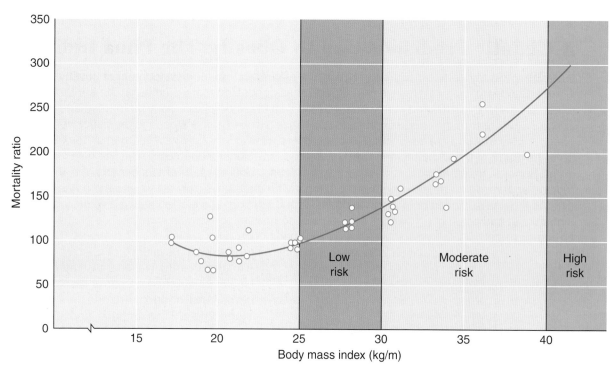

▲ Figure 21.7   The relationship of body mass index to excess mortality. A mortality ratio of 100 represents average mortality. The lowest portion of the curve (body mass indexes under 25) indicates very low risk.

Bray, G.A. "Obesity: Definition, diagnosis and disadvantages." MJA 1985; 142: S2-S8. © Copyright 1985. *The Medical Journal of Australia*—reproduced with permission.

1994, demonstrated that 23.7% of all adults in the United States met the criteria for the metabolic syndrome. In adults over 60 years, the prevalence was greater than 40%. Consistent with the trends in obesity, the prevalence was higher in Mexican-American men and women and in black women.[12]

Furthermore, obesity has been directly related to changes in normal body function, increased risk for certain diseases, detrimental effects on established diseases, and adverse psychological reactions.

Changes in Normal Body Function

The prevalence and extent of changes in body function vary with the individual and with the degree of obesity. Respiratory problems are quite common among people with obesity. These can lead to some other common consequences of obesity, such as lethargy (sluggishness), because of increased carbon dioxide levels in the blood, and polycythemia (increased red blood cell production) in response to lower arterial blood oxygenation. These can lead to abnormal blood clotting (thrombosis), enlarge-

ment of the heart, and congestive heart failure. Those with obesity typically have a lower exercise tolerance because of these respiratory problems and also because of the increased body mass that must be moved during exercise. Additional weight gains further reduce activity levels, and exercise tolerance decreases even more.

Increased Risk for Certain Diseases

An increased risk of developing certain chronic degenerative diseases also is associated with obesity. Both hypertension and atherosclerosis have been directly linked to obesity (see chapter 20). So have various metabolic and endocrine disorders, such as impaired carbohydrate metabolism and diabetes. Obesity is a problem associated particularly with the onset of type 2 (non-insulin-dependent) diabetes.

A major research breakthrough has enabled us to better understand the role of obesity as a risk factor for most of these diseases. Since the 1940s, major sex differences in the way in which fat is stored or patterned on the body have been recognized. As shown in figure 21.8, males tend to store fat in the upper body, particularly the abdominal area, whereas females tend to store fat in the lower body, particularly the hips, buttocks, and thighs. Obesity that follows the male pattern is referred to as **upper body (android) obesity,** or apple-shaped obesity, and the female pattern is referred to as **lower body (gynoid) obesity,** or pear-shaped obesity.

Research beginning in the late 1970s and early 1980s established upper body obesity as a risk factor for the following conditions:[3]

- Coronary artery disease
- Hypertension
- Stroke
- Elevated blood lipids
- Diabetes

Furthermore, upper body obesity appears to be more important than total-body fatness as a risk factor for these diseases. Waist and hip circumference, or girth, measurements can be used to identify people with increased risk. A waist/hip girth ratio greater than 1.0 for men and greater than 0.8 for women indicates increased risk. With upper body obesity, the increased risk may result from visceral fat depots' close proximity to the portal circulatory system (circulation to the liver). Figure 21.9 on page 680 shows a young woman being placed into a computed tomography (CT) scanner (figure 21.9*a*) and CT scans at the level of the fourth lumbar vertebra of two men (figure 21.9, *b* and *c*).[26] Subject B has considerably more visceral (deep) abdominal fat than subcutaneous abdominal fat.

> Obesity places you at significantly increased risk for hypertension, diabetes, coronary artery disease, and metabolic and digestive diseases. The health risks associated with obesity are most likely associated with the manner in which the fat is distributed on the body, with upper body obesity, representing high levels of visceral fat, posing significantly greater risk.

a

b

▲ **Figure 21.8**   (*a*) Upper body (android) obesity; (*b*) lower body (gynoid) obesity.

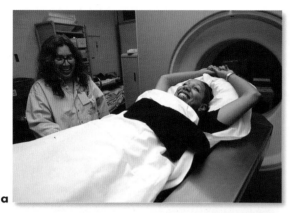

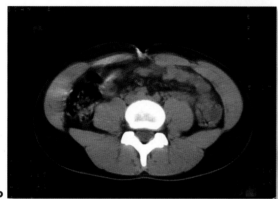

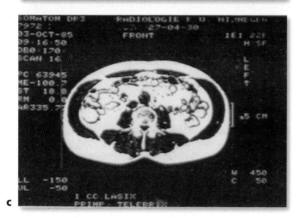

▲ **Figure 21.9** (*a*) A subject in the process of receiving a computed tomography (CT) scan. (*b* and *c*) CT scans at the level of the fourth lumbar vertebra in two people. Subject B (*c*) has considerably more visceral abdominal fat (light areas) than subcutaneous abdominal fat. Subject A (*b*) is leaner but also has considerably less of her fat as visceral fat.

Detrimental Effects on Established Diseases

The effects of obesity on existing diseases are not clear. Obesity can contribute to further development of certain diseases and medical conditions, and weight reduction usually is prescribed as an integral part of treatment.

Conditions that generally benefit from weight reduction include

- angina pectoris,
- hypertension,
- congestive heart disease,
- myocardial infarction (reduced risk of recurrence),
- varicose veins,
- diabetes, and
- orthopedic problems.

Adverse Psychological Reactions

Emotional or psychological problems might be the cause of obesity in a substantial percentage of people with that condition. Furthermore, emotional or psychological problems can arise from the condition itself. In our society, obesity carries a social stigma that contributes substantially to the problems of those who have it. Our media typically glamorize only people with extremely lean bodies. Consequently, some obese people might need professional counseling assistance in their efforts to lose weight.

General Treatment of Obesity

In theory, weight control seems to be a simple matter. The energy consumed by the body in the form of food must equal the total energy expended, which is the sum of the RMR, TEM, and TEA. The body normally maintains a balance between caloric intake and caloric expenditure, but when this balance is upset, weight will be lost or gained. Both weight losses and weight gains appear to depend largely on just two factors: dietary intake and physical activity. This is now recognized as an oversimplification, considering the results of the overfeeding study of monozygotic twins discussed earlier, in which considerable variation in weight gains occurred for the same amount of overfeeding.[5] Not everyone responds to the same intervention in the same way. This difference in response must be considered when one is designing treatment programs for individuals attempting to lose weight, and people trying to lose weight must understand this difference so that those who are low responders won't be discouraged. In the past, we have tended to label the low responders as *noncompliers,* but we know now that this is generally not the case.

Weight loss generally should not exceed 0.45 to 0.9 kg (1-2 lb) per week. Losses greater than this should not be attempted without direct medical supervision. Losing just 0.45 kg (1 lb) of fat a week will result in the loss of 23.4 kg (52 lb) of fat in only a year! Few people become obese that rapidly. Weight loss also should be considered a long-term project. Research and experience have proven that rapid weight losses are usually short lived, and the lost weight is usually quickly regained because rapid weight losses are generally the result of large losses of body water. The body has built-in safety mechanisms to prevent an imbalance in body fluid levels, so the lost water eventually will be replaced. Thus, a person wishing to lose 9 kg (20 lb) of fat is advised to attempt to attain this goal in a minimum of 2-1/2 to 5 months.

Many special diets have achieved popularity over the years, such as the Drinking Man's Diet, the Beverly Hills Diet, the Cambridge Diet, the California Diet, Dr. Stillman's Diet, The Zone, and Dr. Adkin's Diet. Each claims to be the ultimate in effective and comfortable weight loss. Some more recent diets have been developed for use either in the hospital or at home under the supervision of a physician. These are often referred to as very low calorie diets, as they allow only 350 to 500 kcal of food per day. Most of these have been formulated with a certain amount of protein and carbohydrate to minimize the loss of fat-free body mass. Research has shown that many of these are effective, but no single diet has been shown to be more effective than any other. Again, the important factor is the development of a caloric deficit while maintaining a complete, balanced diet that meets the body's vitamin and mineral requirements. The best diet is the one that meets these criteria and is best suited to individual comfort and personality.

Generally, improper eating habits are at least partially responsible for most weight problems, so no diet should be viewed as a quick fix. A person should learn to make permanent changes in dietary habits, especially reducing the intake of fat and simple sugars. For most people, simply eating a low-fat diet will gradually reduce weight to a desirable level. One potential problem with low-fat diets, however, is that people mistakenly assume that low-fat means low-calorie, and this often is not the case. For most people, simply reducing total caloric intake by 250 to 500 kcal per day, combined with a selection of low-fat and low simple sugar foods, would be sufficient to accomplish their desired weight-loss goals.

Hormones and drugs also have been used to assist patients in weight loss by decreasing their appetite or increasing their RMR. Surgical techniques also are used to treat extreme obesity but only as a last resort when other treatment procedures have failed and the obesity is life threatening. Intestinal bypass surgery involves surgically bypassing a large segment of the small intestine, thus reducing food absorption. This procedure is seldom used today because of associated complications. Gastric bypass surgery, which involves surgically reducing the size of the patient's stomach, is also seldom used today because of risks. More recently, gastric stapling to partition off part of the stomach and the insertion of gastric balloons to reduce the capacity of the stomach have been proposed as safer and more effective surgical techniques.

Behavior modification has been proposed as one of the most effective techniques for helping people with weight problems. Major weight losses have been achieved by changing basic behavior patterns associated with eating. Furthermore, these weight losses appear much more permanent. This approach appeals to most people because the techniques seem to make sense and are often easy to incorporate into a normal daily routine. For example, an individual might not have to consciously reduce the amount of food eaten but simply agree that all eating will be done in one location, which often cuts down on snacking. Or an individual might be allowed to take as much food as desired with the first helping, but no second helpings are allowed. Many such simple changes can help regulate eating behavior and result in substantial weight loss.

▶ The etiology of obesity is not simple; it can be caused by any one or a combination of many factors.

▶ Studies of twins indicate that there is a genetic component to obesity. The disorder also has been linked to hormonal imbalances, emotional trauma, homeostatic imbalances, cultural influences, physical inactivity, and improper diets.

▶ Overweight and obesity are associated with increased risk of general excess mortality.

▶ Respiratory problems are quite common among people with obesity. These, in turn, can lead to lethargy and polycythemia.

▶ Obesity increases the risk of certain chronic degenerative diseases. Upper body obesity increases the risk of developing coronary artery disease, hypertension, stroke, elevated blood lipids, diabetes, and the metabolic syndrome. Also, obesity can worsen preexisting health conditions and diseases.

▶ Emotional or psychological problems may contribute to obesity, and the disorder itself, with the stigma it carries, can be psychologically damaging.

▶ In treatment of obesity, it is important to remember that people respond differently to the same intervention.

▶ Weight loss generally should not exceed 0.45 to 0.9 kg (1-2 lb) per week. Simple diet modification, reducing intake of fat and simple sugars, is sufficient to help most people lose weight.

Role of Physical Activity in Weight Control

Inactivity is a major cause of obesity in the United States. In fact, inactivity may be just as important in the development of obesity as overeating! Thus, exercise must be recognized as an essential component in any program of weight reduction or control. Let's examine why physical activity is so important.

Changes in Body Composition With Exercise Training

Physical training can alter body composition. Many people have believed that physical activity has little or no influence on changing body composition and that even vigorous exercise burns too few calories to lead to substantial body fat reductions. Yet research has conclusively demonstrated the effectiveness of exercise training in promoting moderate alterations in body composition.

A person who jogs 3 days a week for 30 min each day at an 11-km/h (7-mph) pace (slightly over 5.4 min/km, or 8.5 min/mi) will expend about 14.5 kcal/min, or 435 kcal for the 30-min run each day. This results in a total expenditure per week of about 1,305 kcal, the equivalent loss of about 0.15 kg (0.33 lb) of adipose tissue (fat plus connective tissue and water) each week just from the exercise period alone. This might lead some people to believe that exercise is a painfully slow way to significantly reduce body fat levels and that there are better and easier ways to lose fat. However, in 52 weeks, providing energy intake remained constant, this person would lose 7.8 kg (17 lb)!

When estimating an activity's energy cost, typically the average or steady-state rate of energy expenditure for that activity is multiplied by the number of minutes that the activity is performed. For example, if the steady-state rate for shoveling snow is 7.5 kcal/min, 1 h of shoveling would require a total of 450 kcal. This would allow an approximate loss of 0.06 kg (0.13 lb) of adipose tissue (450 kcal ÷ 3,500 kcal/0.45 kg of adipose tissue = 0.06 kg, or 450 kcal ÷ 3,500 kcal/lb = 0.13 lb).

But examining the energy expended only during exercise does not give us the full picture. Metabolism remains temporarily elevated after exercise ends. This phenomenon was at one time referred to as the oxygen debt but is now referred to as the **excess postexercise oxygen consumption (EPOC).** Returning the metabolic rate back to its preexercise level can require several minutes following light exercise, such as walking; several hours following very heavy exercise, such as playing a football game; and up to 12 to 24 h or even longer for prolonged, exhaustive exercise, such as running a marathon.

Physical activity is important in both weight maintenance and weight loss. In addition to the calories that are expended during exercise, a substantial expenditure of calories occurs during the postexercise period (EPOC).

The EPOC can require a substantial energy expenditure when considered over the entire recovery period. If, for example, the oxygen consumption following exercise remains elevated by an average of only 0.05 L/min, this will amount to approximately 0.25 kcal/min or 15 kcal/h. If the metabolism remains elevated for 5 h, this would provide an additional expenditure of 75 kcal that would not normally be included in the calculated total energy expenditure for that particular activity. This additional energy expenditure is ignored in most calculations of the energy costs of various activities. The person in this example, by exercising 5 days per week, would expend 375

kcal, or lose the equivalent of about 0.05 kg (0.1 lb) of fat in 1 week, or 0.45 kg (1.0 lb) in 10 weeks, from the additional caloric expenditure during the recovery period alone!

Studies have shown major changes in both weight and body composition with exercise training. One study examined the changes in body composition with diet alone, exercise alone, and a combination of diet and exercise.[41]

Each of three groups of adult women maintained a caloric deficit of 500 kcal per day during a 16-week period of weight loss. The diet-only subjects reduced their daily caloric intake by 500 kcal per day but did not alter their activity level. The exercise-only subjects did not alter their diet but increased their daily activity by 500 kcal. The subjects who combined diet and exercise reduced their daily caloric intake by 250 kcal and increased their daily activity by 250 kcal. Results of this study are illustrated in figure 21.10. Although the

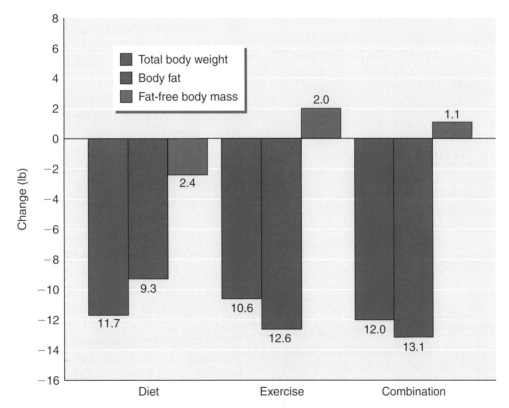

▲ **Figure 21.10**   Changes in total body weight, body fat, and fat-free body mass as a result of using only diet, only exercise, and a combination of diet and exercise.

From W.B. Zuti and L.A. Golding, 1976, "Comparing diet and exercise as weight reduction tools," *Physician and Sportsmedicine* 4: 49-53.[41] Copyright 1976 by McGraw-Hill, Inc. Adapted by permission of McGraw-Hill, Inc.

three groups had similar body weight losses, the two groups who exercised lost substantially more body fat. A major difference was that the two exercise groups gained fat-free body mass, but the diet-only group lost fat-free body mass with weight reduction through diet only.

Most studies, however, have not been able to demonstrate such dramatic changes in weight and body composition with exercise training.[35] But most studies have found similar trends:

- Total weight decreases.
- Fat mass and relative body fat decrease.
- Fat-free mass is either maintained or increased.

Overall, these changes are not large. In a summary of a large number of individual studies that had monitored body composition changes with training, the expected changes from a typical 1-year exercise training program (3 times per week, 30-45 min per day, at 55-75% of $\dot{V}O_2$max) would be as follows: –3.2 kg (–7.1 lb) total body mass, –5.2 kg (–11.5 lb) fat mass, and +2.0 kg (+4.4 lb) fat-free mass.[36] Furthermore, relative body fat would decrease by nearly 6% (e.g., from 30% body fat to 24% body fat).

Although most of these studies have used aerobic training, several other studies have used resistance training and have shown impressive decreases in body fat and increases in fat-free mass. The evidence shows that exercise is an important part of any weight-loss program. But to maximize losses in body weight and body fat, it is necessary to combine exercise with decreased caloric intake.

Attempts to lose weight are much more successful when you lose only 0.45 to 0.9 kg (1-2 lb) per week and when you combine dietary restriction with moderate exercise (300-500 kcal per day). This minimizes loss of fat-free mass and maximizes loss of fat mass.

Mechanisms for Change in Body Weight and Composition

When attempting to explain how exercise causes such changes in body weight and composition, we must consider both sides of the energy-balance equation. Evaluating energy expenditure requires that we consider each of the three components of energy expenditure: RMR, TEM, and TEA. Evaluating energy intake requires that we also consider the energy that is lost in the feces (energy excreted), which is generally less than 5% of the total caloric intake. Keeping this balance in mind, let's examine some of the possible mechanisms through which exercise might affect body weight and body composition.

The energy-balance equation: Energy intake – energy excreted = RMR + TEM + TEA.

Exercise and Appetite   Some believe that exercise stimulates the appetite to such an extent that food intake is unconsciously increased to at least equal that expended during exercise. In 1954, Jean Mayer, a world-famous nutritionist, reported that animals exercising for periods of from 20 min to 1 h per day had a lower food intake than nonexercising control animals.[21] He concluded from this and other studies that when activity is less than a certain minimal level, food intake does not decrease correspondingly and the animal (or human) begins to accumulate body fat. This led to the theory that a certain minimal level of physical activity is necessary for the body to precisely regulate food intake to balance energy expenditure. A sedentary lifestyle may reduce this regulatory ability, resulting in a positive energy balance and weight gain.

Exercise does, in fact, appear to be a mild appetite suppressant, at least for the first few hours following intense exercise training. Furthermore, studies have shown that the total number of calories consumed per day does not change when a person begins a training program. Although some people interpret this as evidence that exercise does not affect appetite, a more accurate conclusion might be that appetite was affected, in fact suppressed, because caloric intake did not increase in proportion to the additional caloric expenditure from the exercise program. In studies conducted on rats, male rats appear to reduce food intake with exercise training, whereas female rats tend to eat the same or even more than nonexercising control rats.[24] There is no obvious explanation for this sex difference. It is unclear whether there is a sex difference in humans.

The decrease in appetite might occur only with intense levels of exercise, in which the resulting increased catecholamine (epinephrine and norepinephrine) levels might suppress the appetite. The increased body temperature that accompanies either high-intensity activity or almost any activity performed under hot and humid conditions also might suppress appetite. We all know from experience that we desire less food when the weather is hot or when our body temperatures are elevated because of illness. This also might explain why a hard running workout results in little or no desire to eat, yet a hard swimming workout elicits a relatively strong craving for food. In the pool, provided that the water temperature is well below body core temperature, the heat generated by exercise is lost very effectively, so core temperature typically is not elevated to the same extent.

> Regular physical activity may assist in better controlling your appetite so that caloric intake balances caloric expenditure.

Exercise and Resting Metabolic Rate   The effects of exercise on the components of energy expenditure became a major topic of interest among researchers in the late 1980s and early 1990s. Of obvious interest is how exercise training might affect the resting metabolic rate, because it represents 60% to 75% of the total calories expended each day. For example, if a 25-year-old man's total daily caloric intake was 2,700 kcal and his RMR accounted for just 60% of that total ($0.60 \times 2,700 = 1,620$ kcal RMR), a mere

1% increase in his RMR would require an extra 16 kcal expenditure each day, or 5,840 kcal per year. This small increase in RMR alone would account for the equivalent of a 0.8-kg (1.7-lb) fat loss per year!

The role of physical training in increasing RMR has not been totally resolved. Several cross-sectional studies have found that highly trained runners have higher RMRs than untrained people of similar age and size. But other studies have not been able to confirm this.[25] Few longitudinal studies have been conducted to determine the change in RMR in untrained people who undergo training for a period of time. Some of these suggest that RMR might increase following training.[7] However, in a very large study of 40 men and women 17 to 62 years of age (HERITAGE Family Study), a 20-week aerobic training program (3 times per week, 35-55 min per day, at 55-75% of $\dot{V}O_2$max) failed to increase RMR even though $\dot{V}O_2$max increased by nearly 18%.[39] Because RMR is closely related to the fat-free mass of the body (fat-free tissue is more metabolically active), interest has increased in the use of resistance training to increase fat-free mass in an attempt to increase RMR.[7]

Exercise and the Thermic Effect of a Meal   Several studies have examined the role of individual bouts of exercise and exercise training in increasing the TEM. A single bout of exercise, either before or after a meal, increases the thermic effect of that meal. Less clear is the role of exercise training on the TEM. Some studies have shown increases, others have shown decreases, and yet others have shown no effect at all. As with measuring changes in RMR accompanying exercise training, measurement of the TEM must be timed carefully with the last exercise bout. When measurements are made within 24 h of the last bout, the TEM is typically lower than it is 3 days afterward.[31]

Exercise and Mobilization of Body Fat   During exercise, fatty acids are freed from their storage sites to be burned for energy. Several studies suggest that human growth hormone may be responsible for this increased fatty acid mobilization.

Growth hormone levels increase sharply with exercise and remain elevated for up to several hours in the recovery period. Other research has suggested that, with exercise, the adipose tissue is more sensitive to either the sympathetic nervous system or the increasing levels of circulating catecholamines. Either situation would increase lipid mobilization. More recent research suggests that this mobilization occurs in response to a specific fat-mobilizing substance that is highly responsive to elevated levels of activity. Thus, we cannot state with certainty which factors are of greatest importance in mediating this response.

Spot Reduction

Many people, including athletes, believe that by exercising a specific area of the body, the fat in that area will be used, reducing the locally stored fat. Results of several early research studies tended to support this concept of spot reduction. But later research suggests that spot reduction is a myth and that exercise, even when localized, draws from almost all of the fat stores of the body, not just from local depots.

One such study used outstanding tennis players, theorizing that they would be ideal subjects for studying spot reduction because they could act as their own controls: Their dominant arms exercise vigorously for several hours every day, whereas their nondominant arms are relatively sedentary.[13] Researchers postulated that if spot reduction is a reality, the nondominant (inactive) arm should have substantially more fat than the dominant (active) arm. The players' dominant arms had substantially greater girths attributable to exercise-induced muscle hypertrophy. But the subcutaneous skinfold fat thicknesses in the active and inactive arms showed absolutely no differences.

Another study examined the localized effects of a 27-day intense sit-up training program. Researchers found no difference in the rate at which fat cell diameter changed in the abdomen, the subscapular region, and the gluteal region.[16] This indicates a lack of specific adaptation at the site of the exercise training (the abdomen). Researchers now theorize that fat is mobilized during exercise either mostly from those areas of highest concentration or equally from all areas, thus negating the spot-reduction theory. Decreases in girth can occur with exercise training, but these result from increased muscle tone, not fat loss.

Low-Intensity Aerobics

As we have discussed in earlier chapters, the higher the exercise intensity, the greater the body's reliance on carbohydrate as an energy source. With high-intensity aerobic exercise, carbohydrate can supply up to 90% or more of the body's energy needs. During the late 1980s, various professional exercise groups promoted low-intensity aerobic exercise to increase the loss of body fat. These groups theorized that low-intensity aerobic training would allow the body to use more fat as the energy source, hastening the loss of body fat. Indeed, the body uses a higher percentage of fat for energy at lower exercise intensities. However, the total calories expended by the body's use of fat does not necessarily change.

> Low-intensity aerobic activity does not necessarily lead to a greater expenditure of calories from fat. More important, the total caloric expenditure for a given period of time is much less than with high-intensity aerobic activity.

This is illustrated in table 21.4. In this hypothetical example, a 23-year-old woman with a $\dot{V}O_2max$ of 3.0 L/min exercises for 30 min at 50% of her $\dot{V}O_2max$ on one day and for 30 min at 75% of her $\dot{V}O_2max$ on another. The total calories from fat do not differ between the low- and high-intensity aerobic workouts: In both cases she burns about 110 kcal of fat during 30 min. Most important, however, for the higher intensity workout she expends about 50% more total calories for the same time period!

Scientists have determined that there is an optimal zone where rates of fat oxidation are at their highest. The Fat_{max} zone, defined as fat oxidation rates within 10% of the peak rate, was found to vary from between 55% and 72% of $\dot{V}O_2max$.[1] This is illustrated in figure 21.11.

Table 21.4

Estimation of Kilocalories Used From Fat and Carbohydrate for a Low- and High-Intensity Aerobic Training Bout

Exercise intensity	Average $\dot{V}O_2$(L/min)	Average RER	% kcal CHO	% kcal fat	kcal for 30 min CHO	kcal for 30 min fat	kcal for 30 min total
Low, 50%	1.50	0.85	50	50	110	110	220
High, 75%	2.25	0.90	67	33	222	110	332

Note. RER = respiratory exchange ratio; CHO = carbohydrate. Subject was a fit but not highly trained 23-year-old woman ($\dot{V}O_2$max = 3.0 L/min).

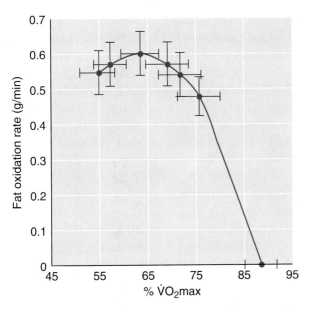

▲ **Figure 21.11**   Rate of fat oxidation at various exercise intensities, expressed as a percentage of $\dot{V}O_2$max.

Reprinted, by permission, from J. Acten, M. Gleeson, and A.E. Jeukendrup, 2002, "Determination of the exercise intensity that elicits maximal fat oxidation," *Medicine and Science in Sports and Exercise* 34: 92-97.

Exercise Gadgets

We seldom get something for nothing. An effortless exercise program would be ideal, of course, but such a program would result in no significant changes in body composition or physical dimensions. With the increasing popularity of exercise, many gimmicks and gadgets have appeared on the market. Some of these are legitimate and effective, but unfortunately many are of no practical value for either exercise conditioning or weight loss. Three such devices were evaluated to determine the legitimacy of their claims: the Mark II Bust Developer, the Astro-Trimmer Exercise Belt, and the Slim-Skins Vacuum Pants. The first device claimed to add 2 to 3 in. (5-8 cm) to the bust within 3 to 7 days, and the last two devices claimed to take inches off of the abdomen, hips, buttocks, and thighs in a matter of minutes. All three devices failed to produce any changes whatsoever when evaluated in tightly controlled scientific studies.[37, 38]

Those who are considering weight reduction often cringe at the thought of increasing their physical activity, and who wouldn't prefer immediate results over waiting for a payoff? But reality must be addressed. To gain the benefits from exercise, it is necessary to actually do the work!

Physical Activity and Health Risk Reduction

An important relationship was discovered during the 1990s that suggests another substantial benefit of fitness and activity. For those who are overweight or obese, their overall risk of death from disease is greatly reduced if they are physically active and fit.[2, 34] This is good news for those who seem destined to remain obese or overweight: An active lifestyle and moderate to high fitness levels can greatly reduce their risk of dying from chronic degenerative diseases, such as coronary artery disease and diabetes.

▶ Inactivity is a major cause of obesity in the United States, perhaps as important as overeating.

▶ The energy expended by activity includes the steady-state rate of energy expenditure during the activity and also the energy expended after the exercise because the metabolic rate remains elevated for some time after the activity ends, a phenomenon known as excess postexercise oxygen consumption (EPOC).

▶ Diet alone causes fat loss, but fat-free mass is also lost. With exercise, either alone or with diet, fat is lost, but fat-free mass is either maintained or increased.

▶ Energy intake – energy excreted = RMR + TEM + TEA when you are in energy balance.

▶ A certain amount of activity appears to be needed for the body to precisely balance energy intake and expenditure.

▶ Research indicates that exercise can suppress appetite.

▶ RMR might increase slightly following training, and even a single bout of exercise increases the TEM.

▶ Exercise increases lipid mobilization from adipose tissue.

▶ Spot reduction is a myth. Low-intensity aerobics burn no more fat than more vigorous exercise, and more total calories are spent in a more strenuous workout.

Diabetes

Diabetes mellitus, also called simply diabetes, is a disorder of carbohydrate metabolism characterized by high blood sugar levels (**hyperglycemia**) and presence of sugar in the urine (**glycosuria**). It develops when there is inadequate production of **insulin** by the pancreas or an inability of insulin to facilitate the transport of glucose into the cells (recall from chapter 5 that insulin is a hormone that reduces the amount of glucose circulating in the blood by facilitating its transport into the cells). We first address the terminology used in defining diabetes and disordered blood sugar control (glycemic control) and then look at the prevalence of diabetes in the United States.

Terminology and Classification

Historically, diabetes mellitus was classified into two major categories: juvenile-onset diabetes (now known as type 1 diabetes) and adult-onset diabetes (now known as type 2 diabetes). This classification was based on the age of onset of diabetes.[20] Unfortunately, there has been an epidemic of type 2 diabetes in children, which largely can be attributed to the increased rates of obesity in children.

Type 1 diabetes is caused by the inability of the pancreas to produce sufficient insulin as a result of failure of the β-cells in the pancreas. Thus, this type is also referred to as **insulin-dependent diabetes mellitus (IDDM)**. Type 1 diabetes accounts for only 5% to 10% of all cases of diabetes.

Type 2 diabetes is the result of the ineffectiveness of insulin to facilitate the transport of glucose into the cells and is a result of insulin resistance. It is also referred to as **non-insulin-dependent diabetes mellitus (NIDDM)**. Type 2 diabetes accounts for 90% to 95% of all cases of diabetes. **Insulin resistance** refers to the condition where a "normal" insulin concentration in the blood produces a less than normal biological response. Insulin's primary function is to facilitate the transport of glucose from the blood into the cell, across the cell membrane. With insulin resistance, your body needs more insulin to transport a given amount of glucose across the cell membrane into the cell. **Insulin sensitivity** is a related term and provides an index of the effectiveness of a given insulin concentration in the blood. As insulin sensitivity increases, insulin resistance decreases.

A third type of diabetes, gestational diabetes, is a form of diabetes that develops in pregnant women and their fetuses in 2% to 5% of all pregnancies. Fortunately, it usually disappears in both mother and baby after delivery. Unfortunately, when gestational diabetes is present, there can be complications during pregnancy.

Another category, **impaired fasting glucose,** refers to those who are borderline diabetic. Diabetes mellitus of either type is diagnosed on the basis of a plasma glucose level of greater than 125 mg/dl following an 8-h fast. Impaired fasting glucose is defined as a plasma glucose level of between 110 and 125 mg/dl, again following an 8-h fast.

Glucose tolerance tests often are used to assist in the diagnosis of diabetes. An oral glucose tolerance test involves drinking a solution in which 75 g of anhydrous glucose is dissolved in water. Plasma glucose levels are measured 2 h later. A glucose value of 200 mg/dl or higher is diagnostic of diabetes. For research studies, an intravenous glucose tolerance test (IVGTT) often is used. A catheter is placed in both arms, with the glucose solution injected into one arm, and blood samples are drawn from the other arm over the course of 3 h. Samples are taken more frequently during the first 15 to 45 min of the test and less frequently from 60 to 180 min. This allows you to develop a curve of both the glucose and insulin responses to the injected glucose load. The IVGTT is more precise than the oral glucose tolerance test.

There are also symptoms of diabetes that can be used to identify those at risk for diabetes. These include

- frequent urination,
- excessive thirst,
- unexplained weight loss,
- extreme hunger,
- sudden vision changes,
- tingling or numbness in hands or feet,
- feeling very tired much of the time,
- very dry skin,
- sores that are slow to heal, and
- more infections than usual.

Prevalence of Diabetes

Approximately 16 million Americans have diabetes. An additional 13.8 million have impaired fasting glucose levels. The prevalence of diabetes increased from 4.9% of the U.S. population in 1990 to 7.3% in 2000—a 49% increase. Between 1990 and 1998, the largest increase (76%) occurred in 30- to 39-year-olds. Among American Indians, ages 45 to 74 years, the prevalence is 43.5% for men and 52.4% for women. The prevalence is higher in the older population, higher in women than men, higher in Mexican-Americans than blacks and whites, and higher in blacks compared with whites. These racial differences in diabetes rates parallel the racial differences in obesity rates discussed earlier in this chapter.

The true prevalence of type 2 diabetes in children has not been established through national epidemiological studies. However, prevalence has been estimated to have increased by as much as 10-fold over the last 20 years. Furthermore, in 10- to 19-year-olds, studies have reported that type 2 diabetes accounted for between 33% and 46% of all diabetes.[20] This is very disturbing, considering that not too long ago we referred to type 2 diabetes as adult-onset diabetes!

Etiology of Diabetes

Heredity appears to play a major role in both type 1 and type 2 diabetes. With type 1 diabetes, the β-cells (insulin-secreting cells) of the pancreas are destroyed. This destruction may be caused by the body's immune system, increased β-cell susceptibility to viruses, or β-cell degeneration.

Type 1 diabetes generally has a sudden onset during childhood or young adulthood. This leads to almost total insulin deficiency, and daily injections of insulin usually are required to control the disease.

In type 2 diabetes, the onset of the disease is more gradual, and the causes are more difficult to establish. Type 2 diabetes often is characterized by any of the following three major metabolic abnormalities: delayed or impaired insulin secretion; impaired insulin action (insulin resistance) in the insulin-responsive tissues of the body, including muscle; or excessive glucose output from the liver.

Obesity plays a major role in the development of type 2 diabetes. With obesity, the β-cells of the pancreas often become less responsive to the stimulation of increased blood glucose concentrations. Furthermore, the target cells throughout the body, including muscle, often undergo a reduction in the number or activation of their insulin receptors, so the insulin in the blood is less effective in transporting glucose into the cell.

Health Problems Associated With Diabetes

Considerable health risks are associated with diabetes. People with this disease have a relatively high mortality rate. Diabetes places a person at increased risk for

- coronary artery disease,
- cerebrovascular disease and stroke,
- hypertension,
- peripheral vascular disease,
- kidney disease,
- eye disorders, including blindness, and
- toxemia during pregnancy.

During the late 1980s, scientists made the important associations among coronary artery disease, hypertension, obesity, and type 2 diabetes. Hyperinsulinemia (high blood levels of insulin) and insulin resistance appear to be the important threads linking these disorders, possibly through insulin-mediated sympathetic nervous system stimulation (increased insulin levels cause increased sympathetic nervous system activity).[9] Again, obesity seems to be the trigger setting off this reaction.

> Diabetes increases your risk for coronary artery disease, cerebrovascular disease and stroke, hypertension, and peripheral vascular disease. Coronary artery disease, hypertension, obesity, and diabetes may be linked through the common pathway of increased insulin levels in the blood or to target cells becoming insulin resistant. However, obesity appears to be the trigger.

General Treatment of Diabetes

The major modes of treatment for type 1 diabetes are insulin administration, diet, and exercise. The dosage of insulin is adjusted to allow normal carbohydrate, fat, and protein metabolism. The type of insulin injected—short-acting or intermediate-acting—and the time of day at which the injections are administered are also individualized to maintain glycemic control throughout the day. For type 2 diabetes, the focus is on three factors: weight loss, diet, and exercise.

A well-balanced diet generally is prescribed for people with diabetes. In the past, patients were prescribed a low-carbohydrate diet to better control blood sugar levels. However, a low-carbohydrate diet necessitates an increase in dietary fat, which can have a major negative effect on blood lipid levels. Because people with diabetes are already at greater risk for coronary artery disease, this is not desirable. Maintenance of proper blood sugar levels is difficult in patients with obesity, so a reduced-calorie diet is necessary for them to achieve major body fat losses. For many people with type 2 diabetes, weight loss alone can bring blood sugar levels back into the normal range. This can be the most important aspect of the treatment plan for the overweight or obese person with diabetes.

Role of Physical Activity in Diabetes

Although no conclusive evidence proves that a physically active lifestyle prevents diabetes, most physicians agree that physical activity is an important part of the treatment plan. Because there is such a disparity between the characteristics and responses of those with type 1 and type 2 diabetes, we discuss each separately.

Type 1 Diabetes

The role of regular exercise and physical training in improving glycemic control (regulation of blood sugar levels) in patients with type 1 diabetes has not been clearly defined and is controversial. The most distinguishing feature differentiating type 1 and type 2 diabetes is that those with type 1 have low blood insulin levels attributable to the inability or reduced ability of the pancreas to produce insulin. Those with type 1 diabetes are prone to hypoglycemia (low blood sugar levels) during and immediately after exercise because the liver fails to release glucose at a rate that can keep up with glucose utilization. For these people, exercise can lead to excessive swings in plasma glucose levels that are unacceptable for the management of the disease. The degree of glycemic control during exercise varies tremendously among individuals with type 1 diabetes. As a result, exercise and exercise training can improve glycemic control in some patients, mainly those who are less prone to hypoglycemia, but not in others.[33]

Although glycemic control is generally not improved by exercise in most people with type 1 diabetes, there are other potential benefits of exercise for these patients. Because these patients have a two to three times greater risk for coronary artery disease, exercise may help reduce this risk. Exercise may also help reduce the risk for cerebrovascular and peripheral arterial diseases.

People with uncomplicated type 1 diabetes do not have to restrict physical activity, providing blood sugar levels are controlled appropriately. A number of athletes who have type 1 diabetes have trained and competed successfully. Monitoring blood sugar levels in an exercising person with type 1 diabetes is important so that diet and insulin dosages can be altered accordingly. Table 21.5 provides guidelines for avoiding hypoglycemia during and after exercise.

Special attention also should be given to the feet of people with diabetes, as it is common for them to experience peripheral neuropathy (diseased nerves) with some loss of sensation in the feet. Peripheral vascular disease is also more common in patients with diabetes, so the circulation to the extremities, especially the feet, often is impaired significantly. Ulcerations and other lesions on the feet account for more than half of all hospitalizations of diabetic patients.[8] Because weight-bearing exercise places additional stress on the feet, the proper selection of footwear and appropriate preventive foot care are important.

Type 2 Diabetes

Exercise plays a major role in glycemic control for people with type 2 diabetes. Insulin

---

Table 21.5

Guidelines for Avoiding Hypoglycemia During and After Exercise in People With Type 1 Diabetes

Consume carbohydrates (15-30 g) for every 30 min of moderate-intensity exercise.

Consume a snack of slowly absorbed carbohydrate following prolonged exercise sessions.

Decrease the insulin dose.

  Intermediate-acting insulin—decrease by 30% to 35 % on the day of exercise.

  Intermediate- and short-acting insulin—omit the dose of short-acting insulin that precedes exercise

  Multiple doses of short-acting insulin—reduce the dose prior to exercise by 30% to 50% and supplement carbohydrates.

  Continuous subcutaneous infusion—eliminate the mealtime bolus or increment that precedes or immediately follows exercise.

Avoid exercising muscle that underlies the injection site of short-acting insulin for 1 h.

Avoid late evening exercise.

From A. Vitug, S.H. Schneider, N.B. Ruderman, 1988, "Exercise and type I diabetes mellitus," *Exercise and Sport Sciences Reviews* 16: 285-304. Copyright 1988 by Williams & Wilkins. Reprinted by permission.[33]

production is generally not of concern in this group, particularly during the early stages of the disease, so the major problem with this form of diabetes is the lack of target cell response to insulin (insulin resistance). Because the cells become resistant to insulin, the hormone cannot perform its function of facilitating glucose transport across the cell membrane. Ivy[15] demonstrated that muscle contraction has an insulin-like effect. Membrane permeability to glucose increases with muscular contraction, possibly attributable to an increase in the number of glucose transporters associated with the plasma membrane. Thus, acute bouts of exercise decrease insulin resistance and increase insulin sensitivity. This reduces the cells' requirements for insulin, which means that people taking insulin must reduce their dosages. This decrease in insulin resistance and increase in insulin sensitivity may primarily be a response to each individual bout of exercise rather than the result of a long-term change associated with training. Some studies have shown that the effect dissipates within 72 h.

Physical activity has many desirable effects for people with diabetes, particularly those with type 2 diabetes. Glycemic control is improved, primarily in people with type 2 diabetes, possibly attributable to the insulin-like effect of muscle contraction on transporting glucose from the plasma into the cell.

> ▶ Diabetes is a disorder of carbohydrate metabolism characterized by hyperglycemia and glycosuria. It develops from inadequate insulin secretion or utilization.
>
> ▶ Type 1 diabetes involves destruction of the β-cells in the pancreas, and it typically has a sudden, early onset. Type 2 diabetes typically involves impaired insulin secretion or action, excessive liver glucose output, or a combination of these.

Diabetes and Exercise

To optimize treatment, whether through insulin, diet, or exercise, people with type 1 diabetes need to monitor their blood sugar levels, allowing them to adjust one or more of these three treatment modalities. Home blood glucose monitoring kits (see figure 21.12) are available that allow the individual to periodically take a finger-stick blood sample to determine his or her blood glucose concentration. This is particularly important when a person with diabetes is contemplating a change in insulin dosage, diet, or exercise patterns. As an example, people often need to reduce their insulin dosage once they start a regular exercise program. Furthermore, this monitoring can occur before, during, and following an exercise bout. This enables an exerciser to know if his or her blood sugar level has decreased to a critical level as a result of the exercise, in which case eating a high-carbohydrate food would be required. Hypoglycemia is a potential risk during exercise, so a person with diabetes should always exercise in the company of another person who is familiar with the disease. A glucose drink or a source of simple sugar, such as jellybeans, should be available if needed to prevent insulin shock.

▲ **Figure 21.12**   A home monitoring kit for checking blood glucose levels.
Photo courtesy of LifeScan, a Johnson & Johnson company.

- ▶ Major modes of treatment for diabetes are insulin administration (if needed), diet, and exercise.

- ▶ In people with type 1 diabetes, glycemic control might or might not be improved with exercise. But these people have a greater risk for coronary artery disease, so exercise certainly can decrease that risk.

- ▶ Blood sugar levels must be carefully monitored when exercising, particularly in people with type 1 diabetes, so diet and insulin dosage can be altered as needed.

- ▶ The feet of people with type 1 diabetes deserve special attention, because peripheral neuropathy causes loss of sensation, and impaired peripheral circulation decreases blood flow. These people might not be aware of injuries to their feet, but these injuries can be very serious.

- ▶ Type 2 diabetes responds well to exercise. Membrane permeability to glucose improves with exercise, which decreases the person's insulin resistance and increases insulin sensitivity.

In Closing . . .

In this chapter, we have concluded our look at the role of physical activity in preventing and treating coronary artery disease, hypertension, obesity, and diabetes. We have seen that exercise can decrease individual risk and also can be an integral part of treatment, improving overall health as well as alleviating some symptoms.

With this chapter, we also conclude our journey to an understanding of exercise and sport physiology. We began this book by reviewing how various body systems function during exercise and how they respond to chronic training. We saw how physical activity and performance are affected by the environment, such as extremes of heat and cold, extremes of pressure, and even the unusual conditions of microgravity. We turned our attention to ways

in which athletes can, or attempt to, optimize their performance. Then we considered the unique differences among older and younger participants and male and female participants in sport and exercise. And finally, we examined the role of exercise in the maintenance of health and the development of fitness.

It has been a long journey from cover to cover, but we hope that you will close this book with a new appreciation of physical activity. Perhaps you will leave this book with a new awareness of how your body performs physical activity. Maybe, if you have not yet done so, you will feel compelled to commit to a personal exercise program. And we hope that you now feel some excitement for exercise and sport physiology, realizing that these areas of study affect so many aspects of our lives.

▷ Key Terms

β-cells
body mass index (BMI)
diabetes mellitus
excess postexercise oxygen consumption (EPOC)
glycosuria
hyperglycemia
hyperinsulinemia
hypoglycemia
impaired-fasting glucose
insulin
insulin dependent diabetes mellitus (IDDM)
insulin resistance
insulin sensitivity
lower body (gynoid) obesity
low-intensity aerobic exercise
non-insulin dependent diabetes mellitus (NIDDM)
obesity
overweight
relative weight
resting metabolic rate (RMR)
spot reduction
thermic effect of activity (TEA)
thermic effect of a meal (TEM)

type 1 diabetes

type 2 diabetes

upper body (android) obesity

▶ Study Questions

1 What is the difference between overweight and obesity?

2. What is ideal body weight, and how is it determined?

3. What is relative body weight? What is its significance?

4. What is body mass index? What is its significance?

5. What is the prevalence of obesity in the United States today? Is there a difference between men and women? Children and adults? Blacks and whites?

6. What are some of the health-related problems associated with obesity?

7. What is the association among obesity, coronary artery disease, hypertension, and diabetes?

8. Describe several methods for treating obesity. Which are the most effective?

9. What role does exercise play in the prevention and treatment of obesity?

10. By what mechanisms might exercise effect losses in total weight and fat weight?

11. How effective is spot reduction? Low-intensity aerobic exercise?

12. Describe the two major types of diabetes. How are they caused?

13. What health risks are associated with diabetes?

14. Describe the role of exercise in treating patients with type 1 diabetes.

15. Describe the role of exercise in treating patients with type 2 diabetes.

▶ References

1. Achten, J., Gleeson, M., & Jeukendrup, A.E. (2002). Determination of the exercise intensity that elicits maximal fat oxidation. *Medicine and Science in Sports and Exercise, 34,* 92-97.

2. Barlow, C.E., Kohl, H.W., III, Gibbons, L.W., & Blair, S.N. (1995). Physical fitness, mortality and obesity. *International Journal of Obesity,* **19**(Suppl. 4), 41-44.

3. Bjorntorp, P., Smith, U., & Lonnroth, P. (Eds.). (1988). *Health implications of regional obesity* (Acta Medica Scandinavica Symposium Series No. 4). Stockholm: Almqvist & Wiksell International.

4. Bouchard, C. (1991). Heredity and the path to overweight and obesity. *Medicine and Science in Sports and Exercise,* **23,** 285-291.

5. Bouchard, C., Tremblay, A., Després, J.-P., Nadeau, A., Lupien, P.J., Theriault, G., Dussault, J., Moorjani, S., Pinault, S., & Fournier, G. (1990). The response to long-term overfeeding in identical twins. *New England Journal of Medicine,* **322,** 1477-1482.

6. Bray, G.A. (1985). Obesity: Definition, diagnosis and disadvantages. *Medical Journal of Australia,* **142,** S2-S8.

7. Broeder, C.E., Burrhus, K.A., Svanevik, L.S., & Wilmore, J.H. (1992). The effects of either high intensity resistance or endurance training on resting metabolic rate. *American Journal of Clinical Nutrition,* **55,** 802-810.

8. Chisholm, D.J. (1992). Diabetes mellitus. In J. Bloomfield, P.A. Fricker, & K.D. Fitch (Eds.), *Textbook of science and medicine in sport* (pp. 555-561). Boston: Blackwell Scientific.

9. Daly, P.A., & Landsberg, L. (1991). Hypertension in obesity and NIDDM: Role of insulin and sympathetic nervous system. *Diabetes Care,* **14,** 240-248.

10. Flegal, K.M., Carroll, M.D., Kuczmarski, R.J., & Johnson, C.L. (1998). Overweight and obesity in the United States: Prevalence and trends, 1960-1994. *International Journal of Obesity,* **22,** 39-47.

11. Flegal, K.M., Carroll, M.D., Ogden, C.L., & Johnson, C.L. (2002). Prevalence and trends in obesity among US adults, 1999-2000. *JAMA,* **288,** 1723-1727.

12. Ford, E.S., Giles, W.H., & Dietz, W.H. (2002). Prevalence of the metabolic syndrome among US adults. *Journal of the American Medical Association,* **287,** 356-359.

13. Gwinup, G., Chelvam, R., & Steinberg, T. (1971). Thickness of subcutaneous fat and activity of underlying muscles. *Annals of Internal Medicine,* **74,** 408-411.

14. Harnack, L.J., Jeffery, R.W., & Boutelle, K.N. (2000). Temporal trends in energy intake in the United States: An ecologic perspective. *American Journal of Clinical Nutrition,* **71,** 1478-1484.

15. Ivy, J.L. (1987). The insulin-like effect of muscle contraction. *Exercise and Sport Sciences Reviews,* **15,** 29-51.

16. Katch, F.I., Clarkson, P.M., Kroll, W., McBride, T., & Wilcox, A. (1984). Effects of sit up exercise training on adipose cell size and adiposity. *Research Quarterly for Exercise and Sport, 55*, 242-247.

17. Keesey, R.E. (1986). A set-point theory of obesity. In K.D. Brownell & J.P. Foreyt (Eds.), *Handbook of eating disorders: Physiology, psychology, and treatment of obesity, anorexia, and bulimia* (pp. 63-87). New York: Basic Books.

18. Keys, A., Brozek, J., Henschel, A., Mickelsen, O., & Taylor, H.L. (1950). *The biology of human starvation.* Minneapolis: University of Minnesota Press.

19. Knowler, W.C., Pettit, D.J., Saad, M.F., Charles, M.A., Nelson, R.G., Howard, B.V., Bogardus, C., & Bennett, P.H. (1991). Obesity in the Pima Indians: Its magnitude and relationship with diabetes. *American Journal of Clinical Nutrition, 53*, 1543S-1551S.

20. Ludwig, D.S., & Ebbeling, C.B. (2001). Type 2 diabetes mellitus in children. *Journal of the American Medical Association, 286*, 1426-1430.

21. Mayer, J., Marshall, N.B., Vitale, J.J., Christensen, J.H., Mashayekhi, M.B., & Stare, F.J. (1954). Exercise, food intake, and body weight in normal rats and genetically obese adult mice. *American Journal of Physiology, 177*, 544-548.

22. National Institutes of Health. (2000). *The practical guide: Identification, evaluation, and treatment of overweight and obesity in adults* (NIH Publication No. 00-4084). Washington, DC: U.S. Department of Health and Human Services.

23. Ogden, C.L., Flegal, K.M., Carroll, M.D., & Johnson, C.L. (2002). Prevalence and trends in overweight among US children and adolescents, 1999-2000. *JAMA, 288*, 1728-1732.

24. Oscai, L.B. (1973). The role of exercise in weight control. *Exercise and Sport Sciences Reviews, 1*, 103-123.

25. Poehlman, E.T. (1989). A review: Exercise and its influence on resting energy metabolism in man. *Medicine and Science in Sports and Exercise, 21*, 515-525.

26. Seidell, J.C., Deurenberg, P., & Hautvast, J.G.A.J. (1987). Obesity and fat distribution in relation to health—Current insights and recommendations. *World Review of Nutrition and Dietetics, 50*, 57-91.

27. Sims, E.A.H. (1976). Experimental obesity, dietary-induced thermogenesis and their clinical implications. *Clinics in Endocrinology and Metabolism, 5*, 377-395.

28. Stunkard, A.J., Foch, T.T., & Hrubec, Z. (1986). A twin study of human obesity. *Journal of the American Medical Association, 256*, 51-54.

29. Stunkard, A.J., Harris, J.R., Pedersen, N.L., & McClearn, G.E. (1990). The body-mass index of twins who have been reared apart. *New England Journal of Medicine, 322*, 1483-1487.

30. Stunkard, A.J., Sørensen, T.I.A., Hanis, C., Teasdale, T.W., Chakraborty, R., Schull, W.J., & Schulsinger, F. (1986). An adoption study of human obesity. *New England Journal of Medicine, 314*, 193-198.

31. Tremblay, A., Nadeau, A., Fournier, G., & Bouchard, C. (1988). Effect of a three-day interruption of exercise-training on resting metabolic rate and glucose-induced thermogenesis in trained individuals. *International Journal of Obesity, 12*, 163-168.

32. Troiano, R.P., Flegal, K.M., Kuczmarski, R.J., Campbell, S.M., & Johnson, C.L. (1995). Overweight prevalence and trends for children and adolescents: The National Health and Nutrition Examination Surveys, 1963 to 1991. *Archives of Pediatric Adolescent Medicine, 149*, 1085-1091.

33. Vitug, A., Schneider, S.H., & Ruderman, N.B. (1988). Exercise and type I diabetes mellitus. *Exercise and Sport Sciences Reviews, 16*, 285-304.

34. Welk, G.J., & Blair, S.N. (2000). Physical activity protects against the health risks of obesity. *Research Digest: President's Council on Physical Fitness and Sports, 3*(12), 1-6.

35. Wilmore, J.H. (1995). Variations in physical activity habits and body composition. *International Journal of Obesity, 19*(Suppl. 4), S107-S112.

36. Wilmore, J.H. (1996). Increasing physical activity: Alterations in body mass and composition. *American Journal of Clinical Nutrition, 63*, 456S-460S.

37. Wilmore, J.H., Atwater, A.E., Maxwell, B.D., Wilmore, D.L., Constable, S.H., & Buono, M.J. (1985). Alterations in body size and composition consequent to Astro-Trimmer and Slim-Skins training programs. *Research Quarterly for Exercise and Sport, 56*, 90-92.

38. Wilmore, J.H., Atwater, A.E., Maxwell, B.D., Wilmore, D.L., Constable, S.H., & Buono, M.J. (1985). Alterations in breast morphology consequent to a 21-day bust developer program. *Medicine and Science in Sports and Exercise, 17*, 106-112.

39. Wilmore, J.H., Stanforth, P.R., Hudspeth, L.A., Gagnon, J., Daw, E.W., Leon, A.S., Rao, D.C., Skinner, J.S., & Bouchard, C. (1998). Alterations in resting metabolic rate as a consequence of 20-wk of endurance training: The HERITAGE Family Study. *American Journal of Clinical Nutrition, 68*, 66-71.

40. World Health Organization. (1998). *Obesity: Preventing and managing the global epidemic. Report of a WHO Consultation on Obesity.* Geneva: WHO.

41. Zuti, W.B., & Golding, L.A. (1976). Comparing diet and exercise as weight reduction tools. *Physician and Sportsmedicine, 4*, 49-53.

▷ Selected Readings

Albright, A., Franz, M., Hornsby, G., Kriska, A., Marrero, D., Ullrich, I., & Verity, L.S. (2000). American College of Sports Medicine Position Stand: Exercise and type 2 diabetes. *Medicine and Science in Sports and Exercise, 32,* 1345-1360.

American College of Sports Medicine and American Diabetes Association Joint Position Stand. (1997). Diabetes mellitus and exercise. *Medicine and Science in Sports and Exercise, 28*(12), i-vi.

Atkinson, R.L. (1989). Low and very low calorie diets. *Medical Clinics of North America, 73,* 203-215.

Ballor, D.L., & Poehlman, E.T. (1994). Exercise-training enhances fat-free mass preservation during diet-induced weight loss: A meta-analytical finding. *International Journal of Obesity, 18,* 35-40.

Bjorntorp, P., & Brodoff, B.N. (Eds.). (1992). *Obesity.* Philadelphia: Lippincott.

Brownell, K.D., & Foreyt, J.P. (Eds.). (1986). *Handbook of eating disorders: Physiology, psychology, and treatment of obesity, anorexia, and bulimia.* New York: Basic Books.

Colberg, S.R., & Swain, D.P. (2000). Exercise and diabetes control. *Physician and Sportsmedicine, 28*(4), 63-81.

Després, J.-P., Lamarche, B., Mauriége, P., Cantin, B., Degenais, G.R., Moorjani, S., & Lupien, P.-J. (1996). Hyperinsulinemia as an independent risk factor for ischemic heart disease. *New England Journal of Medicine, 334,* 952-957.

Eriksson, J.G. (1999). Exercise and the treatment of type 2 diabetes mellitus. *Sports Medicine, 27,* 381-391.

Fulton, J.E., McGuire, M.T., Caspersen, C.J., & Dietz, W.H. (2001). Interventions for weight loss and weight gain prevention among youth. *Sports Medicine, 31,* 153-165.

Gautier, J.F., Scheen, A., & Lefébvre, P.J. (1995). Exercise in the management of non-insulin-dependent (type 2) diabetes mellitus. *International Journal of Obesity, 19*(Suppl. 4), S58-S61.

Glenny, A.-M., O'Meara, S., Melville, A., Sheldon, T.A., & Wilson, C. (1997). The treatment and prevention of obesity: A systematic review of the literature. *International Journal of Obesity, 21,* 715-737.

Grundy, S.M., Blackburn, G., Higgins, M., Lauer, R., Perri, M.G., & Ryan, D. (1999). Physical activity in the prevention and treatment of obesity and its comorbidities: Evidence report of independent panel to assess the role of physical activity in the treatment of obesity and its comorbidities. *Medicine and Science in Sports, 31,* 1493-1500.

Gudat, U., Berger, M., & Lefébvre, P.J. (1994). Physical activity, fitness, and non-insulin-dependent (type II) diabetes mellitus. In C. Bouchard, R.J. Shephard, & T. Stephens (Eds.), *Physical activity, fitness and health* (pp. 669-683). Champaign, IL: Human Kinetics.

Holloszy, J.O., Schultz, J., Kusnierkiewicz, J., Hagberg, J.M., & Ehsani, A.A. (1986). Effects of exercise on glucose tolerance and insulin resistance. *Acta Medica Scandinavica,* (Suppl. 711), 55-65.

Ivy, J.L. (1997). Role of exercise training in the prevention and treatment of insulin resistance and non-insulin-dependent diabetes mellitus. *Sports Medicine, 24*(5), 321-336.

Jakicic, J.M., Clark, K., Coleman, E., Donnelly, J.E., Foreyt, J., Melanson, E., Volek, J., & Volpe, S.L. (2001). American College of Sports Medicine Position Stand: Appropriate intervention strategies for weight loss and prevention of weight regain in adults. *Medicine and Science in Sports and Exercise, 33,* 2145-2156.

Jousilahti, P., Tuomilehto, J., Vartiainen, E., Pekkanen, J., & Puska, P. (1996). Body weight, cardiovascular risk factors, and coronary mortality: 15-yr follow-up of middle-aged men and women in Eastern Finland. *Circulation, 93,* 1372-1379.

Kriska, A. (2000). Physical activity and the prevention of type 2 diabetes mellitus. *Sports Medicine, 29,* 147-151.

Kriska, A.M., Blair, S.N., & Pereira, M.A. (1994). The potential role of physical activity in the prevention of non-insulin-dependent diabetes mellitus: The epidemiological evidence. *Exercise and Sport Sciences Reviews, 22,* 121-143.

Leon, A.S. (1993). Diabetes. In J.S. Skinner (Ed.), *Exercise testing and exercise prescription for special cases* (2nd ed., pp. 153-183). Philadelphia: Lea & Febiger.

Miller, W.C. (2001). Effective diet and exercise treatments for overweight and recommendations for intervention. *Sports Medicine, 31,* 717-724.

Miller, W.C., Koceja, D.M., & Hamilton, E.J. (1997). A meta-analysis of the past 25 years of weight loss research using diet, exercise or diet plus exercise intervention. *International Journal of Obesity, 21,* 941-947.

National Institutes of Health. (1985). Health implications of obesity: National Institutes of Health Consensus Development Conference Statement. *Annals of Internal Medicine, 103,* 1073-1077.

Nieman, D.C. (1998). *The exercise-health connection.* Champaign, IL: Human Kinetics.

Poehlman, E.T., & Horton, E.S. (1990). Regulation of energy expenditure in aging humans. *Annual Reviews in Nutrition, 10,* 255-275.

Ross, R., Freeman, J.A., & Janssen, I. (2000). Exercise alone is an effective strategy for reducing obesity and related comorbidities. *Exercise and Sport Sciences Reviews, 28,* 165-170.

Ross, R., Janssen, I., & Tremblay, A. (2000). Obesity reduction through lifestyle modification. *Canadian Journal of Applied Physiology, 25,* 1-18.

Ryan, A.S. (2000). Insulin resistance with aging. *Sports Medicine, 30,* 327-346.

Schneider, S.H., & Ruderman, N.B. (1990). Exercise and NIDDM. *Diabetes Care, 13,* 785-789.

Stefanick, M.L. (1993). Exercise and weight control. *Exercise and Sport Sciences Reviews, 21,* 363-396.

Strauss, R.S., & Pollack, H.A. (2001). Epidemic increase in childhood overweight, 1986-1998. *Journal of the American Medical Association, 286,* 2845-2848.

Stunkard, A.J., & Wadden, T.A. (Eds.). (1993). *Obesity: Theory and therapy* (2nd ed.). New York: Raven Press.

Tsui, E.Y.L., & Zinman, B. (1995). Exercise and diabetes: New insights and therapeutic goals. *Endocrinologist, 5,* 263-271.

Tudor-Locke, C.E., Bell, R.C, & Myers, A.M. (2000). Revisiting the role of physical activity and exercise in the treatment of type 2 diabetes. *Canadian Journal of Applied Physiology, 25,* 466-491.

Wallberg-Henricksson, H., Rincon, J., & Zierath, J.R. (1998). Exercise in the management of non-insulin-dependent diabetes mellitus. *Sports Medicine, 25*(1), 25-35.

Willett, W.C., Manson, J.E., Stampfer, M.J., Colditz, G.A., Rosner, B., Speizer, F.E., & Hennekens, C.H. (1995). Weight, weight change, and coronary heart disease in women. *Journal of the American Medical Association, 273,* 461-465.

Young, J.C. (1995). Exercise prescription for individuals with metabolic disorders. *Sports Medicine, 19,* 43-54.

1-repetition maximum (1RM)—The maximal amount of weight that can be lifted just one time.

1RM—*See* 1-repetition maximum.

ACE—*See* angiotensin converting enzyme.

acclimation—Physiological adaptation to an environmental stress in a laboratory environment.

acclimatization—Physiological adaptation to an environmental stress in a natural environment.

acetyl CoA—*See* acetyl coenzyme A.

acetyl coenzyme A (acetyl CoA)—The compound that forms the common entry point into the Krebs cycle for the oxidation of carbohydrate and fat.

acetylcholine—A primary neurotransmitter that transmits impulses across the synaptic cleft.

actin—A thin protein filament that acts with myosin filaments to produce muscle action.

action potential—A rapid and substantial depolarization of the membrane of a neuron or muscle cell that is conducted through the cell.

acute altitude (mountain) sickness—Illness characterized by headache, nausea, vomiting, dyspnea, and insomnia. It typically begins 6 to 96 h after one reaches high altitude and lasts several days.

acute muscle soreness—Soreness or pain felt during and immediately after an exercise bout.

acute overload—*See* overload.

acute response—A physiological response to a single bout of exercise.

adenosine diphosphate (ADP)—A high-energy phosphate compound from which ATP is formed.

adenosine triphosphatase (ATPase)—An enzyme that splits the last phosphate group off ATP, releasing a large amount of energy and reducing the ATP to ADP and P_i.

adenosine triphosphate (ATP)—A high-energy phosphate compound from which the body derives its energy.

ADH—*See* antidiuretic hormone.

adolescence—The period of life between the end of childhood and the beginning of adulthood. The onset of puberty marks the beginning of adolescence.

ADP—*See* adenosine diphosphate.

aerobic capacity—*See* maximal oxygen uptake.

aerobic endurance—*See* cardiorespiratory endurance.

aerobic interval training—Repeated, moderate duration, fast-paced exercise bouts with moderate rest intervals between bouts.

aerobic metabolism—A process occurring in the mitochondria that uses oxygen to produce energy (ATP). Also known as cellular respiration.

aerobic training—Training that improves the efficiency of the aerobic energy-producing systems and can improve cardiorespiratory endurance.

afferent division—The sensory division of the peripheral nervous system.

air plethysmography—A procedure for assessing body composition by using air displacement to measure body volume, allowing the calculation of body density.

alcohol—A central nervous system depressant proposed to have ergogenic properties.

aldosterone—A mineralocorticoid hormone secreted by the adrenal cortex that prevents dehydration by promoting renal absorption of sodium.

alveolar capillary membrane—*See* respiratory membrane.

amenorrhea—The absence (primary amenorrhea) or cessation (secondary amenorrhea) of normal menstrual function.

amino acids—The chief components of proteins that are synthesized by living cells or are obtained by the body through the diet.

amphetamine—A central nervous system stimulant proposed to have ergogenic properties.

anabolic steroids—Prescription drugs with the anabolic (growth-stimulating) characteristics of testosterone, taken by some athletes to increase body size, muscle mass, and strength.

anabolism—The building up of body tissue; the constructive phase of metabolism.

anaerobic—In the absence of oxygen.

anaerobic glycolysis—The anaerobic breakdown of glucose to lactic acid in the production of energy (ATP).

anaerobic metabolism—The production of energy (ATP) in the absence of oxygen.

anaerobic threshold—The point at which the metabolic demands of exercise can no longer be met by available aerobic sources and at which an increase in anaerobic metabolism occurs, reflected by an increase in blood lactate concentration.

anaerobic training—Training that improves the efficiency of the anaerobic energy-producing systems and can increase muscular strength and tolerance for acid-base imbalances during high-intensity effort.

angiotensin converting enzyme (ACE)—An enzyme that converts angiotensin I to angiotensin II.

anorexia nervosa—A clinical eating disorder characterized by distorted body image, intense fear of fatness or weight gain, amenorrhea, and refusal to maintain more than the minimal normal weight based on age and height.

antidiuretic hormone (ADH)—A hormone secreted by the pituitary gland that regulates fluid and electrolyte balance in the blood by reducing urine production.

arterial-mixed venous oxygen difference, or a-$\bar{v}O_2$ difference—The difference in oxygen content between arterial and mixed venous blood, which reflects the amount of oxygen removed by the whole body.

arterial-venous oxygen difference, or a-vO_2 difference—The difference in oxygen content between arterial and venous blood at the tissue level.

arteries—Blood vessels that transport blood away from the heart.

arterioles—The smallest arteries that transport blood from larger arteries to the capillaries.

arteriosclerosis—A condition that involves loss of elasticity, thickening, and hardening of the arteries.

atherosclerosis—A form of arteriosclerosis that involves changes in the lining of the arteries and plaque accumulation, leading to progressive narrowing of the arteries.

athlete's heart—A nonpathological enlarged heart, often found in endurance athletes, that results primarily from left ventricular hypertrophy in response to training.

ATP—*See* adenosine triphosphate.

ATPase—*See* adenosine triphosphatase.

ATP-PCr system—The short-term anaerobic energy system that maintains ATP levels. Breakdown of phosphocreatine (PCr) frees P_i, which then combines with ADP to form ATP.

atrioventricular (AV) node—The specialized mass of conducting cells in the heart located at the atrioventricular junction.

atrophy—Loss of size, or mass, of body tissue, such as muscle atrophy with disuse.

autogenic inhibition—Reflex inhibition of a motor neuron in response to excessive tension in the muscle fibers it supplies, as monitored by the Golgi tendon organs.

autoregulation—Local control of blood distribution (through vasodilation) in response to a tissue's changing needs.

a-$\bar{v}O_2$ difference—*See* arterial-mixed venous oxygen difference.

AV node—*See* atrioventricular node.

axon hillock—A part of the neuron, between the cell body and the axon, that controls traffic down the axon through summation of excitatory and inhibitory postsynaptic potentials.

axon terminal—One of numerous branched endings of an axon. Also known as a terminal fibril.

basal metabolic rate (BMR)—The lowest rate of body metabolism (energy use) that can sustain life, measured after an overnight sleep in a laboratory under optimal conditions of quiet, rest, and relaxation and after a 12-h fast. *See also* resting metabolic rate.

BCAA—*See* branched-chain amino acids.

bends—*See* decompression sickness.

β-blockers—A class of drugs that block transmission of neural impulses from the sympathetic nervous system, proposed to have ergogenic properties.

β-cells—Cells in the islets of Langerhans in the pancreas that secrete insulin.

β-oxidation—The first step in fatty acid oxidation, in which fatty acids are broken into separate 2-carbon units of acetic acid, each of which is then converted to acetyl CoA.

bicarbonate loading—Ingesting bicarbonate to elevate blood pH with hopes of delaying fatigue by increasing the capacity to buffer acids.

bioelectric impedance—A procedure for assessing body composition in which an electrical current is passed through the body. The resistance to current flow through the tissues reflects the relative amount of fat present.

blood doping—Any means by which a person's total volume of red blood cells is increased, typically via transfusion of red blood cells or use of erythropoietin.

blood lipids—Blood-borne fats, such as triglycerides and cholesterol.

BMI—*See* body mass index.

BMR—*See* basal metabolic rate.

body build—The morphology (form and structure) of the body.

body composition—The chemical composition of the body. The model used in this book considers two components: fat-free mass and fat mass.

body density (D_{body})—Body weight divided by body volume.

body mass index (BMI)—A measurement of body overweight or obesity determined by dividing weight (in kilograms) by height (in meters) squared. BMI is highly correlated with body composition.

body size—A person's height and mass (weight).

Borg RPE scale—A numerical scale for rating perceived exertion.

bradycardia—A resting heart rate lower than 60 beats/min.

branched-chain amino acids (BCAA)—Specific amino acids—leucine, isoleucine, and valine—that have been postulated to work in combination with L-tryptophan to delay fatigue, primarily through central nervous system mechanisms.

buffer—A substance that combines with either an acid or a base to maintain a constant acid–base (pH) balance.

bulimia nervosa—A clinical eating disorder characterized by recurrent episodes of binge eating, a feeling of lack of control during these binges, and purging behavior, which may include self-induced vomiting and use of laxatives and diuretics. Sometimes the disorder also includes fasting or excessive exercise behaviors.

CAD—*See* coronary artery disease.

caffeine—A central nervous system stimulant believed by some athletes to have ergogenic properties.

calcitonin—A hormone secreted by the thyroid gland that assists in the control of calcium ion concentrations in the blood.

calorie (cal)—A unit of measure of energy in biological systems, where 1.0 calorie is equal to the amount of heat energy needed to raise the temperature of 1.0 g of water 1° C, from 15 to 16° C.

calorimeter—A device for measuring the heat produced by the body (or by specific chemical reactions).

cAMP—*See* cyclic adenosine monophosphate.

capillaries—The smallest vessels transporting blood from the heart to the tissues and the actual sites of exchange between the blood and tissue.

capillary to fiber ratio—The number of capillaries per muscle fiber.

carbohydrate—An organic compound formed from carbon, hydrogen, and oxygen; includes starches, sugars, and cellulose.

cardiac cycle—The period that includes all events between two consecutive heartbeats.

cardiac hypertrophy—Enlargement of the heart by increases in muscle wall thickness or chamber size or both.

cardiac output ($\dot{Q}$)—The volume of blood pumped out by the heart per minute. $\dot{Q}$ = heart rate × stroke volume.

cardiorespiratory endurance—The ability of the body to sustain prolonged exercise.

cardiovascular deconditioning—A decrease in the cardiovascular system's ability to deliver sufficient oxygen and nutrients.

cardiovascular drift—An increase in heart rate during exercise to compensate for a decrease in stroke volume. This compensation helps maintain a constant cardiac output.

cardiovascular endurance training—*See* aerobic training.

catabolism—The tearing down of body tissue; the destructive phase of metabolism.

catecholamines—Biologically active amines (organic compounds derived from ammonia), such as epinephrine and norepinephrine, that have powerful effects similar to those of the sympathetic nervous system.

central nervous system (CNS)—System consisting of the brain and spinal cord.

cerebral infarction—Death of brain tissue that results from insufficient blood supply attributable to blockage or damage of a cerebral vessel. *See also* stroke.

childhood—The period of life between the first birthday and the onset of puberty.

chronic adaptation—A physiological change that occurs when the body is exposed to repeated exercise bouts over weeks or months. These changes generally improve the body's efficiency at rest and during exercise.

chronic fatigue syndrome—A syndrome that appears to involve immune system dysfunction. Patients have incapacitating fatigue, sore throat, muscle tenderness or pain, and cognitive dysfunction; the symptoms may vary in severity over time but generally last for months or years.

chronic hypertrophy—An increase in muscle size that results from repeated long-term resistance training.

CK—*See* creatine kinase.

CNS—*See* central nervous system.

cocaine—A so-called recreational drug that is a central nervous system stimulant which mimics the action of the sympathetic nervous system; it is generally ergolytic.

cold acclimatization—*See* acclimatization.

concentric action—Muscle shortening.

conduction—(1) Transfer of heat or cold through direct molecular contact. (2) Movement of an electrical impulse, such as through a neuron.

congenital heart disease—A heart defect present at birth that occurs from abnormal prenatal development of the heart or associated blood vessels. Also known as congenital heart defect.

congestive heart failure—A clinical condition in which the myocardium becomes too weak to maintain adequate cardiac output to meet the body's oxygen demands; congestive heart failure usually results from the heart being damaged or overworked.

continuous training—Training at a moderate to high intensity without stopping to rest.

contractile velocity (V_o)— The speed of action associated with specific muscle fiber types.

convection—The transfer of heat or cold via the movement of a gas or liquid across an object, such as the body.

coronary artery disease (CAD)—Progressive narrowing of the coronary arteries.

cortisol—A corticosteroid hormone released from the adrenal cortex that stimulates gluconeogenesis, increases mobilization of free fatty acids, decreases use of glucose, and stimulates catabolism of protein. Also known as hydrocortisone.

creatine—A substance found in skeletal muscles most commonly in the form of PCr. Creatine supplements are often used as ergogenic aids because they are theorized to increase PCr levels, thus enhancing the ATP–PCr energy system by better maintaining muscle ATP levels.

creatine kinase (CK)—The enzyme that facilitates the breakdown of PCr to creatine and P_i.

cross-sectional research design—A research design in which a cross section of a population is tested at one specific time and then data from groups within that population are compared.

cross-training—Training for more than one sport at the same time, or training multiple fitness components (such as endurance, strength, and flexibility) within the same period.

cycle ergometer—An exercise device that uses cycling to measure physical work.

cyclic adenosine monophosphate (cAMP)—Intracellular second messenger that mediates hormone action.

DBP—*See* diastolic blood pressure.

decompression sickness (bends)—A condition in which bubbles of nitrogen are trapped in the blood and tissues during a too-rapid ascent from depth during diving, characterized by severe discomfort and pain.

dehydration—Loss of body fluids.

delayed-onset muscle soreness (DOMS)—Muscle soreness that develops a day or two after a heavy bout of exercise and that is associated with actual injury within the muscle.

densitometry—The measurement of body density.

depolarization—A decrease in the electrical potential across a membrane, such as when the inside of a neuron becomes less negative relative to the outside.

detraining—Changes in physiological function in response to a reduction or cessation of regular physical training.

development—Changes that occur in the body starting at conception and continuing through adulthood; differentiation along specialized lines of function, reflecting changes that accompany growth.

diabetes mellitus—A disorder of carbohydrate metabolism characterized by hyperglycemia (high blood sugar levels) and glycosuria (presence of sugar in the urine). The disease develops when there is inadequate production of insulin by the pancreas or inadequate utilization of insulin by the cells.

diastolic blood pressure (DBP)—The lowest arterial pressure, resulting from ventricular diastole (the resting phase).

direct calorimetry—A method that gauges the body's rate and quantity of energy production by direct measurement of the body's heat production.

direct gene activation—The method of action of steroid hormones. They bind to receptors in the cell, and then the hormone-receptor complex enters the nucleus and activates certain genes.

disordered eating—Abnormal eating behavior that ranges from excessive restriction of food intake to pathological behaviors, such as self-induced vomiting and laxative abuse. Disordered eating can lead to clinical eating disorders, such as anorexia nervosa and bulimia nervosa.

diuretics—Substances that promote water excretion.

diurnal variation—Fluctuations in physiological responses that occur during a 24-h period.

DOMS—*See* delayed-onset muscle soreness.

down-regulation—Decreased cellular sensitivity to a hormone, likely the result of a decreased number of cell receptors available to bind with the hormone.

DPB—*See* diastolic blood pressure.

dual-energy X-ray absorptiometry (DXA)—A technique used to assess both regional and total body composition through the use of X-ray absorptiometry.

dynamic action—Any muscle action that produces joint movement.

dyspnea—Labored or difficult breathing.

eating disorders—A group of clinical disorders involving eating. *See* anorexia nervosa, bulimia nervosa.

eccentric action—Any muscle action where muscle lengthens.

eccentric training—Training that involves eccentric action.

ECG—*See* electrocardiogram and exercise electrocardiogram.

EDV—*See* end-diastolic volume.

EF—*See* ejection fraction.

efferent division—The motor division of the peripheral nervous system.

EIAH—*See* exercise-induced arterial hypoxemia.

ejection fraction (EF)—The fraction of blood pumped out of the left ventricle with each contraction, determined by dividing stroke volume by end-diastolic volume and expressed as a percentage.

electrical stimulation training—Stimulation of a muscle by passing an electrical current through it.

electrocardiogram (ECG)—A recording of the heart's electrical activity.

electrocardiograph—A machine used to obtain an electrocardiogram.

electrolyte—A dissolved substance that can conduct an electrical current.

electron transport chain—A series of chemical reactions that convert the hydrogen ion generated by glycolysis and the Krebs cycle into water and produce energy for oxidative phosphorylation.

end branches—Branches coming off the ends of the axons leading to the axon terminals.

end-diastolic volume (EDV)—The volume of blood inside the left ventricle at the end of diastole, just before contraction.

endomysium—A sheath of connective tissue that covers each muscle fiber.

end-systolic volume (ESV)—The volume of blood remaining in the left ventricle at the end of systole, just after contraction.

endurance—The ability to resist fatigue; includes muscular endurance and cardiorespiratory endurance.

energy—The capability of producing force, performing work, or generating heat.

engram—A specific, learned, and memorized motor pattern, stored in both the sensory and motor portions of the brain, that can be replayed on request.

epimysium—The outer connective tissue that surrounds an entire muscle, holding it together.

epinephrine—A catecholamine released from the adrenal medulla that, along with norepinephrine, prepares the body for a fight or flight response. It is also a neurotransmitter. *See* catecholamines.

EPOC—*See* excess postexercise oxygen consumption.

EPSP—*See* excitatory postsynaptic potential.

ergogenic—Able to improve work or performance.

ergogenic aid—A substance or phenomenon that can improve work or athletic performance.

ergolytic—Able to impair work or performance.

ergometer—An exercise device that allows the amount and rate of a person's physical work to be controlled (standardized) and measured.

erythropoietin—The hormone that stimulates erythrocyte (red blood cell) production.

essential amino acids—The eight or nine amino acids necessary for human growth that the body cannot synthesize and are thus essential parts of our diets.

estrogen—A female sex hormone.

ESV—*See* end-systolic volume.

eumenorrhea—Normal menstrual function.

evaporation—Heat loss through the conversion of water (such as in sweat) to vapor.

excessive training—Training in which volume, intensity, or both are too great or are increased too quickly without proper progression.

excess postexercise oxygen consumption (EPOC)—Elevated oxygen consumption above resting levels after exercise; at one time referred to as oxygen debt.

excitatory postsynaptic potential (EPSP)—A depolarization of the postsynaptic membrane caused by an excitatory impulse.

exercise electrocardiogram (ECG)—A recording of the heart's electrical activity during exercise.

exercise-induced arterial hypoxemia (EIAH)—A decline in arterial PO_2 and arterial oxygen saturation during maximal or near-maximal exercise.

exercise physiology—The study of how body structure and function are altered by exposure to acute and chronic bouts of exercise.

exercise prescription—Individualization of the prescription of exercise duration, frequency, intensity, and mode.

exhaustion—Inability to continue exercise.

expiration—The process by which air is forced out of the lungs through relaxation of the inspiratory muscles and elastic recoil of the lung tissue, which increases the pressure in the thorax.

external respiration—The process of bringing air into the lungs and the resulting exchange of gas between the alveoli and the capillary blood.

extracellular fluid—The 35% to 40% of the water in the body that is outside the cells, including interstitial fluid, blood plasma, lymph, cerebrospinal fluid, and other fluids.

extrinsic neural control—Redistribution of blood at the system or body level through neural mechanisms.

fasciculus—A small bundle of muscle fibers wrapped in a connective tissue sheath within a muscle.

fast-twitch (FT) fiber—A type of muscle fiber with a low oxidative capacity and a high glycolytic capacity; associated with speed or power activities.

fat—A class of organic compounds with limited water solubility that exists in the body in many forms, such as triglycerides, free fatty acids, phospholipids, and steroids.

fat-free mass—The mass (weight) of the body that is not fat, including muscle, bone, skin, and organs.

fatigue—General sensations of tiredness and accompanying decrements in muscular performance.

fat mass—The absolute amount or mass of body fat.

fatty streaks—Early lipid deposits within blood vessels.

female athlete triad—Three interrelated disorders—disordered eating, menstrual dysfunction, and bone mineral disorders—to which some female athletes are prone.

$FEV_{1.0}$—*See* forced expiratory volume in 1 s.

FFA—*See* free fatty acids.

fiber hyperplasia—An increase in the number of muscle fibers.

fiber hypertrophy—An increase in the size of existing individual muscle fibers.

fibromyalgia syndrome—A chronic syndrome that includes muscle pain as its dominant symptom but is also characterized by muscle weakness, migraine-type headaches, and depression.

Fick equation—$\dot{V}O_2 = \dot{Q} \times$ a-$\bar{v}O_2$ difference.

force—Strength or energy exerted or brought to bear.

forced expiratory volume in 1 s ($FEV_{1.0}$)—The volume of air exhaled in the first second after maximal inhalation.

Frank–Starling mechanism—The mechanism by which an increased amount of blood in the ventricle causes a stronger ventricular contraction to increase the amount of blood ejected.

free fatty acids—The components of fat that are used by the body for metabolism.

free radicals—Univalent (unpaired) oxygen intermediates that leak out of the electron transport chain during metabolic processes and may damage tissues.

frostbite—Tissue damage that occurs during cold exposure because circulation to the skin decreases, in an attempt to retain body heat, to the point that the tissue receives insufficient oxygen and nutrients.

FT—*See* fast-twitch fiber.

gastric emptying—The movement of food mixed with gastric secretions from the stomach into the duodenum.

gender differences—Any differences between females and males. *See also* sex-specific differences.

glucagon—A hormone released by the pancreas that promotes increased breakdown of liver glycogen to glucose (glycogenolysis) and increased gluconeogenesis.

gluconeogenesis—The conversion of protein or fat into glucose.

glycogen—The form of carbohydrate stored in the body, found predominantly in the muscles and liver.

glycogen loading—The manipulation of exercise and diet to optimize the body's glycogen storage.

glycogenesis—The conversion of glucose to glycogen.

glycogenolysis—The conversion of glycogen to glucose.

glycolysis—The breakdown of glucose to pyruvic acid.

glycolytic enzymes—Enzymes that are specific to the glycolytic energy system.

glycolytic system—A system that produces energy through glycolysis.

glycosuria—The presence of glucose in the urine.

Golgi tendon organ—A sensory receptor in a muscle tendon that monitors tension.

graded exercise test (GXT)—An exercise test in which the rate of work is increased gradually in 1- to 3-min increments, usually to the point of fatigue or exhaustion.

graded potential—A localized change (depolarization or hyperpolarization) in the membrane potential.

growth—An increase in the size of the body or any of its parts.

growth hormone—An anabolic agent that stimulates fat metabolism and promotes muscle growth and hypertrophy by facilitating amino acid transport into the cells.

GXT—*See* graded exercise test.

HACE—*See* high-altitude cerebral edema.

Haldane transformation—An equation allowing you to calculate the inspired air volume from expired air volume, or expired air volume from inspired air volume.

HAPE—*See* high-altitude pulmonary edema.

HDL—*See* high-density lipoprotein.

HDL-C—*See* high-density–lipoprotein cholesterol.

heart rate recovery period—The time it takes for heart rate to return to the resting rate following exercise.

heat acclimatization—*See* acclimatization.

heat cramp—Cramping of the skeletal muscles as a result of excessive dehydration and the associated salt loss.

heat exhaustion—A heat disorder resulting from an inability of the cardiovascular system to meet all the body tissues' needs while also shifting blood to the periphery for cooling, characterized by elevated body temperature, breathlessness, extreme tiredness, dizziness, and rapid pulse.

heat stroke—The most serious heat disorder, resulting from failure of the body's thermoregulatory mechanisms. Heat stroke is characterized by body temperature above 40.5° C (105° F), cessation of sweating, and total confusion or unconsciousness and can lead to death.

hematocrit—The percentage of cells or formed elements in the total blood volume. More than 99% of the cells or formed elements are red blood cells.

hemoconcentration—A relative (not absolute) increase in the cellular content per unit of blood volume, resulting from a reduction in plasma volume.

hemodilution—An increase in blood plasma, resulting in a dilution of the blood's cellular contents.

hemoglobin—The iron-containing pigment in red blood cells that binds oxygen.

hemoglobin saturation—The amount of oxygen bound by each molecule of hemoglobin.

Henry's law—The law stating that gases dissolve in liquids in proportion to their partial pressures, depending also on their solubilities in the specific fluids and on the temperature.

hGH—*See* human growth hormone.

high-altitude cerebral edema (HACE)—A condition of unknown cause in which fluid accumulates in the cranial cavity at altitude, characterized by mental confusion that can progress to coma and death.

high-altitude pulmonary edema (HAPE)—A condition of unknown cause in which fluid accumulates in the lungs at altitude, interfering with ventilation, resulting in shortness of breath and fatigue, and characterized by impaired blood oxygenation, mental confusion, and loss of consciousness.

high-density lipoprotein (HDL)—A cholesterol carrier regarded as a scavenger; theorized to remove cholesterol from the arterial wall and transport it to the liver to be metabolized.

high-density-lipoprotein cholesterol (HDL-C)—The cholesterol carried by HDL.

hormonal agents—A group of hormones proposed to have ergogenic properties.

hormone—A chemical substance produced or released by an endocrine gland and transported by the blood to a specific target tissue.

HRmax—*See* maximum heart rate.

human growth hormone (hGH)—A hormone that promotes anabolism and is believed by some athletes to have ergogenic properties.

hydrocortisone—*See* cortisol.

hydrostatic weighing—A method of measuring body volume in which a person is weighed while submerged underwater. The difference between the scale weight on land and the underwater weight (corrected for water density) equals body volume. This value must be further corrected to account for any air trapped in the lungs and other parts of the body.

hyperbaric environment—An environment, such as that underwater, involving high atmospheric pressure.

hyperglycemia—An elevated blood glucose level.

hyperinsulinemia—High levels of insulin in the blood.

hyperplasia—An increase in the number of cells in a tissue or organ. *See also* fiber hyperplasia.

hyperpolarization—An increase in the electrical potential across a membrane.

hypertension—Abnormally high blood pressure. In adults, hypertension is usually defined as a systolic pressure of 140 mmHg or higher or a diastolic pressure of 90 mmHg or higher.

hyperthermia—Elevated body temperature; anything above a person's normal resting body temperature.

hypertrophy—Increase in the size or mass of an organ or body tissue. *See also* fiber hypertrophy.

hyperventilation—A breathing rate or tidal volume greater than necessary for normal function.

hypobaric environment—An environment, such as that at high altitude, involving low atmospheric pressure.

hypoglycemia—A low blood glucose level.

hyponatremia—A blood sodium concentration below the normal range of 136 to 143 mmol/L.

hypothermia—Low body temperature; anything below that person's normal temperature.

hypoxia—A decreased availability of oxygen to the tissues.

hypoxic vasoconstriction—The constriction of blood vessels in response to low levels of oxygen.

IDDM—*See* insulin-dependent diabetes mellitus.

immune function—The body's normal ability to fight infection and illness with antibodies and lymphocytes.

impaired-fasting glucose—A plasma glucose level of between 110 and 125 mg/dl following an 8-h fast.

indirect calorimetry—A method of estimating energy expenditure by measuring respiratory gases.

infancy—The first year of life.

inhibiting factors—Hormones transmitted from the hypothalamus to the anterior pituitary that inhibit release of some other hormones.

inhibitory postsynaptic potential (IPSP)—A hyperpolarization of the postsynaptic membrane caused by an inhibitory impulse.

inspiration—The active process involving the diaphragm and the external intercostal muscles that expands the thoracic dimensions and thus the lungs. The expansion decreases pressure in the lungs, allowing outside air to rush in.

insulin—A hormone produced by the β-cells in the pancreas that assists glucose entry into cells.

insulin-dependent diabetes mellitus (IDDM)—One of two major categories of diabetes mellitus that is caused by the inability of the pancreas to produce sufficient insulin as a result of failure of the β-cells in the pancreas. This is also known as type 1 diabetes.

insulin resistance—A deficient target cell response to insulin.

insulin sensitivity—An index of the effectiveness of a given insulin concentration on the disposal of glucose.

internal respiration—The exchange of gases between the blood and tissues.

interval training—Repeated, brief, fast-paced exercise bouts with short rest intervals between bouts.

intracellular fluid—The approximately 60% to 65% of total body water that is contained in the cells.

IPSP—*See* inhibitory postsynaptic potential.

ischemia—A temporary deficiency of blood to a specific area of the body.

isometric training—Resistance training involving a static action.

Karvonen method—The calculation of training heart rate by adding a given percentage of the maximal heart rate reserve to the resting heart rate. This method gives an adjusted heart rate that is approximately equivalent to the desired percentage of $\dot{V}O_2$max.

kilocalorie (kcal)—The equivalent of 1,000 calories. *See* calorie.

Krebs cycle—A series of chemical reactions that involve the complete oxidation of acetyl CoA and produce 2 mol of ATP (energy) along with hydrogen and carbon, which combine with oxygen to form H_2O and CO_2.

lactate—A salt formed from lactic acid.

lactate dehydrogenase (LDH)—A key glycolytic enzyme involved in the conversion of pyruvate to lactate.

lactate threshold—The point during exercise of increasing intensity at which blood lactate begins to accumulate above resting levels, where lactate clearance is no longer able to keep up with lactate production.

L-carnitine—A substance important for fatty acid metabolism because it assists in the transfer of fatty acids from the cytosol (the fluid portion of the cytoplasm, exclusive of organelles) across the inner mitochondrial membrane for β-oxidation.

LDH—*See* lactate dehydrogenase.

LDL—*See* low-density lipoprotein.

LDL-C—*See* low-density lipoprotein cholesterol.

lean body mass—The sum of the body's fat-free mass and essential fat. This is not to be confused with fat-free mass.

lipogenesis—The process of converting protein into fatty acids.

lipolysis—The process of breaking down triglyceride to its basic units to be used for energy.

lipoprotein lipase—The enzyme that breaks down triglycerides to free fatty acids and glycerol, allowing the free fatty acids to enter the cells for use as a fuel or for storage.

lipoproteins—The proteins that carry the blood lipids.

longevity—The length of a person's life.

longitudinal research design—A research design in which subjects are tested initially and then one or more times later to directly measure changes over time resulting from a given intervention.

low-density lipoprotein (LDL)—A cholesterol carrier theorized to be responsible for depositing cholesterol in the arterial wall.

low-density–lipoprotein cholesterol (LDL-C)—The cholesterol carried by LDL.

lower body (gynoid) obesity—Obesity that follows the typically female pattern of fat storage, in which fat is stored primarily in the lower body, particularly in the hips, buttocks, and thighs.

low-intensity aerobic exercise—Aerobic exercise performed at low intensity, theoretically to cause the body to burn a higher percentage of fat.

L-tryptophan—An essential amino acid that has been proposed to increase aerobic endurance performance through its effects on the central nervous system. It theoretically acts as an analgesic and delays fatigue.

macrominerals—Those minerals of which the body needs more than 100 mg per day.

MAP—*See* mean arterial pressure.

marijuana—A so-called recreational drug that is generally ergolytic.

maturation—The process by which the body takes on the adult form and becomes fully functional. It is often defined by the system or function being considered.

maximal expiratory ventilation ($\dot{V}_E$max)—The highest ventilation that can be achieved during exhaustive exercise.

maximal heart rate reserve—The difference between maximal heart rate and resting heart rate.

maximal oxygen uptake ($\dot{V}O_2$max)—The maximal capacity for oxygen consumption by the body during maximal exertion. It is also known as aerobic power, maximal oxygen intake, maximal oxygen consumption, and cardiorespiratory endurance capacity.

maximum heart rate (HRmax)—The highest heart rate value attainable during an all-out effort to the point of exhaustion.

mean arterial pressure (MAP)—The average pressure exerted by the blood as it travels through the arteries. It is estimated as follows: MAP = DBP + [0.333 × (SBP – DBP)].

menarche—The onset of menstruation; the first menses.

menses—The menstrual or flow phase of the menstrual cycle.

menstrual cycle—The cycle of uterine changes, averaging 28 days and consisting of the menstrual (flow) phase, the proliferative phase, and the secretory phase.

menstrual dysfunction—Disruption of the normal menstrual cycle; it includes oligomenorrhea, primary amenorrhea, and secondary amenorrhea.

metabolic equivalent (MET)—A unit used to estimate the metabolic cost (oxygen consumption) of physical activity. One MET equals the resting metabolic rate of approximately 3.5 ml of $O_2 \cdot kg^{-1} \cdot min^{-1}$.

metabolic syndrome—A term that has been used to link coronary artery disease, hypertension, type II diabetes, and upper body obesity to insulin resistance and hyperinsulinemia. This syndrome has also been referred to as syndrome X and the civilization syndrome.

microgravity—An environment in which the body experiences a reduced gravitational force.

microminerals (trace elements)—The minerals of which the body needs less than 100 mg per day.

mitochondrial oxidative enzymes—Oxidative enzymes located in the mitochondria.

MK—*See* myokinase.

mode—Type of exercise.

morphology—The form and structure of the body.

motor division—*See* efferent division.

motor reflex—An involuntary motor response to a given stimulus.

motor unit—The motor nerve and the group of muscle fibers it innervates.

mountain sickness—*See* acute altitude sickness.

muscle buffering capacity—The muscles' ability to tolerate the acid that accumulates in them during anaerobic glycolysis.

muscle fiber—An individual muscle cell.

muscle spindle—A sensory receptor located in the muscle that senses how much the muscle is stretched.

muscular endurance—The ability of a muscle to resist fatigue.

myelin sheath—The outer covering of a myelinated nerve fiber, formed by a fatlike substance called myelin.

myelination—The process of acquiring a myelin sheath.

myocardial infarction—Death of heart tissue that results from insufficient blood supply to part of the myocardium.

myocardium—The muscle of the heart.

myofibril—The contractile element of skeletal muscle.

myoglobin—A compound similar to hemoglobin, but found in muscle tissue, that carries oxygen from the cell membrane to the mitochondria.

myokinase (MK)—A key enzyme in the ATP–PCr energy system.

myosin—One of the proteins that forms filaments that produce muscle action.

myosin cross-bridge—The protruding part of a myosin filament. It includes the myosin head, which binds to an active site on an actin filament to produce a power stroke that causes the filaments to slide across each other.

nebulin—A giant protein that coextends with actin and appears to play a regulatory role in mediating actin and myosin interactions.

needs analysis—An assessment of factors that determine the specific training program appropriate for an individual.

negative feedback system—The primary mechanism through which the endocrine system maintains homeostasis. Some body change upsets homeostasis, which triggers release of a hormone to correct the change. Once that correction is accomplished, the hormone is no longer needed, so its secretion decreases.

nerve impulse—The electrical signal conducted along a neuron, which can be transmitted to another neuron or an end organ, such as a group of muscle fibers.

neuromuscular junction—The site at which a motor neuron communicates with a muscle fiber.

neuron—A specialized cell in the nervous system responsible for generating and transmitting nerve impulses.

neurotransmitter—A chemical used for communication between a neuron and another cell.

nicotine—A central nervous system stimulant found in tobacco products that is proposed to have ergogenic properties.

NIDDM—*See* non-insulin-dependent diabetes mellitus.

nitrogen narcosis—A condition caused by breathing air underwater at depths where the partial pressure of nitrogen is elevated, causing the central nervous system to experience a narcotic-like effect and leading to distortions in judgment and sometimes to serious injury or death. Also known as rapture of the deep.

nonessential amino acids—The 11 or 12 amino acids that the body synthesizes.

non-insulin-dependent diabetes mellitus (NIDDM)—One of two major categories of diabetes mellitus that is caused by the ineffectiveness of insulin to facilitate the transport of glucose into the cells and is a result of insulin resistance. This is also known as type 2 diabetes.

nonresponders—Individuals who show little or no improvement compared with others who undergo the same training program.

nonshivering thermogenesis—The stimulation of metabolism by the sympathetic nervous system to generate more metabolic heat.

nonsteroid hormones—Hormones derived from protein, peptides, or amino acids that cannot easily cross cell membranes.

norepinephrine—A catecholamine released from the adrenal medulla that, along with epinephrine, prepares the body for a fight or flight response. It is also a neurotransmitter. *See* catecholamines.

nutritional agents—Nutritional substances proposed to have ergogenic benefits.

obesity—An excessive amount of body fat, generally defined as more than 25% in men and 35% in women; a BMI of 30 or greater.

oligomenorrhea—Abnormally infrequent or scant menstruation.

oral contraceptives—Drugs used for birth control and other medical purposes, which are believed by some female athletes to have ergogenic properties.

osmolarity—The ratio of solutes (such as electrolytes) to fluid.

ossification—The process of bone formation.

osteopenia—The loss of bone mass with aging.

osteoporosis—Decreased bone mineral content that increases bone porosity.

overreaching—A systematic attempt to intentionally overstress the body, allowing the body to adapt even more to the training stimulus, above and beyond that attained during a period of acute overload.

overtraining—The attempt to do more work than can be physically tolerated.

overtraining syndrome—A condition brought on by overtraining and characterized by performance decrements and a general breakdown in physiological function.

overweight—Body weight that exceeds the normal or standard weight for a particular individual based on sex, height, and frame size; a BMI of 25.0 to 29.9.

oxidative capacity of muscle ($\dot{Q}O_2$)—A measure of the muscle's maximal capacity to use oxygen.

oxidative system—The body's most complex energy system, which generates energy by disassembling fuels with the aid of oxygen and has a very high energy yield.

oxygen diffusion capacity—The rate at which oxygen diffuses from one place to another.

oxygen poisoning—A condition caused by breathing concentrated oxygen for a long period, such as during a deep dive, characterized by visual distortion, confusion, rapid and shallow breathing, and convulsions.

oxygen supplementation—The breathing of supplemental oxygen, which is proposed to have ergogenic properties.

oxygen transport system—The components of the cardiovascular and respiratory systems involved in transporting oxygen.

parathyroid hormone (PTH)—The hormone released by the parathyroid gland to regulate plasma calcium concentration and plasma phosphate.

partial pressure—The pressure exerted by an individual gas in a mixture of gases.

partial pressure of oxygen (PO_2)—The pressure exerted by oxygen in a mixture of gases.

pathophysiology—The physiology of a specific disease or disorder.

PCr—*See* phosphocreatine.

PDGF—*See* platelet-derived growth factor.

pericardium—A double-layered outer covering of the heart.

perimysium—The connective tissue sheath surrounding each muscle fasciculus.

periodization—Varying the training stimulus over discrete periods of time to prevent overtraining.

peripheral blood flow—Blood flow to the extremities and the skin.

peripheral nervous system (PNS)—That section of the nervous system through which motor nerve impulses are transmitted from the brain and spinal cord to the periphery and sensory nerve impulses are transmitted from the periphery to the brain and spinal cord.

peripheral vascular disease—Diseases of the systemic arteries and veins, especially those to the extremities, that impede adequate blood flow.

peripheral vasoconstriction—*See* vasoconstriction.

PFK—*See* phosphofructokinase.

pharmacological agents—A group of drugs proposed to have ergogenic properties.

phosphate loading—The practice of ingesting sodium phosphate, which has been proposed to have ergogenic properties.

phosphocreatine (PCr)—An energy-rich compound that plays a critical role in providing energy for muscle action by maintaining ATP concentration.

phosphofructokinase (PFK)—A key rate-limiting enzyme of the anaerobic glycolytic energy system.

phosphorylase—A key enzyme of the anaerobic glycolytic energy system.

physical maturity—The point at which the body has attained the adult physical form.

physiological agents—A group of agents normally present in the body that have been proposed to have ergogenic properties.

placebo—An inactive substance usually provided in a manner identical to an active substance, typically to test for real results produced by the test substance versus equally real results of psychological origin.

placebo effect—An effect produced by the subject's expectations after being administered an inactive substance (placebo).

placebo group—That group in an intervention study which receives a placebo rather than a test substance.

plaque—A buildup of lipids, smooth muscle cells, connective tissue, and debris that forms at the site of injury to an artery.

platelet-derived growth factor (PDGF)—A substance released by blood platelets that promotes the migration of smooth muscle cells from the media of an artery into the intima.

plyometrics—A type of dynamic-action resistance training based on the theory that use of the stretch reflex during jumping will recruit additional motor units.

PNS—*See* peripheral nervous system.

PO$_2$—*See* partial pressure of oxygen.

power—The rate of performing work; the product of force and velocity. The rate of transformation of metabolic potential energy to work or heat.

power stroke—The tilting of the myosin head, caused by a strong intermolecular attraction between the myosin cross bridge and the myosin head, which causes the actin and myosin filaments to slide across each other.

predictive value of an abnormal exercise test—The accuracy with which abnormal test results reflect the presence of a disease.

pregnancy—The state of carrying an embryo or fetus in the body.

premature ventricular contraction (PVC)—A common cardiac arrhythmia that results in the feeling of skipped or extra beats caused by impulses originating outside the SA node.

primary amenorrhea—The absence of menarche (the beginning of menstruation) beyond age 18.

primary risk factors—Risk factors that have been conclusively shown to have a strong association with a certain disease. Primary risk factors for coronary artery disease include smoking, hypertension, high blood lipid levels, obesity, and physical inactivity.

principle of disuse—The theory that a training program must include a maintenance plan to ensure that the gains from training are not lost.

principle of hard/easy—The theory that a training program must alternate high-intensity workouts with low-intensity workouts to help the body recover and achieve optimal training adaptation.

principle of individuality—The theory that any training program must consider the specific needs and abilities of the individual for whom it is designed.

principle of orderly recruitment—The theory that motor units generally are activated on the basis of a fixed order of recruitment, in which the motor units within a given muscle appear to be ranked according to the size of the motor neuron.

principle of periodization—The gradual cycling of specificity, intensity, and volume of training to achieve peak levels of fitness for competition.

principle of progressive overload—The theory that, to maximize the benefits of a training program, the training stimulus must be progressively increased as the body adapts to the current stimulus.

principle of specificity—The theory that a training program must stress the physiological systems critical for optimal performance in a given sport to achieve desired training adaptations in that sport.

progesterone—A hormone secreted by the ovaries that promotes the luteal phase of the menstrual cycle.

prostaglandins—Substances derived from a fatty acid that act as hormones at the local level.

protein—A class of nitrogen-containing compounds formed by amino acids.

PTH—*See* parathyroid hormone.

puberty—The point at which a person becomes physiologically capable of reproduction.

pulmonary diffusion—The exchange of gases between the lungs and the blood.

pulmonary ventilation—The movement of gases into and out of the lungs.

Purkinje fibers—The terminal branches of the AV bundle that transmit impulses through the ventricles six times faster than through the rest of the cardiac conduction system.

PVC—*See* premature ventricular contraction.

$\dot{Q}$—*See* cardiac output.

$\dot{Q}_2$—*See* oxidative capacity of muscle.

radiation—The transfer of heat through electromagnetic waves.

rapture of the deep—*See* nitrogen narcosis.

rating of perceived exertion (RPE)—A person's subjective assessment of how hard he or she is exercising.

recompression—Increasing the pressure exerted on the body, usually in a recompression chamber, to cause nitrogen bubbles in the body to go back into solution. Recompression is used to treat decompression sickness.

rehabilitation programs—Programs designed to reestablish health or fitness following a disability or illness.

relative body fat—The ratio of fat mass to total body mass, expressed as a percentage.

relative weight—The percentage by which an individual is either overweight or underweight, generally determined by dividing the person's weight by the mean weight for the medium frame category for his or her height (from standard weight tables).

releasing factors—Hormones transmitted from the hypothalamus to the anterior pituitary that promote release of some other hormones.

renin—An enzyme formed by the kidneys to convert a plasma protein called angiotensino-

gen into angiotensin II. *See also* renin–angiotensin mechanism.

renin–angiotensin mechanism—The mechanism involved in renal control of blood pressure. The kidneys respond to decreased blood pressure or blood flow by forming renin, which converts angiotensinogen into angiotensin I, which is finally converted to angiotensin II. Angiotensin II constricts arterioles and triggers aldosterone release.

RER—*See* respiratory exchange ratio.

residual volume (RV)—The amount of air that cannot be exhaled from the lungs.

resistance training—Training designed to increase strength, power, and muscular endurance.

respiratory alkalosis—A condition in which increased carbon dioxide clearance allows blood pH to increase.

respiratory centers—Autonomic centers located in the medulla oblongata and the pons that establish breathing rate and depth.

respiratory exchange ratio (RER)—The ratio of carbon dioxide expired to oxygen consumed at the level of the lungs.

respiratory membrane—The membrane separating alveolar air and blood, composed of the alveolar wall, the capillary wall, and their basement membranes.

responders—Individuals who show improvement in response to a training program.

resting heart rate (RHR)—The heart rate at rest, averaging 60 to 80 beats/min.

resting membrane potential (RMP)—The potential difference between the electrical charges inside a cell and outside the cell, caused by a separation of charges across the membrane.

resting metabolic rate (RMR)—The body's metabolic rate early in the morning following an overnight fast and 8 h of sleep. RMR does not require sleeping overnight in a laboratory or clinical facility. *See also* basal metabolic rate.

retraining—Recovery of conditioning after a period of inactivity.

rheumatic heart disease—A form of valvular heart disease involving a streptococcal infection that has caused acute rheumatic fever, typically in children between ages 5 and 15.

RHR—*See* resting heart rate.

RMP—*See* resting membrane potential.

RMR—*See* resting metabolic rate.

RPE—*See* rating of perceived exertion.

RV—*See* residual volume.

saltatory conduction—The means of rapid nerve impulse conduction along myelinated neurons.

SA node—*See* sinoatrial node.

sarcolemma—A muscle fiber's cell membrane.

sarcomere—The basic functional unit of a myofibril.

sarcopenia—The loss of muscle mass associated with aging.

sarcoplasm—The gelatin-like cytoplasm in a muscle fiber.

sarcoplasmic reticulum (SR)—A longitudinal system of tubules that is associated with the myofibrils and stores calcium for muscle action.

SBP—*See* systolic blood pressure.

scuba—Self-contained underwater breathing apparatus.

SDH—*See* succinate dehydrogenase.

secondary amenorrhea—The cessation of menstruation in a woman with previously normal menstrual function.

second messenger—A substance inside a cell that acts as a messenger after a nonsteroid hormone binds to receptors outside the cell.

self-contained underwater breathing apparatus—*See* scuba.

sensitivity—A test's ability to correctly identify subjects who fit the criteria being tested, such as coronary artery disease.

sensory division—*See* afferent division.

sensory–motor integration—The process by which the sensory and motor systems communicate and coordinate with each other.

sex-specific differences—True physiological differences between females and males.

shivering—A rapid, involuntary cycle of contraction and relaxation of skeletal muscles that generates heat.

sinoatrial (SA) node—A group of specialized myocardial cells located in the wall of the right atrium that control the heart's rate of contraction; the pacemaker of the heart.

size principle—The size of the motor neuron dictates the order of motor unit recruitment, with small sized motor neurons being recruited first.

skinfold fat thickness—The most widely applied field technique used to estimate body density, relative body fat, and fat-free mass. It involves measurement with calipers of the skinfold fat at one or more sites.

sliding filament theory—A theory explaining muscle action: A myosin cross bridge attaches to an actin filament, and then the power stroke drags the two filaments past one another.

slow-twitch (ST) fiber—A type of muscle fiber that has a high oxidative and a low glycolytic capacity, associated with endurance-type activities.

sodium–potassium pump—An enzyme called Na^+-K^+-ATPase, which maintains the resting membrane potential in disequilibrium at –70 mV.

specificity—A test's ability to correctly identify subjects who do not fit the criteria being tested.

specificity of training—The principle that physiological adaptations in response to physical training are highly specific to the nature of the training activity. To maximize benefits, training should be carefully matched to an athlete's specific performance needs.

spontaneous pneumothorax—The entrance of air into the pleural cavity, often caused by rupture of alveoli, that can lead to lung collapse.

sport physiology—The application of the concepts of exercise physiology to training athletes and enhancing sport performance.

spot reduction—The practice of exercising a specific area of the body, theoretically to reduce locally stored fat.

sprint training—A form of anaerobic training involving very brief, intense training bouts.

SR—*See* sarcoplasmic reticulum.

ST—*See* slow-twitch fiber.

static-action resistance training—Resistance training that emphasizes static muscle action. Also known as isometric resistance training.

static muscle action—Action in which the muscle contracts without moving, generating force while its length remains static (unchanged). Also known as isometric action.

steady-state heart rate—A heart rate that is maintained constant at submaximal levels of exercise when the rate of work is held constant.

steroid hormones—Hormones with chemical structures similar to cholesterol that are lipid-soluble and diffuse through cell membranes.

strength—The ability of a muscle to exert force—generally the maximal ability.

stroke—A cerebral vascular accident, a condition in which blood supply to some part of the brain is impaired, typically caused by infarction or hemorrhage, so that the tissue is damaged.

stroke volume (SV)—The amount of blood ejected from the left ventricle during contraction; the difference between the end-diastolic volume and the end-systolic volume.

submaximal endurance capacity—The average absolute power output a person can maintain during a fixed period of time on a cycle ergometer, or the average speed or velocity a person can maintain during a fixed period of time. Generally, these tests will last at least 30 min but usually not more than 90 min.

submaximal exercise—All intensities of exercise below maximal exercise intensity.

succinate dehydrogenase (SDH)—A key enzyme of the oxidative enzyme system.

summation—The summing of all individual changes in a neuron's membrane potential.

SV—*See* stroke volume.

swimming flume—A device that uses propeller pumps to circulate water past a swimmer, who attempts to maintain body position by swimming against the current.

synapse—The junction between two neurons.

systolic blood pressure (SBP)—The greatest arterial blood pressure, resulting from systole (the contracting phase of the heart).

T$_3$—*See* triiodothyronine.

T$_4$—*See* thyroxine.

tachycardia—A resting heart rate greater than 100 beats/min.

tapering—A reduction in training intensity prior to a major competition to give the body and mind a break from the rigors of intense training.

taper period—A time during which training intensity is reduced, allowing time for tissue damage from intense training to heal and for the body's energy reserves to be fully replenished.

target cells—Cells that possess specific hormone receptors.

TEA—*See* thermic effect of activity.

TEM—*See* thermic effect of a meal.

teratogenic effects—Effects that cause abnormal fetal development.

testosterone—The predominant male sex hormone.

test specificity—Matching the type of ergometer used in testing to the type of activity an athlete usually performs to ensure the most accurate results.

tethered swimming—A method of monitoring a swimmer in which the swimmer is attached to a harness connected to a rope, a series of pulleys, and a pan that contains weights, which allows the swimmer to swim while maintaining a constant position in the pool.

thermal stress—Stress imposed on the body by external temperature.

thermic effect of activity (TEA)—The energy expended in excess of the resting metabolic rate to accomplish a given task or activity.

thermic effect of a meal (TEM)—The energy expended in excess of resting metabolic rate associated with digestion, absorption, transport, metabolism, and storage of ingested food.

thermoreceptors—Sensory receptors that detect changes in body temperature and external temperature and relay this information to the hypothalamus.

thermoregulation—The process by which the thermoregulatory center, located in the hypothalamus, readjusts body temperature in response to small deviations from the set point.

thermoregulatory center—An autonomic nervous center located in the hypothalamus that is responsible for maintaining normal body temperature.

thirst mechanism—A neural mechanism that triggers thirst in response to dehydration.

THR—*See* training heart rate.

threshold—A minimum amount of stimulus needed to elicit a response. Also, the minimum depolarization required to produce an action potential in neurons.

thyrotropin (TSH)—A hormone secreted by the anterior lobe of the pituitary gland that promotes the release of thyroid hormones.

thyroxine (T$_4$)—A hormone secreted by the thyroid gland that increases the rate of cellular metabolism and the rate and contractility of the heart.

tidal volume—The amount of air inspired or expired during a normal breathing cycle.

titin—A protein that positions the myosin filament to maintain equal spacing between actin filaments.

TLC—*See* total lung capacity.

TLV/RV ratio—The ratio between total lung volume (TLV) and residual volume (RV).

total lung capacity (TLC)—The sum of vital capacity and residual volume.

total peripheral resistance—The resistance to the flow of blood through the entire systemic circulation.

trace elements—*See* microminerals.

training heart rate (THR)—A heart rate goal established by using the heart rate equivalent of a desired percentage of $\dot{V}O_2$max. For example, if a training level of 75% $\dot{V}O_2$max is desired, 75% of $\dot{V}O_2$max is calculated, and the heart rate corresponding to this $\dot{V}O_2$ is selected as the THR.

transient hypertrophy—The "pumping-up" of muscle that happens during a single exercise bout, resulting mainly from fluid accumulation in the interstitial and intracellular spaces of the muscle.

transverse tubules (T tubules)—Extensions of the sarcolemma (plasma membrane) that pass laterally through the muscle fiber, allowing nutrients to be transported and nerve impulses to be transmitted rapidly to individual myofibrils.

treadmill—An ergometer in which a motor and pulley system drive a large belt on which a person can either walk or run.

triglycerides—The body's most concentrated energy source and the form in which most fats are stored in the body.

triiodothyronine (T$_3$)—A hormone released by the thyroid gland that increases the rate of cellular metabolism and the rate and contractility of the heart.

tropomyosin—A tube-shaped protein that twists around actin strands, fitting in the groove between them.

troponin—A complex protein attached at regular intervals to actin strands and tropomyosin.

TSH—*See* thyrotropin.

type 1 diabetes—A type of diabetes mellitus that generally has a sudden onset during childhood or young adulthood and leads to almost total insulin deficiency, usually requiring daily insulin injections. Also known as insulin-dependent diabetes mellitus (IDDM) or juvenile-onset diabetes.

type 2 diabetes—A type of diabetes mellitus in which disease onset is more gradual and the causes are more difficult to establish than in type 1 diabetes. Type 2 diabetes is characterized by impaired insulin secretion, impaired insulin action, or excessive glucose output from the liver. Also known as non-insulin-dependent diabetes mellitus (NIDDM).

undertraining—The type of training an athlete would undertake between competitive seasons or during active rest. Generally, physiological adaptations will be minor, and there will be no improvement in performance.

upper body (android) obesity—Obesity that follows the typically male pattern of fat storage, in which fat is stored primarily in the upper body, particularly in the abdomen.

up-regulation—An increased cellular sensitivity to a hormone, often caused by increased hormone receptors.

Valsalva maneuver—The process of holding the breath and attempting to compress the contents of the abdominal and thoracic cavities, causing increased intra-abdominal and intrathoracic pressure.

valvular heart disease—A disease involving one or more of the heart valves. Rheumatic heart disease is one example.

vasoconstriction—The constriction or narrowing of blood vessels.

vasodilation—The dilation of blood vessels.

VC—See vital capacity.

$\dot{V}CO_2$—The volume of CO_2 produced per minute.

$\dot{V}_E$— The volume of air expired per minute.

$\dot{V}_E$max—*See* maximal expiratory ventilation.

$\dot{V}_E/\dot{V}CO_2$—*See* ventilatory equivalent for carbon dioxide.

$\dot{V}_E/\dot{V}O_2$—*See* ventilatory equivalent for oxygen.

veins—Blood vessels that transport blood back to the heart.

ventilatory breakpoint—The point at which ventilation increases disproportionately compared with oxygen consumption.

ventilatory equivalent for carbon dioxide ($\dot{V}_E/\dot{V}CO_2$)—The ratio of the volume of air ventilated ($\dot{V}_E$) to the amount of carbon dioxide produced ($\dot{V}CO_2$).

ventilatory equivalent for oxygen ($\dot{V}_E/\dot{V}O_2$)—The ratio between the volume of air ventilated ($\dot{V}_E$) and the amount of oxygen consumed ($\dot{V}O_2$); indicates breathing economy.

ventricular fibrillation—A serious cardiac arrhythmia in which the contraction of the ventricular tissue is uncoordinated, affecting the heart's ability to pump blood. *See also* ventricular tachycardia.

ventricular tachycardia—A serious cardiac arrhythmia consisting of three or more consecutive premature ventricular contractions. *See also* premature ventricular contraction and ventricular fibrillation.

venules—Small vessels that transport blood from the capillaries to the veins and then back to the heart.

very low density lipoprotein (VLDL)—A lipoprotein carrier of cholesterol.

very low density lipoprotein cholesterol (VLDL-C)—The cholesterol carried by VLDL.

vital capacity (VC)—The maximal volume of air expelled from the lungs after maximal inhalation.

vitamin—One of a group of unrelated organic compounds that perform specific functions to promote growth and to maintain health. Vitamins act primarily as catalysts in chemical reactions.

VLDL—*See* very low density lipoprotein.

VLDL-C—*See* very low density lipoprotein cholesterol.

$\dot{V}O_2$—The volume of oxygen consumed per minute.

$\dot{V}O_2$ drift—A slow increase in $\dot{V}O_2$ during prolonged submaximal exercise at a constant power output.

$\dot{V}O_2$max—*See* maximal oxygen uptake.

wet bulb globe temperature (WBGT)—A measurement of temperature that simultaneously accounts for conduction, convection, evaporation, and radiation, providing a single temperature reading to estimate the cooling capacity of the surrounding environment. The apparatus for measuring WBGT consists of a dry bulb, a wet bulb, and a black globe.

windchill—A chill factor created by the increase in the rate of heat loss via convection and conduction caused by wind.

work—Force expressed through distance, or a displacement, independent of time.

Note: The letters *f* and *t* after page numbers indicate figures and tables, respectively.

COMMON SCIENTIFIC ABBREVIATIONS AND UNITS AND CONVERSIONS

Common Scientific Abbreviations

Amount of substance

mol (mole)

mmol (millimole)

μmol (micromole)

Distance

km = kilometer

m = meter

cm = centimeter

mm = millimeter

μm = micrometer or micron

in. = inch

ft = foot

yd = yard

mi = mile

Electrical potential difference

V = volt

mV = millivolt

Energy

kcal = kilocalorie or Calorie

cal = calorie

J = joule

kJ = kilojoule

BTU = British Thermal Unit

RQ = respiratory quotient

Force

N = newton

Mass and Weight

kg = kilogram

g = gram

mg = milligram

μg = microgram

lb = pound

oz = ounce

kp = kilopond

Power

W = watt

Pressure

atm = atmosphere

mmHg = millimeters of mercury

Temperature

°C = degrees Celsius or centigrade

°F = degrees Fahrenheit

Time

h = hour

min = minute

s = second

Torque

Nm = Newton meter

Volume

L = liter

ml = milliliter

μl = microliter

gal = gallon

qt = quart

pt = pint

c = cup

tbsp = tablespoon

tsp = teaspoon

International System (SI) Units for quantifying exercise

Mass: kilogram (kg)

Distance: meter (m)

Time: second (s)

Force: newton (N)

Energy: joule (J)

Work: joule (J)

Heat: joule (J)

Power: watt (W)

Velocity: meters per second (m × s^{-1})

Torque: newton-meter (Nm)

Acceleration: meter per second per second
(m × s^{-2})

Angle: radian (rad)

Angular
velocity: radians per second (rad × s^{-1})

Amount
of substance: mole (mol)

Volume: liter (L)

Conversions

Amount of a substance

1 mol = 1,000 mmol

1 mmol = 1,000 μmol

1 mol of a gas = 22.4 L (standard conditions)

1 L of gas (standard conditions) = 44.6 mmol

mol = g/molecular weight

molarity of a solution = mol/L or mol/kg
solvent

mol in a volume = volume (L) × molarity

mmol in a volume = volume (ml) × molarity

Distance

1 m = 100 cm = 1,000 mm = 39.37 in. = 3.28 ft
= 1.09 yd

1 km = 0.62 mi

1 cm = 0.3937 in.

1 μm = 1 micron = 10^{-6} m = 10^{-3} mm

1 in. = 0.0254 m = 2.54 cm = 25.4 mm

1 ft = 12 in. = 0.3048 m = 30.48 cm = 304.8 mm

1 yd = 3 ft = 0.9144 m

1 mi = 5,280 ft = 1,760 yd = 1,609.35 m
= 1.61 km

Energy

1 joule = 0.239 cal

1 kcal = 1,000 cal = 4,184 J = 4.184 kJ

1 BTU = 0.2522 kcal = 1.055 kJ

1 L of oxygen consumed = 5.05 kcal = 21.1 kJ
at an RQ = 1.00

Force

1 N = 0.2248 lb = 0.1020 kg

Heat

1 J = 0.239 cal

1 kcal = 4.184 kJ

Mass and weight

1 kg = 1,000 g = 10^6 mg = 10^9 μg = 2.205 lb

1 g = 1,000 mg = 0.03527 oz

1 mg = 1,000 μg

1 lb = 16 oz = 453.6 g = 0.454 kg

1 oz = 28.35 g

1 μl of water weighs 1 mg

Power

1 W = 1 J/s = 60 J/min = 0.0143 kcal/min

1 W = 60 Nm/min = 6.118 kgm/min
= 6.118 kpm/min

1 kgm/min = 1 kpm/min = 0.1634 W

1 kcal/min = 69.78 W

Pressure

Standard pressure = 1 atm (atmosphere)
= 760 mmHg

Temperature

°C = 0.555 [(°F) – 32]

°F = 1.8 (°C) + 32

Velocity

1 m/s = 3.6 km/h (kph) = 2.237 mi/h (mph)

1 km/h (kph) = 16.7 m/min = 0.28 m/s
= 0.91 ft/s = 0.62 mi/h (mph)

1 mph = 88 ft/min = 1.47 ft/s = 1,609.3 m/h
= 26.8 m/min = 0.447 m/s = 1.6093 kph

Volume

1 L = 1,000 ml = 10^6 μl

1 L = 1.057 qt

1 ml = 1,000 μl

1 qt = 0.9463 L = 946.3 ml = 2 pt = 32 oz

1 gal = 4 qt = 128 oz = 3,785.2 ml = 3.7852 L

1 c = 8 oz = 236.6 ml

1 oz = 29.57 ml = 2 tbsp = 6 tsp

1 tbsp = 3 tsp = approximately 15 ml

1 tsp = approximately 5 ml

Work

1 J = 1 N × 1 m

1 J = 0.102 kgm = 0.102 kp